International Scientific Conference on Efficient Waste Treatment (EWT-2018)

IOP Conference Series: Earth and Environmental Science
Volume 337

St. Petersburg, Russia
13 - 14 December 2018

Editor:

Pietro Zunino

ISBN: 978-1-7138-0136-8
ISSN: 1755-1307

Printed from e-media with permission by:

Curran Associates, Inc.
57 Morehouse Lane
Red Hook, NY 12571

Some format issues inherent in the e-media version may also appear in this print version.

Institute of Physics
Dirac House, Temple Back
Bristol BS1 6BE UK

Phone: 44 1 17 929 7481
Fax: 44 1 17 920 0979

techtracking@iop.org

Additional copies of this publication are available from:

Curran Associates, Inc.
57 Morehouse Lane
Red Hook, NY 12571 USA
Phone: 845-758-0400
Fax: 845-758-2633
Email: curran@proceedings.com
Web: www.proceedings.com

TABLE OF CONTENTS

IOP Conf. Series: Earth and Environmental Science **337** (2019) 011001 doi:10.1088/1755-1315/337/1/011001

Editorial Preface

The international scientific conference "Efficient waste treatment – 2018" (EWT-2018) took place in St. Petersburg from 13th to 14th of December 2018. It was managed by Peter the Great St. Petersburg Polytechnic University. During the conference, multiple progress reports of research projects in the areas of waste recycling and energy efficiency were presented. Professors, scientists and representatives of industry and governmental organizations did those reports. New relationships between scientists and industry representatives were established.

EWT-2018 has collected 87 report by researchers from 10 countries. Presentations by various scientists have provided a review of latest achievements and future prospects in the area of recycling of various wastes for future processing. Informal meetings held during the event allowed some potential partners to establish contact, share new project ideas and start fresh collaborations. Multiple young participants have received certificates for best oral and written presentations, selected by the Program Committee.

Official conference website: http://ewt-conf.ru/

Editor

P. Zunino

IOP Conf. Series: Earth and Environmental Science **337** (2019) 011001 doi:10.1088/1755-1315/337/1/011001

The invited speaker

Pietro Zunino (University of Genoa, Italy)

IOP Conf. Series: Earth and Environmental Science **337** (2019) 011001 doi:10.1088/1755-1315/337/1/011001

Oral Presentations

IOP Conf. Series: Earth and Environmental Science **337** (2019) 011002 doi:10.1088/1755-1315/337/1/011002

Peer review statement

All papers published in this volume of *IOP Conference Series: Earth and Environmental Science* have been peer reviewed through processes administered by the proceedings Editors. Reviews were conducted by expert referees to the professional and scientific standards expected of a proceedings journal published by IOP Publishing.

IOP Conf. Series: Earth and Environmental Science **337** (2019) 012001 doi:10.1088/1755-1315/337/1/012001

Possibilities of municipal waste water treatment by using phototrophic microorganisms under the Moscow climate conditions

K Gorin[1*], D Kolomoitsev[2], A Melnikova[1] and Y Sergeeva[1]

[1]Biotechnology and Bioenergy department, National Research Centre "Kurchatov Institute", Moscow
[2]Agronomy and Biotechnology department, Russian State Agrarian University - Moscow Timiryazev Agricultural Academy

* Corresponding author: gkvbio@gmail.com

Abstract. Application of phototrophic microorganisms for municipal waste water treatment today became promising technology that combines solving of environmental issues and production of valuable biomass. This technology related to biomass cultivation in open ponds by using waste heat, that taken place in the waste water. Also huge volumes of municipal waste water makes economically impossible to apply artificial illumination. Thus application of phototrophic microorganisms for municipal waste water treatment is strongly related to climate conditions. In this paper data concerning phototrophic microorganisms cultivation in Moscow climate conditions are presented. *Euglena gracilis* CCAP 1224/5Z strain was used, cultivation was provided at illumination from 500 – 1200 lux depending on natural illumination. All cultivations were provided on synthetic waste water. It was shown that speed of growth was approximately the same in case synthetic waste water and controls. Content of organic impurities was decreased no less 70% at all rounds of cultivation. Final biomass concentration in case of cultivation on waste water exceeded concentration in the control flasks. Thereby it was demonstrated that it is possible in Moscow climate conditions during spring – autumn period to apply phototrophic microorganism *E. gracilis* CCAP 1224/5Z strain for waste water treatment.

1. Introduction

Lack of water resources and rising of waste water problems is one of the main challenges today. This is related with population growth, urbanization and climate change. Wastewater is contains a lot of organics and nutrients that must be removed before water is returned to natural water bodies to prevent eutrophication. Scientists are currently developing alternative treatment systems for wastewater treatment [1].

One of the most interesting directions of wastewater treatment is based on the microalgae cultivation. This type of system has several advantages over conventional chemical methods, such as high efficiency, lower costs, safety for the environment [2], and also because of their photosynthetic ability to convert solar energy into useful biomass and the use of nutrients such as nitrogen and phosphorus, causing eutrophication [3].

The types of nutrient sources that can be used for microalgae are municipal wastewater, breeding wastewater, industrial wastewater, and anaerobic digestion effluent. Nevertheless, the use of wastewater as nutrients in buildings requires careful monitoring of the harmful chemicals potentially found in the wastewater, such as solvents, paints, and pharmaceuticals, in addition to dirt and mold [4].

The use of microalgae is effective not only for the removal of nitrogen and phosphorus from wastewater [5], but also there is possibility of obtaining biodiesel from biomass [2, 6], including application in remote areas [7]. It should be noted that different types of microalgae have a various ability to remove substances. Therefore, researchers suggest screening to determine the best strain for specific conditions [8, 9].

So in a number of scientific works there are examples of successful cultivation of microalgae in a particular city, belonging to different climatic zones. In the works, the possibility of growing *Chlorella* microalgae under conditions of a temperate climate belt passing to the sea, namely in the city of St. Petersburg, is shown [10]. Wastewater treatment and biomass production for biodiesel production at low temperatures (7-13°C), characteristic of the conditions of the North using cyanobacteria *Synechocystis* and *Chlorella* microalgae [11], cultivation of *Euglena* on wastewater in Tokyo conditions [12].

In the case of the employing of microalgae cultivation for wastewater treatment has another advantage compared with other methods - it is an effective utilization of carbon dioxide, which is also very important for the cities [13].

This study, we examined the possibility of cultivating microalgae *Euglena gracilis* CCAP 1224 / 5Z strain in two types of wastewater, recycling organic and inorganic substances under the Moscow climate conditions.

2. Materials and methods

2.1. Microalgal strain, pre-culture cultivation condition

The culture of *Euglena gracilis* CCAP 1224/5Z from the Culture Collection of Algae and Protozoa (Scotland) was used in the experiments. The microalgae strain were cultured in 250 ml Erlenmeyer flask with 100 ml modified HUT nutrient media [14] at 22 ± 1°C under artificial illumination daylight fluorescent lamps (20 W, 6400K) light intensity 60 μmol PAR m^{-2} s^{-2} under 14:10 (light and dark) photoperiod on a laboratory shaker at 100 rpm.

The composition of HUT nutrient media was as follows mg/l: 20 KH_2PO_4, 288.8 NH_4Cl, 25 $MgSO_4{\cdot}7H_2O$, 400 sodium acetate, 40 Yeast extract and 1 ml of stock solution Pfennig microelements [15]. The composition of stock solution microelements was as follows mg/l: 5000 EDTA, 2000 $FeSO_4{\cdot}7H_2O$, 100 $ZnSO_4{\cdot}7H_2O$, 30 $MnCl_2$, 300 H_3BO_3, 200 $CaCl_2{\cdot}6H_2O$, 10 CuCl2, 20 $NiCl_2{\cdot}2H_2O$, 20 $Na_2MoO_4{\cdot}2H_2O$.

2.2. Experiment conditions

All experiments have been conducted in Erlenmeyer flasks (5000 mL) with 4000 ml of artificial waste water X1 and X1/2 under constant temperature 20 ± 1 °C. Artificial illumination was provided using 20 W daylight normalized fluorescent lamps, whose emission spectrum was in the range of 400–700 nm and a light intensity of approximately 60 μmol PAR $m^{-2}s^{-2}$ under 24 light photoperiod with air bubbling 300 ml/min. The composition of synthetic wastewater referred to as X1 was 1000 mg/L of glucose, 95.5 mg/L NH_4Cl, 56.3 mg/L, urea, 22.6 mg/L KH_2PO_4, 12.6 mg/L, $FeSO_4$*7 H_2O, 309 mg/L $NaHCO_3$ and 35 mg/L yeast extract. The residual water called X1/2 has half the concentration of all chemical species of X1 [1] and the composition of control HUT nutrient media was describe above.

2.3. Methods for the determination of permanganate index

The samples were centrifuged for 10 min at 4000 rpm (centrifuge Awel 20 mf) to eliminate all of the microalgae. Supernatant was used for analysis.

Permanganate index was determined by using method, presented in the GOST R 55684-2013 "Drinking water. Methods for the determination of permanganate index".

In a conical flask with a capacity of 250 cm^3 make 100 cm^3 of a thoroughly mixed sample of the analyzed water, a few pieces of porous porcelain, add 5 cm^3 of sulfuric acid solution, 10 cm^3 of a working solution of potassium permanganate. The contents of the flask are heated on an electric hotplate so that boiling comes no later than 5-7 minutes, and boil for 10 ± 2 minutes, closing with a small conical funnel to reduce evaporation. To the hot solution add 10 cm^3 of oxalic acid working solution. The bleached hot solution is titrated using a burette with a working solution of potassium permanganate until a pale pink color appears, lasting for about 30 s. Record the volume of the working solution of potassium permanganate consumed for titration (V, cm^3).

Permanganate index (I_{Mn}) in terms of atomic oxygen, mgO_2/dm^3, is calculated by the equation:

$$I_{Mn} = (V_3 - V_0)\cdot C\cdot K\cdot 5\cdot M / V_4, \text{ where}$$

V_3 - the volume of the working solution of potassium permanganate consumed for the titration of an aliquot of the sample of the analyzed water;

V_0 - the volume of the working solution of potassium permanganate consumed for titration at idle experiment;

C- concentration of potassium permanganate working solution, 2 mmole/dm^3;

K - coefficient correction for the working solution of potassium permanganate;

5 - stoichiometric coefficient;

M - atomic mass of oxygen for conversion to atomic oxygen equal to 8 g O_2 / mole;

V_4 - titration amount of water analyzed, cm^3.

2.4. Methods for the determination of orthophosphate concentration

The samples were centrifuged for 10 min at 4000 rpm (centrifuge Awel 20 nf) to eliminate all of the microalgae. Supernatant used for analysis.

Orthophosphate concentration was determined by PND F 14.1:2.248-07 "Method of performance measurements mass concentrations of orthophosphates, polyphosphates and general phosphorus drinking, natural and waste water photometric method".

The method is based on the interaction of orthophosphates with ammonium molybdate in an acidic medium with the formation of molybdophosphoric acid, its reduction by ascorbic acid in the presence of antimony chloride, followed by photometric measurement of the stained the blue color of the reduced form of molybdophosphoric acid (molybdenum blue) at wavelength 880 - 890 nm.

9.0 cm^3 of sample (or, when the content of orthophosphates is over 2.0 mg/dm^3 PO_4, its smaller volume diluted to 9.0 cm^3) is poured into a test tube with a screwed stopper, add 0.5 cm^3 mixed molybdenum-acid reagent and leave not less than 2 minutes. Then add 0.5 cm^3 of a solution of ascorbic acid, close the tube with a screw cap and mix. After 15–20 minutes, the measurement of the optical density (concentration, mg/dm^3) of the analyzed sample relative to the blank sample at a wavelength of 880–890 nm (spectrophotometer Thermo scientific Genesys 10S UV-Vis) is carried out. Distilled water is used as an idle sample through the entire course of the analysis.

To determine the concentration of orthophosphates in the sample, a calibration characteristic was established. According to the results, a calibration graph of the dependence of optical density on the concentration of orthophosphate ions (mg/dm^3) was made.

The mass concentration of orthophosphates (mg/dm^3 PO_4) in the analyzed sample is found according to the calibration curve, taking into account the preliminary dilution of the sample according to the equation:

$$X_{PO4} = Cgr\cdot 10 / Vs, \text{ where}$$

X_{PO4} - mass concentration of orthophosphates in the analyzed sample, mg/dm^3 PO_4;

Cgr - mass concentration of phosphates, found on the calibration curve, mg/dm^3;

10 - total volume of solution in a test tube, cm^3;

Vs - volume of water sample analyzed for analysis, cm^3.

2.5. Growth monitoring

Growth was monitored by dry biomass weight. Dry weight was measured according to the APHA method 8111G, [16] with a modified drying temperature (105 °C) and filter (Whatman GF/ C, 1.2 μm).

3. Results and Discussion

Algae growth was measured in terms of increase in biomass concentration (g/dm^3) and is shown in figure 1.

Initially two sample of wastewater X1 and X1/2 and control were inoculated with the same initial cellular concentration of *E. gracilis*. From the first day the culture on X1 and X1/2 medium has shown more rapid grow than control medium. This is consistent with the data obtained in [5]: since the beginning of the cultivation of the mixed indigenous microalgae at the primary effluent with a higher content of COD showed the fastest growth. The growing of *E. gracilis* on a wastewater are higher when control to 10.8 and 27.4 percent for X1/2 and X1 respectively, which is directly related to the richer composition of nutrient media.

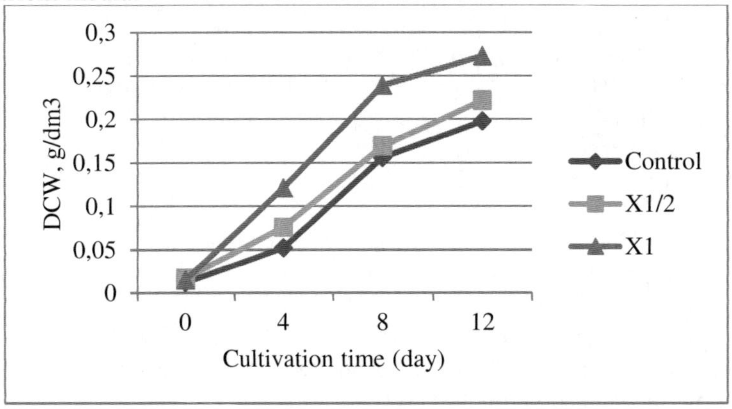

Figure 1. Comparison of culture *E. gracilis* grow throughout different wastewater.

pH was measured for 12 days of culture medium with the purpose of understand its variations figure 2. Initial pH of HUT medium was 7.1 and for artificial wastewater were 8.8 and 9.16 for X1/2 and X1 culture medium respectively. For control medium pH values showed a consistent increase from pH 7.1 to 8.78, what gives accordance with to study [17] about *Euglena* as a source of lipids for potential use as biofuel and wastewater treatment. During cultivation on the X1/2 and X2 no sagnificant change of pH was observed.

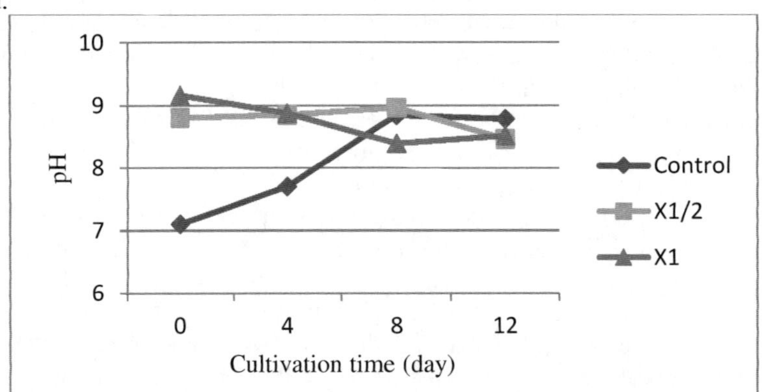

Figure 2. Variations of pH of different types of wastewater.

During this investigation photo-heterotrophic cultivation of *E. gracilis* was performed in order to reduce organic impurities. COD removal efficiency was very similar for both sample of wastewater figure 3. Rapid decrease in COD was observed in first four days and more slowly uptake during next days of cultivation. Significant COD removals were achieved in case of X1 and X1/2 wastewater and compose 73.7% and 72.4%, respectively.

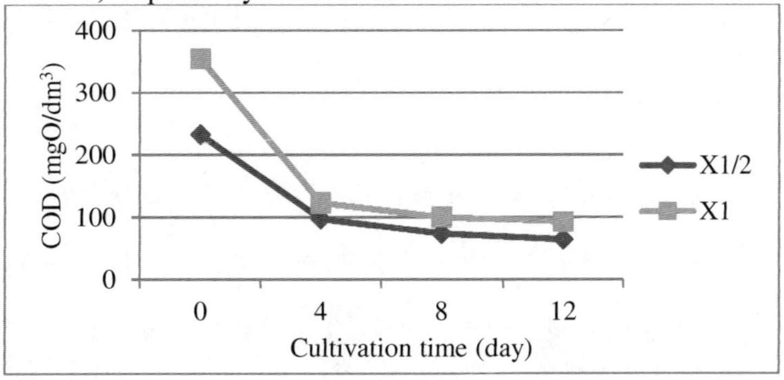

Figure 3. Variations of COD concentration during the experiment for different types of wastewaters.

Phosphorus is one of the main pollutants pose a serious threat to the water quality and cause damage to aquatic life [18]. It is known that the microalgae possess an ability to eliminate metals, nitrates, phosphates. Figure 4 depicts the result for the removal efficiency of orthophosphate in the wastewater samples by microalgae. The concentration of orthophosphate in the wastewater samples X1, X1/2 and control were reduced from 25.1 mg/dm^3 to 7.6 mg/ dm^3, or 69.7% of removal efficiency; from 15.5 mg/dm^3 to 5.7 mg/ dm^3, or 63.2% of removal efficiency; from 23.0 mg/dm^3 to 10.6 mg/ dm^3, or 53.9% of removal efficiency, respectively. In other study process of artificial wastewater treatment more effectively: the phosphate was reduced for 94% by *Chlorella vulgaris* cultivation [19].

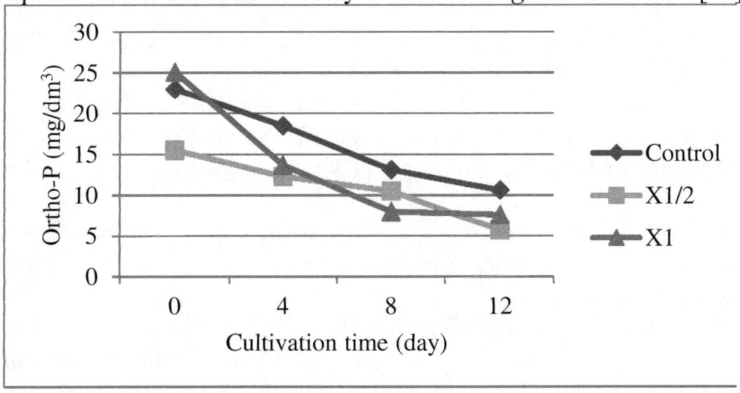

Figure 4. Variation of orthophosphate concentration during the experiment for different wastewaters.

4. Conclusions

In current paper was presented next results:

- Our research demonstrated that *E. gracilis* can effectively growth in the two types of artificial wastewater.
- Microalgae exhibited great ability to eliminate nutrients: after cultivation the decrease COD were more 70% and orthophosphate more 60%;
- Maximum cell biomass after cultivation on a surface medium a bit more than control medium;
- Demonstrated that it is possible in Moscow climate conditions to apply *E. gracilis* microalgae for waste wastewater treatment.

Further research will be focused on optimization of cultivation processes, developing of photobioreactor prototype and determination of lipid content.

Acknowledgements
This work was made with financial support of the Ministry of Science and Higher Education of the Russian Federation. The number of agreement is No. 14.616.21.0079. Unique identification number of applied research is RFMEFI61617/0079

References
[1] Salgueiro J L, Pérez L, Maceiras R, Sánchez Á and Cancela Á 2016 *Int. J. Environ. Res.* **10(3)** 465–70
[2] Lam M K, Yusoff M I, Uemura Y, Lim J W, Khoo C G, Lee K T and Ong H C 2017 *Renew. Energy* **103** 197–207
[3] Abdel-Raouf N, Al-Homaidan A A and Ibraheem I B M 2012 *Saudi J. Biol. Sci.* **19** 257–75
[4] Elrayies G M 2018 *Renew. Sustain. Energy Rev.* **81** 1175–91
[5] Mohamad S, Fares A, Judd S, Bhosale R, Kumar A, Gosh U and Khreisheh M 2017 *IOP Conf. Ser. Earth Environ. Sci.* **67** 012032
[6] Rinna F, Buono S, Cabanelas I T D, Nascimento I A, Sansone G and Barone C M A 2017 *J. Water Process Eng.* **18** 144–149
[7] Gotovtsev P, Vorobiev V, Migalev A, Badranova G, Gorin K, Dyakov A and Reshetilov A 2018 *Robotics* **7** 2
[8] Hoang Nhat P V, Ngo H H, Guo W S, Chang S W, Nguyen D D, Nguyen P D, Bui X T, Zhang X B and Guo J B 2018 *Biores. Tech.* **256** 491–501
[9] Ge S and Champagne P 2017 *Environ. Sci. Technol.* **51** 3558–66
[10] Politaeva N A, Kuznetsova T A, Smyatskaya Y A, Trukhina E V and Atamanyuk I 2018 *Chem. Pet. Eng.* **53** 801–805
[11] Jämsä M, Lynch F, Santana-Sánchez A, Laaksonen P, Zaitsev G, Solovchenko A and Allahverdiyeva Y 2017 *Algal Res.* **26** 65–73
[12] Torihara K and Kishimoto N 2015 *J. Water Environ. Technol.* **13** 195–205
[13] Kawachi M, Ishimoto M, Mori F, Yumoto K, Sato M and Noël H 2013 *MCC-NIES List of Strains, 9th ed.* (Tsukuba: Microbial Culture Collection at National Institute for Environmental Studies)
[14] Gorin K V, Badranova G U, Gotovtsev P M, Shatalova A Y, Grigoriev T E, Krasheninnikov S V, Tihomirov S A, Kondratev O A, Vishnevskaya M V and Vasilov R G 2018 *IOP Conf. Ser. Mater. Sci. Eng.* **292** 012051
[15] Pfennig N, Lippert K D 1966 *Arch. Mikrobiol.* **55** 245-256
[16] A.P.H.A. APHA, A.W. Works 2005 *Standard Methods for the Examination of Water and Wastewater* (Washington, D.C.: APHA-AWWA-WEF)
[17] Mahapatra D M, Chanakya H N and Ramachandra T V 2013 *J. Appl. Phycol.* **25** 855–65
[18] Yadavalli R, Rao C S, Rao R S and Potumarthi R 2014 *Asia-Pacific J. Chem. Eng.* **9** 368–73
[19] Feng Y, Chao L and Zhang D 2011 *Biores. Tech.* **102** 101–5

Membraneless microbial biofuel cell for municipal waste water treatment

M Vishnevskaya[1*], D Gazizova[1,2], A Victorenko[1,2] and I Konova[1]

[1]National Research Centre "Kurchatov Institute", Biotechnology and Bioenergy Department, 123182 Moscow, Russia
[2]Moscow Physical Technical Institute, 117303 Moscow, Russia

* Corresponding author: Vishnevskaya_MV@nrcki.ru

Abstract. Microbial biofuel cell is the promising technology that gives possibilities to generate electricity by degradation of organic contaminations. Thus, this technology became very perspective for municipal waste water treatment to minimize energy consumption of waste water treatment plants. The main limitations of microbial biofuel cell application are related with high cost of membranes and electrodes materials. Absence of membrane and application of cheap materials with low conductivity and low surface area usually leads to decreasing of power output. In this paper we present membraneless microbial fuel cell (MFC) with cheap conductive electrodes that can be applied in future for waste water treatment. Gluconobacter Oxydans VKM V-1280 strain from All-Russian Collection of Microorganisms was used as biocatalyzer for anode. Graphite was applied as electrode material. Immobilization was provided by using mixture of poly (3,4-ethylenedioxythiophene) polystyrene sulfonate (PEDOT:PSS) and poly (ethyleneglycol) diglycidyl ether (PEGDE). Experiments were provided by using synthetic analog of municipal waste water. No additional chemical mediator was applied. Power output was 1.43 mcW/sm2. Electrodes were stable during biofuel cell (BFC) operation. Thus, it was demonstrated that it is possible to apply membraneless MFC for municipal waste water treatment.

1. Introduction

Today, the energy industry is divided into three categories by fuels types: renewable sources, fossil fuels and nuclear sources [1]. The last two are the most popular, however the applying of fossil fuels can harm the human being and environment of the planet generally [2]. For example, greenhouse gases are produced as a result of fossil fuels burning leading to climate changes as well as environmental deterioration [3]. Therefore, the search for new sources of energy now is ongoing.

One of the latest developments in this area is the MFC, which produce energy by oxidizing organic compounds from waste water. Due to this oxidation electrons are emitted and then received on anode, what causes the produced electrical current to be sent to cathode. The anode compartment contains a biocatalyst (microbial cells or enzymes), a mediator and a substrate (oxidizable organics – alcohols, sugars, etc.). Oxidation of the substrate leads to the appearance of a charged state of the anode and the generation of a potential difference between the anode and the cathode. On the external circuit, the electrons are transferred to the cathode by the action of electromotive force. All the described processes occur in the water phase – in the anode compartment there is a physiological buffer that provides the vital activity of microbial cells, in the cathode compartment there is a buffer stabilizing the pH index [4].

Scientists observed the current created by bacteria as early as 10s of the 19th century [5]. But it was only in the 1990s that the MFC began to be considered a promising technology [6]. After the discovery of the fact that the mediator is not a binding part of the MFC, developments in this area have expanded [7].

Research is underway on detection of bacteria strain, which will allow MFC to work with highest efficiency. The selection criteria in this case are bacteria's ability to decompose organic substances and transmit electrons from intracellular space to anode. The most commonly used bacteria for creating of BFC are *Shewanella oneidensis*. However, despite the fact that in the MFC other bacteria are used more often, *Gluconobacter oxydans* that is a promising type of bacteria with which for creation of BFC [8]. It was shown that application of these bacteria can help to increase power output and exclude mediator from anode chamber [9].

The topical issue in new MFC development is search of more cheap material for electrodes. They should provide sufficient conductivity, chemical stability, biocompatibility and high specific surface area [10]. The most common and suitable by electrochemical properties material is graphite. Graphite rods are cheaper than most alternative electrodes. And yet they have acceptable properties for use in microbial biofuel cells [11, 12]. Graphite rods are biocompatible to living organisms and have the ability to transfer electrons. So those cheap materials are still interesting for such applications as a MFC for waste water treatment.

MFC can be used for variety of purposes. For example, it can be used as a biosensor: the MFC with *Gluconobacter oxydans* is capable of providing low-power Internet of Things devices with on board processing, wireless communication and reading functions [13]. MFCs are also used in the field of bioenergy based energy sources for the autonomous mobile robots [14]. Frequently, the MFCs are used to improve the environment, including water treatment [15].

In presented paper we demonstrate possibility of creation of the membranless MFC based on graphite electrodes for waste water treatment with electrical power generation.

2. Experiments

2.1. Reagents

The bacterial strain *Gluconobacter Oxydans* VKM V-1280 was used in this work. Sorbit-D and yeast extract by Dia-M (Russia) were used for preparation of a nutrient [9]. The composition of synthetic wastewater (SWW) was 1000 mg/L of glucose, 95.5 mg/L NH_4Cl, 56.3 mg/L urea, 22.6 mg/L KH_2PO_4, 12.6 mg/L $FeSO_4*7H_2O$, 309 mg/L $NaHCO_3$ and 35 mg/L yeast extract [16]. It was employed as electrolyte. The distilled water was used throughout.

2.2. Test-bed

A chemical chamber (50 ml) with electrodes attached to a 3D printed cap with holes for them was used as membraneless cell for the experiments. Measurements of the current-voltage characteristics (CVC) were carried out on an IPC-Micro potentiostat.

2.3. Obtaining biomass of Gluconobacter Oxydans

The biomass of bacteria *Gluconobacter Oxydans* was obtained from the G.K. Scriabin Institute of Biochemistry and Physiology of Microorganisms RAS from All-Russian Collection of Microorganisms. To further maintain the growth of culture, the nutrient medium was used: sorbit-D (10%), yeast extract (1%), distilled water and agar (2.5%) for dense media. Cultivation (100-150 g) for dispersing the culture was carried out for 24 hours at a temperature of 28 °C. Cultivation was carried out for collecting biomass during 20 hours (centrifugation 15 minutes, 10,000 g). After collecting the biomass, it was washed with cold sterile NC solution (0.9% NaCl, 2 mM $CaCl_2$) [17] and resuspended again in 1 ml of NC solution. The biomass was stored in the refrigerator (+ 4 °C) before using in the experiment.

2.4. Electrode preparation

IOP Conf. Series: Earth and Environmental Science **337** (2019) 012002 doi:10.1088/1755-1315/337/1/012002

Graphite rod electrode (diameter 6 mm, length 61 mm) was used as a cathode. Similar electrode with a mixture of *Gluconobacter Oxydans* biomass with high-conducting polymer PEDOT:PSS and PEGDE was used to better fix the biomaterial on the electrode surface in the ratio 4: 1: 1 as anode. Total mixture (600 μl) was applied to anode and installed to the working area of the electrode equal to 7.3 cm^2. The anode was dried at a temperature of + 27 °C for 2 days.

2.5. CVC measurement
CVC was measured after adding 40 ml of a composition of SWW to the cell. The measurements were carried out on an IPC-Micro potentiostat with a specified voltage from -300 to 300 and back to –300 mV (without interruption) with a step of 50 mV. The time step is about 2 minutes to establish a stable current.

3. Results and discussion
The CVC measurement of BFC was provided. For a visual representation of the experimental results graph of the CVC presented at fig. 1.

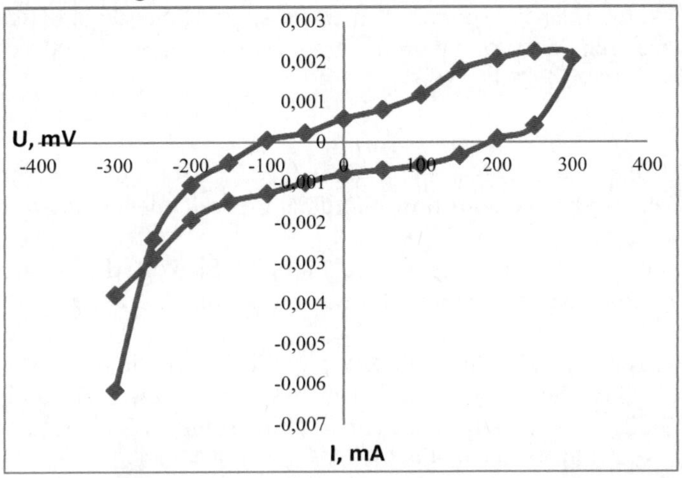

Figure 1. Current dependence of voltage in MFC.

Resulting graph at figure 1 showing that there is a capacity in presented MFC cell. Also it is necessary to mentioned that form of CVC is good correlated with the one presented in other paper for *Gluconobacter oxydans* based MFC (see for example [8]).

The next step was to calculate the various characteristics of the cell. For a complete analysis of the results, it was necessary to obtain the values of the capacity and power of the cell. To calculate the average capacity we used the formula: side.

$$C = I/(U/dt) \tag{1}$$

Where (*C*) is the capacitance, (*U*) is the voltage, I is current and (*dt*) is the measurement time for a particular voltage. The cell capacity made up 5.3 mF average.

The operation process of MFC was provided for consumer and it generated current equal to 0.2 mA and voltage equal to 52.5 mV.

Power generation was calculated by the next formula:

$$P=I*U \tag{2}$$

Where (*P*) is power, (*I*) and (*U*) are current and voltage respectively.

Membraneless microbial biofuel cell showed total power of 10.5 mW. The average value of the specific power is 1.43 mcW/sm^2. Those results demonstrate that this membranless MFC can operate without mediator with respectable power output using waste water as a substrate source. Further research will be focused on optimization of cell itself and optimization of bacterial cell immobilisation method on electrode to obtain higher power output at the same substrate.

4. Conclusion

The obtained results showing that it is possible to develop membranless MFC for municipal waste water treatment applications. Developed cell showing the considerable power output and, it is necessary to mention that there are several ways to enhance efficiency. Further research will be focused to cell design itself and immobilization of bacterial cells efficiency. Also it is necessary to mentioned that this MFC operated without chemical mediator, so makes strong foundation for future practical application.

Finally it is can be mentioned that MFC can be applied in various areas. Its application can enhance municipal waste water and other organics contaminated waters treatment. The development of MFC is an evolving and highly promising direction in nature-like technologies.

5. Acknowledgements

The work was done with the financial support of the Ministry of Science and Higher Education of the Russian Federation in the framework of the grant agreement No. 14.601.21.0017. Unique identifier of applied research (project) RFMEFI60117X0017.

References

[1] Rahimnejad M, Ghoreyshi A, Najafpour G, Jafary T 2011 Power generation from organic substrate in batch and continuous flow microbial fuel cell operations *Appl.Energy* **88** pp 3999-4004

[2] Rahimnejad M, Ghoreyshi A, Najafpour G, Younesi H, Shakeri M 2012 A novel microbial fuel cell stack for continuous production of clean energy *Int. J. of Hydrogen Energy* **37** pp 5992-6000

[3] Feng T, Zhou W, Wu S, Niu Z, Cheng P, Xiong X, Li G 2018 Simulations of summertime fossil fuel CO_2 in the Guanzhong basin *Sci. of The Total Environment* **624** pp 1163-70

[4] Sukhvinder B, Giddey S, Munnings C, Bhatt A, Hollenkamp A 2014 Emerging electrochemical energy conversion and storage technologies *Front. Chem.* **24**

[5] Logan B 2008 *Microbial fuel cells* (Hoboken: John Wiley & Sons)

[6] Allen R, Bennetto H 1993 Microbial fuel cells: Electricity production from carbohydrates *App. Biochem. Biotechnol.* **39** pp 27–40

[7] Mokhtarian N, Ramli D, Rahimnejad M, Najafpour G 2012 Bioelectricity generation in biological fuel cell with and without mediators *World App. Sci. J.* **18** pp 559-67

[8] Alferov S, Minaicheva P, Arlyapov V, Asulyan L, Alferov V, Ponamoreva O, Reshetilov A 2014 Bioanode for a microbial fuel cell based on Gluconobacter oxydans immobilized into a polymer matrix *Appl. Biochem. and Microbiol.* **50** pp 637-43

[9] Reshetilov A, Plekhanova Yu, Tarasov S, Arlyapov V, Kolesov V, Gutorov M, Gotovtsev P, Vasilov R 2017 Effect of some carbon nanomaterials on ethanol oxidation by Gluconobacter oxydans bacterial cells *Appl. Biochem. and Microbiol.* **53** pp 123-9

[10] Alferov V, Vasilov R, Gubin S 2014 The biofuel elements on the basis of the nanocarbon materials *RENSIT* **6** pp 187-208

[11] Ma'arof M, Chala G, Ravichanthiran S, Diasip A 2018 A study on microbial fuel cell (MFC) with graphite electrode to power underwater monitoring devices *IJMET* **9** pp 98-105

[12] Mustakeem 2015 Electrode materials for microbial fuel cells: nanomaterial approach *Materials for Renew. and Sustain. Energy* **4**

[13] Somov A, Gotovtsev P, Dyakov A, Alenicheva A, Plehanova Y, Tarasov S, Reshetilov A 2018 Bacteria to power the smart sensor applications: Biofuel cell for low-power IoT devices *2018 IEEE 4th World Forum on Internet of Things (WF-IoT)* pp 802–6

[14] Gotovtsev P, Vorobiev V, Migalev A, Badranova G, Gorin K, Dyakov A, Reshetilov A 2018

Bioenergy based power sources for mobile autonomous robots *Robotics* **7** pp 2–18
[15] Gotovtsev P, Dyakov A 2016 Biotechnology and Internet of Things for green smart city application *2016 IEEE 3rd World Forum on Internet of Things (WF-IoT)* pp 542–6
[16] Salgueiro J, Pe´rez L, Maceiras R, Sa´nchez A, Cancela A 2018 Semicontinuous culture of Chlorella vulgaris microalgae for wastewater treatment *Int. J. of Env. Res.* **12** pp 765-72
[17] Vostiar I, Ferapontova E, Gorton L 2004 Electrical ‘‘wiring’’ of viable Gluconobacter oxydans cells with a flexible osmium-redox polyelectrolyte *Electrochem. Communications* **6** pp 621–6

IOP Conf. Series: Earth and Environmental Science **337** (2019) 012003 doi:10.1088/1755-1315/337/1/012003

Poultry by-products as a basis of combined meat and vegetable feed for domestic animals

R Khabibullin[1*], D Khrundin[1], R Sabirzyanova[1] and G Ezhkova[1]

[1]Kazan National Research Technological University, Meat and Dairy products Department, 420015 K.Marks str. 68, Kazan, Russia

* Corresponding author: hrustik@yandex.ru

Abstract. The use of poultry by-products remains one of the urgent strategies of the economic efficiency increase for a variety of reasons. The most effective is the by-products processing into the meat and bone meal with a high content of amino acids, fatty acids, vitamins and minerals and consequently with high nutritional value. This meal may be used individually and as a basis of the combined feed for domestic animals. The possibility of using poultry by-products as a basis for feed of domestic cats and dogs was studied. The quantitative and qualitative assessment of carcass after the duck *Anas platyrhynchos domesticus* manual deboning was carried out, and the moisture, proteins and fats content was determined by conventional methods. It was shown that the duck carcass had high nutritional value and can be used as a basis for domestic animals feed. Drawing up a recipe of the combined feed including duck carcass meal, additional protein sources and natural vegetables was carried out. The developed calculation and optimization computer program "Optipet" for the combined feeds recipes optimization was used, which allowed to determine the minimum cost value of the combined feed as required by current law documentation.

1. Introduction

The use of secondary raw material of the poultry processing remains one of the urgent strategy of the economic efficiency improving for a variety of reasons. First of all these by-products are formed annually in huge quantities. Thus, in 2009 more than 55.5 billion broilers with an average weight of 1.8 kg were slaughtered worldwide, while the total production of chicken by-products amounted to 17 million tons per year [1].

Second, by-products not properly treated may cause reproduction of insects, vermin, bacteria and viruses, which may result in water pollution (leaching of pathogenic microorganisms, organic substances and inorganic ions) and air pollution (harmful and foul-smelling gaseous substances) [2].

Third, by-products cannot be disposed of in landfills [3].

Finally, they are a valuable secondary raw material for the heat production by burning in biothermal holes, a source of alternative energy by solid-phase methane digestion or the effective organic fertilizers due to the high content of nutrients N, P, K [4].

However, the most effective is the processing of these by-products into meat and bone meal with a high content of essential amino acids, fatty acids, vitamins and minerals, depending on the composition of feed [1, 5, 6].

Chicken meal has one of the best amino acid profiles among the slaughtering secondary products [7]. Digestibility of raw meal protein is 20-30 % lower than in soybean meal and rapeseed flour [8-11], but higher than in meat and bone meal, keratinous flour, blood flour and corn gluten [12-14].

The sustainability and digestibility of chicken meal protein can be enhanced in various ways, for example, by lactic acid fermentation with molasses or brewer's grain, that reduces pH and stops bacterial and viral development [15]. The same effect is achieved by acid and alkaline treatment [17]. A common method is heat treatment under pressure of acute steam (wet rendering) or heating by dry steam (dry rendering) [2], leading to by-products sterilization and drying up to 8% of moisture. If the fat content is higher than 16% and the rancidity problems may occur during storage, additional extraction with solvents up to 10-12 % fat carried out [18]. Freeze drying also makes it possible to obtain a product without pathogens, but with a high fat content and a lower content of crude protein [19].

Such processing methods make the poultry meal a valuable source of protein for animals. Thus, meal increases the duodenal flow of amino acids in ruminants while ensuring microbial synthesis with sufficient nitrogen amount [20, 21].

Poultry by-product meal can be used as a protein supplement for dairy cows, for calves of Holstein breed and bulls; in the prenatal diet of cows it provides both normal reproductive function and milk production. Daily amounts of 0.5 to 0.7 kg have been recommended [2, 8, 12, 13], it is particularly effective when included in a mixed diet [21].

Poultry by-product meal is a useful and cheap substitute for soybean meal in the diets of lambs, especially after co-extrusion of poultry and soybean meal [2] or with wheat or corn grain [2, 22]. It was used for adult pigs in an amount of up to 10% [2, 23] and up to 100% – for young animals [24].

In the diet of adult poultry by-product meal is used in an amount of up to 7-10%, in the diets of broilers from 4% to 10% and laying hens - from 5 to 7.5 % [25, 26].

In the feeding of angora rabbits chicken meal in the amount of up to 8-10% completely replaced fish meal [27]. With its content of 18%, the growth of rabbits slowed down, but the specific cost of produced meat is minimal [28].

Finally, chicken meal in the amount of 5-25% was effective as an additional feeding for the various fish species aquaculture [29-35].

European legislation is tightening on the use of by-products of slaughter, and since 2002 the EU has introduced a ban on the use of secondary products of poultry processing for animal feeding [36].

The lack of adequate food protein is a problem of feeding pets. The Russian market of pet feed is developing dynamically today – its annual growth averages 12-25% [37]. Most of the market falls on importers – Germany, France, UK, USA and China.

Due to the current political and economic situation in the world, many importers have reduced the volume of supplies. On the other hand, the increase of the Russian population and domestic animals number causes a natural increase in the domestic animals feeding consumption. Thus, now there is a favorable situation for the development and promotion of new feed products [38-42].

The aim of this work was to study the possibility of using duck carcasses for the production of combined meat and vegetable feed for domestic animals.

In order to achieve this goal, we have decided to concentrate on the following tasks:

1. Determination of the fractional and component composition of the duck carcass;
2. Preparation of the combined feed recipe with the duck carcass meal;
3. Optimization of the feed recipe according to biological value and cost indicators utilizing a newly developed computer program;
4. The study of the obtained combined feed component composition.

2. Materials and methods

Duck carcasses after manual deboning were obtained from a local duck processing plant. Pre-differentiation of duck carcass into the three fractions– skin, fleshy part and meat and bone residue - was made. The soft fraction was chopped by cutter and meat and bone residue was pre-treated in a steam heater and also was milled to a homogeneous state. The by-products of poultry processing as well as the combined meat and vegetable feed on its basis were studied with help of accepted analytical methods in accordance with the regulatory documentation [43, 44]. Protein, fats, salt, fibers and moisture content of these three fractions were determined by InfraLum FT-12 Infrared Analyzer.

3. Results and discussion

The mass ratio of individual fractions of the duck carcass frame (Fig.1) and their component composition (Fig.2) were determined.

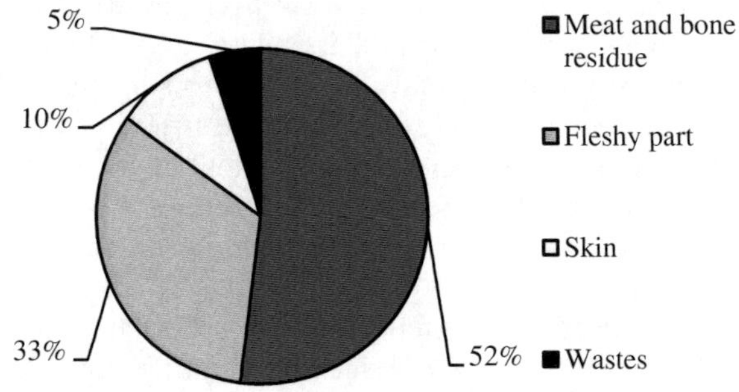

Figure 1. Fractional composition of the duck carcass after manual deboning.

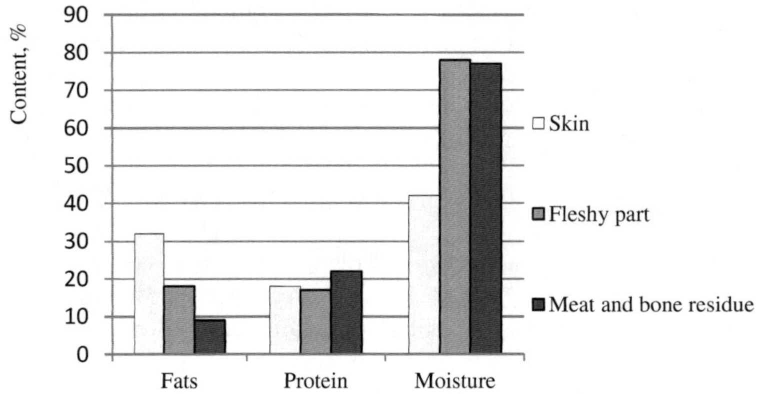

Figure 2. Component composition of the analyzed fractions.

All fractions were characterized by sufficient protein content, especially meat-bone residue - more than 20 %. The fat content is the highest in the fleshy part – 32.2 %, and the lowest in the meat and bone residue – less than 6.0 %. The results showed a high nutritional value of duck carcass and the possibility of its using for pet food.

The recipe of combined meat and vegetable feed for pets based on the duck carcass meal, additional sources of protein were fish meal, non-edible lamb trimmings, nonfat dry milk, additional vegetable ingredients – pumpkin, dried carrots and parsley.

The complexity of the problem of feed composition optimizing is associated with the huge number of options [45-47], so the computing program "Optipet" was developed to calculate and optimize the biological value and feed composition. The optimization criterion was the cost of the feed:

$$\text{Cost} \rightarrow \min \quad (1)$$

Variables were the main components mass fractions, and the boundary conditions determined by the requirements of regulatory and technical documentation and quality assessment considerations. As a result, the optimized formulation and component composition of combined meat and vegetable feed were determined (Tables 1, 2).

Table 1. Optimized recipe of combined meat and vegetable feed.

Components	The composition of components, %					The mass fraction in the feed (%)
	Proteins	Fats	Salt	Moisture	Fibers	
Duck carcass meal	23,10	5,60	0,90	77,20	0,40	47,50

Fish meal	59,40	18,00	0,90	18,00	3,00	4,00
Non-edible lamb trimming	18,75	9,90	0,30	77,00	1,20	10,00
Non-fat dry milk	42,20	0,95	2,40	12,00	0,00	3,00
Pumpkin	2,90	0,10	1,50	89,00	9,00	8,50
Dried parsley	17,00	5,10	1,70	16,00	25,10	3,00
Dried carrots	13,20	3,60	0,80	18,00	22,00	6,00
Broth	24,00	2,60	1,20	98,60	0,00	18,00
The content in the feed	22,36	5,24	0,79	72,32	3,27	100,00

Table 2. Component composition of combined meat and vegetable feed.

Parameter	Calculated value	Actual value
Mass fraction of moisture, %	72,3	74,0±3,3
Mass fraction of proteins, %	22,4	22,3 ±0,9
Mass fraction of fats, %	5,2	5,6±0,4
The content of sodium chloride, %	0,8	1,15±0,03

4. Conclusions

Thus, the most effective way to use poultry by-products is to obtain meat and bone meal and production of combined meat and vegetable feed for domestic animals on its basis.

The developed and optimized formulation of this combined meat and vegetable feed meets the physiological needs of the animal organism and meets the established regulatory requirements.

References

[1] Hertrampf J W, Piedad-Pascual F 2000 *Handbook on ingredients for aquaculture feeds* (Dordrecht: Springer Science & Business Media.)

[2] Heuzé V, Tran G, Chapoutot P, Noblet J, Renaudeau D, Lessire M, Lebas F 2015 Poultry by-product meal *Feedipedia, a programme by INRA, CIRAD, AFZ and FAO* https://www.feedipedia.org/node/214

[3] *Veterinary-sanitary rules for the collection, disposal and destruction of biological waste № 13-7-2/469* 1995 approved by the Chief State Veterinary Inspector of the Russian Federation 4.12.1995 (in Russian)

[4] Salminen E, Rintala J 2002 *Bioresource Technology* **83**

[5] Watson H 2006 *Dogs in Canada*

[6] Dale N, Fancher B, Zumbado M, Villacres A 1993 *J. Appl. Poult. Res.* **2**

[7] Tacon A G J, Metian M, Hasan M R 2009 *FAO Fisheries and Aquaculture technical paper #5402009*

[8] da Silva L D D F, Ezequiel J M B, de Azevedo P S, Barbosa J C, Cattelan J W, de Resende F D, Seixas J R C, do Carmo F R G 1999 *Semina (Londrina)* **20** (1)

[9] de Souza J R S T, Camarao A P, Rego L C 2000 *Braz. J. Vet. Res. Anim. Sci.* **37** (2)

[10] Kamalak A, Canbolat O, Gurbuz Y, Ozay O 2005 *Small Rumin. Res.* **58** (2)

[11] Mustafa A F, Qiao S Y, Thacker P A, McKinnon J J, Christensen D A 2000 *J. Sci. Food Agric.* **80** (11)

[12] Klemesrud M J, Klopfenstein T J, Lewis A J, Shain D H, Herold D W 1998 *J. Anim. Sci.* **75**

[13] Klemesrud M J, Klopfenstein T J, Lewis A J 1998 *J. Anim. Sci.* **76** (7)

[14] Bohnert D W, Larson B T, Bauer M L, Branco A F, McLeod K R, Harmon D L, Mitchell G E 1999 *J. Anim. Sci.* **77** (4)

[15] Cai T, Pancorbo O C, Barnhar H M 1994 *J. Appl. Poult. Res.* **3**

[16] Kim W K, Patterson P H 2003 *J. Dairy Sci.* **86** (10)

[17] Blake J P 2000 *Proceedings XI National Veterinary Medicine Congress*

[18] El Boushy A R Y, van der Poel A F B 2000 *Handbook of poultry feed from waste: processing and use* (Dordrecht: Springer)

[19] Shafer D J, Burgess R P, Conrad K A, Prochaska J F, Carey J B 2001 *J. Poultry Sci* **80** (11)

[20] Bohnert D W, Larson B T, Bauer M L, Branco A F, McLeod K R, Harmon D L, Mitchell G E

1998 *J. Anim. Sci.* **76** (9)
[21] Schroeder J W 1999 *North Dakota State University Extension Service, AS-1180* https: //www.ag.ndsu.edu/ansc/files/annual-reports/copy_of_2012AnnualReportwithAppendix.pdf
[22] Aimiuwu O C, Lilburn M S 2006 *Poult. Sci.* **85** (7)
[23] Rostagno H S, Teixeira A, Donzele J L, Gomes P C, De Oliveira R F M, Lopes D C, Ferreira A J P, Toledo Barreto S L 2005 *Brazilian Tables for Poultry and Swine: composition of feedstuffs and nutritional requirements* (Brazil: Universidade Federal de Viçosa)
[24] Keegan T P, DeRouchey J M, Nelssen J L, Tokach M D, Goodband R D, Dritz S S 2004 *J. Anim. Sci.* **82** (9)
[25] Senkoylu N, Samli H E, Akyurek H, Agma A, Yasar S 2005 *J. Appl. Poult. Res.* **14** (3)
[26] Hosseinzadeh M H, Ebrahimnezhad Y, Janmohammadi H, Ahmadzadeh A R, Sarikhan M 2010 *Int. J. Agric. Biol.* **12** (4)
[27] Ahlawat S S, Khanna N, Sharma D P, Panda P C 2003 *Indian J.Anim.Res.* **37** (2)
[28] Fotso J M, Fomunyam R T, Ndoping B N 2000 *World Rabbit Sci.* **8** (2)
[29] Gunben E M, Senoo S, Yong A, Shapawi R 2014 *Sains Malaysiana* **43** (3)
[30] Fowler L G 1991 *Aquaculture* **99**
[31] Yigit M, Erdem M, Koshio S, Ergun S, Turker A, Karaali B 2006 *Aquacult. Nutr.* **12**
[32] Meeker D L, Hamilton C R 2006 *Essential rendering* ed Meeker D L
[33] Abdel-Warith A, Russell P M, Davies S J 2001 *Aquacult. Res.* **32**
[34] Zhou Q C, Zhao J, Li P, Wang H L, Wang L G 2011 *Aquaculture* pp 322–3
[35] Shapawi R, Ng W K, Mustafa S 2007 *Aquaculture* **273**
[36] Regulation No 1774/2002 of 03.10.2002 *Official Journal of the EC* **45**
[37] Burlakova E 2016 Increased sales of pet food *RBC* https://www.rbc.ru/business/02/02/2016/56af7aa39a7947432ab67cea (in Russian).
[38] Vlasova A Y, Yagodin K V 2015 The Actual problems of management in the economy of the 21st century: *Proceedings of the international scientific-practical conference* (2015) (in Russian)
[39] Antipova L V, Shalimova O A, Senkina T A 2009 *All about the meat* **1** (in Russian)
[40] Antipova L V, Grebenshchikov A V, Kazakov N N 2012 *Bulletin of Voronezh state University of engineering technologies* **1** (in Russian)
[41] Antipova L V, Grebenshchikov A V, Kazakov N N, Selezneva N V 2012 *Izvestiya vuzov: Food technology* **2-3** (in Russian)
[42] Avrorov V A, Luzgin G D, Avrorov G V, Lovchev V V, Poletaeva M A 2016 Condition and prospects of development of modern science: the socio-economic, natural-science studies: *Proceedings of the International scientific-practical conference* (2016) (in Russian)
[43] Russian Federation State Standart R 55453-2013 2014 *Feeding stuffs for nonproductive animals. General specifications* (Moscow: Standartinform) (in Russian)
[44] *Veterinary and sanitary norms and requirements for the quality of feed for non-food animals № 13-7-2 / 10101 of July 5, 1997* (in Russian)
[45] Grebenshchikov A V, Cossacks N N 2012 *The Herald of the Voronezh state Univ. of engin. Tech.* **2** (in Russian)
[46] Khrundin D V, Khabibullin R E, Ezhkova G O 2016 *The Herald of the Voronezh state Univ. of engin. Tech.* **19** (19) (2016) (in Russian)
[47] Zinina O V, Zhakslykova S A 2011 Status, problems and prospects of production and processing of agricultural production: *Proceedings of the international scientific-practical conference* (in Russian)

IOP Conf. Series: Earth and Environmental Science **337** (2019) 012004 doi:10.1088/1755-1315/337/1/012004

Prospects of construction of facilities for processing and storage of radioactive waste of the NPP

A Subbotin[1*], Y Grabilina[1] and D Dneprovskiy[1]

[1] Moscow State University of Civil Engineering, Yaroslavskoe shosse, 26, Moscow, 129337, Russia

* Corresponding author: subbotin-art@mail.ru

Abstract. Use of nuclear fuel on the nuclear power plants (NPP) requires special attention at the further order of radioactive waste (RW). Therefore there is a requirement of innovative and perspective approaches of construction of facilities of processing and storage of RW to protect the environment from a large number of radiation. Relevant the question of prospects and new approaches to construction of these objects arises at design, construction and operation where high requirements of objects of nuclear sector are considered. Nuclear sector, occupies one of key positions in the 21st century. Solvable tasks in the field of the analysis of methods of processing and storage of radioactive waste of the NPP, will allow to make construction of facilities of nuclear power safer and eco-friendly that in the international practice is one of priority tasks for these objects.

1. Introduction

The increasing deficiency of organic fuel and steady growth of energy consumption in the countries of the world community dictate need of development of alternative energy sources, the most real of which is nuclear fuel. The nuclear power occupies one of key positions on obtaining useful energy, both thermal, and electric now. Operation of the nuclear power plants working due to division of kernels leads to the serious problems connected with formation of a huge amount of radioactive waste and has the extreme world importance. By the available estimates so far about 300 thousand tons of the fulfilled nuclear fuel around the world are saved already up, and by 2030 this volume will increase up to 500 thousand tons [1]. In this article the key problem of nuclear power is creation of objects of storage and processing of radioactive waste of the NPP with a possibility of their further use.

In this article key research problems are considered:

- to make the analysis of methods of processing and storage of radioactive waste of the NPP;
- to make comparison of the Russian and international experience in the sphere of construction of these objects;
- to offer relevant ways of storage and processing of radioactive waste of the NPP.

It should be noted that the question of development of theoretical bases and methodical approaches to formation of effective mechanisms at design, construction and operation of objects of nuclear power in the field of storage and processing of Russian joint stock company, is a key position for the world energy companies.

IOP Conf. Series: Earth and Environmental Science **337** (2019) 012004 doi:10.1088/1755-1315/337/1/012004

2. Scientific work

Now there are three fundamental problems defining the relation of society to development of nuclear power and construction of facilities of this appointment as to potentially dangerous technology:

1) risk of severe accidents;

2) treatment of radioactive waste (including with fulfil nuclear fuel);

3) non-proliferation of the sharing materials (risk of global nuclear terrorism).

From the listed problems in public consciousness the problem of treatment of radioactive waste (RW) - the waste containing radioactive isotopes of chemical elements prevails. The nuclear power bears full responsibility for all the waste and completely pays works on their contents and utilization. Radioactive isotopes on the NPP are formed at division ^{235}U and ^{239}Pu and activation by neutrons and protons of constructional materials, impurity of the heat carrier, delay mechanism and fuel. Energy, a half-life period and a type of radiation are important characteristics of degree of danger of isotope [2]. One waste can be diluted and dissipate in the environment, others – to concentrate and be isolated, and the third – only to be isolated and be maintained in storages. Now in the Russian Federation the Federal target program for the address with RW adopted by the resolution of the Government of the Russian Federation in 1995. The program provides: equipment of all NPPs of Russia technical complexes for conditioning of all types of RW; reconstruction and construction on the NPP of storages of the firm and cured RW providing safe storage; development of technology and creation of the equipment for processing of the RW which are formed at a conclusion of operation of power units of the NPP and nuclear facilities; reconstruction acting and a construction in territories of the NPP of the new storages providing compact and safe storage of spent nuclear fuel during all term of operation of the concrete NPP. The processing purpose - extraction from the fulfilled fuel of the sharing nuclides formed during the operation of the reactor (for example, in fuel with burning out of 33 MW • days/kg mass contents ^{235}U makes about 0,83%, and the content of the sharing plutonium isotopes – about 0,68%). Besides, processing of the spent nuclear fuel (SNF) is the intermediate stage on the way of removal of highly active radionuclides from a field of activity of the person [3].

Allocation from SNF of long-living radionuclides allows to make processing products less dangerous and to considerably reduce their volume. As a result of processing receive new nuclear fuel that reduces the need for natural uranium and, besides, simplifies a problem of burial of the received RW as the volume of the cured highly active waste received after processing of SNF is only about 30% of initial volume of SNF [4].

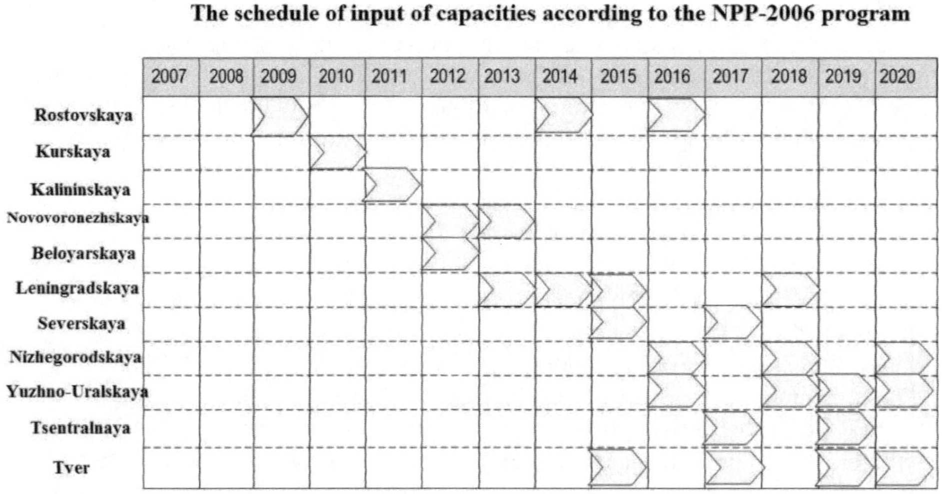

Figure 1. The schedule of input of capacities according to the NPP-2006 program.

Therefore, the tasks presented in a research need to be considered in more detail, including the problems described above.

The analysis of methods of processing and storage of radioactive waste of the NPP consists in the following examples: Kursk NPP-2. Let's consider the main space-planning solutions of the building of processing and storage of Russian joint stock company. The building consists their two main functionally various units and auxiliary facilities:

- block of processing of radioactive waste (BP);
- block of storage of radioactive waste (BS);

and auxiliary constructions:

- reception construction of diesel fuel;
- constructions of a tank of emergency discharge of diesel fuel from the building;
- air-channel [5].

The building of processing and storage of radioactive waste with the air-channel is structurally executed in the form of a monolithic box-shaped design with the external and internal monolithic bearing walls leaning on a base plate. The building consists of several sections of various number of storeys divided by deformation seams. Level of a clean floor of the first floor is taken for relative mark 0,000 that corresponds to absolute mark 158,70. The bearing framework of the building is accepted monolithic reinforced concrete. Need of monolithic reinforced concrete designs is caused by features of technology solutions and requirements of biological protection [6].

The spatial rigidity and stability of the building is provided:

- creation of uniform system, the monolithic reinforced concrete cross, longitudinal walls and monolithic reinforced concrete disks of overlapping which are rigidly connected among themselves;
- tough seal of walls in base plates through anchor cores;
- rigidity of a covering on steel farms, the provided system of communications on the lower and top belts of farms and also fastening of a flooring to runs.

Such constructive scheme of the building provides preservation it from the general progressing destruction at single local refusal of the bearing structural element (for example, parts of a wall). It is also necessary to note that the bearing elements of the constructive scheme have the increased reliability as calculated on extreme natural and emergency technogenic influences taking into account a number of conservative prerequisites:

- small probability of emergence – 10000 years for maximum settlement earthquake;
- use of conservative values of loadings from extreme and emergency influences;
- accounting of wide range of physicomechanical properties of the soil basis;
- use of conservative characteristics of properties of materials (both strength, and dissipative), etc [5].

Besides, in use buildings monitoring of a condition of building constructions (visual, tool) which with guarantee provides impossibility of sudden refusal of the bearing building construction is carried out.

In the building on processing and storage of radioactive waste the complex according to treatment of radioactive waste is placed. The block of processing of RW is intended for reception and processing at the modern level of liquid radioactive environments and waste, and solid radioactive waste (low-active and medium-active) for the purpose of reduction of volume of radioactive waste and their transfer to a form safe at storage, transportation and burial [6]. The block of storage of RW is intended for reception and storage firm and cured (low-active and medium-active) the waste which is formed in ten years of operation of two power units of the NPP (with a possibility of the subsequent expansion of volume of storage for 50 years of operation of nuclear power plant) and also the highly active waste which is formed in 60 years of operation of four power units of the NPP. A reception construction of diesel fuel - the separate construction intended for reception of a tanker truck with diesel fuel. A construction of a tank of emergency discharge of diesel fuel from the building - the buried construction intended for intake of diesel fuel from a tanker truck in the reception capacity and its giving in the system of burning and pyrolysis and return of diesel fuel to emergency capacity in case of accident [7].

IOP Conf. Series: Earth and Environmental Science **337** (2019) 012004 doi:10.1088/1755-1315/337/1/012004

Under the terms of responsibility for radiation safety on AE-5.6 PIN the building of processing and storage of RW belongs to the II category; a reception construction of diesel fuel - to the III category; a construction of a tank of emergency discharge of diesel fuel from the building of processing and storage of RW - to the III category [1].

Space-planning decisions are executed taking into account:

- modular principle of configuration of buildings and rooms;
- rational blocking of groups of rooms by the technological principle;
- climatic conditions of placement NPP;
- reductions of utilities of total structural volumes and building area;
- parameters of external natural and technogenic influences;
- implementation of health and fire-prevention requirements;
- reductions of terms of construction and labor costs;
- conveniences of the organization of construction;
- divisions of buildings and constructions into construction and technological modules from the point of view of their responsibility for safety;
- divisions of buildings and constructions into buildings of the main production and all-station buildings [8].

In a basis of the space-planning solution of the building of processing and storage of radioactive waste with the air-channel the principle of optimum use of volumes of buildings and constructions for ensuring technological processes and also the principle of safety according to normative documents is underlain.

Space-planning decisions are subordinated to purpose of the building [3]. The composite arrangement of buildings and constructions is defined taking into account compact use of the territory of an industrial site, technological communications, fire-prevention and health requirements, conditions of laying of utilities.

Foreign example is the Australian organization in the field of nuclear technologies - ANSTO (the Australian Nuclear Science and Technology Organization). She closely cooperates with the government and the public on the questions concerning use of nuclear technologies. ANSTO has wide experience in treatment of own radioactive waste which it is substantially generated in scientific research and in nuclear medicine. She also maintains close contact with the similar organizations abroad and is a permanent participant of the international forums according to treatment of radioactive waste [9].

ANSTO conducts extensive researches in the field of treatment of radioactive waste. One of developments of the organization is process of neutralization of radioactive waste (RW) with use of technology under the name isostatic pressing (hot-isostatic pressing HIP) under the ANSTO Synroc brand is hotter. Radioactive waste in combination with ceramic material compresses under the high pressure and temperature and hermetically is located in a special ceramic form and then in containers. This process is intended for safe encapsulation of waste on tens of thousands of years [5].

In December, 2009 the U.S. Department of Energy made the decision to use HIP technologies for processing and storage 4400m^3 highly active waste in National laboratory Idaho.

In Australia radioactive waste (RW) is formed as a result of use of radioactive materials in scientific research, the industry, agriculture and medicine [3].

The most part of development of RW falls to the share of ANSTO. Here waste of operation of the OPAL reactor (Open Pool Australian Lightwater), decommissioning of the Moata reactor and, in the future, decommissioning of HIFAR reactors (High Flux Australian Reactor) and OPAL enter. HIFAR was operated within several decades (till 2007) and supported researches and production of radioisotopes for medical and other purposes.

Natural radioactive materials (NORM) are one more source of radioactive waste, they are of particular importance for the mining industry. Also there is a large number of so-called "historical" waste in storages which resulted from activity of various organizations. Researches of radioactive ores, the conducted CSIRO (The Commonwealth Scientific and Industrial Research Organisation), or broad use of radioactive material of radium can serve for creation of the shining dials as examples [10].

So, comparing the Russian and foreign experience of the address of radioactive waste, it is possible to note that the volumes of waste accepted on storage in the countries of the world having nuclear heritage are tens of thousands of cubic meters a year. And volumes of the Russian storages are not comparable with foreign therefore Russia is on the threshold of the solution of problems – and, without having resolved an issue of storages, will not be able to cope with a problem of rehabilitation of the objects and territories polluted during the nuclear arms race. It is necessary to be ready to spend for researches in the field of the address about the RW of means comparable with costs of the leading nuclear countries of these purposes.

Therefore, relevant ways of storage and processing of radioactive waste of the NPP, creation of capacities for storage of the RW, the organizations having experience and the corresponding licenses for the right of treatment of radioactive waste is. It is necessary to be ready to spend for researches in the field of the address about the RW of means comparable with costs of the leading nuclear countries of the world of these purposes. It is also necessary to develop technologies of burning of low-active waste, vitrifications of the RW containing long-living and the sharing substances. It is necessary to increase by 3-4 times reception of RW by storage, and, as a result, volumes of storages. Only the specialized companies can solve these problems with highly skilled personnel. Their creation will allow to improve technologies and methods of the address with RW, will provide system approach to means of radiation and entrance control [2].

3. Conclusion

Proceeding from described above, one may say, that, having considered an object of a research of this work, design and construction of facilities of processing and storage of RW demands special approach and special measures of protection and safety. Let's note that construction of the NPP is the progressing industry, appears more and more power plants using nuclear fuel, so construction of facilities on service of objects of storage and processing of nuclear waste will also not stop but only will increase the need for the future.

The solution of research problems confirm relevance of the chosen subject, on the example of the analysis of methods of processing and storage of radioactive waste; comparisons of the Russian and international experience in the sphere of construction of these objects; offers on relevant ways of storage and processing of RW NPP.

It should be noted that these objects play a large role both in the ecological sphere, and in economic. On the one hand they protect the environment from pollution by radiation, with another – a possibility of recycling of RW not suitable for the NPP, but suitable for other industries, will allow to reduce costs of resources of fuel, and will also reduce the number of the buried RW that well affects ecology.

References

[1] Alekseenko V. A., Ivanov A.B. Geological and hydrological aspects of a problem of burial of liquid radioactive waste in ON "Beacon"//the III International symposium "The atomic Urals: science, industry, life". 4.11. Yekaterinburg, 1995. Page 16-19.
[2] Akhmedzyanov V.R., Lashchenova T. N. Treatment of radioactive waste//Energy, 2008
[3] Korchagin Yu.P. Decrease in receipts of radioactive waste on the NPP and modern technologies of their processing / Yu.P. Korchagin//Treatment of radioactive waste. M.: ENITs All-Russian Scientifically Research Institute NPP, 2002. Page 69.
[4] Skachek M.A. Treatment of spent nuclear fuel and radioactive waste of the NPP: manual for higher education institutions/M. A. Skachek. — M.: MEI publishing house, 2007. — 448 with: silt.
[5] Naumov N.A. The relation of the population of Russia to nuclear power//Bulletin TsOI on NP. 1994. N 7-8. Page 40-44.
[6] Polivoda A.K. The events held by Gosatom of Ukraine on creation of a complex of technological

processes according to treatment of SNF, processing and long storage of RW / A.K. Polivoda//Appendix 2.3 to the Protocol of the fourth meeting of the Working group on removal of the NPP from operation. M.: Interatomenergo, 1996.

[7] Technology and the equipment on the basis of an induction melter with a cold crucible for processing of radioactive metal junks / V.G. Pastushkov, A.V. Molchanov, V.P. Serebryakov, etc.//Treatment of radioactive waste. M.: ENITs All-Russian Scientifically Research Institute NPP, 2002. Page 99.

[8] Egorov N.N. The organization of works on treatment of radioactive waste in Russia//Bulletin TsOI on NP. 1993. N 5. Page 22-26.

[9] Bushuyev N.I. History and technology of nuclear power//Ministry science and education of Russion Federations, Moscow state building university Moscow, 2015. Page 175-192

[10] Kuznetsov SB. Treatment of radioactive waste on the Kursk NPP / SB. Kuznetsov //Treatment of radioactive waste. M.: ENITs All-Russian Scientifically Research Institute NPP, 2002. Page 32.

IOP Conf. Series: Earth and Environmental Science **337** (2019) 012005 doi:10.1088/1755-1315/337/1/012005

The relevance of re-equipment of thermal power plants under solid waste fuel

A Subbotin[1*], I Trufilov[1] and D Vovkogon[1]

[1]Moscow State University of Civil Engineering, Yaroslavskoe shosse, 26, Moscow, 129337, Russia

* Corresponding author: subbotin-art@mail.ru

Abstract. The use of thermal power plants (TPPs) as energy facilities in world practice is the most popular way to generate electricity. Traditional fuel for these objects is used in the form of coal and gas. Also, in world practice, other types of fuel are used for TPPs, namely, municipal solid waste (MSW). The relevance of the research is the proposal for reliable and technological re-equipment of thermal power plants, thereby solving the problem of the environmental situation in the near radius of the construction of thermal power plants for the disposal of solid waste. The object of research is thermal power plants and their ability to re-equip. The main purpose of the research is to analyze the world and domestic experience in the implementation of solid waste as a fuel for thermal power plants and to offer current technological solutions.

1. Introduction

Given the large number of existing thermal power plants in the world, which currently operate on traditional fuel in the form of coal and gas, as well as a limited supply of these resources, it is necessary to note the relevance in choosing another type of fuel for these facilities that could meet the increased safety requirements for the environment. This type of fuel is municipal solid waste, the disposal of which in this way will help solve the problem of landfill reloading. It should be noted that the importance of processing these wastes on a global scale is increasing every year. Every year, only in Russia produces more than 60 million MSW, which, in terms of one person, is about 400 kg [1, 2]. Only about 7–8% of collected MSW is drawn into economic circulation, and only 2% is burned. The low percentage of recycling (use) of solid waste in Russia is associated with insufficient infrastructure development: currently, there are 243 waste disposal complexes, 53 waste sorting complexes, and about 40 incineration plants [3].

Consequently, the object of study is the power stations and their ability to re-equip.

The main tasks solved in the study are:

- to analyze the world and domestic experience in the implementation of solid waste as a fuel for thermal power plants;
- propose current technological solutions in terms of re-equipment of TPPs;
- develop a system of reliable, safe and technological re-equipment of thermal power plants.

On the other hand, the task of developing technical re-equipment and reconstruction of existing thermal generation power plants with the transition to municipal solid waste is to increase the level of environmental protection the environment in terms of waste management.

IOP Conf. Series: Earth and Environmental Science **337** (2019) 012005 doi:10.1088/1755-1315/337/1/012005

2. Scientific work

Considering the re-equipment of TPPs for fuel from MSW, first of all, it is necessary to consider MSW as a new type of fuel.

MSW is a waste from human life, namely food and vegetable waste, paper, magazines, wood, plastics, textiles, glass and various metals. Separate collection, burial and incineration of waste is applied to successfully combat household waste. Today in Russia the vast majority of waste is not sorted. Waste after entering the garbage containers are sent to specially designated landfills, landfills, where they are stored [4]. For example in Russian Federation, Moscow and the Moscow Region, there are more than 50 polygons, 27 of which have exhausted their capacity, and the other 19 are filled to 90%. Dumps of the same garbage more than 1000. Further development of this problem can lead to catastrophic consequences, because the disposal of solid waste causes pollution of the land, pollution of the upper aquifers with toxic substances contained in the filtrate generated at landfills. The release of large volumes of biogas, mainly methane and carbon dioxide, which leads to climate change, as well as to spontaneous combustion and combustion of landfills with garbage [5]. All of that has a negative impact on the environmental situation and the health of the population in nearby communities. In addition, in accordance with the draft federal law No 584399-5 "On Amendments to the Federal Law "On Production and Consumption Wastes", waste reception at landfills will be limited [6]. As a result, it is worth reviewing this technology of waste disposal in favor of more environmentally friendly and innovative methods. One such method is incineration. Accordingly, the most acceptable option may be a thermal power plant for municipal solid waste [7].

Thermal power station (TPP) - a station, a set of installations and equipment necessary for the production of heat and electricity. Subdivided by type of energy produced at condensing power plants. (electrical energy), thermal power plants (both electrical and thermal), CHP (mainly thermal energy). Currently, power plants operate on two types of fuel: coal and gas. In densely populated areas, far from coal deposits, most power plants switched to gas. This is due to the cost of transporting coal and the non-environmentally friendly generation of energy for this type of fuel. Gas power plants are eco-friendly and compact. Fuel flows continuously and in all weather conditions. In this situation - this is the best solution for large cities [8]. But using traditional fuels does not solve urgent environmental problems, because there is a tendency to increase the amount of household waste. In world practice, more than ten technologies of incineration are implemented [9-14]. Important factors when choosing a particular technology is the station's self-sufficiency and the ability to supply electricity and heat to consumers, as well as the fuel itself. For Russian conditions (high humidity of MSW, high content of inert and environmentally hazardous components in MSW), it is preferable to use the technology of layered burning on an inclined-pushing grate.

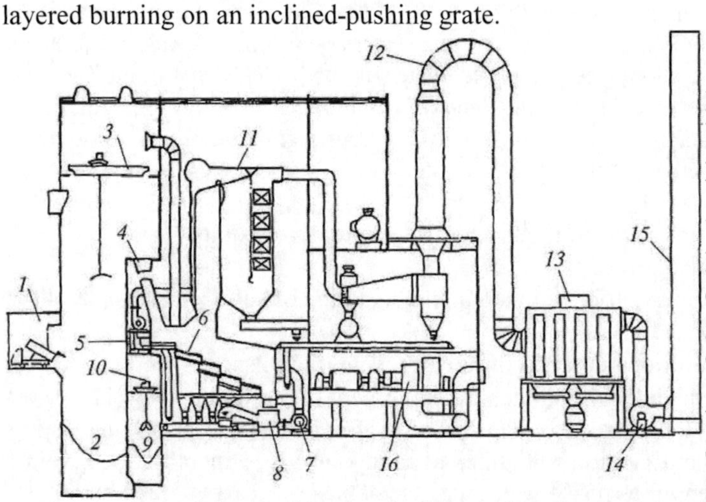

Figure 1. Schematic diagram of layered burning on an oblique-pushing grate.
1 - Tipping hall; 2 - Waste pit; *3* - Waste clamshell crane; *4* - Feed hopper; 5 - Ram feeder; 6 - Pan-dragging grate;7 - Secondary air/flue gas recirculation fan; *8* - Ash conveying system; 9 - Storage of ash; *10* - clamshell crane; *11* - heat-recovery boiler waste heat; *12, 13*- System of cleaning; *14* - Inducted draft fan; *15* - Stack; *16* – Turbogenerator.

Unsorted municipal solid waste is delivered to the receiving compartment using garbage trucks, then they are discharged into a temporary storage facility — an in-depth bunker served by a grapple crane

IOP Conf. Series: Earth and Environmental Science **337** (2019) 012005 doi:10.1088/1755-1315/337/1/012005

(several cranes are usually installed depending on the needs of the units). With the help of a grab crane, waste is moved from the bunker to incineration. At this stage, all oversized garbage is also removed. From the bunker MSW is fed into the feed hopper, which separates the combustion chamber from the environment and prevents unorganized air flow into the furnace. A locking device is installed on the lower part of the receiving funnel - a valve that performs protective functions in the event of an accident. From the receiving funnel, MSW is sent to the firebox to the grate, usually using a hydraulic pusher. The waste incineration temperature is 850 ° C. In the combustion process, slags are formed, which through cooled water fall into a hydraulic slag remover with a pusher device. Under the grate along its entire length, a scraper device is installed to collect spills between the grates. It is important that gas purification and waste incineration technologies are observed, as a result of which they reach the limiting concentrations of organic carbon, carbon oxides, and also perform the process of thermal decomposition of furans and dioxides [15].

Due to the reduced rates of electricity generation during the burning of solid waste, waste sorting and pre-crushing is used to increase the efficiency of fuel of a thermal power plant. Under such conditions, the calorific value (heat engineering indicators of fuel) increases. In addition, it becomes possible to select valuable raw materials for its recycling or composting. A significant increase in the efficiency of using MSW as a fuel for TPPs is achieved by only partial replacement of traditional fuel (coal, gas) with alternative (MSW) [16].

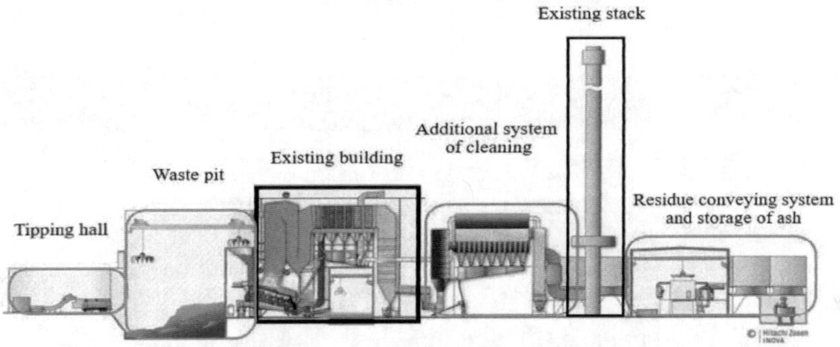

Figure 2. Scheme of re-equipment of the station for fuel from MSW.

Accordingly, in order to effectively implement the re-equipment of TPPs for fuel from MSW, it is necessary to develop a combined layout of TPPs for incineration of both traditional (coal, gas) and MSW, and to have an established waste sorting system [17]. In the Russian Federation, the process of sorting garbage is poorly developed, it is advisable to build a sorting complex directly next to the station. In addition, there must be a waste reception complex. Depending on the type of fuel burned at the station being reconstructed, additional equipment should be used: front furnace for burning solid waste (fuel - coal); installation for gasification of MSW, treatment facilities obtained product gas for subsequent combustion in the boiler furnaces (fuel - natural gas). The refurbished station should include a multi-stage gas cleaning system, ash and slag removal system to prevent the harmful effects on the environment of harmful impurities.

When re-equipping an existing station, there is no need to replace power lines and steam power plants. In addition, unlike other types of fuel, MSW is accepted with a cash surcharge. When re-equipping, it is necessary to make examinations and diagnostics of structures, structures, equipment in order to obtain optimal conditions for generating energy and increasing the reliability of the plant as a whole. In case of detection of defects and damages, it is necessary to work to eliminate them; decommission obsolete and outdated equipment from service. The operating experience of numerous foreign waste recycling stations shows that thermal power plants equipped with modern equipment are environmentally safe, the concentration of emissions of harmful substances from the station itself does not exceed the standard values. For example, in Europe, 28 billion kW / h of electricity and about 70 billion kW / h of thermal energy is already being produced, which prevents emissions of harmful

IOP Conf. Series: Earth and Environmental Science **337** (2019) 012005 doi:10.1088/1755-1315/337/1/012005

substances that could be released during landfill, and also provides significant savings for other types of fuel [18].

In assessing the current situation, we will look at the solution of the key tasks of re-equipping thermal power plants for solid waste fuel. After analyzing the world and domestic experience in the implementation of solid waste as a fuel for thermal power plants, it was revealed that there are problems with the development of infrastructure for the collection and sorting of garbage. And because of this, we do not use this resource again. Based on the above conditions, the use of the combined energy generation scheme is the most rational and efficient. Also, based on the result of the study, a system of reliable, safe and technological re-equipment of TPPs has been developed:

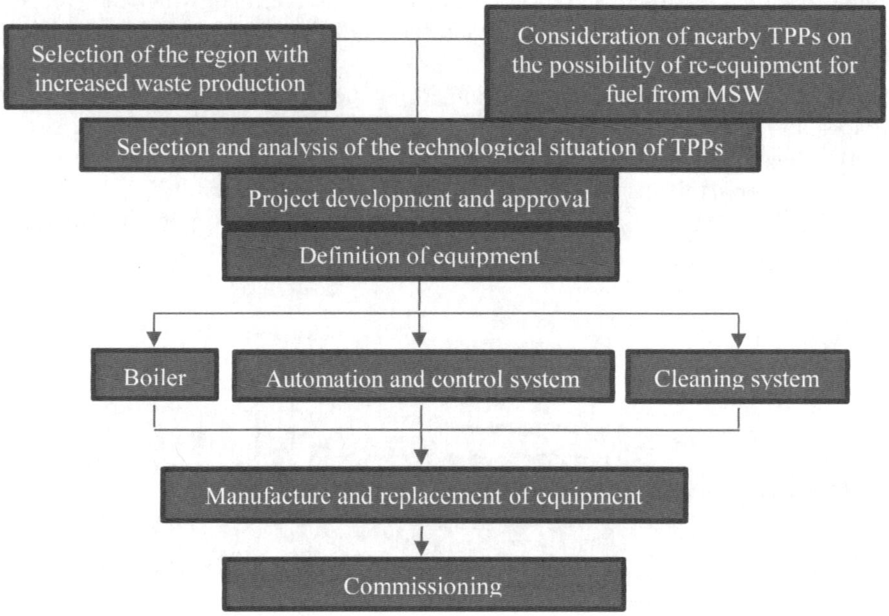

Figure 3. Scheme of the system of reliable, safe and technological re-equipment of TPPs

This scheme demonstrates the sequence of operations for the reconstruction of thermal power plants for fuel from MSW, and also helps to retool effectively and safely, which is the main task of developing thermal generation.

3. Conclusion

The existing waste management system not only in Russia but throughout the world, focused on their disposal, is imperfect and requires modernization, as it leads to environmental pollution. Use of sorted MSW as fuel for TPPs with a combined plant layout is extremely efficient. Incineration of waste will reduce their volume, and, accordingly, their impact on the environment. In addition, MSW is an economically beneficial renewable fuel for thermal power plants, and with an increase in the number of sorting complexes, the relevance of re-equipment of thermal power plants will increase. Stations on this type of fuel have advantages over other technologies for waste disposal: environmental friendliness and efficiency. An analysis of the situation suggests that such stations will become real in Russia in the near future, but they are connected with the adoption of laws aimed at reducing landfill sites.

The tasks we set were completed. An analysis of world and domestic experience in the implementation of solid waste was carried out, and a scheme of reliable technological re-equipment of thermal power plants for Russian conditions was proposed.

References

[1] Kuznetsov V.S. Creating an optimal scheme for the collection and recycling of municipal solid

waste in the city. Report for the Moscow City Duma. M.,(2012)
[2] Asadpoori A, Ankomah C, Asadpoori A, Derevianko O and Shaburov E 2018 Sustained municipal waste management models in Russian megapolicies through utilizing waste-to-energy technologies ed A Mottaeva and B Melović *MATEC Web Conf.* **193** 02039
[3] Comprehensive strategy for the management of municipal solid waste in the Russian Federation (approved by order of the Ministry of Natural Resources and Environment on August 14, 2013 № 298
[4] Tugov A.N., Moskvichev V.F. Use of the energy potential of municipal solid waste in Russia and abroad // Energy: economy, technology, ecology. 2011. №8. P. 32-42
[5] Najman S.M.,Gazeev A.N. Thermal methods for the use of municipal solid waste: a training manual. Kazan: publishing house of Kazan State Technical University (2012)
[6] Aleksashina V.V. Ecology of the city. Incinerators
[7] Tugov A.N. Prospects for the use of solid waste as a secondary energy resources // Power Thermal power generation. 2013. №9. P. 38-48
[8] Akhmetova I G and Chichirova N D 2016 Application of SAW method for multiple-criteria comparative analysis of the reliability of heat supply organizations *Therm. Eng.* **63** 1016–24
[9] Sultanguzin I A, Fedyukhin A V., Kurzanov S Y, Gyulmaliev A M, Stepanova T A, Tumanovsky V A and Titov D P 2015 Prospects for the development of independent power supply systems on the basis of solid fuel thermal conversion technology *Therm. Eng.* **62** 359–64
[10] Fedyukhin A V., Sultanguzin I A, Stepanova T A, Voloshenko E V., Kurzanov S Y and Isaev M V. 2013 Improving solid-fuel pyrolysis and gasification for effective cogeneration *Coke Chem.* **56** 302–6
[11] Fedyukhin A, Sultanguzin I, Gyul'Maliev A and Sergeev V 2017 Biomass pyrolysis and gasification comprehensive modeling for effective power generation at combined cycle power plant *Eurasian Chem. J.* **19** 245–53
[12] Sergeev V V. and Aleshina A S 2011 A gas-generator combined-cycle plant equipped with a high-head heat-recovery boiler *Therm. Eng.* **58** 268–70
[13] Bazarnova J G, Politayeva N A and Bozhuk S G 2017 Innovative technologies secondary use of processed active source Proceedings of the 2017 International Conference "Quality Management, Transport and Information Security, Information Technologies", IT and QM and IS 2017 (IEEE) pp 471–6
[14] Sobgaida N A, Olshanskaya L N and Solodkova A B 2011 Ecological problems of utilization of waste activated sludge of saratov refinery and ways to solve them *Chem. Pet. Eng.* **47** 198–203
[15] Alekseenko S.V.,Basin A.S. Universal technology for the use of municipal solid waste as an unconventional fuel // Energy Saving. 2004. №4. P.42-50
[16] Armisheva G.T., Korotaev V.N., Krivoshein V.G. Reducing the environmental burden in the handling of municipal solid waste through the use of combustible components // Scientific Research and Innovation. 2010. T.4, № 3
[17] Tugov A.N., Tumanovsky A.G., Moskvichev V.F. Experience RTEU in the incineration of solid waste in layer firebox. Report 8th All-Russian Conference with International Participation "Solid Fuel Combustion", November 13-16, 2012
[18] Tugov A.N., Moskvichev V.F. Energy from the garbage can // Energy efficiency and energy saving. 2012. № 4

IOP Conf. Series: Earth and Environmental Science **337** (2019) 012006 doi:10.1088/1755-1315/337/1/012006

Phototrophic microorganisms biomass production with joint utilization of city surface water

K Gorin[1*], V Pojidaev[1], A Borgolov[1] and Y Sergeeva[1]

[1] Biotechnology and Bioenergy department, National Research Centre "Kurchatov Institute", Moscow

* Corresponding author: gkvbio@gmail.com

Abstract. Today surface water especially rain and thaw water became one of the environmental challenges for the cities. Those waters can be contaminated by different impurities including organics, oil products, heavy metals and etc. Utilization and reusing of those types surface water is the point of interest of many city agglomerations. In this paper preliminary results of *Euglena gracilis* CCAP 1224/5Z cultivation at highly organic contaminated surface water are presented. Cultivation was provided at illumination from 500 – 1200 lux depending on natural illumination in Moscow region. It was shown that during experiments that total organic carbon (TOC) content decreased from 41 mg/l to 15 mg/l in five days cultivation period. Final biomass concentration in case of cultivation on waste water exceeded concentration in the control flasks. Also the lipids content has been determined. In the biomass from flasks with surface water lipids content was from 24% to 32% to dry weight, depending on illumination and ambient temperature. Control biomass contains approximately from 28% to 30% in all experiments. Experiments with low TOC close to 10 mg/l showing close results – decreasing of TOC approximately at 50-60% in all flasks. Thereby it was demonstrated that it is possible in Moscow climate conditions during spring – autumn period to apply phototrophic microorganism *E. gracilis* strain for surface water treatment and in this case total lipids content a stay the same as in control cultivations.

1. Introduction

Population growth in urban areas is one of the most significant trends of the 21st century, affecting global economic development, energy consumption, use of natural resources and human well-being. Globally, 3.6 billion people live in urban areas. All of these city dwellers need fresh water, but surprisingly little is known about where large cities get water or the effects of this infrastructure on the global hydrological cycle [1]. The concentration of substances and energy within cities is an inevitable consequence of the dense settlement and the residents' needs for food, clothing and fuel, which in turn greatly affects the distribution of nutrients on the planet [2]. Changes in chemical composition depend on various water sources. Urban runoff is considered to be the main reason responsible for the deterioration of water quality in the catchment areas [3]. Changes in the catchment hydrology associated with urban development also cause canal expansions and purification, as well as the introduction of more suspended and dissolved sediments, such as heavy metals, into water bodies [4].

Despite the fact that the commercial use of microalgae for both large-scale production of biomass, such crops as *Chlorella* and *Dunaliella*, and for wastewater treatment, has about 80 years [5] currently, technologies based on phototrophic microorganisms are only beginning to play a significant role both in the industry of various bioproducts (from fertilizers and biofuels to substances for cosmetics and

IOP Conf. Series: Earth and Environmental Science **337** (2019) 012006 doi:10.1088/1755-1315/337/1/012006

pharmacy), and as self-sufficient systems for normalizing the state of the environment. At the same time, considerable attention is being paid to the introduction of photobioreactors in the urban economy [6, 7].

In this case, microalgae are very attractive object, as they can grow not only in fresh and salt water, but also in the wastewater, using the necessary nutrients for growth, thereby purifying it [8].

There are several concepts of today city water management that can be basis for further integration of phototrophic microorganisms technology [9, 10]. In this regard, great attention is paid to the development of criteria for the compliance of wastewater that can be returned to circulation, for the cultivation of microalgae, as well as the quality of biomass produced and possibility of further biomass processing with receiving of various products [11].

The paper considers the possibility of obtaining biomass of *Euglena* on various types of effluent (supernatant fluid wastewater, suspended solid free-wastewater and suspended solid and microorganisms-free wastewater). The presence of solid suspended particles had a greater negative impact on growth than the presence of microorganisms, therefore, for effective management of the wastewater treatment process, it is necessary to remove solid suspended particles first [12]. Sewage treatment using *Euglena* and obtaining potential raw materials for the production of biofuels: after 8 days of cultivation, the content of ammonium nitrogen decreased by 98%, total nitrogen by 93%, orthophosphate by 85%, 66% of total phosphate and 92% of total organic carbon [13].

In the case of the employing of microalgae cultivation for surface water treatment has another advantage compared with other methods - it is an effective utilization of carbon dioxide, that can be a good point to the zero-carbon discharge city development [14].

So, based on mentioned above information it is possible to say that phototrophic microorganisms can became a good option for treatment of different types of city waste waters. In current work we presented results of highly contaminated by organics surface water treatment by using *E. gracilis* CCAP 1224 / 5Z strain with focus on middle-Russian climate conditions.

2. Materials and methods

2.1. Microalgae strain, pre-culture cultivation condition

The culture of *Euglena gracilis* CCAP 1224/5Z from the Culture Collection of Algae and Protozoa (Scotland) was used in the experiments. The microalgae strain were cultured in 250 ml Erlenmeyer flask with 100 ml modified HUT nutrient media [15] at 20 ± 1°C. Artificial illumination was provided using 20 W daylight normalized fluorescent lamps, whose emission spectrum was in the range of 400–700 nm and a light intensity of approximately 60 μmol PAR m^{-2} s^{-2} under 14:10 (light:dark) photoperiod on a laboratory shaker at 100 rpm.

The composition of HUT nutrient media was as follows mg/l: 20 KH_2PO_4, 288.8 NH4Cl, 25 $MgSO_4{\cdot}7H_2O$, 400 sodium acetate, 40 Yeast extract and 1 ml of stock solution Pfennig microelements [16]. The composition of stock solution microelements was as follows mg/l: 5000 EDTA, 2000 $FeSO_4{\cdot}7H_2O$, 100 $ZnSO_4{\cdot}7H_2O$, 30 $MnCl_2$, 300 H_3BO_3, 200 $CaCl_2{\cdot}6H_2O$, 10 $CuCl_2$, 20 $NiCl_2{\cdot}2H_2O$, 20 $Na_2MoO_4{\cdot}2H_2O$.

2.2. Experiment conditions

Water obtaining from natural water source – "Isetskoe water storage reservoir" (Russia, Sverdlovsk city).

All experiments have been conducted in Erlenmeyer flasks (500 mL) with 300 ml under constant temperature 20 ± 1 °C. Artificial illumination was provided using 20 W daylight normalized fluorescent lamps, whose emission spectrum was in the range of 400–700 nm and a light intensity of approximately approximately 60 μmol PAR $m^{-2}s^{-2}$ under 14:10 (light:dark) photoperiod on a laboratory shaker at 100 rpm.

The composition of nutrient media for control was a modified BBM nutrient media [17] as follows g/l: $NaNO_3$ – 1,25, $MgSO_4$ – 1, KH_2PO_4 – 1,25 and 1 ml of Pfennig stock solution of microelements.

2.3. Extraction of lipids from microalgae biomass and obtaining of fatty acid methyl esters (FAME)
The samples were centrifuged for 10 min at 4000 rpm (centrifuge Awel 20 mf) to sediment biomass. When biomass was freezing at -70°C and next lyophilization (Labconco free zone 2.5) of this biomass were provided.

Lipids from algae dry biomass were extracted with chloroform – methanol solution (2:1) using Bligh and Dyer method. [18] FAMEs were produced by direct methanolysis according to the method described previously [19].

2.4. GLC analysis of FAME
Qualitative analysis of FAMEs was performed with a Bruker 340 GC gas chromatograph equipped with a flame ionization detector and SelectTMBiodiesel for a FAME quartz capillary column (30 m × 0.32 mm × 0.25 μm) under the following linear temperature programming: 190°C (3 min), a temperature increase of 3°C/min, 230°C (5 min). The temperatures of the injection system and detector were 250°C, the linear velocity of the carrier gas (nitrogen) was 20 cm/sec, and the split ratio was 1 : 20. The volume of the injected sample was equal to 1 μL. Fatty acids were identified by means via comparison of the retention time of the analyzed sample with that of a standard sample (FAME mix Supelco–37, Supelco, United States). The content of the individual fatty acids was determined with an internal standard ($C_{17:0}$) [20].

2.5. Determination of total organic and organic dissolved carbon
Determination of total organic carbon (TOC) and dissolved organic carbon (DOC) carried out by using national standard PND F 14.1:2:3:4.279-14 " Method for the determination of organic carbon and total nitrogen in drinking, natural and waste waters by the method of high-temperature oxidation using carbon and nitrogen analyzers".

2.6. Growth monitoring
Growth was monitored by measuring of dry biomass weight. Dry weight was measured according to the APHA method 8111G, [21] with a modified drying temperature (105 °C) and filter (Whatman GF/ C, 1.2 μm).

3. Results and Discussion
Algae growth was measured in terms of increase in biomass concentration (g L^{-1}) and is shown in Fig. 1. Initially two samples of surface water and control were inoculated with the same initial cellular concentration of *E. gracilis*. There was mild initial lag phase observed during the growth for 1 day of *Euglena*. Cultivation in the both cases of control and surface water, from the first day there is a gradual even growth. With the same starting density of 0.033 g L^{-1}, the dry biomass cell weight of the algae cells went up to 0.35 g L^{-1} and 0.37 g L^{-1} on the 5th day for cultivation on control medium and surface water, respectively.

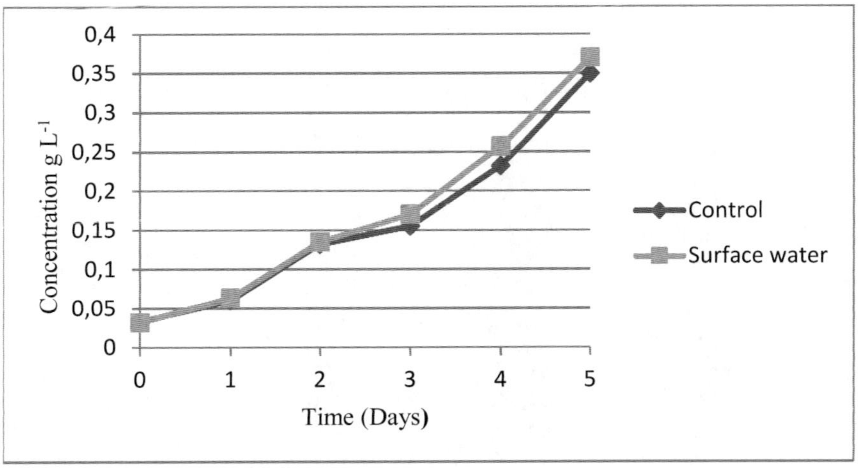

Figure 1. Growth curve of *Euglena* cultivation on control medium and surface water.

Total organic carbon is one of the most important total parameters characterizing organic water pollution. This parameter has its exact definition and is an absolute value representing the content of all organic compounds. In this work surface water was used in which the main part (about 90%) DOC is constituted of TOC. During the experiments investigation photo-heterotrophic cultivation of *E. gracilis* was performed in order to reduce organic impurities and results shown in Fig 2. After cultivation of microalgae the contents of TOC and DOC is decrease for 63.4% and 64.8% respectively. In research high efficiency (about 90%) of reducing TOC for 6 days by cultivating *Euglena* in wastewater is shown [13].

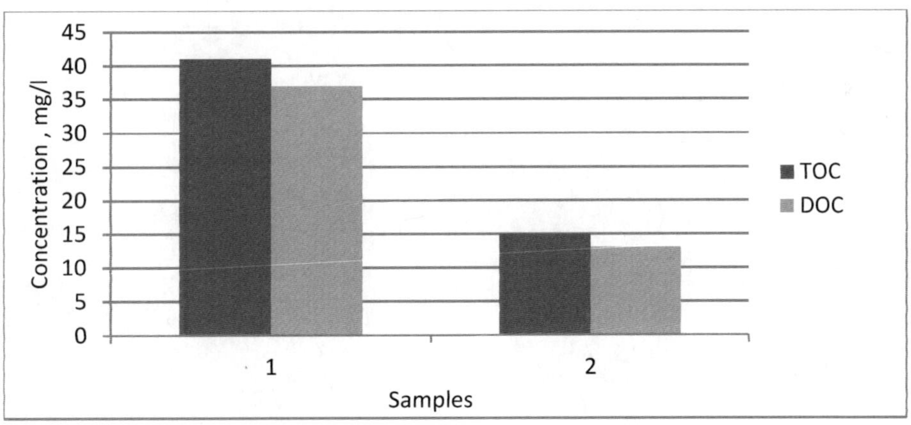

Figure 2. Contents of TOC and DOC in samples: 1 – before and 2 – after cultivation.

As known, the amount and composition of lipids in a cell is affected by the state of the culture [22]. The total lipid content of the samples is presented in Fig. 3. In the biomass from flasks with surface water lipids content was from 24% to 32% to dry weight biomass, depending on illumination and ambient temperature. Control biomass contains approximately from 28% to 30% to dry weight biomass in all experiments. It should be noted the same result 27,2% of total lipid content is was noted earlier by the heterotrophic cultivation of *E. gracilis*, [23].

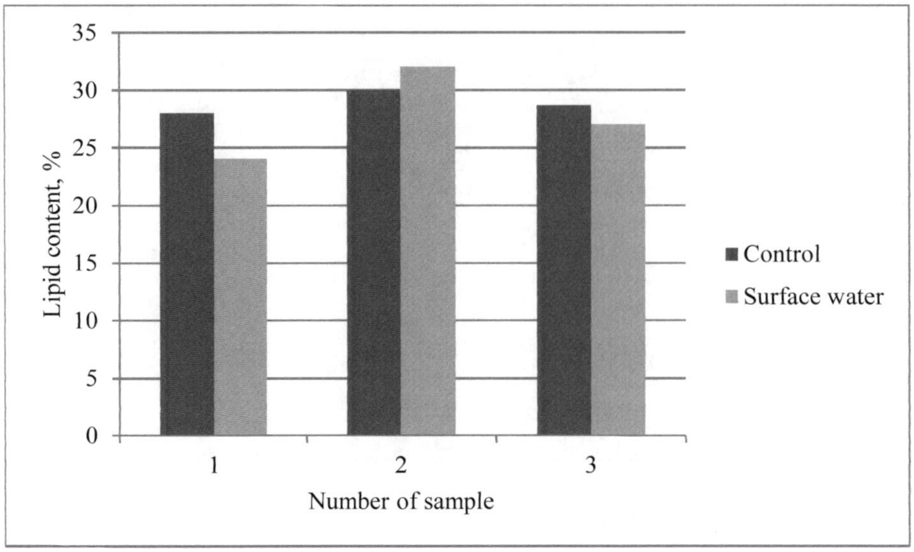

Figure 3. Lipid content by cultivation on control medium and surface water.

The quality of biofuels directly depends on the composition of fatty acids (FA), the main factors are the chain length and the degree of unsaturation. The results of determining the composition of fatty acids are presented in Table 1.

Table 1. Percentage composition of FAME: 1 – cultivation on control medium and 2 – cultivation on surface water.

FAME composition	1	2
$C_{16:0}$	16,19	14,27
$C_{17:0}$	18,06	7,34
$C_{18:0}$	3,32	6,73
$C_{18:1}$	6,06	5,40
$C_{18:2}$	14,75	21,07
$C_{18:3}$	38,49	33,14
$C_{20:0}$	3,14	12,05
	100	100
Saturated FA	40,70	40,39
Unsaturated, including:	59,30	59,61
Monoenoic FA	6,06	5,40
Polyenoic FA	53,24	54,21

The composition of lipids was identified 8 fatty acids. C_{18} FA: stearic acid ($C_{18:0}$), oleic acid ($C_{18:1}$), linoleic acid ($C_{18:2}$) and linolenic acid ($C_{18:3}$) prevailed in both the control and the experimental variants, with more than 50% being polyunsaturated fatty acids (PUFA): linoleic and linolenic. However, the relative content of these acids in the control and experimental variants differ. So, the $C_{18:3}$ /$C_{18:2}$ ratio in the control was 2.61, and in the experimental variant - 1.57, which indicates a decrease in the activity of the desaturation processes in alga cells. Interesting is the fact that the content of saturated fatty acids in both variants is almost the same: 40.70% in the control and 40.39% in the test sample, while the content of margarine ($C_{17:0}$) acid in the experimental variant is almost 2.5 times lower than in the control, and the content of arachic acid ($C_{20:0}$) exceeded almost three times (12.05% and 3.14%, respectively), which may indicate more active elongation processes under the conditions of cultivation of the test sample.

European standards for biodiesel spelled out clear restrictions on the content of PUFAs, the share of which should not exceed 12% for linolenic acid, 1% with 4th and more double bonds. In our study, it was established that the content of linolenic acid in the amount of fatty acids of the lipids of the biomass

of *E. gracilis* exceeded 30%. Similar results on the prevalence of linolenic acid in the cultivation of various microalgae in urban wastewater are presented in [24-26].
This high content of PUFAs causes a high instability of biofuels during storage and, undoubtedly, partial catalytic hydrogenation is necessary to increase its stability when used further as biodiesel [27].

4. Conclusions

- Demonstrated that it is possible in Moscow climate conditions during spring – autumn period to apply *E. gracilis* microalgae for surface water treatment and assess the TOC removal capability: after cultivation of microalgae the contents of TOC was decrease more than 60%;
- Maximum cell biomass after cultivation on a surface medium a bit more than in control medium;
- Evaluate the total lipid content and composition of FA profile of samples are determinates. For biomass samples growing on surface water and control medium content of lipids a stay the same and compose about 30%, but in the fa profile there are differences.

Acknowledgements

This work was made with financial support of the Ministry of Science and Higher Education of the Russian Federation. The number of agreement is No. 14.616.21.0079. Unique identification number of applied research is RFMEFI61617X0079

References

[1] McDonald R I, Weber K, Padowski J, Flörke M, Schneider C, Green P A, Gleeson T, Eckman S, Lehner B, Balk D, Boucher T, Grill G, Montgomery M 2014 *Glob. Environ. Chang.* **27** pp 96–105
[2] Bernhardt E S, Band L E, Walsh C J, Berke P E 2008 *Ann. N. Y. Acad. Sci.* **1134** pp 61–96
[3] Ferreira C S S, Soares D, Ferreira A J D, Costa M L, Steenhuis T S, Coelho C O A 2012 *Geophys. Res. Abstr.* **14** p 3621
[4] Utomo H D, Tan K X D, Choong Z Y D, Yu J J, Ong J J, Lim Z B 2016 *J. Environ. Prot. (Irvine,. Calif)* **07** pp 1547–60
[5] Abdel-Raouf N, Al-Homaidan A A, Ibraheem I B M 2012 *Saudi J. Biol. Sci.* **19** pp 257–75
[6] Elrayies G M 2018 *Renew. Sustain. Energy Rev.* **81** pp 1175–91
[7] Gotovtsev P M, Dyakov A V 2016 *IEEE 3rd World Forum on Internet of Things (WF-IoT)* pp 542–6
[8] Mohammady N G, El-Sayed H S, Taha H M, Fakhry E M, Mahmoud N H, Mohamed J H et al 2015 *Part A. Int J TechnoChem Res.* **3** p 144
[9] Liu L, Jensen M B 2018 *Cities* **74** pp 126–33
[10] Lefebvre O 2018 *Curr. Opin. Environ. Sci. Heal.* **2** pp 26–31
[11] Al-Gheethi A A S, Noman E A, Mohamed R M S R, Apandi N M, Yaakob M A, Pahazri F, Kassim A H M *Management of Greywater in Developing Countries* (Springer, to be published)
[12] Torihara K, Kishimoto N 2015 *J. Water Environ. Technol.* **13** pp 195–205
[13] Mahapatra D M, Chanakya H N, Ramachandra T V 2013 *J. Appl. Phycol.* **25** pp 855–65
[14] Gorin K V, Badranova G U, Gotovtsev P M, Shatalova A Y, Grigoriev T E, Krasheninnikov S V, Tihomirov S A, Kondratev O A, Vishnevskaya M V, Vasilov R G 2018 *IOP Conf. Ser. Mater. Sci. Eng.* **292**
[15] Kawachi M, Ishimoto M, Mori F, Yumoto K, Sato M, Noël H 2013 *MCC-NIES List of Strains, 9th ed.* (Tsukuba Microbial Culture Collection at National Institute for Environmental Studies)
[16] Pfennig N, Lippert K D 1966 *Arch. Mikrobiol.* **55** pp 245-56
[17] Yeh K-L, Chang J-S 2012 *Biores. Technol.* **105** pp 120-7
[18] Bligh E G, Dyer W J 1959 *Can. J. Biochem.Physiol.* **37** pp 911-7
[19] Liu J, Liu Y, Wang H, Xue S 2015 *Biores. Technol.* **176** pp 284–7
[20] Pozhidaev V M, Sergeeva Y E, Gorin K V, Badranova G U, Gotovtsev P M, Borgolov A V,

Vasilov R G 2016 *J. of Analytic.Chem.* **71** 624-9
[21] A.P.H.A. APHA, A.W. Works 2005 *Standard Methods for the Examination of Water and Wastewater*, (Washington, D.C.: APHA-AWWA-WEF) **3**
[22] Solovchenko A E, Khozin-Goldberg I, Didi-cohen S, Cohen Z, Merzlyak M N 2008 *J. Appl. Phycol.* **20** pp 245–51
[23] Constantopoulos G, Bloch K 1967 *J. Biol. Chem.* **242** pp 3538–42
[24] Sturm B S M, Peltier E, Smith V , Denoyelles F 2012 *Environ. Prog. Sustain. Energy* **31** pp 10–6
[25] D'Oca M G M, Viegas C V, Lemoes J S, Miyasaki E K, Moron-Villarreyes J A, Primel E G, Abreu P C 2011 *Biomass Bioenerg.* **35** 1533–8
[26] Hena S, Znad H, Heong K T, Judd S 2018 *Water Research.* **128** 267-77
[27] Sergeeva Y E, Mostova E B, Gorin K V, Komova A V, Konova I A, Pojidaev V M, Gotovtsev P M, Vasilov R G, Sineoky S P 2017 *Appl. Biochem. Microbiol.* **53** pp 807–13

Comparison of Elemental Analysis Techniques for Fly Ash from Municipal Solid Waste Incineration using X-rays and Electron Beams

S An[1,*], B Norlin[1], M Hummelgård[2] and G Thungström[1]

[1]Department of Electronics Design (EKS), Mid Sweden University, Sundsvall, Sweden
[2]Department of Natural Sciences (NAT), Mid Sweden University, Sundsvall, Sweden

*Corresponding author: siwen.an@miun.se

Abstract. With the rapid expansion of the waste incineration business both in Europe and globally, there is a growing need for the elemental analysis for fly ash from municipal solid waste incineration. In this work, samples of washed and unwashed ash from municipal solid waste incineration in Sundsvall are evaluated. Qualitative analysis and semi-quantitative analysis are used to compare two elemental analysis methods, scanning electron microscope with energy dispersive spectroscopy (SEM-EDS) and X-ray fluorescence (XRF) measurement. Both methods are used to retrieve the difference in elemental composition between washed and unwashed fly ash. SEM-EDS accurately detects light elements from well-prepared samples in a vacuum environment, while, for online measurements, XRF is a potential method that analyses hazardous metal content in the fly ash.

1 Introduction

Incineration of the municipal solid waste (MSW) is one of the most efficient ways in waste management, because of the large reduction in volume and the high energy recovery. The EU's rules for waste management, the so-called waste hierarchy, state that the following stages are reuse, material recycling, other recycling – e.g. energy recovery – and finally disposal [1]. In Sweden, a vast majority of this household waste can be recovered or reused, and only four percent is landfilled. Waste incineration provides around 20 percent of all the district heating. The major environmental concern in relation to the short- and long-term impacts of MSW residue disposal is the risk of leaching and subsequent release of potentially harmful substances, particularly inorganic salts and metals trace elements, into the environment [2]. The advantage of municipal solid waste incineration is that many harmful substances break down and the residual substances are bound in ash, which makes them easier to control, handle and recycle. Incineration produces fly ash along with flue gas and heat. The flue gas is cleaned before it is fed out through the chimney, leaving fly ash and sludge from the washing water to be further processed. However, the contents in the fly ash are hazardous to health and the environment, and they thus are commonly stored in storage cells for hazardous waste. Korstaverket in Sundsvall is the first power plant in Sweden, which implements a washing process and allows storage of fly ash on a landfill for non-hazardous waste [3]. The washing process generates a mixture of fly ash and sludge that has low leaching of metal content to the surroundings.

X-ray fluorescence (XRF) technique is a non-destructive method for the elemental analysis, which is fast and fully suitable for simultaneous quantitative determinations. Therefore, it presents some advantages over other spectrometric methods like Neutron Activation Analysis (NAA); which is expensive and not widely available, inductively coupled plasma atomic emission spectroscopy (ICP-AES); which requires dissolution procedures, and are generally time-consuming and can induce losses of some volatile elements [4]. Therefore XRF instrumentation has lower capital cost and is cheaper to use then NAA, is faster than ICP-AES, and is widely available [5], [6].

In this paper, the purpose is to compare two methods to analyse the elemental composition of a random group of fly ash samples: (a) a scanning electron microscope with energy dispersive spectroscopy using electron beams and (b) an X-ray spectrometer using X-rays to irradiate the samples. The fly ash samples are collected both before and after the stabilizing wash process. The major changes affected by the washing process is presented by lowering the Sodium, Potassium, Chloride and Sulfate concentrations. Among the minor constituents increased, e.g. Hg, however, the total concentrations are still low. Final, the washed ash allows to be taken to a class II landfill in Sweden [3].

2 Method

When a material is irradiated by X-rays or highly-focused energetic electron beams, a photon or electron transfers its energy to an inner shell electron in an atom of the material. If the energy of the particle is higher than the binding energy of the electron, it can eject the electron from its atomic orbit. To fill this space, an outer shell electron falls into an empty position of the ejected electron. In this process, the falling electron emits a characteristic fluorescent X-ray photon with an energy which is equal to the difference between the binding energies of those two shells. The energies of the fluorescent photons from a material are characteristic and, therefore, can be used to identify the material [7], [8].

The characteristic X-ray energy lines of the elements of interest are listed in Table 1 [9].

Table 1. Characteristic X-rays.

Element	Z	Fluorescence $k\alpha_1$(keV)	Energy $k\beta_1$(keV)
Sodium (Na)	11	1.040	1.071
Sulphur (S)	16	2.307	2.464
Chlorine (Cl)	17	2.622	2.815
Potassium (K)	18	3.314	3.589
Cadmium (Cd)	48	23.173	26.096

To excite the k-shell electrons of a specific element, the X-ray or electron energy in keV needs to be higher than this energy. The X-ray source used in the XRF measurements is capable of 60 keV while the SEM-EDS source was limited to 20 keV.

2.1 Scanning electron microscopy (SEM)

Two samples, the washed ash and unwashed ash, are prepared. The washed ash is dried in the air and ground manually into fine small grains with the dimension of a few hundred micrometers. The unwashed ash is not concreted by the washing process, and it is thus an already fine and powdery material. Therefore, we do not grind the unwashed ash. A portion of fly ash samples is sprinkled onto double-sided carbon tape mounted on an Aluminum stub (1cm diameter), and then it is sputtered with a conductive coating (Iridium) to avoid charging issues with the electron microscope. This grain mounting enables the analyst to determine the particle morphology, external surface structure and external elemental distribution of individual fly ash particles [10].

SEM is one of the techniques for the chemical and physical characterization of fly ash [11].

SEM operates in vacuum environment, which scans the surface of a conductive sample via a highly-focused energetic electron beam to create an image for detailing the topography and collect the information of the composition of the sample. Three common modes of operation are used in an SEM analysis, namely, backscattered electron imaging (BSE), secondary electron imaging (SEI), and EDS [12]. In this study, we use BSE and EDS to characterize the fly ash samples. The BSE consists of high-energy electrons originating in the electron beam that is reflected or back-scattered out of the specimen interaction volume by elastic scattering interactions with specimen atoms. In the EDS, in order to obtain

IOP Conf. Series: Earth and Environmental Science **337** (2019) 012007 doi:10.1088/1755-1315/337/1/012007

the precise elemental composition of materials, we use a detector (X-MaxN Silicon Drift Detector, Detector Area: 50 mm2, OXFORD instruments) to measure the energy and intensity distribution of X-ray signal generated by a focused electron beam on the specimen.

The elemental analysis is performed in ''spot mode'', in which the beam is localized on a single area manually chosen within the field of view. The location is represented on the provided SEM images. The backscatter rates in heavy elements (high atomic number) are higher than in light elements (low atomic number), resulting in brighter spots shown in the image. Thus, the BSE is used to detect contrast between areas with different elemental compositions. The morphology of a fly ash particle is shown in Figure 1. The SEM images present a poor illumination at the peripheries (vignetting), which is caused by the detector sensitivity dropping towards edges of images at low magnification. The size of the washed ash particles (Figure 1 left) is in a range from less than 10 µm to greater than 200 µm. The unwashed ash particles (Figure 1 right) are down to few nanometers.

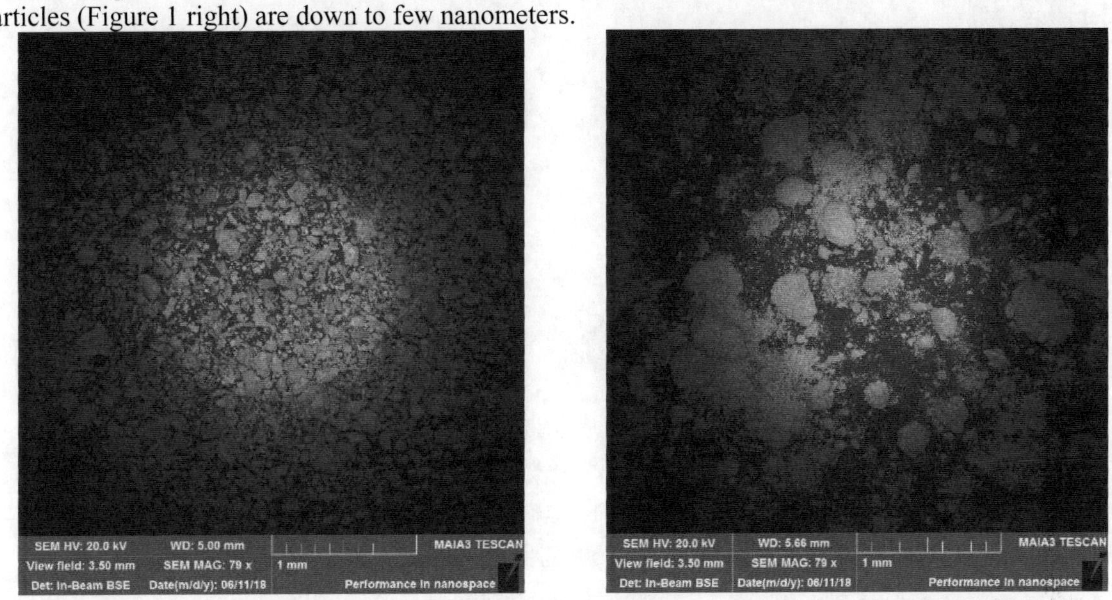

Figure 1. Scanning electron microscopy of (left) Washed fly ash; (right) Unwashed fly ash.

2.2 X-ray fluorescence (XRF)

For loose powders samples, the more finely ground the sample the more likely it is to be an even surface and providing for a better analysis. Irregular sample surfaces hamper X-ray performance. Moreover, they also change the distance from the sample to the X-ray source, which introduces air absorption difference. Pressing powder into pellets is a more rigorous sample preparation than loose powders [13]. The process includes grinding a sample into a fine powder, ideally to a grain size of <75 µm, mixing it with a binding aid, and then pressing the mixture in a die at between 20 000 kg and 30 000 kg to produce a homogenous ground sample pellet. However, pellets sample preparation makes it difficult for an online measurement. In this measurement, the washed sample was roughly ground to up to 3 mm. The exposure area is approximate 4-5 cm2 since the cone angle of the X-ray beam is 48° in this setup. An XRF setup is thus possible to be mounted above a conveyor belt in a plant ash treatment process.

XRF is an analytical method that determines the elemental composition of all kind of materials. It is also capable of determining materials in varies forms such as solid, liquid, and powder [14]. However, the XRF cannot generally make analyses at the small spot sizes typical of SEM work (about 100 micrometres in this project), so it is typically used for the bulk analyses of larger fractions of materials. Because of the relative ease and low cost of sample preparation as well as stability and ease of use of x-ray spectrometers, XRF is used for analysis of major and trace elements in rocks, minerals, and sediment [15].

IOP Conf. Series: Earth and Environmental Science **337** (2019) 012007 doi:10.1088/1755-1315/337/1/012007

An X-ray tube produces a bremsstrahlung continuum extending to low energy, which scatters from the sample. X-rays below the absorption edges do not excite the elements of interest, forming a background that decreases the detection limit of spectrum. If a filter is placed between the tube and sample, X-rays below the K-edge of the filter will be absorbed. It improves the signal to background ratio, improving precision and detection limit [7]. Filters of 0.17 mm Cu and 1 mm Cu are used in this measurements [16]. A 0.3 mm Ge filter also have been considered for these elements below 7 keV.

The XRF measurement setup is shown in Figure 2. An X-ray tube (MOXTEX 60 kV) with Silver (Ag) target is used to irradiate the sample surface. The X-ray fluorescence with discrete energies emitted from the ash sample is detected by an X-ray Spectrometer (Amptek X-123SDD, Detector Area: 25 mm2), which is mounted in an air environment. Due to the large range of fluorescence energies of the materials of interest, two filters and two tube voltages are used to optimize the performance for different elements [17].

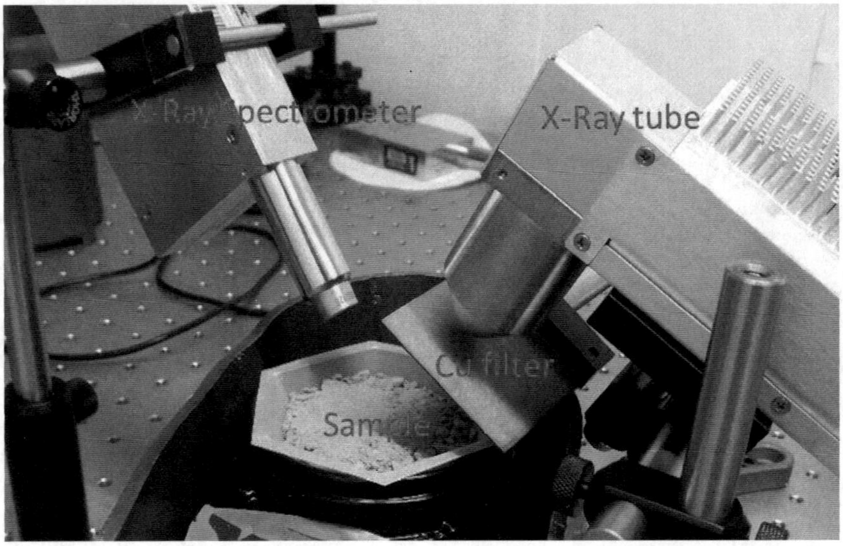

Figure 2. XRF experimental setup, 1 mm Cu filter.

3 Results and discussion

The peak area of each element corresponds to a specific element concentration, and it is influenced by fluorescence yield, sensor absorption, source penetration in the sample, and the escape of fluorescence out of the sample. Grammelis [18] notes that the Na, K, P, Cl, Si and Ca contents determine the ash melting behavior – related to the formation of slag and

fouling. The importance of the quantification of S, Cl and K elements is related to the corrosion of equipment commonly used in energy recovery plants such as heat exchangers. In combustion processes, S, Cl and K partially vaporize in high temperatures and, after cooling, form HCl, KCl, Cl2, alkali chlorides and metal sulfides [18]. A dry/wet scrubber is used to spray fine atomized slurry or lime powder into the hot exhaust gas in order to neutralize the acidic gases such as sulfur oxides and hydrogen chloride [19]. Therefore, the washed ash has higher Ca peak than the unwashed ash.

The SEM-EDS spectra of fly ash are presented in Figure 3. For both washed and unwashed samples, the spectra nicely resolve the peaks from interested elements, Na, S, Cl, K and Cd. For the Na peak, the tail at 0.9 keV shows that it overlaps with the escape peak of Cl. For higher energies elements above 10 keV, the spectra present a poor detection due to the maximum beam energy depends on the energy of the incident electrons. It requires an input energy higher than the detectable peak energy by 30-50%. In order to detect Cadmium $k\alpha_1$ and $k\beta_1$ peaks, an electron beam with a higher energy is needed, which is not involved in this measurement.

IOP Conf. Series: Earth and Environmental Science **337** (2019) 012007 doi:10.1088/1755-1315/337/1/012007

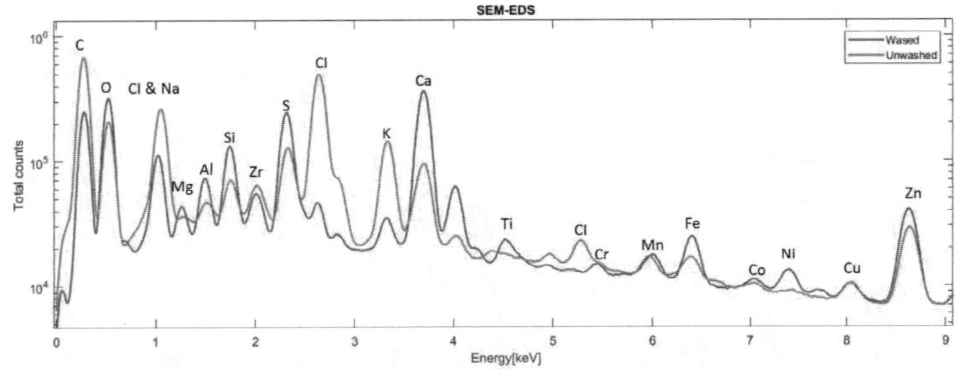

Figure 3. Spectra of washed and unwashed ash sample, SEM-EDS, 20 kV, 5 mins.

XRF spectra of ash sample with 0.17 mm Cu filter are presented in Figure 4. A digital filter (one-dimensional filter) is applied to compute averages and smooth noise data. As same as SEM-EDS spectra shown in Figure 3, the Na peak is disturbed by the Cl escape peak, presenting a merged peak. The light elements below 1 keV are easily absorbed by air, thus it is difficult to detect them in the present setup in the air. Ar kα_1 peak at 2.958 keV and kβ_1 peak 3.190 keV overlap with Cl kβ_1 2.815 keV and K kα_1 3.314 keV, respectively. It should be noted that there is no Ar in the ash, but it presents in the spectra, because the normal atmosphere consists of Ar.

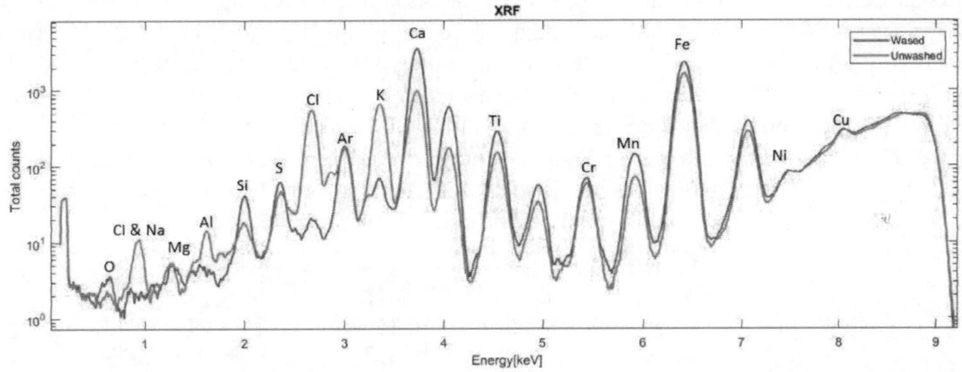

Figure 4. Spectra of washed and unwashed ash, XRF, 0.17 mm Cu filter, 15 kV, 10 mins.

For detecting Cd, a higher X-ray tube energy 40 kV with 1 mm Cu filter is applied. As shown in Figure 5, the Cd peak at 23.173 keV presents a lower peak in the washed ash compared with the spectra of the unwashed ash. The kβ energy of Cd at 26.096 keV is disturbed by the k$\alpha_1$peak of Sb, but the peak presents at 23.173 keV is able to detect Cd. The Ag peak at 22.163 keV is contributed by the X-ray tube target material rather than the ash.

IOP Conf. Series: Earth and Environmental Science **337** (2019) 012007 doi:10.1088/1755-1315/337/1/012007

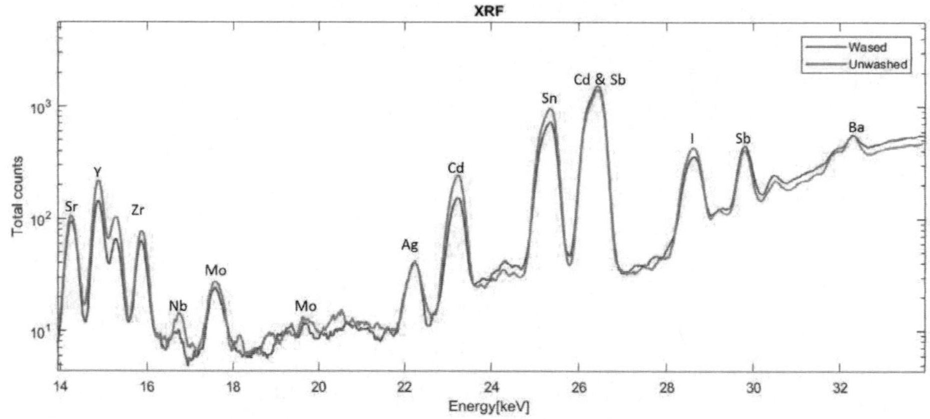

Figure 5. Spectra of washed and unwashed ash, XRF, 1 mm Cu filter, 40 kV, 60 mins.

In order to compare the SEM-EDS and XRF methods on the performance of elemental analysis, we use same the sample (i.e., the ground sample) for both methods. Tab.2 shows the comparison between the SEM-EDS and XRF methods for the elements of interest. Each element is presented as its intensity determined by the area of a peak profile (total counts). By employing a semi-quantitative analysis, we present the relative intensity of the elements of interest. Because the escape peak of the Si sensor in the spectrometer is excited by the Cl peak at 2.62 keV, escape photons has lost 1.74 keV of the original photon energy and shows a peak at 0.88 keV adjacent to the Na peak. Thus, the three W/U ratios (the ratio of washed to unwashed) of Na present different values after the washing process. Although the washing process decreased the sulfate content, both methods do not distinguish between sulfur and sulfate. The results thus shows that the S (sulfur and sulfate) is increased by 12-25 % after the washing process approximately. Cl and K decrease 17-27 % and 42 % for the ground ash sample, respectively.

Table 2. Comparison between SEM-EDS and XRF for the elements of interest in fly ash. The total counts are normalized for each method.

Method	Sample	Na	S	Cl	K	Cd
SEM-EDS (Ground sample, Figure 3)	Washed	0.316	0.611	0.169	0.143	NULL
	Unwashed	0.569	0.329	1	0.337	NULL
	Ra=W/U	0.555	1.857	0.169	0.424	NULL
XRF (Raw sample, Figure 4 and Figure 5)	Washed	3.224e-3	0.094	0.056	0.149	0.312
	Unwashed	3.720e-3	0.082	0.781	1	0.362
	Ra=W/U	0.867	1.146	0.072	0.149	0.862
XRF (Ground sample)	Washed	0.011	0.437	0.208	0.418	NULL
	Unwashed	0.010	0.178	0.779	1	NULL
	Ra=W/U	1.067	2.460	0.267	0.418	NULL

SEM-EDS is a point measurement method and sensitive to the chosen position since the ash samples are inhomogeneous, while XRF measures a surface average of 0.785 cm2 for the ground sample. The effect of the grinding and sample preparation can be evaluated by comparing XRF measurements of raw and ground samples, which is presented in Tab. 3 by comparing ratios of the W/U ratios in Tab. 2.

Table 3. Relative comparison of SEM-EDS and XRF methods and of ground- and raw samples using XRF.

Ratio	Comparison	Na	S	C	K
Rc / Ra	Between XRF and SEM-EDS				
Rc / Rb	Between ground and raw samples				

When comparing the methods in Tab. 3, Rc/Ra = 0.98 (≈1.0) for K indicates that XRF and SEM-EDS show good agreement for detecting K. For the lighter elements, XRF gives a factor between 1.3 and 1.9 higher content in washed ash. This difference might be due to differences in XRF penetration depth for the lighter elements and must be considered. Tab. 3 further shows that sample preparation affects the

XRF measurement to show higher content of all the evaluated elements after the washing process. Calibration of the XRF analysis must therefore consider the ash preparation.

4 Conclusion

XRF can detect Na, S, Cl, K, Cd and other heavier hazardous elements in fly ash. The provided measurements are semi-quantitative and can be calibrated either using dedicated quantitative analysis software for XRF or with other methods like SEM-EDS. The latter provides good spectral resolution for light elements where air absorption of low energy X-rays disturbs the XRF measurement. In samples containing many different elements, the peaks can distort each other. In these fly ash samples, the Na peak is distorted in both XRF and SEM-EDS measurements due to the Silicon escape peak of Cl. The further process for correcting escape peaks, sum peaks, and background removal must be considered [20].

The XRF method is suitable for online measurements since it requires no specific sample preparation and measuring larger fractions of materials. XRF penetrate relatively deep into the ash sample while the SEM-EDS method mainly measures the surface content. This makes the XRF method sensitive to background noise form the container material. An XRF online measurement method thus needs to be calibrated to consider the effect of surrounding elements. Another conclusion is that the calibration of a measurement system must consider the ash preparation.

We conclude that XRF is a potential method for online measurements of hazardous metal content in fly ash from incineration plants. If the metal composition of the fly ash were known on site, it would be possible to adjust the wash process to optimize the content of the washed/stabilized ash. In the future, it is possible to further develop the washing process into the extraction of metals from the fly ash.

References

[1] Dijkgraaf E and Vollebergh H R J 2004 Burn or bury? A social cost comparison of final waste disposal methods *Ecological Economics* **50** (3-4) pp 233-47

[2] Waechter M R, Ionel I and Negrea A 2013 New concept for landfill disposal of municipal solid waste incineration residues *J. of Environmental Protection and Ecology* **14** 3 pp 947–55

[3] Nordmark D, Lindgren E, Vamling M and Lagerkvist A 2011 *Processoptimering av asktvätt, (Optimization of ash washing process)* (Stockholm: Värmeforsk Service AB)

[4] Chand P, Kumar A, Gaur A and Mahna S K 2009 Elemental analysis of ash using X-ray fluorescence technique *Asian j. of Chemistry* **21** 10 pp 220-4

[5] Bettinelli M and Tania P 1990 Spectrometry X-ray Spectrum, 19, 227

[6] Dogan O, Simsek O, Ertugrul M and Kobya M 2001X-ray fluorescence spectrometry analysis of trace elements in fly ash samples of Yenikoy thermal power plants *Instrumentation Science & Technology* **29** 5 pp 433-9

[7] Van Grieken R and Markowicz A 2001 *Handbook of X-Ray Spectrometry*

[8] Reza S, Chang H, Norlin B, Fröjdh C and Thungstrom G 2015 Detecting Cr contamination in water using X-ray fluorescence *In Nuclear Science Symposium and Medical Imaging Conference (NSS/MIC), 2015 IEEE* pp 1-2

[9] Thompson A C, Attwood D T 2001 *X-ray Data Booklet* (California: Lawrence Berkeley National Laboratory) [Online] Available: http://xdb.lbl.gov/xdb.pdf (2001)

[10] Kutchko B G and Kim A G 2006 Fly ash characterization by SEM–EDS *Fuel* **85** (17-18)

[11] Vassilev S V and Vassileva C G 2005 Methods for characterization of composition of fly ashes from coal-fired power stations: A critical overview *Energy Fuels* pp 1084–98

[12] Postek M T, Howard K S, Johnson A H and McMichael K L 1997 *The scanning electron microscope. Handbook of charged particle optics* pp 363-99

[13] Shibata Y, Suyama J, Kitano M and Nakamura T 2009 X-ray fluorescence analysis of Cr, As, Se, Cd, Hg, and Pb in soil using pressed powder pellet and loose powder methods. *X-Ray Spectrometry: An International J.* **38**(5) pp 410-6

[14] Johnson D, Hooper P and Conrey R 1999 XRF Method XRF Analysis of Rocks and Minerals for

Major and Trace Elements on a Single Low Dilution Li-Tetraborate Fused Bead *Adv. X-ray anal* **41** pp 843-67

[15] Brouwer P 2006 Theory of XRF (Almelo, Netherlands: PANalytical BV)

[16] Al-Bataina B A, Maslat A O and Al-Kofahi M M 2003 Element analysis and biological studies on ten oriental spices using XRF and Ames test *J. of Trace Elements in Medicine and Biology* **17**(2) p 85

[17] Norlin B, Reza S and Fröjdh C 2016 X-ray fluorescence measurements of toxic metal content in ash from municipal solid waste incineration *In Nuclear Science Symposium, Medical Imaging Conference and Room-Temperature Semiconductor Detector Workshop (NSS/MIC/RTSD), 2016* pp 1-3

[18] Tiburcio R S, Gomez A C G, Junior M M and Neto A M P 2018 Evaluation of the Elemental Composition of Municipal Solid Waste Bottom Ash: A New Methodology for Sample Preparation *Revista Brasileira de Energias Renováveis* **7**(2)

[19] Lam C H, Ip A W, Barford J P and McKay G 2010 Use of incineration MSW ash: a review. *Sustainability* **2**(7) pp 1943-68

[20] Redus R 2008 *Amptek Application Note XRF-1: XRF Spectra and Spectra Analysis Software*

IOP Conf. Series: Earth and Environmental Science **337** (2019) 012008 doi:10.1088/1755-1315/337/1/012008

Study on the sorption capacity of the adsorbent based on polyurethane and chitin to remove oil spills

Y D T Trang[1*] and L A Zenitova[1]

[1]Department of Synthetic Rubber Technology, Kazan National Research Technological University, Kazan, Republic of Tatarstan, Russian Federation

* Corresponding author: tydtrang@gmail.com

Abstract. Nowadays, oil and its derivatives were considered one of the major energies and they also have been one of the main sources of environmental pollution, especially for water pollution. The tendency in this time of researchers was to find new adsorption, that has high sorption capacity, economical and easy production. By accordance with the present trend, in this study, chitin was used with the dose up to 50% as a filler in the composition of adsorbents basic on polyurethane foam (PUF). The effect of the size (small size: 1-3 mm and large size: 5-10 mm) and the dose (10, 20 and 50%) of filler in the foaming composition to technological parameters of the collected adsorbent, as well as to its absorption capacity were investigated. The results showed that the adsorbents filling with a lower dose of chitin (10 and 20%) were better foaming than the one containing a higher dose of filler. At the same time, the adsorbents containing small chitin also formed the foam better than one filling large size. For 2 types of PUF, the adsorbent from elastic PUF has higher sorption capacity than the semi-elastic adsorbent with the same dose of the filler. For the results of sorption effectivity of the adsorbents presented that the adsorbent with the small size of filler observed greater the absorbency than the one with large size. In which, the elastic adsorbent containing 10% chitin with small size has the highest oil absorption capacity, which reached 13.13 g/g. To compare the sorption capacity of the adsorbent between relation to oil and water, it shown that its oil capacity is significantly higher than the sorption capacity of water.

1. Introduction

Oil and its derivatives were considered one of the major energies; on the other hand, they also had been one of the main sources of environmental pollution, especially water environmental [1-6]. They penetrated into the aquatic environment in different ways, for example: they entered into surroundings from the production, transportation and storage process; or when using oil as fuel for ship engines; either when cleaning storage tanks for oil; or even from wastewater of the refineries or possibly be caused from waterway accidents. Meanwhile, the oil pollution in the water environment has affected the sea and water transport, the economy, the health of people and living organisms. When oil came in contact with water, it has affected the aquatic organisms such as: physiological disorders, diseases, changes biological functions. On the other hands, the decomposition process of the oil in water by the microorganisms occurred very slowly, which leads to the forming the oil layer on the water surface. The oil layers on the water surface reduced the necessary supply air for microorganisms, and it also became the secondary source of pollution for the aquatic environment [7]. So, it is necessary and urgently to remove the oil from the water [8].

The wastewater containing oil and its products have treated by the various methods: mechanical, chemical, biological or sorption method. In which, the sorption method has attracted many scientists because of its simplicity and efficiency. However, the current adsorbents to remove oil spills still had many disadvantages, for instance: expensive, complicated production process, low sorption efficiency and low buoyancy, negative environmental impact [9]. Therefore, it was necessary to create the new adsorbent, which has lower cost, high adsorption and buoyancy ability in the water surface, as well as the possibility of reusability.

Chitin is one of the inexpensive and available materials. It was produced by units of N-acetylglucosamine. This polymer is a popular polysaccharide widely distributed in nature (the second most quantity biopolymers after cellulose); it was contained mainly in the crustacean shell (shrimps, crabs) and the other aquatic animals (coral, jellyfish, squid). Chitin and its derivatives have many valuable properties, such as: antifungal and antimicrobial activity, biodegradable and biocompatible, no toxic for humans and animals. For this reason, it was considered widely renewable resource used in many fields, as an example: wastewater treatment and environmental protection, pharmacy, medicine, agriculture, industry, biotechnology. In recent years, the new functions for chitin were discovered. Hence, its production and consumption were strongly growing [10]. In the wastewater treatment and environment field, chitin was considered a coagulant and flocculating agent for the pollutants in the wastewater such as: the suspended organic compounds and solids, amino acids [11]; it worked as an adsorbent to remove reactive dyes in the wastewater of the textile mills [12]. It also had a good adsorption ability to remove heavy metal ions [11], to remove phenols and chlorophenols in wastewater [12]. Besides, chitin removed quite high the contaminants in refineries wastewater; and so, leaded to improving significantly about the quality of wastewater [13]. In addition, the authors from Brazil reported that chitin and chitosan were able to remove oil spills well in seawater, nevertheless, after sorption process, this biopolymer sunk to the bottom of the sea; causing inconvenient in the removal oil spill response [14]. Moreover, in the previous studies, it can be shown that chitin has sorption capacity to remove oil spills, but after the sorption process the adsorbents not be reused. Thus, its cost was very expensive [11].

Consequently, the aim of the current study created the oil adsorbent based on polyurethane foam (PUF) and chitin, which its adsorption capacity was sufficiently high to remove oil spills. Besides, this adsorbent also had good buoyancy, can be reused and an increase of the amount recover oil after the sorption process. In the limit of this study, the influence of the size and degree of the filler on the foaming and sorption capacity of the adsorbents in two types: the elastic and semi-elastic adsorbent was investigated.

2. Detail experimental

2.1. Materials

Adsorbents - were represented polyurethane foams, were produced by mixing component A filling chitin and with component B; the components were produced by "Dow Izolan", Vladimir, Russia. Component A was used with two types: elastic and hard polyurethane foams. This choice of components PUF due to the structure of the received foams. The elastic foams had interconnection of the cells in their structure, which naturally contributed to the high absorption capacity of this foams. However, they have loosed their buoyancy and sunk when in the saturated state, which led to affect adversely on the living water organisms. On the contrary, the hard foams hard mainly the closed-cell structure, in that the cells were connected; this make difficulty to the penetration of the adsorbate; however, this adsorbent still maintains the buoyancy ability although it was saturated. When the component A with elastic PUF was used, the elastic adsorbent EPUF was obtained; and when component A was used with the hard PUF, the semi-elastic adsorbent SEPUF was obtained. The ratio of components A to B was 1: 0.6.

The chopped chitin as the filler was produced by "Chitosan Vietnam", Ho Chi Minh, Vietnam with two type of the sizes: 1-3 mm (M) and 5-10 mm (B).

a. Chitin (B) b. Chitin (M)

Figure 1. The chitin with different size

2.2. Methods

- Adsorbents (PPU-X) were obtained according to the scheme in figure 2.
- Apparent density was determined according to GOST 409-77 [15].
- The oil sorption capacity of the adsorbents was calculated using the following equation:

$$Q_o = \frac{m_1 - m_0}{m_0} \quad (1)$$

Where: Q_0: the sorption capacity, g/g;
m_1: the weight of adsorbent after process sorption, g;
m_2: the initial dry weight of adsorbent, g;

In the beaker filling with 50 ml of the adsorbate (oil or water) were put about 2 grams of the chopped adsorbents. The sorption process was calculated after 1, 2, 3, 5, 8, 10, 15, 20, 30, 60, 90, 120 minutes of the sorption time [16, 17].

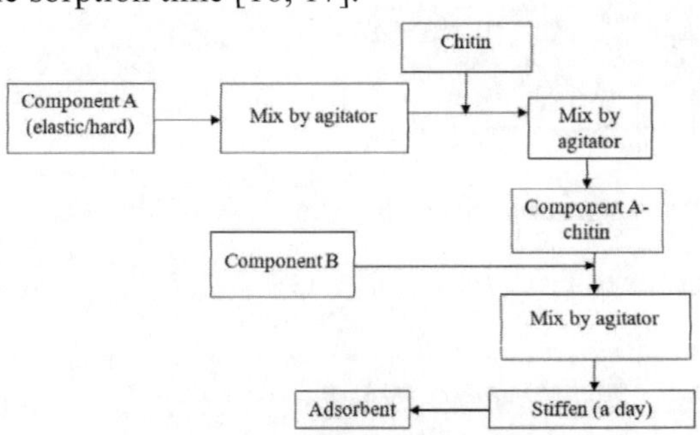

Figure 2. Process flow diagram for the preparation of adsorbent PPU-X [16]

3. Results and discussion

3.1. Technology producing of the adsorbent based on PUF and chitin

3.1.1. The semi-elastic adsorbent

Figure 3 showed the effect of the dose and size of the filler on the foaming process of semi-elastic polyurethane foam (SEPUF). Regardless of the size of the chitin pieces, the more dose of fillers increased, the more the foaming ratio of the foam increased. However, at the dose 50% of filler, the foaming ratio decreased, the foam became denser.

Figure 3. Semi-elastic adsorbents basic on PUF and chitin
a) SEPUF; b) SEPUF-10B; c) SEPUF-20B; d) SEPUF-50B;
e) SEPUF-10M; f) SEPUF-20M; g) SEPUF-50M

The piece size of the filler also affected on the foaming process and the quality of the adsorbent. Smaller pieces of chitin were distributed more evenly in the polyurethane matrix and contribute better dilation of PUF.

For the adsorbent to remove oil spills, its structure was one of the main factors affecting the sorption capacity [18,19]. Form fig. 4 observed that the addition of chitin into the composition of the elastic adsorbent give possible to forming a fairly uniform foam. In which, smaller pieces of chitin were distributed more evenly in the volume of the polyurethane matrix and gave a large number of homogeneous cells.

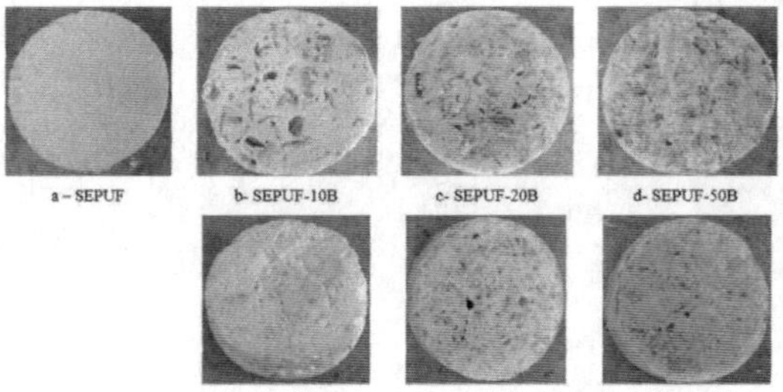

Figure 4. Cross-section of the adsorbents SEPUF and chitin

3.1.2. The elastic adsorbent

Figure 5. The elastic adsorbents based on PUF and chitin
a) EPUF; b) EPUF-10B; c) EPUF-20B; d) EPUF-50B;
e) EPUF-10M; f) EPUF-20M; g) EPUF-50M

The results of the elastic adsorbents showed that the addition of chitin led to decreasing of the foaming. In the comparison in the degree dispersion of the filler, it was noted that in the case with smaller filler, the rise ratio of the foamiing was better. The best rise ratio of the foaming was observed by the adsorbent with 10% dose of filler, for both size of the filler. Fig.5.

The cross-sections of the elastic adsorbents were presented in fig.6. The addition of the chitin increased the pore size in the adsorbent. (Fig. 6b,c,e,f). For the adsorbent containing 50% chitin, it was seen that the pieces of chitin overlap the pores in the foam, led to reduce its absorption ability. In which, the adsorbent filling with the smaller size of chitin gave possibly to form adsorbent with larger and more uniform the number of pores.

It was also observed that EPUF had shrinkage extent less in comparison to SEPUF. For the elastic adsorbents containing 10 and 20% chitin, it was found that the more the amount of filler in the adsorbent, the less their elasticity, which was convenient for chopping. When filling with the dose 50% of chitin, the adsorbent became hard, no elasticity.

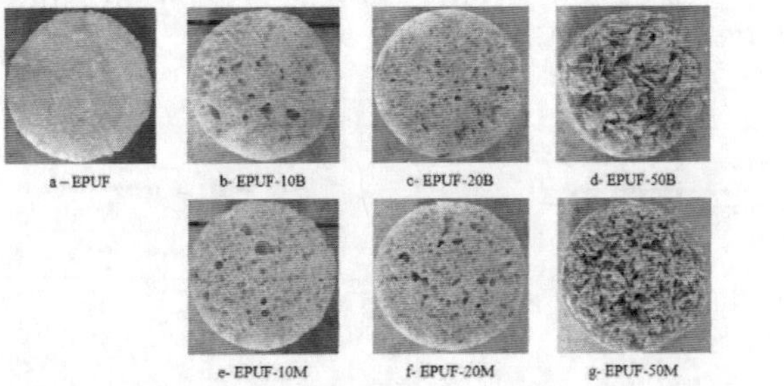

Fig. 6. Cross-section of the elastic adsorbents based on PUF and chitin

3.2. Technological parameters in the foaming of the adsorbents based on PUF-C

For the foaming process of the adsorbent, the main parameters were the "start time" and "rise time" of the foam. The data of table 1 presented that the both "start time" and "rise time" of the foam were higher when the dose of filler in both types of adsorbent increased. Especially, the "start time" and "rise time" of the foam 50% of the filler increased in comparison to the unfilled adsorbents. The influence of the size of filler was evaluated, the results showed that the adsorbent containing filler with smaller size affected less on the foaming parameters; for the dose of filler 10 and 20%, they were commensurate with the initial ones. The large size of the filler impeded the interaction of the components of PUF [20].

For the adsorbent to remove oil spills, the apparent density was one of the main parameters. The higher the apparent density, the heavier the adsorbent, this led to reduced its buoyancy. The purpose of this study created the adsorbent with high buoyancy, which prevented the sinking and ensured to remove it easily at the saturated state. From the data in table 1, it is shown that the addition of chitin into the adsorbent led to the increase of its apparent density. The larger sizes of chitin pieces affected more on increasing the density of adsorbents. Because the higher the quality of foam, the less its apparent density and its structure were more porous.

Table 1. Technological parameters of adsorbents based on PUF-C

Adsorbent	Start time, s	Rise time, s	Apparent density, kg/m3
SEPUF	33	110	64.39
SEPUF-10M	33	112	72.37
SEPUF-20M	34	132	67.53
SEPUF-50M	52	197	141.46
SEPUF-10B	34	138	98.35
SEPUF-20B	40	166	88.69
SEPUF-50B	111	249	148.71
EPUF	29	156	59.10
EPUF-10M	29	156	77.78
EPUF-20M	40	160	91.86
EPUF-50M	87	-	220.29
EPUF-10B	30	161	86.65
EPUF-20B	43	170	132.21
EPUF-50B	90	-	138.30

-: can't determinate

IOP Conf. Series: Earth and Environmental Science **337** (2019) 012008 doi:10.1088/1755-1315/337/1/012008

3.3. Oil adsorption capacity of the adsorbents

Sorption process is the process of movement of absorbed substances through the capillaries in the holes and on the surface of porous solids [6]. To evaluate the effect of the sorption process, the sorption capacity was one of the important factors. The sorption capacity of the adsorbent was the higher, its efficiency for removing oil spills was the greater. The results of the oil sorption capacity of the received adsorbents were shown in fig.7.

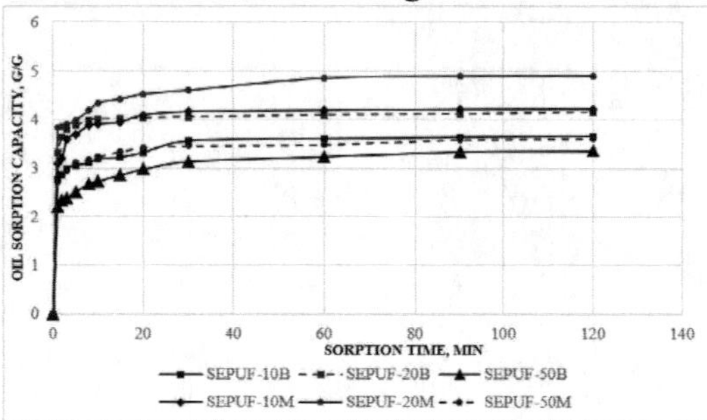

a. For the semi-elastic adsorbent

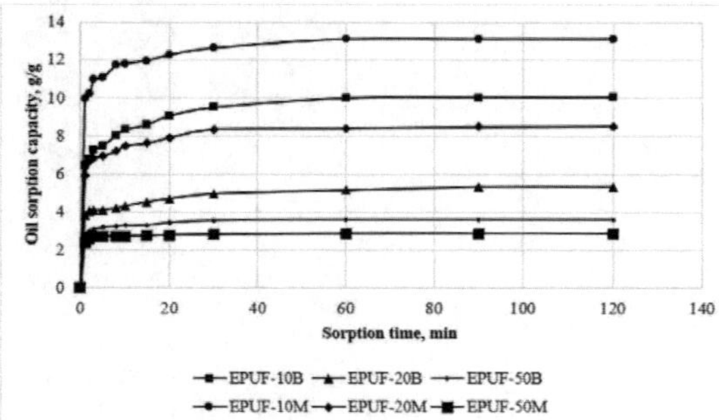

b. For the elastic adsorbent

Figure 7. The relationship between oil absorbency and time sorption of the adsorbent PUF-C

The results from fig.7 presented that the sorption process of oil divided into three main steps:

Step 1: the speed adsorption process increased significantly in the first 15 - 20 minutes of the sorption process. Initially, the porous contained only air, so the oil got into the foam unhindered.

Step 2: the sorption process was slow down in the period from the 20th to 60th minutes.

Step 3: the sorption process came to the balance from the 60th minutes.

3.3.1. The sorption capacity of the adsorbents

a. The semi-elastic adsorbent

The adsorbent containing small chitin had greater oil absorption capacity than the one with the large size of chitin. It was previously noted that the adsorbents containing smaller chitin formed a better foam with the uniform distribution of pores in comparison to the foam containing the larger size of chitin; obviously, this led to the higher sorption capacity of the adsorbent (fig.3-6).

The semi-elastic adsorbent containing 20% chitin reached the highest oil sorption capacity, was 4.9 g/g. Also, the density of this adsorbent was compared to the density of the unfilled foam, it was indicated the higher buoyancy than the unfilled foam.

b. The elastic adsorbent

The oil sorption capacity of the elastic adsorbents also occurred similarly to the semi-elastic adsorbent. Fig.7b. The oil absorbency of the elastic adsorbents containing chitin with small size was presented higher than the elastic adsorbents with the large size of chitin. The adsorbent EPPU-10M with dose 10% of filler reached the greatest absorption ability, was 13.13 g/g, it had the absorption capacity higher when comparing with the analogues containing 20 and 50% chitin.

When comparing the sorption capacity of adsorbents, it was found that the elastic adsorbents had significantly higher oil sorption capacity than the semi-elastic adsorbents (fig.7a,b). This is obvious because the absorption capacity of the adsorbent depended on two factors: the number of pores and their sizes [18]. Meanwhile, the elastic adsorbent contained mainly the open-cells structure in comparison to the semi-elastic adsorbent, in which the number of closed-cells was significantly smaller.

3.4. The water sorption capacity of the adsorbents

For the adsorbents to remove oil spills, the water absorbency was an important factor. The results of the sorption capacity in relation to the water were shown in fig.8.

Similar to the oil adsorption process, the largest amount of water is absorbed in the first 20 minutes. Then, the speed of the sorption process slows down and finally, from 60th minutes of the sorption process, the sorption curve reached the equilibrium status.

In much the same way of the oil sorption process, the adsorbents containing chitin with smaller size, their water absorbency were higher than of adsorbents with a larger size of chitin. However, for the semi-elastic adsorbent filled with 20% chitin, the water absorption reached the highest 4.96 g/g. Meanwhile, the elastic adsorbent with a dose of filling of 10% showed better water adsorption capacity - 5.79 g/g than with the filling of 20 and 50% chitin.

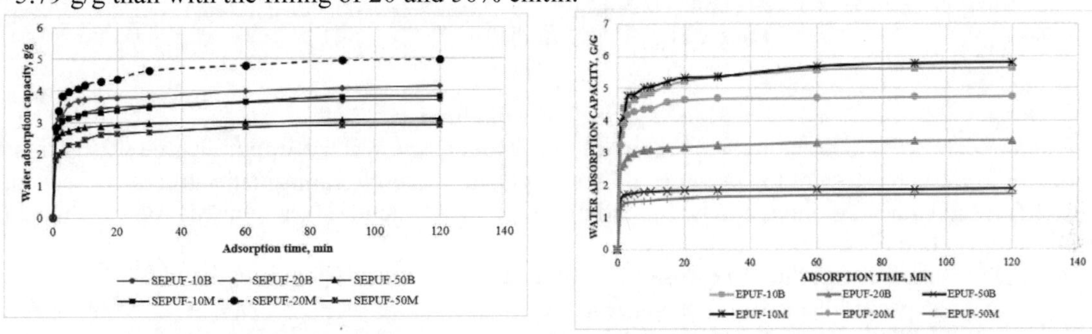

a. For semi-elastic adsorbents b. For elastic adsorbents

Figure 8. The relationship between water absorbency and the adsorption time of the adsorbents PUF-C

Because the removal to oil spills was often carried out on water surfaces, thus the oil adsorption more than water adsorption of the adsorbent was necessary. The selectivity of the adsorption was characterized by the coefficient "*a*" and was calculated by the fomula:

$$a = \frac{Water\ sorption\ capacity}{Oil\ sorption\ capacity} \quad (2)$$

From the formula (2) indicated that if the coefficient "*a*" of the adsorbent was low, then its oil sorption capacity was higher than the water sorption capacity; on the contrary, if the coefficient "*a*" was high, then the water sorption capacity of the adsorbent was higher than its oil sorption capacity. In this study, the coefficient "*a*" of the adsorbents was calculated at sorption times 120 minutes (the table 2).

Table 2. Coefficient "a" of the adsorbents PUF-C

Adsorbents	Coefficient "a"
SEPUF-10M	0.90
SEPUF-20M	1.01
SEPUF-50M	0.80
SEPUF-10B	1.01
SEPUF-20B	0.99
SEPUF-50B	0.92
EPUF-10M	0.44
EPUF-20M	0.56
EPUF-50M	0.60
EPUF-10B	0.56
EPUF-20B	0.63
EPUF-50B	0.52

From the data of table 2 can be observed that the elastic adsorbents had greater oil sorption capacity. And the semi-elastic adsorbents reached absorption capacity related to water higher than to oil.

Chitin was structurally polysaccharide built from chains, which was containing amine N-acetyl-D-glucosamine (β(1–4)-linked) (fig.9). This chains formed a layered structure. Polysaccharide chains are interspersed with layers of protein. Besides, chitin has two hydroxyl groups, one of which at C-3 is secondary, and the second at C-6 is primary. The presence of functional groups in chitin structure on the one hand are the reason of its polarity and hydrophilicity; and on the other hand, the structure of chitin provided the possibility of interaction between functional groups that provide various chemical modifications.

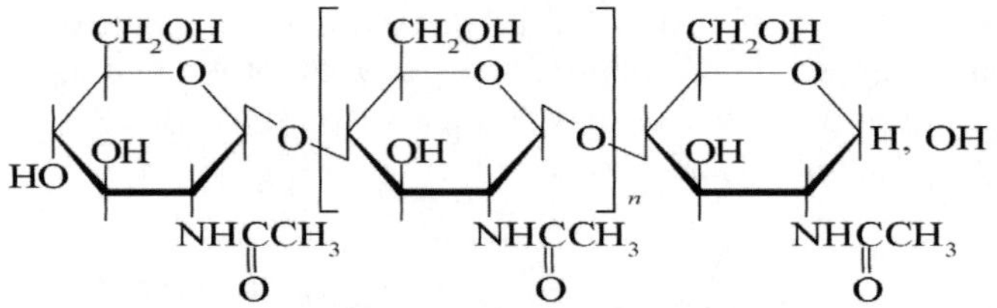

Figure 9. Structure of chitin

It would seem that, due to the hydrophilicity of chitin, led to the water sorption degree was greater than oil sorption. But because of the presence of hydroxyl and N-acetyl groups in the structure of chitin, it caused strong hydrogen interactions, including with the structural components of oil, which contributes to its good oil sorption ability.

In both EPUF and SEPUF, the same dose of chitin is used to filling. Therefore, the PUFF adsorbent have a greater oil capacity than SEPUF adsorbents due to the nature of PUF foam. Apparently, when comparing with the SEPUF adsorbents, the EPUF adsorbents had presence of a larger number of open pores; this affected the increased oil sorption capacity.

Therefore, the use of elastic adsorbents PPU-C was more preferable. The elastic adsorbent with a dose 10% of chitin had the highest oil sorption capacity, reached 13.13 g/g and it also had the highest selectivity "a" equal to 0,44.

Conclusions

The adsorbents based on polyurethane with filling chitin have been developed and investigated, it presented the high sorption efficiency. The elastic adsorbents based on PUF and chitin had higher sorption capacity than the semi-elastic adsorbents. A significant amount of oil was absorbed in the first 15–20 minutes of the sorption process, then the rate of absorption decreased, and after 60 minutes of the sorption process, it occurred saturation state. Adsorbents with chitin pieces size 1-3 mm had greater absorption capacity than the ones with size 5-10 mm of chitin. The highest selectivity of sorption for the water and oil system was presented by the elastic adsorbent based on polyurethane foam containing 10% chitin with size 1-3 mm and its oil sorption capacity reached 13.13 g/g.

References

[1] Aguilera F, Méndez J, Pásaro E, Laffon B 2010 *J. Appl. Toxicol.* **30** p 12
[2] Paulauskienė T, Jucikė I, Jušcˇenko N, Baziukė D 2014 *Water Air Soil Pollut.* **225** p 12
[3] Zhang J, Fan S, Yang J, Du X, Li F, Hou H 2014 *Sci. Total Environ.* **477** (7)
[4] Alonso-Alvarez C, Pérez C, Velando A 2007 *J. Aquat. Toxicol.* **84** (7)
[5] Camacho M, Boada L D, Orós J, Calabuig P, Zumbado M, Luzardo O P 2012 *Mar. Pollut. Bull.* **64** (6)
[6] Moro A M, Brucker N, Charão M F, Sauer E, Freitas F, Durgante J, Bubols G, Campanharo S, Linden R, Souza A P 2015 *J. Environ. Res.* **137** (7)
[7] Wahi R, Chuah L A, Choong T S Y, Ngaini Z, Nourouzi M M 2013 *Purif. Technol.* **113** (12)
[8] Bandura L, Woszuk A, Kołodynska D, Franus W 2017 Minerals, 7(3) p 25
[9] Tri D L M 2012 M S thesis *Chemical engineering, Hanoi national university*
[10] Rangel-Mendez J R, Barrios V A E, Davila-Rodriguez J L *Biopolymers* pp 163-80

[11] Filipkowska U 2006 *Adsorption Sci. & Tech.* **24**(9) p 14
[12] Rhee J-S, Jung M-W, Paeng K-J 1998 *Analytical sciences* **14** (3)
[13] Milhome M A L, de Keukeleire D, Ribeiro J P, Nascimento R F, Carvalho T V, Queiroz D C, 2009 *Nova* **32**(8)
[14] Barros F C D F, Vasconcellos L C G, Carvalho T V, Nascimento R F D 2014 *The Electronic J. of Chem.* **6**(1)
[15] *GOST 409-77* (Moscow: Official publication, and IC Publishing Standards)
[16] Anh K T Q, Ivanova M A, Zenitova L A 2017 *Bulletin of the Technological University* **20**(11)
[17] Ivanova M A, Zenitova L A 2015 *J. Ecology and Industry of Russia* **4**(4)
[18] Kumpaneko I V, Roschin A V, Ivanova N A, Panin E O, Sakharova N A 2015 *Russian j. of phys. chem. B* **9**(2)
[19] El-Din G A, Amer A A, Malsh G, Hussein M 2017 *Alexandria Eng. J. of Egypt*
[20] Sirotkina E E, Novoselova L Yu 2005 *Chemistry for Sustainable Development* **13**

IOP Conf. Series: Earth and Environmental Science **337** (2019) 012009 doi:10.1088/1755-1315/337/1/012009

Waste management system in the brewing industry

K Zhichkin[1*], V Nosov[2] and L Zhichkina[3]

[1]Samara state agricultural academy, Department of Economic Theory and Economics of AIC, 446442, Samara region, Kinel, Uchebnaja str., 2, Russia
[2]K.G. Razumovsky Moscow State University of technologies and management, Department of economics and management 109004, Moscow, Zemlyanoy Val, 73, Russia
[3]Samara state agricultural academy, Department of Land Management, Soil Science and Agrochemistry, 446442, Samara region, Kinel, Uchebnaja str., 2, Russia

* Corresponding author: zskirill@mail.ru

Abstract. Waste management implies such an organization of an enterprise economic activity in which, at the lowest possible level of environmental pollution, an organization creates the prerequisites for efficient utilization, reuse and imparting additional useful properties to waste in order to obtain an additional economic effect. In the Samara region, significant financial resources are spent annually on waste disposal activities, which could be used to increase production efficiency in industry and agriculture. Only current costs of environmental protection from 2011 to 2016 increased from 7.3 to 10.3 billion rubles or by 41.0%. In addition to this, investments in fixed assets related to environmental protection amounted to 2.3 billion rubles in 2011, up to 5.9 billion rubles - in 2014. Among the various types of pollution, solid waste ranks second (about 30%). In the brewing industry, the main type of waste is brewery mash, formed from grain residues. To properly dispose of it, a large area for storing waste is required, and the possibility of use in the future as an organic fertilizer. All this requires significant capital and disposal costs. At the same time, the brewer can be used as a high-quality feed. The article proposes a waste management system that provides for the interaction of the beer industry enterprises with pig-breeding complexes. Activities include the use of brewery mash as a valuable feed for the replacement of concentrated feed. As a replacement result, daily gains increase by 5-7%.

1. Introduction

In the current situation, the role of waste in the country's economy is very large. They are performing a dual function. On the one hand, waste is a brake on the industry and agriculture development, diverting considerable financial resources from the production process, which in another situation was an additional economic and investment reserve [7, 17-20]. For example, only in the Samara region, the current costs of environmental protection in 2016 amounted to a huge amount – 10,2 billion rubles. (table 1). Adding to it targeted investment costs, we get a forced annual reduction in the development potential of the region in the amount of 10 to 15 billion rubles. Among the various types of pollution, solid waste ranks second (about 30%) [1, 2].

Table 1. Current costs and investments in fixed capital aimed at environmental protection in the Samara region, mln. rub.

Indicator	2006	2011	2012	2013	2014	2015	2016
The current costs of environmental protection, total	5647,6	7294,2	7248,9	8306,7	8796,2	9561,5	10283,4
on the treatment of the wastes	722,9	806,4	1016,6	1285,8	2623,2	2990,6	3072,5
In % to the total volume	12,8	11,1	14,0	15,5	29,8	31,3	29,9
Investments in fixed assets, total	985,8	2298,4	2851,4	3592,8	5915,8	4792,1	2358,4
Total	6633,4	9592,6	10100,3	11899,5	14712	14353,6	12641,8

Note: calculated according to the data of the Russian Federation Federal State Statistics Service.

On the other hand, with the right approach, waste plays the role of an additional source of funds, a reserve for reducing the cost of production [9-11]. A striking example is the situation with the waste of the brewing industry. The main type of waste in beer production period - brewery mash. It is formed from the sediment remaining during cooking and sucking off the wort [14]. Consists of the grain remains of barley or other crops [6]. The chemical composition of the brewery mash are given in tables 2, 3, 4 [3-5].

Table 2. The total chemical composition of dry brewery mash

Name	% content
Crude protein	23,44
Crude fiber	14,3
Crude fat	7,73
Ash	2,5
Calcium	0,37
Phosphorus	0,5
Moisture	8,67
N-free extractive substances	43,34

Table 3. Content of basic microelements in dry brewery mash

Name	Contents
Calcium, %	0,37
Phosphorus, %	0,50
Manganese, mg / kg	52,0
Zinc, mg / kg	105,0
Iron, mg / kg	205,0
Copper, mg / kg	15,0

Table 4. Amino acid composition of dry brewery mash

Name	% content
Lysine	0,86
Histidine	0,66
Arginine	1,07
Aspartic acid	1,35
Threonine	0,77
Serine	0,89
glutamic acid	4,57
Proline	2,05

Glycine	0,79
Alanin	0,94
Cystine	0,46
Valin	1,06
Methionine	0,50
Isoleucine	0,79
Leucine	0,57
Tyrosine	0,61
Phenylalanine	1,23

2. Research methods

In the traditional approach, the beer pellet is either disposed of at the waste landfill or on a special site organic fertilizers are prepared from them [8]. To properly dispose of it, a large area for storing waste is required, and the possibility of use in the future as an organic fertilizer [15, 16]. All this requires significant capital and disposal costs. As shown by numerous studies and the existing practice of agricultural enterprises, the brewery mash, with appropriate organization of production and marketing of waste, can be a valuable feed for almost all groups of farm animals [21-26]. The main obstacle for this is the short shelf life of a wet brewery mash [12, 13].

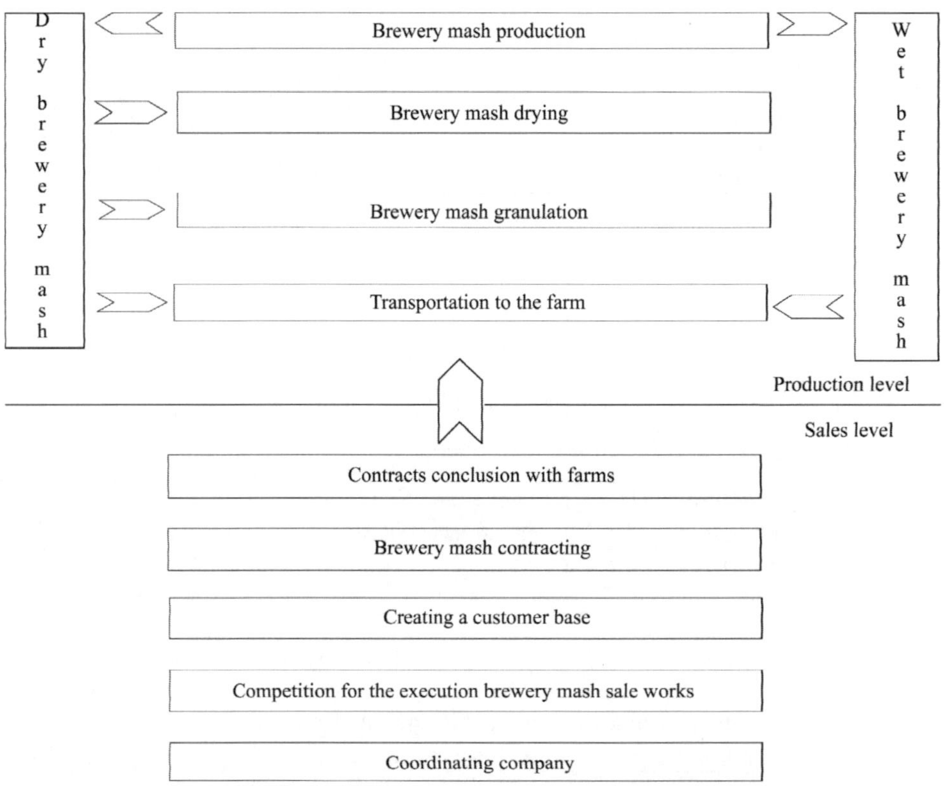

Figure 1. Elements of the waste management system in a brewing company (for example, brewery mash)

In the Samara region, the following waste management scheme is used to reduce the cost of waste disposal and obtaining additional economic effect at breweries. (Fig. 1) The system is based on two main elements: the construction of production for drying and granulation brewery mash, due to which the period of possible brewery mash use increases dramatically, the possibility of transporting the product improves, the possibility of using brewery mash as a component of compound feeds appears (production level); transfer to the outsourcing function for the implementation of the brewery mash, which reduces the risks for the implementation of the product (sales level). The coordinating company is determined by the results of an annual competition, which allows the brewing company to effectively monitor the work and fulfill the contract conditions. Such an organization of the production process and its maintenance ensures the implementation of dry and wet brewery mash with little or no inventory at the enterprise, saving on the construction of warehouses and their maintenance.

One of the possibilities for utilization of waste from the brewing industry is cooperation with pig farms in the region to reduce the fattening pigs cost.

A significant role in the formation of the economic result and cost play feed. At the pig-breeding complex, pigs were fed during the fattening period according to standard recipes provided by the production technology. In the first period of fattening (from 38 to 68 kg), the animals were fed SK-6 feed, in the second period (from 68 to 112 kg) SK-7 feed. 1 kg of such feed contained 1.10-1.17 fodder units, 154-136 g of crude protein and 129-115 g of digestible protein, respectively. In the current environment, the cost of feed in the pork production cost structure is 65-75%. Therefore, it is necessary to look for new types of feed, which will either reduce costs by reducing the cost of 1 fodder unit.

In order to reduce the cost of feed for feeding pigs for fattening, a recipe for feed PC-55 was developed. In 1 kg of feed brand PC-55 contains 0,07 fodder units and 5 g of crude protein less than in the feed CK-6.

The specialists of the pig-breeding industry were given the task of maximally reducing the cost of preparing mixed feeds. This solves the problem of not only reducing costs, but also compliance with the supplied feed formulation. It was decided to work out the replacement of the grain group with a brewery mash, which is formed as a waste of beer production. Fresh brewery mash has a light or slightly chocolate color. It contains mostly insoluble sediment with residues of barley, rice and oatmeal. When fresh, it contains 20,3-20,5% of dry matter and an average of 0,23 fodder units. An experiment was put on replacing part of the feed with a brewery mash in a ratio of 1:10 - 1-2 kg of brewery mash per 100-200 g of feed of recipe PK-55.

Production experience was carried out for two months. For this purpose, four experimental and two control groups of local animals obtained from crossing a yorkshire breed with boars producing landrace breed were formed at the fattening zone of the pig-breeding complex.

The experimental groups I and II and the control group I included animals with a living weight from 45 to 48 kg, the experimental groups III and IV and the second control group included gilt with a living weight of 55-60 kg. The schooling of pigs to the brewery mash was carried out gradually over the course of one week, starting from 0,2 kg per head per day. Feeding pigs of the experimental and control groups was carried out 2 times a day with liquid feed of 75% moisture with a temperature of 30-35 ° C. The control groups were fed according to the norms of feeding for these groups of animals, pigs I and III of the experimental groups 200 g of feed were replaced with 1 kg of fresh brewery mash, II and IV experimental groups instead of 200 g of feed were fed 2 kg of brewery mash. Entering fresh brewery mash into the diet did not have a negative impact on the feed palatability. The remains of brewery mash in the feeders one hour after feeding in all experimental groups was not observed. The state of health in animals of the experimental groups was similar to the controls, and no deviations from normal physiological indications were established. The results are shown in table 5.

Table 5. The results of production experience in the use of brewer mash

Group of animals (feeding rate)	Experimental group ± to control,%	
	1 month	2 months
A group of animals weighing 45-48 kg		
Experienced I (replacement of 200 g of feed for 1 kg of brewery mash)	-0,03	+0,07
Experienced II (replacement of 200 g of feed for 2 kg of brewery mash)	+5,2	+5,2
A group of animals by weight category 55-60 kg		

Experienced III (replacement of 100 g of feed for 1 kg of brewery mash)	+4,8	+3,8
Experienced IV (replacement of 100 g of feed for 2 kg of brewery mash)	+7,4	+6,9

3. Conclusion

As can be seen from the data in the table, a 1:20 replacement provides almost the same gain with the control group. The recommended replacement of 1:10 (100 g per 1 kg of brewery mash or 200 g of feed by 2 kg) increases the monthly weight gain by 5-7%.

The introduction of a new feeding scheme will require minimal cost. To create a point for receiving a brewery mash, storing it and mixing it with a fattening compound feed of a standard recipe will require an investment of 200 thousand rubles.

The creation of the proposed waste management system at the beer industry enterprises will significantly reduce the amount of waste, remove the unpleasant smell from the decaying brewery mash, and increase the profitability of the enterprise as a whole. Attracting a special company for the sale of dry brewery mash will reduce the entrepreneurial risks of the project.

References

[1] Nosov V V, Kozin M N, Gladun T N 2015 *Ecology, Environment and Conservation* **21**(S) pp 103–10
[2] Nosov V V, Kozin M N, Andreev V I, Surzhanskaya I Y, Murzina E A 2016 *Res. J. of Pharm., Biolog. and Chem. Sci.* **7**(6) pp 382–5
[3] Muster-Slawitsch B, Brunner C 2018 *RSC Green Chemistry* **55** pp 430-61
[4] Hegde S, Lodge J S, Trabold T A 2018 *Renew. and Sustain. Energy Rev.* **81** pp 510-23
[5] Bamforth C W 2017 *Annual Rev. of Chem. and Biomolec. Eng.* **8** pp 161-76
[6] Halász A, Lásztity R 2017 *Use of Yeast Biomass in Food Production* pp 1-312
[7] Tippmann J, Becker T 2016 *Chemie-Ingenieur-Technik* **88** (12) pp 1857-68
[8] Djukić-Vuković A, Mladenović D, Radosavljević M, Kocić-Tanackov S, Pejin J, Mojović L 2016 *Waste Management* **48** pp 478-82
[9] Muster-Slawitsch B, Hubmann M, Murkovic M, Brunner C 2014 *Chem. Eng. and Processing: Process Intensification* **84** pp 98-108
[10] Konrád Z, Los J, Fryč J, Kudělka J 2014 *Res.in Agricult. Eng.* **60**(1) pp 10-6
[11] Schnitzenbaumer B, Arendt E K 2013 *European Food Res. and Tech.* **236** (6) pp 1015-25
[12] Arendt E K, Zannini E 2013 *Cereal Grains for the Food and Beverage Ind.* (Woodhead Publishing Limited) pp 1-485
[13] Thomas K 2011 *Liquid Bread: Beer and Brewing in Cross-Cultural Perspective* 7 pp 35-46
[14] Gupta M, Abu-Ghannam N, Gallaghar E 2010 *Comprehensive Reviews in Food Science and Food Safety* **9**(3) pp 318-28
[15] Russell I, Stewart G G 2008 *Biotechnology: Second, Completely Revised Edition* **9-12** pp 419-62
[16] Zanker G, Kepplinger W, Pecher C 2007 *Utilization of By-Products and Treatment of Waste in the Food Industry* pp 273-81
[17] Akhtar A, Sarmah A K 2018 *J. of Cleaner Prod.* **186** pp 262-81
[18] Inglis M 2007 Best Practice and Cost Saving *SB07 New Zeal. Pap.* **057** pp 10-2
[19] Yang W-S, Park J-K, Park S-W, Seo Y-C 2015 *J. Mater. Cycles Waste Manag.* **17** pp 207-17
[20] Barreto V B, Amado F D R, Cruz K V 2014 Engineering Optimization IV *Proceedings of the 4th International Conference on Engineering Optimization, ENGOPT 2014* pp 753-6
[21] Moustsen V, Niemi J, Kyriazakis I, Maes D 2018 *Preventive Veterinary Medicine* **159** pp 106-14
[22] Hansson S O, Åman P, Becker W, De Koning D-J, Lagerkvist C J, Larsson I, Lehrman A, Risérus U, Stymne S 2018 *Trends in Food Sci. and Tech.* **80** pp 131-40
[23] Lopez B I M, Song C, Seo K 2018 *Animal Sci. J.* **89** (10) pp 1381-8
[24] Park H-S, Oh S-H 2018 *J. of Appl. Animal Res.* **46** (1) pp 587-92

[25] Méndez-Palacios N, Méndez-Mendoza M, Vázquez-Flores F, Castro-Colombres J G, Ramírez-Bribiesca J E 2018 *Animal Sci. J.* **89** (7) pp 994-1001
[26] Alves K, Schenkel F S, Brito L F, Robinson A 2018 *J. of Animal Sci* **96** (7) pp 2567-78

IOP Conf. Series: Earth and Environmental Science **337** (2019) 012010 doi:10.1088/1755-1315/337/1/012010

Wastewater treatment of industrial enterprises via adsorption method

O Atamanova[1*], E Tikhomirova[1], A Koshelev[1], M Istrashkina[1], N Politayeva[2], A Podolsky[1]

[1] Yuri Gagarin State Technical University of Saratov, 77 Polytekhnicheskaya St., Saratov, Russia
[2] Peter the Great St. Petersburg Polytechnic University, Polytechnicheskaya 29, St. Petersburg, Russian Federation, Russian Federation

*Corresponding author: O_V_Atamanova@mail.ru

Abstract. The publication justifies the feasibility of adsorption method of industrial wastewater treatment from aromatic compounds. Objects and methods of research are described. Studies of adsorption mechanisms of aromatic compounds on bentonites of various modifications made it possible to establish the most effective sorption rock clay-based materials, as well as filter compositions. Bentonite modified with carbon nanotubes after heat treatment 550 ° C proved to be the most effective with respect to the studied aromatic compounds. The principle of combining sorbents in a single filtering system, based on the sorbent selection for various groups of organic pollutants according to their ability to be ionized in an aqueous medium, has been proposed and investigated. Three most effective multicomponent adsorption filter compositions for sewage treatment from nitro and amino compounds were established. The proposed approach is aimed at designing local water treatment stations for industrial enterprises to deal with aromatic compounds via use of multicomponent adsorption filter compositions.

1. Introduction

The development of improved wastewater treatment methods in sewage systems is a priority area of modern industrial ecology. This procedure is associated with designing highly efficient filtering systems that can provide wastewater treatment for various industries. The most harmful wastewater is produced in the course of industrial production of dyes, chemicals and medicines. Cleaning the effluent of these industries from organic pollutants is especially difficult [1-3]. It has been established [4] that use of adsorption filtering systems is most appropriate for wastewater treatment from organic contaminants. Therefore, the improvement of adsorption technologies and sorbent compositions through their chemical modification plays an important role in creating new, more efficient wastewater treatment methods for enterprises in the field of chemical industry [5].

Development of new methods for adsorption treatment of polluted waters has been conducted at the Department of Ecology, Yuri Gagarin State Technical University of Saratov (SSTU), since 2011 [6-8]. Department scientists focused their efforts on using inexpensive and readily available natural clay materials, primarily bentonites, as sorbents [9-12]. Bentonites are multi-mineral formations consisting predominantly of silicates. Clay rocks are usually highly ion-exchangeable. This property is typical for the minerals with a so-called "sliding" crystal lattice, such as montmorillonite. When this mineral interacts with water, a water molecule is able to enter the gaps between the layers of montmorillonite crystal lattice and push them apart. It is this property of the mineral that explains its high adsorption capacity.

SSTU Department of Ecology scientists developed complex sorbent compositions of a new generation for efficient removal of toxic organic substances, heavy metals and oil refining by-products from industrial wastewater. These compositions were also found effective for water disinfection.

Multi-layer filter compositions are of particular interest, since they provide comprehensive treatment of polluted water from several pollutants simultaneously. Current experimental studies conducted at SSTU Department of Ecology are directed at creating filter compositions for treating wastewater containing various organic toxicants. Special attention is paid to the extraction from the water of benzene (aromatic) compounds that penetrate into natural water bodies with sewage of dye, chemical, pharmaceutical, light, and other types of industry [13]. Benzene compounds are the most toxic and biologically steady pollutants [14, 15].

2. Objects and methods of research

The object of our study was wastewater from industrial enterprises. To solve the problem of wastewater treatment, we proposed various structural modifications of the bentonites: by high-temperature heat treatment (HTHT) at different regimes (with preliminary or direct calcination at a given temperature) and by organic components - carbon nanotubes (CNT) and glycerol, both individually and in combination with each other. Modification of bentonites was carried out jointly with our collaborators from scientific and production enterprise "LISSKON".

Major employed research methods were spot test analysis [16], photometry, titration, etc.

We selected nitro and amino compounds - o-toluidine, o-phenylenediamine and p-dinitrobenzene - as model pollutants for our laboratory studies. Concentrations of these chemical compounds, exceeding MAC, were found in wastewater of textile production in Saratov region [17].

Laboratory studies of various filter compositions using modified bentonites included the following:

1. The study of the bentonite adsorption capacity with respect to aromatic organic compounds for different variants of bentonite modification.
2. The study of multicomponent adsorption filter compositions for water treatment from nitro and amino compounds.
3. Selection of the most effective filter compositions for adsorption filtering systems used for wastewater treatment of industrial enterprises from a complex of organic pollutants. The model water solution to be filtered included the following pollutants: p-dinitrobenzene, m-aminophenol, o-phenylenediamine, p-nitroaniline and o-toluidine. 200 ml of the model solution were filtered through each filter composition.

3. Results and Discussion

The results of our studies on the adsorption capacity of modified bentonites with respect to selected organic pollutants (o-phenylenediamine, o-toluidine and p-dinitrobenzene) revealed that all four modifications of the sorbents selected for research have different adsorption capacity. The highest adsorption capacity was manifested for the bentonite modified with CNT at a firing temperature of 550 ° C (direct calcination); somewhat lower adsorption capacity was shown for the bentonite modified with glycerol and CNT at a firing temperature of 550 ° C (direct calcination); still lower adsorption capacity was found for the bentonite modified with glycerol and CNT at a firing temperature of 550 ° C (calcination with preheating for 30 minutes); finally, the lowest adsorption capacity was discovered for the bentonite without modifiers (direct calcination at a temperature of 550 ° C). Such pattern was observed for all studied pollutants.

Established parameters for o-phenylenediamine, o-toluidine and p-dinitrobenzene adsorption on investigated sorbents in dynamic conditions are included in Table 1.

IOP Conf. Series: Earth and Environmental Science **337** (2019) 012010 doi:10.1088/1755-1315/337/1/012010

Table 1 – Adsorption effectiveness parameters of organic substances on investigated sorbents

Organic substance	Bentonite modification	DEC, mg/g	K_d, mg/dm3	S, %
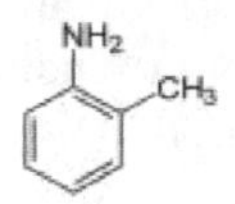	unmodified	3,089±0,011	1,907±0,030	94,600±0,060
	CNT	3,935±0,049	2,623±0,030	95,000±0,110
	glycerol	3,645±0,051	2,433±0,020	94,801±0,100
	glycerol and CNT	3,499±0,010	2,187±0,04	94,667±0,360
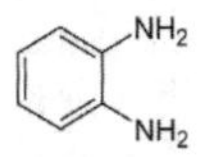	unmodified	2,066±0,013	1,878±0,017	94,500±0,140
	CNT	2,872±0,021	2,872±0,011	95,000±0,350
	glycerol	2,521±0,041	2,480±0,031	94,950±0,250
	glycerol and CNT	2,341±0,014	2,209±0,039	94,700±0,300
	unmodified	1,012±0,051	1,177±0,060	94,267±0,990
	CNT	1,322±0,036	1,762±0,080	95,000±0,430
	glycerol	1,173±0,062	1,563±0,033	95,000±0,120
	glycerol and CNT	1,023±0,010	1,421±0,050	95,200±0,510

DEC – dynamic exchange capacity, K_d – interfacial partition coefficient, S – adsorption rate

In the study of multicomponent adsorption filter compositions, it was established experimentally that the most effective compositions with respect to complex pollution with organic substances were the compositions [18, 19], containing the following sorbents:

1. Bentonite modified with CNT, thermally treated at 550 ° C; AB-17-8 anion exchange resin; ASKG silica gel.
2. Peat; zeolite; organobentonite TS 952752-2000; bentonite modified with heat treatment at 550 ° C; bentonite modified with glycerol after heat treatment at 550 ° C.
3. Zeolite; organobentonite TS 952752-2000; bentonite modified with heat treatment at 650 ° C; bentonite modified with glycerol after heat treatment at 650 ° C.

Figure 1 presents the results of comparative analysis of filter compositions.

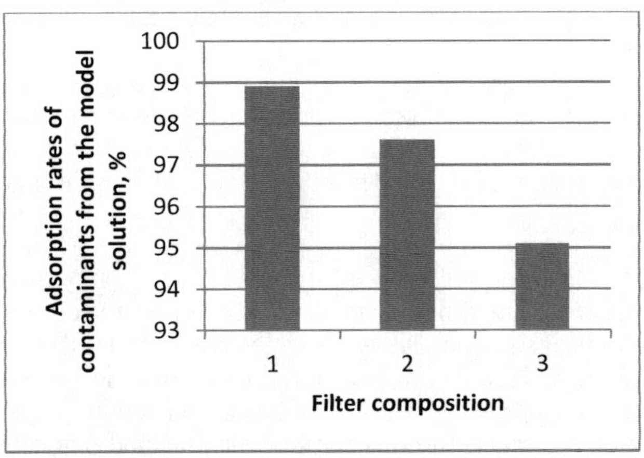

Figure 1. Adsorption rates (%) of nitro and amino compounds from the model solution

The effectiveness of water treatment from amino and nitro compounds for each of three listed filter compositions exceeded 95% [20].

4. Conclusions

Studies of the adsorption capacity of modified bentonites with respect to o-phenylenediamine, o-toluidine and p-dinitrobenzene established that model solutions filtered through bentonite modified with CNT (direct calcination at 550 ° C) stably reached a negative reaction faster than in case of other sorbents. It was also found that the highest efficiency of water treatment from the listed pollutants was observed for the bentonite modified with CNT (direct calcination at 550 ° C, no preheating).

The principle of combining different sorbents in the structure of a single filtering system, based on sorbent selection for various groups of organic substances according to their ability to be ionized in an aqueous medium, has been proposed and investigated. In relation to the amino and nitro compounds, the most effective filter composition contained the following sorbents: bentonite modified with CNT, thermally treated at a temperature of 550 ° C (direct calcination); AB-17-8 anion exchange resin; ASKG silica gel.

5. Acknowledgements and Conflict of Interests

The authors are grateful to the scientific-industrial enterprise LISSKON located in the city of Saratov, Russia, for using their laboratory space and equipment for conducting the research. The study was supported by the grant No. 5.3922.2017/PCh from the Russian Federation Ministry of Education. The authors state no conflict of interests involved.

References

[1] Lu G, Wang C, Tang Z, Guo X 2007 *Ecotoxicology* **16**(7) pp 485–90
[2] Valero E, González-Sánchez M-I, Pérez-Prior M-T 2014 *Env., Energy and Climate Change I* **32** pp 317-39
[3] Agladze G, Tsurtsumia G, Jung B-I, Kim J-S, Gorelishvili G 2007 J. of Appl. Electrochem. **37**(9) pp 985–90
[4] Koganovskiy A, Klimenko N, Levchenko T 1983 Ochistka i ispolzovaniye stochnykh vod v promyshlennom vodosnabzhenii (Leningrad: Chemistry) (Russian)
[5] Dytnersky Yu I 1995 *Processes and equipment of chemical technology: textbook for universities. 2 nded. Part 2. Mass exchange processes and equipment* (Moscow: Chemistry)
[6] Minzatu V, Negrea P, Negrea A, Ciopec M, Bumbu B, Golban A, Vasile A 2016 *WIT Transactions on Ecology and the Env.* **209** pp 185-95
[7] Istrashkina M, Atamanova O, Tikhomirova E 2016 *Izvestiya Samarskogo nauchnogo tsentra RAN* **18** 2(2) pp 381-4 (Russian)
[8] Istrashkina M, Tikhomirova E, Atamanova O, Vedeneyeva N 2016 *Izvestiya Samarskogo nauchnogo tsentra RAN* **18** 2(3) pp 687-91 (Russian)
[9] Adeyemo A, Adeoye I, Bello O 2017 *Appl. Water Sci.* **7** (2) p 543
[10] McKay G, Otterburn M, Aga J 1985 *Water, Air, and Soil Pollution* **24**(3) pp 307–22
[11] Lagaly G 2007 *Surfactants and Colloids in the Environment* **95** pp 61-72
[12] Golubev V, Garibyants A 1971 *Heterogeneous Processes of Geochemical Migration* (Moscow: Plenum Publishing Corporation) pp 45-60
[13] Vasilenko T, Mishina A 2018 *Solid State Phenomena* **284** pp 719-25
[14] Neun D, Penn A, Snyder C 1994 *Archives of Toxicology* **68**(9) pp 535–40
[15] Branton P, Bradley R 2011 *Adsorption* **17** (2) pp 293–301
[16] Faygl F 1962 *Kapelnyi analiz organicheskikh veshchestv: per. s angl.* ed Kuznetsova V (Moscow: Goskhimizdat) (Russian)
[17] *Sovershenstvovaniye sorbtsionnykh metodov ochistki zagryaznennykh prirodnykh I stochnykh vod* ed Tikhomirovoy E (Saratov: SGTU imeni Gagarina Yu.A.) (Russian)
[18] Wu H, Li X, Zhao W, Meng Z, Zhang W, Qian L 2017 *Materials Science Forum* **898** pp 1923-8
[19] Kharlamova T, Gorokhova L 1980 *Voprosy okhrany okruzhayushchey sredy na predpriyatiyakh sluzhby byta* (Moscow: MTI) pp 89-97 (Russian)
[20] Istrashkina M, Tikhomirova E, Podolsky A, Atamanova O 2017 *Proceedings of the Sixth International Environmental Congress (Eighth International Scientific-Technical Conference) "Ecology and Life Protection of Industrial-Transport Complexes" ELPIT-17 20-24 September. Samara-Togliatti, Russia: Edition ELPIT* (Samara: Publishing House of Samara Scientific Centre) pp 1007

Hydrothermal liquefaction-isomerization of biomass for biofuel production

M Kotelev[1], I Tiunov[1], Evgenii Ivanov[1] and Z Namsaraev[2*]

[1]Gubkin Russian State University of Oil and Gaz, 119991 Moscow, Russia
[2]NRC Kurchatov Institute, 123182 Moscow, Russia

* Corresponding author: zorigto@gmail.com

Abstract. During the last years, a lot of research activity has been devoted to a biomass conversion into liquid fuels. In order to produce fuels of the same quality as those obtained from crude oil, an appropriate conversion technology is required. Hydrothermal liquefaction (HTL) of biomass is a catalytic process that is considered to be a very promising technology for biofuel production. HTL products have a relatively high content of gasoline and diesel fractions, low oxygen content and high heating value. Moreover, HTL does not require a drying of the raw materials prior to conversion that allows a significant reduction of the price of the biofuel. We developed a new catalytic system for HTL, consisting of halloysite nanotubes as aluminosilicate carrier and strontium/titanium/tin oxide as an isomerizing component. Due to the sufficiently high specific surface area of halloysite nanotubes, a higher content of gasoline fraction in biocrude products was achieved. In addition, the introduction of the isomerizing component significantly increased the octane number of the gasoline fraction.

1. Introduction

Currently, many methods have been proposed for the conversion of biomass into liquid hydrocarbons. Two main approaches can be distinguished: 1) processing of individual components of biomass (in particular, hydrolysis of cellulose followed by fermentation to obtain bioethanol, transesterification of triacylglycerides); 2) conversion of biomass as it is. The second approach includes pyrolysis that can be carried out either in the presence of hydrogen (hydropyrolysis) or without it, and hydrothermal liquefaction (HTL) to produce bio-oil. Of all the methods, the HTL process is the most economically feasible, since its implementation does not require energy-intensive deep drying of biomass, as in the case of pyrolysis. At the same time, raw materials with moisture content of 50–90% by weight are quite suitable for processing. Thus, the biomass of green plants, aquatic vegetation or microalgae can be converted directly into bio-oil [1]. During the HTL process, biomass is subjected to rapid heating in a closed volume to a temperature of 250-600 °C, and then it is kept at the maximum temperature for a while. Then the reaction mass is cooled, and fractioned.

The main products of the HTL process are: a mixture of methane and C2-C3-hydrocarbon gases, a mixture of liquid hydrocarbons or bio-oil, aqueous phase containing ammonium nitrogen, carboxylic acids and other water-soluble organic substances, and solid residues. The heating rate, maximum temperature and exposure time can be varied during the process. The most valuable component is bio-oil, so the HTL process is usually carried out under such conditions that maximize its yield. In particular, the use of catalysts increases the yield of bio-oil and usually helps to reduce the process temperature, and in some cases affects the quality of the product as well.

To assess the quality of bio-oil, such indicators as fractional composition (gasoline and diesel fraction content) oxidative stability of individual fractions, and density are used. In addition, the following indicators are of particular interest from practical point of view: the octane number of the gasoline fraction and the low-temperature properties of the diesel fraction.

For example, in patent [2] authors describe a heterogeneous catalyst for hydrothermal liquefaction of biomass consisting of a hydrogenating component (ruthenium) and porous carrier with pore volume of 30-60 %. However, the use of the noble metal as a hydrogenating component greatly increases the total cost of the catalyst and the entire technological process.

Homogeneous catalysts are also used for hydrothermal liquefaction of microalgae biomass [3]. The increase in bio-oil yield for various catalysts occurs in the following order: Na2CO3 > CH3COOH > KOH > HCOOH. However, usage of organic acids as catalysts leads to increased nitrogen content in the bio-oil, compared to the bio-oil obtained with alkali catalysts. One of the disadvantages of using homogeneous catalysts is that they are consumed during the process and cannot be regenerated, which leads to low economic efficiency of the process.

Biller and coauthors [4] studied three heterogeneous catalyst of the following composition: $Co/Mo/Al_2O_3$, Pt/Al_2O_3 and Ni/Al_2O_3. A 10 % increase in low heating value of the bio-oil and degrease of oxygen content in the products was observed, while there was no significant yield change detected. It was found that $Co/Mo/Al_2O_3$ and Pt/Al_2O_3 catalysts have a greater effect on carbohydrates and proteins of biomass, while Ni/Al_2O_3 reduces the oxygen content in the products of lipids decomposition.

Saber and coauthors [5] studied effect of various catalysts (nano-Ni/SiO_2, zeolite and Na2CO3) on the yield and quality of bio-oil at low temperatures (210 °C, 230 °C, 250 °C) produced from microalgae *Nannochloropsis sp.* It was established that the use of catalysts leads to a decrease in the content of oxygen and nitrogen in bio-oil while increasing the yield; bio-oil yield decreases in the series: nano-Ni/SiO_2 > zeolite> Na_2CO_3. The maximum yield was achieved with nano-Ni/SiO_2 catalyst is achieved at a temperature of 250 °C and was 30%.

The efficiency of using HZSM-5 and Ce/HZSM-5 catalysts was demonstrated by Xu and coauthors on *Chlorella pyrenoidosa* microalgae [6].

Another heterogeneous catalyst for the HTL process is described in patent [7]. The catalyst consists of nickel oxide on the surface of powdered clay. The surface area of the catalyst is relatively low (10-15 m^2/g) which hinders transport of biomass components towards catalyst's active centres, that leads to increased content of high-boiling components in the product. In addition, the absence of isomerization activity of the used catalyst leads to a low quality of the product (octane number of the gasoline fraction of the bio-oil does not exceed 66).

Table 1 shows summarized literature data on the processes of hydrothermal liquefaction of microalgae biomass.

Table 1. Literature data.

Microalgae strain	T, °C	Exposure time, min	Catalyst	Yield, %	Source
Cyanobacteria sp.	325	45-60	-	21.0	[8]
Bacillariophyta sp.	325	45-60	-	18.2	[8]
Chlorella pyrenoidosa	280	120	-	39.4	[9]
Aurantiochytrium sp. K RS101	400	10	-	51.2	[10]
Nannochloropsis sp.	600	1	-	67.0	[11]
Nannochloropsis sp.	250	60	nano-Ni/SiO2	30.0	[5]
C. pyrenoidosa	300	20	HZSM-5	34.0	[6]
C. pyrenoidosa	300	20	Ce/HZSM-5	49.9	[6]
Spirulina / Chlorella	350	60	1 M Na_2CO_3	20.0/27.3	[12]
Spirulina / Chlorella	350	60	1 M KOH	15.2/22.4	[12]
Spirulina / Chlorella	350	60	1 M HCOOH	14.2/19.1	[12]

Spirulina / Chlorella	350	60	1 M CH3COOH	25.3/204	[12]
Nannochloropsis oc. / Chlorella vulgaris	350	60	Co/Mo	25.5/38.7	[13]
Nannochloropsis oc. / Chlorella vulgaris	350	60	Ni/Al	18.1/30	[13]
Nannochloropsis oc. / Chlorella vulgaris	350	60	Pt/Al	30.2/38.9	[13]

As can be seen from Table 1 in modern works, the yield of bio-oil reaches 50-60% wt. of the original microalgae biomass. A large fraction of carbon is transformed in water-soluble form. The yield of gaseous products is about 20 %. The dominant component in gaseous products is CO_2. In addition, gaseous products may contain small amounts of N_2, CH_4, and C_2-hydrocarbons.

The use of catalysts, both homogeneous and heterogeneous, is considered one of the promising ways to improve the method of HTL which is currently not well developed [14]. An alternative to catalytic HTL may be the catalytic upgrading of bio-oil in a separate process, following hydrothermal liquefaction [15].

The goal of this work is to developed a new catalyst in order to increase the efficiency of hydrothermal liquefaction of biomass. We used Al_2O_3, kaolin and halloysite as carrier for the catalyst. Halloysite is a natural alumosilicate clay consisting of nanosized tubes. Its unique structure and low cost make it of particular interest for using as a catalyst component. Halloysite TEM micrograph is shown in Figure 1.

In this paper, we investigated not only the yield of hydrothermal liquefaction products, but also the properties of the resulting gasoline and diesel fraction. In particular, an attempt was made to increase the catalyst isomerizing activity by adding oxides of strontium, titanium and zirconium. In our opinion, this should have contributed to an increase in the content of branched hydrocarbons in the products, which in turn should contribute to the increase of octane number of the gasoline fraction and improvement of the low-temperature properties of the diesel fraction.

Figure 1. TEM micrograph of halloysite nanotubes.

2. Methods

The catalyst was prepared as follows. 18 g of finely dispersed halloysite clay nanotubes or Al_2O_3 or kaolin and 1 g of isomerization agent (tin oxide, titanium oxide, zirconium oxide) were grinded and mixed. 150 g of 10 % solution of polyvinyl alcohol with molar mass of 6000 (pore-forming component) and 1.5 g of 85 % solution of orthophosphoric acid (binder component) were added to the mixture. Then the mixture was pushed through an extrusion device, dried for 5 h at 140 °C and then tempered at 400 °C for 10 h. Then the mixture was soaked in 50 g of 10 % solution of sulfuric acid for 1 h followed by drying at 250 °C for 5 h.

The catalyst's activity was evaluated using 150 g of cyanobacterium Anabaena variabilis biomass. The biomass and 5 g of the catalyst were placed in an autoclave, heated up to 300 °C at a rate of 70 °C/min under vigorous stirring and kept at this temperature for 30 minutes. Then the autoclave was cooled to the room temperature and the products were separated into liquid and solid phases by filtration followed by liquid phase separation into aqueous and organic phases in a separating funnel. The weight of organic phase was measured and bio-oil yield was calculated.

To estimate bio-oil fraction composition it was dehydrated and distilled. The following fractions were collected: initial boiling point – 180 °C (gasoline fraction) and 180 °C – 360 °C (diesel fraction). The hydrocarbon composition of the gasoline fraction was analyzed using a gas chromatograph, and the octane number of the fraction is calculated from the analysis results. To assess the low-temperature properties of the diesel fraction, its pour point is measured.

3. Results

The yield of bio-oil was 37 % of the dry weight of the biomass taken. Fraction analysis of the bio-oil showed that the content of gasoline fraction was 54 % wt., and 39 % wt. for diesel fraction. The residues accounted for the rest.

The octane number of the gasoline fraction, calculated from the results of chromatographic analysis, was 86. The pour point of the diesel fraction was minus 7 °C.

The results of experiments with various catalysts are shown in Table 2.

Table 2. HTL process results with various catalysts.

Isomerization agent	Finely dispersed component	Organic phase yield, %	Gasoline fraction content, %	Research octane number	Diesel fraction content, %	Pour point of the diesel fraction
SnO		37	54	86	39	- 7
TiO	Al_2O_3	36	55	88	38	- 10
ZrO		34	45	86	52	- 10
SnO		26	15	60	41	0
TiO	kaolin	16	28	64	40	–
ZrO		20	32	73	64	- 6
SnO	halloysite	26	13	65	72	5
TiO	clay	21	14	54	65	2
ZrO	nanotubes	30	15	66	68	- 1

As can be seen from the presented results, the use of all three isomerizing agents in combination with finely dispersed aluminum oxide leads to an increased yield of liquid products of hydrothermal liquefaction, as well as increase in the content of high-octane hydrocarbons in the gasoline fraction, and, in addition, a significant decrease in the pour point of the diesel fraction. It should be noted that the diesel fraction with a pour point in the range from -5 to -10 °C is quite suitable for the production of winter-grade diesel fuel. The average octane number of the gasoline fraction in the resulting products is 87 points, which is also a good result that exceeds a gasoline fraction obtained without a catalyst by 10-20 points [7, 16, 17].

When using kaolin and halloysite as a fine and porous component the yield of liquid products does not increases. Moreover, the content of high-boiling components in the bio-oil significantly decreases. This fact suggests that, Al_2O_3, also exhibits significant cracking activity.

The increased yield of bio-oil can be explained by the activity of the catalyst due to its greater specific surface area, and the presence of isomerization component. At the same time, a high value of the specific surface of the catalyst leads to an increased content of the gasoline component, while the isomerizing component is responsible for increase in the octane number of the gasoline fraction. Moreover, the described catalyst also improves low-temperature properties of the diesel component of the bio-oil due to the increased content of branched hydrocarbons in the latter.

4. Acknowledgements
This work was supported by the Ministry of Education and Science of the Russian Federation, grant 14.574.21.0137 (unique project identifier RFMEFI57417X0137).

References

[1] Elliott D C 2016 *Algal Research* **13**
[2] *EP2904070* 2015
[3] Biller P, Sharma B K, Kunwar B, Ross A B 2015 *Fuel* **159**
[4] Biller P, Friedman C, Ross A B 2013 *Biores. Tech.* **136**
[5] Saber M, Golzary A, Hosseinpour M, Takahashi F, Yoshikawa K 2016 *Appl. Energy* **183**
[6] Xu Y, Zheng X, Yu H, Hu X 2014 *Biores. Tech.* **156**
[7] *CN105797730* 2016
[8] Huang Y, Chen Y, Xie J, Liu H, Yin X, Wu C 2016 *Fuel* **183**
[9] Yu G, Zhang Y, Schideman L, Funk T L, Wang Z 2011 *Transactions of the ASABE* **54** (1)
[10] Vo T K, Lee O K, Lee E Y, Kim C H, Seo J-W, Kim J, Kim S-S 2016 *Chem. Eng. J.* **306**
[11] Faeth J L, Savage P E 2016 *Biores. Tech.* **206**
[12] Ross A B, Biller P, Kubacki M L, Li H, Lea-Langton A, Jones J M 2010 *Fuel* **89** (9)
[13] Biller P, Riley R, Ross A B 2011 *Biores. Tech.* **102** (7)
[14] Barreiro D L, Prins W, Ronsse F, Brilman W 2013 *Biomass and Bioenergy* **53**
[15] Barreiro D L, Gómez B R, Ronsse F, Hornung U, Kruse A, Prins W 2016 *Fuel Proces. Tech.* **148**
[16] Baker E G, Elliott D C 1988 *Research in Thermochemical Biomass Conversion* (Netherlands: Springer)
[17] Biller P, Ross A B 2011 *Biores. Tech.* **102** (1)

Using the method of biotesting to assess the toxicity of waste medical and biological practices containing nanomaterials

T Shulgina[1*], O Nechaeva[2], A Torgashova[2] and N Darin[3]

[1]Saratov Research Institute of Traumatology, Orthopedics and Neurosurgery, 148 Chernyshevskystreet, Saratov, Russia;
[2]Saratov State Technical University, 77 Politechnicheskayastreet, Saratov, Russia;
[3]LLC "M9", 10 Komzina street, Togliatti, Russia

* Corresponding author: tshylgina2012@yandex.ru

Abstract. One of the current modern scientific directions is the research and introduction of products obtained using nanotechnologies into various spheres of human activity. Among the variety of nanostructures, metal nanoparticles are found to be the most widely used, which, due to the uniqueness of their properties and small size, are considered as the most promising agents in the composition of bactericidal agents. They are used as specific carriers of drugs and as components in the composition of antibacterial agents. In addition, it has been proven that the bacteriolytic and bacteriostatic properties of some metals are greatly enhanced with decreasing size. Nanoparticles of various metals, among which silver nanoparticles are the most widely used, can be used independently; however, their high ability to aggregate requires the choice of stabilizers, which usually use natural and synthetic polymers. However, their use is limited due to the high toxicity of metallic nanoparticles. Therefore, the search for the safest options for metal nanoparticles is relevant. This article presents data on the toxicity study of prototypes of silver nanoparticles stabilized by synthetic and natural polymers on biological test objects.

1. Introduction

Today, the pharmaceutical industry is the main component in maintaining the health of the population, which leads to an increase in the quality and duration of life. But at the same time, there is a downside – it is the waste of the pharmaceutical industry, which have an impact on the environment. Pharmaceutical contamination of wildlife is closely associated with increased consumption of medicines, which, in turn, contributes to demographic aging, the increasing prevalence of chronic diseases, the availability of low-cost treatment and the emergence of new drugs [1]. Currently, about 4000 active drugs have been registered [2, 3].

Pharmaceutical substances entering the environment can vary in a certain way under the influence of both biotic and abiotic factors. However, a number of compounds are not only resistant to natural destruction, but also able to change their activity with the formation of physicochemical modifications that pose a potential danger to living organisms and humans [3, 4]. These include, in particular, nitrogen-containing heterocyclic compounds exhibiting antimicrobial activity, some analgesics, as well as drugs used in sleep disorders, epilepsy and other diseases [2].

Nanomaterials are considered as the basis for the formation of a new technological element. Their essence consists of the purposeful manipulation of material objects and the creation of structures in the range of sizes from units to tens of nanometers. They are characterized by unique physicochemical

IOP Conf. Series: Earth and Environmental Science **337** (2019) 012012 doi:10.1088/1755-1315/337/1/012012

properties associated with a high ratio of their surface to volume, which indicates a high efficiency of their action [5-9]. Nanotechnologies are used to create biocompatible materials for surgery, antiseptic dressings, for targeted drug delivery to tissues [10-21].

Effective vaccines of a new generation are being developed, which are hybrid nanostructures formed by nanometers of artificial origin together with recombinant DNA molecules and virus-like particles [22, 23]. Therefore, nanoparticles should be carefully investigated not only for the toxicity of a particular nanoparticle, but also for the degree of exposure to its compounds, as well as their effect depending on the concentration of the substance in the medium and the ability to react with other compounds. The main risk factor for the use of nanotechnology in medicine is the lack of information about the interaction of specific nanoparticles with the human body [24-26].

At the same time, to assess the safety of nanomaterials, it is necessary to develop new toxicological approaches, since the inapplicability of traditional dosimetryto them [27, 28].

At present, it is necessary to change the previously established views on the problem of utilization of pharmacological waste in connection with anthropogenic changes occurring in the environment. There are four main ways of utilization of pharmaceutical trades, regulated by 87 and 155 instructions of the Ministry of Environment: discharge into industrial sewage system; burning; accommodation on specially equipped sanitary landfills; splitting up. None of the ways is environmentally friendly. The development of a unified integrated approach to the search for environmentally friendly methods of disposal and recycling of waste today is an important issue that requires a consistent and focused decision [29-32].

One of the methodological techniques that allows you to evaluate the effect of toxic environmental factors on the organism as a whole, or on its individual function is biotesting. Using biotesting methods, it is possible to model the effects of a particular factor with a general biological effect on a living organism, which contributes to a rapid assessment of the toxicity of the studied factors.

Due to the relevance of the above stated material, the purpose of the study was to evaluate the acute toxicity indicators of silver nanoparticles stabilized by various polymeric compounds on the model of a biotest object *Daphnia magna* Straus.

2. Objects and methods of research

The acute toxic effect of prototypes of silver nanoparticles stabilized by natural and synthetic polymers was estimated by the mortality of Daphnia magna Straus after 48 hours of exposure according to the standard method [33].

As prototypes, silver nanoparticles solutions stabilized with natural polymers were used: carboxymethyl cellulose (Ag / CMC), sodium oleate (Ag / OleNa) and synthetic polymer: dioxyl sodium sulfosuccinate AOT (AgW), sodium dodecyl sulfate (Ag / SDS), polyazolidine ammonium, hygrochloride, sodium dodecyl sulfate (Ag / SDS), polyazolidine ammonium, hydrochloride, sodium dodecyl sulfate (Ag / SDS), polyazolidine ammonium, polymerase, sodium dodecyl sulfate (AgW) hydrate ions of iodine (Ag / PG-m), polyvinyl alcohol (Ag / PVA). The working concentrations of the test samples were 3%, 2%, 1%, 0,5%, 0,25% and 0,125%.

According to the method, the assessment of the impact of experimental samples of silver nanoparticles presented in the test medium was carried out according to the daphnia mortality rates (*D. magna* Straus) in comparison with the control culture in samples that did not contain the dispersions studied (control). The criterion of the acute toxic effect of the compound was the death of 50% or more of the biotest objects for 48 hours in the test sample, as compared with the control samples in which all crustaceans maintained their viability.

Daphnids mortality counts were taken every 24 hours in experimental and control samples. The experiment was stopped if more than 50% of the crustaceans were observed to die within 24 hours in all the test tubes. Motionless individuals were considered dead if they did not begin to move for 15 seconds after lightly rocking the tube.

IOP Conf. Series: Earth and Environmental Science **337** (2019) 012012 doi:10.1088/1755-1315/337/1/012012

The number of dropped efippiums was also taken into account, since it was established that as a result of the reaction to the toxicant, daphnids molt and ephippium drops.

In experiments to determine the acute toxic effect of experimental samples of silver nanoparticles were established:

- lethal concentration causing death of 100% test organisms (LC_{100-48});
- the average lethal concentration causing death of 50 % and more test organisms (LC_{50-48});
- the harmless concentration, causing the death of not more than 10% of the test objects (HC_{10-48});
- the minimum concentration of a harmless compounds, when were not observed death of organisms (HC_{0-48}).

Statistical processing of the data was performed in the program "Statistica 6.0".

3. Results and Discussion

In the course of the research, it was found that according to the calculation of the percentage of dead daphnia relative to the control, all nanoparticle samples showed a toxic effect on the biotest-object under study (table 1).

Based on the data obtained, the acute toxicity indicators for the investigated aqueous dispersions of metal nanoparticles were determined, which are presented in Table 2.

As a result of the research, it was found that solutions of silver nanoparticles were characterized by different levels of toxicity with respect to test objects, which depended on the stabilizer content in the sample.

The study of the toxicity indicators of silver nanoparticles stabilized by carboxymethyl cellulose made it possible to establish an acute toxic effect of the drug against daphnia in the concentration range of 0,5–3%, for which the toxicity criteria of 0.125% and 0.25% toxicity were exceeded, had no toxic effect.

Table 1. Daphnia survival rate in test samples of silver nanoparticles under study (exposure 48 hours)

Laboratory codes of prototypes	Daphniadeathrate A, % Compound concentrations, %						control
	3	2	1	0,5	0,25	0,125	
AgW	100	80	50	20	10	0	0
Ag/CMC	90	90	70	40	25	5	0
Ag/OleNa	80	80	70	70	50	40	0
Ag/PG-m	90	80	50	40	15	0	0
Ag/PVA	90	70	70	50	40	25	0
Ag/SDS	90	90	75	70	50	15	0

Table 2. Indicators of the acute toxicity of silver nanoparticles (%)

Prototypes	LC_{100-48}	LC_{50-48}	HC_{10-48}	HC_{10-48}
AgW	3	1-2	0,25	0,125
Ag/CMC	-	3-0,5	0,25	0,125
Ag/OleNa	-	3-0,25	0,125	-
Ag/PG-m	-	3-1	0,5-0,25	0,125
Ag/PVA	-	3-0,25	-	0,125
Ag/SDS	-	3-0,25	-	0,125

A high level of toxicity was established for a solution of silver nanoparticles stabilized by sodium oleate, since the toxicity criteria were exceeded for five working concentrations (3%, 2%, 1%, 0,5%, 0,25%). Acute toxic effect on the biotest object was absent only when the drug concentration was 0.125%.

An aqueous solution of silver nanoparticles exerted a toxic effect in the concentration range of 0,5–3%. Harmless for test objects were working concentrations of 0,25% and 0,125%.

Evaluation of the effect of working dilutions of silver nanoparticles stabilized by sodium dodecyl sulfate showed that concentrations from 0.25% to 3% exceeded the toxicity criteria for the biotest object and only the concentration of 0.125% was harmless.

When used as a stabilizer of metal nanoparticles of polyazolidine ammonium modified with iodine hydrate ions, it was found that a solution of silver nanoparticles had an acute toxic effect on biotest objects in the concentration range of 1-3%. For the remaining working dilutions, the toxicity criteria were not exceeded.

The effect of silver nanoparticles stabilized by polyvinyl alcohol, in the concentration range of 0,5–3% on daphnia,was characterized as acute toxic, since the toxicity criteria were exceeded for them. Concentrations of 0,25% and 0,5% were harmless when exposed to a biotest object.

The absence of toxic action for most working dilutions of silver nanoparticles solutions, stabilized by polyazolidine ammonium, modified by iodine hydrate ions, makes it possible to consider this stabilizer as the most promising.

4. Conclusions

Thus, it was shown that silver nanoparticles, stabilized by various versions of polymers, had an acute toxic effect, which depended on their concentration. The use of biotest objects makes it possible to reliably estimate the degree of nanomaterials toxicity, in particular, silver nanoparticles. The conducted studies allowed us to establish basically a similar pattern of manifestation metals nanoparticles of toxicity, stabilized by various polymeric compounds, with respect to *Daphnia magna* Straus.

References

[1] Felicity T 2017 *Public Health Panorama* **3(1)** pp 1-140
[2] Mukhutdinova A N, Rychkova M I, Tyumina E A, Vihareva E V 2015 *Bulletin of Perm University* **1** pp 65-76
[3] Barenboim G M, Chiganova M A 2012 *Water: chemistry and ecology* **10** pp 40-6
[4] Study on the environmental risks of medicinal products. FinalReport.ExecutiveAgencyforHealthandConsumers (2013)
[5] Baytukalov T A, Author. dis. Cand. Chemicalofsciences 2006 p 20
[6] Andrusishina I N 2011 *Toxicologist problems* **3** pp 5-14
[7] Ershov B G 2001 *Russian Chemical Journal* **XLV 3** pp 20-30
[8] Toropova Y G, Golovkin A S, Malashicheva A B, Korolev D V, Gorshkov A N, Gareev K G, Afonin M V, Galagudza M M 2017 International Journal of Nanomedicine **12** pp 593-603
[9] Sirotkina A L, Fedorovich E D, Sergeev V V 2018 *High Temperature* **56(5)** pp 732-7
[10] Singh R, Smitha M S, Singh S P *J NanosciNanotechnol* 2014 **14 (7)** pp 4745-56
[11] Sosenkova L S, Egorova E M 2011 *Journal of Physical Chemistry* **85 (2)** pp 317-26
[12] Ayala-Núñez N V, Lara H H, Ixtepan L, Rodríguez C 2009 *Nanobiotechnology* **5** pp 2-9
[13] Kolesnikov I E, Golyeva E V, Kalinichev A A, Kurochkin M A, Lähderanta E, Mikhailov M D 2017 *Sensors and Actuators, B: Chemical* **243** pp 338-45
[14] Timin A S, Muslimov A R, Lepik K V, Saprykina N N, Sergeev V S, Afanasyev B V, Vilesov A D, Sukhorukov G B 2016 *Journal of Materials Chemistry B* **4(45)** pp 7270-82
[15] Prikhozhdenko E S, Lengert E V, Parakhonskiy B V, Sukhorukov G B, Yashchenok A M 2016 *Acta Physica Polonica A* **129(2)** pp 247-9
[16] Salova A V, Belyaeva T N, Leontieva E A, Zlobina M V, Kharchenko M V, Kornilova E S 2016 *Oncotarget* **7(5)** pp 6029-47
[17] Dobrovolskaya I P, Yudin V E, Popryadukhin P V, Ivan'kova E M, Morganti P 2016 *Journal of Applied Cosmetology* **34(1-2)** pp 1-8
[18] Nepomnyashchaya E K, Velichko E N, Pleshakov I V, Aksenov E T, Savchenko E A 2017

Journal of Physics: Conference Series **841(1)** 012020
[19] Valueva S V, Vylegzhanina M E, Alekseeva P E, Sukhanova T E 2018 *Technical Physics* **63(9)** pp 1248-53
[20] Bazarnova Y, Politaeva N, Lyskova N 2018 *Journal of Biosciences* **73(7-8)** pp 291-6
[21] Maksimova Y G, Vasil'ev D M, Zorina A S, Ovechkina G V, Maksimov A Y 2018 *Applied Biochemistry and Microbiology* **54(2)** pp 173-8
[22] Petrova-Brodskaya A V, Bondarenko A B, Timin A S, Plotnikova M A, Afanas'ev M V, Semenova A A, Lebedev K I, Gorshkov A N, Gorshkova M Yu, Egorov V V, Klotchenko S A, Vasin A V 2017 *Voprosy Virusologii* **62(6)** pp 259-65
[23] Timin A S, Muslimov A R, Petrova A V, Lepik K V, Okilova M V, Vasin A V, Afanasyev B V, Sukhorukov G B 2017 *Scientific Reports* **7(1)** p 102
[24] Tereshchenko V P, Cartel T N 2010 Monograph p 240
[25] Tamayo L A, Zapata P A, Rabagliati F M [et al.] 2015 *Journal of Materials Science: Materials in Medicine* **26 (3)** pp 516-24
[26] Gliga A R, Skoglund S, Wallinder I O [et al.] 2014 *Particle and Fibre Toxicology* **76 (3)** pp 12-24
[27] Prodanchuk N G, Balan G M 2009 *Modern Problems of Toxicology* **3 (4)** pp 4-18
[28] Arora S, Rajwade J M, Paknikar K M 2012 *Toxicology and Applied Pharmacology* **258** pp 151-165
[29] Tkachenko T V, Bezryadina A S 2017 *International Student Scientific Herald* **4** pp 619-21
[30] Khaliullin T O, Zalyalov R R 2009 *Kazan Medical Journal* **4** pp 578–84
[31] Onishchenko G G, Tutelyan V A, Gmoshinsky I V, Khotimchenko S A 2013 *Hygiene and Sanitation* **1** pp 4-11
[32] The impact of nanomaterials on the environment [Electronic resource]URL: http: // www. Nanonewsnet
[33] Methods for determining the toxicity of water extracts from the soil, sewage sludge and waste, drinking, waste and natural water mortality test *Daphnia magna* Straus 2006

Functional beverages containing pectin from different raw material

N Limareva[1*], L Donchenko[2], V Malaknov[1], E Semenova[1]

[1]North-Caucasus Federal University 355009, Stavropol, Russia
[2]Kuban State Agrarian University, 350044, Krasnodar, Russia

* Corresponding author: nlimareva@pfncfu.ru

Abstract. The article presents the results of experimental studies on the development of new types of pectin-containing beverages. Vegetable and berry juice, liquid pectin and natural flavor are selected as a recipe basis of drink. The results of study of chemical composition of the developed beverages showed that the recipe ratio of the selected raw materials provides an optimal ratio of physiologically active components, giving them functional properties. Such data provides a basis for their inclusion in the human diet to reduce the number of alimentary diseases.

1. Introduction

During almost the entire period of human civilization, food was mainly considered as a means of satisfying hunger, appetite and taste needs. However, in recent decades, it has been established that unbalanced nutrition is the cause of the increase in the number of chronic diseases. Thus, the incidence of cardiovascular disease has increased 8-12 times, endocrine disorders 5 times. Diabetes mellitus ranks third in the world among all diseases [1].

According to domestic and foreign studies of scientists and specialists of the world health organization (who), the health of the population is more than 50% dependent on the way and conditions of life; in 10 – 20% it is due to the influence of heredity (genetic factors); approximately the same effect is exerted by the external, human environment, and only about 8 – 10% is accounted for the work of practical health care[2].

In view of the above, the organization of rational nutrition is relevant in modern society. It is possible on the basis of expansion of the range and increase in volumes of production of functional food.

The concept of functional nutrition as an independent scientific and applied direction in the field of healthy nutrition in modern terminology was developed in the early 90s. From the modern point of view, the term "functional food products" means such food products that are intended for systematic use in the composition of food rations by all age groups of the healthy population in order to reduce the risk of diseases associated with nutrition, preserve and improve health through the presence in their composition of physiologically functional food ingredients [3].

No less important in modern environmental conditions is the presence of components with detoxification properties in food. According to the Corte-Dubininin scale to assess the toxicity of pollutants recommended by the world health organization (WHO), toxic metals are ranked first in terms of the degree of negative impact on human health (135 points) [4].

Such natural polysaccharides as pectin substances have the highest detoxification properties in relation to toxic metals [5]. In addition, pectin is a physiological component. Thus, in accordance with

EU432/2012, pectins are recommended for lowering cholesterol and glucose levels in the blood by 6 and 10 g/day respectively [6].

The most physiologically effective food is a drink. It should be noted that the market of soft drinks is one of the most dynamically developing amid other food products. Manufacturers do not get tired to offer new drinks, preparing new items for the beginning of the summer season.

According to various sources, the world market of non-carbonated beverages amounted to about 80 million liters in 2018. This segment has been developing steadily over the past 4 years, showing an average global growth of 4.7% per year. At the end of 2018, taking into account the planned growth of the category of 5.4%, the world market of non-carbonated beverages should exceed 85 million liters. Analysts agree that the growth rate of the global market of non-carbonated beverages will not only continue over the next few years, but even slightly accelerate to 5.7% by 2019-2020.

One of the main trends in the European beverage market is health care. Consumers are beginning to pay more attention to certain useful properties of certain drinks. At the same time, manufacturers, respectively, by all means and possible methods try to emphasize the usefulness of the product for health. No less popular trend is the functionality. Consumers want to have the magic drink. Categories of functional drinks, sports drinks, brain improvement and other activities continue their world ascent to Olympus.

With this in mind, the aim of our study was to develop recipes of juice-based drinks with the addition of pectin substances as prolongators of biologically active compounds, as well as detoxicants of toxic metals.

To achieve this goal, the following research objectives are set:

- to study the chemical composition of the main industrial raw materials to determine their physiologically active ingredients;
- perform simulations of beverages recipes with balanced composition based on their functional orientation;
- experimentally confirm the recipes of new types of functional drinks.

2. Theoretical and experimental research methods

As an experimental site we have chosen 2 regions in the South of Russia, in particular Krasnodar and Stavropol territories.

Krasnodar region is located in the South of Russia, in the South-Western part of the North Caucasus. The region takes the 3rd place among regions of the Russian Federation, in number of inhabitants — after Moscow and the Moscow region.

Stavropol region is located in the Central part of Ciscaucasia and on the Northern slope of the greater Caucasus. The area of the region is larger than, for example, the Netherlands. In the West and South-West of the Stavropol territory is bordered by Krasnodar territiry.

This geographical location determines the features in the chemical composition of plant materials. In addition, the agro-industrial sector and the resort and sanatorium complex are traditionally developed in these regions. This, in turn, leads to the development of the beverage market, both for indigenous people and vacationers.

Varieties of pumpkins cultivated on an industrial scale, pumpkin juice and puree, pectin concentrated extracts of apple, grape pomace, forage watermelon and beet pulp, juice of BlackBerry fruits, rose hips were selected as the objects of our research.

Pumpkin varieties. There are three types of pumpkin:

Large-fruited pumpkin (Cucurbita maksima),

Butternut Pumpkin (Cucurbita moschata),

Hard or ordinary Pumpkin (Cucurbita pepo), and its varieties are zucchini and Patisson.

We have selected the following varieties of pumpkin for the study:

Volga grey 92. Middle-maturing type, resistant to disease, belongs to large-fruited pumpkin. Powerful plant forms rounded-flat large (weighing 10-12 kg) fruit of greenish-gray color. Rind is hard and thin. The flesh is dense, bright orange, taste is very good.

Mozoleevskaya 47 belongs to a hard-core or ordinary pumpkin. Middle-maturing type. The elongated fruits with striped rind. Flesh is dense, orange. Fruit weight up to 5 kg. Long shelf life.

Marble. It belongs to the Muscat varieties of pumpkin. The variety is late-maturing (about 135 days); fruits are average, weighing 5-6 kg, flattened, the surface is wrinkled-segmented, lumpy, gray and dark gray bark, with specks of marble type. The flesh is intense orange, thick, firm, crispy, very sweet and tasty. Shelf life of 8-9 months.

Ker-2. New product of Kuban breeders. The fruit is orange in color, weight of of 1.17 to 2.45 kg. Rind of Mature fruit is a thin (1 mm), plastic. The subcortical layer 1.5-2.0 mm is light orange with yellow tint. Ripe pumpkins have a dense sweet flesh with no grassy taste and look attractive.

The chemical composition of pumpkin fruits and juices used to improve organoleptic characteristics and increase nutritional value was determined by methods generally accepted in the biochemistry of plant products: soluble dry substances – by refractometric method; fractional composition of sugars and acids, mineral content – by capillary electrophoresis; vitamin content – by HPLC and fluorometry.

The method of quantitative determination of pectin substances in plant raw materials is based on the extraction of pectin from plant raw materials and its transfer to the dissolved state [3]. The basis of the study of the extracts of hydropectin and protopectin is the calcium-pectat method and precipitation with ethanol. In addition, the content of free and esterified carboxyl groups in pectin preparations and the degree of esterification by conductometric method were studied. Complexing ability was determined by trilonometric titration according to the degree of binding of lead ions.

For the correct assessment of the chemical composition of the studied varieties of pumpkin fruit were selected in harvest maturity.

3. Results and discussion

Pumpkin, like other vegetables, consists of water, insoluble and soluble substances. Pumpkin fruit contain about 70-90% water, 1.3-3.3% insoluble substances (cellulose, lignin, kutin, protopectin, insoluble proteins, etc.), 7.5 – 23% and higher soluble substances (sugars, organic acids, nitrogenous soluble substances, hydratopectin, water-soluble vitamins, coloring, vitamins, minerals, etc.).

In our objects of research humidity ranged from 81,3±0,5% (grade Ker-2) to 84,9±1,2% (grade Marble).

The most important part of pumpkin fruit are sugars and organic acids.

The results showed that the total amount of sugar in the varieties of pumpkins ranged from 5.7 (sort Mozoleevskaya) to 8.1% (Marble).

Pumpkin fruit contained three types of sugar: fructose, glucose and sucrose. At the same time, sucrose prevailed in all varieties. The glucose content is 8.6 times less in the Marble variety (0.12%) than in the Ker – 2 variety (1.03%). At the same time, the content of fructose in the Marble variety is 31.5% less than in the Volga gray variety 92.

According to the content of organic acids, the studied samples do not differ significantly (figure 1.)

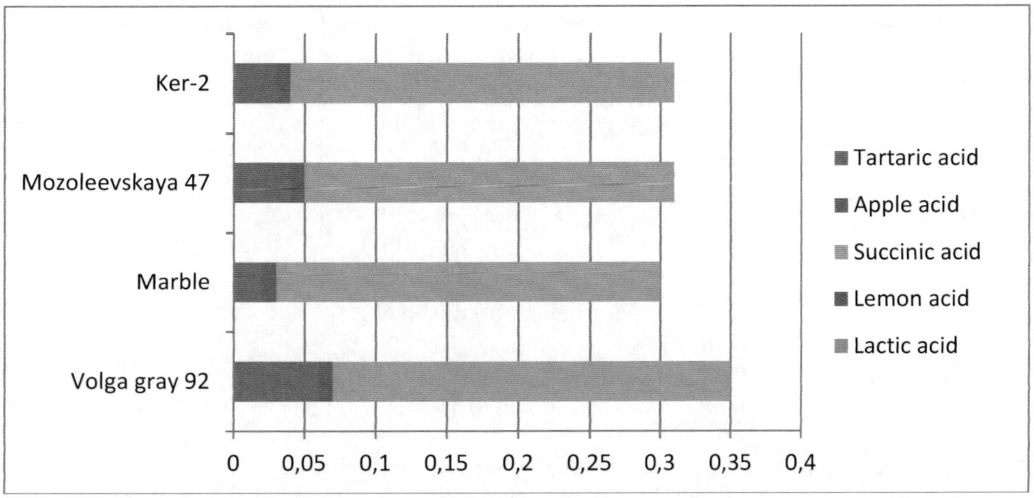

Figure 1. Fractional composition of organic acids in the studied pumpkin samples

However, by fractional composition in the type of the Volga gray 92 there is 2.5 – 3 times more tartaric acid. It is known that natural tartaric acid has antioxidant properties and has a beneficial effect on metabolic and digestive processes in the body. Citric acid and lactic acid in the studied varieties are absent, and by the content of malic acid, they do not differ (0.01%) and its content is very low.

Relatively high content of succinic acid was found in all varieties (0.26 – 0.28%). The results of previous studies have shown that succinic acid is an adaptogen (increases the body's resistance to adverse environmental factors), which is a positive factor in the development of functional products.

Vitamins play an important role in the prevention of various diseases. It is known that pumpkin fruits are rich, first of all, in carotene. Therefore, we have evaluated the pumpkin fruit according to the content of this vitamin. It was found that the studied samples are indeed characterized by a high content of carotene. Its contents amounted to 440 mg/100 g in sample of the pumpkin varieties of the Volga grey 92, and 390 mg/100g – of Marble. In other varieties the content of carotene was in the range of 370 mg/100 g (cultivar Mozoleevskaya 47) and 300 mg/100 g (in the type of Ker-2). The mass fraction of water-soluble vitamins-C and Niacin – is relatively high and practically does not differ among varieties, and is 14 and 0.1 mg/100 g of fruit pulp, respectively. Vitamin B6 content-low (0.02 mg / 100 g of fruit pulp). There are no significant differences between the varieties.

It should be noted that the daily requirement of the human body in vitamin C is 50 mg with this in mind, it is clear that the fruit of pumpkin can be reasonably considered as a functional source of this vitamin to meet the physiological needs (the degree of providing daily needs is 28%).

Taking into account the recommended norms of consumption of the considered vitamins, the degree of safety in the consumption of 100 g of pumpkin fruit will be: in niacin – 50%, in vitamin B_6– 1.0%.

Thus, pumpkin fruit in accordance with the requirements for food products are a functional source of vitamin C, niacin and carotene.

The nutritional value of food is determined by the mineral composition. Minerals are essential components of food.

In this regard, we have analyzed the studied pumpkin fruit for the content of such basic macronutrients as potassium, calcium, magnesium, sodium, phosphorus and trace elements as iron, zinc, copper.

Data on the mineral composition of the studied varieties of pumpkin fruit is presented in table 1.

Table 1. The content of minerals in the studied samples of pumpkin fruit

Substance	Varieties of Pumpkin fruit			
	Volga grey 9292	Mozoleevskaya 47	Marble	Ker-2
	Macroelements, mg/100 g			
K	84,1	90,0	80,4	89,1
Ca	26,5	25,0	27,2	24,0
Mg	10,0	12,0	14,0	13,5
Na	3,5	4,0	4,1	3,9
P	14,7	15,0	15,5	16,0
	Microelement, mg /100 g			
Fe	0,40	0,40	0,5	3,7
Zn	0,20	0,24	0,24	0,23
Cu	0,20	0,18	0,19	0,18

From the tabular data it follows that by the degree of daily provision of macronutrients calcium is higher. Safety range is 3,3-4,7%. The degree of daily provision of the human body with potassium while eating pumpkin fruit is insignificant-about 2%. A low degree of daily provision is observed in phosphorus-only 1.2%.

The degree of daily supply of iron, copper and zinc is low in the range of 4.5 – 12%.

Taking into account the requirements for functional food test samples can not be considered as a source of minerals.

A very important component of pumpkin fruit are pectin substances.

The results of the study of the total content and fractional composition of pectin substances of the selected objects are shown in figure 2.

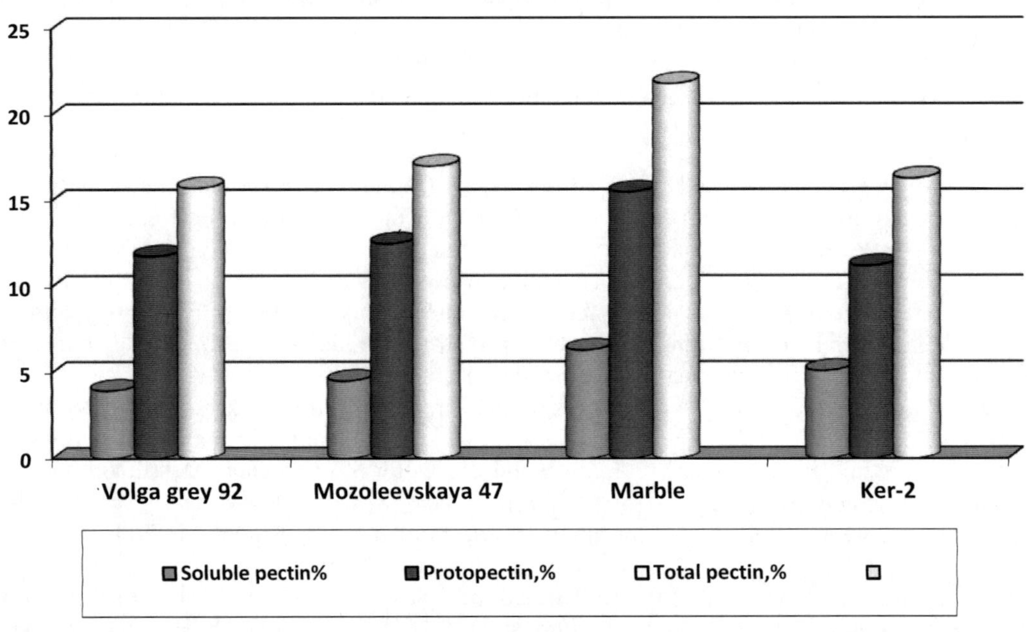

Figure 2. Fractional composition and total content of pectin in the selected objects of research, % to absolutely dry substance

From the presented data it can be seen that by the total content of pectin substances Muscat Marble variety (21.8%) is higher.

In all studied samples fractional composition of pectin substances is represented by protopectin and soluble pectin (hydropectin). In this case, the content of protopectin prevails over the soluble fraction, which is natural for vegetable raw materials.

The results of studies of the analytical characteristics of the pectin taken from the studied pumpkin fruit showed that the content of free carboxyl groups of pectin samples is practically the same.

At the same time, the lowest degree of esterification was observed in pectin of the pumpkin of the Volga gray variety 92 (37.94%), the largest – in pectin of the pumpkin of the marble variety (43.27%).

Thus, the taken pectins should be attributed to low esterified.

It is known that the complexing properties of pectin substances depend on the content of free carboxyl groups, i.e. the degree of etherification of carboxyl groups with methanol. The degree of esterification determines the linear charge density of the macromolecule, and, consequently, the strength and method of communication of cations.

With a decrease in the degree of esterification, that is, with an increase in the charge of the macromolecule, the connection of pectin substances with cations increases. Therefore, we can predict a greater complexing ability of the variety Volga grey 92.

Thus, the results of the study give grounds for the conclusion that the fruit of pumpkin should be considered a recipe component as a source of low-esterified pectins and carotene. This is consistent with the results of our earlier study [8]. This is preferable when processing large-fruited pumpkin with extracting juice from it, in particular variety of the Volga grey 92.

We further consider BlackBerry juice and rosehip fruit as possible recipe components in order to expand the functional orientation. This choice is due to their unique chemical composition and availability.

Thus, to improve the organoleptic properties and enrichment of functional drinks by P-active substances, vitamins and organic acids BlackBerry juice is of greatest interest as a recipe component.

It was found that the juice of BlackBerry berries contains glucose (2.8-3.6%), fructose (3.1-3.2%), sucrose (0.35-0.58%), organic acids (up to 2.2%), in particular malic and citric acid.

BlackBerry is also rich in P-active substances-up to 1500 mg %. It should be noted that the study of substances with p-vitamin activity has recently become increasingly important. Substances of vegetable origin, often with different chemical properties, but with commonality of biological action contain vitamin P.

The main representatives of P-active substances are flavonoids (catechins, leucoanthocyanins, flavonols, anthocyanins and copolymerized forms of these compounds). The BlackBerry juice from the above group mainly contains leucoanthocyanins, flavonols, and anthocyanins, which have not only capillary-strengthening action, but also antitoxic. The latter property is manifested in the binding by complexation of heavy metal ions, which is relevant in modern conditions. To prevent diseases and maintain physical and mental activity R-active connections required for 100 – 200 mg per day, to provide therapeutic action daily intake increases by 5 – 10 times, i.e. up to 1 – 2 g.

In the study of BlackBerry juice from different manufacturers, the amount of P-active substances amounted to 598 and 708 mg/100g. The content of flavonols was in the range of 111.6 – 129,1; anthocyanins – 309,6 – 465,0; leucoanthocyanins – 116,0 – to 105.3 mg/100 g.

Thus, judging by the experimental data, BlackBerry juice can be reasonably considered as a functional source of P-active compounds.

To enhance the antioxidant properties it is necessary to use ascorbic acid. Ascorbic acid regulates energy metabolism, reduces oxidative stress, stimulates the formation of collagen and procollagen, participates in the metabolism of folic acid and minerals, activates the synthesis of some steroid hormones, catecholamines. Wild rose berries can be considered such a natural source of raw materials.

It is known that rosehip berries contain a whole complex of essential minerals and vitamins, including the highest concentration of vitamin C (higher than in lemons, oranges, black currants), which makes them valuable for medicine and healthy eating. Rosehip fruit contain up to 17% vitamin C, 12% carotene, vitamins B2, E, K, P, flavonoids, 18% sugar, 3.7% pectin, 4.5% tannins, organic acids, trace elements — iron, manganese, phosphorus, potassium, magnesium, molybdenum, cobalt, chromium, copper, a significant amount of potassium salts. The use of rosehip helps to strengthen the immune system, increase the body's resistance to bacteria, helps to improve the digestive system, helps to slow down the aging process.

Liquid pectin in the form of pectin concentrate from Apple pomace with a mass content of pectin substances 4% is produced in industrial volumes by "SunLand" (Hungary), from beet pulp, feed watermelon and grape pomace – by experimental production of the Kuban state agrarian University.

In the preparation of recipes we used the method of profiling with the construction of profilograms. Based on the analysis of sensor profiles, the optimal ratio of components was chosen.

When optimizing the beverage recipes with the help of mathematical modeling, in order to obtain a product that combines a balanced micronutrient composition, functional activity and favorable taste, first were determined the optimal ratio of fruit and vegetable fillers and pectin substances that provide favorable taste qualities of beverages.

For this purpose, three-factor simplex-centroid plans were used. The model samples prepared according to the planning matrix were tasted, assessing them by five indicators (signs): color, sweetness, consistency, aroma, fullness of taste. These signs for the created a beverage are chosen as preferred that best reflect the type of the future drink.

The results were processed using the methods of statistical and graphical analysis by building ternary graphs in the program "STATISTIKA 7.0", which allowed to determine the most appropriate in terms of organoleptic characteristics of the final product. On the basis of the tasting assessment on a 10-point scale implemented in the non – alcoholic food industry, the following ratios of the mass content of recipe components in the drink were established: berry juice – 15%, pumpkin pectin – containing juice – 30%, liquid pectin from various raw materials with a pectin content of 4% - 40%, infusion of rose hips-15%.

The pectin part of the drink was made up of liquid Apple, beet and grape pectin, liquid pectin from feed watermelon. This choice is due to the difference in the properties of pectins. Thus, Apple liquid pectin and liquid pectin from the feed watermelon play the role of thickeners, which gives the full taste of drinks. Beet and grape liquid pectin are good complexing agents in relation to toxic metals. Complexing ability is from 40 (for the wine), 75 mg of Pb2+/g of dry pectin (of beet).

The results of experimental studies have also shown that the use of Apple and grape liquid pectin may increase the optimal content of the beverage pectin to 50 – 60 %, which increases its detoxification properties.

Physical and chemical parameters of the studied beverages, including pumpkin and BlackBerry juice, rosehip infusion and liquid pectin of different types are presented in table 1.

Table 2. Physical and chemical parameters of pectin-containing beverages

Pectin extract	Dry substance, %	Total pectin, %	Acidity, %
apple	17,2	2,06	3,8
beet	14,6	1,98	2,0
forage watermelon	13,6	1,67	2,6
grape	15,5	1,25	3,2

The content of pectin substances in the developed beverages based on pectin concentrates from various types of plant raw materials is from 1.27 to 2.06 %.

In order to determine the functional properties, we conducted research to determine the complexing ability of the developed beverages. The data obtained is presented in figure 3.

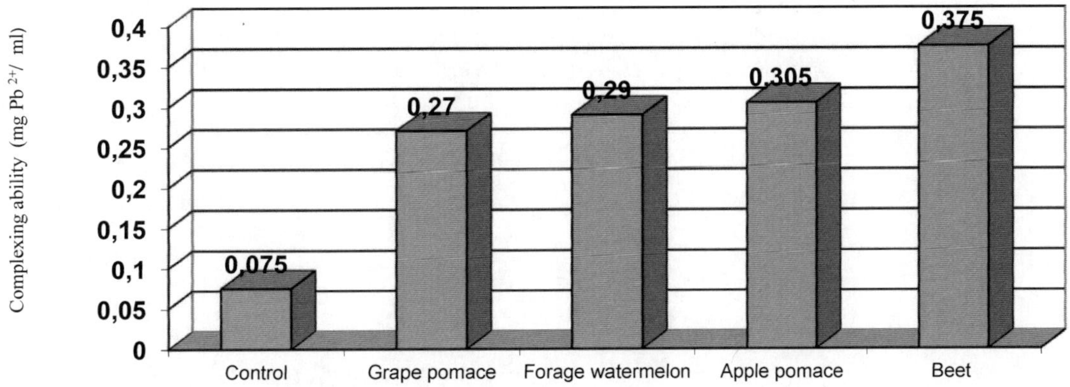

Figure 3. Complexing ability of drinks

It follows from figure 3 that as a result of the use of liquid pectins in the recipe, the complexing ability of the developed beverages increases. The juices with beet pectin have the greatest complexing ability-5 times more in comparison with the control sample. Comparative analysis of beverages using pectin products obtained from different raw materials showed that drinks with pectin from the fruit of forage watermelon by complexing ability are not inferior to drinks with Apple pectin and surpass drinks with grape pectin.

4. Summary

Based on the above, it can be concluded that the recipe ratio of the selected raw materials provides good organoleptic characteristics and optimal ratio of physiologically active components, giving them functional properties.

At the same time, drinks have high detoxification and antioxidant properties and are a functional daily source of vitamins (C, niacin, carotene), P-active compounds, pectin and minerals for modern human. This provides a basis for their inclusion in the human diet to provide a rational and balanced diet.

In addition, these drinks can be classified as specialized food products and recommended for hard labour workers (mining, chemical, nuclear, metallurgical and other industries) and the population living in ecologically polluted areas.

5. Conclusion

Thus, the results of experimental studies give grounds for the conclusion that it is advisable to consider the selected raw sources – pumpkin and BlackBerry juice, infusion of rose hips, liquid pectin – as a basis for the development of functional food drinks.

6. Conflict of interest

Authors declare no conflict of interest.

References

[1] World health organisation 2016 *Global report on diabetes*
[2] Donchenko L V, Nady`kta V D 2018 *Food safety*
[3] Donchenko L V et al 2018 *Functional food technology* (Moscow: Urait)
[4] Dubinin N P 1997 *Some problems of modern genetics*
[5] Il'ina I A, Donchenko L V, Zemskova Z G 2003 Technology features of pectins with high-prolonging properties *Vestnik Rossijskoj akademii sel'skohozyajstvennyh nauk* **3** pp 8–10
[6] EU 432/2012. Commission Regulation *official Journal of European Union 2012*
[7] Zhirenchina Z U, Kizatova M Z, Donchenko L V, Donchenko E V, Kurasova L A 2016

Comparative characteristics of the chemical structure of apples for the elaboration of functional food products *Res. J. of Pharm., Biolog. and Chem. Sci.* **7**(3) pp 2360-70
[8] Limareva N S, Donchenko L V 2014 Innovative drinks based on functional vegetable juices. *Polythematic network electronic scientific journal of Kuban state agrarian University* **95** pp 511–40
[9] Bush P L 2014 Pectin chemical properties, uses and health benefits ed Bush P L (New York: Nova Science Publishers)

Application development for accidental pollution assessment on chemical manufacturers (pollution from chemical waste)

V Nosov[1], M Tindova[2*], K Zhichkin[3], M Mirgorodskaya[4]

[1]K.G. Razumovsky Moscow State University of technologies and management, Department of economics and management 109004, Moscow, Zemlyanoy Val, 73, Russia
[2]Yuri Gagarin Saratov State Technical University, Economics, Applied Mathematics and System Analysis Department, 410054 Saratov, Polytechnicheskaya street, 77, Russia
[3]Samara state agricultural academy, Department of Economic Theory and Economics of AIC, 446442, Samara region, Kinel, Uchebnaja str., 2, Russia
[4]K.G. Razumovsky Moscow State University of technologies and management, Department of economics and management 109004, Moscow, Zemlyanoy Val, 73, Russia

* Corresponding author: mtindova@mail.ru

Abstract. Considering that in recent decades there has been a dramatic increase in the amount of industrial disasters, among which accidents on chemical facilities retain a significant place, forecasting tasks and accident consequence assessment on hazardous chemical facilities are analyzed in this paper. The task of the emergency area size determination, as well as the determination of building destruction level, and the facility personnel and public expenses. The objective of the paper is to review the existing methodologies of forecasting and create a software system for accident consequence forecasting on hazardous chemical facilities. **Methodology.** The analysis of different methods of damage estimation, conducted by the author, allowed for the general scheme of damage from accidents on potentially hazardous facilities estimation to be made. Furthermore, the author analyzed potential objects of chemical pollution in the Saratov region. The paper demonstrates phases of theoretical development and implementation of software program for accident consequence assessment and modeling on hazardous chemical facilities. The developed software system has a connection with Google Maps GMap.NET, which allows to visualize the obtained assessment results. **Results.** The result of practical implementation of the application developed by the author is an accident model for a facility, using chlorine substances, the amount of pollution measurement and territory pollution determination. **Conclusion.** The paper shows that the implementation of the program system developed in the work of organizations will help the heads to minimize possible losses of working personnel, public and organization financial losses.. The author notes that the application developed is free from the main disadvantages of existing programs, more specifically, the absence of the possibility to estimate the physical damage caused by the accident.

IOP Conf. Series: Earth and Environmental Science **337** (2019) 012014 doi:10.1088/1755-1315/337/1/012014

1. Introduction

In recent decades there has been a dramatic increase in the amount of industrial disasters, among which accidents on chemical facilities retain a significant place. Accidents on chemical facilities may not only lead to inventory and human losses, but also to irreparable and inestimable damage for the environment. This is amply reflected in the Bhopal gas tragedy of 1984 and numerous accidents that have occurred across the world before or since [1]. After the Bhopal tragedy similar tragedies have also occurred. For example, the Buncefield disaster of 2005 has caused estimated losses of $500 billion, and the BPL refinery disaster of 2010 has harmed larger areas of marine environment than any other past accidents [2; 3; 4]. Unfortunately, there are many similar cases. Ever so often one accident triggers another accident or a series of accidents, thereby escalating the disaster by "domino effect» [5; 6; 7]. This has required development of mechanisms for accidents on chemical facilities protection, accident consequence assessment, and also for accidents forecasting. Many methods have been improved, and new methods have also been developed [8; 9; 10; 11]. The main task of accident consequence assessment and forecasting on hazardous chemical facilities is to determine the size of the emergency area, the level of building destruction, and the facility personnel and public expenses [12; 13].

For implementation of forecasting it is necessary to develop a software and design a three-phase mathematical model for analysis and estimation.

In the first phase forecasting of results of possible natural and man-caused emergencies is conducted, using average conditions (average weather conditions per year; average amount of people in buildings and on streets; average population density, etc.) This phase is conducted until emergencies.

In the second phase forecasting and analysis of the emergency are conducted immediately after the source of emergency is revealed according to updated information (emergency period, weather conditions during this period, etc.)

In the next phase results of forecasting and real situation according to emergency response service data are clarified [14].

The objective of the paper is to review the existing methodologies of forecasting and create a software system for accident consequence forecasting on hazardous chemical facilities.

Hazardous chemical facilities (HCF) include [15]:

1) chemical industry companies, as well as chemical equipment (units) and plants, manufacturing and using emergency and hazardous chemical substances (EHCS);

2) petroleum refineries

3) railway stations, ports, terminals and warehouses at the terminal (intermediate) points of EHCS movement;

4) production in other industrial sector using EHCS

5) vehicles (containers and tank cars, tank trucks, river and sea tankers, pipeline transport, etc.)

Chemical production facilities and facilities using toxic substances in their production process seem to be of the greatest danger.

2. Methodology (hazardous chemical environment identification methods)

The chemical environment refers to the extent and degree of chemical contamination of the air (in the country), which affects human activity and industrial objects work [16].

The chemical damage zone may have one or several chemical contamination areas, defined as populated area in a chemical damage zone.

To identify the chemical environment means to determine chemical contamination zones and to map them (or to locate them on a diagram or on a plan). The chemical environment assessment is conducted using the forecasting method (using civil defense headquarters), and chemical intelligence data (commanders of paramilitary forces of civil defense).

Based on the assessment of the magnitude of a real danger that depends not only on the toxicity of substances, but also on the size of their reserves and their spread in the atmosphere, the list of EHCS from which protection should be ensured, includes nine basic substances: Chlorine, Ammonia,

IOP Conf. Series: Earth and Environmental Science **337** (2019) 012014 doi:10.1088/1755-1315/337/1/012014

Phosgene, Sulfur dioxide, Hydrogen cyanide, Hydrogen sulfide, Carbon disulfide, Hydrogen fluoride, and Acrylonitrile [17].

It should be noted that in Saratov Region there are facilities using hazardous chemicals in their production: "Saratovgassynthesis" in Saratov (Hydrogen cyanide, Acrylonitrile, Sodium cyanide, etc.), "VK-3" in Saratov (Chlorine), and OJSC "Appatit" in Balakovo (Sodium fluorosilicate, Sulfuric acid and Phosphoric acid).

To assess the chemical environment means to identify the chemical damage scale (the depth, the chemical contamination area), and to define it on the location map.

Two types of the chemical environment can be distinguished: actual chemical environment, identified by the chemical intelligence using such instruments for control as gas analyzers, and potential chemical environment, identified by the forecasting, based on mathematical calculation and modeling and mapping the obtained situation.

The actual situation is identified after the accident and formation of the damage zone. The chemical environment, identified by the forecasting method, only conditionally describes the chemical damage.

The general scheme for definition of damage from accidents on hazardous facilities is presented in Figure 1 [18]. As we analyze this scheme we can see the necessity of using the software in the accidents in the chemical industry damage assessment process due to the need to quickly process large amounts of information.

3. Analysis of software programs for chemical pollution assessment existing

Today there are a few software programs for pollution assessment on the Russian market that are in the public domain. The CJSC "Scientific Engineering Research Center" 2011 software program "Risk Study and Safety Assessment" is the most commonly used (judging by the number of downloads). This program unit allows forecasting of contamination zones sizes in case of accidents on reservoirs and storages during transportation by railway, pipeline and other means of transport, and also in case of destruction of chemically hazardous facilities. It also allows mapping the zones of possible contamination and sectors of EHCS spreading depending on the speed and direction of wind, and to generate extended reports with the displaying of the calculations performing. Inability of estimating the accident damage and choosing the statistical data for affected communities estimation is the main disadvantage of this software program (data must be entered manually).

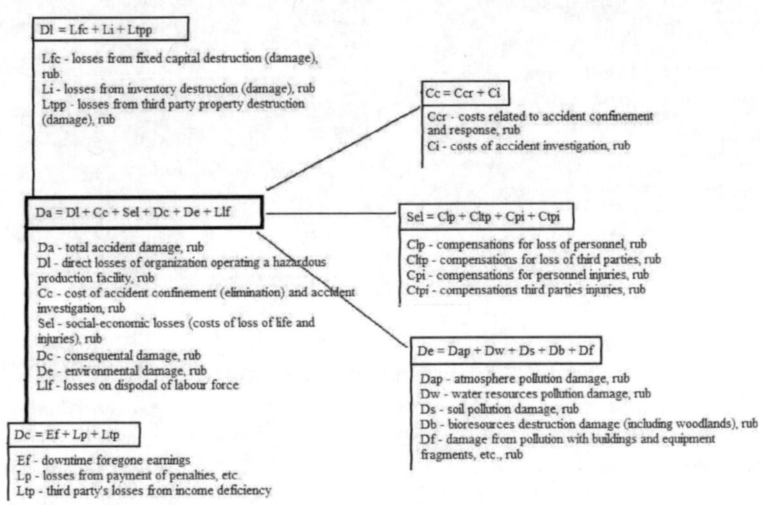

Figure 1. Scheme for definition of damage on hazardous facilities.

Another software program is the RD 52.04.253-90 project, developed by the LLC Scientific Engineering Research and Production Enterprise "Aviainstrument" based on the 1990 standard scale of

IOP Conf. Series: Earth and Environmental Science **337** (2019) 012014 doi:10.1088/1755-1315/337/1/012014

contamination forecasting procedure. This program can be applied to accidents and scenario calculations (potential accidents and their consequences modeling) in cases of spreading of highly hazardous chemical (HHC) in the atmosphere in gaseous or aerosol state. It allows forecasting of contamination zones sizes in case of accidents on reservoirs and storages during transportation by railway, pipeline and other means of transport, and also in case of destruction of chemically hazardous facilities. This software has the same disadvantages as the first one, moreover, the methodology of accident forecasting for accident estimation is quite old-fashioned.

Therefore, the main disadvantage of both software products is the lack of damage definition. This function will help the heads of economic entities to define potential losses in case of emergency, and to allocate financial resources correctly for depreciation charges calculation.

4. Practical application design for chemical pollution assessment

Based on the fact that the most popular operating system nowadays is Windows, it would be appropriate to develop the program for it on the first step, and, having considered the leading programming languages and their advantages and disadvantages, the C# language was chosen as the programming language. The programming environment, developed by the founding company of the chosen programming language - Microsoft, and their product - Visual Studio 2017, were chosen as the development environment [19].

The next step was the development of the program system structure, which, according to the ultimate goal - the potential damage area forecasting - would consist of units, each of which would be implemented in separate forms. It should be noted that EHCS can be chosen from the list which contains 33 substances (Fig. 2B).

The first unit is responsible for the possibility of data entry into the form (Fig. 2A). In the form it is also necessary to choose the state of matter for the substance, the amount of EHCS in the environment, type of spill, weather conditions at the time of the accident and statistical data, as well as to choose the amount of people suffered from the accidents on the economic entity or outside.

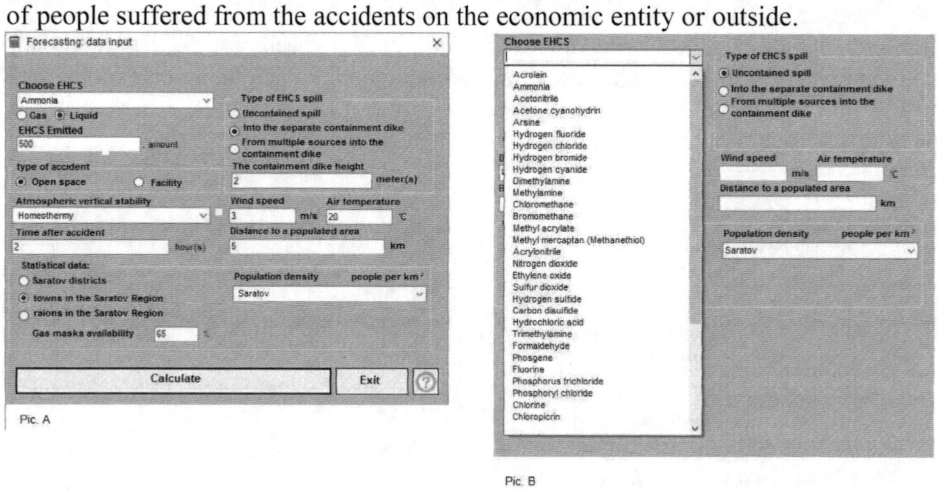

Figure 2. A) Data-entry form; B) The list of substances in the box

This form includes three additional buttons. When clicking "Calculate", all the attributes chosen are considered, clear structure of mathematical calculation is created and implemented, and the new form with the calculation data is opened. The "Exit" button is responsible for the program shutdown and deleting it from the operating system processes. When clicking the question mark button the online help is opened.

Moreover, for all text boxes the event handler was created, which does not allow incorrect data to be entered. In this case the program does not block the "dot", "comma" and figures entry, the entry of other symbols is not available. If the user tries to enter non-existent data, then after clicking "Calculate" a message window will be opened, showing the line where the mistake was made.

The next unit (form) demonstrates the outcome with some source data for the better calculation reliability analysis (Fig 3A).

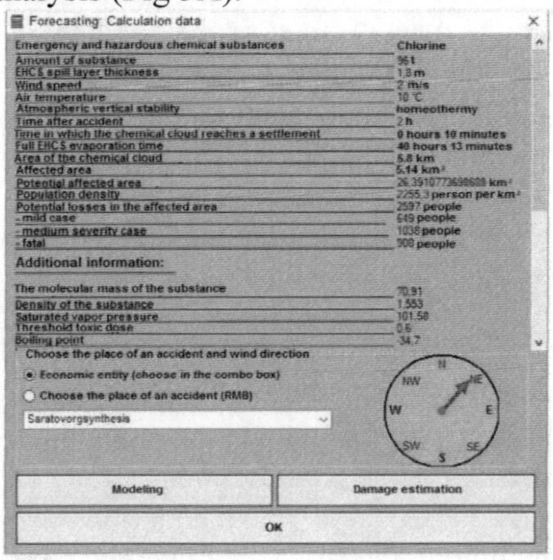

Pic A

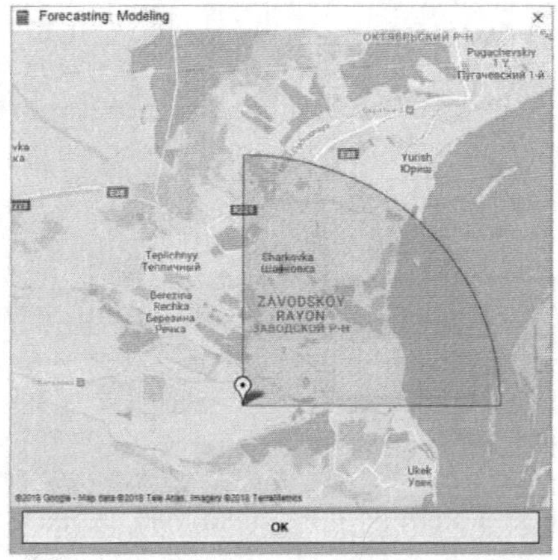

Pic. B

Figure 3. A) Calculation data form; B) Area affected by the accident map modeling

It is further proposed that the user sets additional parameters for the accident consequences map modeling. For this the accident location must be chosen. If the user wants to choose from the list of standard economic entities in the Saratov Region ("Saratovgassynthesis", "VK-3" and OJSC "Appatit"), the "Economic Entity" button should be clicked. When choosing the place of an accident, launch coordinates are the central part of Saratov.

The next parameter needed for modeling implementation is the wind direction measurement, which can be done by a compass (Fig. 3A) [20].

The Google GMap.NET framework was used in our program for the map rendering. This library is free, cross-platform has an open source code, and allows to define objects geolocations. Fig. 3B illustrates the example of the damage zone visual display. The yellow marker shows the accident area and defines the contaminated air cloud route direction; the potential EHCS contamination zone is displayed as a red circle, semicircle or sector depending on the speed of wind in the ground air; the potential chemical contamination zone is shaded semi-transparent red.

5. Results

Let us consider the developed program system operation using the example of an accident on the industrial facility using chlorine substances in their production activities. For this the area of hazardous chemicals contamination in case of an accident on a hazardous chemical facility should be identified, as well as the time in which the chemical cloud reaches the populated area given the following source data: the EHCS type - Chlorine; the amount of EHCS (Q0) - 96 tons; the EHCS storage conditions - compressed fluid; the bund wall height (H) - 2 meters; time after the accident (N) - 4 hours; weather conditions - homeothermy; air temperature (T) - 10°C; the wind speed (V) - 2 meters per second; distance from the EHCS emission source to populated area (X) - 2 km; statistical data - cities of the Saratov Region; gas masks availability for population - 30%. The result is shown in the Fig. 3A.

To define the affected area it is necessary to add some data to the form with the obtained result: "Saratovgassynthesis" is chosen as the accident scene; the wind direction is north-east (NE). The result is shown in the Fig. 3B.

6. Conclusion

The software developed is not cross-platform and operates only on Windows 7, 8, 10 operating systems, the minimum system requirements for a computer comply with the system requirements for the operating systems. The 4.0 .NET Framework is also required for the operation. In conclusion it should be noted that the application developed for the accidents on chemical facilities pollution assessment lacks the main disadvantage of the similar programs on the market, specifically the lack of possibility to assess the physical damage at the facility where the accident took place, and in the polluted territory. The implementation of the program system developed in the work of organizations will help the heads to minimize possible losses of working personnel, public and organization financial losses.

References

[1] Abbasi T, Abbasi T, Abbasi S A 2018 Role of HAZOP in assessing risk of accidents in chemical process industries: capability and lacunae *Intern. J. of Engineering, Science and Math* **7** pp 482–7
[2] Abbasi T, Pompapathi V, Tauseef S M, Abbasi S A 2017 An assessment of the prevailing codes/standards and models for determining safe spacing between two or more hazardous storage tanks *Intern. J. of Engineering, Science and Math* **6** pp 255-68
[3] Abbasi T, Tauseef S M, Suganya R, Abbasi S A 2017 Types of accidents occurring in chemical process industries and approaches to their modeling *Intern. J. of Engineering, Science and Math* **6** pp 424-55
[4] Tauseef S M, Abbasi T, Thiruselvi, Abbasi S A 2017 The risk of domino effect associated with the storage of liquefied petroleum gas (LPG) and the safety codes for accident prevention *J. of Engin., Science and Math.* **6** pp 456–82
[5] Abdolhamidzadeh B, Abbasi T, Rashtchian D, Abbasi S A 2010 A new method for assessing domino effect in chemical process industry *J. of Hazardous Materials* **182** pp 416–26
[6] Abdolhamidzadeh B, Rashtchian D, Abbasi T, Abbasi S A 2011 Domino effect in process-industry accidents - An inventory of past events and identification of some patterns J. of Loss Prevention in Process Industries **24** pp 575–93
[7] Khan F, Abbasi S A 2011 An assessment of the likelihood of occurrence, and the damage potential of domino effect (chain of accidents) in a typical cluster of industries *J. Loss Prevent* **41** pp 283-306
[8] Nosov V V, Kozin M N, Gladun T N 2015 Optimization of the farm production structure taking into account weather, economic and environmental conditions *Ecology, Environment and Conservation* **21** pp 103–10
[9] Nosov V V, Kozin M N, Andreev V I, Surzhanskaya I Y, Murzina E A 2016 Increasing The Efficiency of Land Resources Use for an Agricultural Enterprise *Res. J. of Pharmac., Biol. and Chem. Sci.* **7** pp 382–5
[10] Kaur M, Nagpal A K 2017 Evaluation of air pollution tolerance index and anticipated performance index of plants and their application in development of green space along the urban areas *Env. Sci. and Poll. Res.* **24** pp 18881-95
[11] Duan W, Bin H 2015 Emergency Response System for Pollution Accidents in Chemical Industrial Parks, China *Intern. J. of Envir. Research and Public Health* **12 pp** 7868-85
[12] Salzano E, Basco A, Busini V, Cozzani V, Marzo E, Rota R, Spadoni G 2013 Public awareness promoting new or emerging risks: Industrial accidents triggered by natural hazards (NaTech) *J. of Risk Res.* **16** pp 469-85
[13] Kumar V, Pandey S, Tauseef S M, Abbasi T, Abbasi S A 2018 Pool Fires in Chemical Process Industries: Occurrence, Mechanism, Management *J. of Failure Analysis and Prevention* **18** pp 1224-61
[14] Akimov V A, Vorobiev Y L, Faleev M I 2012 *Natural and Man-Caused Emergencies Safety* (Moscow: Abris)
[15] Bezborodova O E 2014 *Forecasting of Accidents Consequences on Hazardous Production Facility and Their Impact on the Environment* (Penza: PSU)

[16] Kapustin S Y, Malahov V I, Bashkov A P 2001 *Technical Guide on Emergency Chemical Environment Forecasting and Assessment* (Ivanovo: Ministry of Education of the RF)
[17] Tindova M G 2012 Environmental Danger Economic Assessment Fuzzy Model *Econ.: Yesterday, Today, Tomorrow* **3-4** pp 129-39
[18] Tindova M G, Maximov A A 2016 Fuzzy Data Analysis on Damage Definition in Cases of Inappropriate Use of Lands *Agrifood Economics* **4** pp 59-64
[19] Integrated Development Environment of Visual Studio, New Possibilities of Visual Studio, (https://www.visualstudio.com/ru/vs)
[20] Tindova M G 2005 Possibility of GIS-Technologies Using in Real-Estate Evaluation Problems Solving *Economic Analysis: Theory and Practice* **4** pp 60-4

Technology of regeneration and recycling of frying fats by using a new microgranulated nanosorbent in the catering

S Eliseeva[1]*, I Simakova[2], R Perkel[1], J Kostenko[2] and A Smolentceva[1]

[1]Peter the Great St. Petersburg Polytechnic University, St. Petersburg, Polytechnicheskaya st., 29, St. Petersburg, 195251, Russia
[2]Vavilov Saratov State Agrarian University, Saratov, Theater Square, 1, 410012, Russia

* Corresponding author: eliseeva_sa@spbstu.ru

Abstract. The article considers the technological capabilities and prospects for regeneration of frying fats in the catering enterprises using active filtration by a micro-granulated nanosorbent with subsequent recycling. The offered recycling technology involves the restoration of the quality of degraded frying fat with the purpose of its reuse and/or subsequent utilization. It is known that degraded frying fats contain products of oxidative and hydrolytic decomposition which vary according to their composition and structure. Actually, there is no universal technology that would provide the quality and safety indicators of frying fat for its further usage both in catering or for recycling in the market of secondary raw materials. Development of new microgranulated nanosorbent is based on fundamentally new technologies for modifying the surface of highly dispersed clay minerals of the Russian raw materials market using carbon nanomaterials. The granular sorbent is more preferable for use in catering, it allows easy separation of oil from the adsorbent without significant effort spent on filtration. The granular sorbent is lighter, more voluminous, and it significantly increases the adsorption efficiency even by contact method. The authors propose the technology of regeneration and recycling of frying fats with using new microgranulated nanosorbent in the catering enterprises.

1. Introduction

Despite the relevance of a wholesome diet, due to the tendency of modern man to lead a healthy lifestyle, there exists an obvious popularity of fast food, especially among young people, for which the food fries is dominated [1 - 4]. However, the systematic consumption of deep fried foods causes many alimentary diseases [5 - 8]. In addition, growth in production of fried potatoes is accompanied by an increase of waste of the fourth class danger to the environment which is degraded frying fats. The problem of its utilization and recycling has not yet been solved in Russia.

Nowadays there arises an issue concerning extension the shelf life of the used fried fats. For this purpose many fast food companies practice its daily passive filtering in order to remove charred particles of fried foods that accelerate the thermal oxidation of the used fat. However, the only filtration cannot remove fat-soluble polar compounds that accumulate during the frying process. One of the rational ways of saving food resources is periodic adsorption cleaning, as a stage of the process of deep frying, during the whole cycle of frying and at its final stage [9 - 17].

Regeneration of used fat is also a solution to the problem of their disposal. As a result, the regenerated fat is sold in the secondary market, where the demand for it is large enough, since the cost of the processed fat is 60% lower than the cost of it while it is fresh. The regeneration and cleaning of waste fat are the only correct and legitimate alternative for recycling of solid wastes of food industry and catering [18].

Currently, despite a large variety of sorbents for water purification [19-29], there is no industrial production of special sorbents in Russia, which would provide effective purification of the used fat from oxidation products. The existing technology is difficult to implement in the conditions of work of public catering enterprises. Also it do not meet all the requirements of quality, allowing one to reduce to the maximum permissible values only certain indicators of safety of used fats. So, there is a problem of development new "green technologies" aimed at ensuring the safety of fast foods fried in deep frying and solving the problem of utilization of used fat as solid waste in the food industry.

The solution proposed by the authors, involves using of simple technology, applicable to the catering enterprises and is based on the removal of products of oxidation of frying fats by adsorption purification method with the use of a high dosage nanosorbent. It is a special effective and relatively cheap product, made on the basis of mineral raw materials, which are mined in Russia.

The aim of this work was to develop a prototype of microgranulated nanosorbent and technology of regeneration and recycling of deep used fat for using in the food industry and catering.

The technical task of the work was to develop a prototype of microgranulated nanosorbent based on Russian natural materials. The efficiency of the sorbent was evaluated by a number of toxic secondary oxidation products removed from the purified fat, the speed of the purification process, organoleptic indicators of the quality of the purified fat, the possibility of using the purified fat in the process immediately after purification without additional operations or compliance with the quality indicators of the purified fat indicators acceptable for processing and disposal.

2. Objects and methods of research

The objects of this research were: nanosorbent based on natural materials; samples of deep fat (palm oil) initial and after heat treatment; the frying fats (palm oil), regenerated after treatment with a nanosorbent.

In order to develop and study the nanosorbent, we used laser diffraction analyzer of particle size SALD-2201 from SHIMADZU (Japan). Study of the porous structure of the nanosorbent samples was carried out by low-temperature nitrogen adsorption using the high-speed gas sorption analyzer Quantachrome NOVA (USA). Determination of structural characteristics was performed using the Brunauer-Emmett-Taylor technique. The desorption or adsorption branch of the isotherm in the pressure range 0.967 - 0.4 p/po was used as the initial data for calculations by the BJH method.

The granulated material was produced on the basis of the natural mineral raw components: flask, dolomite, magnesium silicate and montmorillonite. Usage of montmorillonite (Al_2O_3•$4SiO_2$•H_2O) is governed by the presence of all properties which are characteristic for natural nanoscale particles with a layered expanding cell.

Preparation of natural raw components for nanosorbent production included the stages of washing and drying. After bringing the particles of the composition of the natural materials to an equal size (from 0.5 to 10 μm) in a ball planetary mill, a pilot batch of adsorbents was produced by vortex rolling.

The composition of the modified natural mineral raw materials (flask, dolomite, magnesium silicate and montmorillonite) in certain ratios was gradually added into the container of the device intended for mixing bulk materials with water. In the rotating drum, the complex of adsorption materials was wetted with an aqueous solution, which was fed at a given speed. As a result of uniform mixing, a ductile mass with a moisture content of 15-30% and a density of 1 to 2 g/cm^3 was obtained. To impart strength or obtain sorbents with improved properties at the time of mixing there were added carbon nanomaterials.

In the special cylinder, absorbing the liquid phase, the sorbent particles fall on the inner surface of the cylinder due to centrifugal forces and are rolled into regular spherical granules. At the end of the granulation process heat treatment is carried out.

The optimum heat treatment temperature is defined as the minimum of the temperature range for the granulate 2 hour-holding time (with rotation) with the condition of compliance with the requirements for chemical and mechanical resistance of the granules. The experimentally determined optimum annealing temperature was 550°C.

The obtained granules are sent to a conveyor moving at a low speed from 0.01 to 0.05 m/s, which is blown with hot air from a heat gun at a temperature from 50 to 90°C. The granules of the adsorbent composition are dried to a humidity of 5-8% in a muffle furnace at a temperature of 450-700°C for 50 - 60 minutes or in a microwave oven for 5-8 minutes. The radiation frequency of the microwave device is 2450 MHz, its power is from 1 to 10 kW. The remaining in the granules moisture evaporates, forming a porous structure of the granules, which contributes to an increase in its specific surface area, and hence the sorption capacity. At the end of heat treatment, the granules are cooled to ambient temperature by blowing in a directed air flow. After cooling, the granules are sieved by fractions.

The quantitative analysis of the elemental composition of the nanosorbent was carried out using the energy dispersive x-ray fluorescence spectrometer Rayny EDX-720 SHIMADZU by calibration curves of fundamental parameters.

The acid number (Q.p.), the peroxide number (P.P.) and the total content of copolymers insoluble in petroleum ether (CIPE, %) were determined in all samples of used fats.

3. Results and discussion

The indicators of the elemental composition of the nanosorbent in mass fractions (%) are shown in Figure 1.

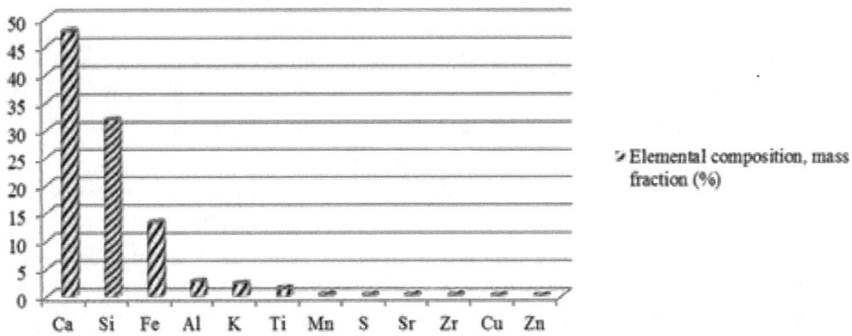

Figure 1. Indicators of the elemental composition of the nanosorbent, mass fractions (%).

In accordance with the required indicators of appearance and technical characteristics, the complex granulated nanosorbents should comply with the requirements specified in Table 1.

Table 1. Technical characteristics of the nanosorbent.

Indicators	Units	Characteristics
Appearance	Irregular gray granules 	
Particle size	mm	0.25-0.5

IOP Conf. Series: Earth and Environmental Science **337** (2019) 012015 doi:10.1088/1755-1315/337/1/012015

Humidity	%	3.0
The sieve residue:		
Sieve 60		17.0
Sieve 50	%	
Sieve 36		
Sieve 10		
Abrasion resistance	%	0.2± 0.2
Bulk density	g/dm3	1.2

The data on specific surface, total pore volume and distribution of the pores of the nanosorbents by its diameter are shown in Table 2

Table 2. Specific surface, total pore volume and distribution of the nanosorbent pores by diameter.

Nanosorbent	Specific surface, m^2/g	Total pore volume, cm^3/g	Average radius of micropores, Å	Distribution of pore diameters (nm), % 1.5-2.0	2.0-5.5	5.0-11.0	More than 11
	510	0.27	16.097	2.80	32.39	26.79	38.47

The main tasks of this method were to simplify the technology of cleaning and separating the nanosorbent from fat, increase the amount of toxic products removed (CIPE), improve organoleptic characteristics of cleaned fat, and consider the possibility of introducing this technology into the production process in food industry and catering.

Regeneration of degraded fat was carried out on a laboratory setup in the Department of "Food Technology" of Vavilov Saratov State Agricultural University.

The oxidized fat, depending on the stage of the technological process, is heated or cooled to 70°C, nanosorbent is introduced in the amount of 5-10% from weight of fat, all the components are stirred for 15-30 minutes at 50-70°C, and then filtered.

According to this study, the content of CIPE in the purified palm oil when using the microgranulated nanosorbent is reduced by more than 70%, which significantly exceeds the degree of purification of used fats achieved by imported synthetic adsorbents based on magnesium silicate - Magnesol XL and Dalsorb.

In order to study the efficiency of the developed technology, we performed the quality assessment of products fried in the refined fat using safety and organoleptic indicators (Figures 2-4).

Under the conditions of a real technological process, the product from flour (Chak-Chak) was fried at 150°C. The palm oil was used for 10 hours, during the entire heating time, the samples were taken to study the safety performance and sensory properties of the product. The kinetics of peroxide accumulation during the operation of purified fat is presented in Figure 2.

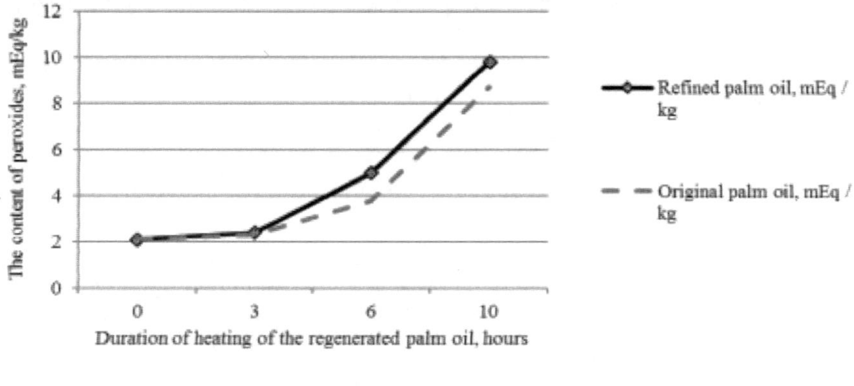

Figure 2. The accumulation of peroxides in purified palm oil used for 10 hours in the actual production process.

When using the palm oil as frying fat, cleaned with a nanosorbent, the kinetics of changes of the peroxide number both in the purified or fresh fat are identical, that characterizes its sufficient thermal stability. After 10 hours of continuous heating of the palm oil, its peroxide number approaches a limit value of 10 mEq of active oxygen/kg.

The peroxide number of the palm oil extracted from the product is close to the peroxide number of the oil in which the product was fried, Figure 3.

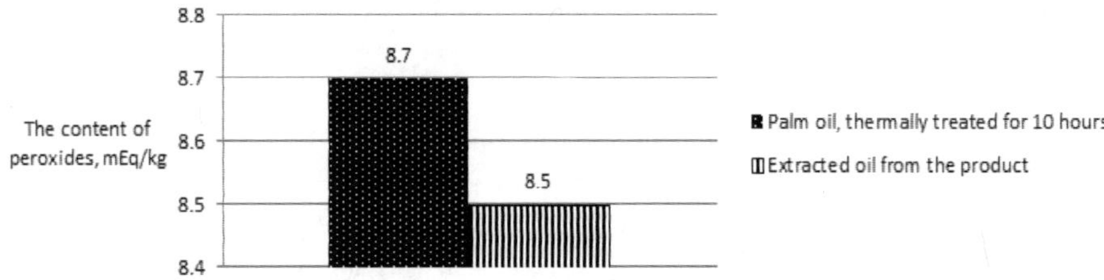

Figure 3. Comparative evaluation of the content of peroxides in palm oil extracted from the product and palm oil, in which the product was fried.

Figure 4 shows a behavior graph for content of toxic oxidation products of CIPE in palm oil without and with usage the purification technology.

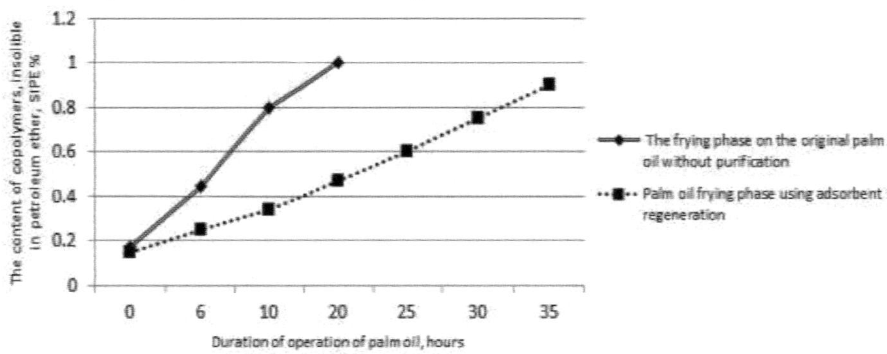

Figure 4. Changes in the content of CIPE, % in palm oil without cleaning and using cleaning with a nanosorbent.

As we can see from Figure 4, the application of the proposed cleaning technology allows one to increase the life of using of the palm oil to 10 hours, taking into account its indicators of safety.

Thus, the offered technology allows one to improve the indicators of safety of thermally oxidized frying fat in order to prolong its time-of-use up to 10 hours and to reuse it in the technological cycle after cleaning, as well as for subsequent disposal of used frying fat in terms of oxidative damage.

The organoleptic indicators of quality in purified oil corresponded to the norm.

4. Conclusion

As a result of the work, we developed the prototype microgranulated nanosorbent and the technology of regeneration and recycling of frying fat for applying in the food industry and catering enterprises. The advantages of the microgranulated nanosorbent consist in: a significant decrease in the concentration of CIPE in the cleaned fat by 70%; ease of applying in the practical process; simplicity of separation of the adsorbent from the fat; use of domestic raw materials for the production of sorbent.

The developed method of regeneration of frying fats allows use of purified fat immediately after cleaning, without subsequent deodorization, since the listed adsorbents have got a neutral or basic reaction without any acid.

The technology of purification of frying fats by nanosorbent developed by the authors is universal and the areas of its application are extensive. The authors have considered the possibility of using this technology on the example of the production of fast food products, but there is a huge potential for using nanosorbent in the technology of purifying vegetable oils from oxidation products during storage or preparation for recycling.

References

[1] Sahin S, Gulum Sumnu S 2009 *Advances in deep-fat frying of foods* (Boca Raton: CRC Press, Taylor & Francis Group)
[2] Billek G 2000 *Eur. J. Lipid Sci. Technol* **102** pp 587-93
[3] Simakova I V, Perkel R L, Volovey A G, Kutkina M N 2013 *Bulletin of the Vavilov Saratov State Agrarian University* **10** pp 47-50
[4] Simakova I V, Nosov A S, Makarova A N 2013 *Bulletin of the Vavilov Saratov State Agrarian University* **8** pp 59-62
[5] Blumenthal M M 1991 *Food Technology* **45** (2) pp 68–74
[6] Cohn J 2002 *Curr. Opin. Lipidol.* **13** pp 19-24
[7] Simakova I V, Makarova A N, Perkel R L 2011 *Sci. and Tech. J "Technology and Commodity Innovative Food Products"* **3** (8) pp 67-74
[8] Simakova I V, Nosov A S, Kotukova N A 2012 *Storage and processing of agricultural products* **5 pp** 42-44
[9] Boki K, Wada T, Ohno S 1991 *J. Amer. Oil Chem. Soc* **68** (8) pp 561-5
[10] Bouchon P, Pyle D L 2005 *Food and Bioproducts Processing* **83** pp 253–60
[11] Brian S 2006 Adsorbent Treatment of Frying Oil: Commercial Frying Case Study. *Abstracts of World Conference and Exhibition on Oilseed and Vegetable Oil Utilization, 14-16 August 2006, Istanbul, Turkey* p 14
[12] Buckley H A, Eatson A J, Johnson L R 1984 *Miner. Mag* **48** (346) pp 119-26
[13] Yates R A, Caldwell J D 1992 *JAOCS* **69** Sep
[14] Yates R A, Caldwell J D 1993 *JAOCS* **70** May
[15] Proctor A 2006 Vishal Jain. Production of CLA Rich Soy Oil. *Abstracts of World Conference and Exhibition on Oilseed and Vegetable Oil Utilization, 14-16 August 2006, Istanbul, Turkey* p 9
[16] Christie W W, Dobson G 2000 *Eur. J. of Lipid Sci. and Tech.* **102** pp 515–20
[17] Claxson A W D, G E Hawkes, D P Richardson, D P Naughton 1994 *FEBS Lett* **335** pp 81–90
[18] Kutlu B, Atac O, Turkay S 2006 Pretreatment of Used Frying Oil for Biodiesel Production *Abstracts of World Conference and Exhibition on Oilseed and Vegetable Oil Utilization, 14-16 August 2006, Istanbul, Turkey* p 12
[19] Leshchinskaya A P, Ezhova N M, Pisarev O A 2016 Synthesis and characterization of 2-hydroxyethyl methacrylate-ethylene glycol dimethacrylate polymeric granules intended for selective removal of uric acid *Reactive and Functional Polymers* **102** pp. 101-9
[20] Chechevichkin A V, Vatin N I, Samonin V V, Grekov M A 2017 Purification of hot water by zeolite modified with manganese dioxide *Magazine of Civil Engineering* **76(8)** pp. 201-213
[21] Politaeva N A, Slugin V V, Taranovskaya E A, Alferov I N, Soloviev M A, Zakharevich A M 2017 Granulated sorption materials for waste waters purufucation from zink ions (Zn2+) *Izvestiya Vysshikh Uchebnykh Zavedenii, Seriya Khimiya i Khimicheskaya Tekhnologiya* **60(7)** pp. 85-90
[22] Smyatskaya Y, Kosheleva A, Taranovskaya E 2018 Sorption properties of materials based on residual biomass *MATEC Web of Conferences* **245** 18005
[23] Kosheleva A, Atamaniuk I, Politaeva N, Kuchta K 2018 Adsorption of rare earth elements using bio-based sorbents *MATEC Web of Conferences* **245** 18001
[24] Garkushina I S, Polyakova I V, Pisarev O A 2017 Frontal dynamics of erythromycin sorption on monolithic molecularly imprinted polymer sorbents *Russian Journal of Physical Chemistry*

91(11) pp. 2225-29

[25] Groshikova A R, Medvedev R Y, Panarin E F 2017 Synthesis of organic–inorganic sorbent containing phenylboronic acid as glucose-binding ligand *Russian Journal of General Chemistry* **87(10)** pp. 2376-79

[26] Zhenisbekovna U S, Satayev M I, Viktorovich S V 2016 Production of active carbons from apricot pit shells by thermal activation in the mixture of carbon dioxide and water vapors *Biosciences Biotechnology Research Asia* **13(3)** pp. 1319-25

[27] Taranovskaya E A, Sobgayda N A, Markina D V 2016 Absorbent materials based on chitosan for sewage treatment from heavy metal ions *Ecology and Industry of Russia* **20(5)** pp. 34-39

[28] Marchenko G N, Akhmetova I G, Mindubaeva L M and Moiseeva M A 2006 Efficiency of use of granulated chitosan in the complex process of water conditioning at thermal power plants *Power Technol. Eng.* **40** 185–9

[29] Sobgaida N A, Ol'shanskaya L N, Makarova Y A 2009 Removing heavy-metal ions from effluents by means of sorbent formed from wood-working and agribusiness wastes *Chemical and Petroleum Engineering* **45(9-10)** pp. 580-584

IOP Conf. Series: Earth and Environmental Science **337** (2019) 012016 doi:10.1088/1755-1315/337/1/012016

Silica extraction from integrated refinery of rice straw for production of value-added products

Cuong Thai Dinh[1*], Hung Nguyen Duy[2], Dien Le Quang[1], Lubov A. Zenitova[2] and Thanh Nguyen Trung[1]

[1]Hanoi University of Science and Technology, Hanoi, Vietnam
[2]Kazan National Research Technological University, Kazan, Russsia

* Corresponding author: dien.lequang@hust.edu.vn

Abstract. Rice straw is an agricultural residue from rice production, which can be a potential lignocellulosic feedstock for various value-added products. In this study, silica was extracted from the rice straw. The biomass was pre-treated with ethanol and was cooked with 18% w/w sodium hydroxide to extract lignin and silica. The silica was precipitated from the alkaline solution using 30% sulfuric acid to pH 11.5 and was recovered through centrifugation. The obtained solid was washed and then combusted at 575 °C for 6 hours. The obtained ash was dissolved in sodium hydroxide solution at pH 8.0 and then re-precipitated by sulfuric acid to form silica powder, which obtained a yield of about 6.7% w/w dry rice straw. The characteristics of amorphous disk-formed silica, determined using XRD, SEM, EDS, and FT-IR.

1. Introduction

Rice straws used to be burned in the field for having no known use, which causes environmental pollution. This spurred interests to utilize them in the production of biofuels, chemicals, and new materials. In many rice-producing countries, rice straw is burned for energy production, which when combined can provide 14% of the world's energy [2,19,20]. In this study, extraction of the high-value added product, silica, from rice straw is achieved.

Rice straw has a high content of inorganic substances, ranging between 10-17% w/w dry material. Several chemical analyses of rice straw and rice husk from different rice varieties grown in Vietnam and other countries have confirmed the abundant presence of amorphous silica, which has several applications [4,18,21]. Compared with other agricultural residues such as bamboo and sugarcane bagasse, rice straw cellulose has higher silica content. The ash produced from the combustion of rice straw contains about 75% silicon [1].

Extraction of the bio-silica involves burning followed by alkaline and acid treatment [5-7, 13-18]. However, methods for silica recovery from the biomass has major limitations and drawbacks [8], such as low extraction efficiency and low purity because other organic compounds are extracted as well. Furthermore, bio-treatment of rice straw for enhanced silicon recovery requires complex pre-treatment [9], which may not be financially viable for large-scale production due to high costs and use of uncommon microorganisms.

Rice straw has a high content of waxes [10], which interferes with silica recovery, but has a high value and market. Therefore, an integrated production of silica, waxes, lignin, and cellulose can be applied to boost the feasibility of rice straw processing, compared with the method of Prehydrolysis-Kraft Pulping [11].

IOP Conf. Series: Earth and Environmental Science **337** (2019) 012016 doi:10.1088/1755-1315/337/1/012016

The study utilized the whole rice straw biomass by integrated refining of all its components. The products obtained were cellulose, lignin, extractives, and organic silica. Without pre-treatment, alkaline extraction of silica was more difficult. The study showed that removal of waxes using ethanol, prior alkaline extraction, resulted in higher silica recovery efficiency.

2. Materials and methods

2.1. Material

Oryza sativa L. rice straw, harvested on October 2017 from the Red River Delta of Vietnam, was used for the experiments. The dry rice straw was grinded and screened to prepare the suitable sample that was retained on 24 mesh screen for experiments. The suitable particle sizes for rice straw pretreatment should be smaller than 0.7 cm and bigger than 0.07 cm. The prepared biomass was stored in plastic bags at room temperature. It was not washed before ethanol extraction and alkaline treatment.

Rice straw was prepared by TAPPI method T264 cm-97 for analysis of chemical composition, which was determined as follows: cellulose content 39.7% (TAPPI T17 wd-70); lignin content 17.2% (TAPPI T222 om-98); pentosane 24.3% (TAPPI T223); extractives in ethanol 4.2% (TAPPI T204); ash 13.5% (TAPPI T211 om-93); and 1% sodium hydroxide solubility of Wood and Pulp (TAPPI T212) and hot water solubility 18.3% (TAPPI T207).

2.2. Ethanol extraction of Rice Straw

For each experiment, about 300 grams of prepared rice straw was extracted by ethanol at 50°C for 120 minutes, with a solid/liquid ratio of 1/12 (w/v). It was done in a 5-L stainless reactor, which was heated in a water tank. Pre-extracted rice straw was washed, dried, and used for silica preparation by alkaline treatment.

2.3. Silica preparation from extracted Rice Straw

Fifty grams of ethanol-extracted rice straw was treated with sodium hydroxide solution with a solid/liquid ratio of 1/10 in a 1-L stainless steel tank which was heated in a glycerine tank. Sodium hydroxide dosage, temperature, and processing time were adjusted according to the aims of each experiment. Alkaline-treated rice straw was pulped for 5 minutes and then filtered to collect the black liquor (alkaline solution) and cellulose.

The alkaline solution was then treated with 30% sulfuric acid to pH 11.5 for silica precipitation, and then to pH 3.0 for lignin precipitation. The precipitates were then filtered, washed, dried and analyzed. The silica powder recovered was burned at 575°C to obtain the ash, which was then dissolved using sodium hydroxide solution to pH 8 under heat, and filtered to remove the insoluble components. The filtered solution was diluted using sulfuric acid to precipitate silica. Silica suspension was centrifuged, and the pellet was washed and dried for the determination of yield and analysis of product properties.

Amorphous structure of silica dioxide was determined using XRD and FT-IR. The chemical composition of silica powder was determined by EDS. A form of silica particles was studied by FE-SEM JEOL JSM-7600F.

3. Results and Discussion

As known, rice straws are covered by a layer of lipophilic extractives [10], such as waxes, which are insoluble in water. During chemical processing, these hydrophobic layers make it hard for alkaline or acidic solutions to penetrate inside the plant cell walls. Meanwhile, silica compounds can glue with these compounds, forming a layer on the surface of the straw. Therefore, the compounds of straw are hard to separate. These properties were key to the separation of these compounds. To improve the efficiency of silica recovery, the rice straw was pre-treated with ethanol prior extraction. Besides being able to obtain extractives, the cell walls were easier to penetrate, improving silica removal from straw. It is a new method, different from the previous studies.

IOP Conf. Series: Earth and Environmental Science **337** (2019) 012016 doi:10.1088/1755-1315/337/1/012016

The objective of this study is to determine the appropriate conditions for separation and purification of silica dioxide from straw. The extractives and cellulose are valuable by-products of the process.

3.1. Alkaline treatment of Rice Straw

The pre-treated rice straw was treated with 5-25% w/w sodium hydroxide solution at 60°C for 180 minutes. Residual solid was washed, dried, and combusted, and the ash was collected for analysis. The alkaline solutions were treated with H_2SO_4 30% to pH 11.0 to obtain the inorganic substances.

The determination of the ash content of alkaline-treated rice straw showed that to significantly separate the inorganic matter from the straw, the sodium hydroxide solution must be more than 15% w/w (Figure 1). The maximum yield for extracted inorganic substances was achieved using an alkaline concentration of 18% (w/w), obtaining 82% of ash. The residual solid collected in this setup was 48.3% w/w, which indicated that lignin and carbohydrate were dissolved in the alkaline solution. Compositional analysis of the residual solid showed that 72.4% were lignin and 26.6% were carbohydrates.

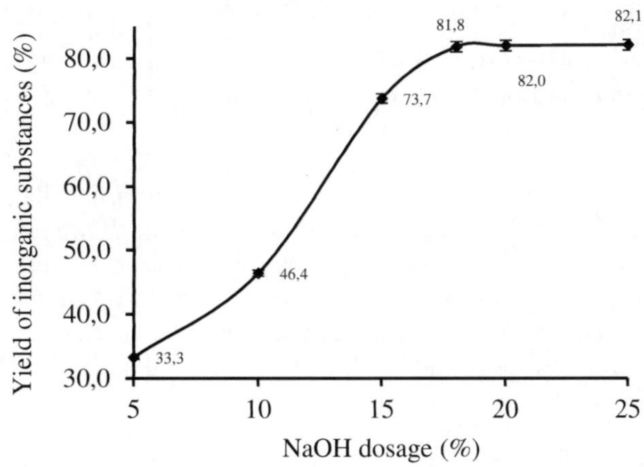

Figure 1. Effect of alkaline dosage on the yield of dissolved inorganic substances. Conditions: treatment time of 180 min and treatment temperature of 60°C)

The effect of temperature on the yield of inorganic substances indicated that with the same time span and alkaline concentration, treatments at temperatures greater than 60°C were ineffective (Figure 2). On the other hand, treatments at temperatures less than 60°C gave a lower yield of inorganic substances. For instance, samples subjected to 50°C and 40°C only yielded 40% and 30% inorganic substances, respectively. This showed the significant effect of temperature on the degradation and dissolution of components during alkaline treatment.

IOP Conf. Series: Earth and Environmental Science **337** (2019) 012016 doi:10.1088/1755-1315/337/1/012016

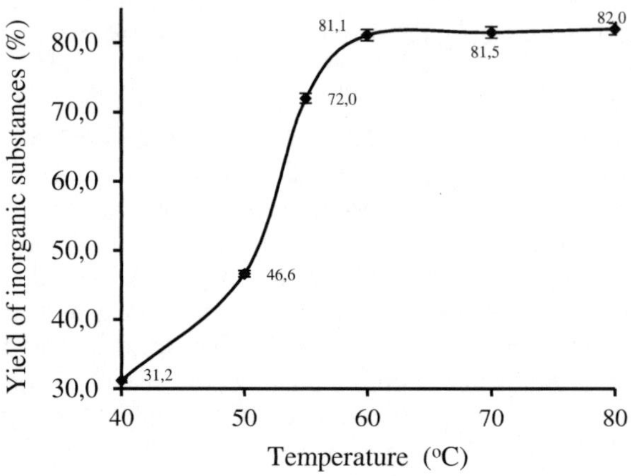

Figure 2. Effect of temperature on the yield of dissolved inorganic substances.Conditions: NaOH dosage of 18% (w/w), treatment time of 180 min.

Dissolution of inorganic substances and other organic components from rice straw into the alkaline solution requires some time. However, longer treatment time does not always translate to being more effective. From the experiments, it was observed that within the first 90 minutes, 40% of the total inorganic substances was dissolved. This amount doubled in the next 60 minutes before it reached the "threshold" (Figure 3). Sufficient treatment time is required to swell the straw, for the alkaline solution to seep through the cell walls via osmosis, and for the inorganic substances to diffuse outwards through the cell wall.

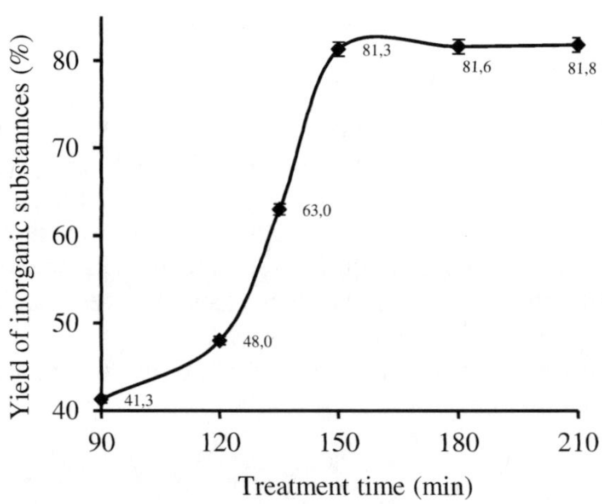

Figure 3. Effect of treatment time on yield of dissolved inorganic substances. Conditions: NaOH dosage of 18% (w/w), temperature of 60°C.

The suitable conditions of alkaline treatment for the extraction of silica, containing some inorganic substances, from rice straw were fixed using a solid/liquid ratio of 1:10, sodium hydroxide concentration of 18% w/w dry rice straw, temperature of 60°C, and treatment time of 150 mins.

IOP Conf. Series: Earth and Environmental Science **337** (2019) 012016 doi:10.1088/1755-1315/337/1/012016

3.2. Silica Preparation

For silica preparation, 1.0 kg of ethanol-pre-extracted rice straw was treated with sodium hydroxide at suitable conditions to obtain about 7.2 liters of silica containing alkaline liquor. Afterward, it was treated with sulfuric acid 30% at 60°C to pH 11.5-11.6. At this pH value, lignin has not precipitated yet. These conditions are required for the effective preparation of silica. The precipitate was then filtered, washed, dried and burned at 575°C for 3 hours. The grey-white ash produced was dissolved in sodium hydroxide solution at 100°C for 3 hours, then treated by sulfuric acid 25% at pH 6.0. The solution was centrifuged, and the pellet produced was dried to obtain the amorphous silica. These conditions for silica preparation have been optimized by a series of experiments.

Preparation of silica from non-pre-extracted rice straw has been carried out following the same method, except pre-treatment. A higher yield of 6.7% w/w was obtained with the ethanol-pre-treated rice straw compared to the 2.2% of the non-pre-treated samples.

The X-ray diffraction of silica (Figure 4) showed a broad peak between 20° and 30°, centered at 22.5°, typical of the amorphous silica [12]. EDS and XRD data gave clear evidence of the high purity of the collected amorphous silica, which was extracted from pre-treated rice straw.

EDS analysis confirmed that the as-derived product contained Si and O at a (52%:38%) mass ratio. These amorphous mesoporous silica disks have great potential to be prepared and utilized as molecular sieves, catalyst preparation, absorbents, etc.

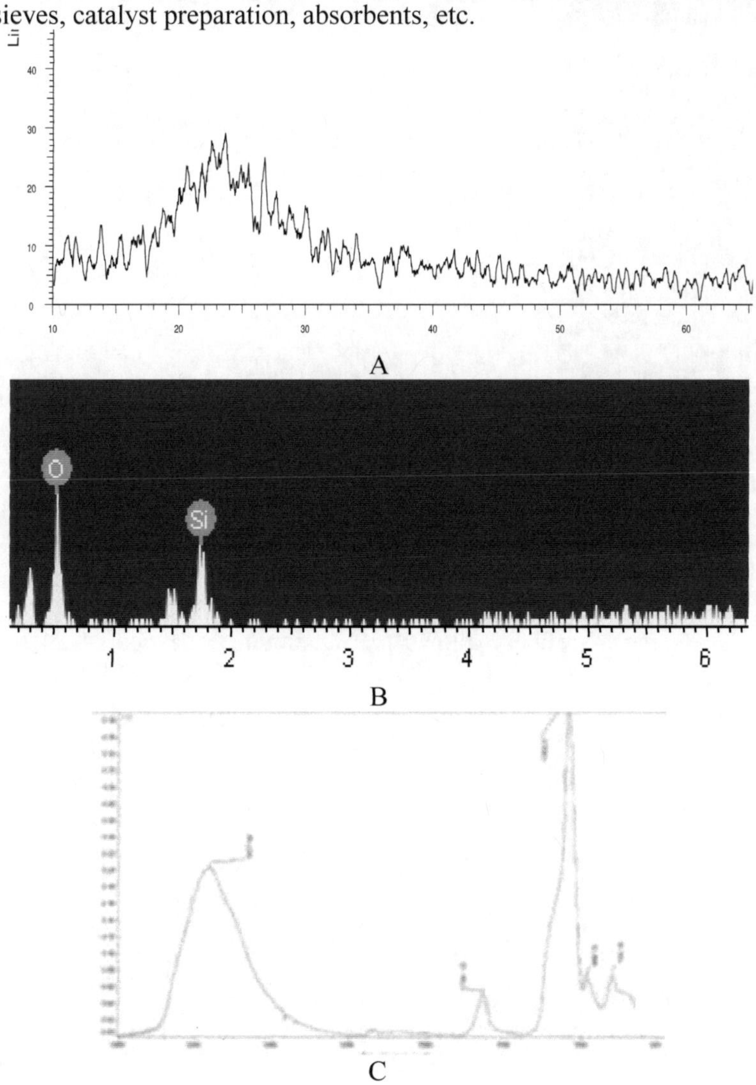

Figure 4. Characterization of silica powder: XRD (A), EDS (*B*), and FTIR (*C*)

IOP Conf. Series: Earth and Environmental Science **337** (2019) 012016 doi:10.1088/1755-1315/337/1/012016

Scanning Electronic Microscopy (Figure 5) showed that the silica powder agglomerates were of the size 6-8 micrometer. Due to the nonconductive nature of silica, the charges quickly accumulate on the powder surfaces, even with gold sputtering, showing visible artifacts and limiting resolution.

Figure 5. SEM micrographs of silica from pre-extracted rice straw

Figure 6 shows the flow chart of the overall process of isolation of cellulose and by-products from the rice straw integrated refinery.

The ethanol-dissoluble extractives are interesting compounds, which can be used as a preservative for wood and other applications. Unbleached cellulose is used for bioethanol and nanocellulose production. Lignin can be precipitated by sulfuric acid after silica.

Compared to other methods, this scheme of refinery allows for maximum utilization of all the components of rice straw.

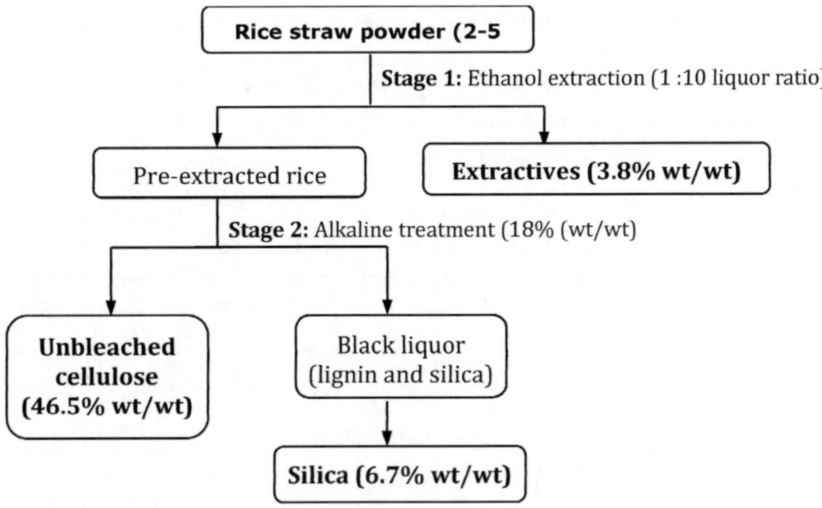

Figure 6. The schematic representation of integrated refinery of rice straw

4. Conclusion

The modern method of lignocellulosic refineries strives to tackle the problem of agricultural wastes by the effective utilization of all the components. This study has identified ethanol pre-extraction as the best method for the removal of inorganic substances in rice straw cell walls, improving the recovery of silica, through sodium hydroxide treatment.

The sequential extraction method in alkaline solution is an efficient process for the isolation and purification of silica and lignin from rice straw.

Recovery of silica from the black liquor obtained from rice straw alkaline pulping can be considered as one of the achievements in the field of agricultural residues refining. The successful extraction of purified silica confirmed the viability of the method presented herewith in obtaining this value-added product for the rice straw refinery.

References

[1] Kargbo F R, Xing J, Zhang Y 2010 *Agricult. and Biology J. of North America* **1** (3) pp 195-200
[2] Matsumura Y, Minowa T, Yamamoto H 2005 *Biomass and Bioenergy* **29** p 347
[3] Isikgor F H, Becer C R 2015 *Polymer Chem.* **6** p 4497
[4] Dien L Q, Phuong N T M, Hoa D T, Hoang P H 2015 *Appl. Biochem. Biotech.* **175** (3) p 1536
[5] Chen H, Wang F, Zhang C, Shi Y, Jin G, Yuan S 2010 *J. of Non-Crystalline Solids* **356** p 2781
[6] Zaky R R, Hessien M M, El-Midany A A, Khedr M H, Abdel-Aal E A, El-Barawy K A 2008 *Powder Tech.* **185** p 31
[7] Hessien M M, Rashad M M, Zaky R R, Abdel-Aal E A, El-Barawy K A 2009 *Materials Sci. and Eng.* **162** p 14
[8] Lu P, Hsieh Y L 2012 *Powder Tech.* **225** p 149
[9] Zemnukhova L A, Panasenko A E, Artem'yanov A P, Tsoy E A 2015 *Biores.* **10** (2) p 3713
[10] Xiao B, Sun X F, Sun R C 2001 *J. of Wood chem. and Tech.* **21** (4) p 397
[11] Dien L Q, Cuong T D, Hoang P H, Hoa D T, Thanh L T 2015 *Int. J. of Materials Chem. and Physics* **1** (3) p 359
[12] Lu P, Hsieh Y L 2012 *Powder Tech.* **225** p 149
[13] Adam F, Chew T S, Andas J 2011 *J. Sol–Gel Sci Technol* **59** p 580
[14] Jal P K, Sudarshan M, Saha A 2004 *Colloids Surf. Physicochem. Eng. Aspect* **240** p 173
[15] Zaky R R, Hessien M M, El-Midany A A, Khedr M H, Abdel-Aal E A, El-Barawy K A 2005 *Powder Technology* **185** p 31

[16] Hessien M M, Rashad M M, Zaky R R, Abdel-Aal E A, El-Barawy K A 2009 *Materials Sci. and Eng.* **162** p 14
[17] Le V H, Thuc C N H, Thuc H H 2013 *Nanoscale Res.Letters* **8** p 58-68
[18] Fernandes I J, Calheiro D, Sánchez F A, Camacho A L D, Rocha T L A D C, Moraes C A M, Sousa V C D 2017 *Materials Res*. **20** p 512
[19] Lim J S, Manan Z A, Alwi S R W, Hashim H 2012 *Renew. and Sustain. Energy Reviews* **16** p 3084
[20] Delivand M K, Barz M, Gheewala S H 2011 *Energy* **36** (3) p 1435
Phuong N T M, Hoang P H, Hoa D T 2017 *Clean Tech. and Env. Policy* **19** p 1313
[21] Abraham A, Mathew A K, Sindhu R, Pandey A, Binod P 2016 *Biores. Tech.* **215** p 29

IOP Conf. Series: Earth and Environmental Science **337** (2019) 012017 doi:10.1088/1755-1315/337/1/012017

Glycerol modifier influence on interfacial activity of bentonite during the adsorption of o-toluidine

A V Kosarev[1]*, E I Tikhomirova[1], N A Politaeva[2] and A L Podolsky[1]

[1]Gagarin State Technical University of Saratov, School of Urbanistics, Civil Engineering and Architecture, Polytechnicheskaya, 177, Saratov, 410054, Russian Federation
[2]Peter the Great St. Petersburg Polytechnic University, Higher School of Biotechnology and Food Technology, Polytechnicheskaya, 29, St.-Petersburg, 195251, Russian Federation

* Corresponding author: aleteia@inbox.ru

Abstract. This paper is devoted to the study of changes in the surface activity of systems containing unmodified bentonite and bentonite modified with glycerol additives upon adsorption of o-toluidine. A relationship between surface activity and the value of o-toluidine adsorption on modified aluminosilicate was obtained. It is shown that the presence of a modifier-glycerol in the adsorption system enhances the process of lowering the surface tension.

1. Adsorption activity of modified aluminosilicates

The purification of industrial wastewater from organic compounds plays an important role for applied ecology today. Meanwhile, models of the relationship between the structural and adsorption characteristics of aluminosilicate systems under conditions of dynamic adsorption have not been sufficiently developed. Clay materials are one of the most effective types of adsorbents. It should be noted montmorillonite, which is the main component of natural bentonite. This was established in [1] the influence of the structural factor in the modification of montmorrilonite by intercalation, which leads to an increase in the proportion of active adsorption centers on the surface of the aluminosilicate. It is also shown that this leads to an increase in the efficiency of adsorption of heavy metals on them. The authors [2] showed that aluminosilicates adsorption activity is determined by the structural characteristics of the system: the interplanar distance, the distribution of the adsorption centers along the surface of the adsorbent, the diameter of the adsorptive molecule. In [3] it was found that the nature of formaldehyde adsorption on aluminosilicate adsorbents depends on the type of modification of the adsorbent: acid modifiers change the specific surface of the adsorbent more efficiently than the main ones.

The authors of [4] showed that the processes of adsorption of zinc ions from solutions on natural and modified montmorillonites can be described using Langmuir and S-shaped isotherms, the shape of which is determined by both the surface properties of the sorbent and the forms of zinc ions in aqueous solutions. In [5] it was established that cationic dyes molecules can aggregate with each other as a result of aluminosilicates surface adsorption, which leads to an increase in the efficiency of their extraction from water. The authors [6] showed that for the adsorption of methylene blue and crystal violet at low values of the adsorption capacity the endothermic effect prevails, and at high values the exothermic effect prevails.

The authors [7-8] note that the adsorption of cationic forms of dyes on aluminosilicates has a monomolecular mechanism that follows the Langmuir adsorption isotherm and is an endothermic process, during which adsorbent-adsorbate layers with low orderliness are formed. A technology for purifying water from the Orange II dye was proposed in [9]. Processes of preliminary extraction of cationic surfactants from waste and the subsequent use of organomodified bentonite were combined in this technology. It was shown in [10] that anionic dyes are also effectively removed from water using aluminosilicates, modified hescamethylenediamine. The authors [11] note that clinoptilolite can be a very effective adsorbent for removing organic dyes, such as safranin T, from wastewater.

Adsorption mechanism obeys the Langmuir isotherm in this case. At the same time, the adsorption of toduidine blue on a given zeolite is described by a second-order kinetics model and obeys isotherms of both Langmuir and Freundlich [12]. The kinetics of main green and basic violet ions adsorption on the surface of montmorillonite was simulated using the fractal approach. Electrostatic interactions are not dominant in the interaction "zeolite - the main dye" [14]. The authors of [15] showed that montmorillonite, modified in an alkaline medium with polyethyleneimine, is associated with it through electrostatic interactions. The mechanism of modifying aluminosilicates with the addition of surface-active polymer modifiers is discussed in article [16]. Three types of composite materials are distinguished in depending on the structure of the components and the conditions of synthesis. traditional microcomposite is formed in the case when the modifier molecules do not penetrate into the layer structure of the aluminosilicate. The second modification variant corresponds to the intercalated composite, corresponds to the case when the modifier molecules penetrate into the interlayer space of the aluminosilicate and cause an increase in this distance and changing the adsorption activity of the system. The third variant of the modification, exfoliated nanocomposite, corresponds to the case when the modifier molecules completely separate the particles of aluminosilicate, so that they completely lose their orderliness. [17,18]. It was shown in [19] the adsorption of cationic dyes on aluminosilicate modified by cationic surfactants is due to the hydrophobic interaction of the adsorbate with the modified surface, as well as due to ion exchange.

2. Adsorption process modeling

The ratio between surface tension coefficient σ and the adsorption characteristics of the adsorbate associated with the aluminosilicate is given by the Gibbs equation:

$$\frac{\Gamma_0 RT}{C} = -\frac{\partial\sigma}{\partial C} \tag{1}$$

where G_0 - the value of adsorption of o-toluidine on the modified aluminosilicate; C is the equilibrium concentration of the adsorbtive in the aqueous macrophase of the solution; R is the universal gas constant; T-temperature. The ratio (-∂σ / ∂c) characterizes the surface activity of the adsorbing substance. For bentonite modified with non-ionic modifier additives, the equation (1) is:

$$\frac{\Gamma_m RT}{C} = -\frac{\partial\sigma}{\partial C} \tag{2}$$

The surface tension of a system containing modified aluminosilicate σ_m is related to the surface tension σ of a system containing unmodified aluminosilicate by the ratio

$$\sigma_m = k\sigma, \tag{3}$$

where k is the coefficient, showing how many times the adsorption of this component on the modified adsorbent is higher than on the non-modified one:

$$k = \frac{\Gamma_{m\infty}}{\Gamma_\infty}, \tag{4}$$

where $\Gamma_{m\infty}$ and Γ_∞ - values of the limiting adsorption capacity in relation to this adsorbate modified and non-modified aluminosilates respectively. From equations (1) and (2) taking into account (3), the:

$$(\Gamma_m + \Gamma_0)\frac{RT}{C} = -\frac{\partial\sigma}{\partial C}(1+k) \tag{5}$$

It follows that the amount of surface activity caused by the modification of aluminosilicate can be found in the following way:

$$-\frac{\partial\sigma}{\partial C}=\frac{(\Gamma_m+\Gamma_0)}{(1+k)}\frac{RT}{C} \tag{6}$$

3. Experiment results

We have experimentally studied the adsorption of o-toluidine on pure bentolit, and on the bentonite, modified by glycerol additives. Laboratory investigations are performed on the basis of the scientific and educational center "Industrial ecology" of Gagarin State Technical University of Saratov with the use of the verified test and auxiliary equipment.

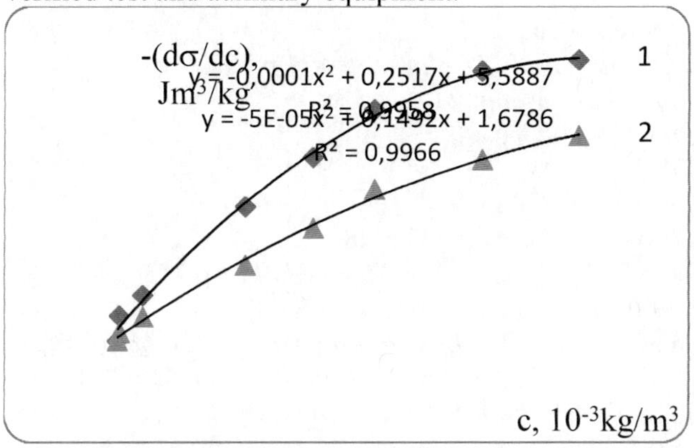

Figure 1. Change of surface activity (- ∂σ/∂c) on bentonite, modified glycerol (curve 1) and non-modified bentonite (curve 2) from the equilibrium concentration of o-toluidine in solution.

Experimental determination of the amount of adsorption capacity is based on determination of residual concentration of o-toluidine in the aqueous phase of the solution containing adsorbent. We had an experiment to determine this value. We dissolved 1.5 g of 2-naphthene-3,6-disulfonic acid sodium salt in 50 ml of distilled water. Then we added the obtained solution Na2CO3 to the resulting system and diluted with water to the volume of 100 ml. We mixed 1 ml of the solution with 1 ml 2 N hydrochloric acid and added 1 ml of 0.25% solution NaNO2 for the reaction of diazotization. We poured this solution in 15 minutes with 10 ml of 1 M solution Na2CO3 and 0.5 ml of reagent solution, and then diluted with water to a volume of 25 ml. After 1 h we photometric the resulting solution on a two-beam scanning UV-type Spectrophotomometer «Promekolab» PE-6100 UV at a wavelength of 485 nm [20].

Increase of the size (- ∂σ/∂c) at increasing of concentration o-toluidine is caused by decrease of a superficial tension at adsorption (Fig. 1). The presence of the modifier-glycerol leads to an increase in the interlayer distance in the structure of adsorbent, as well as the of micro-environment polarity reduction of the molecules o-toluidine. It is strengthens the process of lowering the surface tension.

This work is topical for consideration of industrial ecology problems on improvement of adsorption purification of sewage from organic compounds by modification of aluminosilicate adsorbents.

4. Conclusions

The work shows that the presence of the modifier-glycerol leads to increase in the dependence of the surface activity of the system (- ∂σ/∂c) from the concentration of o-toluidine in the solution. It is can be explained by the fact that presence of the modifier-glycerol leads to an increase in the interlayer distance in the structure of adsorbent, as well as the of micro-environment polarity reduction of the molecules o-toluidine. It is strengthens the process of lowering the surface tension.

The given work is the result of carrying out of research work within the framework of grant for winners of competitive selection of the projects executed by scientific collectives of research centers and (or) scientific laboratories of educational Organizations of higher education (state task of the Ministry of Education and Science of the Russian Federation, project 5.3922.2017/PCh).

IOP Conf. Series: Earth and Environmental Science **337** (2019) 012017 doi:10.1088/1755-1315/337/1/012017

References

[1] Yurchenko V V, Sviridov A V, Sviridov V V, Nikiforov A F, Pryanichnikov S V 2017 *Sorption and chromatographic processes* **17** p 506

[2] Ivanova E S, Gavronskaya Yu Yu, Stozharov V M, Pak V N 2014 *J. general chem* **84** p 185

[3] Nazimova L A, Strelnikova O Yu, Khodosova N A, Belchinskaya L I, Roessner F 2014 *Sorption and chromatographic processes* **14** p 185

[4] Ganebnykh E V, Sviridov A V, Maltsev G I 2015 *Chemistry for sustainable development interes.* **1** p 89

[5] Miyamoto N, Kawai R 2000 *Applied Clay Science* **16** p 161

[6] Rytwo G, Ruiz-Hitzky E 2003 *J. of Thermal Analysis and Calorimetry* **71** p 751

[7] Insuwan W, Rangsriwatananon K 2014 *J.of Porous Mater.* **21** p 345

[8] Khenifi A, Bouberka Z, Sekrane F 2007 *Adsorption* **13** p 149

[9] L Zhu 2008 *J. Ma Chem. Eng. J.* **139** p 503

[10] Alver E, Metin A U 2012 *Chemical Engineering J.* **200** p 59

[11] Qiua M, Qian Ch, Xub J, Wu J, Wang G 2009 *Desalination* **243** p 286

[12] Alpat S K, Ozbayrak O, Alpat S, Akçay H 2008 *J. of Hazardous Materials* **151** p 213

[13] Wang C C, Juang L C, Hsu T C, Lee C K 2004 *J. Colloid Interf. Sci.* **273** p 80

[14] Hu Q H, Qiao S Z, Haghseresht F, Wilson M A, Lu G Q 2006 *Ind. Eng. Chem. Res.* **45** p 733

[15] Pshinko G N, Puzyrnaya L N, Kosorukov A A, Goncharuk V V 2010 *Radiochem.* **52** p 291

[16] Chvalun S N, Novokshonova L A, Korobko A P, Brevnov P N 2008 *Russian Chem. J.* **LII** p 52

[17] Hussain F, Hojjati M, Okamoto M, Gorga R E 2006 *J. of Composite Material.* **40** p 1512

[18] Alix S, Follain N, Tenn N, Alexandre B 2012 *J. Phys. Chem.* **116** p 4937

[19] Tarasevich Y I, Trifonova M Y, Bondarenko S V, Dolenko S A 2010 *Colloid J.* **72** p 565

[20] Kosarev A V, Atamanova O V, Tikhomirova E I, Istrashkina M V 2018 *Water and ecology: problems and solutions* **3**(75) p 24

IOP Conf. Series: Earth and Environmental Science **337** (2019) 012018 doi:10.1088/1755-1315/337/1/012018

The use of clays materials of the Tyumen region in the treatment of natural and waste water

A Zagorskaya[1]* and L Pimneva[1]

[1]Tyumen Industrial University, Volodarskogo str., 38, Tyumen, 625000, Russia

* Corresponding author: zagorskajaaa@tyuiu.ru

Abstract. An assessment was made of the ecological status of water bodies in the South of the Tyumen Region, which play a significant role in the region's water use and are sources of drinking and industrial water supply. The hydrochemical characteristics of natural waters and the water use conditions that determine the indicators of water quality in water bodies and streams are considered. Critical levels of pollution have been established for zinc, copper, manganese, iron, cadmium, chromium, nickel and aluminum in all water bodies of the Tyumen region. The reasons for optimizing the purification processes at the existing water supply and wastewater treatment facilities in the region are given. The structure and chemical composition of clay materials of the Tyumen region was studied. The efficiency of extraction of nickel ions of kaolin, montmorillonite and mixed kaolin and montmorillonite clays has been investigated. The quantitative characteristics of the process of nickel ion adsorption on all the presented natural clays are obtained. It is established that the mixed natural clay has higher values of exchange capacity in relation to nickel ions. It is noted that the description of the adsorption equilibrium corresponds to the Langmuir model. The possibility of using kaolinite, montmorillonite and mixed clays as a reagent for purifying natural and waste waters at the existing treatment facilities of the Tyumen region is presented.

1. Introduction

According to the classification of the Federal Service for Hydrometeorology and Environmental Monitoring of the Russian Federation, water bodies of the Tyumen region belong to the 4th class - "dirty". The unsatisfactory state of water bodies is associated with the transit formation of the main water masses and pollution in the overlying areas (Sverdlovsk, Kurgan, Omsk regions and the Republic of Kazakhstan). Statistics show that industrial growth in the Tyumen region also makes a significant contribution to water pollution. In total, about 130 large industrial and agricultural facilities operate in the region with a total consumption of 315.4 million tons of water per year. Of the effluents entering the rivers and lakes of the Tyumen region, only 69.4% are sufficiently cleaned [1]. The dynamics of pollution in the water bodies of the region is presented in Figure 1 and 2.

IOP Conf. Series: Earth and Environmental Science **337** (2019) 012018 doi:10.1088/1755-1315/337/1/012018

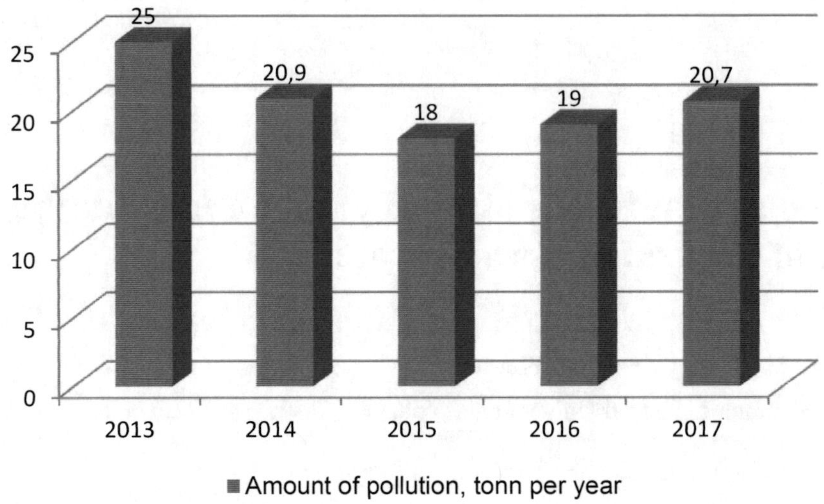

Figure 1. Mass of pollution discharged into water bodies over the past 5 years

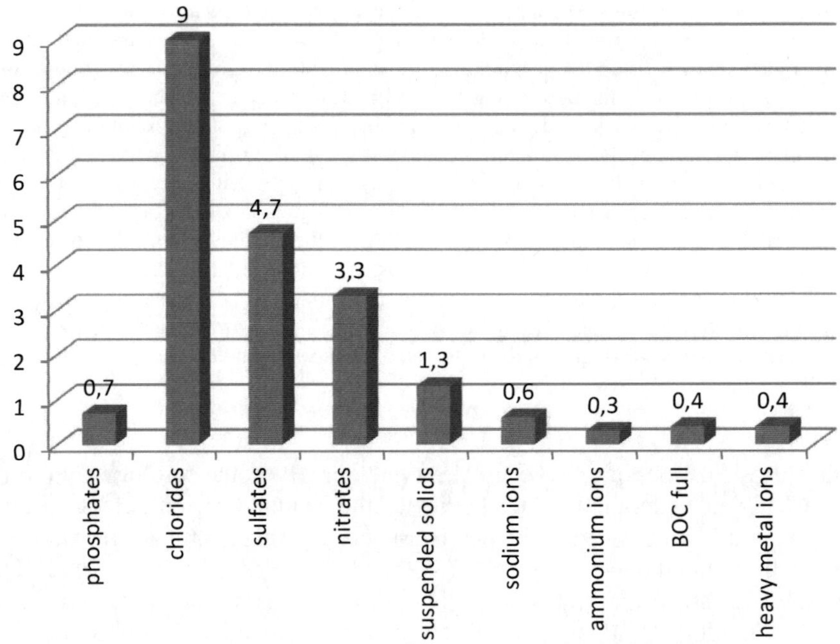

Figure 2. Kinds of pollutions in ton per year

The specificity of water pollution in the Tyumen region is related to the natural background for a number of pollution: difficult and easily oxidizable organic substances (COC, BOC), compounds of iron, copper, manganese [1, 2]. However, for the last 8 years there has been a steady accumulation of heavy metal ions in natural components of the environment [2-4]. The concentrations of heavy metal ions in the water bodies of the city are presented in Figure 3. Special attention should be paid to pollution first and second hazard classes: cadmium, lead, mercury and nickel. Reducing their concentration, blocking the transfer of the food chain is important for the health of the population of the region as a whole [5].

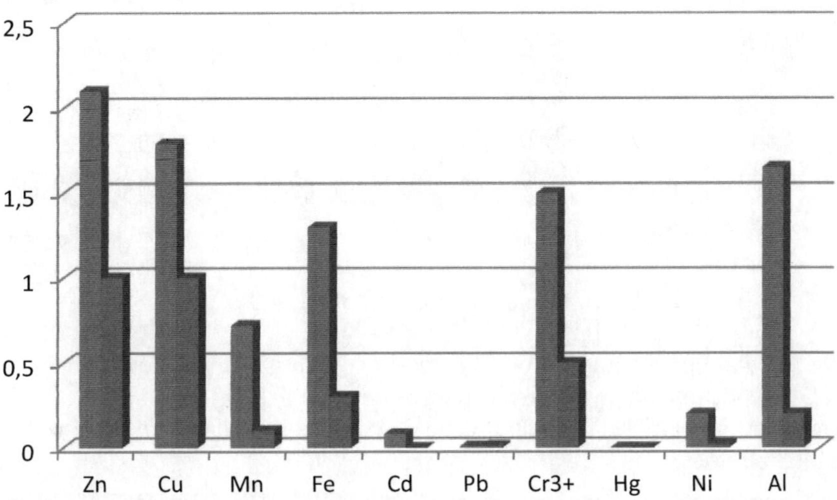

Figure 3. According to the concentration of heavy metal ions (mg / dm3) MAC of water bodies (mg / dm3)

In this regard, the technologies of purification of natural and waste waters are acquiring particular relevance, which make it possible to extract heavy metal ions as efficiently as possible with minimal economic costs, and a number of research studies are now being carried out [6-13]. The presence in the Tyumen region of clay deposits allows us to consider them as available material, which can easily be integrated into the existing scheme of cleaning any treatment facilities in the area.

2. Subjects and methods

The purpose of this work is to study the adsorption properties of natural clay minerals of the Tyumen region in relation to nickel ions and the possibility of using them for the purification of natural and waste waters.

In the work were used natural clay minerals of the Tyumen region. Analysis of the chemical composition of clays was determined by a scanning electron microscope (SEM).

According to the data obtained, its main components are oxides in the amounts presented in Table 1.

Table 1. - The chemical composition of natural clays, mass. %

Clays	SiO_2	Al_2O_3	Na_2O	K_2O	CaO	Fe_2O_3	TiO_2	MgO	Ni
			Before adsorption						
MMT	50.24	20.78	0.73	3.37	0.60	8.52	0.91	1.79	-
K	49.22	46.87	0.96	0.80	0.86	1.12	0.94	-	-
K + MMT	53.62	20.29	0.41	3.88	1.53	13.13	1.71	1.63	-
			After adsorbtion						
MMT	48.62	16.59	0.39	0.37	0.33	7.49	0.52	1.64	1.52
K	50.20	45.77	0.06	0.44	0.41	1.38	1.08	-	0.008
K + MMT	52.18	19.57	0.34	3.81	0.52	16.66	1.84	1.28	0.008

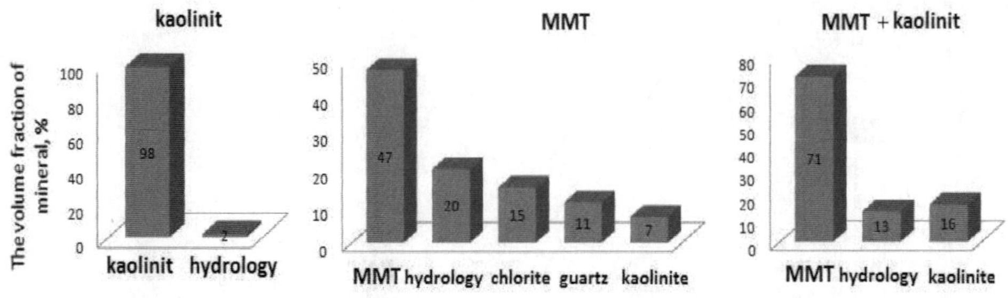

Figure 4. Phase composition of natural clays

The mineralogical and phase composition of natural clays (Fig. 3.) was determined by X-ray phase analysis using a DRON diffractometer with a copper anode (λ = 1.5406 Å).

Figure 4 shows that the particles of natural clays have sizes less than 1 micron in the form of thin leaves with irregular outlines.

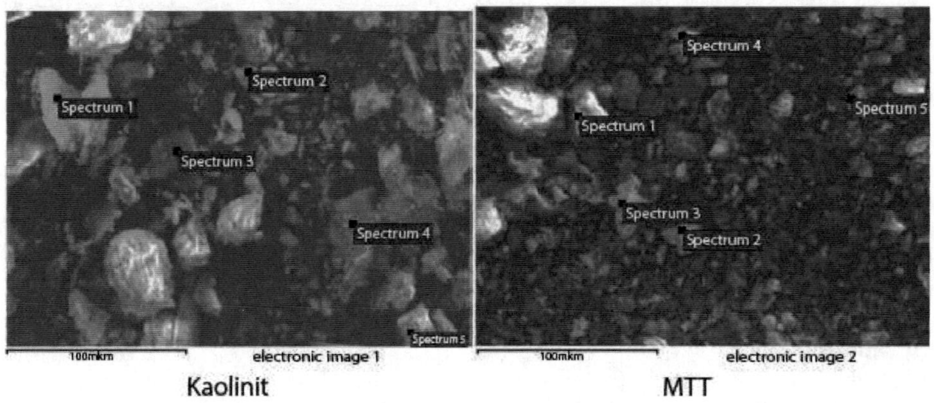

Figure 5. SEM micrograph

The adsorption activity of natural clays is determined by the fact that they contain the mineral montmorillonite [14]. The crystal lattice of montmorillonite is a three-layer packet in which between the layers of silicon-oxygen tetrahedra there is a layer of aluminum-octahedra The distance between the packages can vary from 0.4 to 2.0 nm, which promote to the ion exchange of such clays.

The objects of study of the adsorption properties of natural clays were model nickel nitrate aqueous solutions. Dilution methods were used to prepare solutions with various concentrations of nickel ions from nickel nitrate solutions. The adsorption process was studied under static conditions. The ion concentration is nickel-plated before and after the adsorption process by the standard method using the trilonometric method [15]. The adsorption capacity of clay minerals was calculated as a function of the concentration of the solutions at a constant temperature. The amount of adsorption is determined by the excess of the substance at the phase boundary compared with the equilibrium amount of this substance. The calculation of adsorption was carried out according to the equation:

$$A = \frac{(C_0 - C_P) \cdot V}{m} \tag{1}$$

where C_0 - the initial concentration of nickel ions in solution, mol l; C_P - the equilibrium concentration of nickel ions in solutions after the adsorption process, mol / l; V – adsorbent mass, g , m – adsorption, mol/l;

The results of the analyzes were processed with the calculation of the arithmetic average value of the adsorption value for each of the studied concentrations from three experiments performed in parallel.

3. Results

As a result of the experimental studies shown in Figure 6, that with an increase in the concentration of nickel ions the magnitude of adsorption increases.

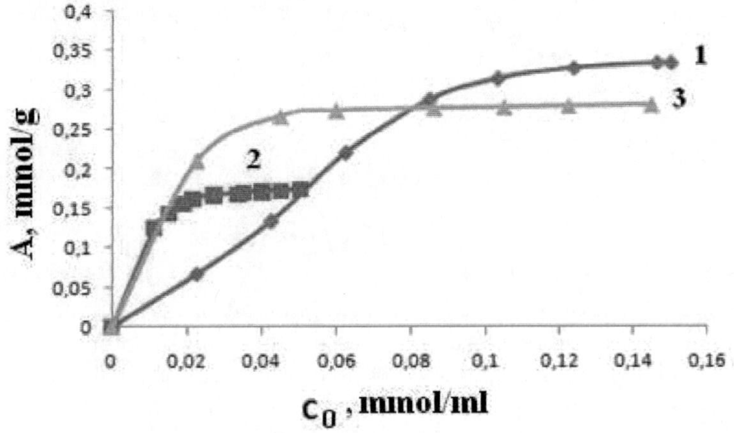

Figure 6. Adsorption isotherms of nickel (II) ions depending on the concentration of the initial solution on clays: kaolinite (1), montmorillonite (2) and mixed (3)

According to the BET classification, the isotherms obtained on three types of natural clays are a type I isotherm, which reflects monomolecular adsorption [3]. The lower part of the isotherms is convex (Fig. 6, curves 2 and 3) can be associated with the presence of micro- and macropores in the adsorbent. In the clay adsorbent with a mixed composition (Fig. 6, curve 1), the convex lower part of the isotherm is absent, adsorption does not occur due to micropores.

The type of isotherms shows the degree of affinity of the adsorbed ions to the adsorbent. The adsorption isotherm allows to determine the maximum exchange capacity of the adsorbent. To describe the adsorption process, the Langmuir model was used:

$$A = A_{\infty} \cdot \frac{k_L \cdot C_P}{1 + k_L \cdot C_P} \qquad (2)$$

or

$$\frac{C_P}{A} = \frac{C_P}{A_{\infty}} + \frac{1}{A_{\infty} \cdot k_L} \qquad (3)$$

where A - the amount of sorbed copper and $A\infty$ - the limiting sorption value (mmol / g),

C_p - the equilibrium concentration of copper ions in solution (mmol / ml),

K_L- the concentration constant of sorption equilibrium characterizing the intensity of the sorption process, ml / mmol.

Figure 7 shows the Langmuir adsorption isotherm in linear form.

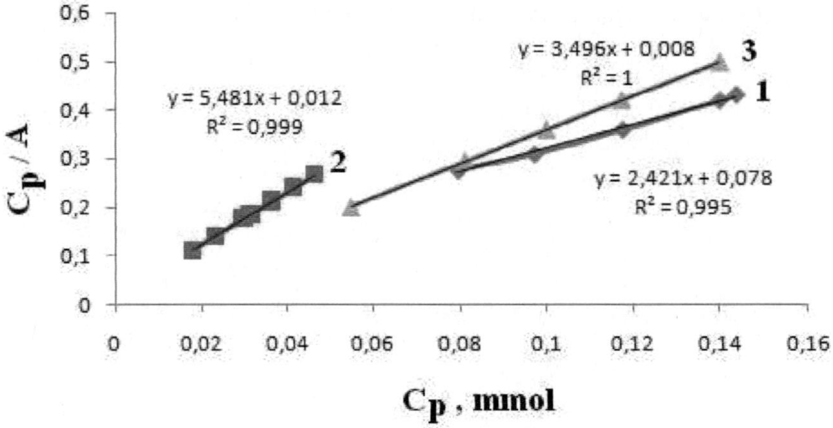

Figure 7. Isotherm of nickel ion adsorption on clays: kaolinite (1), montmorillonite (2) and mixed (3)

Linear treatment of nickel ion adsorption isotherms from aqueous solutions on natural clays: are presented in Table 2.

Table 2. Presentation of nickel (II) ion adsorption isotherms using the Langmuir model

Const.	MMT clay	Kaolin clay	K + MMT
A_∞, mmol/g	83.33	12.82	125
k_L, ml/mmol	0.002	0.032	0.002
R^2	0.999	0.995	1

The obtained calculations show that the maximum adsorption increases in the series:

Kaolinite> montmorillonite> mixed (K + MMT) clays

According to the obtained results, it can be concluded that clays containing montmorillonite in their composition have a higher adsorption capacity with respect to nickel ions as compared to kaolin clay.

4. Conclusion

In the course of the research, it was found that the sorption capacity of montmorillonite clays with respect to nickel ions is 122 mg Ni / g (montmorillonite), 18.8 mg Ni / g (kaolinite) and 183.38 mg Ni / g (mixed K + MMT) clay This confirms the possibility of using this type of clay in the processes of purification of natural and waste waters.

References

[1] *Report on the environmental situation in the Tyumen region* 2018 Administration of the Tyumen region

[2] Guzeeva S A 2014 Environmental state of surface water and bottom sediments of lakes tyumen *Bulletin of the Krasnoyarsk State Agrarian University* **8** (95) pp 134-9

[3] Skipin L, Gaevaya E, Zaharova E, Petukhova V, Sidorova K 2016 Biogeochemistry of heavy metals in trophic chain in terms of the south of tumen region *Procedia Engineering* **165** pp 860-8

[4] Gashev S N, Kurkhinen Yu P 2015 Dynamic processes in the fauna of the spine of the western siberia and their causes *Bulletin of Tyumen State University. Ecology and nature management* **1** (1) pp 80-9

[5] *Report of the health department of the Tyumen region* 2018 Administration of the Tyumen region

[6] Maksimova Y G, Vasil'ev D M, Zorina A S, Ovechkina G V, Maksimov A Y 2018 Acrylamide and Acrylic Acid Biodegradation by Alcaligenes faecalis 2 Planktonic Cells and Biofilms *Applied Biochemistry and Microbiology* **54(2)** pp. 173-8

[7] Politaeva N, Bazarnova Y, Smyatskaya Y, Slugin V, Prokhorov V 2017 Impact of carbon dopants on sorption properties of chitosan-based materials *Journal of Industrial Pollution Control* **33(2)** pp. 1617-21
[8] Sobgaida N A, Olshanskaya L N, Makarova Y A 2010 Cleaning petroleum products from waste water with composite filters based on waste products *Chemical and Petroleum Engineering* **46(3)** pp. 171-7
[9] Chechevichkin A V, Vatin N I, Samonin V V, Grekov M A 2017 Purification of hot water by zeolite modified with manganese dioxide *Magazine of Civil Engineering* **76(8)** pp. 201-13
[10] Barashkova P S, Molodkina L M 2018 Track-etched membranes back-flushing and regeneration during the natural water purification *Magazine of Civil Engineering* **77(1)** pp. 112-20
[11] Matveyeva A N, Pakhomov N A, Murzin D Y2016 Recycling of Wastes from the Production of Alumina-Based Catalyst Carriers *Industrial and Engineering Chemistry Research* **55(34)** pp. 9101-08
[12] Smyatskaya Y, Kosheleva A, Taranovskaya E 2018 Sorption properties of materials based on residual biomass *MATEC Web of Conferences* **245** 18005
[13] Andrianova M, Bondarenko E 2016 Changing of Cations Concentrations in Waters of Polluted Urban River *MATEC Web of Conferences* **53** 01037
[14] Meunier A 2005 *Clays* (Heidelberg: Springer)
[15] Schwarzenbach G, Flaška G 1970 *Complexometric titration* (Moscow: Chemistry)
Voyutsky S S 1976 *The course of colloid chemistry* (Moscow: Chemistry)

Silicone fire-resistant compositive material modification

R Sadykov[1]*, A Rakhmatullina, V Voilochnikov[2] and D Taramasova[2]

[1]Kazan National Research Technological University, 420015, Russia, Kazan, Karl Marks Str., 68.
[2]«Vesto» LTD, Russia, Kazan, Gagarin Str., 26-35.

* Corresponding author: gonzik22007@yandex.ru

Abstract. The effect of the thermally expanding flame retardant dosage as well as its combination with the modifier - the product of silicone rubber waste destruction - on the properties of cold-cured compositions based on low-molecular-weight silicone rubber was studied. It was shown that equal mass substitution of aluminum hydroxide for thermally expanding graphite in an amount of 5% by weight results in sealants with a swelling ratio of 5.8, which is sufficient for fire retardant systems used in nuclear and thermal power plants. It was established that the use of an optimal amount of destructive material (5% by weight) in combination with thermally expanding graphite leads to an increase of the conditional strength at break of silicone flame-resistant cold-curing compositions by 35%. The use of thermally expanding graphite and the silicone waste destruction product in the formulation of a silicone sealant reduces the cost of the composites obtained by 6% in terms of raw materials.

1. Introduction

Composite materials based on low molecular weight silicone rubbers have a complex of unique properties: heat resistance (from -60 to +250°C), bio inertness, resistance to radiation and ozone exposure. This is due to the nature of their chemical bonds of the main molecular chain of the polymer. Thus, the Si – O bond energy in silicone polymers is 450 kJ / mol, while the bond energy of fluorinated carbon atoms in fluor elastomers is only 356 kJ / mol, the C – C bond in hydrocarbons is 336 kJ / mol [1].

Nonetheless, the each year growing fire safety requirements are forcing new ways to improve the flame retardant properties of silicone composites. These requirements are particularly high in the nuclear industry, aerospace, construction, automotive and shipbuilding.

One of the solutions is the creation of silicone fire-resistant sealants that form mechanically strong vulcanite foam (swelling effect) under the action of high temperatures (fires), which can significantly reduce the thermal conductivity of the coating [2-4].

The aim of the work is to increase the flame retardant properties of silicone composite materials (sealants, mastics) by modifying them with intumescent additives.

2. Research methods

Characteristics of the components used in the preparation of the prototypes of silicone flame-retardant sealant:

1. Low-molecular silicone rubber «107 RTV» - viscosity 20000 mm^2/s (at 25°C), $\rho = 0.975$ g / cm^3 (at 25°C), volatile matters 0.67% (at 150°C, 3 hours) (China).

2. Polymethylsiloxane 100 - a clear, colorless liquid, viscosity 100 mm^2/s (at 25°C), volatile matters 0.65% (at 150°C, 3 hours), the refractive index 1.4030, the flash point 315°C (China).

3. Aluminum hydroxide special «FRAMIAL» - white powder, $Al(OH)_3$ content = 99.5%, ρ = 2.4 g/cm^3, Mohs hardness 2.5-3.5, refractive index 1.56 -1.58, humidity <0.3%, electrical permeability (ISO 787-14) <100 μS/cm, pH = 8-10, loss on ignition > 34% (800°C, 1 hour), average particle size 15 microns. 1711-015-40705684-2006 Standart (Russia).

4. Thermo-expanding graphite - carbon content 99%, expansion ratio 350 ml/g, average particle size 177 microns, volatile matters content <10%, pH = 5-7 (China).

5. Diethylaminomethyltriethoxysilane (ADE-3) - a clear, colorless or yellowish liquid, the content of the basic substance is 90.02% (China).

6. Ethyl silicate-32 - a clear, colorless liquid, the SiO_2 content 32.47% wt., ρ = 0.9810 g/cm^3, the content of chlorine ions is 0.001% wt. (China).

7. Silicone rubber waste destruction product - a grey suspension with a typical smell, pH = 6-7, the proportion of the solid phase is 50% wt, the chemical composition of the liquid fraction is 30% wt. oligodimethylsiloxane, 70% wt. oligoethoxysiloxane [5,6].

Standard physio-mechanical tests of sealants were carried out in accordance with State Standards: 21751-76, 21981-76.

The coefficient of expansion of vulcanizates was determined according to the method of VSK-04-2001 (Ministry of Emergency Situations of Ukraine). Two steel plates (50x50x2 mm) were uniformly applied with a sample of fire-resistant sealant. After complete vulcanization, the plates with the vulcanizer were thermostatically controlled for 3 hours at T = 70°C. Then the thickness of the applied coating of both plates was measured (L_{k1}, L_{k2}). Next the sample was placed in a muffle furnace for 30 minutes at T = 340°C, after which the thickness of the expanded coating was determined (L_{t1}, L_{t2}). The coefficient of expansion of the vulcanizates was determined by the formula:

$$K=0,5*(L_{m1}/ L_{\kappa1}+L_{m2}/ L_{\kappa2}) \quad (1)$$

3. Experimental results and their discussion

There is a huge amount of flame-retardant additives in the world, which can be classified as follows (Figure) [7-13].

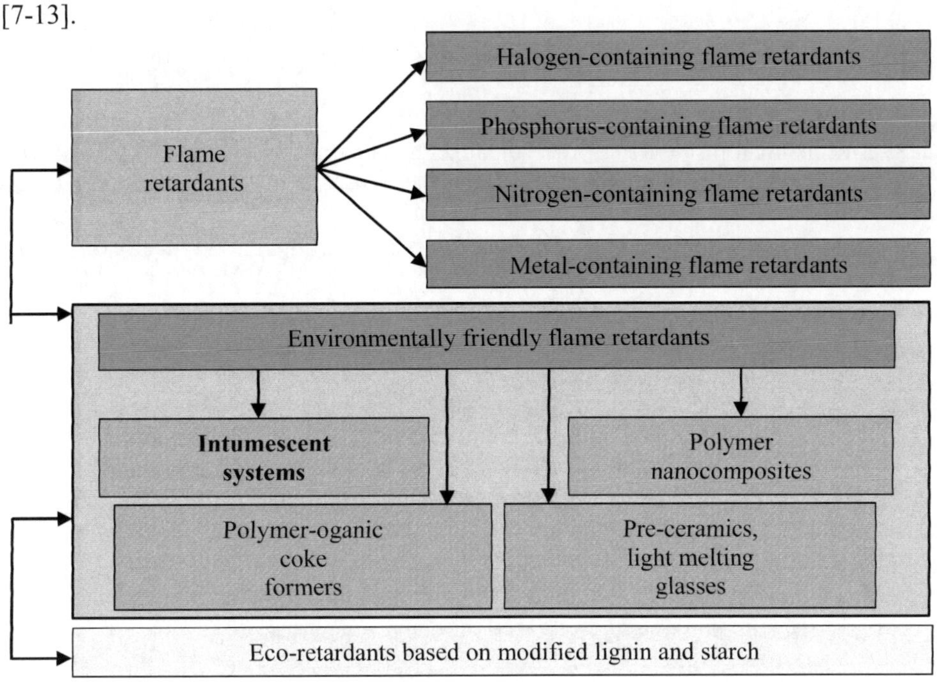

Figure 1. Main types of fire retardants

In order to give the silicone sealants the effect of expansion under the action of high temperatures (fires), thermally expanded (thermo-oxidized) graphite was introduced into the silicone sealant formulation. The choice of this intumescent additive is determined by the lower cost (160 rubles/kg) and simplicity of introduction into the system compared to, for example, a system consisting of ammonium polyphosphate (165 rubles/kg) and pentaerythritol (130 rubles/kg). It should be taken into the consideration that the amount of thermally expanding graphite in the composite, to achieve the desired coefficient of expansion, is less by 50-60%.

A series of physico-mechanical and fire tests made it possible to determine the optimal dosage of thermally expanding graphite in the composition of the sealant - 5.0% wt. (Table 1). This optimum is determined based on the fact that the required coefficient of expansion for silicone fire-resistant sealants used in fire-retardant systems at nuclear and thermal power plants is at least 5.

Table 1. Composition and technical characteristics of silicone flame-retardant sealant samples with different content of thermally expanding graphite

Components	**Content,% wt.**							
Low molecular weight silicone rubber	30,0							
Polymethylsiloxane	15,0							
Aluminum hydroxide	55,0	54,0	53,0	52,0	51,0	**50,0**	49,0	48,0
Thermally expanding graphite	0	1,0	2,0	3,0	4,0	**5,0**	6,0	7,0
Curing composition (based on the finished paste): ADE-3/Ethyl silicate-32	3,0 / 3,2							
The name of indicators	**Test results**							
Shore A hardness	35	35	35	35	35	**37**	38	40
Conditional strength at break, MPa	2,25	2,0	1,85	1,8	1,65	**1,5**	1,35	1,25
Elongation at break, %	210	210	185	160	180	**165**	170	175
Swelling coefficient	1,3	2,3	3,1	3,7	4,6	**5,8**	7,7	8,9

However, the replacement of part of aluminum hydroxide by graphite led to a significant decrease in the physical and mechanical properties of the vulcanizate. This may be due to the poor distribution of the intumescent additive in the composition due to the larger volume of graphite compared to aluminum hydroxide.

The next step was to use the silicone rubber waste destruction product upon receipt of silicone sealant to improve the strength characteristics of the composite [14].

An efficient and cost-effective technology for processing silicone rubber wastes, developed at the «Vesto» LTD [15], allows to obtain liquid products of destruction, which can later be used as a water-repellent additives in water-dispersion paints [16]. In the course of further destruction products processing, which consists in their vacuum degassing [17], an «oligomeric» fraction is formed. It is used at «Vesto» LTD as a polyfunctional additive in the production of one-component industrial and household silicone sealants [14], as well as two-component potting and building mastic.

A complex of physical and mechanical tests showed that the optimal content of «oligomeric» fraction in the formulation was 5% wt. (table 2). The increase of the strength characteristics with the introduction of a destruction product into the system is primarily associated with the filler nanoparticles contained in it (particle size from 1 to 70 nm [5]), which, as is known [18,19], significantly increases the physical and mechanical properties of silicone rubber based products. The presence of oligodimethylsiloxane in the amount of 30% wt. in the liquid fraction [5,6] has a plasticizing effect on the process of obtaining a silicone flame-retardant sealant, allowing to replace part of the expensive imported plasticizer -

IOP Conf. Series: Earth and Environmental Science **337** (2019) 012019 doi:10.1088/1755-1315/337/1/012019

polymethylsiloxane, which ultimately will lead to a decrease in the cost of the final product (Table 3-4).

Table 2. Composition and technical characteristics of silicone flame-retardant sealant samples with different contents of the silicone rubber waste destruction product

Components	**Content,% wt.**							
Low molecular weight silicone rubber	30,0							
Polymethylsiloxane	15,0	14,0	13,0	12,0	11,0	10,0	9,0	8,0
Destruction product	0	1,0	2,0	3,0	4,0	5,0	6,0	7,0
Aluminium hydroxide	50,0							
Thermally expanding graphite	5,0							
Curing composition (based on the finished paste): ADE-3/Ethyl silicate-32	3,0 / 3,2							
The name of indicators	**Test results**							
Shore A hardness	37							
Conditional strength at break, MPa	1,5	1,7	1,8	2,0	2,2	2,3	2,15	2,1
Elongation at break, %	165	165	160	160	155	155	155	150
Swelling coefficient	5,8							

The use of a destruction product as a technological additive does not change the production technology of silicone flame-retardant sealant.

A comparative analysis of tables 3-4 shows that an optimized formulation will reduce the cost of sealant production by 16,500 rubles (6%) per 1000 kg of finished product.

Table 3. Cost calculation of silicone fire-resistant sealant obtained by the control recipe

Raw materials name	**Raw materials amount, % wt.**	**1 kg raw materials cost, rubles**	**1 t of product price, rubles**
Low molecular weight silicone rubber	30,0	390,0	117000,0
Polymethylsiloxane	15,0	400,0	60000,0
Destruction product	0	70,0	0
Aluminium hydroxide	50,0	60,0	30000,0
Thermally expanding graphite	5,0	160,0	8000,0
Curing composition (based on the finished paste): ADE-3/Ethyl silicate-32	3,0/3,2	1800,0/190,0	60080,0
Total			275080,0

Table 4. Cost calculation of silicone fire-resistant sealant obtained by the experienced recipe

Raw materials name	**Raw materials amount, % wt.**	**1 kg raw materials cost, rubles**	**1 t of product price, rubles**
Low molecular weight silicone rubber	30,0	390,0	117000,0
Polymethylsiloxane	10,0	400,0	40000,0
Destruction product	5,0	70,0	3500,0
Aluminium hydroxide	50,0	60,0	30000,0
Thermally expanding graphite	5,0	160,0	8000,0
Curing composition (based on the finished paste): ADE-3/Ethyl silicate-32	3,0/3,2	1800,0/190,0	60080,0
Total			258580,0

IOP Conf. Series: Earth and Environmental Science **337** (2019) 012019 doi:10.1088/1755-1315/337/1/012019

4. Conclusion

Thermally expanding graphite introduction to the basic formulation of silicone flame-retardant cold-curing sealant in an amount of 5% by weight allows to get composites with a high temperature exposure coefficient of expansion equal to 5.8. At the same time there is a decrease in physical and mechanical properties. However, the combination of thermally expanding graphite in the same dosage with the silicone rubber waste destruction product leads to leveling of the strength properties, reaching the level of the indicator of tensile strength at break for the basic formulation at a dosage of modifier 5% by weight. Equal-mass replacement of expensive and scarce polymethylsiloxane liquid to the silicone rubber waste destruction product in the formulation of a sealant leads to a reduction in the cost of the finished product by 6% (in terms of raw materials).

References

[1] Harmonov I V 1976 *Synthetic rubber* (Leningrad: Chemistry) p 462

[2] Ustinov A, Zybina O, Tomakhova A, Pavlov S 2018 The enhancement of operating properties of intumescent fire-protective compositions *MATEC Web of Conferences* **245** 11008

[3] Nikitina M, Ustinov A, Kiseleva V, Babikov I 2018 New fire retardant compositions for fire-resistant automatic curtains *MATEC Web of Conferences* **245** 11004

[4] Zybina O, Marina G, Andrey P 2017 The research of influence polymeric compounds on the effectiveness of intumescent coatings for the fire-protection of construction structures *IOP Conference Series: Earth and Environmental Science* **90(1)** 012206

[5] Sadykov R A, Rakhmatullina A P, Beskrovny D V, Voilochnikov V M 2016 *Bull. of the Tech. Univ.* **19** (21) pp 45-8

[6] Sadykov R A, Rakhmatullina A P, Voilochnikov V M, Gizzatullina L I 2017 *INEOS J.* pp 193-4

[7] Varfolomeev S D, Lomakin S M, Sakharov P A 2010 *Chem. J.* pp 42-5

[8] Zybina O, Gravit M, Stein Y 2017 Influence of carbon additives on operational properties of the intumescent coatings for the fire protection of building constructions *IOP Conference Series: Earth and Environmental Science* **90(1)** 012227

[9] Krivtcov A, Gravit M, Zimin S, Nedryshkin O, Pershakov V 2016 Calculation of Limits of Fire Resistance for Structures with Fire Retardant Coating *MATEC Web of Conferences* **53** 01032

[10] Gravit M V, Golub E V, Antonov S P 2018 Fire protective dry plaster composition for structures in hydrocarbon fire *Magazine of Civil Engineering* **79(3)** pp. 86-94

[11] Sirotkina A, Fedorovich E, Sergeev V 2017 Model of formation and roughness calculation of the porous layer on the heated surface during nanofluids boiling *Psychology of Learning and Motivation - Advances in Research and Theory* **67** pp. 101-6

[12] Snegirev A Y, Talalov V A, Stepanov V V, Korobeinichev O P, Gerasimov I E, Shmakov A G 2017 Autocatalysis in thermal decomposition of polymers *Polymer Degradation and Stability* **137** pp. 151-61

[13] Korotkov A S, Gravit M 2017 3D-map modelling for the melting points prediction of intumescent flame-retardant coatings *SAR and QSAR in Environmental Research* **28(8)** pp. 677-89

[14] Sadykov R A, Rakhmatullina A P, Voilochnikov V M 2015 *Bull. of the Tech. Univ.* **18** (8) pp 104-6

[15] *Russian Federation Patent 2412219* 2009

[16] Sautina N V, Yezhov M V, Taramasova D R, Barabanov V P, Sitnov S A 2013 *Paints and varnishes and their app.* **3** pp 22-4

[17] *Russian Federation Patent 2572786* 2016

[18] Kononenko A S 2012 *Bull. of the BSAA* **1** pp 29-32

[19] Gusev A I 2005 *Nanomaterials, nanostructures, nanotechnologies* (Moscow: FizMatLit)

IOP Conf. Series: Earth and Environmental Science **337** (2019) 012020 doi:10.1088/1755-1315/337/1/012020

Liquidation of oil spills using a sorbent based on chitosan

Quyen Thi Quynh Anh[1]*and L A Zenitova[1,]*

[1]Department of Synthetic Rubber Technology, Kazan National Research Technological University, Kazan, Republic of Tatarstan, Russian Federation.

* Corresponding authors: quynhanhmoitruong@gmail.com, liubov_zenitova@mail.ru

Abstract. The development of a sorbent for the oil spill treatment based on polyurethane foam and chitosan. Chitosan is a polysaccharide – an insoluble fiber type – derive from chitin which is found in the shells of some marine life such as shrimp, lobster and crab. It is also found in some types of edible fungi and insects.
Carry out a research into the technology of polyurethane foam filling with chitosan in the amount of 20% and 50% by mass. Elastic and semi-elastic polyurethane foam was used as a polymer matrix of the sorbent. Sorbents with 20% chitosan by mass have the best oil absorbent capacity. Absorbent capacity of the sorbent based on elastic polyurethane foam, full filled with chitosan, soluble in acid, is ~ 12, 65 g / g, and the semi-elastic sorbent, full filled with chitosan, soluble in acid, is ~ 6.12 g / g. Regardless of the polymer matrix type, the sorbent has a high buoyancy (does not sink in the carbonated state), and the full saturation of the sorbent with oil occurs almost for the first 15 to 20 minutes

1. Introduction

Crude oil is a natural fluid which consists of a complex mixture of organic molecules, mainly hydrocarbon, with various chemical and physical properties. Besides that, it is also the main raw material for the production of petrochemical products, such as solvents, chemical fertilizers, plastics, pesticides, asphalt, fuel, the remaining 12% is used for petrochemistry.

More than 50% of the world's crude oil is produced in the Gulf of Arabia from onshore and offshore wells. The number of oil spills registered in the Arabian Gulf was 550 cases, a total of 14,000 barrels from 1995 to 1999, and 11,000 barrels spilled from 2000 to 2003 [1].

There are various solutions to realize the removal of oil from sea surface and to prevent the coastline pollution from oil, for example, mechanical extraction and using of dispersants. However, the goal of any spill treatment is to minimize marine pollution and protect coastline from oil. For this reason, the elimination must be carried out extremely quickly in order to remove oil from the sea surface [2]. When oil spills, and penetrates the surface, consequences will extend over for a long time [3].

The main methods for the elimination of oil pollution on the water surface include 4 groups:

- The elimination of booms (booms allow you to move oil to any direction for collection);
- Chemical methods (dissolution in water or oil of surface-active substances that change the ratio of the surface energy of the interfacial boundaries in the oil-water system);
- Using special sorbents;
- Biological methods (use of destructor microorganisms) [4].

Adsorption, an efficient separation method that is widely used as an effective and economical solution to remove pollutants. Published researches have shown that chitosan is effective in removing compounds of various toxic metals, such as cadmium, mercury, molybdenum, uranium, vanadium, platinum and palladium [5, 6]. Chitosan, a hydrophilic natural polymer obtained by alkaline dissolvent of chitin, which is contained mainly in the shells of shrimp and crab.

In addition, chitosan has promising applications such as textile manufacturing, high-quality paper production, agriculture, and even the nuclear industry [7-11].

On the other hand, polyurethane foam is also able to absorb oil and oil products [12,13]. However, it is not economically feasible solution.

The purpose of this study is to develop technologies for the production of petroleum sorbent based on polyurethane foam and chitosan as well as the study of its properties.

2. Experimental part

2.1. Materials

The object of study was a sorbent obtained on the basis of polyurethane foam and chitosan.

The sorbent components were used:

- Component A (el) manufactured by ELASTOKAM LLC, Nizhnekamsk (TU 2226-068-10480596-07) based on a mixture of polyether, catalysts, surfactants and water; Dow Izolan, Vladimir, Russia for the production of flexible polyurethane foam - PPUel, as well as the Apge component for the production of semi-rigid polyurethane foam.
- Component B for flexible polyurethane foam produced by Dow Izolan, Vladimir, Russia
- Chitosan - water-soluble chitosan (HV), acid-soluble chitosan (HC), Chitosan, Vietnam.

2.2. Methods

2.2.1. Determine the foaming technological parameters

The experiment of sorbent composition according to the technological flowsheet with free foaming was carried out in a 500ml beaker; determine the "start time" PUF (Tc, s) and "rise time" PUF (τp, s).

Determination of apparent density was carried out according to GOST 409-77.

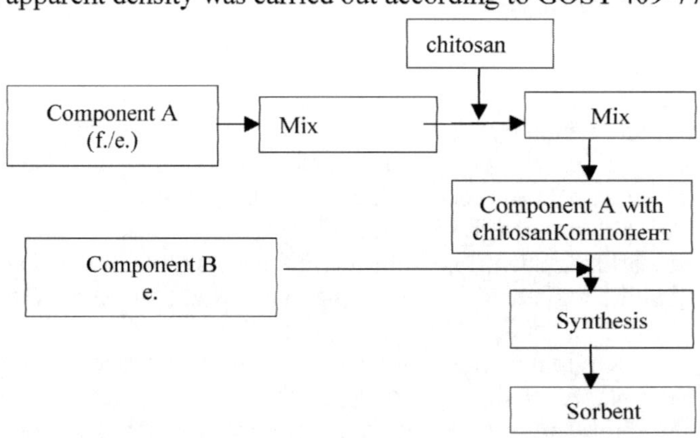

Figure 1. Technological scheme of obtaining oil sorbent

2.2.2. Obtain sorbent

To obtain a petroleum sorbent, there are two methods as below:

The first method is to obtain a sorbent based on elastic PPU (PPUel) from component A (el) and component B for flexible polyurethane foam.

The second method is to obtain a sorbent based on semi-rigid PPU (PPUpr) from component A (r) and component B for flexible polyurethane foam.

There is an important factor for sorbents is buoyancy. Elastic foams have interconnected cells in their structure, which naturally contributes to the high absorption capacity. However, in a saturated state, they can lose buoyant property and sink, which adversely affects marine life. Rigid foams, in the other hand, have a closed-cell structure, the cells in the foam are not communicated, so it is difficult for sorbate to penetrate into the foam, besides that, a sorbent retains its buoyant property in saturated state. Therefore, two types of sorbents were used, if PPUel tended to be "flooded," PPUr would avoid this drawback.

Chitosan was used as filler in the amount of 20 and 50% by mass of the sum of components A and B.

The ratio of components: A (el / r.): B (el.) = 100: 60. After manufacturing the sorbent, it was crushed to a particle size of Du = 0.125 cm^3.

To determine the oil capacity, 50 ml of oil and a certain amount of sorbent were added to the plastic form. Oil capacity was determined by the mass difference of the saturated and initial sorbents through 1; 1.5; 2; 3; 5; 8; 10; 15; 20; 30; 60; 90 and 120 minutes in oil. For the oil intensity, the maximum saturation of the sorbent with oil was taken, when the oil intensity curve — the absorption time reached the plateau [14,15] according to the formula:

$$\text{oil capacity} = \frac{m_2 - m_1}{m_1} \quad (1)$$

Where: m_1 – dry sorbent mass; (g)

m_2 - sorbent mass after oil absorption through 1, 2 …, 120 minutes (g).

3. Results

3.1. Determination of foaming technological parameters.

Table 1. Technological parameters of foaming PPU, depending on the composition

Sorbent	ts, c	tp, c	Apparent density, kg / m^3
PPUpr	30	116	0,0644
PPUpr - 20-HC	32	121	0,6234
PPUpr -50-HC	44	139	0,8380
PPUpr -20-HV	29	198	0,0785
PPUpr -50-HV	53	341	0,1182
PPUel	27	150	0,0591
PPUel -20- HC	28	162	0,0930
PPUel -50-HC	54	173	0,3656
PPUel -20-HV	92	2075	0,1193
PPUel -50-HV	-	-	0,3846

It is seen that the temporal parameters of foaming depend both on the amount of filler, the type of PU foam and filler. Naturally, the more filler, the "harder" composition foamed. Both the "start time" and the "rise time" are lower in the case of foam filling is lower degree.

In addition, the "start time" of PPUel comes a little earlier than for PPUp sorbent. The dependence is similar for the "rise time" indicator. High temporal parameters of foaming compared to traditional – unfilled polyurethane foam – are due to the presence in composition of "heavy" filler with a large amount. The filler in large quantities makes the sorbent heavier. So in the case of unfilled PPU-sorbent, the density is lower than that of PPUel and PPUpr with filler.

The indicator "start time" for the sorbent, PPUpr-50-HC comes a little earlier than for PPUpr-50-HV. The same pattern is observed for PPUel-20-HV and PPUel-20-HC. A similar pattern is observed for the indicator "time rise" foam. This result is reached probably due to the composition of chitosan.

It is obvious that HV's structure components influent the process of foam formation. Indeed, chitosan contains methylol and hydroxyl groups capable of interacting with the isocyanate groups in the component B. Figure 1.

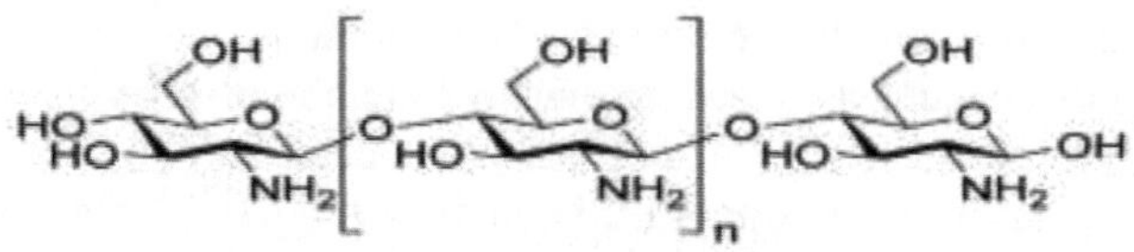

Figure 2. Structural formula of chitosan.

During the synthesis of polyurethane foam, there are series of reactions:

$$OCN-R-NCO + HO-OH \rightarrow -O-\underset{\underset{O}{\|}}{C}-NH-R- \quad (1)$$

$$-NCO + H_2O \rightarrow -NH_2 + CO_2 \quad (2)$$

$$-NCO + -NH_2 \rightarrow -NH-\underset{\underset{O}{\|}}{C}-NH- \quad (3)$$

$$-NH-\underset{\underset{O}{\|}}{C}-NH- + -NCO \rightarrow -\underset{\underset{CONH-}{|}}{N}-\overset{\overset{O}{|}}{C}-NH- + -NCO \quad (4)$$

$$-O-\underset{\underset{O}{\|}}{C}-NH- + -NCO \rightarrow -\underset{\underset{CONH-}{|}}{N}-COO- \quad (5)$$

The main reaction is urethane formation between the hydroxyl-containing component (component A) and the isocyanate groups of the isocyanate component (component B) (reaction 1). Since water is present in component A, the isocyanate groups of component B will react with water to create carbon dioxide, which is a chemical foam blowing agent (reaction 2). In addition, the isocyanate groups of component B react with amines presented in component A to create urea groups (reaction 3), then, interacting with the isocyanate groups of component B, produce biuret groups (reaction 4). These components are responsible for branching and stitching polymer chains latter. Urethane groups formed by reaction 1, when interacting with the isocyanate groups of component B, according to reaction 5, create allophanate groups, as like as the biuret groups, cause branching and cross-linking of polymer chains. In the case of the prevalent reaction rate of polymer chains formation (reaction 1.3-5), the foaming process by reaction 2 is difficult or does not occur. In the case of the predominant interaction rate of isocyanate groups with water, the foam grows rapidly, but at a certain moment it coalesces without creating the polymer network. In order to equalize the speed of these reactions, catalysts are given into the PUF system, namely in component A, so that the elongation and cross-linking of the polymer chains occur simultaneously.

Water-soluble chitosan in the process of obtaining PPU may partially dissolve in the water of component A and is able to react with the isocyanate of component B, competing with the hydroxyl groups of component A. Thus, component B becomes insufficient for full foaming of the PPU composition.

For acid-soluble chitosan, this process is impossible due to its water resistance.

3.2. The effect of exposure time on its oil capacity

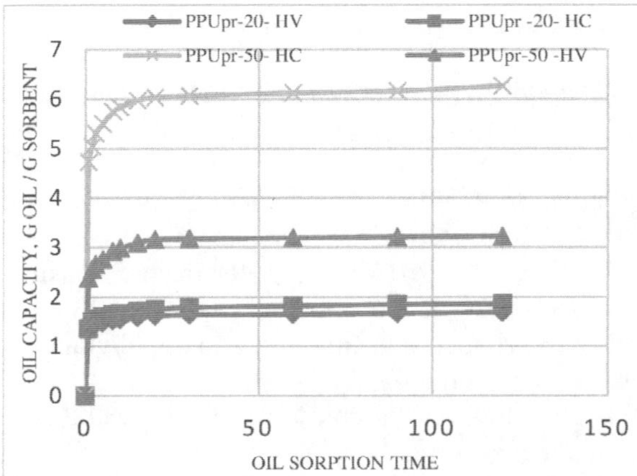

Figure 3. The PPUpzh sorbent's oil capacity dependence on its composition and exposure to oil time.

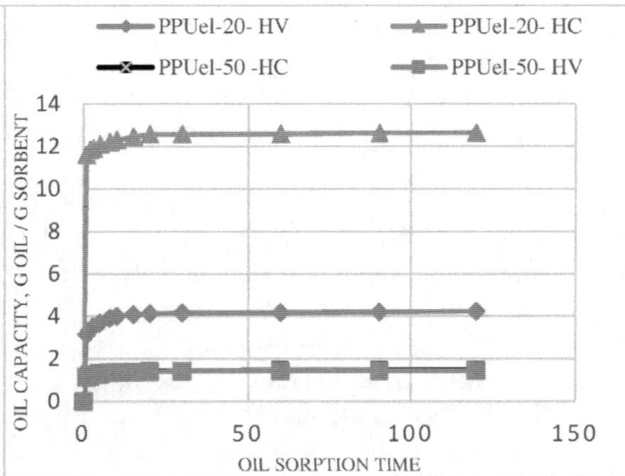

Figure 4. The PPUel sorbent's oil capacity dependence on its composition and exposure to oil time.

From the shown in Figures 3 and 4, it is clear that in the first 15-20 minutes there is a rapid absorption of oil for all sorbents. After 30 minutes the sorption PPUel reaches an equilibrium state. The balance of sorption PPUpr is achieved in 60 minutes.

The most effective among PPUel is PPUel-20-HC sorbent. Its absorption capacity in relation to oil is ~ 12.65 g / g. Figure 3

Among the sorbents PPUpr the most effective is PPUpr-50-HC sorbent. Its absorption capacity in relation to oil is ~ 6.12 g / g. Figure 2

When comparing chitosan HC and HV, it was found that chitosan, soluble in acid, has a greater oil capacity than chitosan, which is soluble in water.

Both types of sorbent retain their buoyancy in the saturated state.

4. Conclusions

Thus, an effective sorbent is the one based on flexible polyurethane foam, filled with 20% by mass of chitosan, soluble in acid. Its absorption capacity in relation to oil is ~ 12.65 g / g. The advantages of this sorbent are that it has a high rate of retrieval per litre of absorbed oil, does not sink even in saturated state, has a high sorption rate (90% absorption in 15-20 minutes), easy to operate, allows to remove the absorbed substance, non-toxic, be manufactured on the basis of available renewable materials.

References

[1] De Pagter R H, Whiddon D J 1994 *SPE Paper 27173, presented at the 2nd Int. Conf. on Health, Safety & Environment in Oil & Gas Exploration & Production, Jakarta, Indonesia, 25-27 January*

[2] Velikhov E H 1996 *Environmental protection at oil-producing facilities in modern condition* (Moscow: Oil industry) pp 47-9

[3] Shorina T S 2009 Effect of oil and oil products on various components of environment *Bulletin of Orenburg State University* **6** pp 642-5

[4] Ratliff M D 1994 *SPE Paper 27170, presented at the 2nd Int. Conf. on Health, Safety & Environment in Oil & Gas Exploration & Production, Jakarta, Indonesia, 25- 27 January.*

[5] Orlov D S, Sadovnikova L K, Lozanovskaya I N 2002 *Ecology and protection of the biosphere during chemical pollution* (Moscow: Vysshaia shkola)

[6] Bunich P G 1978 Resource and environmental problems of the oceans *Problems of Sea ecology* **7** (Kiev: Institute of economics/ Sciences academy of Ukraine) pp 19-28

[7] Yrum K, Pekdemir T, Copur M 2004 Surfactants treatment of crude oil contaminated soils *J. of Colloid and Interface Sci.* **276** pp 456-64

[8] Mazlova E A, Shagarova L B 2001 *Environmental solutions in the oil and gas complex* (Moscow: Tekhnika)
[9] N S Chikina 2010 *Oil and petroleum product spill treatment using a sorbent based on polyurethane foam and grain waste* dissertation C.Sc. Kazan Station Tecnological Universitet Kazan
[10] Gerente C, Lee V K C, Cloirec P L, McKay G 2007 *Crit. Rev. Env. Sci. Technol.* **37** (41)
[11] Chang Y H, Huang C F, Hsu W J, Chang F C 2007 *J. Appl. Polym. Sci.* **104** p 2896
[12] Taranovskaya E A, Sobgayda N A, Alferov I P, Morev P A 2015 Wastewater treatment using chitosan *Bulletin of Orenburg State University* **10** (85) pp 322-6
[13] Taranovskaya E A, Sobgayda N A, Markina D V 2016 Production technology and use of granular sorbents based on chitosan *Chemical and Oil&Gas Engineering* **5** pp 42-5
[14] Abdullin V F, Artemenko S E, Arzamastsev O S 2008 Features of chitin biopolymer extraction processes from the shell of crustaceans *Chemical Fibers* **6** pp 21-4
[15] Abdullin V F 2006 *Technology and properties of chitosan biopolymer from crayfish shell* dissertation C.Sc. (Saratov)
[16] Arzamastsev O S, Artemenko S E, Abdullin V F 2011 Intensification of the producing chitosan films process *Vestnik Saratovsky State Technical University* **4** (60) 2 p 112–4
[17] Ivanova M A, Zenitova L A 2011 Recovery of absorbed oil from PPU-OZK sorbent *Actual problems of polymer science* pp 75-6
[18] Chikina N S, Mukhamedshin A V, Ankudinova A V, Zenitova L A, Shirotkin A S, Garabagiu A V 2009 Reducing the environmental burden of oil and petroleum products spills using a sorbent based on polyurethane foam and grain waste *Bulletin of Kazan Technological University* **6** pp 184-92
[19] Ivanova M A, Zenitova L A, Chikina N S, Murtazina R T, Yanov V V 2011 "OILFORMSORB" sorbent for oil and petroleum products spills treatment *Modern problems of life safety: theory and practice. Proceedings of the International Scientific and Practical Conference. Part II*
[20] Ivanova M A, Zenitova L A, Chikina N S, Murtazina R T, Yanov V V 2011 "Regeneration of absorbed products by the sorbent PPU-OZK *Chemical working of Russian Conference "Butlerov Heritage – 2011"*

IOP Conf. Series: Earth and Environmental Science **337** (2019) 012021 doi:10.1088/1755-1315/337/1/012021

Deactivation of the catalytic complex of obtaining alkylbenzenes by oxygenate compounds

E Bayguskarova[1]*, R Mukhamedzyanov[1], E Vasilyeva[1] and R Akhmedyanova[1]

[1]KNRTU, Synthetic rubber technology, 72 Karl Marx Street, Kazan, Russia.

* Corresponding author: bayguskarova90@mail.ru

Abstract. The possibility of environmentally safe and energy-saving deactivation of the catalytic complex based on aluminum chloride used in the process of alkylation of benzene with olefins, oxygenate compounds – epoxy soybean oil, propylene oxide and isopentane oxidation products is shown.It has been established that the deactivating ability increases in the series: oxygenates → propylene oxide → epoxidized soybean oil (ESBO).

1. Introduction

The sources of many environmental problems associated with the formation of wastewater are industrial enterprises [1-8].

Production of ethyl- and isopropyl benzene is one of the most large-scale commercial processes in terms of consumption and processing of benzene. Almost 75% of the petrochemical benzene produced in the world is consumed in the production of ethylbenzene and isopropyl benzene. Almost all ethylbenzene is used in the production of styrene, the most important monomer for polymers production[9].

The technology of alkylation of benzene with ethylene (propylene) in the presence of liquid-phase acid catalysts has been introduced at many petrochemical enterprises in Russia and in the world. Aluminum chloride is widely used as an active component of the catalyst system because of its following advantages : use of a smaller molar excess of benzene relative to olefin; ability to catalyze both alkylation and transalkylation reactions, mild conditions of the process[10-14]. However, the use of aluminum chloride has a number of disadvantages: equipment corrosion; formation of a large amount of wastewater at the stage of the catalytic complex deactivation [15-19].

Traditionally, deactivation of $AlCl_3$-based catalyst complexes is carried out by aqueous-alkaline treatment of the reaction products. This produces a large amount of wastewater. In the production of 1 ton of ethylbenzene, 10 tons of wastewater is formed. Disposing 1 ton of wastewater costs 6000 rubles. If the company produces 100 thousand tons of ethylbenzene per year, then the approximate cost of wastewater disposal will be 6 billion rubles. Thus, the technology of producing alkyl benzenes in the presence of $AlCl_3$is energy-intensive and environmentally hazardous. Solving the problem of reducing the amount and even eliminating the formation of polluted wastewater is an urgent task.

In recent years, the use of epoxy compounds, in particular, olefin oxides [20], for example, propylene oxide (PO) for deactivation of catalyst systems based on metal chlorides has been described.

Along with non-technological PO, more high-boiling epoxy compounds, such as epoxy resins, are used. [21-23].

IOP Conf. Series: Earth and Environmental Science **337** (2019) 012021 doi:10.1088/1755-1315/337/1/012021

It is worth noting that the use of epoxy compounds for the deactivation of catalyst systems based on Lewis acids is described only for the polymeric petroleum resin synthesis processes, but for deactivation of catalytic systems in alkylation processes, with the exception of our published results [24], there is no information available. Oxygen-containing organic compounds formed during the oxidation of saturated hydrocarbons, in particular isopentane, were used as deactivating agents for the first time. [25-29].

The purpose of the study - the development of environmentally friendly process of deactivating the catalyst complex used in industrial processes for the alkylation of benzene with ethylene or propylene with oxygenate compounds.

To assess the possibility of deactivating the catalyst complex used in industrial processes for the alkylation of benzene with ethylene or propylene with oxygenate compounds, the model reaction of benzene alkylation with hexene-1was chosen. Gustavson liquid complex, obtained on the basis of aluminum chloride, propyl chloride and benzene, was used as a catalyst($AlCl_3$:C_3H_7Cl:C_6H_6 = 1:0.5:2mol.). As a deactivating agent, propylene oxide, epoxidized soybean oil, and products of liquid-phase oxidation of isopentane were used.

2. Experimental part

The physical and chemical characteristics of the epoxidized soybean oil are presented in table 1.

Table 1. Characteristics of epoxidized soybean oil, manufacturer Hairima, China

Indicator	Value
Colorsaturation (Pt-Co)	120
Density, g/cm^3	0.993
Epoxy oxygen content, % wt.	6.2
Acid index, mg KOH/g	0.5
Iodine index, g I_2/100 g	1.6
Flash point, °C	300
Refraction index, 20 °C	1.47

Isopentane oxidation is carried out in the liquid phase at a batch plant, in the metal reactor, equipped with a tank jacket, a cooler, systems for hydrocarbon dosing and loading, and a cooled sampling instrument. A porous ceramic plate is installed at the bottom of the reactor to form a stream of small gas bubbles. The process is carried out at a temperature of 120 °C and a pressure of 30 kgf/cm^2.

The composition of oxidation products is given in table 2.

Table 2. Isopentane oxidation products

S No.	Components	Content,% mass
1	Isopentane	85.48
2	Alcohols	4.29
3	Ketones	6.85
4	Esters	1.71
5	Ethers	0.15
6	Heterocycles	1.17
7	Diols	0.29
8	Acids	0.07
Total		100

Propylene oxide is a colorless liquid with an etheric odor. It is mixed with water in a volume ratio of 1:1.5 (20 °C), ether and alcohol.

The physical and chemical characteristics of the propylene oxide are presented in table 3.

The catalyst complex is prepared in a vessel with a hermetically-sealed plug to prevent the ingress of moisture. The calculated amounts of toluene and aluminum chloride are loaded, and then propyl chloride is added. The resulting mixture is kept for at least 6 hours at a temperature of 60°C.The resulting catalyst complex is a homogeneous system.

Alkylation is carried in a three-neck glass flask equipped with a mixing device and a back flow condenser. The flask is loaded with feed materials – benzene and catalyst complex. Then, hexane-1 is added via the addition funnel while stirring intensively. Alkylation reaction is carried out at a temperature of 60°C for 3 hours. The concentration of the aluminum chloride catalyst complex is 2% of the mass.

Table 3.The physical and chemical characteristics of the propylene oxide

Indicator	Value
Melting point, T_M,°C	104.4
Boiling point, T_B, °C,	34.5 - 34.9
Flashpoint, °C, higher than	-30
Density, 20 °C, ρ, g/cm^3	0.859
Refraction index, n^{20}_D	1.3667

The process of the catalyst complex deactivation is carried out by adding the calculated amount of the deactivating agent to the alkylation reaction mass. During the deactivation process, samples of reaction mass, in which the amount of free $AlCl_3$and the degree of the catalyst complex deactivation are determined, is taken according to the methodology given in the work [10].

IR spectra were obtained by FTIR spectroscopy on a Thermo Scientific Nicolet iS10 spectrometer.

The aluminum content in the bottom product was determined by burning samples of the bottom product in a muffle furnace at a temperature of 800 ° C until constant mass was reached[30].

3. Results and its discussion

Since aluminum chloride is particularly dangerous in terms of the release of aggressive HCl, deactivation can be carried out by transferring it to a compound that does not have acidic properties and is not capable of producing HCl.

In general, the scheme of a catalyst complex deactivation with an epoxy compound can be represented as follows [11]:

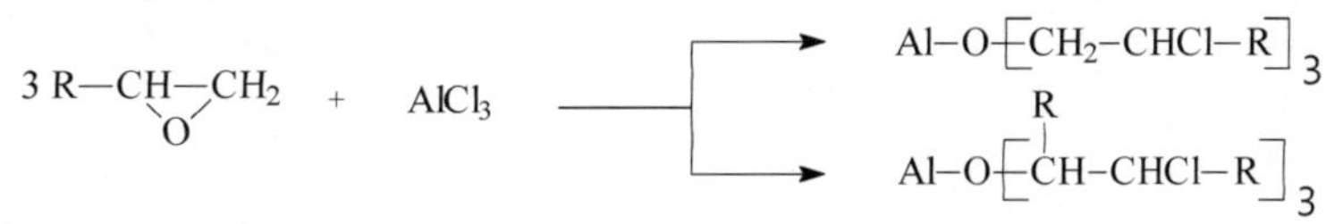

(1)

Thus, during the deactivation reaction, the termination no free aluminum chloride occurs with the formation of a stable organic compound and thereby the release of free hydrogen chloride is prevented.

Initially, the deactivation process was studied on a model system, which was a solution of a catalyst complex in toluene. Epoxidized soybean oil was used as a deactivating agent.

In the course of the work, the influence of temperature, the molar ratio of aluminum chloride and epoxy groups (EG) [$AlCl_3$]:[EG], and the nature of the deactivating agent on the degree of deactivation was studied.

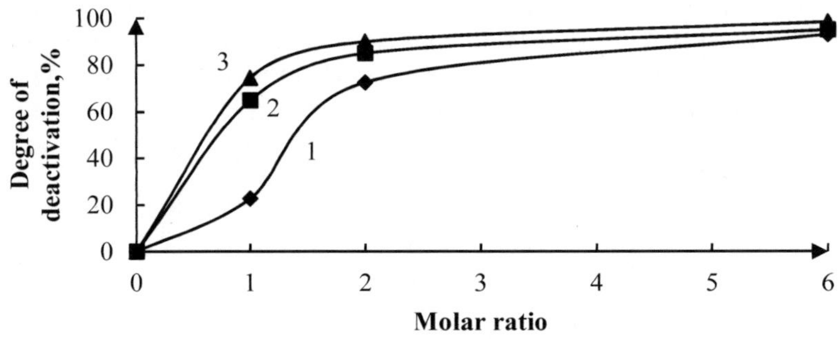

Figure 1. Dependence of $AlCl_3$-containing catalyst complex degree of deactivation on molar ratio of [$AlCl_3$]:[EG] in the reaction mass at temperature T= 50 °C (1), 60°C (2), 70°C (3).

As the figure shows, the degree of deactivation increases with the increase of molar ratio and temperature of deactivation .A further increase in the molar ratio and temperature was impractical, since no increase in the degree of deactivation of the catalyst complex was observed.

Thus, the optimal conditions for the deactivation of the catalyst complex were selected:molar ratio[$AlCl_3$]:[EG]= 1:3; temperature70°C; time 45 minutes, at which the degree of deactivation reaches 98%.

Then the deactivation process was studied on a model reaction of benzene alkylation with hexene-1.

As deactivating agents, compounds containing oxiranerings, which are propylene oxide and epoxidized soybean oil (ESBO), and oxygenates were used. Their composition is given in table 2.

Deactivation was carried out for 60 minutes at a temperature of 70 °C and a molar ratio of [$AlCl_3$]:[EG] = 1:3.

Study in the process of the $AlCl_3$-containing catalyst complex deactivation in the reaction mass of benzene alkylation with hexene-1 by time (Fig. 2) showed that the most effective deactivating agent was ESBO. In this case, the degree of the catalyst complex deactivation (the degree of $AlCl_3$conversion to a neutral compound) was 98%.The products ofisopentane oxidation turned out to be the least effective, which may be due to the fact that they contain oxygen-containing groups of a different nature.

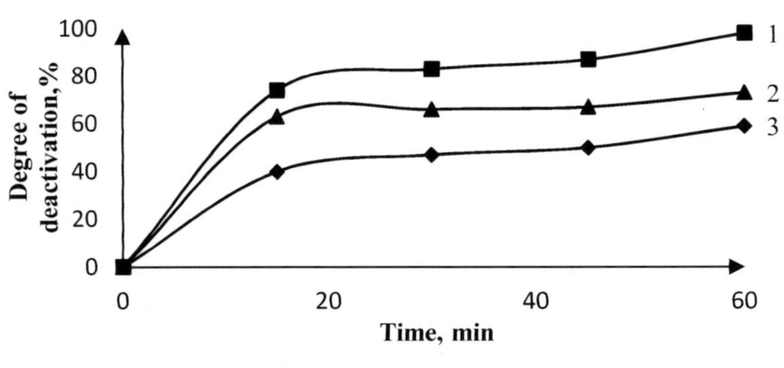

Figure 2. Dependence of $AlCl_3$-containing catalyst complex degree of deactivation on time of deactivation of benzene alkylation with hexane-1. 1 – ESBO, 2 – propylene oxide, 3 – isopentane oxidation products.

When an excess of epoxidized soybean oil is used for catalyst complex deactivation, a sufficiently large amount of the product of its interaction with $AlCl_3$may be formed. As a result, the problem of finding ways for its qualified use arises. After deactivation, the reaction mass was subjected to vacuum

IOP Conf. Series: Earth and Environmental Science **337** (2019) 012021 doi:10.1088/1755-1315/337/1/012021

distillation, and after the fractionation in the temperature range of 90-100 °C and a vacuum of 15 Mmhg the bottom product was obtained, which was a gel-like homogeneous mass of a dark yellow color.

According to IR spectroscopy, the transmission band in the region of 820 cm^{-1}, typical for epoxy groups and present on the spectrum of epoxidized soybean oil (Fig. 2, curve 2), is absent in the spectrum of the bottom product (Fig. 2, curve 1).This indicates that the deactivation process is carried out with the opening of the epoxy ring. In addition, a wide band is observed in the spectrum of the bottom product in the region of 3400 cm^{-1}, which may be due to the presence of hydroxyl groups in it. There is also a wide band in the region of 1600 cm^{-1}, which may indicate the presence of aromatic compounds with unsaturated group [30].

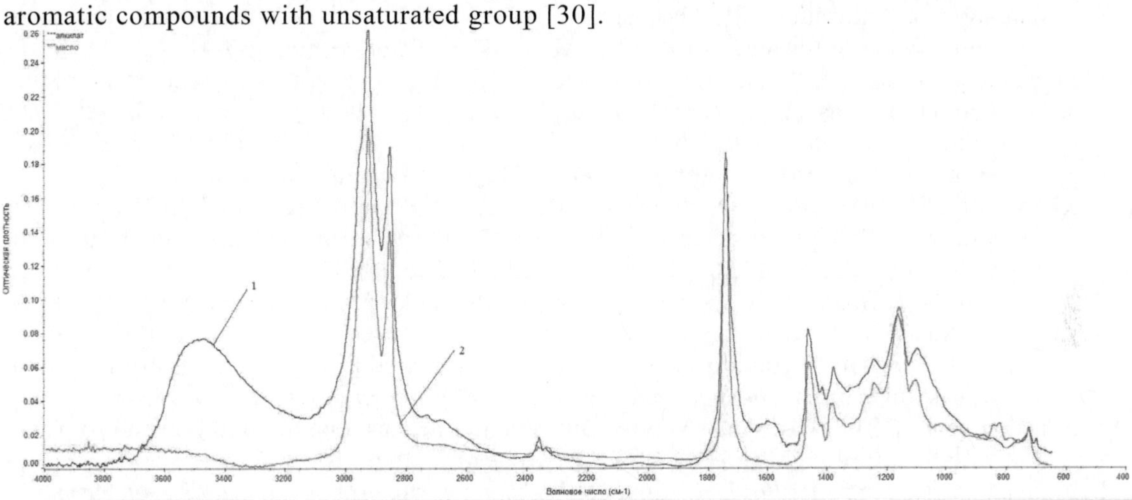

Figure 3. IR spectra of the bottom product obtained during the deactivation of aluminum chloride by epoxidized soybean oil (1), epoxidized soybean oil (2).

The fact that aluminum passes into a bound state and remains in the bottom product after the process of deactivation can determine further uses of the bottom product.

Table 4. Aluminum content in the bottom product

Molar ratio	Alcontent in the bottom product, %	
[ESBO] :[$AlCl_3$]	Theoretical	Experiment
6:1	1.18	0.50
6:1	1.18	0.55
3:1	1.50	0.60
3:1	1.50	0.60

As the table shows, the Al content in the bottom product is 0.5-0.6% by weight.

The presence of various functional groups and aluminum in the composition of the bottom product allows us to consider the possibility of using it as a modifying, multifunctional additive to polymers.

4. Conclusion

For the first time, the laws of deactivation of an $AlCl_3$-containing catalyst system for the alkylation of benzene with olefins with oxygen-containing compounds have been established. It was established that the deactivating ability increased in the series: oxygenates C_5→ propylene oxide → ESBO. The degree of deactivation increases with increasing molar ratio [$AlCl_3$]: [EG], temperature, and process duration.

Using the technology of the $AlCl_3$-based catalyst system deactivation with oxygenate compounds gives a positive environmental effect and reduces waste disposal costs.

References

[1] S Gerzeliyev, S Myachin 2009 Ethylbenzene *Russian. The Chem. J.*
[2] Politayeva N A, Smyatskaya Y A, Slugin V V 2018 Wastewater cleaning in a composite filter after magnetic treatment *Comptes Rendus de L'Academie Bulgare des Sciences* **71(6)** pp. 766-71
[3] Sobgaida N A, Ol'shanskaya L N, Nikitina I V 2008 Fiber and carbon materials for removing oil products from effluent *Chemical and Petroleum Engineering* **44(1-2)** pp. 41-4
[4] Alshabab M S, Andrianova M, Alsalloum D 2016 Modification of Wastewater Treatment Technology at Cottonseed Oil Plant *MATEC Web of Conferences* **53** 01040
[5] Smyatskaya Y, Kosheleva A, Taranovskaya E 2018 Sorption properties of materials based on residual biomass *MATEC Web of Conferences* **245** 18005
[6] Ivanchenko O, Khabibullin R, Bhat R 2018 Wastewaters of meat-processing enterprise: Assessment of genotoxic potential *MATEC Web of Conferences* **245** 18002
[7] Khabibullin R, Ivanchenko O, Petrov A, Bhat R 2018 Optimization of the process of anaerobic-aerobic purification of waste waters of food production using the spatial separation of stages *MATEC Web of Conferences* **245** 18003
[8] Arakcheev E N, Brunman V E, Volkov A N, D'yachenko V A, Petkova A P, Sorokin A V, Brunman M V, Konyashin A V 2016 Complex Electrolytic Unit for Producing Anode Liquor and Ferrate to Treat Waters Contaminated by Oil-Refining Products and Harmful Admixtures *Chemical and Petroleum Engineering* **52(1-2)** pp. 143-8
[9] Ethylbenzene (EB): 2016 World Market Outlook and Forecast up to 2020 [Online source]. – Available at: http://mcgroup.co.uk/researches/ethylbenzeneeb
[10] Lipovich V G 1985 *Alkilirovanie aromaticheskikh uglevodorodov [Alkylation of Aromatic Hydrocarbons]* (Moscow: Chemistry)
[11] Topicheva A V, Zavgorodniy S V, Kryuchkova V G 1962 *Reaktsii alkilirovaniya organicheskikh soyedineniy olefinami [Organic compounds alkylation with olefins]* (Moscow: Chemistry)
[12] Polubentsova M F, Duganova V V, Mikhaylenko G A 1996 *Gen. chem. J.* **66** pp 630-4
[13] Kosyakov D S, Ipatova E V, Krutov S M, Ul'yanovskii N V, Pikovskoi I I 2017 Study of Products of the Alkaline Decomposition of Hydrolysis Lignin by Atmospheric Pressure Photoionization High-Resolution Mass Spectrometry *Journal of Analytical Chemistry* **72(14)** pp. 1396-1403
[14] Andreeva L N, Nekrasova T N, Nazarova O V, Bezrukova M A, Zolotova J I, Lebedeva E , Tsvetkov N V, Panarin E F 2016 Polyelectrolyte behavior of copolymers of 2-deoxy-2-methacrylamido-D-glucose with cationic comonomers in water and dimethylsulfoxide solutions *European Polymer Journal* **83** pp. 22-34
[15] Kolesnikov I M, Babin E P 1980 *Alkilirovaniye benzola v prisutstvii alyumosilikatnykh katalizatorov [Alkylation of benzene in the presence of aliminosilicate catalysts]* (Kiev: Visha shkola)
[16] Lipovich V G 1979 *Synthesis and transformations of alkyl aromatic compounds* (Irkutsk: IGU Publ)
[17] Lesment T N, Feingold S I, Liiv T K 1969 *Petrochem.J.* **9** pp 379-85
[18] Olbright L F, Golsby A R 1982 *Alkilirovaniye Issledovaniye I promyshlennoye oformleniye protsessa [Alkylation. Research and industrial design of the process]* (Moscow: Chemistry)
[19] Baguyskarova E S, Akhmedyanova R A, Miloslavskiy D G 2015 *Russian J. of Appl. Chem. pp* 44-6
[20] Manamkova A A, Bondaletov V G, Belousova A S, Lyapkov A A 2010 *Bulletin of the Polzunovksiy J.* **4** pp 271-4
[21] Akhmedyanova R A, Miloslavskiy D G, Liakumovich A G, Plaksin A S, Portnaya A A 2011 *Bulletin of the Techn. Univ. J.* **18** pp 254-9

[22] Bondaletov V G, Fiterer E P, Tolmacheva V Y, Bondaletova L I 2005 *Izvestiyavusov Publ.* **48** pp 73-6

[23] Bondaletov V G, Tolmacheva V Y, Fiterer E P, Toryan A A, Manakova A A, Petrenko T V 2009 *Bulletin of the Tomsk Polytechnic Univ.J.* **314** pp 98-100

[24] Baguyskarova E S, Akhmedyanova R A, Miloslavskiy D G 2017 Energy-saving environmentally friendly technology for the production of polymeric petroleum resins based on the bottoms of styrene distillation *IX vserossiyskaya Karginskaya konferentsiya «Polimery 2017» [IX All-Russian Kargin conference "Polymers-2017"]* (In Russian).

[25] Reich H J 1978 *Oxidation in Organic Chemistry* (New York: Academic Press)

[26] Emanuel N M 1974 *Teoriya I praktika zhidkofaznogo okisleniya [Theory and practice of liquid-phase oxidation]* (Moscow: Science)

[27] Galimov R A, Gayfullin A A, Harlampidi H E 2007 *Okislenie alkanov do sinteticheskikh zhirnykh kislot* [The oxidation of alkanes to synthetic fatty acids] (Kazan: KSTU)

[28] Emerson W S 1948 The liquid phase oxidation of ethylbenzene *J. of the American Chem. Soc.* **70** pp 3764-7

[29] State Standard 1461-75 Petroleum and petroleum products. Method of ash test 1976 (Moscow: Standartinform Publ.) (In Russian)

[30] Kazitsina L A 1971 *Ispol'zovanie UF, IK I YAMR spektroskopii v organicheskoy khimii[The use of UV, IR and NMR spectroscopy in organic chemistry]* (Moscow: Vysshaya shkola)

IOP Conf. Series: Earth and Environmental Science **337** (2019) 012022 doi:10.1088/1755-1315/337/1/012022

Components of solid municipal waste in construction compositions

N Fomina[1]* and M Polyanskiy[1]

[1]Yuri Gagarin State Technical University of Saratov, Department "Building Materials and Technologies", Polytechnic St., 77, Saratov, Russia.

* Corresponding author: fominanani@yandex.ru

Abstract. The effectiveness of chemical, thermal, mechanical and complex methods of disposal of polymer waste was analyzed. The results of studies aimed at the development of the composition of a thermoplastic binder using polymeric municipal waste, taking into account the sample sizes of various components in waste sorting complexes, are presented. The binder is filled with mineral powders and shredded wood waste. Received compositions are used for the manufacture of construction products with a long life cycle. Standard methods for determining the properties of products for constructional purpose, the parameters of the fillers were used. Compositions for the manufacture of polymer-mineral composite building materials are offered. The optimal parameters of filling and dosage of functional additives are established.

1. Introduction

Such positive tendencies of the XX-XXI centuries as the growth of industrial production, an increase in population and life expectancy, an improvement in the quality of life also have their negative consequences. This includes the rapid increase in the amount of industrial and household wastes. The practice of handling municipal waste most acceptable today for the Russian Federation is transporting them by municipal services to waste sorting complexes [1-6]. The main task of such complexes is to reduce the volume of garbage. The efficiency of the complexes increases due to the extraction of useful components from the garbage, such as: glass, metal, paper and cardboard, textiles, wood, plastic. The number of extracted components is not so great. This number is about 10% of the total mass of garbage. The remaining 90% is reduced in volume by grinding. The resulting mass is to be disposed of at special landfills.

It is obvious that there are various plastic such as different packaging, broken household items, toys, etc in a large volume of municipal waste. If we consider plastic by type of polymer, then polyethylene terephthalate (PET) is most successfully removed by waste sorting complexes. PET beverage bottles are easily identified and removed separately - transparent, green, brown, white; PET bottles are also separated from the oil. Polyolefins (low and high pressure polyethylene (HDPE, LDPE), polypropylene (PP)) are also successfully extracted in the form of containers from household chemicals, canisters, containers from food products, disposable tableware, transparent and colored bags. Communal waste also contains quite a lot of polystyrene (PS), because this polymer in foam form (press-free - EPS) is widely used as protective packaging for household appliances, various goods and products. EPS is also one of the most popular types of thermal insulation, and accordingly it is a component of construction waste. However, the sampling of EPS from waste is difficult, and it is extracted in small quantities. Thermosetting polymers and polyvinyl chloride (PVC) are difficult to identify, and therefore are almost completely disposed of. Such a variant of polymers, like rubber in automobile tires due to the increased hazard class, is not allowed on waste sorting complexes, and should be disposed of by specialized methods. In this way PET plastic and plastics as HDPE, LDPE, PP are a mass commercial products of waste sorting complexes. According to Waste Management JSC, PET is about 70% of all recoverable plastic and 30% is polyolefins.

The effectiveness and applicability of various methods for the processing of polymeric waste, depending on the type and properties of the polymers, the volume of education, the degree of pollution and other specific factors, are reflected in domestic and foreign surveys and monographs [7-10], as well as in numerous Russian studies (Karpov Institute of Non-Ferrous Metallurgy [11], K. Berkunov State University named after Kh. M. Berbekov [12], VGU named after Stoletov [13], KuzGTU named after TF Gorbachev [14], MSTU named after G.I. Nosov [15, 16], Peter the Great St. Petersburg Polytechnic University [17], SSTU named after Gagarin Yu. A. [18, 19]), European [20], American [21, 22] and other [23-25] scientific schools.

It is known that chemical methods for processing PET waste, such as transesterification [26], hydrolysis [13, 14, 26, 27], glycolysis [25, 21, 24, 28], algolysis [26], methanolysis [13, 21, 29], and others, with obtaining valuable low-molecular products, or destructive decomposition to produce sorbents [15, 18, 22], are complex in technological design, durable and often energy-intensive. Methods for processing PET waste based on thermal exposure to produce low-molecular products (pyrolysis, thermolysis [30]) are also associated with technological difficulties and significant energy costs. Thermal processing, which is aimed at obtaining energy, is ineffective, because the calorific value of polymers is commensurable with that of, for example, wood or coke [31], and the cost of polymer fuel is much higher (the reason for this is the cost of collecting and sorting waste).

Due to the relative simplicity and low energy consumption thermomechanical methods of processing polymer wastes are the most possible. Polymer waste is ground by cutting in high-performance shredders. The resulting crushed material is heated to the softening temperature or to the temperature of melting and then on the special equipment loose agglomerate is received. More common technology is in melting polymers and extruding a melt, followed by cutting the material into bulk granules. The technologies of mechanical processing of PET-waste into flexs are also quite common. The resulting semi-finished products - agglomerate, granulate, flex - are subject to further processing into products and can serve as a substitute for the primary polymeric raw materials.

Examples of successful processing of polymeric waste in building products include the production experience of the companies of OOO Prompolymer (Kirov), Skolkovo Mechanical Plant (Samara Region), Pstile (Bashkortostan), and others.

Using polymer semi-finished products in the form of granules and flexes for the manufacture of building products, having a cost lower than the cost of primary raw materials (approximately 20-30%), but still quite tangible, leads to increased prices for products. Often polymer-mineral building products are not in demand, due to the fact that they are competing with similar products based on cement, which are inferior in properties, but significantly cheaper than polymer-based products.

Therefore, the aim of the work was to develop compositions for the manufacture of construction products based on polymeric waste, using waste without prior cleaning and processing into semi-finished products.

2. Materials and Methods

Polymeric waste PET, HDPE, LDPE, PP, obtained by sorting municipal waste, were studied. Wastes were crushed into particles up to 5 cm.

Mineral powder materials obtained by drying, crushing and grinding samples of regional rocks (quartz sand, limestone, chalk, clay, and flasks) were investigated as fillers. In order to reduce the density of products, as a filler we studied wood waste obtained by grinding wood components of municipal waste, such as elements of old furniture and construction products made of wood that were out of circulation. Crushed wood waste was pretreated with sodium silicate solution, and then dried.

Standard methods for determining the properties of products for constructional purpose, the parameters of the fillers were used.

3. Results and Discussion

The possibility of manufacturing building products based on PET waste by heating at 270 °C for 1 hour in forms of compacted composition from a mixture of PET-waste with particle sizes up to 1 cm and quartz sand, without using melting and mixing equipment is shown (Table 1).

Table 1. Properties of samples made by pressing mixtures % with following heating based on PET waste.

The content of the PET, % by weight of the composition	Average density, kg/m^3	Compressive strength, MPa	Bending strength, MPa	Water absorption, % by weight
15	1890	11.8	1.6	10.0
20	1920	18.0	2.0	5.0
30	1970	22.4	1.6	3.0

The resulting products have satisfactory compression strength at 20-30% binder content in the composition. It is necessary to increase the time for keeping samples in the mold at a temperature of 270 °C with an increase of PET waste in the composition. But this leads to an increase in technological energy consumption. It should be noted that the samples obtained by this method have a low flexural strength, the noticeable surface porosity and low decorativeness of the samples.

The possibility of manufacturing building products based on PET waste with the preparation of a mixture of PET waste with particle sizes up to 5 cm and quartz sand in a melting and mixing unit at a temperature of 250-270 °C is also shown. The technology involves laying the hot mixture into molds and pressing the mixture for 2 minutes, holding it for 5 minutes with following remove the sample from the mold (Table 2).

The strength of the samples obtained by this method is significantly higher. The highest value of the strength parameters is observed at 30% content of PET waste in the composition.

In this way to obtain high-quality products, it is necessary to completely melt the PET waste.

Table 2. Properties of the Samples Made by Pressing of Hot Mixtures Based on PET-Waste.

The content of the PET, % by weight of the composition	Average density, kg/m^3	Compressive strength, MPa	Bending strength, MPa	Water absorption, % by weight
15	1813	10.0	4.0	7.0
20	1935	22.5	5.6	2.7
30	1961	76.4	17.4	0.3
40	1838	65.8	13.9	0.3
50	1720	61.7	9.0	0.3

Additionally, samples were made according to the technological variant of the preparation of the composition in the melting-mixing unit, with the only difference that other mineral fillers of various degrees of dispersion were used. It was established that with the introduction of coarse carbonate chips (limestone with a particle size of 0.315-5 mm) into the composition, the properties of the samples become close to the properties of samples filled with quartz sand. The usage of finely dispersed fillers with a specific surface area in the range from 1000 to 8000 cm^2/g, regardless of the chemical and mineralogical composition of the filler (finely divided flasks, montmorillonite clays, limestone, chalk), decreases in physical and mechanical properties of the samples.

This indicated the need to change the composition of the binder, the need to introduce polyolefins into the composition. It is also necessary to use modern additives-bearers in the composition. This modern additives facilitate the processes of mixing thermoplastics among themselves and with mineral fillers. Also they contribute to enhancing the adhesive interaction at the level of the interfacial layers. Compositions have been developed [19, 32]. The binder is a complex of PET-waste, waste of polyolefins (preferably PP), additive additives (copolymer of ethylene and vinyl acetate Etatilen EVA-g-GMA), filled with fine mineral powders, to obtain high-quality durable building products with strength at compression up to 70 MPa and frost resistance over 200 cycles.

The possibility of introducing wood waste into the compositions was considered. On the one hand, such a filler will reduce the density of products, on the other hand, it will effectively utilize the wood components of municipal waste. The properties of samples made by pressing from hot mixtures based on a binder (including PET-waste, PP-waste, the addition of Ethatilene EVA in a ratio of 70% : 25% : 5% by weight) filled with mineral powders and chopped wood waste are investigated. The binder content in compositions, filled with mineral powders only, was 70% by weight. Shredded wood waste was introduced by replacing the volume of mineral filler with an equivalent volume of wood filler. Volumetric parts of the fillers for the convenience of analysis transferred to the mass parts (Table 3).

The decrease in the density of the obtained samples with an increase in the content in the compositions of shredded wood waste leads to a regular decrease in strength. However, the resulting strength indicators are at a fairly high level. This allows the usage of the developed compositions for the manufacture of construction products for structural purposes.

Table 3. Properties of Samples Made by Pressing from Hot Mixtures, Based on a Binder (from PET-Waste, PP-Waste, Auxiliary Additive), Filled with Mineral Powders and Crushed Wood-Waste.

Binder content, parts by weight	Filler content parts by weight		Average density, kg/m3	Compressive strength, MPa	Bending strength, MPa	Water absorption, % by weight
	mineral	wood				
100	233	-	1900	39.6	10.3	0.2
100	160	8	1610	35.9	16.1	0.5
100	120	14	1572	34.7	14.7	0.5
100	40	25	1389	34.2	9.4	0.5
100	-	48	1110	25.1	6.2	2.7

4. Conclusions

Studies have shown the possibility of manufacturing construction products based on polymer waste, using waste without prior cleaning and processing into semi-finished products. It is advisable to combine in one composition PET-waste and waste polyolefins, preferably PP-waste. The optimal content of the polymeric binder in the filled composition was established. It is 30% by weight of the composition. The possibility and expediency of introducing wood components of municipal waste into the compositions is shown.

5. Acknowledgments

The studies were carried out with the support of an individual grant Of the Foundation for the promotion of small enterprises in the scientific and technical sphere within the framework of the program "UMNIK-2017».

References

[1] Orlova T, Melnichuk A, Klimenko K, Vitvitskaya V, Popovych V, Dunaieva I, Terleev V, Nikonorov A, Togo I, Volkova Y, Mirschel W, Garmanov V 2017 Reclamation of landfills and dumps of municipal solid waste in a energy efficient waste management system: Methodology and practice *IOP Conference Series: Earth and Environmental Science* **90(1)** 012110

[2] Asadpoori A, Ankomah C, Asadpoori A, Derevianko O, Shaburov E 2018 Sustained municipal waste management models in Russian megapolicies through utilizing waste-to-energy technologies *MATEC Web of Conferences* **193** 02039

[3] Chusov A, Neguliaeva E, Romanov M 2018 Optimization of the Solid Waste Management System in Saint-Petersburg Based on the Morphological Composition Study *MATEC Web of Conferences* **193** 02037

[4] Kholodiakov A, Bondal A, Zubkova M, Chusov A, Stroganov A 2016 Methane component of the modern WWTP biogas as an alternative energy source *Solid State Phenomena* 871 pp. 208-15

[5] Payan L F B, Lopez D C 2018 Small-scale steam turbine and electric generation from municipal solid waste *International Journal of Environment and Waste Management* **22(1-4)** pp. 74-86

[6] Badenko V, Kurtener D, Yakushev V, Torbert A, Badenko G 2016 Evaluation of current state of agricultural land using problem-oriented fuzzy indicators in GIS environment *Lecture Notes in Computer Science* (including subseries Lecture Notes in Artificial Intelligence and Lecture Notes in Bioinformatics) **9788** pp. 57-69

[7] Malik N, Kumar P, Shrivastava S, Ghosh S 2017 *Int. J. of Plastics Tech.* **21**(1) pp 1-24

[8] Pol D R, Baknell K B 2009 *Polimernye smesi* (Saint-Petersburg: Scientific foundations and technologies)

[9] Bobovich B B, Devyatkin V V 2000 *Pererabotka othodov proizvodstva i potrebleniya* (Moscow: Intermet Inzhiniring)

[10] La Mantia F 2006 *Vtorichnaya pererabotka plastmass* (Saint-Petersburg: Professiya)

[11] Bedanokov A Yu, Beshtoev B Z, Mikitaev M A, Mikitaev A K, Sazonov V V 2009 *Ecological problems of our time: materials of scientific-practical seminar* (Maykop: IP Magarin O.G.)

[12] Beeva D A, Mikitaev A K, Barokova E B, Beev A A, Borisov V A, Yakokutova A A 2013 *Fundamental research* **10** pp 2878-81

[13] V N Shahova, A A Vorob'eva, I A Vitkalova, A S Torlova, E S Pikalov 2016 *Modern high technologies* **11** pp 320-5

[14] Teryaeva T N, Lyadov S S 2015 *Quality management of education, products and environment: proceedings of the 9th all-Russian scientific and practical conference* (Biysk: Polzunov Altai State Technical University)

[15] Bod'yan L A, Varlamova I A, Girevaya H Ya, Kalugina N L, Medyanik N L 2014 *Modern problems of science and education* **2** pp 700

[16] Bod'yan L A, Varlamova I A, Girevaya H Ya, Kalugina N L, Girevoj T A 2015 *Modern high technologies* **2** pp 15-8

[17] Ogorodov L I, Lustina O V 2017 *Magazine of Civil Engineering* **6** (74) pp 17–32

[18] E A Buharova 2016 *Sorbcionnye materialy na osnove othodov poliehtilentereftalata i soedinenij grafita dlya ochistki stochnyh vod*, PhD dissertation (Saratov)

[19] Fomina N N, Ivashchenko Yu G, Polyanskij M M *Patent RU 2623754* 2017

[20] Blanken T C, Van Helden A K, Majoor J C J E, Mooiweer H H, Reynhout M J, Wisse W *Patent EP 1395527 B1* 2004

[21] Mendivil-Escalante J M, Gomez-Soberon J M, Almaral-Sanchez J L, Corral-Higuera R, Arredondo-Rea S P, Castro-Beltran A, Cabrera-Covarrubias F G 2015 *Int. J. Structural Analysis & Design* **2** (1) pp 105-9

[22] Strain I N, Wu Q, Pourrahimi A M, Hedenqvist M S, Olsson R T, Andersson R L 2015 *J. Mater. Chem* **3** pp 1632-40

[23] Elamri A, Abid K, Harzallah O, Lallam A 2015 Nanocomposites Coating and Manufacturing *Am. J. of Nano Research and Application* **3** (4-1) pp 11-16

[24] Kim J, Jeong D, Son Ch, Lee Y, Kim T, Moon Il 2007 *Korean J. Chem. Eng* **24** (6) pp 1076-83

[25] K B Prabhu, G J Chomal, S Mkulkarni 2013 *Education and Green Sky - Materials Technology for a Better World: International Technical Conference* (Long Beach) pp 1256-1270

[26] L S Kalinina, M A Motorina, N I Nikitina, N A Hachapuridze 1984 *Analiz kondensacionnyh polimerov* (Moscow: Chemistry)

[27] Ivanov A M, Harichkin A S 2008 *Int. polym. science and technology* **6** pp 9-11

[28] Ruamcharoen P, Phetphaisit C W , Bumee R, Ruamcharoen J, Niyomdecha B, Nillawat S 2012 *Advanced Materials Research: 2nd International Conference on Key Engineering Materials* **488-489** pp 648-653 (Singapore)
[29] Evmenov S D, Silinina E B, Smirnov A V 2004 *Journal KuzSTU* **5** pp 62-5
[30] Gogol Eh V, Mingazetdinov I H, Gumerova G I, Egorova O S, Malceva S A, Grigoreva I G, Tunakova Yu A 2013 *Vestnik KNRTU* **16** (10) pp 163-8
[31] Shapovalov V, Tartakovskij Z, Kudyan S 2012 *The Science and Innovations* **9** (115) pp 9-11
[32] Fomina N N, Ivashchenko Yu G, Polyanskij M M, Pavlova I L 2018 *J. Phys: Conference Series* **365** (3) 03

IOP Conf. Series: Earth and Environmental Science **337** (2019) 012023 doi:10.1088/1755-1315/337/1/012023

Selection of bacteria capable of biodegradation of organic nitrogen-containing compounds and metagenomic analysis of their living environment

A Maksimov[1], E Vladimirova[2*], Yulia Pavlova[1] and Anna Shilova[1]

[1]Institute of Ecology and Genetics of Microorganisms UB RAS, 614990 Perm, Russia
[2]Peter the Great St.Petersburg Polytechnic University, St.Petersburg, Russia

* Corresponding author: vladymyrovaliza18@mail.ru

Abstract. Samples of karst lakes, rivers, industrial effluents, and natural soil were studied for selection of bacteria for biodegradation. A metagenomic analysis of the used natural and man-modified media was carried out. It has been established that the most unique are the compositions of the metagenomes of natural karst reservoirs, as well as microcenoses of industrial effluents. Most of the detected bacteria belong to Proteobacteria. As a result of selection, strains of microorganisms with high activity of nitrocellulose biodegradation, aromatic nitrates, quinoline derivatives and amides were obtained. In particular, Pseudomonas strains and R. erythropolis, effectively utilizing nitrobenzenes, nitrocellulose, quinolines, which were isolated from industrial effluents. At the same time, Pseudomonas strains were isolated from the natural karst lake, which quickly utilized the nitrocellulose. It was also found that natural soils contain a large number of prokaryotes that can utilize organic nitrates. The results confirm the view that natural environments, due to high metabolic diversity, are the richest source of producers of a wide variety of enzymes. An obvious reason for the proliferation of the metabolic systems of technogenic pollutants in natural environments is the presence in natural environments of some quantities of their structural analogues - nitrogen-containing plant metabolites and soil microflora, as well as lignin decomposition products.

1. Introduction

Synthetic nitrogen-containing compounds, such as organic nitrates, quinolines, amides, etc., are widespread environmental pollutants. The most common of them are organic nitrates, such as nitrocelluloses, nitrobenzenes, nitrotoluenes, nitrophenols, are widely used in industry for plastics, paints, varnishes, adhesives, pesticides, dyes, medicines, explosives, fungicides, analytical reagents, etc.

The danger of nitro compounds for human health is due to their high toxicity, as well as the possible formation of mutagenic metabolites. Various physical and chemical methods have been developed for the degradation of nitrogen-containing organic compounds. The most common are incineration and chemical processing. However, these methods are not environmentally friendly and are applicable only to dry concentrated waste, while much of the damage to the environment is caused by industrial runoff and soil pollution.

Many organic nitrates are resistant to biodegradation due to their high toxicity for most microorganisms (nitrophenols) or low availability (water-insoluble nitrocellulose, 2,4,6-trinitrotoluene, etc.). Because of this, they accumulate in the environment.

Due to the high relevance of research in the field of biodegradation of organic nitrates have developed over several decades [1]. The ability to biotransformation and biodegradation of nitro compounds is known in the strains of the species *Pseudomonas fluorescens*, *Micobacterium*, *R. erythropolis* [2-6]. Aerobic and anaerobic processes of such decomposition are known. The main class of enzymes involved in the transformation of trinitrotoluene are nitroreductases; enzymes of the family of flavoproteins, oxidase, hydrogenase, and peroxidase can also participate in degradation [2].

In the last decade there has been a significant expansion of knowledge about the biotransformation of TNT at the genetic and biochemical level and the development of the application of these processes in the field of biotechnology. Research in the field of biotransformation of nitro compounds has received a new impetus through the use of modern molecular methods, including metagenomic studies [7-10].

Despite the long history of studying the processes of biotransformation of nitroorganic compounds, this trend remains relevant as a result of the practical relevance and sustainability of such substrates to biological and abiogenic degradation.

Actual are the selection of new cultures of microorganisms - active biodestructors, the identification of new genes and enzymes of metabolism of organic nitrates, enzymological analysis [2,8,11].

2. Materials and research methods

2.1. Objects of study

Samples from karst lakes, bottom sediments of rivers, industrial effluent and natural soils of Perm region were used for isolate microorganisms-biodestructors of nitrogen-containing organic compounds.

With a low level of anthropogenic chemical load:

- Molebnoe Karst Lake and the failure of Kishertsky karst area);
- Bottom sediments of the river Suz'va and its tributaries, Nytvensky district of the Perm region;
- Sod-meadow soil, Nytvensky district of the Perm region.
- Bottom sediments of the Yazovaya River, Perm (industrial effluents of OAO Kamteks-Khimprom).

2.2. Substrates, media and culture conditions

The isolation of communities (consortia) and pure cultures transforming organic nitrogen-containing compounds was carried out on a mineral medium with the addition of a nitrogen-containing compound as the sole source of carbon and / or nitrogen.

As substrates nitrobenzen- and 1,3-dinitrobenzenes, 2,4–dinitrophenol, dinitrocellulose (colloxylin), quinoline, and 8-nitroquinoline were used.

Cultures were grown in liquid mineral medium N of the following composition, g / l: K_2HPO_4 - 1.0; KH_2PO_4 - 1.0; NaCl - 0.5 A trace element solution was added to the medium, g / l: $MgSO_4 \times 7H_2O$ - 0.5; $CaCl_2$ - 0.005; $CoCl_2 \times 6H_2O$ - 0.01; $FeSO_4 \times 7H_2O$ - 0.005 [12]. Ammonium chloride at a final concentration of 10 mM was used as a standard source of nitrogen, and glucose at a concentration of 0.1% was used as a source of carbon and energy. Pure cultures were isolated by plating on LB agar medium (Luria Bertani) containing (g / l): tryptone 10.0, yeast extract 5.0, NaCl 10, agar (Sigma) 15. Agar media were prepared by adding purified agar for microbiological media (Sigma) to the final concentration 1.5%.

Cultures were grown under conditions of periodic cultivation in Erlenmeyer flasks in 50 ml of nutrient medium on an orbital shaker with a rotation speed of 150 rpm at a temperature of 30 ° C.

Isolates were isolated by the method of cumulative culture. After 30 days, they were seeded on Petri dishes with agar N medium containing selective substrates at a concentration of 10 mM. On days 7-14 of growth, the resulting colonies were selected, transferred to Petri dishes with fresh selective medium and with full LB medium.

The density of the culture was evaluated by the change in the optical density of the cell suspension at $\lambda = 540$ nm, considering the dilution.

Minimum inhibitory concentrations of substrates were determined by the growth intensity of the culture in the Luria – Bertani medium (LB) with different concentrations of chloroquine by serial dilution in a 96-well polystyrene plate (Eppendorf). The plates were incubated in a thermostat at a temperature of 30° C for three days and the optical density at 540 nm was measured on an Infinity M1000 plate spectrophotometer, Tecan, Switzerland. For the minimum inhibitory concentration took the highest concentration at which there was no significant increase in the density of the culture.

2.3. Identification of selected cultures

Taxonomic analysis of bacteria was carried out on the basis of standard cultural-morphological, biochemical, and chemotaxonomic characteristics according to the Berge bacterium determinant [13], and also using specialized guidelines. Also, for species identification, amplification of a fragment of the 16S rRNA gene with universal primers 27F (50-AGAGTTTGTCCTGGCTCAG-30) and 1492R (50-TACCTTGTTACGACTT-30), sequencing of the obtained PCR product and comparison of sequences deposited in the Ezio database of data has been used in the Ezio data database, has been used. www.ezbiocloud.net/). A metagenomic analysis of the samples studied was carried out on the 16S rRNA genes on the MiSeq platform (Illumina) [14].

2.4. Molecular genetics methods

DNA sequencing was performed using an Applied Biosystem 3500XL system using the Big Dye Terminator Cycle Sequencing Kit reagents according to the manufacturer's instructions. Comparison of gene sequences was carried out using the software packages VectorNTI11, ClustalW, as well as the original application YACWGUI 1.2.

DNA extraction was carried out by the alkaline method (heating): the material of the colony was transferred with a microbiological loop in 0.5 ml of 0.01 M NaOH, mixed on a vortex, kept in a solid-state thermostat at 97 ° C for 10 minutes, then in a freezer at - 18°C 10 minutes. Then the warm-up and cooling cycle was repeated twice more.

2.5. Biotransformation and analysis

Biotransformation products were determined by GC-MS methods on an 689 / 573T MSD apparatus (Agilent-Hewlett Packard), under the following conditions: extraction of samples with ethyl acetate; capillary column HP 5MS, column initial temperature - 60 ° C (3 min), heating to 315 ° at a rate of 7 ° / min; evaporator temperature 280 ° C, carrier gas helium, as well as using HPLC LC10, Shimadzu on a column Luna C18 (250 x 4.6 mm). The flow rate was 1 ml / min at 25 ° C. Detection was performed at 230 and 340 nm.

The amount of water-insoluble aromatic nitrates and nitrocellulose was determined gravimetrically on an analytical balance after drying to constant weight.

Mathematical processing of experimental data was performed using MS Office Excel and Statistica 5.0.

3. Results and discussion

3.1. Isolation of bacteria communities

It is known that among nitrogen-containing organic compounds of one series, amines, amides and nitriles are most easily subjected to biological utilization, while nitro compounds and nitrogen heterocycles are generally resistant to biodegradation due to lower bioavailability and, in some cases, high toxicity.

Among cultures of microorganisms capable of biodegradation of nitro compounds, anaerobic bacteria attract the particular attention of researchers. The reduction of nitro groups to nitroso derivatives, hydroxylamines or amines is catalyzed by nitroreductases through the sequential addition of pairs of electrons whose donors are co-substrates. It is believed that the degradation of most polynitroaromatic compounds is possible only under anaerobic conditions [10,11,15].

At the same time, the possibility of using aerobic processes attracts a higher rate of metabolism and growth of aerobic cultures, as well as the potentially greater ease of controlling aerobic processes. In this regard, in our work we used microorganisms capable of aerobic growth and transformation of substrates under aeration conditions.

The method of cumulative culture obtained on the basis of nitrogen-free salt medium N with the addition of nitrocellulose and nitroaromatic compounds (nitrobenzene, p-nitrophenol, quinoline, 8-nitroquinoline) as selective substrates, as a source of carbon and / or nitrogen, obtained communities of microorganisms capable of using nitro groups of organic compounds as a source of nitrogen. Investigated the activity of the obtained microbial communities in the process of biodegradation of a number of organic nitrates. The degree of biodegradation of the substrate, expressed as a percentage of the amount introduced into the medium and the growth of bacteria, given in units of optical density at a wavelength of 540 nm, was evaluated. The high activity of communities isolated from industrial effluents of Kamteks-Khimprom, in particular, during the decomposition of dinitrocellulose (Table 1) is shown.

From the literature, a similar result is known in the anaerobic process using sulfate-reducing bacteria of the genus *Desulfovibrio* [16,17], however, in the case of the consortium studied, utilization was observed in a wide range of physicochemical conditions.

The ability of selected microorganisms to utilize organic nitrates has been investigated. Cultures were grown on mineral medium containing up to 0.25% of aromatic nitrates as the sole source of nitrogen. It is established that the most optimal growth agent under these conditions is nitrobenzene, which provided the maximum biomass growth. At the same time, it was found that a consortium of bacteria isolated from industrial wastewater degrades mononitrophenol in a weakly alkaline medium (pH 7.8) at a concentration of more than 1 mmol / l in 5 hours, with the formation of aliphatic carboxylic acids, which significantly exceeds the rate of biodegradation, described in the literature for strains *Arthrobacter*, *Burkholderia*, and others [2,3,12,18]. For example, for the strain *Burkholderia* sp. FDS-1 - 9 hours at a concentration of 0.6 mmol / l MNF, pH7 - transformation to hydroquinone [18].

Table 1. Disposal of substrates introduced into the medium as a source of carbon and nitrogen or a source of nitrogen

Substrate	**Substrate biodegradation,% / Culture density, OD540** Sample (community) number* 1	2	3
0,1% p-nitrophenol	2,23±0,24/ 0,33±0,16	3,50±0,85/ 0,32±0,17	22,12±3,01/ 1,12±0,23
0,1% p-nitrophenol + 0,1% glucose	2,43±0,43/ 1,43±0,41	4,11±0,85/ 2,47±1,22	24,32±2,99/ 2,12±0,54
0,1% nitrobenzene	6,35±0,53/ 1,21±0,42	6,90±0,53/ 0,99±0,43	27,31±3,15/ 3,00±0,81
0,1% nitrobenzene + 0,1% glucose	7,54±0,76/ 1,99±0,35	8,31±0,99/ 2,99±0,85	37,79±3,21/ 4,93±1,27
1% quinoline	4,43±0,21/ 0,87±0,23	5,83±0,75/ 1,31±0,09	23,25±3,14/ 3,08±0,46
1% quinoline + 0,1% glucose	5,36±1,01/ 1,34±0,35	6,96±0,78/ 3,20±0,07	54,12±5,25/ 6,46±0,93
0,1% 8- nitroquinoline	3,53±0,42/ 0,63±0,14	4,99±0,55/ 0,98±0,12	23,57±2,67/ 1,00±0,81
0,1% 8- nitroquinoline + 0,1% glucose	4,42±0,75/ 1,234±0,43	6,87±0,79/ 2,93±0,35	27,35±2,81/ 3,93±1,27
1% dinitrocellulose	4,33±0,85/	6,62±0,85/	49,45±3,22/

IOP Conf. Series: Earth and Environmental Science **337** (2019) 012023 doi:10.1088/1755-1315/337/1/012023

	1,12±0,31	1,01±0,46	2,89±0,85
1% dinitrocellulose + 0,1% glucose	5,74±0,75/ 1,71±0,45	8,62±0,85/ 1,68±0,57	64,12±6,99/ 5,56±0,85

* The community numbers correspond to the samples from which they were isolated:

1 - Sod-meadow soil, Nytvensky district of the Perm region;

2 - Bottom sediments of the Syuzyva River;

3 - Bottom sediments, industrial effluents of Kamteks-Khimprom JSC.

3.2. Metagenomic analysis

A metagenomic analysis of the media from which the active crops were selected: the technogenically altered environment (bottom sediments and industrial effluents of Kamteks-Khimprom JSC), as well as natural environments — karst lakes and rivers that do not have significant chemical pollutants (Suzyva, Nytvensky region), natural soil.

It has been established that the most unique is the composition of the metagenome of natural karst reservoirs, as well as microcenoses of industrial effluents containing "nitrotel". As a result of metagenomic analysis, it was found that most of the detected bacteria in natural reservoirs belong to the *Proteobacteria* fillet - 61% of the total number of reads, mostly represented by the *Betaproteobacteria* class (37.5%), the *Alphaproteobacteria* class (10.3%), the *Deltaproteobacteria* class (5%), class *Gammaproteobacteria* (2%), class *Epsilonproteobacteria* (0.1%). the class *Betaproteobacteria* is represented by the orders of *Burkholderiales* (8%), the family *Comamonadaceae* (5.6%); order of *Gallionellales* (7.2%), the family *Gallionellaceae* (7.2%) and order *Rhodocyclales* (5%), the family *Rhodocyclaceae* (5%). A significant proportion of identified prokaryotes belonged to the phyla: *Bacteroidetes* (14.7%), *Cyanobacteria* (0.4%), *Acidobacteria* (1.1%), *Chlorobi* (0.6%). Sulfate-reducing bacteria belonged to the class *Deltaproteobacteria* (5%), represented by the orders *Desulfuromonadales* (0.6%), *Desulfovibrionales* (0.5%), Desulfobacterales (0.5%). *Desulfuromonadales* order is represented by the family of Geobacteraceae 0.3% (of the total number. The *Desulfovibrionales* family (0.5%) belongs to the *Desulfovibrionales* order (0.5%). The *Desulfobacterales* order is represented by the families: *Desulfobulbaceae* (0.2%), *Desulfobulbaceae* (0.2%).

3.3. Pure cultures

Active strains isolated using subculturing were identified by polyphasic taxonomy and analysis of 16S RNA genes as representatives of the proteobacteria *Pseudomonas fluorescens* (4 strains) and *Pseudomonas denitrificans* (4) and Pseudomonas sp. (one); actinobacteria of the genera *Rhodococcus* (15), in particular, species of *Rhodococcus erythropolis* (9 strains), *Rhodococcus ruber* (2), *Rhodococcus rodnii* (4), genera *Arthrobacter* (7), *Gordona* (1). In particular, strains *Pseudomonas fluorescens* N-17, N19 and *Pseudomonas putida* N-26, as well as *R. erythropolis* N-4 and N-14, which efficiently utilize nitrobenzenes and nitrocellulose, were isolated from the effluent of nitrocellulose. At the same time, the strains *Pseudomonas* sp. N52, N53, quickly disposed of colloxylin. It has also been established that natural soil contains a large number of diverse cultures of prokaryotes capable of utilizing aromatic nitrates, albeit at a slower rate compared with isolates from industrial effluents. Despite the fact that the majority of organic nitro compounds — environmental pollutants — are of technogenic origin, a large number of bacteria and their communities capable of using these substances were sown from natural soils and waters that are not in contact with anthropogenic organic nitrates. This may be due in part to the fact that in open systems it is possible to transfer microorganisms from media that have industrial pollution, as well as the possibility of transferring trace amounts of nitroorganic compounds.

An obvious reason for the spread of such metabolic systems is the presence in natural environments of certain amounts of structural analogs of man-made pollutants, such as various nitrogen-containing plant metabolites and soil microflora, as well as lignin decomposition products, including its abiogenic destruction. And, as is well known, natural environments, including rich soils, due to the highest metabolic diversity, are the richest source of producers of a wide variety of enzymes.

It has been shown that most of the studied actinobacteria are able to use the nitrogen of the nitro groups of organic nitroesters, as is known for other microorganisms and fungi [16-20] . It was established that the majority of bacterial cultures - active transformants had nitroreductase activity. Nitroesterase activity was detected only in two isolates, which did not show a high rate of biodegradation under the conditions of model experiments.

As a result of screening studies, genes characteristic of some representatives of the genus *Pseudomonas* were detected in 6 studied *Pseudomonas* cultures using the Ps_nfsA-F/R primers.

Analysis of the genes encoding the biodegradation enzymes of other nitrogen-containing compounds, in particular, amidases, was performed as previously described. Phylogenetic analysis of amidase genes was obtained, as well as the results of studies of the activity, thermal and pH stability of controlled enzymatic activities in the transformation of nitrocellulose and aromatic nitrates (nitrobenzenes, nitrotoluenes).

It has been shown that most of the studied actinobacteria are able to use the nitrogen of the nitro groups of organic nitroesters, as is known for other microorganisms and fungi [18, 20-23] .

3.4. Minimal inhibitory concentrations and cultivation

The complexity of the utilization of aromatic nitro compounds is caused by the influence of the following factors: the lower availability of nitro groups for enzymes caused by steric hindrances (nitro groups are associated with an aromatic ring); hydrophobicity, low solubility of aromatic nitriles in the aqueous phase; the difficulty of diffusion of biotransformation products and substrates into and out of the cell; toxicity of aromatic nitriles for cells.

To assess the effect of the latter factor on transforming cells, minimal inhibitory concentrations of a number of aromatic nitro compounds for the isolated bacteria were determined. It was found that, in general, strains of the genus *Rhodococcus* were the most resistant to the presented substrates. The most toxic for bacteria were nitrophenols and nitroquinoline (MIC for *Pseudomonas* strains - 31.3, for *Rhodococcus* - 62.5. Toluene nitrates were less toxic, apparently due to low solubility in aqueous medium. Colloxylin did not inhibit the growth of bacteria.

The ability of the isolated strains to grow and transform nitro compounds as the sole substrate (source of carbon and nitrogen) and with the addition of glucose (the aromatic substrate is the source of only nitrogen) was investigated.

It was established that the most active biotransformation of *Rhodococcus* strains was when using as a carbon source 0.1% acetate and 0.25% nitroaromatic compound or 1-5% colloxylin at an ambient temperature of 28-30 ° C and slightly alkaline pH values of 7.2-8.

It was shown that highly active cultures of the strains *Pseudomonas* sp. N-52, *P. fluorescens* N-17, N19, *P. putida* N-26, *R. erythropolis* N-5 and N-14 in optimal conditions of 7.7 mg O_2 / l, pH 7.8–8.0, 25°C are able to transform colloxylin c from high speed - up to 0.5 g / 1 g dry weight of cells / day.

4. Conclusions

Thus, microbial communities and pure cultures capable of using the nitro groups of man-made organic compounds as a nitrogen source were selected. In particular, strains of *P. fluorescens*, *P. putida* and *R. erythropolis*, which effectively utilize nitrobenzenes, nitrocellulose, substituted quinolines and amides, were isolated.

It is also shown that microorganisms which possess metabolic systems for the disposal of nitroorganic compounds inhabitant not only in the environment, polluting chemicals, but also in the environment of karst lakes, rivers, soils, remote from industrial centers and not tangible man-made stress. At the same time, the representatives of bacteria, active in the process of biodegradation of nitroorganic compounds, do not belong to the dominant genera identified as a result of metagenomic analysis. The results confirm the view that natural environments, as the best metabolic diversity, are the richest experience of producing a wide variety of enzymes. An obvious reason for the propagation, of the metabolic systems of technogenic pollutants in natural environments is the presence in natural environments of some quantities of their structural analogues - nitrogen-containing plant metabolites and soil microflora, as well as lignin decomposition products.

Highly active cultures of *Pseudomonas* sp. N-52, *P. fluorescens* N-17, N19, *P. putida* N-26, *R. erythropolis* N-5 and N-14, capable of utilizing the nitro-organic connections at high speed, are promising for use in local wastewater treatment of chemical plants.

5. Acknowledgments

This work was supported by RFBR research project No. 16-44-590359.

References

[1] Cartwright N J, Cain R B 1958 *Biochemical J.* **71** pp 248–61
[2] Yu G Maksimova, A Yu Maksimov, V A Demakov 2018 *Appl. Biochem. Microbiol.* **34** pp 9-23
[3] Ju K-S, Parales R E 2010 *Microbiol. Mol. Biol. Rev.* **74** pp 250–72
[4] Li T, Zhang T C, He L 2017 *J. Biotechnol.* **264** pp 8-16
[5] Qiu L, Wang H, Wang X 2017 *Water Environ Res.* **89** pp1970-80
[6] Min J, Wang J, Chen W, Hu X 2018 *AMB Express.* **8** p 43
[7] Auer N, Hedger J N, Evans C S 2005 *Biodegradation.* **16** pp 229-36
[8] Kao C-M, Lin B-H, Chen S-C et al 2016 *Bioremed. J.* **20** pp 165–73
[9] Muter O, Potapova K, Limane B et al 2012 *J. Environ. Manage.* **98** pp 51–5
[10] Shilova A V, Maksimov A Yu, Maksimova Yu G 2018 *Biomiks.* **10** pp 24-7
[11] Saratovskikh E A, Avdeeva L V, Yarullin R N et al 2018 *J. Therm. Anal. Calorim.* **134** pp 653-64
[12] Maksimov A Yu, Kuznetsova M V, Ovechkina G V et al 2003 *Appl. Biochem. Microbiol.* **39** pp 55-9
[13] Garrity G 2005 *Bergey's Manual of Systematic Bacteriology (2nd ed.)* (New York: Springer)
[14] Gilbert J A, Dupont C L 2011 Microbial Metagenomics: Beyond the Genome *Annual Rev. Marine Science.* **3** pp 347–71
[15] Kuo D T F, Simini M, Allen H E 2018 *Water Research.* **128** pp 138-47
[16] Giacomucci L, Toja F, Sanmartin P, Toniolo L et al 2012 *Biodegradation.* **23** pp 705–16
[17] N B Tarasova, O V Gorshkov, O E Petrova 2009 *Mikrobiologiia.* **78** pp 192-6
[18] Zhang Z, Hong Q, Xu J, Zhang X, Li S 2006 *Biodegradation.* **17** pp 275-83
[19] Scherbakov V A, Saratovskikh E A, Yarullin R N 2018 *Appl. Biochem. Microbiol.* **54** pp 55-62
[20] Freedman D L, Cashwell J M, Kim B J 2002 *Waste Management.* **22** pp 283-92
[21] Sundaram S T, Zhang Y Z, Sharma A, Ng K, Brodman B W 1995 *J. Appl. Polymer Sci.* **58** pp 2287-91
[22] Zhang M, Liu G-H, Song K et al 2015 *Chem. Eng. J.* **259** pp 876–84
[23] Subashchandrabose S R, Venkateswarlu K, Krishnan K et al 2018 *J. Hazard Mater.* **347** pp 176-83

IOP Conf. Series: Earth and Environmental Science **337** (2019) 012024 doi:10.1088/1755-1315/337/1/012024

Increasing the Waste Management Efficiency in the Arctic Zone of Russia through the Projects of Eco-Industrial Parks' Development

M Liubarskaia[1]*, M Tsurkan[2] and A Artemiev[3]

[1]Saint-Petersburg State University of Economics, Department of State and Territorial Management, 21 Sadovaya str., 121023 Saint-Petersburg, Russia
[2]Tver State University, Department of State Management, 33 Zhelyabova str., 170100 Tver, Russia
[3]Tver State Technical University, Department of Geodesy and Cadastre, 22 A. Nikitin emb., 170026 Tver, Russia

* Corresponding author: lioubarskaya@mail.ru

Abstract. The relevance of the study is due to the waste management and environmental safety issues at the territory of the Arctic zone of Russia. Some of these problems could be solved in the process of creating and operating of eco-industrial parks on the basis of a project approach. The purpose of the article is to systematize the ideas on the possibilities of applying project management for the development of eco-industrial parks in the Arctic zone of Russia. The article clarifies the concept of "eco-industrial park" in the framework of relevant Russian legislation and practice. The classification of eco-industrial parks has been developed taking into account different characteristics, including content of main waste flow and type of processing technology. The main challenges for the implementation of the project approach on the territory of the Arctic zone of Russia are indicated. The results of the research are of practical importance for public authorities, the governing body and residents of eco-industrial parks, specialists in the field of waste management, scientists studying the environmental and economic aspects of the Arctic zone of Russia.

1. Introduction

At present, at the territory of the Arctic zone, one of the most important areas of sustainable development is the provision of effective waste management. Preservation of unique Arctic ecosystems and elimination of accumulated pollution in the Arctic regions are the key tasks of state policy and fulfillment of Russia's international obligations [1,2].

The Arctic coast of Russia from Murmansk to Pevek and many islands of the Arctic zone of the Russian Federation are littered with piles of metal debris, including rusted ships and machinery, cars, containers, endless deposits of two-hundred barrels and huge tanks for oil products. According to some estimates, in the coastal zone of the Arctic Ocean there are up to 4 million tons of industrial and construction debris, as well as from 4 to 12 million iron barrels.

Enlarged, in the regions of the Arctic zone, there are five main environmental problems:

- the negative consequences and threats of the ongoing global climate change;
- land degradation and violation of land using conditions;
- change of biological diversity and reduction of bio resources reserves;

- deterioration of the habitat of the indigenous population of the Arctic regions of Russia and the conditions of their traditional nature using.

The most significant accumulation of pollutants and violations of Arctic natural landscapes occurred in the 1930s and 1980s, when intensive industrialization and extensive extraction of natural resources in the regions of the North of Russia was underway. By the end of that time frame, about 70% of enterprises of the industry of extraction and processing of natural resources contributed to the formation of a significant amount of solid, liquid and gaseous waste [3].

Sources of pollution in the regions of the Arctic zone of Russia are presented in Table 1.

Table 1. Main industries contributing to pollutions in the regions of the Arctic zone of Russia.

Region	Industry/Object
Kola	Non-ferrous metallurgy, mining industry, nuclear power plant (NPP), thermal power plants (TPP), objects of radioactive waste generation
Severodvinsk (Archangelsk Region)	Pulp-and-paper industry, military facilities, thermal power plants (TPP), objects of radioactive waste generation
Islands of the Arctic zone of Russia	Military facilities, scientific stations and hydro-meteorological services
Novozemelsky	Military facilities, flooding of nuclear installations and other radioactive waste in the Kara Sea
Timano-Pechora	Extraction and transportation of hydrocarbons
Vorkuta	Mining industry, thermal power plants (TPP)
Norilsk	Mining and metallurgical industry
Yano-Indigirsky	Mining industry
Chukchi	Mining industry, nuclear power plant (NPP), thermal power plants (TPP), objects of radioactive waste generation

Source: compiled by authors based on [3]

For example, on Belyi Island there are 42 sources of pollution. The largest source of waste is building timber, including demolition residues (99%). The other sources include barrels (1%), scrap and waste of ferrous metals (0.1%), household waste (less than 0.1%), lead batteries (less than 0.1%), construction and demolition waste (less than 0.1%), etc.

Solving these problems requires the following measures:

- implementation of new technologies, including technologies for cleaning the territory of islands, coastal zones and water areas of the Arctic seas from anthropogenic pollution;
- conducting scientific research on the assessment of nomenclature and quantitative characteristics of negative impacts (risks) on the environment throughout the Arctic;
- conducting predictive assessments of the types and volumes of hazardous wastes generation, emissions of harmful substances;
- creation of technology for collection and processing of accumulated hazardous wastes taking into account climatic conditions. [3]

The institutional environment for the implementation of these measures can be created through special clusters, or Arctic eco-industrial parks. The best foreign practices of eco-industrial parks have been created in Denmark, Austria, Finland, Great Britain, China, and Sweden. Their functioning showed that 100% of the waste could be gained 25% of certified fuel for the production of electricity and heat, 5% of secondary material resources, 40% of compost and man-made soil, and only 20% is a residue that can be sent in compressed form to landfills [4].

IOP Conf. Series: Earth and Environmental Science **337** (2019) 012024 doi:10.1088/1755-1315/337/1/012024

The problem of energy supply is relevant for the Arctic zone of Russia. The Arctic regions are characterized by high energy intensity and low efficiency of natural resource extraction, costs of northern production in the absence of effective compensation mechanisms, low labor productivity, and underdevelopment of the energy system, as well as irrational structure of generating capacities, high production costs of generation and transportation of electricity. A high level of economically justified tariff for the production of electric energy is 600-2000 rubles per kW per hour.

Undoubtedly, the Arctic regions have special climate challenges. Even under not favourable climate conditions, the first experience of creating an eco-industrial park has already been obtained in the Murmansk region. The mechanism of functioning of this park could be improved within the framework of the project approach.

At the same time, increase in the number of eco-industrial parks on the territory of the Arctic zone of Russia could be gained through implementation of related projects.

So it's equally important to use project approach for creation and development of eco-industrial parks, but both aspects remain practically unexplored in the scientific literature and will be presented in this study.

2. The concept and classification of eco-industrial parks

In the world practice, the term "eco-industrial park" means the group of industrial enterprises increasing the efficiency of using natural resources through cooperation and interaction based on a systematic approach.

The purpose of creating eco-industrial parks is to achieve economic, environmental and social benefits while reducing the consumption of primary material and energy resources, involving secondary resources in the production turnover, reducing the level of environmental impact of existing industrial facilities and eliminating the accumulated environmental damage [5].

The presence of an industrial component in the definition of the concept of "eco-industrial park" does not allow it to be identified with the concept of "industrial techno park", which, of course, is broader.

Criteria for classifying an industrial techno park as an eco-industrial park are following:

- availability of exchange of material and energy resources both within the park and with remote partners;
- organization of a cascade system of water supply and integrated wastewater treatment in the eco-industrial park with possible involvement of municipal wastewater from closely located settlements;
- organization of a common system for managing the material and energy resources of the eco-industrial park (through the identification of a focal point and (or) "anchor tenant");
- use of the general infrastructure of the park, including administrative, transport, energy and water supply systems, waste management of the eco-industrial park.

The coordinating centre of the eco-industrial park is the governing body of the eco-industrial park with functions of searching and developing the opportunities for implementation of network exchanges of material and energy resources between the enterprises within the eco-industrial park and beyond. The approximate organizational structure of the eco-industrial park is shown in Figure 1.

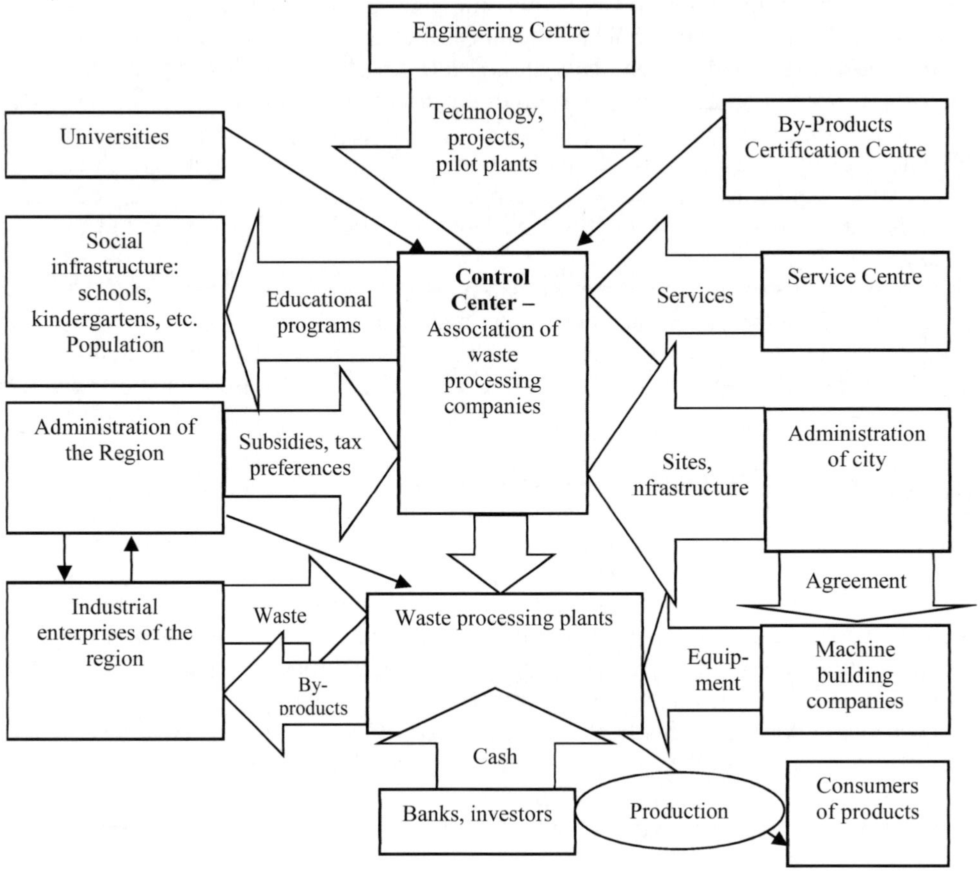

Fig. 1. Organizational structure of Eco-industrial Park. Source: compiled by authors.

Anchor resident of the eco-industrial park is the key organization in the eco-industrial park, which provides most of the material and energy flows for industrial symbiosis:

- a physical anchor tenant (for example, an energy company, a large industrial enterprise);
- an institutional anchor tenant (for example, a scientific and technical university).

By the authors' opinion, the key difference between the techno park and the eco-industrial park (EIP) is the functioning of the EIP on the basis of the principles of industrial symbiosis, which implies:

- reuse of by-products and (or) waste;
- exchange of product-specific materials between two or more companies to replace the use of commercial products or raw materials);
- joint use of public services (infrastructure facilities) and joint management of such resources as energy, water, electricity and heat, as well as joint operation of water treatment facilities and gas cleaning equipment;
- joint provision of services, satisfying the general needs of companies for the implementation of auxiliary activities such as fire safety, transportation and others.

In addition, there is an approach that allows us to identify the concept of "eco-industrial park" with the notion of "techno-ecosystem". According to it, the creation of an eco-industrial park is the development and transformation of an existing industrial technological system into a techno-ecosystem. That means the integration of the technosystem and ecosystem based on the ecosystem waste management mechanism.

IOP Conf. Series: Earth and Environmental Science **337** (2019) 012024 doi:10.1088/1755-1315/337/1/012024

Technosystem (enterprise, city) is a spatially defined set of production components and living organisms, united by a single environment of existence.

Ecosystem is a spatially defined set of living organisms and their habitat, united by a real-energy and information interaction.

Ecosystem waste management is the management of material and energy flows through information interaction between ecosystem components and the environment, ensuring self-organization and sustainable development of the ecosystem [6].

The authors' approach, which is also used by domestic practitioners, unites all the above definitions and allows us to propose the following interpretation of the concept of "eco-industrial park". It is a complex of technological solutions, equipment, facilities that provide a unified technological process with ultimate goal is manufacturing goods from recycled materials.

In order to systematize the idea of potential eco-industrial parks that at the territory of Russia within the framework of the current legislation, the classification can be proposed with the following criteria:

- orientation to the type of waste;
- sources of formation of flows of material and energy resources;
- availability of a key organization that generates most of the flows of material and energy resources;
- level of localization.

Depending on the orientation of the type of waste EIP can be:

1. oriented on utilization and processing of solid municipal waste components;
2. oriented on mixed utilization, neutralization and processing of municipal solid waste, industrial, medical and biological waste.

Depending on the sources of material and energy resources, eco-industrial parks can be divided into:

1. EIP with the flows of material and energy resources focused on solid municipal waste;
2. EIP with the flows of material and energy resources focused on industrial waste and by-products;
3. EIP of combined type.

Depending on the availability of a key organization that generates most of the flows of material and energy resources, eco-industrial parks can be divided into:

1. EIP without anchor resident;
2. EIP with an anchor resident.

Depending on the level of localization, eco-industrial parks can be divided into:

1. Federal EIP focused on the production of final products from secondary raw materials, supplied from regional eco-industrial parks with a total capacity of at least 1.5 million tons.
2. Regional EIP focused on minimization of waste disposal by maximizing the selection of useful fractions, disposing of certain types of waste and neutralizing organic, medical, biological and other waste depending on the specifics of the subject (the proportion of decontamination and disposal is 60-80%). It is a union of operators for processing, disposal, neutralization, disposal of waste., and also an area of activity of the regional operator.
3. Corporate or urban EIP, primarily focused on addressing the relevant problems of the urban community.

3. Increasing waste management efficiency through the implementation of the project approach for the development of eco-industrial parks in the Arctic zone of Russia

As it was noted in the introduction, use of the project approach for the development of eco-industrial parks in the Arctic zone of Russia has prospective both at the stages of EIP creation and EIP functioning improvement (Figure 2).

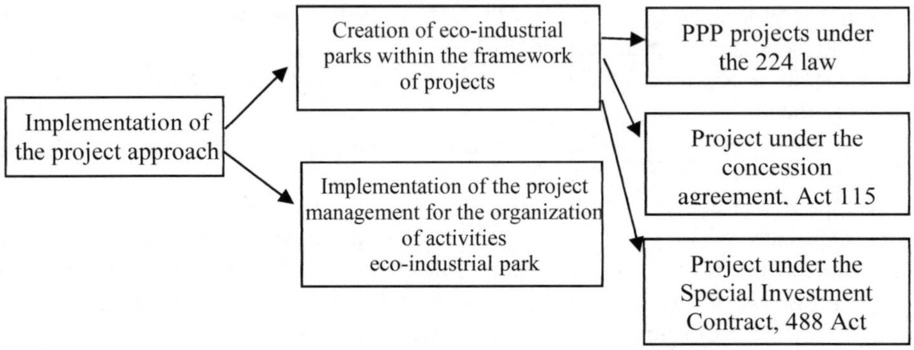

Figure 2. Directions for the implementation of the project approach for increasing environmental safety through the development of eco-industrial parks in the Arctic zone of Russia. Source: compiled by authors.

The creation of eco-industrial parks on the territory of the Arctic zone of Russia can be implemented within the framework of projects of public-private partnership, concession projects, special investment contracts.

Thus, the total volume of investments in the sphere of functioning of eco-industrial parks is 12%, taking into account the need to provide a logistical relationship between the objects of the park at the level of 20%.

At the same time, only a few concession agreements with regional authorities have been concluded in the Arctic zone of Russia, and the territory of implementation of these agreements is limited by Murmansk and Arkhangelsk regions.

Murmansk region has the most favorable investment climate, and the best practice of applying the project approach for the development of the region's infrastructure. Only in 2017, the region's authorities concluded 4 concession agreements and 1 special investment contract. By now, there are 16 agreements and contracts in the region with 107 billion rubles of total investments and 18,000 of new jobs created.

The project for development of eco-industrial park in the Murmansk region involves the territory of Murmansk city, municipalities of Severomorsk, Aleksandrovsk, Vidyayevo, Zaozersk, and Kola and Pechenga districts. The Government of the Murmansk region, as the concessionaire, provides the facilities created for the project to JSC "Waste Management" for 40 years.

The project includes formation by the end of 2018 a landfill with capacity of 250 thousand tons per year, waste sorting complex with capacity of 180 thousand tons per year, and waste processing stations with total capacity of 70 thousand tons per year. By now, in the framework of the project, 1.86 billion rubles were invested in solving environmental problems in the region.

As a result of the project, it is planned to close 5 landfills, launch a mandatory waste treatment system, including stages from collection to processing and disposal, and monitoring of illegal disposal of solid waste.

The facilities in the framework of the project are fully comply with the requirements of the Russian legislation in the field of environmental protection and ensure minimal impact on the ecosystem of the Murmansk region. The project uses Russian technologies and equipment that take into account the severe climatic characteristics of the Murmansk region.

To calculate the environmental and economic efficiency of project's implementation, the following formula can be used within the framework of the mentioned mechanisms:

$$BE = \sum_{i=1}^{t} \frac{D_k^t + D_p^t - RB}{(1+Sr)^t} \tag{1}$$

where Dk – expected revenues of the regional budget of the Arctic zone of Russia from the functioning of the eco-industrial park for each year of the calculation period, thousand rubles;

Dp – savings of the budget of the region of the Arctic zone of Russia from the functioning of the eco-industrial park for each year of the calculation period, thousand rubles;

RB – expenditures of the budget of the region of the Arctic zone of Russia from the functioning of the eco-industrial park for each year of the calculation period, thousand rubles;

Sr – relevant rate of refinancing established by the Central Bank of the Russian Federation;

t – the estimated period for the budgetary effect, from tmin = 1 year till tmax = planned period of operation and maintenance of the eco-industrial park within the framework of a specific project implemented in the Arctic region of Russia.

Project of creation or development of eco-industrial park at the territory of the Arctic zone of Russia could be considered effective if the result of calculation of the budgetary efficiency is equal or more than 1.

In this context, the project is a set of interrelated activities aimed at recycling or processing a certain type of waste under time and resource constraints to obtain a measurable result by the project team (temporary organizational structure).

Each project is considered as an object of management, which requires planning, organization and control of the labor, financial and logistical resources, used to effectively achieve the project objectives [8].

The use of project management for the development of eco-industrial parks is seen as the most effective way of organizing the EIP activities since each project is an innovative business unit [9, 10]. The project team achieves the results through the application of modern methods, instruments and technologies. Thus, the implementation of each project is a transition to a new qualitative level of EIP development.

The project team is formed from the number of participants of the eco-industrial park. EIP participants are residents (tenants), as well as representatives of local (regional) authorities, expert communities, industry associations, unions, research institutes and other interested parties.

A resident (tenant) of an eco-industrial park is an enterprise involved in industrial symbiosis and (or) using the integrated services of an eco-industrial park (administrative, logistic, etc. infrastructure). If there is an anchor resident in the eco-industrial park, it's of a great use to include the manager of this enterprise into the project team. The head of the project team could be the manager of an anchor resident or any other resident (enterprise). For each project it's determined individually.

The implementation of the project approach for the development of eco-industrial parks presupposes the presence of a project office in EIP structure, as a unit responsible for methodological and organizational support of project management in the eco-industrial park, planning and monitoring the project portfolio, implementation and development of the information system for collection, accumulation of the data and reporting on projects.

The implementation of the project management for EIP development is presented in Figure 3.

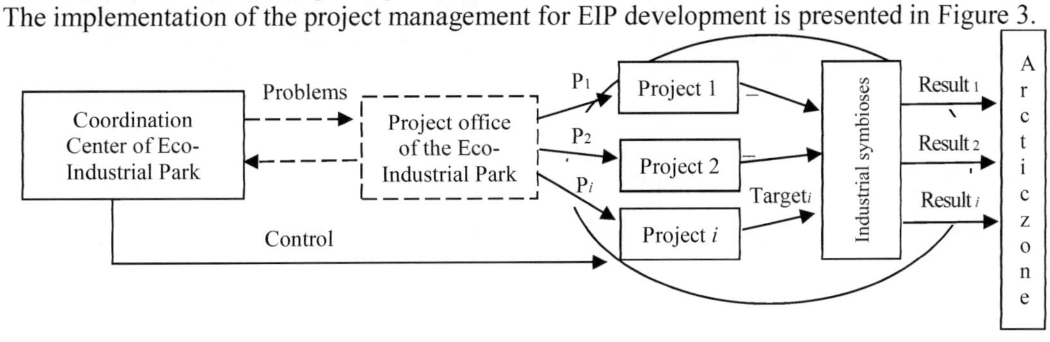

Figure 3. The Concept of Project Management Implementation for Eco-industrial Parks Development. Source: compiled by authors.

IOP Conf. Series: Earth and Environmental Science **337** (2019) 012024 doi:10.1088/1755-1315/337/1/012024

Projects focused on solving of similar or interrelated tasks can be combined into a program, a set of projects and form the portfolio of the eco-industrial park.

4. Challenges of the implementation of the project approach to increasing environmental safety in the Arctic zone of Russia based on eco-industrial parks

The main challenges for the implementation of the project approach for the development of eco-industrial parks on the territory of the Arctic zone of Russia are the following (Table 2):

1. Severe climate conditions;
2. Demographic features;
3. Overpriced resources, lack of supply of resources;
4. Remoteness from the main production and distribution markets;
5. Lack of institutional environment in the regions for the implementation of projects for the EIP creation.

Table 2. The main challenges of the implementation of the project approach to increasing environmental safety in the Arctic zone of Russia.

Challenge	Description	Methods of overcoming
Severe climate conditions affecting the requirements for projects and increasing the amount of investments	Low temperatures Large amount of precipitation, especially snow Short construction season Difficulties in selecting a land plot for a landfill that complies with legislation	State co-financing of capital-intensive projects and using of special funds
Geography affecting the accessibility of project sites	High percentage of island territories Large distances between settlements Vast uninhabited territories	Legislative support for the using waste as secondary material and energy resources
Conditions of scarcity and overpricing on resources affecting the accessibility of materials and labour	Decreasing population Lack of qualified personnel High investment and operating costs Higher prices for fuels, payroll, additional social benefits	Tax privileges and preferences for state procurement for enterprises using secondary material and energy resources (VAT, property tax)
Remoteness from the main production and distribution markets	Limited access to markets for by-products High transport leverage and unprofitable high rates for waste processing	Implementation of green tariffs for transporting waste to disposal sites
Lack of institutional environment in the regions for the implementation of projects for the creation of eco-industrial parks	The authorized body in the sphere of PPP, concessions and special investment contracts is not defined The specialized information resource in the sphere of PPP, concessions and special investment contracts has not been created	Expanding the institutional environment established in the Murmansk region to all Arctic zone of Russia
Qualification requirements for participants in the implementation of the project approach	Lack of specialists with project competencies (the type of professional competencies that allow to implement project management in accordance with the role in the project)	Recruiting external specialists Training under supplementary education programs

Source: compiled by authors based on [11]

IOP Conf. Series: Earth and Environmental Science **337** (2019) 012024 doi:10.1088/1755-1315/337/1/012024

5. Conclusions

Currently, the development of eco-industrial parks in the Arctic regions of Russia is at the stage of formation. The expected results of their functioning include the following:

- improving environmental situation;
- increasing share of recycled waste and using by-products;
- growing number of jobs for qualified specialists;
- increasing revenue side of the budgets of the Arctic regions.

To improve the manageability and efficiency of the eco-industrial parks, their functioning can be advanced within the framework of the project approach. The positive experience of implementation of this concept was obtained in the eco-industrial park of the Murmansk region.

At the same time, the distribution of this type of management under study to eco-industrial parks may face a number of difficulties, including climate and location challenges, absence of project competencies among potential project participants (regional authorities and EIP residents' employees).

References

[1] Perez E, Yaneva Z 2016 *Pol. Scien. J.* **10** pp 441-9
[2] Werner K et al 2016 *Pol. Scien. J.* **10** pp 364-73
[3] Sokolov I 2017 *Ecol. Econ. J.* **2** (10) pp 18-27
[4] Eco-industrial parks of Russia 2018 *Mun. Sol. Waste J.* **1** (55)
[5] Mariev V, Smirnova T 2017 *Mun. Sol. Waste J.* **2** pp 14–7
[6] Eco-industrial parks of Russia 2018 *Mun. Sol. Waste J.* **1** (56)
[7] Borisov V, Pochukaeva O 2017 *Studies on Russian Econ. Devel. J.*
[8] Mariev V, Smirnova T 2017 *Mun. Sol. Waste J.* **3** pp 21–3
[9] Ernstson H, Sorlin S 2013 *Ecol. Econ. J.* **86** pp 274-84
[10] Gibbs D, Deuts P 2007 *Clean. Prod. J.* **15, 17** pp 1683-95
[11] Hintsala H, Niemela S, Tervonen P 2016 *Pol. Scien. J.* **10** pp 450-7

IOP Conf. Series: Earth and Environmental Science **337** (2019) 012025 doi:10.1088/1755-1315/337/1/012025

A study of the chemical composition and antioxidant properties of products of wild berries processing

L Nilova[1]*, S Malyutenkova[1] and R Ikramov[1]

[1]Institute of Industrial Management, Economics and Trade, Peter the Great St. Petersburg Polytechnic University, 195251, Saint-Petersburg, Russia

* Corresponding author: nilova_l_p@mail.ru

Abstract. The article presents the results of studies of the chemical composition and composition of individual antioxidants of bilberries and lingonberries and their marc, which are waste products after squeezing the juice from the berries. The yield of juice from berries averaged 64%. Dry substances, acidity, composition of sugars, pectin substances and individual antioxidants - total phenolic compounds, total flavonoids, total anthocyanins were determined in berries and in the marc. During squeezing of the juice, a redistribution of chemical components was occurred with a predominance of total phenolic compounds, total flavonoids and total anthocyanins in the marc. As a result, the marc showed higher antioxidant properties than juices, which were investigated using the DPPH and FRAP methods. The antioxidant properties of the marc are due to the transition in them of anthocyanins, which make up 64% and 59%, respectively, of their total amount in bilberries and lingonberries. The correlation between DPPH and FRAP tests for anthocyanins was more than 0.952.

1. Introduction

Wild berries growing in northern latitudes can be an additional source of biologically active substances in human nutrition. Bilberry, lingonberry, blueberry, cranberry, and cloudberry grow in the forests of Russia, Finland, Norway, and Sweden. Recent studies have shown that these berries contain significant amounts of polyphenols, anthocyanins, hydroxycinnamic and hydroxybenzoic acids, which are natural antioxidants [1-5]. Their qualitative and quantitative composition varies widely depending on the type of berries, place of growth, cropping period, storage time, and processing - freezing and thermal heating. Wild berries contain more biologically active substances than their cultivated forms [6-8]. The quantitative composition of phenolic compounds between freshly harvested berries and berries sold in retail can vary by about 10 times [2, 4, 6, 9]. Thus, in wild berries sold in the consumer market in Sweden, the content of total phenolic compounds is 253.6 mg/100 g DW in bilberry, and 219.7 mg 100 g DW in lingonberry, while in freshly picked berries it can reach 3050 and 1350 mg/100 g DW, respectively [2, 4, 10, 11]. Processing berries in contact with oxygen leads to the loss of total phenolic compounds and flavonoids, while the anthocyanins are more stable [10, 12, 13]. Heat treatment of food using wild berries and their derivative products can lead to an increase in the antioxidant properties of the product being developed due to the synergism of biologically active substances and melanoidins [14]. To preserve the biologically active substances of extracts of berries, innovative encapsulation systems using natural whey proteins have been proposed [15]. A high content of biologically active substances is characteristic not only of berries, but also of leaves, stems, and sprouts, which are used medicinally [4, 11, 16].

It has been established that consumption in the diet of wild berries with a high content of fats contributes to their better metabolism by gut microflora, and prevents lipid peroxidation [17]. Due to the presence of natural antioxidants, especially anthocyanins, the consumption of fresh berries or dry preparations from them contributes to the prevention of alimentary dependent diseases, the main of which being cardiovascular and oncological ones [3, 18-20]. The pharmacological action of bilberry is the most widely studied. To obtain pharmacological preparations, bilberry extracts are used [13]. However, the usual extraction does not allow for the complete transfer of biologically active substances contained in the peel and seeds into the extract. For this purpose, cavitation, fermentation, and production of CO_2-extracts are used for extraction [3, 21-23]. Fermentation of berries with saccharomycete allows obtaining additives from them with a high content of sugars while enhancing antimicrobial properties [9, 21]. It has been established that phenolic compounds and ellagotannin extracts of wild berries inhibit the development of Staphylococcus aureus and Bacillus cereus [6].

Seasonality of berry production has led to the growing need to obtain food from them with a long shelf life. In addition to traditional jams, confitures, fruit drinks, juices are obtained from wild berries, which are used as one of the components in fermented dairy products and mixed juices. The marc, remaining expression of juices is not used in food production.

The aim of the research was to study the chemical composition and antioxidant properties of the wild berries marc that remained after expressing the juice.

2. Materials and methods

2.1. Research objects

The most common wild berries of the northwestern federal district of Russia of the heather bloodline of the Vaccínium genus — blueberry and lingonberry — were chosen as objects of research. Berries were collected in the territory of the Vyborg district of the Leningrad region. The juice was obtained from the berries using an electromechanical-fy machine for expressing juice EKM-3, Electrosila JSC, St. Petersburg, Russia. The obtained juice was filtered to separate the pulp. Berries and marc after expressing the juice were used for research..

2.2. Research methods

2.2.1. Chemical composition research methods

The mass fraction of moisture was determined by the complete desiccation method at a temperature of 60 °C; titrated acids - by a titration method after preliminary extraction on a water bath; pectic substances - by the Metlitz method, the mass fraction of fiber - by the gravimetric method after preliminary hydrolysis [24].

The mass fraction of the sugars composition was determined on an Agilent 1260 Infinity II liquid chromatograph (USA). Capillary column - Bio-Rad Carbohydrates 300 mm x 7.8 mm. The mobile phase is bidistilled water, the flow rate of the mobile phase was 0.6 ml/min; the column temperature was 80 °C; the detection was performed using a refractometric detector; isocratic mode., the volume of the injected sample is 1 ml. Solutions of sucrose, glucose, and fructose in bidistilled water were internal standard. The results were processed using the Agilent 1260 Infinity II tool software.

2.2.2. Research methods for individual antioxidants

The determination of ascorbic acid was performed by a titrimetric method with a solution of sodium 2,6-dichlorophenolindophenolate, after preliminary extraction of ascorbic acid with a 2% hydrochloric acid solution [24].

Total phenols assay by Folin–Ciocalteau reagent. Ethanol extracts of the marc were kept in the dark with Folin – Ciocalteau reagent for 30 minutes at room temperature. Optical density was measured on a SHIMADZU 1240 spectrophotometer (SHIMADZU, Japan) at a wavelength of 735 nm. The results obtained are expressed in mg of gallic acid (GA) [4].

IOP Conf. Series: Earth and Environmental Science **337** (2019) 012025 doi:10.1088/1755-1315/337/1/012025

Total flavonoids assay was determined spectrophotometrically by reaction with aluminum chloride. Extraction of flavonoids was carried out with 60% ethyl alcohol. Optical density measurement was carried out after 30 minutes on a SHIMADZU 1240 spectrophotometer (SHIMADZU, Japan) at a wavelength of 420 nm. The results obtained were expressed in mg of rutin [24].

Total anthocyanins assay expressed in terms of cyanidine was determined by pH-differential spectrophotometry of samples at pH 1.0 and 4.5 at a wavelength of 700 nm and maximum values in the visible area on a SHIMADZU 1240 spectrophotometer (SHIMADZU, Japan).

2.2.3. Antioxidant properties research methods

Antiradical activity was determined by Glewind [24]. Extraction was carried out with a 50% ethanol solution. AOA was determined using a diphenylpycrylhydrazyl (DPPH) stable free radical, which is recovered in the reaction with the antioxidants of the product for 30 minutes in the dark at room temperature. The light absorption of the original solution of DPPH is 0.5 relative units. Optical density was measured at a wavelength of 517 nm on a SHIMADZU 1240 spectrophotometer (SHIMADZU, Japan). The calibration curve was built by ascorbic acid (AC).

Antioxidant activity (chelating capacity) was determined by the FRAP method with ferric chloride, *o*-phenanthroline and triton X 100 on a SHIMADZU 1240 spectrophotometer at a wavelength of 505 nm [24]. The calibration curve was built by AC.

3. Results and discussion

The yield of juice from bilberry and lingonberry was almost the same and was about 64%. The share of marc accounted for about 20%, while the rest was pulp (Figure 1). Bilberry and lingonberry are real berries, so the juice with pulp contained seeds with a more noticeable amount of them in the bilberry pulp.

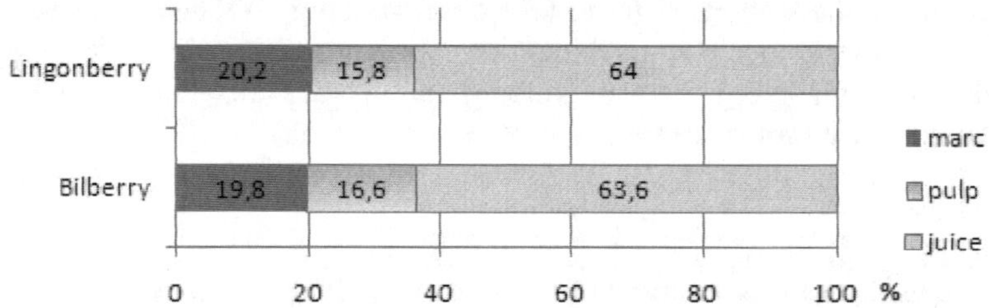

Figure 1. The ratio of constituent parts of wild berries, %

Further research was carried out in whole berries and marc after juice expression. The obtained marc contained 2.6-2.7 times dry solids than the juice (Table 1). Dry solids of the marc predominantly contained cellulose, reducing sugars and titrated acids. Pectic substances passed into the juice with pulp, while the marc contained their trace quantities.

The total reducing sugars assay in bilberry was 1.8 times higher than in lingonberry. In berries, regardless of their type, fructose and glucose prevailed. The amount of sucrose was not material and did not exceed 1.0% of dry solids. When processing berries, the distribution of sugars in the constituent parts depended on their type. In bilberry marc, as well as in berries, fructose and glucose prevailed. In lingonberry marc, the content of glucose and sucrose increased, and fructose decreased quantitatively, which indicates its transition to juice and pulp. As a result, glucose was the dominant sugar. The distribution of titrated acids was also uneven. In the marc of bilberry and lingonberry, the titrated acidity decreased by almost 2 times. At the same time, titrated acidity was 1.5 times higher in the lingonberry marc than in bilberry marc.

Table 1. The chemical composition of berries and marc after juice expression

Indicators	Bilberry		Lingonberry	
	whole berry	marc	whole berry	marc
Dry matter, %	18.2±0.4	49.7±0.6	18.8±0.5	48.9±0.5
Sucrose, % DM	1.00±0.02	0.5±0.02	0.87±0.01	1.29±0.02
Glucose, % DM	11.50±0.10	6.9±0.20	5.67±0.20	10.77±0.30
Fructose, % DM	13.00±0.42	10.0±0.30	7.5±0.32	6.96±0.30
Acidity, % on citric acid	1.50±0.01	0.76±0.01	2,1±0.01	1,1±0.03
Pectic substances, % DM	3.74±0.12	0.01±0.01	3.35±0.10	0.01±0.01
Cellulose, % DM	8.62±0.28	34.9±0.45	8.99±0.23	36.9±0.52

Wild berries are considered as sources of antioxidants containing a complex of compounds of phenolic nature [2, 4, 10, 11]. The study of the composition of individual antioxidants showed that bilberry and lingonberry contain total phenolic compounds, common flavonoids, total anthocyanins and vitamin C in their composition, with their predominance in bilberry (Table 2). And while the amount of total phenolic compounds and flavonoids in bilberry is 1.4 and 1.2 times more than in lingonberry, respectively, the amount of anthocyanins is more than 2 times greater in bilberry, causing its dark purple color. But the berries of lingonberry contained more vitamin C.

Table 2. Individual antioxidants assay

Indicators	Bilberry		Lingonberry	
	whole berry	marc	whole berry	marc
Total phenolic compounds, mg GA/ 100 g	588.9±12.6	682.4±10.9	425.5±14.5	508.1±10.1
Total flavonoids, mg rutin / 100 g	465.0±11.4	510.2±10.5	371.2±11.8	450.4±7.3
Total anthocyanins, mg cyanidin / 100 g	313.0±8.8	514.8±8.5	182.5±9.0	207.4±5.3
Vitamin C, mg / 100 g	18.34±0.62	6.88±0.53	22.10±0.60	7.25±0.55

When expressing juices, a significant part of the individual antioxidants was transferred into the marc. Thus, the amount of total phenolic compounds in marc increased by 15.9 and 19.4%, respectively, of their content in bilberry and lingonberry. Marc contained less total flavonoids than total phenolic compounds, especially the bilberry marc, which was only 9.7% more. Total anthocyanins, which are glycosides, often associated with the matrix of plant materials, have passed into the marc most of all. In the bilberry marc, their number was 64.5% more than in berries, which was more than 2.6 times than in the lingonberry marc. The content of vitamin C both in the berries and marc was insignificant, regardless of the type of berries.

The presence of individual antioxidants determined their antioxidant properties, which were determined by their antiradical capacity with respect to the DPPH-radical and antioxidant capacity by their chelating capacity using the FRAP method (Figure 2).

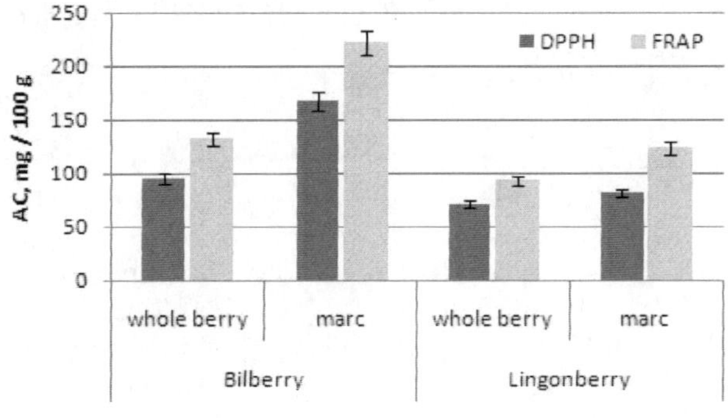

Figure 2. Antioxidant capacity of wild berries and marc (mg AC/ 100 g)

Bilberry and lingonberry marc had greater antioxidant properties compared with berries. The antiradical activity of bilberry marc, determined by the DPRH-test and the chelation capacity by the FRAP method, was 1.7-1.8 times higher than in berries. In lingonberry, the differences between berries and the marc were less significant. Antiradical activity differed only by 14%, and chelating capacity - by 33.6%.

The greatest influence on the antioxidant properties of berries and marc was made by common anthocyanins. The correlation between the DPRH-test and the content of anthocyanins approached to 1, amounting to 0.977 with an acceptance probability of 0.95. The number of anthocyanins influenced the chelating capacity with a high degree of probability (Table 3). A similar dependence of the transition of anthocyanins and their effect on antioxidant properties was also obtained by other authors [7, 8], both with whole berries from different regions and with different fractions of berries obtained by extraction.

Table 3. Correlation (R^2) between total phenolic compounds, total flavonoids and total anthocyanins of antioxidant capacity DPPH and FRAP

Indicators	Total phenolic compounds	Total flavonoids	Total anthocyanins	P-value
DPPH	0.829	0.682	0.977	$< 0{,}05$
FRAP	0.869	0.777	0.952	$< 0{,}05$

The effect of total phenolic compounds on the antioxidant properties of berries and marc also turned out to be great and ranged from 0.829 to 0.869, but was more associated with chelating capacity. Common flavonoids had the least effect on antioxidant properties, especially on antiradical activity. The correlation coefficient, in this case, was the lowest - 0.682.

4. Conclusion

Wild bilberry and lingonberry are natural sources of antioxidants that can be used to prevent oxidative stress when used fresh or processed. In the process of juice production, about 20% of marc is formed, into which antioxidants transfer, forming their antiradical activity and chelating capacity. The total phenolic compounds, total flavonoids and total anthocyanins assay in the marc is higher than in whole berries. Total anthocyanins transfer to the marc most of all, the amount of them in marc is 64.5% more than in the berries causing dark purple color. Lindonberry marc contains 2.6 times less common anthocyanins than bilberry marc. The more bilberry and lingonberry and the products of their processing contained anthocyanins, the higher their antioxidant properties were. The correlation between DPPH and FRAP- tests for anthocyanins was 0.977 and 0.952, respectively.

References

[1] Häkkinen S, Heinonen M, Kärenlampi S, Mykkänen H, Ruuskanen J, Törrönen R 1999 *Food Research International* **32** pp 345-53
[2] Hajazimi E, Landberg R, Zamaratskaia G 2016 *LWT - Food Science and Technology* **74** pp 128-34
[3] Fan Z-L, Wang Z-Y, Liu J-R 2011 *Food Chemistry* **129** pp 402-7
[4] Bujor O-C, Ginies Ch, Popa V I, Dufour C 2018 *Food Chemistry* **252** pp 356-65
[5] Ancillotti C, Ciofi L, Pucci D, Sagona E, Giordani E, Biricolti S, Gori M, Petrucci W A, Giardi F, Bartoletti R, Chiuminatto U, Orlandini S, Mosti S, Del Bubba M 2016 *Food Chemistry* **204** pp 176–84
[6] Tian Ye, Puganen A, Alakomi H-L, Uusitupa A, Saarela M 2018 *Food Research International* **106** pp 291–303
[7] Colak N, Torun H, Gruz J, Strnad M, Hermosin-Gutierrez I, Hayirlioglu-Ayaz S, Ayaz F A 2016 *Food Chemistry* **201** pp 339–49
[8] Feng Ch, Su Sh, Wang L, Wu J, Tang Z, Xu Ya, Shu Q, Wang L 2016 *Food Chemistry* **204** pp 150-8

[9] Puišo Ju, Jonkuvienė D, Mačionienė I, Šalomskienė J 2014 Colloids and Surfaces B *Biointerfaces* **121** pp 214–21
[10] Bornšek Š M, Polak T, Skrt M, Demšar L, Ulrih N P, Abram V 2015 *Food Chemistry* **173** pp 61–9
[11] Bujor O-C, Le Bourvellec C, Volf I, Popa V I, Dufour C 2016 *Food Chemistry* **213** pp 58–68
[12] Nilova L, Malyutenkova S 2017 *Agronomy Research* **16** pp 1444-56
[13] Śaponjac V T, Čanadanović-Brunet Ja, Ćetkovic G, Djilas S, Cetojević-Simin D 2015 *LWT - Food Science and Technology* **61** pp 615-21
[14] Nilova L P, Pilipenko T V 2016 *Problems of Nutrition* **85** (6) pp 39-47
[15] Betz M, Steiner B, Schantz M, Oidtmann J, Mader K, Richling E, Kulozik U 2012 *Food Research International* **47** pp 51–7
[16] Feng Ch-Yo, Wang W-W, Ye J-F, Li Sh-Sh, Wu Q, Yin D-D, Li B, Xu Ya-Ju, Wang L-Sh 2017 *Food Chemistry* **219** pp 490–5
[17] Lehtonen H M, Lindstedt A, Jarvinen R, Sinkkonen Ja, Graca G, Viitanen M, Kallio H, Gil A M 2013 *Food Chemistry* **138** pp 982-90
[18] Wang S, Meckling K A, Marcone M F, Kakuda Yu, Tsao R 2011 *Food Research International* **44** pp 2545–54
[19] Kivima A S, Ehlers P I, Turpeinen A M, Vapaatalo H, Korpela R 2011 *J. of Functional Foods* 3 pp 267-74
[20] Leon-Gonzalez A J, Sharif T, Auger C, Abbas M, Fuhrmann G, Schini-Kerth V B 2018 *J. of Functional Foods* **44** pp 227–34
[21] Viljanen K, Heinio R-L, Juvonen R, Kosso T, Puupponen-Pimia R 2014 *Food Chemistry* **157** pp 148-56
[22] Fatkullin R, Popova N, Kalinina I, Botvinnikova V 2017 *Agronomy Research* **16** pp 1295-303
[23] Babova O, Occhipinti A, Capuzzo A, Maffei M E 2016 *J. of Supercritical Fluids* **107** pp 358–63
[24] Rogozhin V V, Rogozhina T V 2015 *Workshop on biochemistry of agricultural production* (St. Petersburg: GIORD)
[25] Salishcheva O V, Donya D V 2013 *Foods and Raw Materials* **1** (2) pp 76–84

IOP Conf. Series: Earth and Environmental Science **337** (2019) 012026 doi:10.1088/1755-1315/337/1/012026

Development of territorial scheme of municipal solid wastes treatment in Tyumen region

E Gaevaya[1], S Tarasova[1*] and O Udartseva[1]

[1]FSBEI of Higher Education «Tyumen industrial university», Lunacharskogo St., 2, Tyumen, 625001, Russia

* Corresponding author: tarasovass@tyuiu.ru

Abstract. At the present time the majority of wastes have negative influence on the environment. At the same time, about 70% of wastes can be used repeatedly, as raw material, transforming them into new products and materials. Russia possesses the existing and commissioned waste recycling and waste burning plants. But only 3-15% of municipal solid waste are utilized. Moreover, many of these plants are not loaded at full capacity or have to buy the wastes abroad. This is due to the fact that Russia still doesn't have effective system of wastes treatment – separate collection and following utilization or waste neutralization. To prevent the increase of areas of used municipal solid waste fields and unauthorized dumps, it's required to track the full chain of wastes movement from the place of their formation to the place of their dumping.

1. Introduction

Industrial and consumption wastes (hereinafter - wastes) are the substances or subjects which are formed in the process of production, works and services or in the process of consumption which are removed, are intended for removal or subjected to removal according to current Federal law [1].

Depending on the degree of negative influence on the environment, the wastes are divided according to the criterion established by federal body of legislative authority, carrying out state regulation in the sphere of environment protection. They are divided into five classes of hazards [1,2].

Storage of industrial and consumption wastes is allowed on: the production territory of the main producers (manufacturers) of wastes; reception centers for collection of secondary raw materials; territory and in the buildings of socialized plants on recycling and neutralization of toxic wastes on open specially equipped platforms [2].

Collection of wastes on the production territory is intended for selective collection and accumulation of separate types of wastes; using of wastes in the following technological process with a purpose of neutralization, partial or full, and utilization on the auxiliary productions [1,3].

Collection of wastes is carried out according to the requirements of Sanitary Rules and Norms 2.1.7.1322 "Hygienic requirements to the placement and neutralization of industrial and consumption wastes" [4]. For the waste collection it's provided the installation of container-type waste bins with covers mounted on specially equipped site with solid artificial waterproof coating (reinforced-concrete slabs) and efficient protection from wind and atmospheric precipitations with the observance of unhindered vehicular access for their loading and removal to the object of placement and utilization. Storage of wastes on the unprotected ground is not allowed [5,6,7].

Processes of waste treatment (wastes life cycle) include the following stages: formation, accumulation and storage, primary processing (sorting, dehydration, neutralization, pressing, calibration etc.), transportation, secondary recycling (neutralization, modification, utilization, using as secondary raw material), storage, burial and burning [8,9,10].

Waste burial is applied to noncombustible refuse and such wastes which evolve toxic substances during the burning process.

Wastes field (municipal solid waste) is not an ordinary dump. Modern fields for utilization are complex engineering structures equipped with systems of fight with underground waters pollution and atmospheric air [11,12,13]. Some fields can process the gas formed in the process of wastes decomposition into electric energy and heat. Unfortunately, today this mostly refers to European countries, since in Russia very small percentage of fields corresponds to these characteristics [14,15,16].

The main minus of traditional wastes burial consists in that even using numerous systems of cleaning and filtration, this utilization type doesn't give possibility to get rid of such negative effects of wastes decomposition as putrefaction and fermentation which pollute air and water [17,18].

Seriousness of influence of waste processing and burial on the environment depends on the volume of produced wastes, their composition, amount of illegally buried wastes, number of wastes placed on the dump and standards on the plants of wastes processing. Future influence of waste management process will depend on how the indicated factors change. The final processing of wastes means the burial on the dump or burning. And these two types of final processing have an influence but it's negative in both cases [19].

Placement of wastes on the dumps leads to the methane emanation – one of the greenhouse gases and dangerous chemical substances which have harmful influence on the environment.

With the course of time, while the wastes decompose, they sag. At the same time shallow cavities appear. The water accumulates in them and the whole plot turns into marsh with poisonous water. For the periodic control for the quality of ground waters, monitoring wells are made along the dump perimeter [21].

2. Methods and conditions

For the development of territorial scheme there were taken into account the volumes of municipal solid wastes formation for 26 municipal regions and urban districts of Tyumen region, presented in the Figure 1.

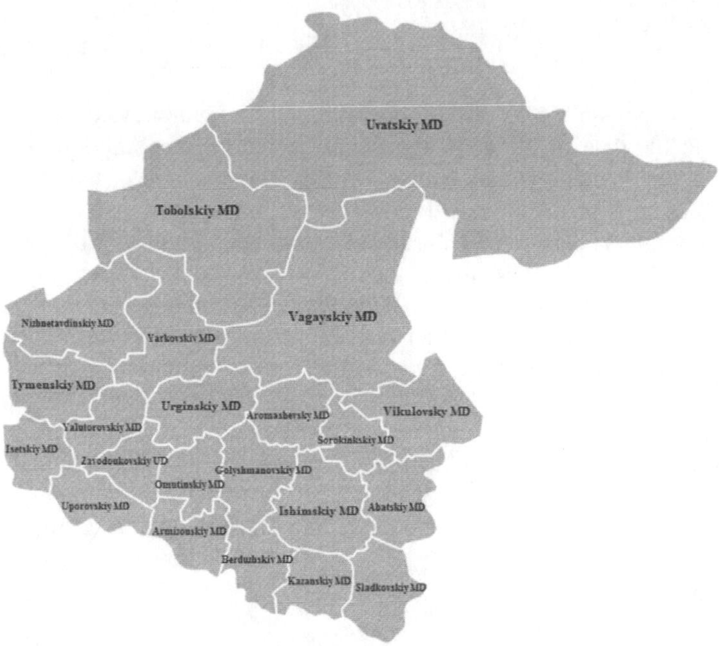

Figure 1. Map of municipal districts and urban districts of Tyumen region

IOP Conf. Series: Earth and Environmental Science **337** (2019) 012026 doi:10.1088/1755-1315/337/1/012026

To study the dynamics of change of recycled wastes volumes, there were used actual data about the volumes of municipal solid wastes formation, appeared as the result of daily activities of the population for the following types of wastes: wastes from houses with joint collection; unsorted wastes from houses (except large-scale ones); rubbish and street sweeping; wastes (rubbish) from building and repair works; wastes of paper and cardboard from office activity and paper work; large-scale wastes from houses; wastes of unpolluted polypropylene package [22,23].

Places of disposal of municipal solid wastes in Tyumen region were presented by the following objects: MSW fields; authorized dumps; unauthorized dumps.

In the studied municipal regions there were observed unauthorized dumps, except Abatskiy, Vikulovskiy, Isentsiy and Yalutorovskiy municipal districts and Tyumen.

To determine the dynamics of municipal solid wastes formation in every residential area of municipal regions and urban districts there was calculated demographic population forecast on which the required index depends, it's determined by the formula 1:

$$P_t = P_0 \times \Delta t \qquad (1)$$

where P_t –forecasted level of population number; P_0 –basic level of population number; Δ – average annual increase of population; t – forecasting period.

According to the forecast of municipal solid wastes formation by residential areas for the period from 2016 to 2026, some municipal districts observe decrease of population that leads to the decrease of municipal solid wastes formation. Graphical data is given in the figure 2.

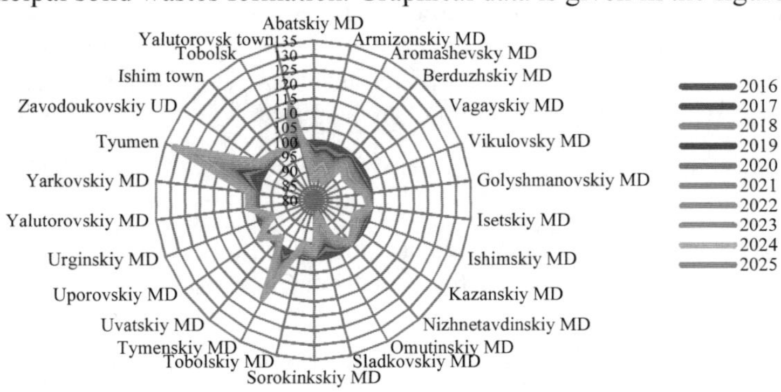

Figure 2. Dynamics of population decrease and formation of municipal solid wastes in municipal districts of Tyumen region

Maximal decrease of population by 2016 will take place in Sladkovskiy municipal district – by 14.90% with regard to the data of 2016. Insignificant decrease of population size and formation of municipal solid wastes will be observed in Tobolsk by 0.05%.

Significant dynamics of population growth and formation of municipal solid wastes in municipal districts of Tyumen region by 2026 is observed in Tyumen municipal district and makes up 18.36% regarding the same indices of 2016. The lowest growth is observed in Ishim town and it will be 1.61% by 2026. Maximal value 31.80% is forecasted in Tyumen.

For the collection, accumulation and further operations on treatment with municipal solid wastes, there will be built 19 waste transfer stations and 10 sites for accumulation with solid waste compactor by the end of 2026 in Tyumen region.

At the present time, territory of Tyumen region possesses 21 objects of municipal solid wastes disposal which correspond to the requirements of effective legislation and are included in SRWDS.

During 2018:

1) complete transfer of municipal wastes to the waste recycling plants is planned in the following regions:

- Abatsky municipal district;
- Ishimsky municipal district;
- Ishim town;
- Nizhnetavdinsky municipal district;

- Tobolsk municipal district;
- Tobolsk;
- Tyumen Municipal District;
- Tyumen;
- Yarkovsky municipal district;
- Yalutorovsk;
- Yalutorovsk municipal district

2) partial transfer of municipal solid plants and fields of MSW is planned in the following regions:

- Vagayskiy municipal district;
- Zavodoukovskiy urban district.

3) total transfer of municipal solid wastes to the MSW fields will be continued in the following regions:

- Armizonsky municipal district;
- Aromashevsky municipal district;
- Berdyuzhsky municipal district;
- Vikulovsky municipal district;
- Golyshmanovsky municipal district;
- Isetsky municipal district;
- Kazan municipal district;
- Omutinsky municipal district;
- Sladkovsky municipal district;
- Sorokinsky municipal district;
- Uvatsky municipal district;
- Uporovsky municipal district;
- Yurginsky municipal district.

In Uvatskiy municipal district transfer of wastes will be carried out to three WSM fields in Uvat village, Demyanskoe village and Pershino. Percentage ratio of residential areas transferring MSW to the fields.

33 villages transfer their wastes to the MSW fields of Uvatskiy municipal region. Amon them: 21 residential areas transfer MSW to the field of the village Uvat, 5 village settlements transfer wastes to the village Demyanskoe, 7 village settlements – to Pershino. This dynamics of distribution of the wastes transfer by the residential areas to MSW fields is determined by the geographical location relative to each other, that contributed to the transportation of wastes.

In Yarkovskiy municipal district transfer of wastes will be carried out to 3 waste recycling plants. Percentage ratio of residential areas transferring MSW to the waste recycling plants is given in figure 3.

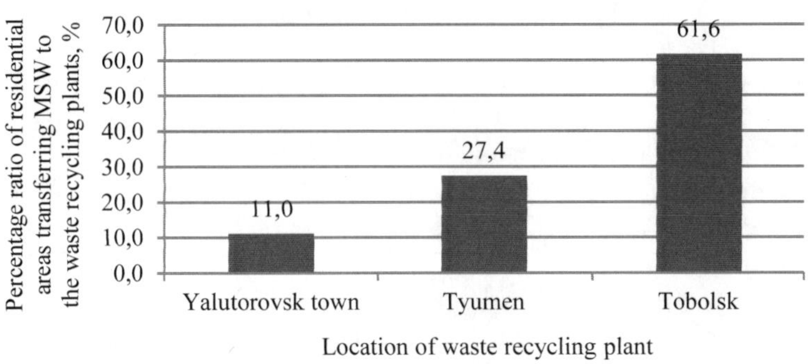

Figure 3. Percentage ratio of residential areas transferring MSW to the waste recycling plants in Yarkovskiy municipal district

To the waste recycling plant in Ylutorovsk there will be transferred MSW from 8 village settlements; to Tyumen – from 20 village settlements; to the plant in Tobolsk – from 45 residential areas. This

IOP Conf. Series: Earth and Environmental Science **337** (2019) 012026 doi:10.1088/1755-1315/337/1/012026

gradation of distribution of MSW transfer by residential areas to the waste recycling plants is stipulated by more rational transportation depending on their geographical location relative to each other.

In Vagayskiy municipal district the transfer of wastes by residential areas is carried out to the field of MSW located in Vagay village and waste recycling plant in Tobolsk. Percentage ratio is given in the figure 4.

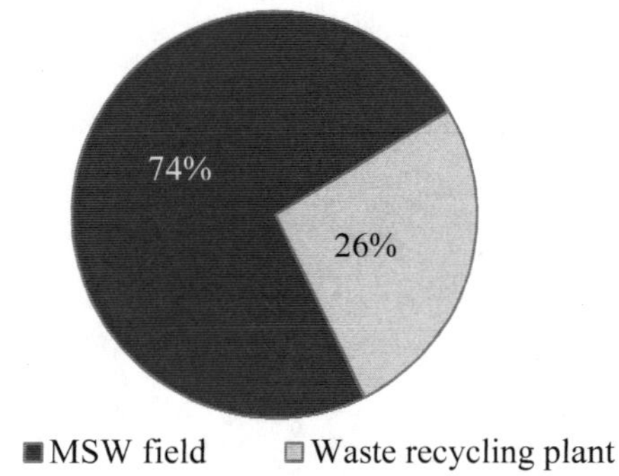

Figure 4. Percentage ratio of residential areas carried out wastes transfer to MSW field and to the waste recycling plants in Vagayskiy municipal district

Wastes from 84 residential areas are transferred to MSW field, and to the waste recycling plant – from 30 residential areas.

Analogous situation is observed in Zavodoukovskiy municipal district that is given in the figure 5.

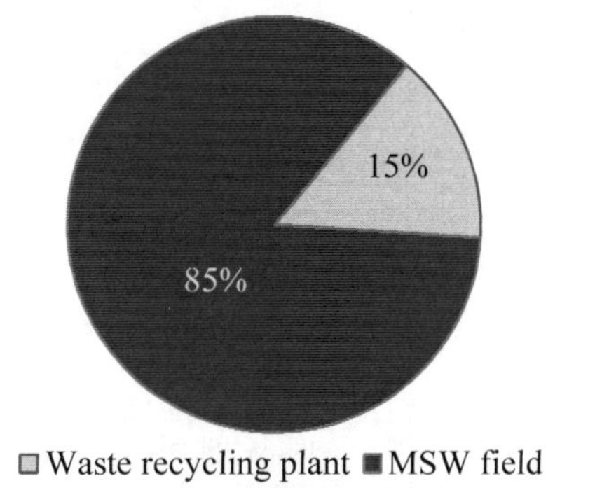

Figure 5. Percentage ratio of residential areas carrying out transfer of wastes to the MSW field and to the waste recycling plants in Zavodoukovskiy urban district.

3. Conclusions

From 47 settlements, 7 settlements transfer waste to the waste recycling plant, the remaining settlements - to the MSW field. Waste transportation to the waste recycling plant from village settlements is stipulated by their favorable geographic location relative to the plant, a small figure may indicate that the waste recycling plant has not been put into full operational capacity and cannot accept more waste for recycling.

When calculating the population for the period from 2016-2026, the number of generated MSW will be 499865.54 tons.

The objects of accumulation and disposal of municipal solid waste are identified:

- fields of MSW (14 municipal districts);
- waste recycling plants (Ishim, Yalutorovsk; Tyumen, Tobolsk);
- waste transfer stations (19 municipalities);
- sites with the presence of a press packer (7 municipal districts).

The developed territorial scheme will allow timely collection of transportation and processing of MSW, in order to improve the quality of the environment.

References

[1] *Russian Federation. Laws. On production and consumption waste* 1998 (Eksmo)

IOP Conf. Series: Earth and Environmental Science **337** (2019) 012026 doi:10.1088/1755-1315/337/1/012026

[2] *Russian Federation. Orders. On the federal waste classification Catalog* 2015
[3] *GOST 30772-2001 Resource saving. Waste management. Terms and definitions* (Standards Publishing, 2002)
[4] *Sanitary Rules and Norms 2.1.7.1322-03 Hygienic requirements for the disposal and disposal of production and consumption waste* 2002 (Standardinform)
[5] *Russian Federation. Laws. On licensing of certain types of activities* 1998 (Eksmo)
[6] Bobovich B B 2000 *Recycling of production and consumption waste* (Moscow: Intermet Engineering)
[7] Smetanin V I 2000 *Protecting the environment from industrial and consumer waste* (Moscow: Kolos)
[8] *Russian Federation. Laws. On environmental protection* 2001 (Eksmo)
[9] *Russian Federation. Laws. On licensing activities for the collection, transportation, treatment, utilization, disposal, placement of waste of I-IV hazard classes: Resolution* 2015
[10] *Russian Federation. Laws. On the sanitary-epidemiological well-being of the population* 1999 (Eksmo)
[11] *Russian Federation. Laws. On the inclusion of waste disposal facilities in the state register of waste disposal facilities* 2014
[12] Akimova T A, Kuzmin A P, Haskin V V 2001 *Ecology* (Moscow: UNITI)
[13] Arustamov E A, Levakova I V, Barkalova N V 2001 *Ecological basics of nature management* (Moscow: Dashkov I Ko)
[14] Belkov V M *Methods, technologies and concepts for utilization of carbon-containing industrial and solid household waste*
[15] Gutenev V V 2007 *Industrial Ecology* (Moscow: MarT)
[16] Palgunov P P 2000 *Utilization of industrial waste* (Moscow: Science)
[17] Grinin A S, Novikov V N 2002 *Industrial and household waste: Storage, utilization, processing* (Moscow: FAIR-PRESS)
[18] Hotuntsev Yu L 2004 *Ecology and environmental safety* (Moscow: Academy)
[19] *Sanitary Rules and Norms No. 4015-85 Limit content of toxic compounds in industrial wastes and storage tank located outside the territory of an enterprise (organization)* 1985 (Standardinform)
[20] Muravia L A 2003 *Ecology and life safety* (Moscow: UNITI)
[21] Denisova V V 2006 *Ecology* (Moscow: ICC "MarT")
[22] *Ecological state, use of natural resources, environmental protection of Tyumen Region* 2006 Department of subsoil use and ecology of the Tyumen region
[23] *Ecological state, use of natural resources, environmental protection of Tyumen Region* 2006 State committee for environmental protection of Tyumen region

IOP Conf. Series: Earth and Environmental Science **337** (2019) 012027 doi:10.1088/1755-1315/337/1/012027

Justification of methodological approaches to standardisation of irrigation as an element of resource saving and minimization of the anthropogenic load on agrobiocenosis

G Olgarenko[1], V Olgarenko[2*], I Olgarenko[2] and V I Olgarenko[3]

[1]Federal State Research Institution All-Russia Scientific and Research Institute for Irrigation and Farming Water Supply Systems "Raduga", Russia
[2]Novotcherkassk Institute on Engineering and Land Reclamation named after A K Kortunov – a branch of FGBOU VO «State Agriculture University of the Don», Russia
[3]Federal state budget scientific establishment «The Russian scientific research institute of land improvement problems», Russia

* Corresponding author: olgarenko_vi@mail.ru

Abstract. The analysis shows that the value of agricultural crops' evapotranspiration is influenced by many non-linearly changing factors in the closed system "soil - plant - atmosphere". Therefore, the method of determining evapotranspiration is complemented by computational models, with the help of which evapotranspiration is determined with a sufficient degree of accuracy from the results of actual observations, including the hydrometeorological parameters of the corresponding irrigated areas, followed by comparison of these values with data obtained when determining potential evapotranspiration on the existing networks of hydrometeorological and water balance stations. As a result of research, empirical dependencies of multivariate determination of evapotranspiration, as well as on moisture reserves in the calculated soil layer and potential evapotranspiration, were established, taking into account the growth and development phases and the microclimatic coefficient of the summer planting period. The above methodological approach substantiates the new water and energy saving methods of forecasting, planning of irrigation processes of agricultural crops and operation-al management based on the spatial and temporal variability of hydrometeo parameters and actual moisture reserves in the calculated soil layer for the conditions of use of livestock waste.

1. Introduction

Evapotranspiration of agricultural crops has significant spatiotemporal variability, therefore the establishment of its optimal norm should be considered as a methodology for determining forecast indicators [1, 2]. They are established by retrospective calculation of evapotranspiration on the basis of observations of meteorological parameters during the required time period and an appropriate statistical analysis. In this way, the most probable values are obtained both for the near and long term [3-5]. The methodology for predicting evapotranspiration is based on the improvement of the bioclimatic method. It is necessary to know its actual value, which provides [6, 7]:

- calculation of evapotranspiration taking into account the dynamics of soil moisture in the calculated layer, as well as the variability of hydrometeorological parameters of a particular year;

- taking into account the effect of groundwater recharge on the indicators of evapotranspiration;
- determination of the amount of irrigation water infiltration beyond the limits of the design layer;
- establishment of the dynamics of evapotranspiration deficit and irrigation regimes for years of different water availability, taking into account the spatial and temporal variability of hydrometeorological parameters and actual moisture reserves in the calculated soil layer.

2. Materials and methods

The data obtained during water balance and agrometeorological studies by scientists of the Federal state budget scientific establishment «The Russian scientific research institute of land improvement problems» and Don state agrarian university for the period from 1990 to 2016, as well as materials of the Federal state budget scientific establishment «The Russian scientific research institute of land improvement problems» researches were used for 1960–1980 [8-11] to assess the influence of hydrometeorological factors and soil moisture variability on the evapotranspiration of various crops,.

The studies, which were based on the dependences of the bioclimatic coefficients were carried out over a five-year period in Agroenterprise Bessergenevskoye LLC, the village of Bessergenevskaya of the Oktyabrsky district of the Rostov region [12, 13].

The ameliorative system is also filled with treated animal waste, therefore, constant monitoring of changes in the salt composition of the water used for irrigation is carried out (Table 1).

Table 1. The salt composition of water used for irrigation (per $1/dm^3$)

Salt composition	Cl^-	SO_4^{2-}	HCO_3^-	Ca^{2+}	Mg^{2+}	Na^+	K^+	Ion sum	Dry residue	pH
gr	0,16	0,22	0,265	0,087	0,048	0,131		0,93	0,94	7,9
mmol	4,9	4,77	4,33	4,5	3,86	5,66		–	–	

3. Results and discussion

The authors established paired coefficients of their correlation with evaporation (E_ω), air temperature (t), lack of air humidity (d_φ), precipitation (P), lack of natural humidification ($E_\omega - P$), evapotranspiration (ET) and irrigation th rule (M), which is based on many years of research by scientists of the Federal state budget scientific establishment «The Russian scientific research institute of land improvement problems» and Don state agrarian university to study the dynamics of evapotranspiration of various crops depending on changes in actual meteorological parameters measured at the corresponding meteorological stations for the conditions of specific irrigated arrays. (table 2) [14, 15].

Table 2. Paired coefficients of multiyear evapotranspiration and irrigation norm with main meteorological factors

Indicator	E_ω	t	d_φ	P	$E_\omega - P$	ET	M
E_ω	1,00	0,90	0,94	0,81	0,83	0,96	0,74
t	0,34	1,00	0,80	0,82	0,90	0,73	0,58
d_φ	0,94	0,89	1,00	0,74	0,89	0,82	0,72
P	0,83	0,84	0,74	1,00	0,92	0,20	0,65
$E_\omega - P$	0,84	0,92	0,89	0,93	1,00	0,93	0,98
ET	0,95	0,74	0,85	0,78	0,91	1,00	0,82
M	0,74	0,58	0,71	0,65	0,96	0,78	1,00
Standard deviation	83,0	178,3	178,0	57,8	97,9	77,2	112,0
The coefficient of variation V, %	10,4	6,4	7,3	30,6	12,0	12,1	33,0

Analysis of the data shows that the lack of natural moisture $(E_\omega - P)$, which characterises the heat and moisture supply of the calculated soil layer during the growing season, has a significant impact on the amount of evapotranspiration. The interrelation between evapotranspiration and potential evapotranspiration is the strongest, it's a complex hydrometeorological characteristic, the coefficient of variation is 0.96.

The magnitude of evapotranspiration is based on determining of the bioclimatic coefficients, as well as the characteristics of their components based on scientific and analytical studies and statistical analysis of long-term experimental data on the cultivation of summer planting potatoes in the conditions of the floodplain's Lower Don, aimed at creating new design models that ensure the rational use of water and energy resources, high productivity of potato cultivation, conservation and improvement of soil fertility, creating a favorable ecological situation in the irrigated agricultural landscape [16–22].

Data on the accumulated active temperature, the sum of the deficits in air humidity and the magnitude of evapotranspiration are presented in Table 3.

Table 3. Data on temperatures, deficit of air humidity and evapotranspiration on the main phases of the potatoes' development for the summer planting period for the survey years under study on average

Characteristic	Growth and development phase						During the growing season
	Young growth	Budding	Blossom	Termination of foliage growth	Wilting foliage	Technical ripeness	
Evapotranspiration (*ET*) ,mm	41,4	109,8	49,6	78,0	61,2	36,2	376,2
The sum of active temperatures (Σt), °C	254,6	448,6	178,6	357,6	347,0	213,8	1800,2
The amount of deficiency of air humidity (Σd_φ), mb	161,1	330,6	130,9	267,0	247,8	164,6	1335,6

According to A. M. Alpatyev, linear bioclimatic patterns depending on the radiation balance (R), sums of average daily deficits of air humidity (Σd_φ) and air temperature (Σt) have an accuracy of up to 90%. Currently, curvilinear dependencies are used, which allows to increase the accuracy of determining the parameters under study, but it is proposed to use several hydrometeorological characteristics to build more accurate dependencies. One-factor and multivariate correlation analysis was conducted, the reliability of the relationship between the values considered was determined to assess the closeness of the relationship between the components of the bioclimatic coefficients' formula: evapotranspiration (*ET*, mm), the sum of active temperatures (Σt ° C) and the amounts of air humidity deficits (Σd_φ, mb).

An analysis of the reliability of the connection's tightness of the characteristics under consideration showed that the temperature has a correlation coefficient of 0.88; air humidity deficit - 0.74; making it possible to characterize their connection with evapotranspiration as very close and close, respectively. It should be noted that during multifactor correlation analysis, the variation of evapotranspiration by 79% (determination (r^2) - 0.79) is related to the effect of the studied factors - the sum of active air temperatures and the sum of air humidity deficits and can be mathematically explained this data.

The dynamics of evapotranspiration value (*ET*, mm) change from the sum of active temperatures (Σt ° C) and the amounts of air humidity deficits (Σd_φ, mb) is presented in Figure 1 and dependence (1)

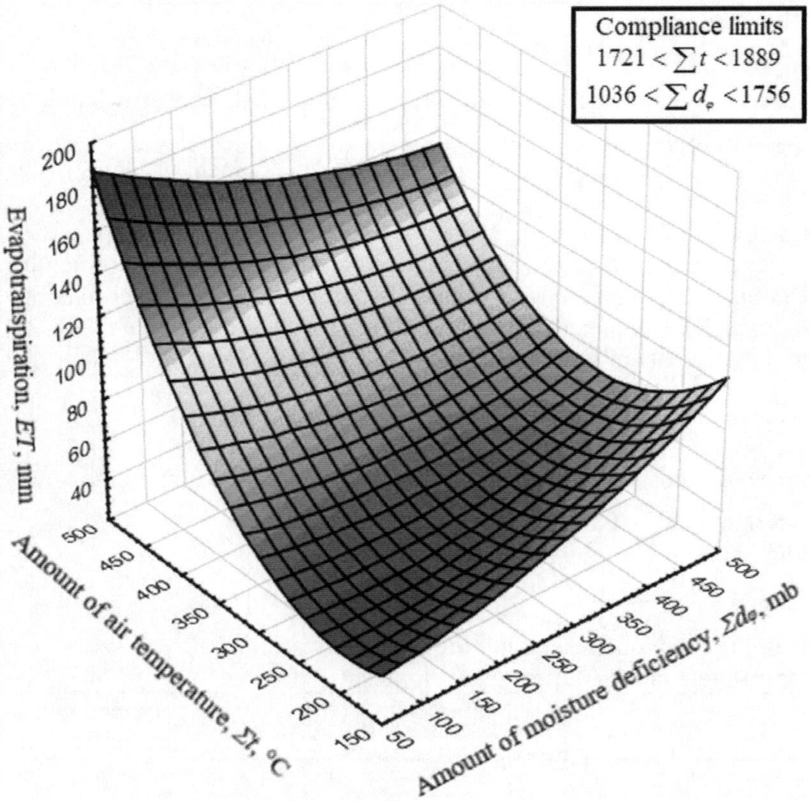

Figure 1. Dynamics of evapotranspiration changing which depends on temperature and humidity deficit

$$ET = 92{,}183 + 0.13 \cdot \sum d_{\varphi} - 0{,}573 \cdot \sum t + \left(0{,}2 \cdot \sum d_{\varphi}{}^{2} - 0{,}7 \cdot \sum t \cdot \sum d_{\varphi} + 1{,}6 \cdot \sum t^{2}\right) \cdot 10^{-3} \quad (1)$$

where ET - evapotranspiration, mm;
$\sum t$ - the amount of air temperatures, °C;
$\sum d_{\varphi}$ - the amount of deficiency of air humidity, mb.

The actual value of potato evapotranspiration (ET) is determined by determining the elements of the water balance equation of the irrigated field according to:

$$ET = K_{\omega} \cdot E_{\omega} \cdot K_{0}, \quad (2)$$

where ET - evapotranspiration, mm;
K_{ω} - bioclimatic coefficient;
E_{ω} - potential evapotranspiration, the value of which is obtained from the weather station, mm;
K_{0} - microclimatic coefficient taking into account the change of microclimate of an agricultural field under the influence of irrigation.

For the conditions of the Lower Don's floodplain as a whole for the irrigated area under consideration $K_{0} = 0{,}87$.

Bioclimatic coefficients of potatoes are determined by the empirical equations of the authors depending on the biological characteristics of potato plants and the effect of actual moisture reserves in the calculated soil layer, taking into account the development phases. The equations have a general form:

$$K_{\omega} = A_{1} \cdot \left(\frac{W_{H} + W_{K}}{2 \cdot W_{HB}}\right)^{2} + A_{2} \cdot \left(\frac{W_{H} + W_{K}}{2 \cdot W_{HB}}\right) + A_{0}, \quad (3)$$

where K_ω - bioclimatic coefficient;

A_1, A_2, A_0 - equation parameters;

W_H, W_K, W_{HB} – moisture reserves in the calculated soil layer: initial, final, and at a humidity corresponding to the lowest water capacity.

The parameters of equation (4) for the various phases of growth and development of plants of the potato are given in table 4.

An evaporator is installed to determine the amount of evapotranspiration in time in the crop rotation area, the indications of which are compared with the corresponding data of the nearest meteorological station. In the absence of evaporators, the magnitude of potential evapotranspiration is determined by regional dependencies on bridges, which are obtained by the authors as a result of modifying the model of V. P. Ostapchik, which is based on the data of weather stations in the Rostov region:

$$E_\omega = A \cdot (d_\varphi)^{b \cdot t}, \tag{4}$$

where A, b – empirical parameters which are obtained by the authors for the conditions of the Lower Don's floodplain. A = 1,220; b = 0,003; R^2 = 0,84;

d_φ – deficiency of air humidity, mbar;

t – air temperature, °C.

Table 4. Parameters of the equation for determining the bioclimatic coefficient for different phases of potato development

Growth and development phases	Equation parameter A_1	A_2	A_3	Approximation reliability
Young growth	–0,114	1,747	0,025	0,98
Budding	–0,068	0,229	0,601	0,97
Blossom	–0,082	0,605	0,672	0,99
Termination of foliage growth	–0,073	0,550	1,395	0,97
Wilting foliage	–0,295	0,741	1,027	0,96
Technical ripeness	–0,027	0,181	0,645	0,92

The performed experiments and the obtained data allowed us to establish a complex empirical relationship that allows us to determine the actual moisture reserves of the soil in the conditions of the Lower Don's floodplain:

$$ET = 92{,}183 + 0.13 \cdot \sum d_\varphi - 0{,}573 \cdot \sum t + \left(0{,}2 \cdot \sum {d_\varphi}^2 - 0{,}7 \cdot \sum t \cdot \sum d_\varphi + 1{,}6 \cdot \sum t^2\right) \cdot 10^{-3} =$$
$$= \left(A_1 \cdot \left(\frac{W_H + W_K}{2 \cdot W_{HB}} \right)^2 + A_2 \cdot \left(\frac{W_H + W_K}{2 \cdot W_{HB}} \right) + A_0 \right) \cdot 1{,}22 \cdot {d_\varphi}^{0{,}003 \cdot t} \cdot 0{,}87 \tag{5}$$

so, we get:

$$A_1 \cdot \left(\frac{W_H + W_K}{2 \cdot W_{HB}} \right)^2 + A_2 \cdot \left(\frac{W_H + W_K}{2 \cdot W_{HB}} \right) + A_0 = \frac{92{,}183 + 0.13 \cdot \sum d_\varphi - 0{,}573 \cdot \sum t + \left(0{,}2 \cdot \sum {d_\varphi}^2 - 0{,}7 \cdot \sum t \cdot \sum d_\varphi + 1{,}6 \cdot \sum t^2 \right) \cdot 10^{-3}}{1{,}0614 \cdot {d_\varphi}^{0{,}003 \cdot t}} \tag{6}$$

The deviations of the actual values of meteorological parameters from their mean multiyear values should be used to improve the accuracy of the calculation of the evapotranspiration of the irrigated field.

Deviations of potential evapotranspiration and evapotranspiration from their mean multiyear values, according to the corresponding weather stations and experimental results, are calculated by the following modular coefficients:

$$K_{\Delta E} = \frac{E_\omega}{\overline{E_\omega}}, \tag{7}$$

$$K_{\Delta ET} = \frac{ET}{\overline{ET}}, \tag{8}$$

where $K_{\Delta E}$ and $K_{\Delta ET}$ - modular deviations of potential evapotranspiration and evapotranspiration values;

ET and E_ω - actual values of evapotranspiration, potential evapotranspiration, which are obtained from weather station data, and transpiration, mm;

$\overline{ET}$, $\overline{E_\omega}$ – mean annual values of potential evapotranspiration and evapotranspiration, mm.

- Based on the analysis of experimental and average annual data, a nonlinear dependence of the relative deviations of evapotranspiration on potential evapotranspiration (Figure 2), which is described by a second-degree polynomial type equation, is obtained:

$$K_{\Delta ET} = f(K_{\Delta E}) = 1{,}51 \cdot \left(\frac{E_\omega}{\overline{E_\omega}}\right)^2 - 1{,}48 \cdot \left(\frac{E_\omega}{\overline{E_\omega}}\right) + 0{,}9, \tag{9}$$

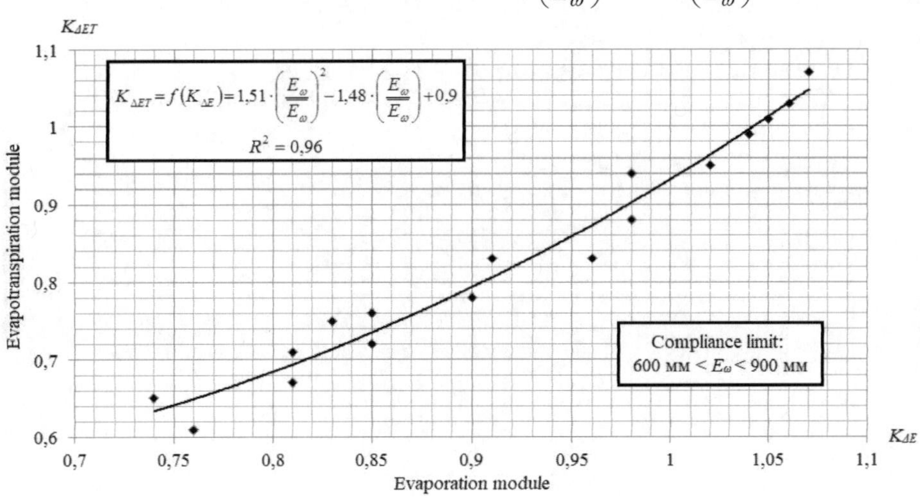

Figure 2. Empirical dependence of the relative deviations' dynamics of evapotranspiration on potential evapotranspiration

In order to avoid excessive accumulation of errors, it is recommended to periodically correct the values ET and E_ω by comparing them with actual measurement data on a specific field. Such measurements should be carried out 1-2 times during the growing season.

4. Findings

The conducted multifactor correlation analysis allows us to conclude that the variation of the evapotranspiration rate by 79% is related to the effect of the studied factors, and the total correlation coefficient is 0.89. As a result, the empirical dependences of evapotranspiration on the sum of active air temperatures, the sum of air humidity deficits and their multifactorial dependence were obtained. Empirical dependences of bioclimatic coefficients on the actual moisture reserves in the calculated soil layer are established, and a complex empirical dependence is established on the establishment of actual soil moisture reserves.

IOP Conf. Series: Earth and Environmental Science **337** (2019) 012027 doi:10.1088/1755-1315/337/1/012027

The results of the studies allowed us to establish a significant increase in the accuracy of determining evapotranspiration and potential evapotranspiration according to the models developed in comparison with actual data, which ensured a significant increase in the efficiency of water and energy resources used in the implementation of irrigation regimes of potato plantations in the conditions of the Lower Don's floodplain.

References

[1] Vasilyev S 2017 *S. J. of RSRILIP* **27** 18
[2] Vasilyev S, Olgarenko V, Olgarenko I 2017 *S. J. of RSRILIP* **28** 13
[3] Balakay G, Vasilyev S, Babichev A 2017 *S. J. of RSRILIP* **26** 19
[4] Shedrin V, Vasilyev S 2017 *SWoDAIC* **1** 3
[5] Olgarenko V, Yurchenko L, Olgarenko V, Olgarenko I 2017 *NLVAC* **45** 7
[6] Helman D, Osem Y, Yakir D, Lensky I 2017 *AaFM* **232** 12
[7] Olgarenko G, Turapin S 2016 *MaWM* **6** 3
[8] Grigorov S, Malyuga A, Korotkov A 2007 *RAS* **33** 7
[9] Vlasov M, Vasilyev S, Grigorov S 2018 *S. J. of RSRILIP* **31** 16
[10] Babichev A, Olgarenko V 2016 *S. J. of RSRILIP* **22** 19
[11] Babichev A, Balakay G, Monastyrskiy V 2015 *Fertility* **86** 3
[12] Vasilyev S, Vinnilov D 2016 *S. J. of RSRILIP* **23** 17
[13] Olgarenko V, Olgarenko I, Olgarenko V 2017 *MTED (RA)* **4** 6
[14] Olgarenko G, Savushkin S 2007 *YAA* **3** 4
[15] Olgarenko G 2010 *Ways of impr. the eff. of irrigate. agricult.* **35** 8
[16] Jansons V, Jurenoks V 2012 *Economics and Business* **22** 6
[17] Hirano T, Suzuki K, Hirata R 2017 *AaFM* **232** 12
[18] Bazan G, Marino P, Guarino R, Schicchi R 2015 *ABF* **52** 19
[19] Fraga H, Malheiro A, Moutinho-Pereira J, Jones G, Alves F, Pinto J, Santos J 2014 *REC* **14** 12
[20] Yoo S, Lee S, Choi J, Im J 2016 *PaWE* **14** 11
[21] Hirano T, Suzuki K, Hirata R 2017 *AaFM* **232** 12
[22] Olgarenko V, Olgarenko I, Olgarenko V 2017 *iWoSD* **9** 16

IOP Conf. Series: Earth and Environmental Science **337** (2019) 012028 doi:10.1088/1755-1315/337/1/012028

The role of waste banks in the reduction of solid waste sent to landfill in Semarang, Central Java, Indonesia

M A Budihardjo[1,*], I F S Wahyuningrum[2], F I Muhammad[1], R Pardede[1]

[1]Department of Environmental Engineering, Diponegoro University, Indonesia
[2]Department of Accounting, Semarang State University, Indonesia

*Corresponding author: m.budihardjo@ft.undip.ac.id

Abstract. Approximately 1,000 tonnes of solid waste are generated in Semarang every day, only 850 tonnes of which are transferred to the Jatibarang Landfill. Solid waste that is not transferred to the landfill can trigger environmental and health problems. With the help of community groups, reduction of solid waste is achievable using 'the three Rs' of waste reduction and the implementation of waste banks. Waste banks could act as mediums for depositing waste, improve the socio-economic status of local residents, and develop community connections. Waste banks improve the volume of waste reduction and reduce landfilled waste, helping to extend the operational life of landfills. This study examined the role of waste banks in Semarang and the potential to improve waste management based on the sampling of solid waste generation and composition. This topic was investigated with a purposive sampling method. Results showed that the existing system of waste banks in Semarang is ineffective in reducing waste. However, the system could be improved by increasing waste bank usage, including for various types of valuable waste in Semarang.

1 Introduction

According to the Regional Secretary of Semarang, in 2017, 1.6 million people lived in Semarang's 16 sub-districts. Each day, approximately 1,000 tonnes of solid waste are generated in Semarang, only 850 tonnes of which are transferred to the Jatibarang Landfill. The remainder, solid wastes which are not transferred to the landfill, can trigger environmental and health problems. Amid increasing amounts of solid waste generated in the city, a well-organized system of solid waste management is needed to reduce waste.

UU no. 18 of 2018, a law that administers solid waste management in Indonesia, explains the need to change from conventional to advanced waste reduction methods, focusing on minimisation and treatment of solid waste. Minimisation can be achieved by reducing, reusing, or recycling waste, commonly known as the 'three Rs' of waste reduction [1].

Certain types of waste have economical value if sold or reformed into other things, as raw materials or finished goods. If Semarang locals, as generators of solid waste in the city, are to play an important role in Semarang's solid waste management, they will require a physical location for waste management operational processes. Semarang's waste banks could act as mediums for depositing waste, improve the socio-economic status of local residents, and develop community connections in the city [2]. An effective system of waste banks may improve the city's solid waste management, increasing the amount of waste deposited into the waste banks. This system would improve the volume of waste reduction and landfilled waste in the city while extending the operational life of Semarang's landfill [3]. Moreover, if these systems function effectively, the amount of residual waste in the city will be diminished, its environment will become cleaner, and health outcomes could be improved [4].

2 Methodology

The study was conducted over four months, from September to December 2018, in Semarang, Central Java, Indonesia. Solid waste generation and composition calculation were conducted using SNI 19-3964-1994 [5] in three stages. The first stage was solid waste sampling in domestic and non-domestic areas of

Semarang over eight consecutive days. Sample points were selected from various locations around Semarang to represent the city's solid waste generation and composition.

In this stage of the study, Semarang was divided into five sections: Tugu, Tembalang, Central Semarang, Gajahmungkur, and Gayamsari. During the study, a trash bag was distributed to the city's waste sources the day before weighing of waste commenced. The next stage was solid waste weighing, which was conducted using tools such as gloves, trash bags, shovels, and weight scales. Weighing steps were conducted twice to measure the combined weight of all waste types and separate weight of each waste type to understand the city's waste composition. The last stage of this study was a purposive sampling-based questionnaire, which was targeted at waste bank developers and consumers in Semarang. The number of respondents was calculated with Slovin's formula to identify the correct value of the sample in certain populations and apply this number to a larger population

$$n = \frac{N}{N(d)^2+1} \tag{1}$$

Annotation:

n = number of samples
N = number of populations
d = degree of freedom (ex: 0.1, 0.05, 0.01)

The number of respondents for this study's purposive sampling questionnaire is shown in Table 1 below:

Table 1. Number of respondents.

Sub-District	Population (Houses)	Respondents (Houses)
Mijen	13.958	4
Gunungpati	21.889	6
Banyumanik	31.199	9
Gajahmungkur	11.992	3
South Semarang	13.997	4
Candisari	15.189	4
Tembalang	38.846	11
Pedurungan	40.777	12
Genuk	22.157	6
Gayamsari	15.879	5
East Semarang	14.972	4
North Semarang	23.632	7
Central Semarang	12.075	3
West Semarang	32.211	9
Tugu	6.408	2
Ngaliyan	30.769	9
Total	345.950	100

Besides the questionnaire, an interview was conducted at waste banks in Banyumanik and Pedurungan to understand the quantity and type of waste withdrawn and deposited in the waste bank each month. This interview helped the authors to review the performance of existing waste banks in several areas of Semarang.

3 Results

The city's solid waste generation and composition were calculated based on the weight of waste. This calculation was used because the reduction rate could be determined by the difference between weight with and without implementation of waste banks. In addition, weight is a common unit of measurement used by Semarang's waste scavengers and collectors in their trading processes. For this reason, the use of weight in this calculation made the study easier to understand and more relevant to real-world conditions.

Table 2. Total solid waste generation in Semarang.

Domestic Facility	Number of People	Solid Waste Generation Rate (kg/day/person)	Total Solid Waste Generation (kg/day)
Housing	1,701,172	0.43	732,002.91
	Sub-Total		732,002.91

Non-Domestic Facility	Number of Units	Solid Waste Generation Rate (kg/day/unit)	Total Solid Waste Generation (kg/day)
Elementary School	591	13.77	8,137.04
Junior High School	220	21.86	4,808.78
Senior High School	189	27.38	5,174.47
Shop	681	3.12	2,124.61
Restaurant	4,352	6.72	29,230.05
Health clinic	37	8.3	306.92
Hospital	26	239.33	6,222.57
Market	47	928.09	43,620
Sub-Total			99,624.43
Total Solid Waste Generation of Semarang			831,627.35

Table 2 illustrates the total solid waste generation from domestic and non-domestic facilities in Semarang. In total, 732,002.91 kilograms of solid waste were generated from homes used as domestic facilities each day. Non-domestic facilities such as schools, markets, and hospitals produced 99,624.43 kilograms of solid waste each day. Together, both types of facilities generated 831,627.35 kilograms of solid waste each day. This table shows that individual domestic facilities generated more solid waste than individual non-domestic facilities. However, domestic facilities generated less solid waste than non-domestic facilities as a group.

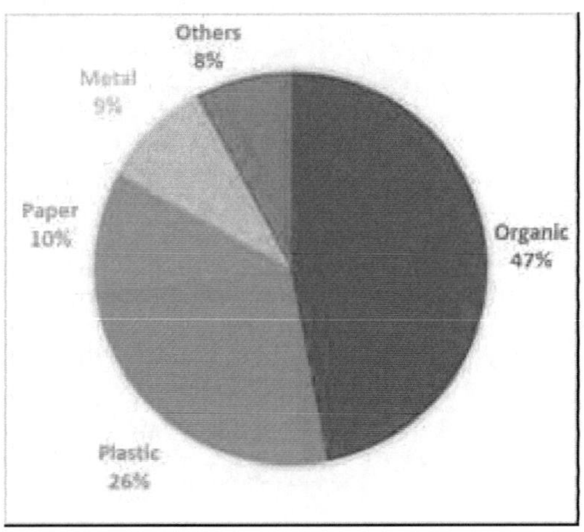

Figure 1. Solid waste composition of domestic facilities.

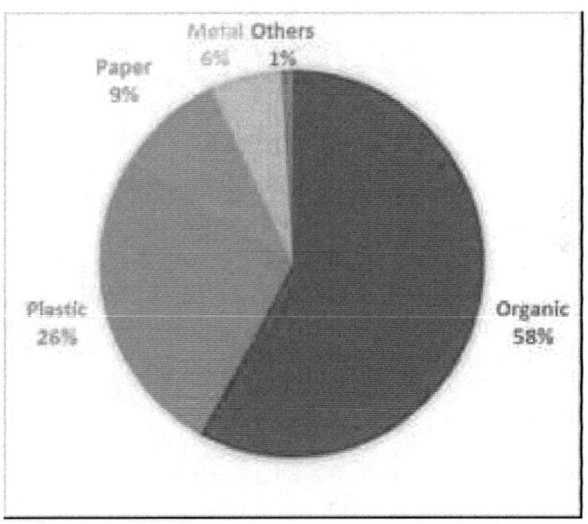

Figure 2. Solid waste composition of non-domestic facilities.

Figure 1 and 2 demonstrate the percentage of composition for each type of waste in domestic and non-domestic facilities respectively. Solid waste composition is differentiated into five different type of waste: organic, plastic, paper, metal, and others. Both domestic and non-domestic facilities produced mainly organic waste (47% and 58% respectively), followed by plastic (26% for both types of facility). Three types of waste accounted for no more than 10% of solid waste from domestic and non-domestic facilities: paper, metal, and others (including glass, rubber, or cloth).

Table 3. Solid waste generation for each type of waste.

Type of Waste	Domestic Facilities (kg/day)	Non-Domestic Facilities (kg/day)	Total (kg/day)	Percentage (%)
Plastic	1916000.69	25,636.74	216,637.43	26.05
Paper	72,000.46	8,809.03	80,809.49	9.72

IOP Conf. Series: Earth and Environmental Science **337** (2019) 012028 doi:10.1088/1755-1315/337/1/012028

Metal	68,000.82	5,884.4	73,885.22	8.88
Organic	345,000.08	57,569.31	402,569.39	48.41
Others	56,000.86	1,724.96	57,725.82	6.94
Total	732,002.91	99,624.43	831,627.35	100.00

Table 3 shows the amount of solid waste generated by each type of waste in domestic and non-domestic facilities. Organic waste accounted for 48.41%, or almost the half, the city's total waste. Plastic accounted for 26.05%, while waste types like paper (9.72%), metal (8.88%), and others (6.94%) were less significant. Overall, this table illustrates that the people of Semarang mainly consume products made from organic materials and plastics, rather than paper, metal, or other materials (such as glass, rubber, or cloth). In addition, the city's high percentage of organic waste may be caused by the enormous amount of food waste and plant matter litter in Semarang.

Results of this study's purposive sampling via questionnaire and interview revealed that the city's waste bank customers usually deposited their waste each month. Although most deposits were organic waste, customers also deposited their inorganic waste. However, most customers collected organic waste until they accumulated a quantity large enough to be sold at the waste bank.

Table 4. Valuable waste in all sub-districts of Semarang.

Sub-District	Number of Customers (Houses)	Valuable Waste (kg/month)							Total (kg/month)
		Paper		Plastic		Metal		Other	
		Plain Paper	Card-Board	Bottle	Cup	Iron	Alumi nium		
Mijen	82	49.4	54.6	20.8	13	46.8	31.2	44.2	260
Gunungpati	245	131.1	144.9	55.2	34.5	124.2	82.8	117.3	690
Banyumanik	405	157.7	174.3	66.4	41.5	149.4	99.6	141.1	830
Gajahmungkur	237	89.3	98.7	37.6	23.5	84.6	56.4	79.9	470
South Semarang	110	60.8	67.2	25.6	16	57.6	38.4	54.4	320
Candisari	92	53.2	58.8	22.4	14	50.4	33.6	47.6	280
Tembalang	256	121.6	134.4	51.2	32	115.2	76.8	108.8	640
Pedurungan	370	165.3	182.7	69.6	43.5	156.6	104.4	147.9	870
Genuk	84	49.4	54.6	20.8	13	46.8	31.2	44.2	260
Gayamsari	180	96.9	107.1	40.8	25.5	91.8	61.2	86.7	510
East Semarang	72	53.2	58.8	22.4	4	50.4	33.6	47.6	280
North Semarang	67	36.1	39.9	15.2	9.5	34.2	22.8	32.3	190
Central Semarang	320	148.2	163.8	62.4	39	140.4	93.6	132.6	780
West Semarang	423	169.1	186.9	71.2	44.5	160.2	106.8	151.3	890
Tugu	78	53.2	58.8	22.4	14	50.4	33.6	47.6	280
Ngaliyan	350	138.7	153.3	58.4	36.5	131.4	87.6	124.1	730
Total	3,371	1,573	1,738	662	414	1,490	993	1,407	8,280
Average	210.69	98.33	108.68	41.40	25.88	93.15	62.10	87.98	517,5
Percentage		19%	21%	8%	5%	18%	12%	17%	

Table 3 demonstrates the amount of valuable wastes that could be deposited in waste banks in all sub-districts of Semarang. During this study, there were 8,280 kilograms of valuable waste in total. The sub-district with the highest number of waste bank customers was West Semarang, with 423 houses. North Semarang had the lowest number, with only 67 houses. West Semarang households also produced the highest amounts of valuable waste (890 kilograms per month), with Banyumanik households in second place, producing 830 kilograms per month.

According to the Environmental Agency of Semarang (DLH), there are only 62 active waste banks in Semarang. With the city's average total deposited waste rate at 517.5 kilograms per month, approximately 1,069.5 kilograms of waste are deposited in waste banks all around Semarang each day. In addition, the current percentage of waste reduction

caused by residents depositing waste in Semarang's waste banks is shown in the following calculation:

- Solid waste generation in Semarang = 831,627.35 kilograms per day
- Existing deposited waste = 1,069.5 kilograms per day

IOP Conf. Series: Earth and Environmental Science **337** (2019) 012028 doi:10.1088/1755-1315/337/1/012028

– Reduction percentage = (1,069.5 / 831,627.35) kilograms per day x 100%
= 0.13%

From this calculation, it is shown that only 0.13% of waste reduction in Semarang can be attributed to the implementation of waste banks. This percentage may increase if more customers use the waste banks in the upcoming year. However, this number remains far below the amount of valuable waste that could be deposited in the waste banks, as shown in the Table 3. Semarang's waste banks could reduce the amount of solid waste transported to Jatibarang Landfill significantly. However, this outcome is possible only if a greater percentage of the city's valuable waste is deposited to the waste banks so that only invaluable waste is deposited in the landfill.

4 Conclusion

At present, Semarang's waste banks are operated by community groups with the help of the city's government. In total, 1.069,5 kilograms of solid waste are deposited into the city's waste banks each day. This figure has contributed to a 0.13% reduction of solid waste generation in Semarang. This number is small compared with the amount of valuable waste produced across Semarang. If more of the city's valuable waste is deposited in waste banks, solid waste generation will be reduced significantly, along with landfilled waste. In summary, effective implementation of waste banks may reduce solid waste generation in Semarang and extend the operational life of the city's Jatibarang Landfill.

References

[1] Aryenti A 2011 Improvement of Community Role through Waste Bank Deposit in Babakan Surabaya, Kiaracondong Bandung *J. Permukiman*
[2] Wijayanti D R, Suryani S 2015 Waste bank as community-based environmental governance: a lesson learned from Surabaya *Procedia-Social and Behavioral Sciences*
[3] Raharjo S, Matsumoto T, Ihsan T, Rachman I and Gustin L 2017 Community-based solid waste bank program for municipal solid waste management improvement in Indonesia: a case study of Padang city *J. of Material Cycles and Waste Management*
[4] Ulhasanah N and Goto N 2018 Assessment of citizens' environmental behavior toward municipal solid waste management for a better and appropriate system in Indonesia: a case study of Padang City *J. of Material Cycles and Waste Management*
[5] National Standardization Agency 1994 *Collecting and Measuring Method of Municipal Solid Waste Generation and Composition.* SNI 19-3964-1994

IOP Conf. Series: Earth and Environmental Science **337** (2019) 012029 doi:10.1088/1755-1315/337/1/012029

Effectiveness of use of phosphogypsum on the solonetzic soils of Siberia and Ural

L Scipin[1*], D Scipin[2], E Zakharova[1] and I Kustysheva[1]

[1]Tyumen Industrial University, 38, Volodarskogo, Tyumen, 625001, Russia
[2]Tyumen State University, 6, Volodarskogo, Tyumen, 625003, Russia

* Corresponding author: sav.vs83@rambler.ru

Abstract. The problem of rational use of saline soils as applied to the specific region becomes one of the main problems in Russia as well as abroad. The article describes the potential possibility to include these soils for rational land utilization in agro-industrial complex system. This will allow us to broaden the degree of their utilization on ploughed fields, hayfields and pastures in Siberia and Ural. Area of these lands as applied to this territory is more than 10 million ha. Obligatory chemical melioration is required for using of these soils on the level of climatic soil type. Its carrying out using gypsum, phosphogypsum and other ameliorants-coagulants allows us to increase fundamentally the fertility of solonetzic soils with their following implementation as yielding agricultural fields. It's important to note that influence of ameliorants-coagulants takes place up to 40 and more years and their dose is up to 20 t/ha for perennial grasses and 40 t/ha for crops. The mentioned regions possess mineral, productive and scientific resources for the solution of such an important problem in conditions of Siberia and Ural.

1. Introduction

At present moment the problem of use of saline soils is one of the main in the land utilization system. Saline soils include such soils which contain ready soluble soils in the amounts that are toxic for plants. The salt source for these soils are rock formation, product of volcanic eruption, eolian transport of salts, atmospheric precipitation ground waters, spray-type irrigative ground-waters, vegetation (salt warts). The most widespread among the saline soils are solonetzic soils containing in the occluded state an increased amount of exchangeable sodium in the illuvial horizon (more than 15% of the cation exchange capacity) [10].

Solonetzic soils possess some negative hydrophysical and chemical properties: high dispersion, swelling ability, weak water permeability and filterability, ash structure, slow process of physical maturity of the soil, adhesiveness in wet state, low water mobility, decreased air exchange, strong compression of soil material in case of drying up, solidity and high resistance, alkaline reaction of the media, salts toxicity, deficit of calcium and excess of sodium [1-5].

It's suffice to note that the area of indicated soils in complex with climatic type soils in Ural and Western Siberia is more than 10 million ha (table 1) [6, 7].

Table 1. Structure of solonetz complexes in Ural and Western Siberia, thousands of ha.

Region, territory	Total	Presence of solonetz, %		
		10-30	30-50	>50
Orenburg	1431	720	99	606

Chelyabinsk	1365	721	92	551
Kurgan	2203	636	190	1377
Sverdlovsk	631	225	50	302
Tyumen	389	176	30	182
Omsk	3364	485	678	1812
Novosibirsk	4317	2311	233	2258
Kemerovo	146	1826	44	21
Altai territory	1798	1183	143	472
	15254	6108	1562	7581

Territory including Orenburg, Chelyabinsk, Kurgan, Tyumen, Omsk, Novosibirsk regions and Altai territory is characterized by large heterogeneity of natural conditions. Here there are landscapes of dry steppe with kastanozems (Kulunda village, Orenburg region), landscapes of steppe with south and ordinary chernozems (Omsk, Chelyabinsk, Orenburg, Novosibirsk regions, Altai territory), landscapes of outlier with ordinary and solonetz chernozems (Chelyabinsk, Kurgan, Omsk regions) and landscapes of northern forest-steppe with chernozem-like meadow soils, lixiviated and gray forest soils (Tyumen, Kurgan, Chelyabinsk, Omsk, Novosibirsk regions and Altai territory) [8].

2. Materials and methods

During the ameliorative diagnostics of solonetzic soils a series of important characteristics is taken into account, in particular, soil complexes where among the climate type soils, the solonetzic soils occupy 10-30%, 30-50 and more than 50%. Fertile land with less than 10% of solonetzic soils is not taken into account in the ameliorative fund. In the complex soils with 10-30 to 50% of solonetzic soils, selective chemical amelioration is reasonable (if the solonetzic soils are located as small spots), and above 50% - continuous melioration. Hydrodynamic conditions determine numerous ameliorative properties of solonetzic soils. Type of these conditions are established according to the level of groundwater occurrence taking into account content of soil-forming and basement rocks.

The expedience of amelioration depends on the type of hydrological condition. Character of of solonetz use is determined according to this condition. Meadow solonetzic soils can be masters when the mineralization of ground waters doesn't increase critical values. Melioration of meadow-steppe and steppe solonetzic soils doesn't depend on condition of the ground waters (table 2) [7].

Table 2. Critical mineralization of ground waters depending on the depth of their occurrence and rock composition.

Depth of groundwater occurrence, m			
gault clay, light loam	Medium loam	medium clay, heavy loam	Mineralization, g/l
<1.0	<1.2	<1.5	<1.5
1.0-1.2	1.2-1.5	1.5-2.0	1.5-3.0
1.2-1.5	1.5-2.0	2.0-2.5	3.0-5.0
1.5-2.0	2.0-2.5	2.5-3.0	5.0-10.0
2.0-2.5	2.5-3.0	3.0-3.5	>10.0
>2.5	>3.0	>3.5	-

Salinization character of solonetz is depth of salt horizon occurrence, chemism and salinization degree. The most laborious solonetzic soils in a view of amelioration are those with high level of location of maximum of salts and with increased soda content.

Degree of solonetzicity of soils depends on exchangeable sodium and plays crucial role in calculation of chemical ameliorant dose. By the content of exchangeable sodium, solonetzic soils are divided into residual (less than 10% of base exchange capacity), soils with low sodium content (10-20%), soils with medium sodium content (20-40%) and soils with high sodium content (more than 40%). Methods of

IOP Conf. Series: Earth and Environmental Science **337** (2019) 012029 doi:10.1088/1755-1315/337/1/012029

solonetzic soils amelioration should be strictly differentiated in connection with quality and quantity variety [4].

During the amelioration of solonetzic soils (improvement of physical properties, creation of thick fertile layer, change of exchangeable sodium with calcium, removal of excess of water-soluble salts of root habitable layer) for the increasing of productiveness of growing cultures it's recommended to apply two main methods: chemical and self-amelioration method. Chemical amelioration method provides for introduction of chemical ameliorants from the outside. This method is the only way to increase the fertility of solonetzic soils with deep occurrence of carbonates and gypsum (deeper than 40-50 cm). It's reasonable to combine gypsuming with watering. In case of high occurrence of deep plantage plowing and step processing [7].

Texts carried out in Western Siberia and in Ural have proved high efficiency of chemical amelioration. Term of aftereffect continues more than 40-50 years [1-3, 5, 7, 8, 10].

Local gypsum deposits are located in Altai territory where there were found about 20 deposits. The most perspective for using is Zhira deposit (Zhira lake of Kuludinsk region), reserves of which are enough for complete amelioration of all solonetzic soils of Altai territory and neighboring regions of other territories. Its efficiency is checked and proved on the territory of West-Siberian region of Ural and Republic of Kazakhstan [4].

Enormous reserves of natural gypsum are located in Perm territory (Kungur town), carnallite deposit is near Solikamsk town. Carnallite has proved itself as good coagulant.

However, it's worth noting that waste of chemical industry – phosphogypsum is rather promising as ameliorant. Beside gypsum, it contains 0,5-1,5% of labile phosphorus (P_2O_5). Its enormous reserves are in Sverdlovsk region in Revda town.

Similar effect on soil is made by green vitriol – waste of paint and coatings industry. It's applied in Chelyabinsk region and is small grain powder with good flowability and weak caking ability, its dose should be increased in comparison with gypsum by 1.6 times.

As applied to Tyumen region, natural reserves of diatomite, zeolite and mud from the cleaning of Velijansk water intake artesian waters can be considered as important reserves of ameliorant-coagulant. It's worth noting that diatomite reserves in northern regions amounts to billions of tons. Preliminary laboratory tests show that the largest efficiency is demonstrated by heat-treated diatomite. The effectiveness of ordinary diatomite is very low. Zeolite and surface and artesian waters cleaning mud wastes influence selectively, taking into account chemism of soil salinization. This proves the necessity of industrial tests on their solonetzic soils of neutral and soda salinization.

It's important to note that the theoretical and practical issues connected with mastering of solonetzic soils received basis in the works by scientists from Siberia and other regions. However, taking into account genetic features of solonetzic soils in different zones and even within the limits of certain region, the constant improvement of applied methods of chemical and agrobiological amelioration and their complex use as applied to specific agro-ameliorative conditions is required.

The goal of studies is to reveal long-term consequence of calciferous ameliorants for hydrophysical and physical and chemical properties of solonetzic soils, their efficiency as applied to West-Siberian region.

Study objectives: to give theoretical and practical underpinning of dose of introduced ameliorant, taking into account the rate of exchange reactions outflow; to track the change of quantity and quality content of salts in soil under the influence of gypsuming with different doses, to establish possibilities of exchangeable sodium displacement from the absorbing complex for the long ameliorative period, to determine the response of crops-phytoameliorants to the long-term consequence of chemical amelioration.

Soil cover of test station in the northern forest-steppe (state farm "Vagayskiy" of Omutinskiy region) consists of long-fallow crusted and fine solonetz. According to the classification, it belongs to the meadow multisodium, high-carbonate, deep-gypsum and heavy-loam soils. An amount of exchangeable sodium in solonetzic horizon is 40-72% of exchange capacity. Maximal content of water-soluble salts (0,6%) is related to 10-40 cm layer.

Study methods: experiments were conducted on hydromorphic chernozem-meadow solonchak sulfate-soda medium-saline deep-carbonate deep-gypsum crusted multisodium column solonetz of the Tyumen region. Experiments on chemical amelioration were started in 1972, the area of plots is 400 m2, repetition - 4 times. Exchange sodium was determined by the Gedroyts method. Analysis of the aqueous extract was carried out according to the following methods: alkalinity from soluble carbonates - potentiometrically, chloride ions - complexometricly, potassium and sodium by difference.

The studies have shown that during observational years, the level of weakly-mineralized ground waters varied from 1.2 to 2.4 m, its maximum was reached after complete snow melting. As control, long-fallow solonetz was used. Salt condition of long-fallow solonetz has shown that total content of salts in meter layer varied within the limits of 48-58 t/ha and their largest amount was reached during dry years. In the composition of salts, soda dominated - 28.2 t / ha, chlorides accounted for up to 7 t/ha, sulphates - 5.3 t/ha. The dominance of soda in solonetz gave them sulphate-soda salinization, which led to an increase in their alkalinity and toxicity. The introduction of phosphogypsum in 1972 significantly changed the qualitative and quantitative composition of the aqueous extract over the entire soil profile (Table 3).

Table 3. Anion content (t/ha) in water extract of crusted solonetz of sulphate-soda salinization after 34-year amelioration, 1972-2006.

Layer, cm	CO^{2}_{3}+HCO_3	Cl^-	SO^{2-}_{4}	Σ anions
		Control, without ameliorant		
0-10	1.1	0.6	0.5	2.2
0-40	5.7	2.9	2.5	11.1
0-100	28.2	7.0	5.3	40.5
		Phosphogypsum, 21 t/ha		
0-10	0.3	0.2	1.0	1.5
0-40	1.9	1.1	5.0	8.0
0-100	7.9	4.6	16.7	29.2
		Phosphogypsum, 43 t/ha		
0-10	0.1	0.1	0.9	1.1
0-40	0.7	0.7	4.9	6.3
0-100	5.2	3.1	17.2	25.5

With a full dose of phosphogypsum, the removal of soda from the soil profile was more intense, while the content of products of exchange reactions in the form of sulfates increased by 2 times. This phenomenon did not lead to an increase in the intensity of the salt regime, since the toxicity of sulfates is several orders of magnitude lower than that of soda. Thus, the amount of salts in meter layer has decreased in comparison with control variant by 15-20 t/ha. At the same time the essential decrease was noted at the expense of soda decrease. For the study years its content in this layer has dropped by 3.6-5.4 times. Under the complete dose of phosphogypsum, soda removal from the soil profile took place more intensive, at the same time content of products of the exchange reactions in a form of sulphates has increased by 2 times. This phenomenon didn't lead to the strengthening of salt condition intensity, since toxicity of sulphates is few times lower of that of soda.

The data of table 4 shows that introduction of ameliorants in full and half doses led to the significant weakening of solonetzicity. Thus, initially the content of exchangeable sodium in crusted solonetz (0-40 cm layer) was on the level of medium-sodium to multi-sodium (28.8-53.4% of exchange capacity), then after the gypsuming in the mentioned doses the level of solonetzicity decreased to the residual and low-sodium level, i.e, 2.2-19%. Accepted earlier provision on the short duration or total absence of ameliorative process in conditions of close location of the ground waters wasn't confirmed. It's important to note that phosphogypsum dose for 0-15 cm layer (21 t/ha) could be compared well with full calculated dose of a layer for 0-30 cm (43 t/ha) by its duration. Amount of exchangeable sodium maintained in 0-20 cm layer was on the equal layer under the half as well full dose of ameliorant.

Table 4. Content of exchange cations in meadow solonetz of sulphate-soda salinization for the ameliortive period of 1972-2006.

Depth, cm	%, of exchange capacity			
	Na	K	Ca	Mg
		Control, without ameliorant		
0-10	28.8	8.5	8.6	24.6
10-20	45.9	5.4	7.4	23.9
20-40	53.4	7.2	4.8	24.6
		Phosphogypsum, 21 t/ha		
0-10	4.4	7.2	18.3	18.4
10-20	4.4	3.4	16.4	16.4
20-40	19.0	5.8	26.3	26.3
		Phosphogypsum, 43 t/ha		
0-10	2.2	5.9	28.2	20.2
10-20	3.9	4.6	29.7	18.9
20-40	16.6	5.1	22.4	21.5

The efficiency of dry mass of perennial grass herbage during the study years by half-dose, was 21.2 centner/ha, in case of full dose – 24.7 centner/ha. It didn't exceed 3.2 centner/ha on the long-fallow land (control).

3. Conclusion

Use of chemical industry waste – phosphogypsum on the crusted solonetzic soils allows us to obtain stable efficiency of perennial grasses for more than 34 years under the significant improvement of the main chemical properties of these soils. For the meadow formation on the ameliorated solonetz, it's enough to introduce full dose of phosphogypsum for 0-15 cm layer, calculated according to Gedroyts method. In the field crop rotation the dose of ameliorant should be calculated for the layer of 0-30 cm.

References

[1] Fedotkin V 1993 *Solonetzic soils of Siberia and the Urals* (Novosibirsk: Science)
[2] Kovda V 1937 *Solonchak and solonetzic* (Moscow: USSR Academy of Sciences)
[3] Berezin L 2006 *Amelioration and the use of solonetzic soils in Siberia: monograph* (Omsk: FGOU VPO OmGAU)
[4] Semendyaeva N 2014 *Dynamics of salt composition of Baraba solonetzes during the 27–32-year-old action of gypsum* (Novosibirsk: NGAU)
[5] Novikova A 2004 *Saline soils, their distribution in the world, the improvement of land and environmental issues* (Kharkov: HNAU)
[6] Yeremchenko O 1997 *Natural-anthropogenic changes in solonetzic soils in the Southern Trans-Urals* (Perm: Perm University)
[7] Skipin L 2000 *Siberian solonetzic soils: environmental aspects of development* (Yalutorovsk)
[8] Khusainov A 2012 *Opportunities for reclamation and development of solonetzic soils of Western Siberia*
[9] Reclamation of solonetz of the Southern Urals and Western Siberia 1976
[10] Oborin A 1958 *On melioration and development of solonetzes in conditions of irrigated agriculture of the chernozem strip of Western Siberia Problems of melioration of solonetzes* (Moscow: USSR Academy of Sciences)

IOP Conf. Series: Earth and Environmental Science **337** (2019) 012030 doi:10.1088/1755-1315/337/1/012030

Efficiency of chemical ameliorants for reclamation of bore mud and solonetzic soils of Siberia and Ural

L Scipin[1], D Scipin[2], V Petukhova[1*] and Z Monakhova[1]

[1]Tyumen Industrial University, 38, Volodarskogo, Tyumen, 625001, Russia
[2]Tyumen State University, 6, Volodarskogo, Tyumen, 625003, Russia

* Corresponding author: sav.vs83@rambler.ru

Abstract. The problem of solonetzic soils fertility is still one of the most important in the country. The areas of Siberia and Ural make up about 10 million hectares. To improve chemical properties of solonetz, chemical method of amelioration is applied. It's possible to use this method on the bore muds. Number of mud storage pits only in Khanty-Mansisk is about three thousand. On Yamal the situation is the same. As ameliorants-coagulants in the conditions of Siberia and Ural their natural reserves, wastes of mining and chemical industry can be used. The method of tubes was used for the comparative analysis of the studied ameliorants. Equal soil and bore mud samples with increasing doses of coagulants allow to assess maximal filterability of solonetz, bore mud and reveal their optimal dose. The studies have shown that choice of ameliorant-coagulant is mostly determined by the chemism of soil and bore mud salinization, and also content of exchangeable sodium in soil adsorption complex. As promising ameliorants-coagulants we should consider the wastes of chemical industry – phosphogypsum, and as raw material of mining industry – gypsum, carnallite, calcium chloride.

1. Introduction

The saline soil including solonetz-like soils are widely spread in numerous countries. According to V.A. Kovda, their area in the world is about 950 million hectares. Within Russia Federation they occupy about 30 million hectares, including 10 million hectares in Western Siberia. The scientists of Russia and Western Siberia in particular have accumulated rich experience on amelioration of these soils [1-4].

The difficulty in mastering of solonetzic soils is connected with the variety of their properties within the limits of each soil and climatic zone. The designing of solonetz amelioration depends on hydrological condition, chemism and degree of salinization, depth of water-soluble salts occurrence, exchangeable sodium content in solonetz horizon, availability of gypsum and carbonates in the layer of 0-40 cm, power of humus horizon etc.

During the amelioration of solonetzic soils it's recommended to apply two main methods: chemical and self-amelioration. Chemical amelioration is based on the process when chemical substances, ameliorants, are added to soil and after the interaction with solonetz the fundamental changes of its chemical and physical properties take place. Gypsum, phosphosypsum, sulphuric acid, acid wastes of industry, green vitriol etc. can be such ameliorants. Under their influence an alkalinity is neutralized, sodium is removed from soil and soil material transfers into aggregate state. Solonetz becomes mellow, it is able to absorb and let precipitation wetness and meltwater through into deep horizons, its efficiency is able to increase up to the level of zonal soils [5-7].

2. Materials and methods

The obligatory condition of chemical amelioration is availability of materials, corresponding characteristics of solonetzic soils properties for every filed massif. The scientists of Siberia and Ural have accumulated experience of amelioration of solonetz with gypsum, phosphogypsum, sulphuric acid, green vitriol, artificial aggregate stabilizer etc.

It's important to note that in conditions of Western Siberia and Ural there are all potential possibilities to actively master these lands. Thus, Altai Territory possesses local reserves of gypsum (about 20 deposits), its large reserves in Ural are located in Perm Territory (Kungur town), reserves of carnallite are in Salikamsk, diatomite – in Kamyshlov (Sverdlovskiy region) and northern districts of Tyumen region. In conditions of Polar Ural there is zeolite deposit. This region uses as ameliorants-coagulants wastes of chemical industry in form of phosphogypsum (Revda town), wastes of paint and coatings industry in form of green vitriol (Chelyabinsk region), carbonic calcium gypsum (industrial waste, Ekaterinburg town), muds of surface and artesian waters cleaning (Tyumen etc). Mining of calcium, magnesium and iron-bearing materials are often used in building sphere and their wastes can be utilized on scientific basis for the ameliorative mastering [8-12].

Taking into account properties of salinized soils even within the limits of one land-user, efficiency of different coagulants will be different. In connection with this, we have carried out comparative study of influence of different ameliorants-coagulants, represented by the wastes of chemical and mining industry.

Goal and objectives of the studies – to determine comparative efficiency of different ameliorants-coagulants on chemical and physical properties of solonetz and bore mud.

The object of study is meadow solonetz. In the first case the studied solonetz refer to multisodium, even its surface can contain up to 40% of exchange capacity and solonetz and sub-solonetz horizons – up to 56-72%. Degree of soil salinization of tested plot is very high. Maximum of salts refers to solonetz and sub-solonetz horizons (table 1).

By anion content, solonetz refers to sulphate-sodic, by cation content – to magnesium-sodium, by the occurrence depth of upper border of water-soluble salts it's saline. Response of soil solution of crusted solonetz changes from alkaline (layer 0-10 cm) to strongly alkaline in the soil stratum of 10-60 cm.

Table 1. Salt composition of water extract of meadow crusted solonetz of sulphate-sodic salinization, Tyumen region

Depth, cm	Salts sum, %	% / Mg-eq/100 g of soil					
		HCO_3^-	Cl^-	SO_4^{2-}	Ca^{2+}	Mg^{2+}	Na^+
0-10	0.284	0.138 / 2.26	0.007 / 0.2	0.029 / 0.6	0.006 / 0.3	0.003 / 0.24	0.0058 / 2.58
10-20	0.389	0.246 / 4.04	0.007 / 0.2	0.029 / 0.6	0.008 / 0.4	0.004 / 0.3	0.025 / 4.14
20-40	0.416	0.253 / 4.14	0.008 / 0.24	0.038 / 0.8	0.003 / 0.16	0.002 / 0.14	0.112 / 4.88
40-60	0.373	0.243 / 3.98	0.008 / 0.24	0.019 / 0.4	0.003 / 0.14	0.001 / 0.12	0.099 / 4.32
60-80	0.341	0.224 / 3.68	0.008 / 0.22	0.014 / 0.3	0.003 / 0.16	0.002 / 0.14	0.09 / 3.9
80-120	0.333	0.209 / 3.42	0.007 / 0.2	0.024 / 0.5	0.002 / 0.1	0.001 / 0.1	0.09 / 3.92

The second object of studies was presented by meadow solonetz which refers to chloride-sulphate (neutral) type of salinization. By cation composition it's sodium or sodium-calcium one (Table 2). Content of water-soluble salts in horizons A_{nax} (0-22 cm) and B_2 (22-41 cm) is 0,385; 0,492%, correspondingly. Ca^{2+} and Mg^{2+} prevail in the content of absorbed cations. Exchange capacity in the layer 0-30 cm is 33,7-35,5 mg-eq/100 g of soil.

Table 2. Analysis of water extract of meadow neutral solonetz of Tyumen region.

Layer depth, cm	Salts sum, %	% / Mg-eq/100 g of soil					
		HCO_3^-	Cl^-	SO_4^{2-}	Ca^{2+}	Mg^{2+}	Na^+
0-10	0.362	0.012	0.082	0.149	0.034	0.006	0.079
		0.20	2.34	3.10	1.72	0.50	3.42
10-20	0.408	0.048	0.118	0.110	0.023	0.015	0.096
		0.78	3.38	2.30	1.04	1.24	4.18
20-40	0.492	0.03	0.165	0.139	0.018	0.017	0.130
		0.38	4.72	2.90	0.92	1.44	5.64
40-60	0.572	0.040	0.173	0.168	0.021	0.016	0.154
		0.66	4.94	1.04	1.04	1.34	6.70
60-80	0.626	0.034	0.158	0.226	0.023	0.015	0.170
		0.56	4.50	4.70	1.14	1.22	7.40
80-100	0.519	0.035	0.154	0.158	0.021	0.017	0.134
		0.58	4.40	3.30	1.04	1.42	5.82

As the third ameliorative object there are bore muds of northern district of Tyumen region. Chlorides and hydrocarbonates prevail in the composition of their water extract. Samples of the bore mud have as a rule chloride-sodic salinization by anion composition and calcium-sodium salinization by cation composition. The source of chlorides is sea sedimentary deposits. The source of soda is caustic and soda salt which is added to the content of bore mud to facilitate the drilling process (table 3.).

Table 3. Results of water extracts of the bore mud, 2006.

BM samples	Water extract % / Mg-eq/100 g of soil								Dry remains, %
	potassium K^+	sodium Na^+	calcium Ca^{2+}	magnesium Mg^{2+}	Chloride Cl^-	sulphates $S0_4^{2-}$	carbonates $C0_2^{2-}$	bicarbonates HCO_3^{2-}	
1	0.002	0.052	0.02	0.006	0.071	0.0001	0.006	0.079	0.21
	0.061	2.244	0.499	0.251	2.00	0.0011	0.100	1.300	
2	0.003	0,076	0,01	0.003	0.107	0.00008	0.006	0.085	0.26
	0.077	3.319	0.249	0.126	3.00	0.00085	0.100	1.400	
3	0.004	0.047	0.02	0.006	0.053	0.00007	0.0	0.061	0.17
	0.096	2.036	0.499	0.251	1.500	0.00075	0.0	1.00	
4	0.005	0.088	0.02	0.006	0.036	0.00028	0.0	0.073	0.20
	0.130	3.819	0.499	0.251	1.00	0.0029	0.0	1.200	
5	0.723	0.150	0.05	0.015	1.314	0.00002	0.006	0.134	2.42
	18.48	6.510	1.247	0.628	37.00	0.00022	0.100	2.200	

The content of exchangeable sodium in the bore mud varies from low to medium sodium. Considering that the chemical and physical properties of bore mud are identical to saline soils, and the number of mud storage pit in Khanty-Mansiisk autonomous district reaches only about three thousand, the situation is similar in the YNAO, this suggests similar approaches to their remediation. A number of coagulant ameliorants used on solonetz according to the results of our research can be used on the bore muds.

Methodology of studies. We determined the effectiveness of coagulants in laboratory conditions by tube method. Combination of similar samples in the tests with neutral and sulphate-sodic solonetz under the increasing samples of different ameliorants with equal gradation and following water filling with

IOP Conf. Series: Earth and Environmental Science **337** (2019) 012030 doi:10.1088/1755-1315/337/1/012030

constant norm provided demonstration of different filterability. This is stipulated by the activity of coagulation and displacement of exchangeable sodium from the soil absorbing complex. Comparative action of all tested ameliorants was assessed according to the water filtration. As a standard we used traditional coagulant – gypsum, maximal amount of filtrate was taken as one. Analogous dose was compared with all other ameliorants, including bore muds.

Preliminary analysis of water extract was carried out according to the following methods: alkalinity, from the soluble carbonates (CO_3) - by potentiometry, by chloride-ions argentometry according to Mohr, sulphate-ions in the presence of indicator nithromazo – according to Aidinyan, calcium and magnesium – by chelatometry, potassium – by difference or by flame-photometry.

Table 4. Scale of efficiency of ameliorated substances on solonetzic soils and bore muds by the filterability.

Ameliorants, including industrial wastes	Sulphate-sodic solonetz	Neutral solonetz	Bore mud
Without ameliorant (control)	0	0	0
Gypsum ($CaSO_4 \cdot 2H_2O$)(st)	1.00	1.00	1.00
Calcium chloride ($CaCl_2 \cdot 2H_2O$)	1.20	1.10	1.10
Quick lime	0.97	-	1.00
Slack lime	-	-	0.20
Ferric sulphate ($FeSO_4 \cdot 7H_2O$)	0.84	1.10	1.70
Aluminum sulphate ($Al_2(SO_4)_3 \cdot 18H_2O$	0.97	1.10	1.10
Diatomite	-	-	0.08
Carnallite	0.20	1.10	0.80
Mud of artesian waters cleaning (heat-treated)	-	-	0.02
Mud of surface waters cleaning (heat-treated)	-	-	0.02
Phosphogypsum	1.10	0.80	1.20

Note: maximal filtration on solonetz of sulphate-sodic salinization made up 75 ml/day by the gypsum, 80 ml/day on neutral and 10 ml/day – on the bore mud.

The data of table 4 shows that the main part of studied minerals and their wastes after the mining and chemical industry are able to fundamentally change physical properties of saline soils and bore mud. It's important to note that without their introduction, there was almost no filterability of these soils and bore muds. This indicates the impossibility of fundamental improvement of these objects without chemical amelioration, The studies have proved the main provision that the ameliorant dose on solonetzic soils depends on the content of exchangeable sodium in them. Neutral solonetz with low level of sodium required 2 times less amount of ameliorant for maximal filtration then high-sodium solonetz of sulphate-sodic salinization.

Preliminary studies have demonstrated wide possibility to use mining and chemical industry wastes for amelioration of solonetz of sulphate-sodic salinization. Filterability varied within 0.84 – 1.20 in relation to the gypsum effect (taken as unit). Influence of carnallite wastes was very weak, under the index 0.2 in relation to gypsum. The range of use of ameliorants on neutral solonetzic could be compared with influence of the standard (gypsum). Range of use of ameliorants on neutral solonetzic soils was wider. All studied ameliorants can be used to master them.

Reclamation of the bore muds in conditions of northern regions of Tyumen region is connected with bad hydrophysical and chemical properties, such as: ash structure, crust formation, high hydrophilic nature, very weak water permeability, alkalinity, toxicity etc. Application of coagulants fundamentally illuminates the series of their negative properties.

It's reasonable to use presented ameliorants, excluding slack lime, on the stage of biological reclamation of the bore muds.

3. Conclusion.

To master the saline soils of different chemism together with traditional ameliorant – gypsum, we should actively use wastes of mining and chemical industry in a form of calcium chloride, ferric sulphate and aluminum, phosphogypsum and quick lime. Carnallite wastes proved themselves well on low-level sodium solonetzic soils of neutral salinization and the bore muds.

References

[1] Kovda V 1937 *Solonchak and solonetzic* (Moscow: USSR Academy of Sciences)
[2] Novikova A 2004 *Saline soils, their distribution in the world, the improvement of land and environmental issues* (Kharkov: HNAU)
[3] Berezin L 2006 *Amelioration and the use of solonetzic soils in Siberia: monograph* (Omsk: FGOU VPO OmGAU)
[4] Semendyaeva N 2014 *Dynamics of salt composition of Baraba solonetzes during the 27–32-year-old action of gypsum* (Novosibirsk: NGAU)
[5] Khaustov A 2006 *Environmental protection in oil production* (Moscow: Delo)
[6] Skipin *L* 2000 *Siberian solonetzic soils: environmental aspects of development* (Yalutorovsk)
[7] Fedotkin V 1993 *Solonetzic soils of Siberia and the Urals* (Novosibirsk: Science)
[8] Petukhova V 2016 *Environmental protection in oil production*
[9] Gaevaya E et al 2017 Possibilities of disposal of drilling waste in the formation of a soil-like environment *Proceedings of higher educational institutions* **2** pp 82-89
[10] Safarov A 2016 Using waste oil and gas industry in the production of environmentally friendly road mixes *Ural Eco. Herald* **1** pp 19-23
[11] Solodovnikov A 2015 Experience disposal of drilling waste in the Khanty-Mansiysk Autonomous Okrug Ugra *Env. protection in oil and gas complex* **12** pp 44-8
[12] Vadyunina A *1961 Methods for determination of physical properties of the soil in field and laboratory* (Moscow: Vysshaya shkola)

IOP Conf. Series: Earth and Environmental Science **337** (2019) 012031 doi:10.1088/1755-1315/337/1/012031

Ecological, biotechnological and economic aspects of wheat grain processing

I Jarkova[1*], J Slepokurova[1], N Alekhina[1] and A Samokhvalov[1]

[1]FSBEI VO "Voronezh State University of Engineering Technologies", Revolution Avenue, 194039, Voronezh, Russia

* Corresponding author: zharir@mail.ru

Abstract. The article is devoted to the consideration of issues related with finding ways how to reduce the environmental burden in the wheat grain processing. Based on the analysis of the current state and prospects for the development of the grain processing, flour and bakery industries, the prospect of finding technological solutions for the processing of wheat grain, precluding the formation of bran as a by-product, is substantiated. It has been shown that, due to its chemical composition, wheat bran is a valuable ingredient from a biomedical point of view for inclusion in the diet of a modern person. It is noted that currently used in the baking industry for the enrichment of high-grade wheat flour with dietary fiber and bran can not be called the best from an economic point of view, since they lead to a significant increase in the cost of the final product for the consumer. The authors proposed the most rational from an environmental and nutritional points of view, the method of processing wheat grain, the technology of grain bread of high nutritional value from bioactivated wheat. Due to the process of bioactivation, the content of phytin in the grain is reduced by 75% compared with the original, that allows increasing the bioavailability of the mineral substances of bread for the human body to assimilate. Economic calculations have shown the advantage of the proposed technology of “Bioamarant” bread compared to a wheat flour product containing wheat diet bran (“Dairy-bran bread”): the cost is lower by 19.54%. When producing bread from bioactivated wheat in the amount of 22.5 tons / day, 13.4 tons of wheat should be spent. That is, the absence of the process of grinding this amount of wheat grain prevents the formation of 2.5 tons of bran per day, as well as the formation of emissions into the atmosphere in the amount of 3.2-6.5 tons (in terms of a year) of harmful substances, which will certainly allow reduce environmental stress.

1. Introduction

The main grain crop in the world is wheat. According to FAO, its global production in 2014 was 729.0 million tons. According to forecasts of the Organization of Economic Cooperation and Development and the USDA, the world wheat harvest in 2024/2025 will be 776.2 million tons. At the same time, among the world leaders in terms of wheat cultivation the third place belongs to Russia [1-3].

Russia exports a third part of the grown grain, about 11% is reserved to the seed fund, 10% is feed grain, and the main part (almost half of the crop) is sent for processing (Figure 1), which results are in a variety of main products. At the same time, the main production is accompanied by the accumulation of by-products (Figure 2).

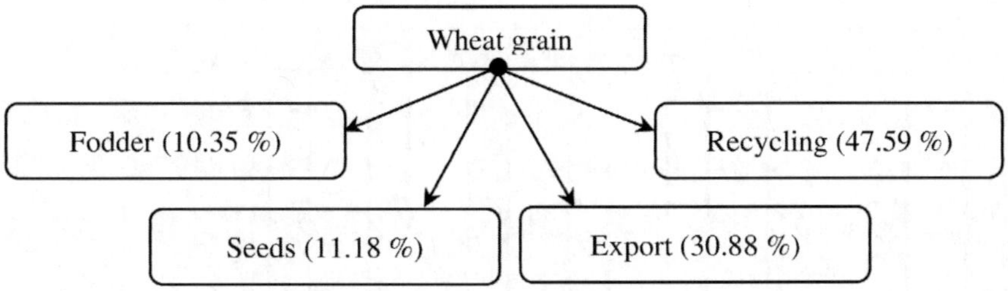

Figure 1. The main uses of grain [4]

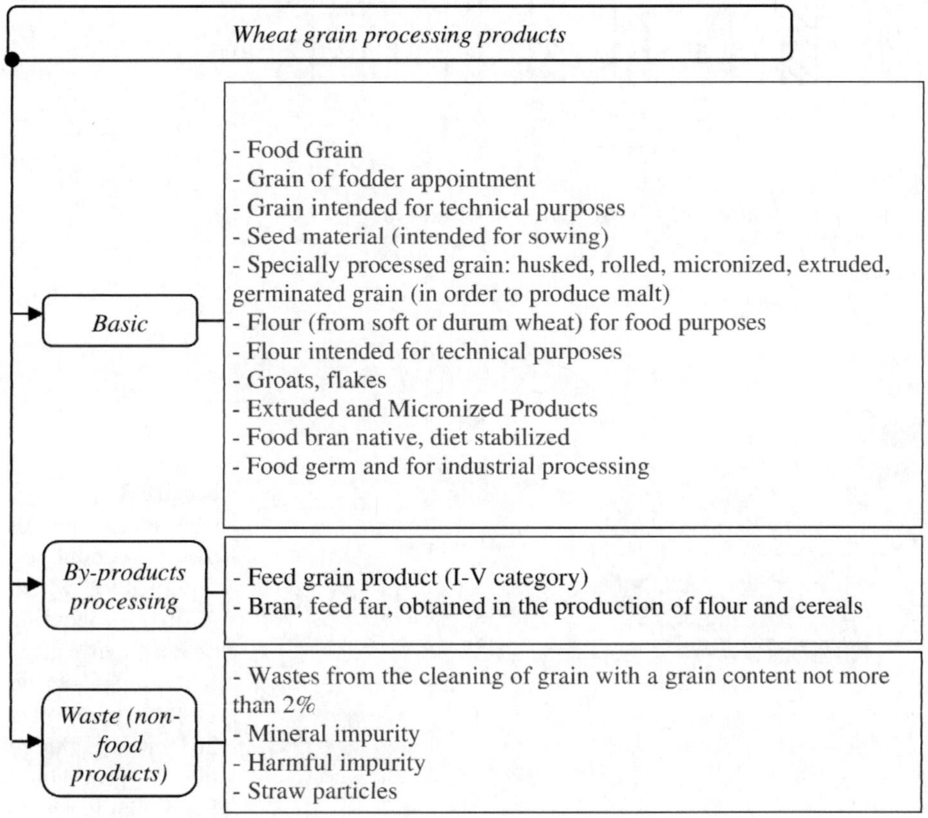

Figure 2. Classification of wheat grain processing products [5]

According to the data provided by the Expert-Analytical Center of Agribusiness [6], wheat flour is produced mainly from the highest grade wheat flour (Fig. 3, a): its annual production for the period 2010-2018. Exceeded the production of first-grade flour by 2.3-3.2 times, and second-grade flour and wheat-rye - by 6.9-11.4 times. The average annual production of wheat flour of the highest grade for the entire period under consideration exceeded the same indicator of the production of first grade flour and the total volume of second grade wheat flour and wheat rye flour, respectively, in 2.7 and 9.6 times (Fig. 3, b).

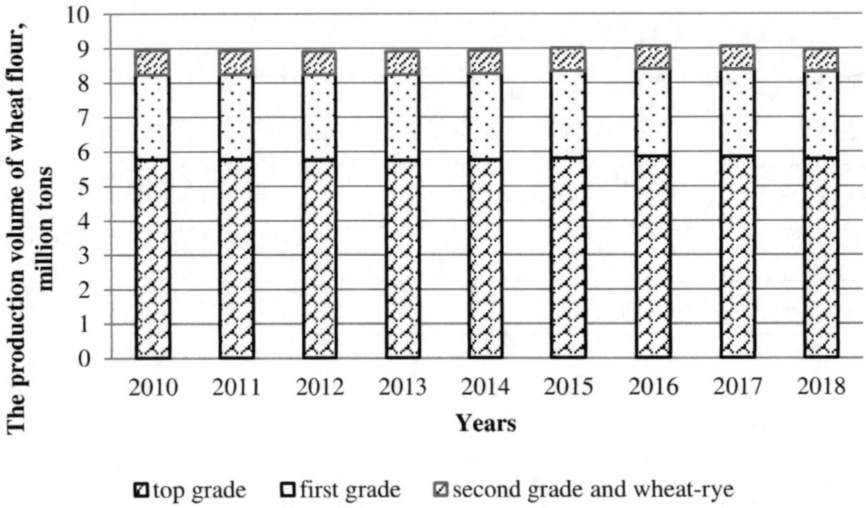

a) the production volume, million tons

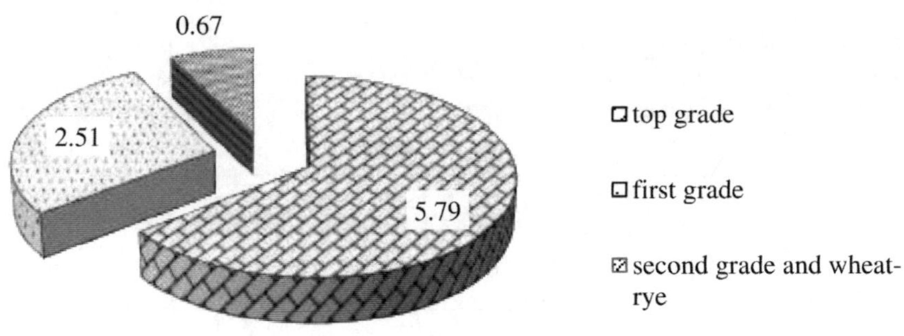

b) the average annual production for the period 2010-2018, million tons

Figure 3. Characteristics of the production volumes of wheat grain processing products in Russia for the period 2010-2018 (in 2018 it is included the production from January to August)

In this way, the statistical data show that at the present stage the process of processing grain into final (commercial) products is accompanied by the receipt of by-products, in particular bran, in the most technologically possible amounts. It is known that during the production of wheat flour of the highest grade, 18.5-21.5% (according to [7] - 12.1-19.1%) bran are formed in parallel, that means the production of 8.402 million tons of wheat flour in 2017 was accompanied by the accumulation of 2.334 million tons of bran. According to [8], in 2007 in Lithuania, the production of 0.402 million tons of flour and cereals was accompanied by the formation of 65.594 million tons of bran. According to the news agency APK-Inform [9], in 2014 the production of wheat flour in Russia amounted to 8.900 million tons, and wheat bran - 2.500 million tons.

According to the Russian Research Institute of grain and its processed products (RRIG), the volume of bran produced at Russian grain processing enterprises can reach 4,500–5,000 tons [10].

Existing volumes and methods of processing wheat grain contribute to the exacerbation of the problem of by-products and waste using and utilization. It should be noted that at the present time bran that is formed in large volumes is not rationally used [11]. Due to the large volumes of this by-product, a negative impact on the environment can occur, such as soil contamination (dumps), water (wastewater from washing machines) [10].

The main method of wheat bran utilization is currently the production of animal feed and poultry feed [12, 13]. This method is economically least expensive and therefore is beneficial for milling enterprises. This is confirmed by the data from the analysis of the products sold range by large milling enterprises, including: OJSC Buturlinovsky Milling Plant (Buturlinovka, Voronezh Region), OJSC Melkombinat (Tver), OJSC Voronezhsky MK, Voronezh , LLC Tambovkrakhmal Agrokombinat (the village of Umet of the Tambov Region)), Kursk Bread Products Plant CJSC (Kursk), Rybinsk Milling Plant JSC (Rybinsk, Yaroslavl Region).

Flour mills sell bran in mound or in compacted (granulated) form. Bran using in unconsolidated form (mound) has significant drawbacks: high costs for their storage and transportation; it takes a lot of space and requires special conditions for storage and transportation; strongly dusty, that increases the risk of fire [14].

Production of dietary bran, which can be used as a raw material for the food industry, is possible. However, the production of dietary wheat bran is insignificant. This is primarily due to necessity of significant financial costs on the part of the enterprise not only for the installation of additional equipment, which makes it possible to obtain a product that meets safety standards requirements of regulatory documents, but also for measures to ensure environmental safety of production associated with necessity of cleaning large volumes of wastewater [8]. The discharge of wastewater from grain processing operations usually requires a permission. Compliance with new legal requirements for environmental protection, health, safety and hygiene may require capital investments in new equipment [15].

Experts note the low technological level of the Russian grain processing industry and the almost complete absence of enterprises that integrate deep grain processing (the production of flour from grain is not deep processing) [16]

RRIG scientists believe that deep processing of grain, including wheat, is one of the promising areas of the grain industry development, the main task of which is the effective all components of grain raw materials using. In the classical sense, this is the process of the grain separating into its components, as a result of which it can be isolated: separate protein fractions, for example, gluten, concentrates and protein isolates are obtained; fractions of starch (A, B, C); soluble and insoluble dietary fiber; some other components [1]. The authors [17] proposes a method of wheat bran utilization by carrying out the enzymatic hydrolysis of starch, followed by the production of bioethanol. A biotechnological method has been developed to process wheat bran to produce a product that can be used to prevent mold growth when storing wet fodder grain and vegetables; as an effective fertilizer while growing organic vegetables and berries; to combat mold while growing food mushrooms [18].

However, these developments of large-scale industrial implementation are currently not received.

Analysis of the scientific and technical information on the problem under consideration has shown that the search for ways how to reduce the environmental burden from the by-product (wheat bran) generated at the flour mills, which are justified from an economic and biomedical point of view, is relevant.

In this regard, we can formulate the objectives of the study:

1. To assess the significance of the wheat bran composition from biomedical point of view.
2. To analyze the existing technologies of using wheat bran in the baking industry.
3. To develop the technology of grain bread with high nutritional value made from bioactivated wheat.
4. To substantiate the possibility of reducing the environmental burden of wheat bran storage and utilization and the economic efficiency of the developed technology for grain bread.

2. Methods

The assessment of significance of the wheat bran composition from biomedical point of view and ways how to use them for reducing the environmental burden was carried out by means of available scientific meta-analysis , technical, statistical and patent information. The assessment of the reduction of environmental stress in the implementation of the proposed method for the production of grain bread was carried out empirically. In addition, analytical and graphical methods, including logical and process approaches, were applied.

The bioactivation of wheat grain was carried out as follows: the wheat grain was preliminarily cleaned of weed and grain impurities, washed. Next, wheat was kept for 20-24 hours in water at a temperature of 18-20 ºC, and then germinated for 10-12 hours. In the process of bioactivation of the grain, the content of phytin was determined by its extraction with hydrochloric acid, followed by precipitation with ferric chloride and the formation of iron phytate insoluble in dilute acid. After the ashing, phosphorus phytin was determined by a colorimetric method (according to Fiske-Subbarou).

The calculation of the cost of the proposed grain bread was performed by the calculation method.

3. Results and Discussion

The problem of grain bran utilization for food purposes was considered by researchers for a long time, and if in the 40s of the last century (immediately after the end of the Second World War) bran acted as an accessible source of protein [19], then at the beginning of the XXI century the focus shifted to the possibility of isolated extraction of these, the protein component [20], as well as the increasing importance of bran, is becoming a source of dietary fiber [21].

The recommended consumption rate of wheat bran is up to 40 g per person with an average weight of 70 kg [22]. This is due to the high content in the bran of biologically valuable substances (table 1), including dietary fiber, protein, minerals, vitamins (B6, PP, E, etc.), phenolic substances with antioxidant properties [22]. It is generally recognized the benefits and feasibility of the inclusion of bran in the formulation of products intended to improve the performance of carbohydrate and lipid metabolism in patients with diabetes mellitus [23-29] for the prevention of cardiovascular diseases [22, 25, 26, 28, 29], normalization of body weight [26, 27], which have antioxidant properties [22].

Table 1. Nutritional Information on Wheat Bran

Indicators	Wheat bran diet GOST R 53496	Wheat bran
Mass fraction of protein, %	16.0	13.2-18.4[1]
Fat content, %	3.8	3.5-3.9[1]
Mass fraction of starch, %	16.6	13.8-24.9[1]
Mass fraction of dietary fiber, %	47.0	43.2
Mass fraction of soluble dietary fiber, %	no standard	9.82[1]
Mass fraction of insoluble dietary fiber, %		33.38[1]
Ash content, %	5.5	3.4-8.1[1]
Biologically active components		
Phytic acid, mg%		2180-5220[1]
Ferulic acid, mg%		500-1500[1]
Vitamins, mg / 100 g:	no standard	
betaine		1000-1300[1]
choline		47-100[1]
niacin		14-18[1]
nicotinic acid	15.5	no information
tocopherols	7.0	1.4[1]
pantothenic acid		2.2-3.9[1]
riboflavin		0.39-0.75[1]
biotin	no standard	0.048[1]
thiamine		0.54[1]
pyridoxine		1-1.3[1]

Mineral substances, mg / 100 g	
Na	25.2±8.74[2]
K	405.5±50.8[2]
Ca	41.5±10.8[2]
Mg	190.6±43.4[2]
Fe	11[1]; 5.68±0.95[2]
Zn	7.3[1]; 3.58±0.55[2]
Se	78[1]

[1] – according to the source [25]; [2] – according to the source [22]

The dietary fibers that make up the bran base have a high water-absorbing capacity, which can contribute to faster saturation and speed up intestinal motility [25]. In addition, dietary fibers can serve as a substrate for the development in the large intestine of microorganisms that synthesize metabolites that improve human health [30, 31].

In the food industry can be used bran made according to GOST R 53496-2009 and corresponding in terms of safety requirements of the TP TC 021/2011 TR.

One of the trends in the production of "healthy" foods is the production of wheat flour with a high content of dietary fiber, which is achieved by adding bran and / or dietary fiber to the highest-grade wheat flour [25, 26, 32]. There are various methods of pretreatment of bran before mixing them with flour (grinding, extrusion, heat treatment, biological treatment), aimed at minimizing their negative impact on the technological properties of flour and, as a consequence, preserving the high quality of bread produced from it [25, 33]. When dosing bran to 15% by weight of flour, the use of improvers, in particular distilled monoglyceride Myverol ™, is recommended to correct the rheological characteristics of the dough and the quality of the bread [26].

Technologies for the use of bran have been developed, subjected to thermoplastic extrusion in the production of cakes and products from shortcrust [34], the production of extruded products [27].

It is noted that the biological value of proteins in a mixture of the highest grade of wheat flour and wheat bran is higher than in the highest grade of wheat flour [19]. This is due to the introduction of bran lysine and tryptophan.

Technologies for the production of bakery products from whole-ground flour (without the selection of bran) have been developed [35, 36].

Nutritionists advise to minimize the use of refined products, to enrich the diet with ballast substances, including dietary fiber, "use unmilled grain", i.e. with shells [21, 28, 37].

The technology for the production of whole-grain bread in which the process of fine grinding of grain and the production of whole-grain mixture occurs using a rotary pulsation hydromechanical homogenizer [38] has been developed.

The main issue requiring attention when consuming whole grains is the presence of phytin in its membranes, which reduces the bioavailability of the mineral substances present in the product and, above all, calcium. We have proposed a method for the production of bread from bioactivated wheat (Fig. 4). It was established that during the swelling of wheat grain within 24 hours in drinking water and its subsequent germination within 12 hours, the content of phytin is reduced by 75% compared with the original grain [39].

The advantage of bread making technology is also a reduction in the cost of its production compared to traditional bakery making technologies: a significant reduction in cost due to the difference in the price of flour and grain, an increase in bread yield [40, 41].

It should be noted that the tendency of reducing the demand for the bran production of mixed feeds (as the main way of their processing) is forming, since in 2016 there is a decrease in the price of feed grain, which is more preferable for this industry in terms of the usefulness of the composition [9].

The use of bran as a fuel is also known [42]. However, it should be noted that the burning of bran is accompanied by the slagging of heating surfaces, sintering of the ash residue, complicates the operation of blower fans and reduces the efficiency of combustion, which together leads to low plant efficiency and high operating costs [43].

To confirm the effectiveness of reducing the environmental burden in the implementation of our proposed method for the production of grain bread, we will carry out a calculation using the example of Voronezh - city of 1 million and 200 thousand people. Suppose that grain bread from bioactivated wheat grain will be consumed by every fourth inhabitant of a vegetable garden in the amount of 50 g per day, that means daily production of such bread should be 22.5 tons. To ensure such a volume of production, 13.4 tons of wheat must be spent (from which 10.9 tons of flour can be obtained). Since we use whole grains, this technological method allows us to avoid the formation and the need to utilize 2.5 tons of bran per day, as well as the formation of emissions into the atmosphere, accompanying the process of grinding the grain. As a reference, it should be noted that according to the methodology "Determination of gross and specific emissions to the atmosphere for grain processing enterprises and elevators" (Annex No. 37 to the order of the Minister of Environmental Protection No. 298 of November 29, 2010), the amount of harmful substances emitted into the atmosphere is 1-2 kg per ton of flour produced, which will reduce the environmental burden in the amount of 3.2-6.5 tons of harmful substances per year.

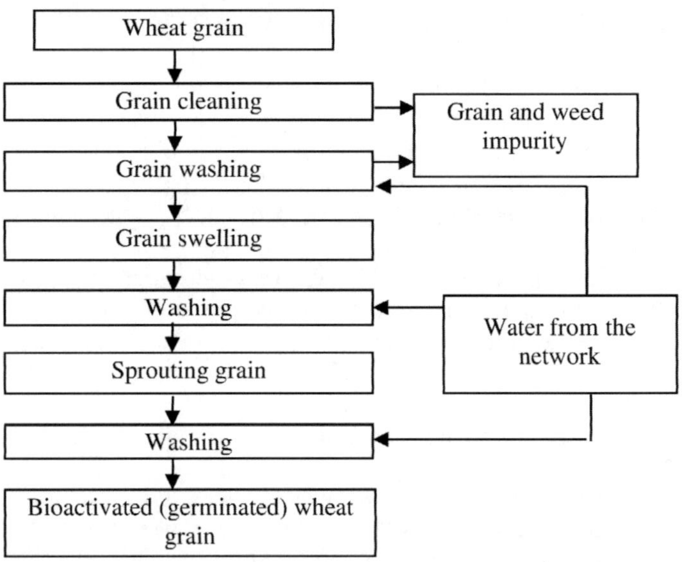

Figure 4. Functional diagram of the production of bioactivated wheat grain

The advantage of the technology of making grain bread from bioactivated wheat is a reduction in the cost of its production compared to traditional technologies for preparing bakery products: a significant reduction in cost due to the difference in the price of flour and grain, reduction in energy costs, increase in bread yield. Confirmation of the economic feasibility of using the proposed technology is the calculation of production costs for traditional and developed technologies, respectively (Table 2).

Table 2. Calculation of the planned cost estimate and the project of the wholesale price of products

Costing items	The cost of 1 ton of bread, rubles:	
	"Milk-bran" (control)	«Bio-amaranth»
Raw materials, basic&supporting materials	23,030	18,530
Transportation and procurement costs	2,300	1,850
Fuel	2,300	1,850
Electricity	2,300	1,850
Basic and additional salary	4,610	3,710
Social insurance contributions	1,390	1,120
Equipment maintenance costs	1,150	0,930
General running costs	1,150	0,930

Production cost	38,240	30,760
Selling expenses	0,290	0,240
Total cost	38,530	31,000
Profitability,%	20	20
Price per unit of product (weight 0.5 kg), rubles	25.4	20.4

The calculated indicators show that the feasibility of using new technology, as the cost of 1 ton of «Bio-amaranth» bread is on 19.54% lower than the control product («Milk-bran» bread), and the price of 0.5 kg of the control and experienced products is 25.4 rub. and 20.4 rub. respectively.

4. Conclusions

In this way, based on the completed research, we can make the following conclusions:

1. Due to its chemical composition, wheat bran is a valuable medico-biological ingredient for inclusion in the modern person diet.

2. The technologies used in the baking industry for the enrichment of high-grade wheat flour with dietary fiber and bran are not the best from the economic point of view, since they lead to a significant increase in the cost of the final product for the consumer. This is largely due to the need to solve environmental problems arising from the organization of the grinding of grain into flour.

3. We have proposed a method for processing wheat grain, precluding the formation of bran (Fig. 5) - the technology of grain bread of enhanced nutritional value from bioactivated wheat grain was developed. Due to the process of bioactivation, the content of phytin in the grain is reduced by 75% compared to the original, which allows increasing the bioavailability of the mineral substances of bread for the human body to digest.

4. Economic calculations showed the advantage of the proposed technology of «Bio-amaranth» bread compared to a wheat flour product containing wheat diet bran (dairy bran bread): the cost price is lower by 19.54%.

5. When producing bread from bioactivated wheat in the amount of 22.5 tons / day, 13.4 tons of wheat should be spent. That is, the absence of the process of grinding this amount of wheat grain prevents the formation of 2.5 tons of bran per day, as well as the formation of emissions into the atmosphere in the amount of 3.2-6.5 tons (in terms of a year) of harmful substances, which will certainly allow reduce environmental stress.

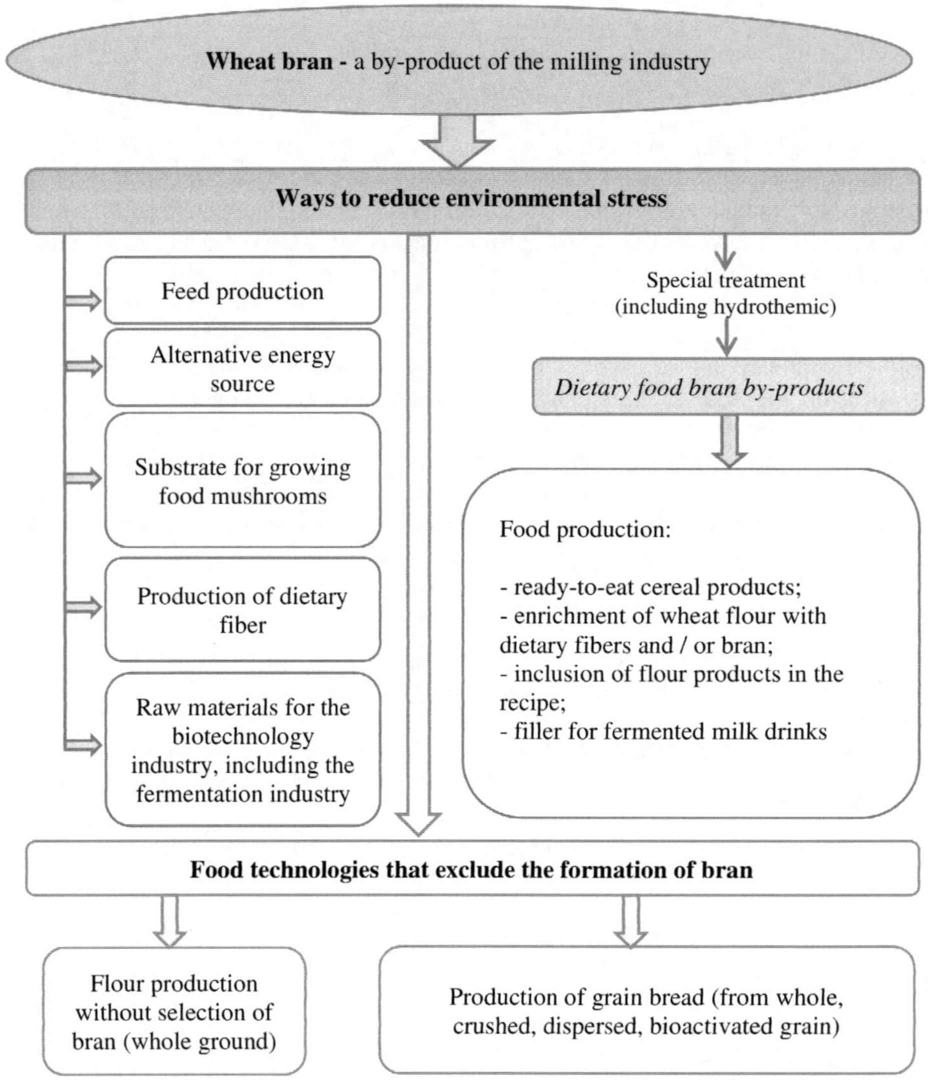

Figure 5. The main ways to reduce the environmental load from the by-product of flour by-products - wheat bran

5. Acknowledgement
This work was executed with financial support of applied researches by the Ministry of Education and Science of the Russian Federation in the framework of the Federal target program "Research and development on priority directions of scientific-technological complex development of Russia for 2014-2020" with the agreement on grant No. 14.577.21.0256 of 26 September 2017, the Unique identifier, PYAR RFMEFI57717X0256.

References
[1] Meleshkina E P *2018 Modern methods, means and standards in the field of evaluation of the quality of grain and grain products, Collection of materials of the 15th All-Russian Scientific and Practical Conference (June 4-8, 2018, Anapa)*
[2] http://www.tsenovik.ru/news/Novosti-APK/Korma/V-sezone201718-Rossiya-stanetuverennym-liderom-mirovogo-eksporta-pshenitsy/?sphrase_id=1363264
[3] The expert-analytical center of agribusiness "AB-Center" *Production of wheat in the world. Wheat producing countries* http://ab-centre.ru/page/proizvodstvo-pshenicy-v-mire-strany-proizvoditeli-pshenicy

[4] Beregantova E V 2016 *The market for products of deep processing of grain in the Russian Federation: state, prospects* (Moscow: NRU HSE)
[5] *Order of the Russian State Bread Inspectorate from 05.09.03 № 49* 2003
[6] http://ab-centre.ru/chart/muka-pshenichnaya-proizvodstvo-v-rossii-po-sortam-po-mesyacam
[7] Golubev I G, Shvanskay I A, Konovalenko L Yu, Lopatnikov M V 2011 *Waste Recycling in the Agricultural Sector: A Handbook* (Moscow: FSBI Rosinformagrotekh)
[8] Kliopova I, Petraškienė V 2009 *Environmental Research, Engineering and Management J.* **3** (49) pp 44-55
[9] Kalaida P *Russian wheat bran market: price trends 2013/14* https://www.apk-inform.com/ru/exclusive/topic/1030820#.W8ib0bsVSUk
[10] Solovieva J P, Meleshkina E P 2018 *Modern methods, means and standards in the field of evaluation of the quality of grain and grain products, Collection of materials of the 15th All-Russian Scientific and Practical Conference (June 4-8, 2018, Anapa)* pp 86-8
[11] Khusid S B, Zholobova I S, Nikolaevich D S, Nesterenko E E 2014 *Scientific journal KubGAU* **98** (04) pp 1-9
[12] Antimonov S V, Solovich S Y, Ganin E V 2008 *News of universities. Food technology* **4** pp 8-10
[13] Solovieva Zh P, Solovieva E V 2018 *Modern methods, means and standards in the field of evaluation of the quality of grain and grain products, Collection of materials of the 15th All-Russian Scientific and Practical Conference (June 4-8, 2018, Anapa)* pp 86-88
[14] https://www.openbusiness.ru/biz/business/svoy-biznes-proizvodstvo-otrubey/
[15] European Bank for Reconstruction and Development (EBRD) 2009 *Environmental and Social Guidelines by Industry: Products from Grain-Processing Plants* 2009 14 p. https://www.ebrd.com/downloads/about/sustainability/grainr.pdf
[16] Gluhih S 2010 *Feed J.* **2** pp 56-8
[17] Ohtemenko I N, Sartakova O Yu, Sartakova E A 2009 *Polzunovsky messenger J.* **3** 340-4
[18] Machihina L 2010 *Bakery products J.* **2** pp 36-9
[19] Chick H, Cutting M E M, Martin C J, Slack E B 1947 *British J. of Nutrition*, **1** (2-3) pp 161-82
[20] Kolpakova V V, Zaytseva L V, Smirnov E A 2005 *Pat RF 2250026 A23J1 / 12 A method of obtaining a protein product from bran*
[21] Global strategy on nutrition, physical activity and health. World Health Organization http://www.who.int/diabetes/ru/
[22] Rubio C, González-Weller D, Caballero J M, Romano A R, Paz S, Hardisson A, Gutiérrez Á J, Revert C 2018 *Journal of Functional Foods* **48** pp 558-65
[23] Sharafetdinov H H 2008 *Nutrition issues J.* **2** (77) pp 23–31
[24] Krivonosova O G, Petrova V A, G I 2014 *Nutrition issues J.* **3** (83) p 57
[25] Zhang H, Wang H, Cao X 2018 *J. Food Res Int.* **113** pp 24-35
[26] Saccotelli M A, Conte A, Burrafato K R, Calligaris S, Manzocco L 2016 *Food Sci Nutr.* **5** (3) pp 689-95
[27] Oladiran D A, Emmambux M N, de Kock H L 2018 *LWT - Food Science and Technology* **97** pp 778-86
[28] U.S. Department of Health and Human Services and U.S. Department of Agriculture 2015 *2015 – 2020 Dietary Guidelines for Americans. 8th Edition* https://health.gov/dietaryguidelines/2015/guidelines/.
[29] Kazakov E D 1997 News of universities *Food technology J.* **4-5** pp 9-11
[30] Flint H J, Duncan S H, Louis P 2017 *Current Opinion in Microbiology J.* **38** pp 59–65
[31] Holscher H D 2017 Dietary fiber and prebiotics and the gastrointestinal microbiota *Gut Microbes* **8** (2) pp 172–84
[32] Hutchinson J B, Martin H F 1970 *J Sci Food Agric.* **21** (3) pp 148-51
[33] Lai C S, Hoseney R C, Davis A B 1989 *Cereal Chem.* **66** (3) pp 217-9
[34] Kuscheva N B 1994 *The use of structured wheat bran in the production of culinary products (abstract of Ph.D.) (St. Petersburg: Rotaprint SPTEI)*

[35] Sayed-Ahmad B, Talou Th, Straumite E 2018 Evaluation of Nutritional and Technological Attributes of Whole Wheat Based Bread Fortified with Chia Flour *Foods* **7** (9)
[36] Rebellato A P, Klein B, Wagner R 2018 *Food Sci Technol. J.* **55** (9) pp 3575-83
[37] Mellette T, Yerxa K, Therrien M, Camire M E 2018 *Foods J.* **7** (6)
[38] Nemenuschaya L A 2018 *Modern methods, means and standards in the field of evaluation of the quality of grain and grain products, Collection of materials of the 15th All-Russian Scientific and Practical Conference (June 4-8, 2018, Anapa)* pp 89-92
[39] Alehina N N, Ponomareva E I, Nikiforova I V 2018 *The change in the course of the technological process of the content of fitin in cereal crops, Collection: Technologies and equipment of the chemical, biotechnological and food industry Materials of the XI All-Russian Scientific and Practical Conference of Students, Postgraduates and Young Scientists with International Participation* pp 556-557
[40] Alehina N N 2007 *Development of accelerated technology of bread of increased nutritional value from bioactivated wheat (dis. Cand. Tek.)* (Voronezh: VGTA)
[41] Sanina T V, Magomedov G O, Alehina N N 2008 *Bread from bioactivated wheat (monograph)* (Voronezh: VGTA)
[42] Degtyareva D 2018 *In the world of science J.* **5-6**
[43] Gaidabrus M A, Valter K A, Zaitsev A I *2017 Possible directions for the use of bran in heat and power engineering, Collection of materials of the All-Russian scientific-practical conference of young scientists "Young Russia" (April 18-21, 2017 Kemerovo)*

IOP Conf. Series: Earth and Environmental Science **337** (2019) 012032 doi:10.1088/1755-1315/337/1/012032

From solid waste management towards the circular economy and digital driven symbiosis

H Ruohomaa[1*] and N Ivanova[2*]

[1]Häme University of Applied Sciences, Hämeenlinna, Finland
[2]Petersburg State Transport University, Visiting Prof. of HAMK University

* Corresponding authors: heikki.ruohomaa@hamk.fi, nivanova@pgups.ru

Abstract. This article is about regional development based on the principles of circular economy and the industrial symbiosis framework. Nowadays tremendous amount of solid waste from household, industrial and agricultural materials accumulate in the territories of different countries around whole world. Finland has experience in waste techno parks design and in the development of waste branch around the whole country. Finland has considered be one of the leading country in Circular Economy and its government strategic goal is to be the leading country in the world in Circular Economy until 2025. The Finnish experience shows that there has to be close co-operation between companies, universities and authorities, but also common goal and understanding.
The main questions: which are the elements that could be adapted from the Finnish Circular Economy and the resource efficiency thinking, and how to implement ideas into other cultural and industrial environments in Russia. Our researchers together could discuss and decide which are the key elements that can be used and which of them contradict to using industrial symbiosis development in Russia.

1. Introduction

The World Economic Forum report [1] has termed this period of accelerating innovation in science and technology – the transformative change in data and technology capabilities combined with a merging of digital, physical and biological realms and its consequences on society as the Fourth Industrial Revolution. It is not only transforming social networks, scientific research and whole industries; it is also radically reshaping biological and material science innovations. Harnessing these opportunities and proactively managing the risks manifest by the rapid evolution of new science and technologies will inevitably require more creativity and agility in current governance frameworks and financing arrangements.

A crucial role in market evolution is happening by exploring how businesses can use the circular economy to drive arbitrage opportunities across complex, global supply chains [2]. A circular economy is restorative and regenerative by design and aims to keep products, components, and materials at their highest utility and value at all times. It questions whether with creativity and innovation we can build a restorative economy.

To be successful on new challenges of Industry 4.0 development, enterprise-university partnerships has to be intense and main objective should be a shared learning [3]. Long-term co-operation creates a background for new co-innovation and co- evolution. The technologies of the Fourth Industrial Revolution have generated enormous excitement about the opportunities they offer as well as concern about governance, regulation and ethics [1].

There are as well exciting possibilities: The convergence of new technologies is creating unprecedented opportunities in all aspects, from business-to-business commerce to humanitarian intervention. The melding of artificial intelligence (AI) with big data capabilities – not to mention the actual exponential accumulation of data itself – has created a fascinating world of communications, collaboration and interaction, not just between people but also between machines and between people and machines. Through digital transformation, the use of new technologies like big data, open data, cloud, IoT, platforms, artificial intelligence, and social networks with increasing intelligence and automation enterprises can capitalize on new opportunities and optimize existing operations to achieve significant business improvement [4].

2. Theoretical background

The new approach is the revised legislative proposal on waste. It covers long-term recycling targets for municipal and packaging waste, measures to limit landfilling, and incentives for Member States to use economic instruments at the national level [5]. It also aims to promote cooperation between industries, with waste from one process becoming secondary materials for others, through a simplified legal framework for by-products and end-of-waste status, creating more certainty for operators in these markets. That leads us for an opportunity to build value networks from the point of side flows of waste. It leads to a business opportunity for industrial symbiosis.

The key notice on sustainability is what extent it will be impact to specific business sector and influences on creation of circular economy business and industrial symbiosis. It is an industry-changing paradigm integrating innovation, differentiation and transformation. Tajani [6] has started, that there will be no sustainability without competitiveness, and there will be no long-lasting competitiveness without sustainability and there will be neither of them without a quantum leap in innovation.

Industry 4.0 describes the organization of production processes based on technology and devices autonomously communicating with each other along the value chain. Industry 4.0 architecture takes account of the increased digitalization of various industries where physical objects are seamlessly integrated into the information network, allowing for decentralized production and real-time adaptation in the future [7]. What is characteristics to Industry 4.0 is that it involves connecting products to each other. Industry 4.0 is a closely linked to Cyber-Physical Systems (CPS) [8] and can be defined as transformative technologies, which manage interconnected systems between its physical assets and computational capabilities [9]. To be successful on new challenges of Industry 4.0 development, enterprise-university partnerships has to be intense and main objective should be a shared learning. Long-term co-operation creates a background for new co-innovation and co- evolution. Adapting Industry 4.0 framework as a basis for development activities has expected to provide an opportunity for remarkable competitive advantage for businesses but also for regions [10].

Nidumolu, Prahalad and Rangaswami, [11] explain widely why sustainability is now the key driver of innovation. The biggest benefit comes from generating and processing big data. Information technology helps make large amounts of data available for different purposes in the public or private sector with minimum effort and costs [12]. Conradie & Choenni [13] have recognized that the most reliable method how to identify the datasets with big potential is a demand-based approach, which means a close cooperation with potential end users. Every participant in ecosystem has an opportunity to participate on creation of innovations and value, which generally is produced on the boundaries of ecosystem stakeholders [14]. Business ecosystem can be born as group activity or a single company can create it by collaborative activities [15].

3. Research questions and research approach

Nations are struggling with big global challenges like climate change, lack of resources, waste problems etc., but nations and companies are also facing changes because of digitalization at the stage of the fourth industrial devolution.

The role of circular economy and digitalization as a business driver is growing in waste management and as digital symbiosis in industrial innovation platform. The main research questions are

a) What are the future key factors to solve waste problem on a sustainable way?

b) How to manage change towards circular economy from waste management?
c) How to evaluate the profitability of circular economy development investment?

This article introduces a concept model for utilizing circular economy and digitalization as a business and innovation driver to facilitate the transition from waste management towards circular economy and digital symbiosis.

4. Circular economy framework

The circular economy refers to an industrial economy that is restorative by intention; aims to rely on renewable energy; minimizes, tracks, and hopefully eliminates the use of toxic chemicals; and eradicates waste through careful design [2]. The system diagram in **figure 1** illustrates the continuous flow of technical and biological materials through the value network.

The term goes beyond the mechanics of production and consumption of goods and services, in the areas that it seeks to redefine (examples include rebuilding capital including social and natural, and the shift from consumer to user). The concept of the circular economy is grounded in the study of non-linear, particularly living systems [2]. This ensures enhanced flows of goods and services.

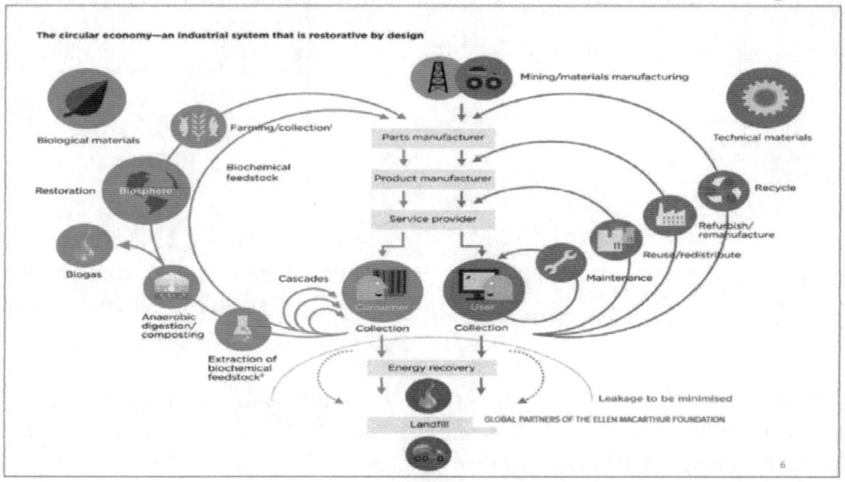

Figure 1. The circular economy - an industrial system that is restorative by design [2].

5. New business via industrial symbiosis

Changing economic structures, intensifying environmental problems and increasing demand-related requirements all mean that companies need to come up with more resource-intensive solutions. These can be achieved through co-operation between companies operating in various sectors, and alongside local authorities and municipalities.

Industrial symbiosis is an industrial ecosystem in which unused or residual resources of one company are used by another. This results in mutual economic, social and environmental benefits. It is a process involving several companies – firms that complement one another provide mutual added value through efficient use of raw materials, technology, services and energy. Joint development of industrial symbiosis provides an efficient way of thinking up, developing and testing product and service innovations. This, in turn, engenders new competencies, on which new business operations can utilize.

Business ecosystems based on industrial symbiosis provide more added value by using fewer natural resources than traditional industrial value chains. This means more efficient energy and water consumption, and reduced waste. In industrial symbiosis, industrial operations, energy production, primary production, waste processing and the services supporting these combine to form an entity that provides products and services for end-user needs, while optimizing resource use. [19]

6. Challenges and opportunities in Russia

21st century is characterized as the first Urban Century in the history of humankind.

In Russia waste management was a challenge already in Soviet times, but since the beginning of 2000ies it has become one of the most critical economic and environmental problems in Russia. For many years, the Russian government's priorities focused on oil and gas, defence and heavy industries, while waste management was considered to be an issue for future generations and continuously neglected.

Russia has a legal and regulatory framework in the environmental area, but responsibilities are distributed between many bodies that enforce and control the regulations poorly. As a result, environmental protection in general and waste management in particular, is an increasing concern.

With recent changes in lifestyle, increasing waste volumes to such high levels that current facilities and outdated technologies are insufficient, the problem becomes definitely crucial. Experts believe that poor environmental situation contribute to a also short life expectancy at birth in Russian Federation [18].

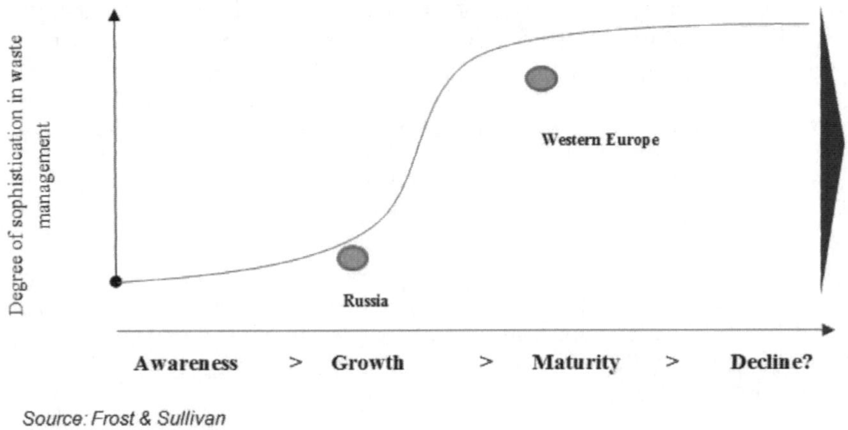

Figure 2. There is huge potential to develop circular economy in Russia [18].

Anyhow, comparison to western European development can be misleading, because of size of country or the structure of industry or the consumer habits. Eg. in case of Finland, all the waste is not collected and treated in a proper way because of long distances.

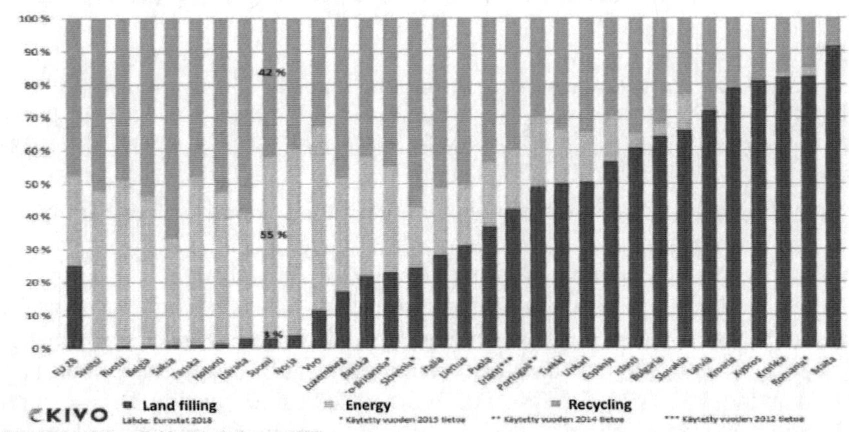

Figure 3. Treatment of Household waste in European countries in year 2016 [19]..

Circular Economy concept has also a very strong strategic approach for many countries. Since some materials, which are needed in high-tech products, are running short and availability from the ground is decreasing. Therefore, those strategic materials, which are needed, are mainly in the products, which we use now.

In case of Finland, circular economy is a part of Finnish innovation policy and Finland wants be the leading country in circular economy in the year 2025.

Experiences coming from Finland show that joint development of industrial symbiosis provides an efficient way of thinking, developing and testing product and service innovations. This, in turn, engenders new competencies, on which new business operations can be based.

The co-operation between government, enterprise and universities is essential to succeed in co-evolution, when building up cumulative competence in creation of solutions for circular economy by benefiting digitalization in it. It is also essential to have a common vision to direct the local operation and funding. Otherwise, the activities can splinter as small pieces and do not form parts of the whole vision.

In rapidly changing operational environment, it a clear and commonly understood vision is required.

7. Forming triangle - Industrial symbiosis, digitalization and legislation

On the above mentioned, we can suppose that triangle consists of the industrial symbiosis, digitalization and legislation. This industrial symbiosis can involve stakeholders at all levels (companies, authorities of different levels and municipalities, smart and digital technologies and legislation agencies and experts) to develop a methodology and software platform to facilitate the exchange of information that can support industrial symbiosis networks, create and support pilot projects and replicate their results at local and regional levels.

Such model has to be designed on the triangle presented on the Figure 4 and, considering Economical Environmental and social issues during whole companies' lifecycle, it can lead to use fewer natural resources and more benefit. Applying these models can create new business and new challenges for companies and economy.

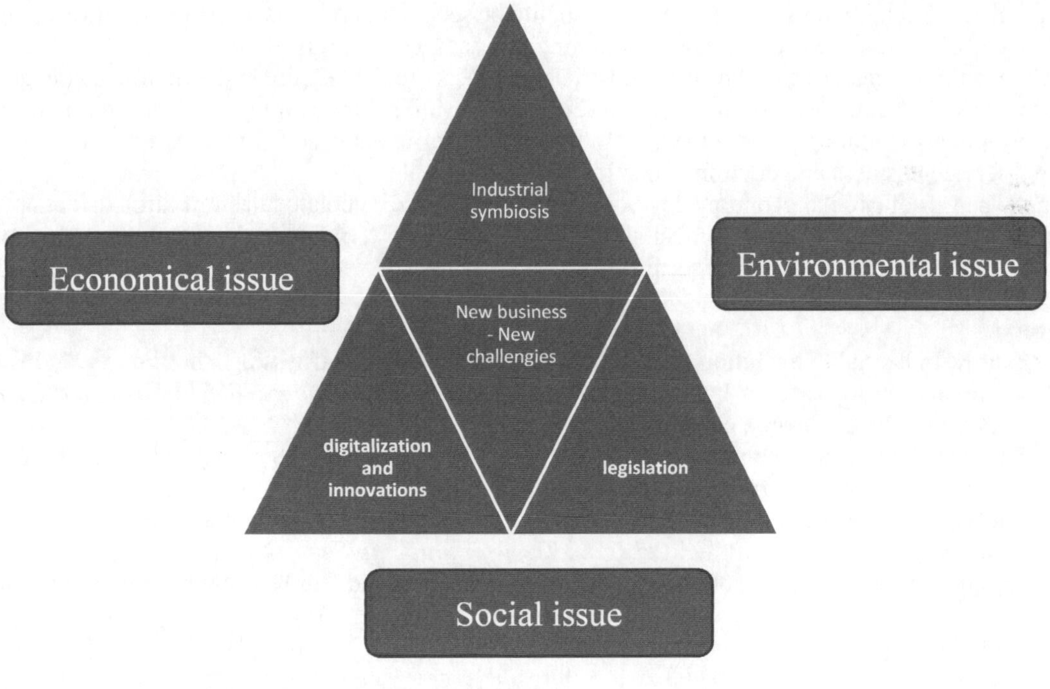

Figure 4. Industrial symbiosis, digitalization and legislation as forming triangle

8. Towards the opportunities of new material sources

Megacities are facing challenges in its waste management. On the other hand megacities have not yet been active to take use the latest concepts in circular economy thinking. Including this, in the fast development of digitalisation, there will be new technical solutions in market, which can provide new possibilities and business models.

Therefore, there are new tools and concepts to build up the future approach and strategy to solve waste problem or, in the other words, take benefit to circulate material and use the waste for the materials of new products or turn into energy. One proposed approach is to take full use of the concept of circular economy and learn the best practices from other countries. Anyhow, learning the best practices does not mean simply coping ideas from other countries and bringing copies. In these cases, we should see learning so that we are able to implement the best practices, so that they fit into local environment and traditions.

In this case, the approach will be related to circular economy and digitalisation, there will be need for large and long-lasting cooperation between government, universities and enterprises to be able to manage the change. Anyhow, this change has to be seen so that it creates huge amount new innovation and build-up new business models and business.

The expenses of circular economy development should not be calculated in short turn business profit basis but it should be seen as long term strategic investment which will bring long term benefits to different sectors of society.

9. Discussion and Conclusions

Circular economy and industrial symbiosis should not be seen only the way to solve the waste management problem, but it can also be seen as the part of social and economic development. The change towards circular economy takes time and it needs actions from different sectors (universities, government and private sector).

Digital ecosystems and industrial symbiosis should be seen as continuous developing entirety, which will change and renew themselves thus maintaining their competiveness.

Digitalization, circular economy and industrial symbiosis are the key drivers, which will change the business environments, a way to act and create innovation. Innovations will need "real life" (transdisciplinary) piloting environments. Both, resource efficiency and digitalization are directing towards service business and environments.

The expenses of circular economy development should not be calculated in short turn business profit basis but circular economy development should be seen as long term strategic investment which will bring long term benefits to different sectors of society.

References

[1] Fourth Industrial Revolution for the Earth Series 2018 Harnessing the Fourth Industrial Revolution for Life on Land *Towards an Inclusive Bio-Economy. World Economic Forum, January 2018, Geneva, Switzerland*

[2] Towards the Circular Economy 2014 Accelerating the scale-up across global supply chains World Economic Forum Report, February 2014, Geneva, Switzerland

[3] Ruohomaa H, Mäntyneva M, Salminen V 2018 Renewing a University to Support Smart Manufacturing within a Region *Digital Transformation in Smart Manufacturing*

[4] Salminen V, Kantola J, Ruohomaa H 2016 Digitalization and Big Data Supporting Responsible Business Co-Evolution *2nd International Co- Evolute Conference on Human Factors, Business Management and Society (Inside AHFE 2016), Orlando, USA July 27-31, 2016.*

[5] European Commission 02.03.2016 https://kumu.io/ellenmacarthurfoundation/educational-resources#circular-economy-educational-resources/key-for-general-resources-map/intro-to-the-circular-economy

[6] EU Environment 2016 Taking Europe closer to the circular economy *Magazine Environment for Europeans*

[7] European parliament, briefing (Sept 2015) 4.0 Industry digitalization for productivity and growth. http://www.europarl.europa.eu/thinktank (internet)

[8] Deloitte (2015) Industry 4.0 challenge: Challenges and solutions for the digital transformation

and use of exponential technologies
[9] PwC (2016) 4.0 Industry: Building the digital enterprise. Global Industry Survey
[10] Ruohomaa H, Kantola J, Salminen V 2017 *Value Network Development in Industry 4.0 Environment Advances in Human Factors, Business Management and Leadership* (Cham: Springer International Publishing AG)
[11] Nidumolu R, Prahalad C K, Rangaswami M R 2009 Why Sustainability is Now the Key Driver of Innovation *Harward Business Review*
[12] Kitchin R 2014 Big Data, new epistemologies and paradigm shifts *Big Data and Society, April–June*
[13] Conradie P, Choenni S 2014 On the Barriers for Local Government Releasing Open Data *Government Information Quarterly* **31** https://www.researchgate.net/publication/261989071 (March 1,2018)
[14] Iansiti M, Levien R 2004 *The keystone advantage: what the new dynamics of business ecosystems mean for strategy, innovation, and sustainability* (Boston: Harvard Business School Press)
[15] Järvi K et al 2013 Ecosystem architecture design: Endogenous and exogenous structural properties *Acta Universitatis Lappeenrantaensis*
[16] 2018 Industrial Symbiosis - One man's waste is another man's raw material. *Sitra* https://www.sitra.fi/en/topics/industrial-symbiosis/#what-is-it-about
[17] Future of waste management in Russia 2018 Tekes, Helsinki, Finland Eurostad 2018, graph by CEWEP

Utilization of wastes from grain processing industries

S Efremova[1*], Ye Kulikova [2], V Konovalov[3] and N A Politaeva[4]

[1]Department of biotechnology and tehnosfernaja safety, Penza State Technological University, Penza, 1A/11 proyezd Baydukova/Gagarina street, Penza 440039 Russia,
[2]Department of plant breeding, seed production and plant biology, Penza State Agrarian University, Penza State Agrarian University, Penza, 30 Batanichtskaj street, 440014 Russia,
[3]Department of Machine Building Technology, Penza State Technological University, 1A/11 proyezd Baydukova/Gagarina street, Penza, 440039 Russia,
[4]Saint Petersburg State Polytechnical University named after Peter the Great, Saint Petersburg, Russia

* Corresponding author: efremova_s15@mail.ru

Abstract. Waste materials resulting from the grain processing industries are harmless and easily amenable to enzymatic and microbial bioconversions. Basidiomycetous mushrooms grown in vitro are promising producers of protein and essential amino acids. Mushroom biomass production under appropriate cultivation conditions using agricultural by-products is a cheap and affordable way to obtain a high-quality food product. The present work explores the possibility of using grain wastes for the cultivation of golden oyster mushrooms under deep cultivation conditions. For these purposes, cornmeal was used as a source of carbon waste and gluten as a source of nitrogen. A high substrate colonization rate was observed. The fruiting cycle lasted about 26 to 28 days on the studied substrates. The most prolonged cycles were obtained on the medium supplemented with sawdust. The maximum and the minimum yields were obtained on wheat straw substrate (171.2 g per kg of substrate) and a grain waste substrate (113.4 g per kg) respectively. On birch sawdust subtrate and on a substrate of grain waste and birch sawdust mixture, the yields reached 143.2 g and 130.3 g in this order. The average yielding crop attained 113.4 g of fresh mushrooms grown on 1 kg of substrate under a relative humidity of 70-75%. This promising work stream not only proposes a waste disposal method, but also, a significant source of income for the grain processing enterprises.

1. Introduction

The increasing economic growth rate of the agriculture and food industry has aggravated the problem of waste management, including grain wastes disposal [1, 2].

Waste materials resulting from the grain processing industries are harmless and easily amenable to enzymatic and microbial bioconversions as well as to various types of preprocessing [3, 4]. Therefore, it would be relevant to continue exploring this research area.

Bioconversion is a natural way to utilize cellulosic waste. It is based on the destruction of an organic substrate by microorganisms and fungi. In this way, it allows to solve two main tasks: the creation of an economically advantageous production process of a target product and, a method of waste disposal for potential environmental pollutants [5 -7].

According to the study of mushroom growing techniques worldwide, different types of wastes are used for substrate preparation, among which: corn stalks, cabbage stalks, rice straw, sunflower husks, coconut shells, cotton and wood processing waste. For the growth of oyster mushrooms, mixtures of straw, soybeans, flax, potato peel, cocoa bean processing waste, sugar cane, coffee, tobacco and grapes are also used as a substrate [8 - 12]. In previous studies, industrial waste has been successfully used for the cultivation of blue oyster mushrooms. In fact, the use of cardboard as a substrate not only increased the growth yields, but also provided greater biological efficiency correlated with a greater productivity of mushroom with larger cap diameter [13].

Under the same cultivation conditions, the spawn rate of oyster mushrooms on substrate dry weight basis will differ depending on the used substrate [14]. In previous studies, winter wheat straw provided a yield of 64.6% of fruiting bodies, corncobs - 46.7%; cotton waste - 68.4%; cotton waste and straw (4:1) mixtrure - 75.8% [15-18].

The main substrate that could be used for the cultivation of oyster mushrooms are wastes such as vegetable wastes from the agriculture industry. However, in order to expand the resource base of industrial production of mycelium of higher fungi and to increase the yield of a biomass enriched with protein, deep cultivation on starch containing plant substrates, including grain waste, is proposed. Yet, grain waste requires a mandatory pretreatment using acid or enzymatic hydrolysis [19].

The pretreatment of raw materials contributes directly to an increase in the yield of mycelium and its protein content. Enzymatic hydrolysis is an environmentally friendly process, since there is no threat to human health. On the other hand, acid hydrolysis requires the use of strong inorganic acids that emit harmful emanations. In this regard, grain processing wastes were pretreated using "Alfalad" and "Glyukolad" enzyme preparations produced by Ladyzhensky factory.

Basidial mushrooms grown under artificial conditions (including Pleurotus ostreatus) are promising producers of environmentally friendly proteins and essential amino acids. Mushroom biomass production using deep cultivation conditions with the optimal growth environment and favorable nutritional medium, is a cheap and affordable way to obtain high-quality food products [19]. The mycelium of the Pleurotus genus cultivated on liquid nutritional medium contains 40-50% of crude protein and 30-40% of true protein. These proteins are considered valuable biological compounds and are characterized by a good digestibility, and the mycelium contains all the essential amino acids. The digestibility of proteins of mushroom biomass obtained by deep cultivation represents 59.3-79.4%.

Thus, one of the potential plant substrates for mushroom cultivation is starch-containing grain waste from the grain processing industry. Therefore, the purpose of this work is to study the potential use of grain waste for the cultivation of golden oyster mushroom (Pleurotus citrinopileatus).

2. Methods

The study of the use of grain waste for the cultivation of golden oyster mushroom (Pleurotus citrinopileatus) was carried out in laboratory conditions with the following experiment scheme: 1. Control - wheat straw; 2. Grain waste; 3. Grain waste + birch sawdust (in equal proportions); 4. Birch sawdust. The experiment was performed in triplicate.

The studied substrates were filled into cotton bags and placed in fermentation tanks at a temperature of 60–70° C for 23–24 hours while basifying water with quicklime at a rate of 0.7–1 kg per 100 liters of water. After that, the substrates were autoclaved. Oyster mushroom mycelium was inoculated after cooling the substrate to 28° C [19]. The object of this study was the mycelial culture of golden oyster mushroom.

The grain spawn planting was performed by uniform mixing with the substrate (moistened to 70%) in the amount of 3% by weight. Then, polyethylene bags were tightly packed, tied and placed in a thermostat environment at + 25° C to colonize the substrate with mycelium. After that, 6 cuts of 4-5 cm longs were made with a sterile scalpel in a staggered manner along the entire length of the bags. The bags overgrown with mycelium were placed in a room where humidity was maintained at 85-90% and the temperature between + 18° and 19° C under natural light. Water tanks and a fan were used in order to maintain the high humidity in the room. At the beginning of the mushroom fruiting, the actively

IOP Conf. Series: Earth and Environmental Science **337** (2019) 012033 doi:10.1088/1755-1315/337/1/012033

liberated carbon dioxide was removed from the room by aerating and turning on the fan. During the month, we monitored The growth, fruiting bodies formation and the yield of goldenoyster mushroom Pleurotus citrinopileatus were monitored for a month [15–17].

If infected sites appeared in the bags, they were removed and treated with a solution of sodium chloride at 250 g/1 liter of hot water [22].

3. Results and observations

Wastes of the grain processing industry include: bran, wastes associated with the cleaning and sorting of grain mass (grain waste), grain weed impurity, injured grains, feeble and germinated grains, seeds of wild plants, substandard grains [21].

During the analysis of the studied grain processing wastes, we isolated two fractions: a fraction with a predominant starch content and another with a predominant fiber content (Table 1). The starch-containing fraction represented 58.2% in 1 kg of waste, and, the fraction containing a predominance of fiber - 41.8% by weight in 1 kg of waste.

Table 1. Physical composition of wastes

Fraction	Fraction composition	Content in 1 kg of waste, %
Predominantly starchy	Grain waste, damaged grains, puny and sprouted grains, wild plants seeds, substandard grains	58,2
With a predominance of fiber	Bran, chaff, remnants of straw and weeds	41,8

Substrates for mushroom cultivation must have a number of properties, most important of which are biotechnological - first of all, the ability of the substrate to meet the nutritional needs of the mushroom.

Easily available substances are less significant for oyster mushrooms than for its competitor, mold fungi; especially in the first days after grain spawn planting. Oyster mushroom refers to fungi that are equally capable of destruction, both cellulose and lignin, which are highly resistant to biodegradation. Among the competitors of the oyster mushroom, a special place is given to such fungi as Trichoderma green mold. In the course of its vital activity, it can utilize cellulose and very quickly develop on substrates in the presence of easily accessible nutrients.

The selectivity of the substrate is one of the most important biological properties. It is determined by the chemical composition of the raw material and accordingly, by the activity of the beneficial microflora located on its surface. The biological selectivity of a substrate is the ability of a thermo-treated substrate to acquire the most favorable conditions for the development of oyster mushrooms and to put the competitive microflora into an inactive state.

In various degrees, any type of vegetal raw material may acquire these properties that are of key importance for the production process. The biochemical mechanism of the acquisition of selective properties by substrates is based on a change in the substrate content in readily available sugars. The effect high-temperature on the substrate leads, upon prolonged exposure, to the chemical hydrolysis of polysaccharides and the accumulation of readily available substances and principally sugars.

When selecting the appropriate substrate for the cultivation of basidiomycetes, a great attention is paid to the carbon and nitrogen contents. The most suitable substrates are hay clover, wheat straw and sunflower husk. Grain waste requires the conversion of its polysaccharide complexes into easily digestible di- and monosaccharides in order to increase its protein content. This is achievable for example, by acid or enzymatic hydrolysis [6] as well as thermal or chemical treatments of the substrate for the suppression of the microflora development.

A comparative analysis of the chemical composition of traditionally used vegetable substrates in the mushroom production industry (Table 2) shows that this raw material (grain waste) contains 12.5% protein, 3.0% of fat; fiber -7.2% and therefore can be used for the cultivation of mushrooms.

Table 2. The chemical composition of grain waste and other vegetable substrates, % of dry weight

Substrate	Proteins	Lipids	Fibers	Ca	P	N	K
Grain waste	12,5	3,0	8,2	0,7	0,4	0,7	1,6
Burclover, hay	14,8	22,8	28,9	1,5	0,2	2,4	2,0
Burclover, straw	8,8	1,5	40,4	2,2	0,1	1,4	2,0
Barley, straw	3,7	1,6	37,7	0,3	0,11	0,6	1,3
Soy, straw	6,1	1,4	41,1	1,7	0,1	1,0	1,0
Soy, hay	14,1	2,4	27,2	1,3	0,2	2,4	0,8
Clover, hay	11,7	3,4	29,2	1,5	0,2	1,9	2,0
Corn stalk	2,3	0,4	32,1	0,2	0,02	0,4	0,4
Oats, straw	4,1	2,2	36,1	0,1	0,1	0,7	1,3
Wheat, straw	3,9	1,5	36,9	0,2	0,1	0,6	0,8
Sunflower husk	19,6	1,1	35,9	0,2	0,1	3,1	0,5

The rate of substrate coverage strongly depends on the seeding rate of oyster mushrooms. The higher it is, the faster the biofouling of the substrate. The development of mycelium also depends on the quality of the substrate, since competitive organisms inhibit and sometimes completely inhibit the growth of mycelium. In the first week of incubation, the mycelium grows in thin threads covering the entire substrate block- this is the stage of colonization. At the end of this stage, a small thickened cushion of mycelium can already be seen around the perforations: the primordia mushrooms will be formed on this roller in the future. During the second week of incubation, the nutrients of the substrate are being consumed and the mycelium network starts developing and thickens - the block becomes whiter and whiter, firmer to the touch. The best time to transfer the substrate blocks to the mushroom fruiting workshop is when the whole substrate is covered by mycelium, but fungi in the form of stroma has not yet formed a strong seal and the mushrooms from perforation did not appear yet. The dynamics of mushroom biofouling of the substrate are presented in table 3.

Table 3. Dynamics of mushroom biofouling of the substrate

Substrate type	Date of occurence			
	Inoculation	Primordium formation	1st mushroom harvest	2nd mushroom harvest
Control - wheat straw	2.12	15.12	21.12	28.12
Grain waste	2.12	19.12	23.12	29.12
Grain waste + birch sawdust	2.12	17.12	24.12	30.12
Birch Sawdust	2.12	18.12	24.12	30.12

Full biofouling of the substrate is determined visually. If necessary, cross-sections of the substrate into units help determine the uniformity of the growth of mycelium in the thickness of the substrate. The overexposure of blocks or reincubation cause either the appearance of stroma or the appearance of oyster mushroom primordia.

In our study, the golden oyster mushroom showed a high colonization rate of the substrate. Primordia began to form at 13-17 day. The first substrate completely colonized was the wheat straw substrate, and the last was the grain waste substrate on day 17. A possible reason for this is a relative decrease in the intensity of gas exchange in the volume of the starch-enriched substrate, which led to a slower formation of primordia and later, of fruiting bodies and accordingly, an increase in the consumption of the substrate to support the vital processes of the mycelium [22].

Oyster mushroom under high levels of carbon dioxide (as, for example, in the incubation chamber) forms a larger number of fungal primordia than with an excess of fresh air. When overexposing primordia for 1-2 days, deformed and damaged fruiting bodies appear.

IOP Conf. Series: Earth and Environmental Science **337** (2019) 012033 doi:10.1088/1755-1315/337/1/012033

The yield of the oyster mushroom Pleurotus citrinopileatus grown in laboratory conditions is presented in Figure 1.

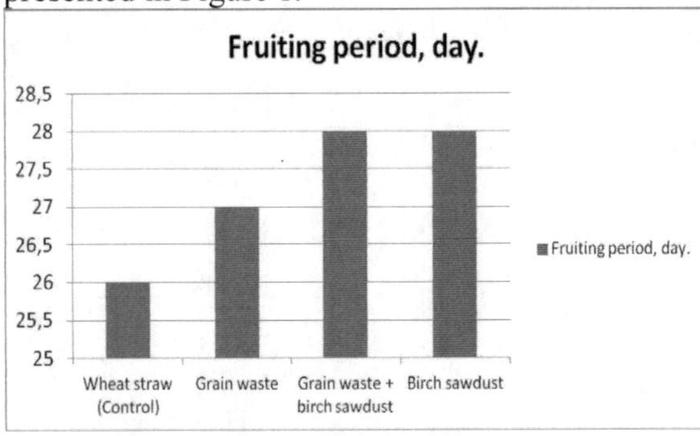

a)

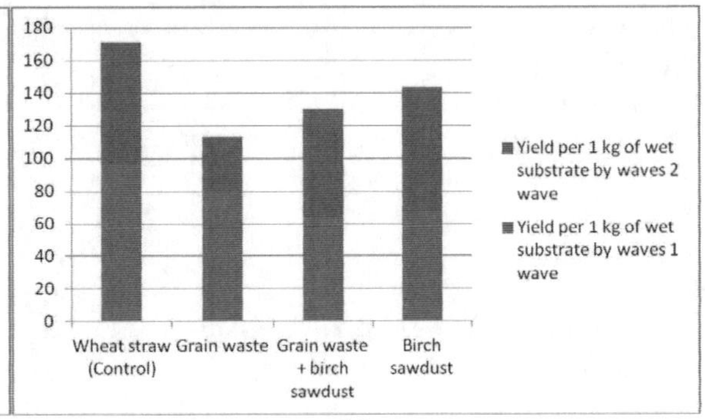

b)

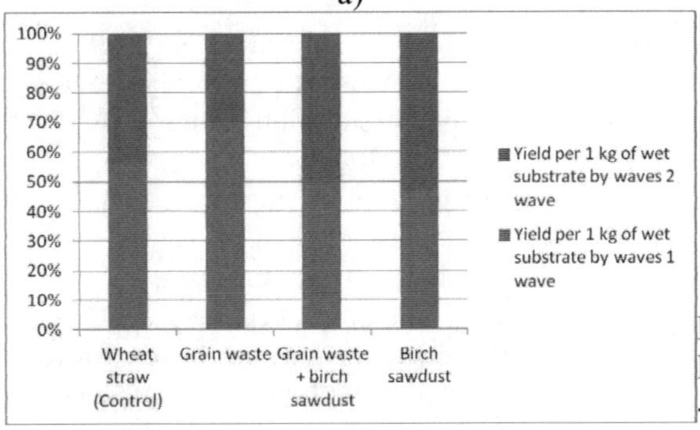

c)

Figure 1. Yield of golden oyster mushroom Pleurotus citrinopileatus (a - fruiting period, days; b - yield referred to 1 kg of wet substrate,%; c - yield referred to 1 kg of wet substrate, g)

According to the obtained data, the fruiting period of golden oyster mushrooms on the studied substrates represented 26-28 days. This duration was more stretched on the variant auditioned of sawdust. The maximum and the minimum yields were obtained on wheat straw substrate (171.2 g per kg of substrate) and grain waste substrate (113.4 g per kg) respectively. On birch sawdust substrate and on the substrate of grain waste and birch sawdust mixture, the yields reached 143.2 g and 130.3 g in this order.

The most intense strain recovery was reached during the first wave of fruiting on grain waste and wheat straw with 70.0 and 56.7% of the total yield, respectively. Such a high product recovery yield obtained on the substrate of grain wastes is probably associated with a high content of available starch. The product recovery was the lowest on the birch sawdust substrate (47.3%) due to its high cellulose content. The mushroom yield in the first and second waves was approximately the same on the substrate with on the substrate of grain waste and birch sawdust mixture (49.3 - 50.7%).

Under industrial conditions of oyster mushroom production, the most profitable is the two-week harvest window, which shortens the production cycle duration and makes it possible to increase the number of crop turns per year.

In the course of the experiment, the initial dry masses of all organic substrates were taken into account; at the end, the mass of the fruiting bodies and the amount of undigested (residual) substrate were taken into consideration (Fig. 2).

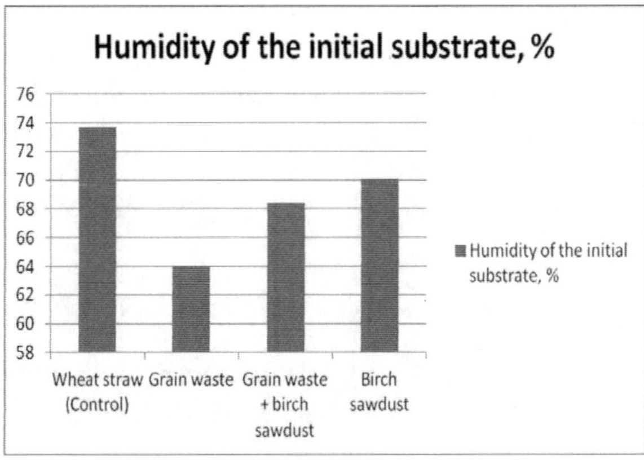

a)

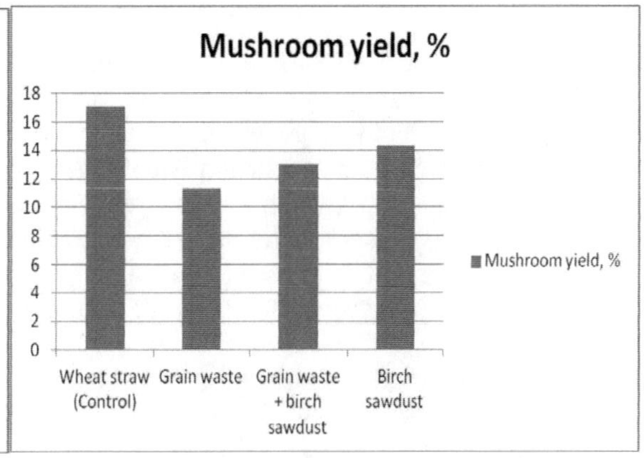

b)

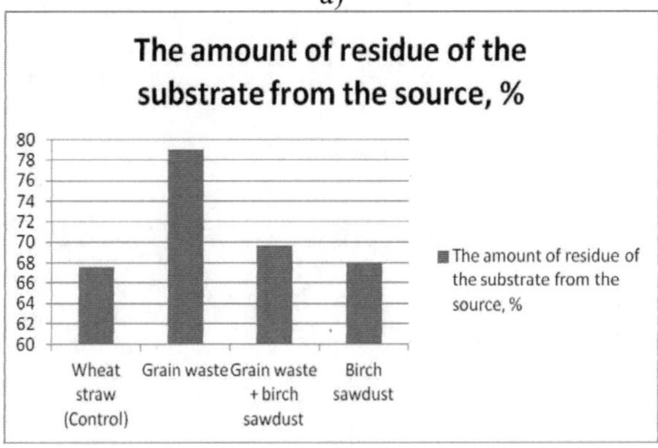

c)

Figure 2. Substrate conversion rates of Pleurotus citrinopileatus mushroom (a: moisture content of the initial substrate%; b: mushroom yield,%; c: percentage of the substrate residue referred to its initial amount,%)

The largest amount of undigested substrate residue was obtained when using grain waste in its pure form - 79.0%, which was related to a mushroom yield of 11.3%. When using grain wastes with the addition of birch sawdust, the degree of digested substrate increased to 69.6%, and the mushroom yield increased to 13.0%, which is lower than the control by 2.9 and 23.9% respectively.

In this way, it can be stated that increasing the nutritional value of the substrate by adding grain processing waste results in a limited increase in the mushroom yield of Pleurotus citrinopileatus, since at the same time, the gas exchange intensity decreases as well as the degree of substrate conversion. Therefore, cellulose-containing components must be added to the substrate with grain waste. Thermal or chemical treatments are also needed to suppress the development of microflora.

4. Conclusions

Grain wastes and grain processing wastes are harmless and easily amenable to enzymatic and microbiological bioconversion. Golden oyster mushroom (Pleurotus citrinopileatus) grown under laboratory condition on grain processing wastes demonstrated a high rate of substrate colonization. The average yielding crop attained 113.4 g of fresh mushrooms grown on 1 kg of substrate under a relative humidity of 70-75%. This promising work stream not only proposes a waste disposal method, but also, a significant source of income for the grain processing enterprises.

References

[1] Kulikova E G, Efremova S Yu 2014 *XXI vek: itogi proshlogo i problemy nastoyashchego plyus* **1(17)** pp 103-9

[2] Kulikova E G 2013 *Innovacionnye tekhnologii v APK: teoriya i praktika. Sbornik statej Vserossijskoj nauchno-prakticheskoj konferencii* pp 110-3

[3] Neverova O A, Gorelikova G A, Poznyakovskij V M Pishchevaya biotekhnologiya produktov iz

syr'ya rastitel'nogo proiskhozhdeniya, Novosibirsk p 416
[4] SHapavalov L V 1988 Zernovye kul'tury **2** pp 23-4
[5] Kulikova E G, Logvina O A 2015 Vliyanie spichechnogo proizvodstva na sostoyanie ehkosistem i biokonversiya ego othodov veshenkoj limonnoshlyapkovoj, XXI vek: itogi proshlogo i problemy nastoyashchego plyus **5(27)** pp 32-9
[6] Zubareva I M, Lyapustina O V, Lebedeva O M, Smetanin V T 2010 Nedelya ecologa tezisy dokladov Mezhdunarodnogo nauchnogo simpoziuma pp 78-81
[7] Sobgaida N A, Olshanskaja L N, Makarova Y A 2009 *Khim. Neftegaz. Mashinstr* **9** pp 36-9
[8] Gambato G, Todescato K, Pavao E, Scortegagna A, Fontana R, Salvador M, Camassola M 2016 *Bioresource Technology* **207** pp 46-51
[9] Owaid M, Abed I, Al-Saeedi P 2015 *Emirates Journal of Food and Agriculture* **27(7)** pp 556-61
[10] Jeyanthi Rebecca L, Seshiah C, Kowsalya E, Sharmila P 2015 *International Journal of Pharmacy and Technology* **7(2)** pp 8887-88
[11] Sharmila P, Jeyanthi Rebecca L, Tissopi T, Kowsalya E 2015 *Der Pharmacia Lettre* **7(8)** pp 193-6
[12] Gambato G, Todescato K, Pavao M, Salvador M, Camassola M 2016 *Bioresource Technology* **207** pp 46-51
[13] Owaid M N, Abed A M, Nassar B M 2015 *Emirates Journal of Food and Agriculture* **27(7)** pp. 537-41
[14] Dawidowicz L, Jasinska A, Siwulski M 2018 *Notulae Botanicae Horti Agrobotanici Cluj-Napoca* **46(1)** pp 156-60
[15] Bis'ko N A Biologiya i kul'tivirovanie s"edobnyh gribov roda veshenka1987 p 148
[16] Dudka I A, Vasser S P Griby: Spravochnik mikologa i gribnika1987 p 535
[17] Morozov A I Griby: Rukovodstvo k razvedeniyu 2000 p 304
[18] Perstieva E V, Dreveckaya V L Vyrashchivanie gribov roda Pleurotus na tverdyh othodah// EHkologicheskie problemy biodegeneracii promylennyh, storoitel'nyh materialov i othodov proizvodstv: mat. vseros. konf. - Penza, 1998 pp 97-98.
[19] Kiseleva O V Tekhnologiya vyrashchivaniya biomassy miceliya serno-zheltogo trutovika Laetiporus sulphureus v usloviyah glubinnogo kul'tivirovaniya s cel'yu polucheniya belkovyh pishchevyh dobavok // avtoref dis…kand tekhn. Nauk. – Krasnoyarsk 2013 p 20
[20] Ternova K G Agrotekhnologicheskoe obosnovanie kul'tivirovaniya veshenki obyknovennoj na kostre l'na // avtoref dis…kand s.-h. nauk. – Moskva 2006 p 20
[21] Sazonova I A 2012 *Ecologicheskaya biotekhnologiya* p 106
[22] Manukovskij N S Kinetika biokonversii lignocellyuloz 1990 p 112

IOP Conf. Series: Earth and Environmental Science **337** (2019) 012034 doi:10.1088/1755-1315/337/1/012034

Composite systems using ash from burning production waste

S Y Efremova[1*], N I Zubrev[2], M I Panfilova[3] and T V Matveyeva[2]

[1]Department of biotechnology and tehnosfernaja safety, Penza State Technological University, Penza, 1A/11 proyezd Baydukova/Gagarina street, Penza 440039 Russia,
[2]Department of Technosphere Safety, Russian University of Transport, Russian Open Academy of Transport, 22/2 Chasovaya St., 125190, Moscow, Russia
[3]Department of Physics and Construction Aerodynamics, National Research Moscow State University of Civil Engineering, Moscow, Yaroslavskoe highway, 26,

* Corresponding author: efremova_s15@mail.ru

Abstract. This article proposes a method of obtaining composite solutions using ash from industrial wastes as a structuring additive. Their introduction into the technology of composite solutions not only as modifiers of the structure and properties of composite binders, but also as fine additives, can significantly reduce the cement consumption and prevent waste placement, as well as, improve the performance properties of the latter and obtain new promising types of materials.

During this study, the chemical and mineral composition of the ash was determined and the hazard class was calculated. Structure formation in composite solutions was studied when cement was replaced with different amounts of ash. The possibility of replacing 10% of cement with ash without loss of the technological parameters of the system has been established.

1. Introduction

The environmental strategy is considering the involvement of wastes as sources of raw materials in the field of circular economic. One of the promising streams in this regard is the creation of new technologies for composite solutions used to strengthen the soil in the construction and repair of tunnels. Introducing effective structure modifiers and composite binders to the technology of composite solutions, and more concretely, fine additives can significantly reduce the cement consumption and prevent the disposal of waste on already crowded rubbish dumps and authorized landfills. Such wastes may include ashes from fuel combustion and wastes, which can improve performance properties and produce new promising types of materials.

At present, it has been proven that replacing of 25–30% of Portland cement with fly ash in concrete is effective. Such a replacement presents no significant differences of concrete strength if compared with the concrete without the ash addition [1–4]. This was confirmed during the construction of the dam of the Bratsk hydroelectric station and the Dnieper hydroelectric complex [5]. In Germany, about 4 million tons of fly ash are used annually in the production of building mortars, concrete products and other materials containing cement [6, 7]. However, their use is limited due to the different origins of the initial fuel [8–13] and the variable chemical composition of the ashes obtained from combustion invarious thermal installations.

There are other known examples of the use of cement slurries with partial replacement of cement with ash resulting from coal burning for the storage and annular space of collector tunnels, as well as ash from the burning of railway sleepers [14-17].

At the enterprises of railway transport, the Forsazh-2 installation is often used for the incineration of oily waste. After temporary storage, the resulting ash is placed in the landfill. This waste can be used combined with ashes of a different origin to obtain composite systems for the transport construction industry.

Composite solutions, as a rule, consist of cement, bentonite and liquid glass. However, the main component in the formulation of composite solutions is cement. The ashes from the combustion of fuel oil and oily waste can be used as an additive in composite solutions, allowing to reduce the consumption of cement, due to its partial replacement.

The purpose of this study is to experimentally confirm the possibility of using ash obtained from the incineration of oily waste as a structuring additive to composite solutions. This will solve environmental problems and help saving cement during the building process without the deterioration the strength characteristics of the material.

2. Material and method

Composite solutions, including bentonite, cement-ashes in different ratios and liquid glass, were studied in this work [5]. The purpose of this work was to study the effect of ash addition on the structure and properties of cement stone. The composition of the ash resulting from the combustion of oily waste includes, mass. %: silicon dioxide - 57.36; aluminum oxide - 23.59; magnesium oxide - 0.93; calcium oxide - 4.07; sodium oxide - 0,51; manganese oxide - 0.035; potassium oxide - 1.07; phosphorus pentoxide - 1.13; titanium dioxide - 1.28; iron oxide - 4.33; sulfur oxide - 2.92; chlorine - 0.08; total - 97,305. Heavy metals - 2,695%. The total amount is 100%. The concentration of heavy metals was, respectively, in ppm: As - 35; Ni — 96; Cu - 161; Sn - 15; Cr - 122; Co - 30; Mo - 11; V - 152; Ni — 96; Zn — 192; Rb - 23; Sr - 1697; Zr - 313; Ba - 2125; U - 5; Th - 30; F - 668; Y - 63; Nb - 30; Pb - 73; La - 122; Ce - 231; Nd - 101; Sc - 25; Ga - 35.

According to the chemical composition of the ash, the environment and health hazard classes have been calculated. It turned out that it belongs to the third class of human health hazard and fourth of the environment hazard class [18].

Based on the X-ray powder diffraction obtained by using the X-ray diffractometer X-Pert PRO MPD (PANalytical, Netherlands), the ash powder was found to contain: coesite, graphite agidride, sulfur plagioclase, jarosite and cristobalite.

During the study, soda-modified bentonite clay of the P2T2A brand, cement of the brand M500 and liquid glass (density - 1.46 g / cm^3, silicate module - 2.7-3.4%) were used.

Bentonite clay was added to the composite solution to increase its plasticity and viscosity and to retain its resistance to delamination. It consists of oxides of silicon, aluminum and calcium, which are the basis for hydraulic binders and have chemical affinity with cement.

The composition of bentonite is established by the method of semi-quantitative X-ray phase analysis (wt.%): 75-80 montmorillonite, 15-17 quartz, 1-2 kaolinite and muscovite type hydromica 1-2. The sorption capacity of bentonite is 113.3 mg · eq / 100g.

3. Results and discussion

The choice of formulation of the composite solutions in relation to engineering and geological conditions is based on the assessment of their rheological and technological properties. Among the most important technological properties of injection solutions is the time of sedimentation or setting, since the solution in a more or less homogeneous state must be delivered (injected) at the right time interval. Equally important is the ultimate strength and durability of the hardened mortar. Therefore, when developing the technology of production of composite solutions containing ashes, two time intervals are to be considered. The first one is the necessary time of curing the solutions, that is, before its transition to a solid state. The second time interval corresponds to the study of the increase in strength during long-term storage.

The effect of the addition of ash on the change in the rheological properties of cement was determined using bentonite solutions during the curing process at a water / cement ratio of 2: 1 in the presence of

5% liquid glass. At the same time, cement was replaced with ash at an amount which varied from 5 to 20%. In this regard, ash was injected into the previously soaked 5% suspension of bentonite, after which the calculated amount of cement and liquid glass were introduced while stirring (Table 1).

Table 1. Changes in plastic strength of the composite solution containing different ash content at different storage durations

Structure formation time, τ, hours	The ash content in the mixture,%				
	0	5	10	15	20
	Strength of the alumina-cement mixture Pm, 10-3 MPa				
1	1,887	1,351	1,211	0,801	0,543
2	3,654	2,716	2,214	1,723	1,253
3	6,757	6,245	5,284	5,011	4,352

Figure 1 shows that an increase in the ash concentration in the composite solution induces an initial strength of the composite solution that decreases with time.

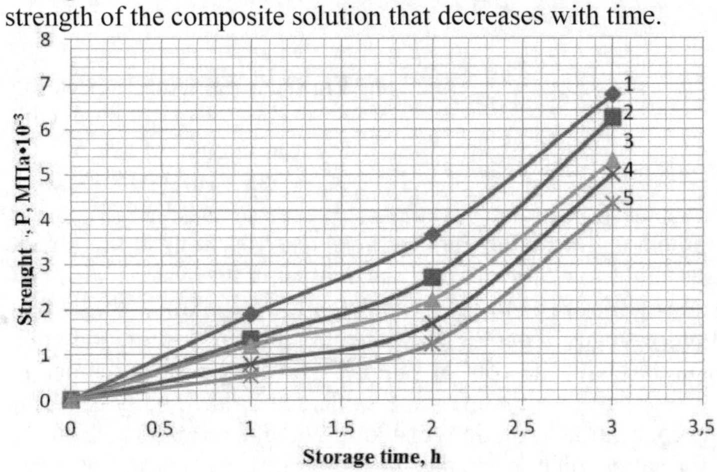

Figure 1. Structuring a composite solution with the introduction of ash,%: 1 - 0; 2 - 5; 3 - 10; 4 - 15, 5 – 20.

A graphical method was used to determine the exact time of reaching loss of fluidity of the solution. For these pruposes, in fig. 1, a straight line was drawn parallel to the abscissa axis, corresponding to the strength of an alumina-cement solution of 2.5 • 10-3 MPa. Perpendiculars dropped from the intersection of the line with curves 1 - 5 indicate the curing time of the solution. The obtained data are presented in Figure 2 showing that with the increase of the cement replacement with ash, the curing time of the solution increases.

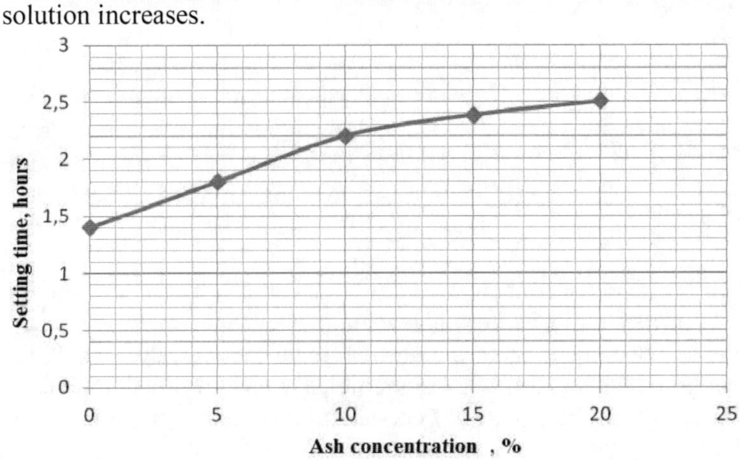

Figure 2. Setting time of the composite solution at different ash concentrations.

The strength of the hardened composite solutions was determined within 28 days on an IP-100 hydraulic press - at 7, 14 and 28 days. A non-linear change in the strength of composite systems with an increase in the ash content in the mixture was established. During storage, after 7, 14 and 28 days, different pattern of changes in the strength properties of composite mixtures prepared with different replacements of cement with ash can be observed (Figure 3).

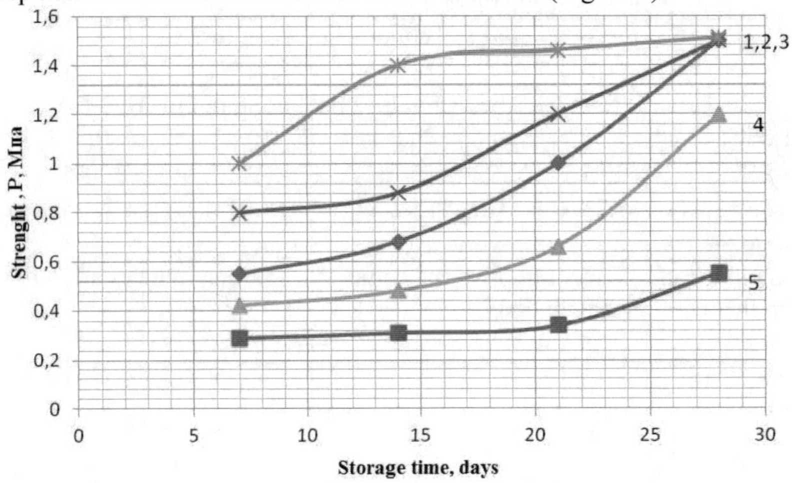

Figure 3. Structural formation of a composite solution with the introduction of ash, %: 1 – 0; 2 - 5; 3 - 10; 4 - 15, 5-20.

Fig. 4 shows that increasing the ash content in the mixture induces a decrease in strength. Moreover, after 7 and 14 days, the strength decreases slightly with increasing ash content in the system. This fact can be explained by the low hydratation rate of ash and cement. 28 days after the preparation, the pattern formation nature of the composite solutions changes. After 28 days, the strength of the solution with a 10% ash content in the cement is comparable with the control sample. The change in the structure formation of composite solutions with different ash additives can be explained by the ability of sulfur of ash to disproportionate, which occurs in concentrated solutions of alkalis with the formation of soluble sulfides and sulfites. In addition, the formation of homochains in concentrated solutions of basic sulphide is possible: Na2S + (n-1) S = Na2Sn, in the presence of which the structure formation of the system apparently accelerates [19].

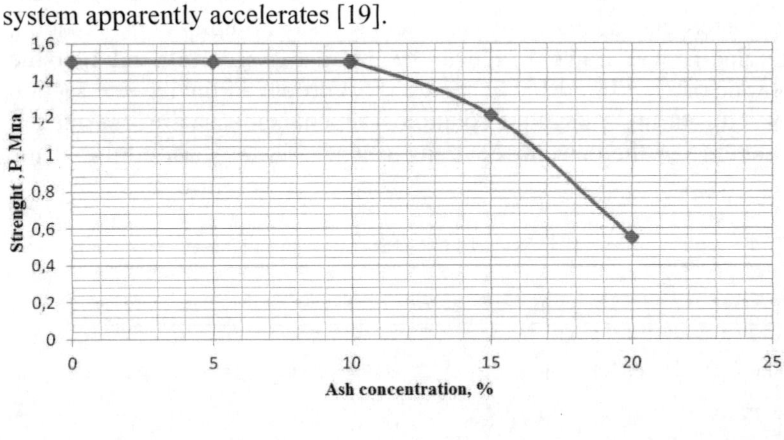

Figure 4. Change in the strength of the composite solution at different ash concentrations in cement after 28 days of storage.

It has been established that when bentonite and cement are mixed with water, an alkaline medium of pH = 12 is formed in the mixture. It turned out that during leaching, a change in the chemical composition of the ash occurs, with the sulfur content decreasing to its greatest extent [20, 21]. Thus, sulfur that has passed into the mixing water falls in the form of nanoparticles, which eventually increase in size to highly dispersed colloidal sizes and fill the cavities of crystallizing cement stone, forming a monolith.

The sample with a 10% replacement of cement with ashes achieved the strength of the control sample for 28 days shows that even such a composition can be recommended for practical application of these composite solutions in the construction and operation of transport infrastructure facilities.

The comparison of the strength analysis results obtained using the sample containing 10% ash and the control sample at day 28 shows that such this composite material composition can be recommended for practical applications in the construction of transport infrastructure facilities.

4. Conclusions.

The work investigated the possibility of neutralization and reuse of ash from the incineration of oily waste for the partial replacement of cement in injection solutions used in the construction of transport facilities.

It is established that with an increase in the concentration of ash in the composite solution, the increase in the strength of the solution decreases with time. It has been established that it is possible to replace 10% of cement with ash has since there is no decrease in the strength characteristics of concrete at after 28 days of storage which was comparable with the control sample. This makes it possible to save cement. It has been proven that using secondary raw materials such as ash to reduce the consumption of expensive materials, and improve the environment is possible.

According to the obtained results, it can be concluded that the use of such composite solutions is promising.

References

[1] Dvorkin L I and Dvorkin O L 2007 Stroitel'nye materialy iz othodov promyshlennosti: Uch.-sprav. Posobie (Rostov-na-Donu: Feniks)

[2] Ignatova O A, Berdov G I and Fomenko V V 2011 Osobennosti processov tverdeniya zolo – cementnyh vyazhushchih Sovremennye naukoemkie tekhnologii **1** pp 80-2.

[3] Rubcov I V, Mitrakov V I and Rubcov O I 2007 Zakreplenie gruntov zemlyanogo polotna avtomobil'nyh i zheleznyh dorog (Moscow: ASV)

[4] Fursov L F and Yakovleva I I 1981 Ispol'zovanie zoly TEHS dlya in"ekcionnyh rabot Izv. VNIIG im. B.E. Vedeneeva **150** pp 77-80.

[5] Nekrasov A S and Sinyuk Yu V 2007 Perspektivy razvitiya toplivno-ehnergeticheskogo kompleksa Rossii na period do 2030 goda Problemy prognozirovaniya **4** pp 21-53.

[6] Vatin N I, Petrosov D V, Kalachev A I and Lahtinen P 2011 Use of ashes and ash-and-slad wastes in construction Magazine of Civil Engineering **4**

[7] Putilin E I and Cvetkov V S 2003 Obzornaya informaciya otechestvennogo i zarubezhnogo opyta primeneniya othodov ot szhiganiya tverdogo topliva na TEHS (Moscow: Soyuzdonii).

[8] Kokubu I and Yamasada D 1976 Cementy s dobavkoj zoly unosa Shestoj mezhdunarodnyj kongress po himii cementa (Moskva. 1974) **3** pp 83-93

[9] CBNTI Minavtodora RSFSR 1977 Opyt ispol'zovaniya aktivnyh i neaktivnyh zol unosa TEHS Francii v dorozhnom stroitel'stve (po materialam zarubezhnogo opyta) EHkspress-informaciya Povyshenie ehffektivnosti proizvodstva i kachestva dorozhnyh rabot **18**

[10] Elks A D and Redman G T R 1965 Site control and the construction of embanknents using pulverised fuel ash from Lagoons The Surveyor and Municipal Engineer

[11] Shuster Dzh S and Hansen R L 1972 Zola unosa, kak stroitel'nyj material dlya nasypnyh sooruzhenij. American Society of Civil Engineers Gournal of Power Division

[12] Haga N, Ohkawa V, Kawamoto T, Konno M and Mizoguchi J 1981 Utilisation of blast furnace and steel slags in road construction NipponSteel Techn.Rept. **17**.

[13] Sobgaida N A and Solodkova A B 2015 Recycling Spent Activated Sludge Int. J. of Environmental Problems **1** (1) pp 64-74.

[14] Lyapidevskij B V, Nikitin A V, Rodina G P and Badamshin S O 2008 In"ekcionnye sostavy dlya zablochnogo i zatrubnogo prostranstva kollektornyh tonnelej Nauka – moskovskomu stroitel'stvu. Sbornik tekhnicheskoj informacii **2** pp 19-21.

[15] Zubrev N I and Ustinova M V 2013 Recikling zoly ot szhiganiya otrabotannyh derevyannyh shpal *HKHI vek: itogi proshlogo i problemy nastoyashchego plyus: Periodicheskoe nauchnoe izdanie* **1** 09(13) pp 140-4.

[16] Ustinova M V 2012 *Razrabotka tekhnologii kompozitnyh cementno-bentonitovyh sistem s dobavkoj zoly ot szhiganiya shpal i primenenie ih pri stroitel'stve i ehkspluatacii ob"ektov infrastruktury zheleznodorozhnogo transporta*. Dissertaciya na soiskanie uchenoj stepeni kandidata tekhnicheskih nauk. (Moskva)

[17] Sanitarnye pravila po opredeleniyu klassa opasnosti toksichnyh othodov proizvodstva i potrebleniya SP 2.1.7.1386-03 (zaregistrirovany v Minyuste 19 iyunya 2003 goda № 4755)

[18] Massalimov I A, Husainov A N, Abdrakipova L F and Mustafin A G 2009 Vydelenie nano chastic sery iz rastvorov polisul'fidov kal'ciya i natriya *Zhurnal prikladnoj himii* **82** 12 pp 1946-51.

[19] Zubrev N I, Matveeva T V, Ustinova M V et al Patent №2645691 «Sposob polucheniya tamponazhnogo rastvora»

[20] Matveeva T V, Zubrev N I and Ustinova M V 2015 Ispol'zovanie zoly ot szhiganiya mazuta v kompozitnyh rastvorah dlya stroitel'stva *Mezhdunarodnyj nauchno-issledovatel'skij zhurnal* **9** (40) pp 61-3

IOP Conf. Series: Earth and Environmental Science **337** (2019) 012035 doi:10.1088/1755-1315/337/1/012035

Substantiation of the expediency of using the effect of electric fields on the process of moisture removal from porous waste structures

A Kholudeneva[1*]

[1]Penza state technological university, 440039, Penza, Baidukov passage / st. Gagarin, 1a / 11 Penza, Russia

* Corresponding author: linblondy@mail.ru

Abstract. During this study, information was collected and analyzed in the perspective of producing an electroosmotic installation, in other words, a converter of colloidal polydisperse systems into capillary-porous by electroosmotic dehydration in order to reduce the residual moisture of sludge from sewage and the pulp and paper industry. Exponents were taken as models expressing the mass of the pulp in function of the drying time and the value of electric current. According to the coefficients of determination, the models were consistent with the initial experimental data. Exponential models were used to further optimize the process of converting colloidal polydisperse systems into capillary -porous.

1. Introduction

The purpose of this study is to collect and analize information in the perspective of producing an electroosmotic installation, in other words, a converter of colloidal polydisperse systems into capillary-porous by electroosmotic dehydration in order to reduce the residual moisture of sludge from sewage and the pulp and paper industry. The possibility of producing insulation materials using dehydrated osprey as a raw material was analyzed in order to increase the productivity of technological innovations in the environmental safety field of industrial enterprises [1]. The relevance of the study is confirmed by the urgent need for innovative approaches to solve the environmental issue by modern industrial enterprises. In addition, high quality indicators of imported products create a competitive environment for developing local enterprises. A solution to this situation may be the active introduction of modern technologies into the local production.

During various technological processes a significant amount of wet waste and residues is formed. In the pulp and paper industry (PPI), a large number of osprey is formed. Pulp is a multi-tonnage, environmentally friendly treatment plant waste of the paper industry, which is a composite material containing environmentally friendly compounds of the raw material mixture for paper making[2].

2. Methods

The wet pulp is initially presented in a colloid polydisperse system. However, with a decrease in moisture content, wet waste takes on the typical properties of capillary-porous materials. For further energy-efficient disposal of wet waste, the possibility of producing a model for converting colloidal polydisperse systems into capillary-porous, necessary for pressing and electrodynamic separation of moisture from waste structures is considered.

IOP Conf. Series: Earth and Environmental Science **337** (2019) 012035 doi:10.1088/1755-1315/337/1/012035

The principle of the model for converting colloidal polydisperse systems into capillary-porous for the purpose of further electroosmotic dehydration is to achieve the maximum energy-efficient effect from the dehydration process. Figure 1 shows schematically an optimized device of an electroosmotic installation for the dehydration process of waste paper production, where: 1- body made of insulating material; 2 - mobile electrode; 3 - stationary electrode with perforation; 4- screw cap, 5- creating (regulating) the initial pressure (torque wrench) of the compression of the osprey (6). Electrodes (2) and (3) are supplied with wires, (7) and (8) with an electrical voltage (electric current) that provides the necessary electric field strength, causing particle velocity variation, 9 - perforation in the housing along the entire perimeter, providing additional convection dehydration after the completion of the electroosmosis process when the electrode is raised. (2), 10 is an external casing angle regulator that provides convection dehydration of the lower layers of the mass after the electroosmosis process is completed, 11- isolation pallet for collecting the separated moisture.

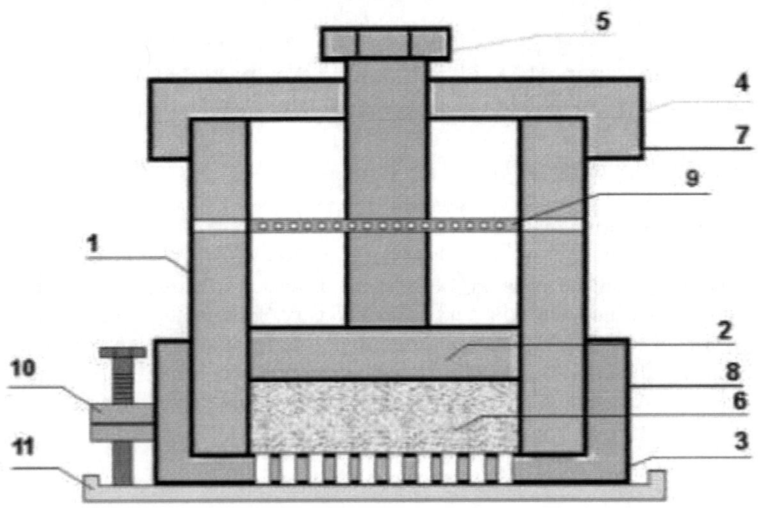

Figure 1. Laboratory electroosmotic installation

This technical solution allows the most energy-efficient electroosmotic dewatering process if compared with the already existent technology, since the variation of direct and alternative current (for example 8, 16, 24 V) at an output current of the voltage source of 20A. This produces a Joule heat effect. The released heat energy in the process of dehydration is given in Table 1, contributing to dehydration without additional energy costs [3].

Table 1. The energy released in the osprey during the osmosis.

t^*,h	0	0,08	0,16	0,25	0,33	0,42	
W_0, W	0	3,919	6,96223	9,5808	11,3588	12,888	
t^*,h	0,5	0,58	0,66	0,75	0,83	0,91	1
W_0, W	13,9274	14,7340	15,360	15,8993	16,265	16,5494	16,7939

Thus, the technical solution contributes to the additional separation of moisture under the action of gravity and partial convection drying, thanks to the perforated cathode and the housing of the installation.

To increase the dehydration and increase the energy efficiency of the system, the structure of the pulp solid phase can be modified. It is possible to achieve this in several ways: by coagulation of the pulp with various chemical reagents, by flocculation, by introducing certain filler materials, by thermal conditioning (sudden changes in temperature) — by freezing (or thawing), as well as by magnetic or electromagnetic treatment. As a result, a significant enlargement of the particles of the pulp occurs and the average surface area of the dispersed phase and the dispersion medium decreases. The average surface energy decreases and the forces of moisture binding to solid particles significantly weakens. The redistribution of types (forms) of moisture is actively carried out, and the content of “free water” increases due to a decrease in the total amount of “the bound water”. Moreover, all the above methods are quite costly [4].

The research results show that the method of electroosmosis can remove a significant amount of liquid from the volume of pulp without resorting to the use of coagulants and at the same time the energy costs can be less than when using thermal drying. It was established, as a result of a series of experiments, that the method of electroosmosis removes moisture from wet pulp quite effectively.

According to the analysis, the electrokinetic phenomenon of moisture transfer through the pore space of capillary-porous media, caused by an external electric field imposed on the diaphragm, was one of the first to be observed by Professor F.F. Reuss. The flight, which he called "water expelling force." Subsequently, this phenomenon called "electroosmosis" refers to the movement of fluid relative to the solid skeleton of a porous material under the action of an electric field in the direction determined by the sign of the electrokinetic potential. [5] When an electroosmosis process is carried out, the divergence of the phase charges leads to the displacement of moving counter ions in a constant electric field together with the liquid phase to the corresponding pole of the current source. The electroosmotic transfer of fluid through the pore space of a capillary-porous body is determined by the electrokinetic potential and the structure of the electrical double layer at the interface of the phase.

The linear electroosmotic velocity of the fluid is determined by the formula:

$$V_l = V_g + \frac{I\xi\varsigma}{4\pi^2\eta\chi r^2}, \tag{1}$$

and the volumetric electroosmotic velocity of the fluid will be equal to:

$$V_o = \frac{I\xi\varsigma}{4\pi\eta\chi}, \tag{2}$$

where V_l – the electroosmotic linear speed;

V_g – hydrostatic speed;

V_o – electroosmotic volumetric speed;

I – current intensity;

ξ – dielectric constant;

χ – the electrical conductivity of the fluid;

η – fluid viscosity;

r – radius of the capillary;

ς – electrokinetic potential. [6]

The earliest known model of the structure of a double electric layer, which allowed to explain not only the existence of electrokinetic phenomena, but also to quantify the value of the capacitance of a double electric layer, was proposed by Helmholtz [7].

Applying innovative technologies such as paper waste dehydration (pulp), the end result is not only the production of unique products (insulation), but also the achievement of environmental and social impact (more than 2000 m3 of waste does not enter the soil factories with average power), while reducing the negative impact on the human health, due to the decay and decomposition of waste [8].

The results of the experiments on the study of electroosmosis obtained in the developed installation, are presented in Table. 2. The reflect of the process of reducing mass m_j (Figure 2) and the process of reducing current i_j (Figure 3).

Table 2. Experimental data on the study of electroosmosis.

t_j	0	0,16	0,33	0,5	0,66	0,83	1
m_j	70	48	38	35	32	29	27
ij	2,5	1,63	1,21	0,8	0,53	0,31	0,09

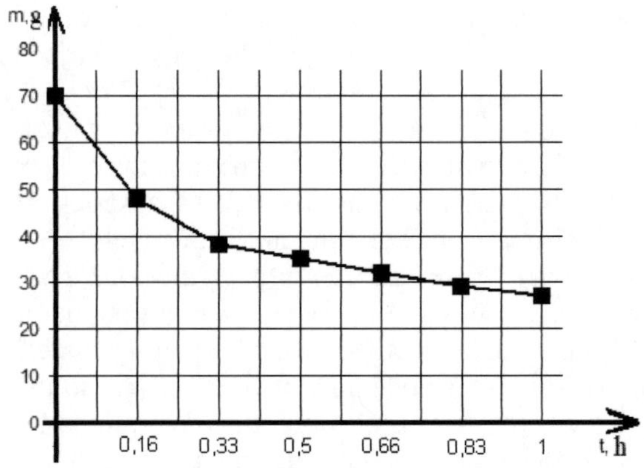

Figure 2. Experimental graph of mass loss in the process of electroosmotic dehydration of the pulp

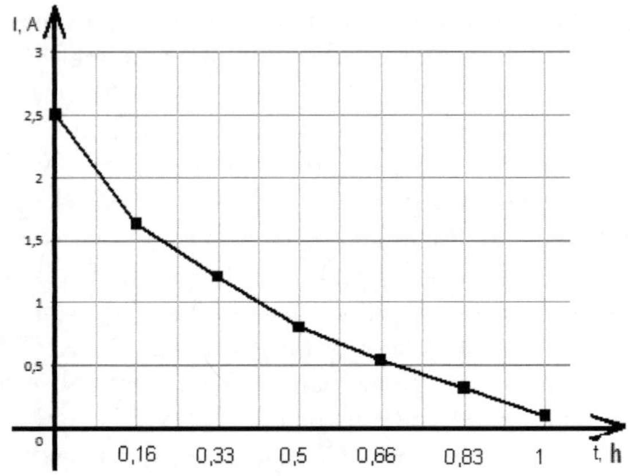

Figure 3. Experimental graph of current reduction in the process of dehydration (drying)

From these data, it follows that as models mass expression of the pulp, in function of the drying time and the value of electric current (under certain initial conditions), we can take the exponents:
to change the mass of the pulp:

$$m(t_j) = m_0 \cdot e^{-\frac{t_j}{\tau_m}}, \tag{3}$$

where m_0 - initial value of the mass of the osprey, loaded into the installation ;

τ_m - time constant of weight reduction

t_j - moment of mass control $t_j \in \{0; t_1; ...t_m\}$,

And to change the current:

$$i(t_j) = I_0 \cdot e^{-\frac{t_j}{\tau_i}}, \tag{4}$$

where I_0 - initial value of the electric current flowing through the scopa;

τ_i - time constant of the process of reducing the current when on the plates (electrodes) of the installation.

while $U = const$ on the plates (electrodes) of the installation.

Observations have shown that, when the process of controlling the effect of electroosmosis, started, the pulp dries to a certain extent by the convective method. This loss was excluded during the processing of experimental data and in the further study of electroosmosis. This made it possible to estimatemore accurately the relationship between the electric current flowing through the pulp and the decrease in its mass caused by electroosmosis.

To avoid the influence of oscillations of the initial mass of the pulp that occurs when the installation is loaded, we move from the subject scales of mass m (t) and current i (t) to universal ones by following the formulas

$$\delta_m(t_j) = \frac{m_{oc}(t_j)}{m_o} = e^{-\frac{t}{\tau_m}} \; ; \tag{5}$$

$$\delta_i(t_j) = \frac{i_{oc}(t_j)}{I_o} = e^{-\frac{t}{\tau_i}} \tag{6}$$

The data presented in table 9 are in units of universal scales.

Table 3. Experimental data on the study of electroosmosis in universal scales.

$j_ч$	0	0,16	0,33	0,5	0,66	0,83	1
$\delta_m(t_j)$	1	0,69	0,54	0,5	0,46	0,41	0,39
$\delta_i(t_j)$	1	0,65	0,48	0,32	0,12	0,09	0,03

The main parameters in models (5) and (6) are time constants. For their estimation, the model is subjected to linearization by logarithmization.

$$\ln \delta_m(t_j) \cdot \tau_m = -t_j \; ; \tag{7}$$

$$\ln \delta_i(t_j) \cdot \tau_i = -t_j \tag{8}$$

and to the known rules of the method of least squares (OLS).

Taking into account (7) and (8) the sums of squares for (S^2) and i can be written as:

$$S_m^2 = \sum (t_j - \ln \delta_m(t_j) \cdot \tau_m)^2 \; ; \tag{9}$$

$$S_i^2 = \sum (t_j - \ln \delta_i(t_j) \cdot \tau_i)^2 \; , \tag{10}$$

Where $t_j, \delta_m(t_j), \delta_i(t_j)$ are data of the experiment.

$$\frac{\partial S_m^2}{\partial \tau_m} = \sum_{j=0}^{n} \left[t_j + \ln \delta_m(t_j) \cdot \tau_m \right] \cdot \ln \delta_m(t_j) = 0 \; ; \tag{11}$$

$$\frac{\partial S_i^2}{\partial \tau_i} = \sum_{j=0}^{n} \left[t_j + \ln \tilde{\delta}_i(t_j) \cdot \tau_i \right] \cdot \ln \tilde{\delta}_i(t_j) = 0 \tag{12}$$

After simple transformations, we get:

$$\tau_m = \frac{\sum_{j=0}^{n} (t_j \cdot \ln \delta_m(t_j))}{\sum_{j=0}^{n} (\ln \delta_m(t_j))^2} \; ; \tag{13}$$

$$\tau_i = \frac{\sum_{j=0}^{n}(t_j \cdot \ln\delta_i(t_j))}{\sum_{j=0}^{n}(\ln\delta_i(t_j))^2} \quad = \quad (14)$$

According to the results of the experiment:

τ_m^* =0,856 h; $\tau_i^* = 0{,}323$ h.

Here, we check the conformity of the models to the original experimental statistics by estimating the coefficients of determination.

$$n_{\tilde{\delta}_m} = \sqrt{1 - \frac{\sum_i(\delta_m(t_j) - \bar{\delta}_m\ (t_j))^2}{\sum_i(\delta_m(t_j) - \tilde{\delta}_m\ (t_j))^2}}\ ; \quad (15)$$

$$n_{\tilde{\delta}_i} = \sqrt{1 - \frac{\sum_i(\delta_i(t_j) - \bar{\delta}_i(t_j))^2}{\sum_i(\delta_i(t_j) - \tilde{\delta}_i(t_j))^2}}\ ; \quad (16)$$

$$n_{\tilde{\delta}_m} = 0{,}905;$$

$$n_{\tilde{\delta}_i} = 0{,}764$$

According to the coefficients of determination, it can be concluded that the models correspond to the initial experimental data.

From figures (2) and (3), it follows that $m(t)$ and $i(t)$ are partially interrelated, and to a large extent, have differences according to the calculations (by 2 to 4 times). This is the evidence that the process of drying by electroosmosis is also influenced by another factor: it is likely that it concerns the effect of the Joule heat accumulated in the pulp. This process is specific and requires additional research, which will be conducted in a separate article.

As a result, it can be concluded that the parameters of models can vary. The latter will be implemented by the variation of the electric field at the electrodes of the installation, but this fact is a detailed implementation of the technological process [9].

The results of repeated experiments show that the mode of drying of electroosmosis and the mode of electrophoresis can be represented by exponents, but the exponents have different time constants.

It should be noted here that the processes of electroosmosis and electrophoresis can take place in an experimental dosage of pulp simultaneously: with a residual volume (mass) of water and technological dosage of a binder (liquid glass).

As the indirect measurements showed, the essence of the method is as follows: $\Delta_1, \Delta_2, \ldots$ in this case, the pulp dries to a certain extent in a convective way [10]. This loss must be eliminated in the study of electroosmosis. This procedure will allow a more accurate assessment of the correlation between the electric current flowing through the pulp and the "loss" of its mass caused by electroosmosis.

After the conversion (deduction of mass loss, $\Delta_1, \Delta_2 \ldots$), the graph takes the form shown in figure 5:

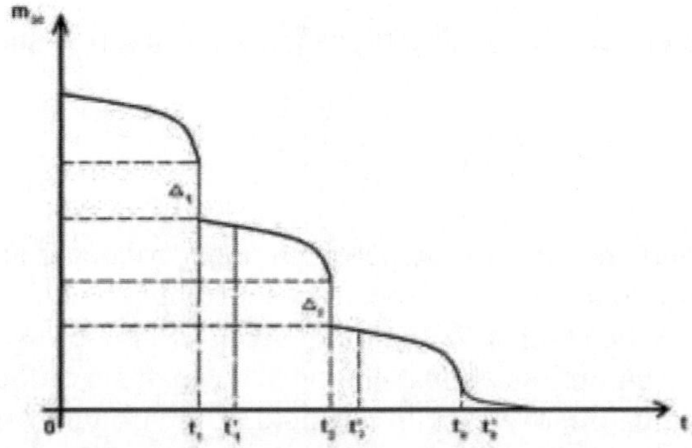

Figure 4. The results of the experiments on the pulp drying by electroosmosis and electrophoresis, taking into account the mass loss of the pulp during reload

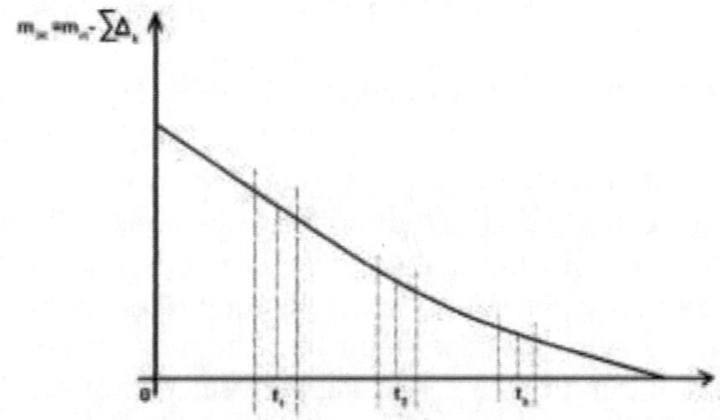

Figure 5. The results of the experiments on the osprey drying by electroosmosis and electrophoresis, taking into account the mass loss of the pulp during a reboot after conversion

$$\begin{cases} m(\bar{t}_1) = \dfrac{m(t_1) + m(t_1')}{2} \\ m(\bar{t}_2) = \dfrac{m(t_2) + m(t_2')}{2} \\ \ldots \end{cases} \qquad (17)$$

The parameter model is estimated according to (17). As its analytic transformation, we take the exponent: $m_{oc}(t) = m_{oc}^{e^{-\frac{t}{\tau}}}(t_0)$ or $\delta_{\text{нос}}(t) = \dfrac{m_{oc}(t)}{m_{oc}(t_0)} = e^{-t/\tau}$,

where τ — unknown parameter.

Similar calculations should be repeated to estimate the parameters of the electroosmosis current and electrophoresis (in the last cases, it is necessary to assess more accurately the changes in the mass of the pulp - increase or decrease).

Later, we get $\delta_{\text{нос}}(t_i)$ and $i_{oc}(t_i)$. With this in mind, we find estimates for the correlation functions.

$$r(\delta_{\text{нос}} i_{oc}) = \frac{k(\delta_{\text{нос}} i_{oc})}{\delta(\delta_{\text{нос}})\delta(i_{oc})} = \frac{\sum \delta_{\text{нос}}(t_i) - \frac{m_{\text{нос}}(t_i)}{m_{\text{нос}}(t_0)})(i_{oc}(t_i) - I_{oc}(t_i))}{\sqrt{\frac{1}{\text{n-1}} \sum_1^n \left[(\delta_{\text{нос}}(t_i) - \frac{m_{\text{нос}}(t_i)}{m_{\text{нос}}(t_0)} \right]^2} \cdot \sqrt{\frac{1}{\text{n-1}} \sum_1^n (i_{oc}(t_i) - I_{oc}(t_i))^2}} \qquad (18)$$

The data of indirect measurements and their analysis allows us to conclude that the process of electroosmosis and electrophoresis is manifested together. Their differentiation is difficult. Therefore, it will be necessary in the future to take advantage of the phenomenological estimates (the time constant and the time of censoring experimental data) based on experimental data [11]. In further experiments, we will measure the mass of the pulp (m_j) at different corresponding times (t_j), where j is the number of the time section: $(j \in \{0,1,2,...n...\})$

3. Conclusions

In the course of this study, information was collected and analyzed in the perspective of producing an electroosmotic installation, in other words, a converter of colloidal polydisperse systems into capillary-porous by electroosmotic dehydration in order to reduce the residual moisture of sludge from sewage and the pulp and paper industry. The proposed model is an optimized installation. The analysis of the possibile improvement of the performance of technological innovations in the field of environmental safety in industrial enterprises. Exponents were taken as models expressing the mass of the pulp as a function of drying time and the value of the electric current. According to the obtained coefficients of determination, the models were consistent with the initial experimental data. Exponential models were used to further optimize the process of converting colloidal polydisperse systems in capillary -porous. The studied model of the wet waste conversion into raw materials intended for being used as building materials according to the obtained experimental data can significantly reduce the costs of enterprises for recycling and increase the ecological level of the pulp and paper industry.

References

[1] Ryzhakov V V, Holudeneva A O and Ryzhakov M V 2015 Issledovaniya korrelyacionnyh svyazej processov ehlektroosmosa i soputstvuyushchih *Mezhotraslevoj nauchno - tekhnicheskij zhurnal «Oboronnyj kompleks – nauchno - tekhnicheskomu progressu Rossii»* **3** (127) pp 41-3.

[2] Ryzhakov V V and Holudeneva A O 2013 Rezul'taty issledovanij osobennostej rynka uteplitelej i ih ispol'zovanie pri organizacii proizvodstva «EHkoplit» *Mezhotraslevoj nauchno - tekhnicheskij zhurnal «Oboronnyj kompleks – nauchno - tekhnicheskomu progressu Rossii»* **4** pp 60-3

[3] Ryzhakov V V, Holudeneva A O and Ryzhakov M V 2014 Faktory, vliyayushchie na ehkologicheskuyu bezopasnost' i ehffektivnost' processa izgotovleniya ehkoplit *Nauchno-metodicheskij zhurnal «XXI vek: itogi proshlogo i problemy nastoyashchego plyus»: Periodicheskoe nauchnoe izdanie, seriya «Ehkologiya»* **5**(21) pp 104-8

[4] Ryzhakov V V, Holudeneva A O and Ryzhakov M V 2013 Menedzhment kachestva produkcii na osnove sootnosheniya «stoimost' - kachestvo» v prilozheniyah *Ezhekvartal'nyj nauchnyj zhurnal «Vektor nauki Tol'yattinskogo gosudarstvennogo universiteta»* **3** pp 251-5

[5] Ryzhakov V V, Kupryashin V A and Holudeneva A O 2013 Osobennosti tekhnologii utilizacii othodov bumazhnogo proizvodstva na osnove ispol'zovaniya ehlektroosmosa *Nauchno-tekhnicheskij zhurnal «Oboronnyj kompleks – nauchno-tekhnicheskomu progressu Rossii»* **1** pp 40-3

[6] Chernyahovskaya T N 2008 *Marketingovaya deyatel'nost' predpriyatiya. Teoriya i praktika* (Moscow: Vysshee obrazovanie)

[7] Koncepciya dolgosrochnogo social'no-ehkonomicheskogo razvitiya Rossijskoj Federacii na period do 2020 g. (utverzhdena rasporyazheniem Pravitel'stva Rossijskoj Federacii ot 8 avgusta 2009 goda N 1121-r.). EHlektronnyj fond pravovoj i nauchno-tekhnicheskoj dokumentacii http://docs.cntd.ru/document/902130343

[8] Perechen' prioritetnyh napravlenij razvitiya nauki, tekhnologij i tekhniki Rossijskoj Federacii (utverzhden Ukazom Prezidenta Rossijskoj Federacii ot 7 iyulya 2011 g. №899) EHlektronnyj fond pravovoj i nauchno-tekhnicheskoj dokumentacii. http://docs.cntd.ru/document/902287707

[9] Pegat A 2009 *Nechetkoe modelirovanie i upravlenie.* (Moscow: Binom. Laboratoriya znanij)

[10] Mosyagin V I 2008 *Ohrana prirody v otraslyah himicheskoj pererabotki drevesiny* (Moscow:

EHkologiya)
[11] Klark J D A 2003 *Tekhnologiya cellyulozy* (Moscow: Lesnaya promyshlennost').
[12] Goryunova V V 2010 Avtomatizirovannoe proektirovanie processov tekhnicheskogo obsluzhivaniya *Informacionnye tekhnologii v proektirovanii i proizvodstve* **2** pp 60-5.
[13] Matyunin I E 2015 *Optimizaciya transportno-gruzovyh processov na promyshlennyh predpriyatiyah* (Minsk: Vysshaya shkola).
[14] Ryzhakov V V, Holudeneva A O and Rudyuk M Yu 2016 *Logistika i CALS-tekhnologii CHast' 1 Logistika. Uchebnoe posobie*(Penza: Penz. Gos. Tekhnol. Un-t).
[15] Dudyshev V D, Semashko V D, Rohmistrov O V and Litikov A P 2006 Ustrojstvo udaleniya vlagi iz poristyh materialov. Patent Rossijskoj Federacii. RU 52735 U1. Opubl. 27.04.2006, Byul. № 12
[16] Holudeneva A O, Zhigalin S N, Ryzhakov V V and Kupryashin V A Razrabotka ehkologicheski chistoj tekhnologii utilizacii othodov proizvodstva bumagi. *Sbornik. dokl. mezhdunar. nauchno - prakt. konf. «Molodezh' i nauka: modernizaciya i innovacionnoe razvitie strany». CHast' 3* pp 248 –50.
[17] Tihomolova K P 2014 *Ehlektroosmos* (Moscow: «Himiya»).
[18] Holudeneva A O 2016 Teoreticheskie obosnovaniya vozmozhnosti vnedreniya avtomatizirovannoj informacionno-izmeritel'noj sistemy upravleniya processom ehlektroosmosa *Sbornik nauchnyh trudov mezhdunarodnoj nauchno - prakticheskoj konferencii «Informacionnye tekhnologii v ehkonomicheskih i tekhnicheskih zadachah»* pp 197-200
[19] Kholudeneva A O 2016 The research of the electroosmotic dehydration features of porous waste taking into account the osmosis physical model influence and the voltage values dynamics *Int. J. of Applied Engineering Research* **11** (19) p 9842
[20] Kholudeneva A O and Ryzhakov V V Processes of drying of moisture-containing materials: problems of ecology and choice of theoretical, schematic and experimental directions for their resolution *Int. J. of Applied Engineering Research*

Technology of biscuit semi-finished products using recycled pumpkin seeds

E Moskvicheva[1*], I Timoshenkova[1,*], D Chernikova[1], E Fedinishina[1] and M Bernavskaya[1]

[1]Peter the Great St. Petersburg Polytechnic University, Graduate School of Biotechnology and Food Technology, 194021 Novorossiyskaya st., 48, Russian Federation

* Corresponding author: itimoshenkova@spbstu.ru

Abstract. A possibility and expediency of pumpkin seeds flour utilization for production of biscuit semi-finished product was studied; the influence of the flour on consumer properties of the product was investigated. Studies were conducted on organoleptic and physicochemical indicators of the finished biscuit semi-finished products. Grounding on the implemented studies a new production technology of pumpkin seeds flour based biscuits was developed, shelf life, biological, and nutrition value were identified. It was established that the optimal ratio of pumpkin seeds flour and starch is 90.5±0.5% to 9.5±0.5%. It was established that after consuming 100 g of the biscuit with a complete substitution of wheat flour to the gluten-free mixture 44% of daily need in magnesium, 50% - in selenium, 39% - in copper, 48% - in phosphorus, 46% - in manganese, 36% - in vitamin B9, and 36% - in linoleic acid is satisfied.

1. Introduction

Flour confectionery remains an integral part of any cuisine and plays a great role in nutrition of humanity.

Most flour confectionery is characterized by poor contents of protein, essential amino acids, macro- and microelements, vitamins, and other vital nutrients, as well as by a high caloric content. Regarding current consumer requirements flour confectionery needs to not only be a mere calorie carrier but to provide a specific functionality.

A substantial part of food products is taken by products devoid of grain crop protein which causes an autoimmune chronic Gee's disease [2].

Due to that the search for raw materials, which could allow creating gluten-free products of high quality possessing a balanced chemical content and high nutritional value, comes to the forefront. Ways of production of gluten-free confectionery were developed with the use of various flour types, such as triticale, rice, buckwheat, corn, fat-free flour from grape, watermelon, thistle fruit seeds; vegetable, fruit and vegetable and berry purees, pastes and powders; grain and groats conversion products [3-12]. Currently pumpkin flour, which is produced during the processing of its seeds, is of special interest. Chemical contents of pumpkin seeds are characterized by a high composition of vitamins, macro- and microelements, essential amino acids and polyunsaturated fat acids.

The purpose of this paper is to develop a technology for a biscuit semi-finished product with a complete substitution of wheat flour to pumpkin seeds flour.

In compliance with the mentioned purpose the following tasks were identified:

- to investigate technological characteristics of pumpkin seeds flour;
- to investigate organoleptic and physicochemical indicators of biscuit baked products with a complete substitution of wheat flour to pumpkin seeds flour with varying ratio of starch from a dry mixture mass and to identify optimal ratio of pumpkin seeds flour and starch from dry mixture mass;
- to develop a technology of a biscuit with a complete substitution of wheat flour to a gluten-free mixture;
- to conduct a complex evaluation of the quality of new biscuit baked products.

Pumpkin seeds flour and biscuit of recipe #1 according to a collection of flour confectionery and bun products [13-17] were chosen as the object of investigation.

2. Methods of investigation

Sampling and their preparation for analysis was conducted in compliance with a State Standard 5904; pumpkin seeds flour humidity – State Standard 9404; autolytic activity – State Standard 27495; amount and quality of gluten – State Standard 27839; acidity of the flour - State Standard 27493; mass fraction of moisture in biscuit semi-finished products – State Standard 21094; organoleptic indicators – State Standard 50763; porosity of biscuit semi-finished products – State Standard 5669.

According to commonly known methods the following physicochemical indicator were identified:

- Gelation temperature – by heating the flour suspension to the temperature of 55-95°C with a 10°C step and subsequent centrifuging of a cooled sample during 10 minutes with a 3000 revolutions per minute speed.
- Density of baked gluten-free products – as a ratio of mass to the volume of a baked semi-finished product's fragment.
- Baking losses of gluten-free products – by a ratio of difference between test preparation mass and baked semi-finished product to test preparation mass;
- Swelling capacity of crumb of baked gluten-free products – according to a method developed by Saint-Petersburg State Science and Research Institute of Baking Industry;
- Water activity of baked gluten-free products using a device Aqulab Pawkit.

A generalized organoleptic indicator was identified by (1).

$$\sum_{\iota=1}^{\nu} \pi_{\iota} \kappa_{\iota} , \tag{1}$$

where p_i is a value of i organoleptic indicator;

k_i is a value of ponderability coefficient of i organoleptic indicator.

Coefficients of ponderability were established depending on significance of an organoleptic indicator (table 1). The sum of all the coefficients was set as 20. This way, with a 5-point scale of organoleptic indicators evaluation, the maximum value of a generalized organoleptic indicator amounts to 100.

Table 1. Coefficients of ponderability of organoleptic indicators of gluten-free biscuit semi-finished products.

Organoleptic indicator	Appearance	Color	Texture	Aroma	Taste
Ponderability coefficient	5.0	2.0	4.0	4.0	5.0

Organoleptic analysis of gluten-free biscuit products was conducted according to a 5-point scale. For a comparative organoleptic evaluation a system of indicators was developed which includes appearance, color, texture, aroma, and taste (table 2).

Pumpkin seeds flour is a loose powder of a light-green color, no crunch is felt during chewing, which speaks for the absence of mineral admixture, taste and aroma are typical for pumpkin seeds.

IOP Conf. Series: Earth and Environmental Science **337** (2019) 012036 doi:10.1088/1755-1315/337/1/012036

Table 2. Organoleptic evaluation criteria.

Organoleptic indicator	Verbal characteristic of points 5	4	3	2	1
Appearance	Correct form, complies with form in which baked products were produced earlier, with somewhat salient upper crust without cracks and tears, smooth surface, thin and soft upper crust	Correct form, insignificant tears, insufficient extent of rise, smooth surface, thin upper crust	Correct form, with small defects of surface (tears, cracks, hollows), thicker upper crust	Incorrect form of the biscuit, flat upper crust with significant tears and cracks and hollows, thick crust	Incorrect form of the biscuit, absence of rise, thick and flat crust with significant tears and cracks
Color	Light-brown crusts, light green crumb that complies with the color of pumpkin seeds, even color	Light-brown crusts, light green or yellowish crumb, even color	Dark-brown crusts, yellowish crumb, even color	Dark-brown crusts and crumb, uneven color	Dark color of crusts and crumb, uneven color, burned parts
Texture	Well-developed porosity, even pore distribution, soft and elastic crumb	Uneven pore distribution, soft and elastic crumb	Weakly expressed porosity, unevenly distributed pores, friable crumb	Not expressed porosity, quite firm crumb	Nonporous biscuit, teared crumb
Aroma	Well expressed, complies with pumpkin seeds smell, without foreign hints	Well expressed, complies with pumpkin seeds smell, without foreign hints	Typical to the type of product with a small hint of pumpkin seeds smell	Not expressed, presence of foreign smells	Aroma untypical to the studied type of products, presence of foreign hints
Taste	Well expressed, complies with pumpkin seeds smell, without foreign tinges	Well expressed, complies with pumpkin seeds smell, without foreign tinges	Weakly expressed, presence of bitter tinge	Not expressed, presence of foreign tinges	Taste untypical to the studied type of products, presence of foreign tinges

3. Results and screenings

Results of investigation of physicochemical parameters are demonstrated in a table 3.

Table 3. Physicochemical indicators of pumpkin seeds flour quality.

Indicators	Premium grade wheat flour	Pumpkin seeds flour
Humidity, %	14.5±0.2	7.6±0.2
Autolytic activity, %	30.0±0.5	24.0±0.5
Amount of gluten, %	83.0±0.1	-
Acidity, ° T	4.0±0.2	22.6±0.2

The autolytic activity indicator of pumpkin seeds flour complies with the indicator of premium grade wheat flour. An increased acidity of pumpkin seeds flour (5 and a half times more than of wheat flour) is explained by the presence of a great amount of organic and polyunsaturated fat acids, which will contribute to intensification of sucrose hydrolase process during biscuit dough formation.

Thus, pumpkin seeds flour does not possess baking properties of wheat flour, however, it may be recommended to use for confectionery industry as the main and additional component of gluten-free mixture, during creation of functional products, for flour confectionery products that do not require a high content of gluten, like biscuit.

Gluten-free mixture based on pumpkin seeds flour and method of adding eggs were taken as factors that influence on quality of biscuit semi-finished products and especially on porosity of the finished products.

Gluten-free mixture used for production of biscuit semi-finished products consists of pumpkin seeds flour and starch.

Investigation of organoleptic indicators and humidity of biscuit semi-finished products was conducted based on gluten-free mixture samples varying ratio of starch from 0 to 20% with a 2% step.

On the grounds of experimental data regression equations were obtained characterizing dependency of generalized organoleptic indicator (2) and humidity of biscuit semi-finished products (3) on starch content (x, ratio) in a gluten-free mixture.

$$\Psi_1=\frac{1}{0{,}01+0{,}15\xi^2} \quad (2)$$

$$\Psi_2=\frac{1}{0{,}03+0{,}14\xi} \quad (3)$$

Correlation coefficients of obtained regression equations are 0.86 and 0.99 correspondingly, which allows speaking about functional dependency of generalized organoleptic indicator and humidity of biscuit semi-finished product on starch content in a gluten-free mixture. Determination coefficients of obtained regression equations are 0.75 and 0.98 correspondingly, corrected determination coefficients are 0.72 and 0.97 correspondingly which proves significance of the chosen factors.

Graphic interpretation of equations (2) and (3) are illustrated at Fig. 1 and Figure 2.

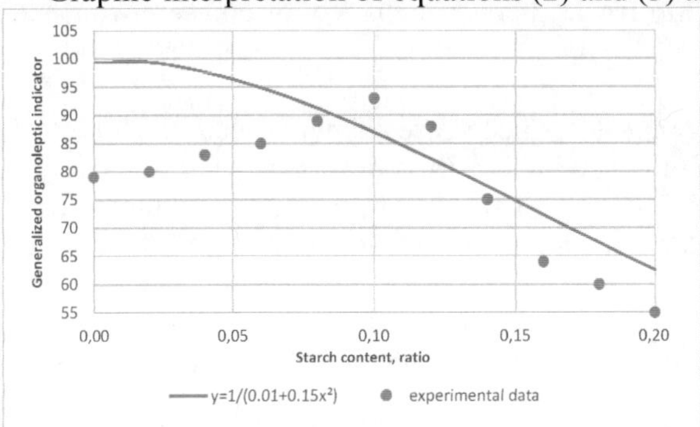

Figure 1. Dependency graph of generalized organoleptic indicator of biscuit semi-finished products on starch content in a gluten-free mixture

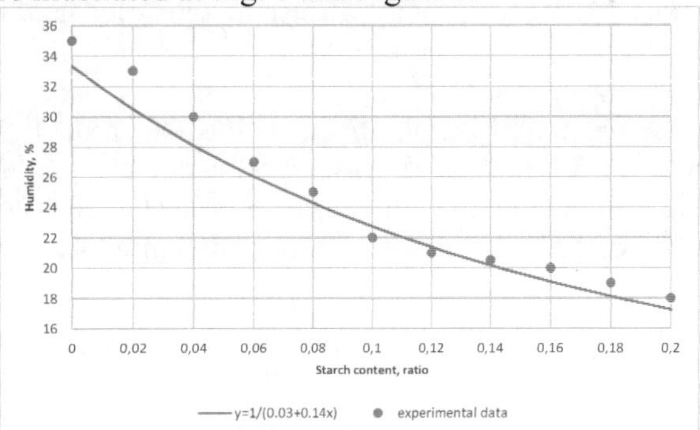

Figure 2. Dependency graph of humidity of biscuit semi-finished products on starch content in a gluten-free mixture

Analysis of revealed dependencies showed that ration of pumpkin seeds flour and starch 90.5±0.5% to 9.5±0.5 % is close to optimal and provides satisfactory quality of biscuit semi-finished products.

Sectional view of pumpkin seeds flour based biscuit sample is demonstrated at Fig. 3. It is evident that the baked biscuit sample had a small rise and sank after baking, which speaks for a weak stability of biscuit dough.

Stored gluten-free biscuit semi-finished products were evaluated according to generalized organoleptic indicator, water activity and mass as well as organoleptic indicators grounding on which shelf-life period of baked products can be revealed.

Organoleptic indicators of the second sample during storing were substantially superior over those of the first sample due to finely-porous and elastic structure of the biscuit.

On the basis of experimental data regression equations were obtained characterizing dependency of water activity indicator (4) and mass (5) on storing duration (days).

$$\Psi_3 = \sqrt{0.56 + 0.61/\xi} \tag{4}$$

$$\Psi_4 = 26.18 + 3.2/\xi \tag{5}$$

Correlation coefficients of obtained regression equations are 0.994 and 0.992 correspondingly, which allows speak about functional dependency of water activity indicator and mass of biscuit semi-finished product on storing duration. Determination coefficients of obtained regression equations are 0.985 and 0.983.

Investigation results of water activity of baked biscuit semi-finished products during storing are demonstrated at Figure 3, changes in mass of biscuit semi-finished products – at Figure 4.

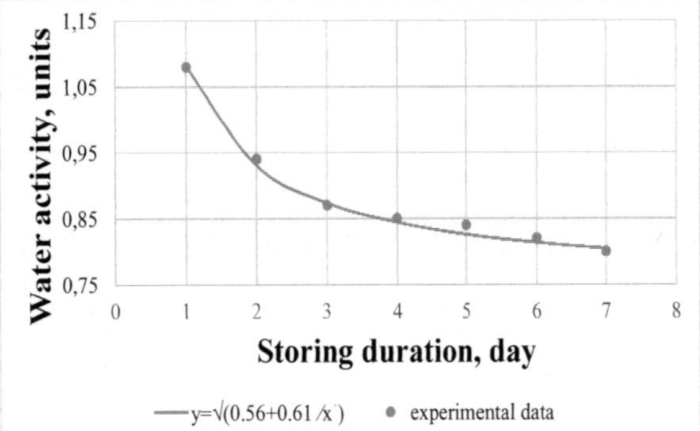

Figure 3. Water activity of biscuit semi-finished products

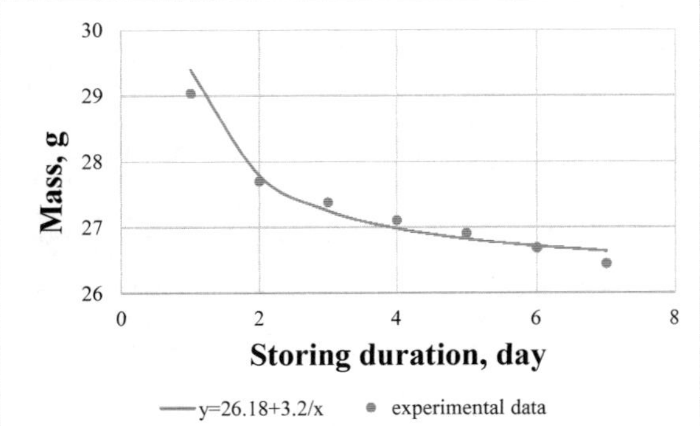

Figure 4. Changes in mass of biscuit semi-finished products

Water activity of biscuit semi-finished products during storing decreases hyperbolically down to the value of 0,8, the mass of biscuit semi-finished product made with a method of separate addition of eggs and boiling protein with hot syrup insubstantially decreases after baking. At the same time, the process of staking that is related to retrogradation of starch is almost not observed, the products stays soft and elastic.

Biscuit semi-finished products based on pumpkin seeds flour satisfies the human need for tryptophan amino acid, score of which equals 100%. Limiting acids in the studied semi-finished product are lysine and threonine.

Expediency of pumpkin seeds flour utilization in a biscuit semi-finished product technology was proved by investigation of its nutritional value (table 4) and mineral elements, vitamins and polyunsaturated fat acids composition.

Table 4. Nutritional value of biscuit semi-finished products.

Indicator	Biscuit semi-finished product
Protein content, g/100 g of product	11
Lipid content, g/100 g of product	8
Carbohydrate content, g/100 g of product	13
Energy value, kcal	150

4. Conclusion

Technological characteristics of pumpkin seeds flour were investigated, on the basis of which the studies were conducted on the main organoleptic and physicochemical indicators of biscuit semi-finished products with a complete substitution of wheat flour to pumpkin seeds flour with variation of starch ratio.

It was established that the optimal ratio of pumpkin seeds flour and starch is 90.5±0.5% to 9.5±0.5%.

After consuming 100 g of the biscuit with a complete substitution of wheat flour to the gluten-free mixture 44.48% of daily need in magnesium, 49.56% - in selenium, 38.95% - in copper, 48.29% - in phosphorus, 46.40% - in manganese, 35.74% - in vitamin B9, and 36.04% - in linoleic acid is satisfied.

References

[1] Locmanov A S , Nazimova G I and Romanov A S 2011 *Techniques and technologies of food industries* **3** pp 71-77
[2] Revnova M O 2006 *Life without gluten* **3** pp 4-63
[3] Docenko V F 2014 *East-European J. of Advanced technologies* **10** pp 64-9
[4] Lukin A A, Merenkova S P and Fomina T U 2016 *Young scientist* **10** pp 263-6
[5] Matveeva T V and Koryachkina S Y 2016 *Scientific basis, technologies, recipes 360* (Spb: GIORD)
[6] Matveeva T V 2009 *Commodity researcher of Food Products* **7** pp 17-21
[7] Moskvicheva E V, Safonova E E and Timoshenkova I A 2017 *Int. science and research J.* **8** pp 46-50
[8] Moskvicheva E V, Timoshenkova I A and Chernikova D A 2017 *SpbPU: material of scientific conf.* pp 95-8
[9] Perfilova O V 2009 *Storing and recycling of agricultural materials* **9** pp 52-4
[10] Bazarnova Yu G and Ivanchenko O B 2016 *Voprosy Pitaniia* **5** pp 100-7
[11] Ji Y 2007 *Food Chemistry* **100** pp 53-8
[12] Sae–Eaw A 2007 *J. of Food Science* **72** 33 92-7
[13] Popov V S, Sergeyeva S S and Barsukova N V 2016 *Bulletin of the University of technology* **19** pp 147-51
[14] Pankina I A and Barsukov A V 2015 *III international scientific and practical Conf.* pp 127-31
[15] Barsukova N, Moskvicheva E, Timoshenkova I and Tsopa E 2018 *SHS Web of Conf.* **44** p 8
[16] Batista J E R, Braga L P, de Oliveira R C, Silva E P and Damiani C 2018 *Food Science and Technology* **38** pp 250-4
[17] Costa L L, Tomé P H F, Jardim F B B, Damasceno K A and Campagnol P C B 2018 *Int. Food Research J.* pp 1489-96

IOP Conf. Series: Earth and Environmental Science **337** (2019) 012037 doi:10.1088/1755-1315/337/1/012037

Intensification of pectin obtaining from apple pomace

O Ivanchenko[1*], E Aronova[1], P Balanov[2], I Smotraeva[2]

[1]Higher School of Biotechnology and Food Technology, Peter the Great Saint Petersburg State Polytechnical University, 199005, Politekhnicheskaya Str., 29, St.Petersburg, Russia
[2]Saint Petersburg National Research University of Information Technologies, Mechanics and Optics (ITMO), Kronverksky Pr. 49, Saint Petersburg, Russia

* Corresponding author: obivanchenko@yandex.ru

Abstract. This study is devoted to the problem of improving the methods of pectin obtaining from apple pomace. The relevance of the study is dictated by the fact of formation of a large amount of waste when receiving apple juice. The output of apple juice at the enterprises of the processing industry is from 50 to 75%, therefore, 25-50% is waste. In our studies we used 42 KHz frequency of ultrasonic vibrations. The powers of ultrasound waves were 30 and 50W. The output of pectin depends on the sample exposure time, and with time passing this growth decreases and eventually stops, which may indicate that a certain amount of destruction has been reached at a given power. The maximum increase in the output of pectic substances is observed when the ultrasonic treatment time is 15 minutes and it is 26%. Capacity also affects pectin output. With power increasing from 30 to 50 W, the amount of pectin released increases by an average of 10%. It is shown that the use of separation at the stage of solution partitioning from pectin allows one to increase the speed of the process by a factor of two.

1. Introduction

The main raw materials for pectins are waste of a large number of food productions. The citrus pectin is obtained from orange peels, which is most popular in the world due to its low cost. The peel of oranges contains a large amount of pectinesterase enzyme, which promotes the deetherification of pectin [1]. In addition, pectin is obtained from mandarin peels, lime, lemons, sunflower baskets, sugar beets [2], black currants [3]. In South America pectin is obtained from some species of cacti [4].

Pectins are among the polysaccharides that can be used in various sectors of food industry as thickeners, gelling agents, emulsifiers, stabilizers and therapeutic and prophylactic additives [5]. The most important property of pectin, from a medical point of view, is its ability to bind with cholesterol, radionuclides, heavy metal ions [6].

The degree of etherification plays a significant role when using pectins. High-methoxyl pectin (HM-pectin, HE-pectin) possesses a weak charge and is less sensitive to charged molecules, such as proteins. HM-pectin increases the colloidal stability of protein molecules, thereby stabilizing acidic beverages. In this case, the interaction mechanism, apparently, is the electrostatic interaction between the protein and pectin, and not the adsorption of pectin on the surface of the protein molecule [7].

In dairy industry and in production of juices from fruits, citrus pectins are used, and the pectin from lime has the best quality in the technology of these products. Pectin from oranges and grapefruit has satisfactory characteristics, and pectin from mandarins does not suit at all [8].

IOP Conf. Series: Earth and Environmental Science **337** (2019) 012037 doi:10.1088/1755-1315/337/1/012037

Usage of pectic substances in baking has several reasons. In bakery production, pectin is used to increase volume and water retention during storage, which allows one to obtain softer and more porous products, as well as improve structural and mechanical properties of crumb and to increase its humidity. These valuable properties of pectins are due to its high water-holding capacity, which allows one to improve the quality indicators of bread. High hydrocolonic properties are related with the fact that pectins during the process of water absorption compete with flour starch and reduce its hydration, and pectins in baking processes are able to release moisture, causing starch gelatinization [2].

Due to the ability to form jelly, pectin is widely used for jelly production: marmalade, jelly fillings for sweets [9].

Pectin from sugar beet pulp is used in dietetic and pharmaceutical product technologies, since this pectin has the property of binding heavy metals. Apple and tomato beverages based on beet pectin have been created, which can be recommended to megacities residents, as well as to people working in hazardous industries, such as metallurgy, chemical industry. Pectin from sunflower baskets is used in cosmetics production [10].

To obtain pectin, it is very expedient to use pomace obtained from apples as a result of their processing into juice [11]. The output of apple juice is from 50 to 75%, therefore, the remaining part is pomace. For companies that produce juices, this is waste and its competent disposal is an absolute part of costs, which ultimately affects the cost of production. In this regard, usage of waste from one production as a raw material for another enterprise is already a strong resource-saving factor.

Apple pomace has its drawbacks. Since this is a plant product and, moreover, it is very juicy (its humidity is 70–90%), it is susceptible to microbiological damage, which limits its applicability in its native form. On the other hand, if pectin production is carried out near enterprises of apples primary processing, this problem is partially removed. In addition, it is possible to store and transport the pomace in dried form. However, in this case, significant energy costs are inevitable. It is important to understand that in order to obtain pectin, using one way or another, dried pomace still needs to be moistened and the evaporated moisture (plus associated costs) inevitably will increase the cost of the product and complicate the process chain. An interesting option is the pomace preservation with the help of chemical reagents. The cost of this technique is less than for drying the product, but there still remains the question about the separation of the preservative from the semi-finished product during pectin production.

2. Experimental part

Technology for pectin production from apple pomace exists and is implemented in the world with varying degrees of success [12-14].

A typical technological scheme for pectin producing includes several steps:

1. Obtaining the pomace.
2. Drying and subsequent grinding.
3. Extraction of soluble substances with cold water and its separation.
4. Extraction of pectin components of pomace using hot water (temperature up to 92 °C) at pH from 3.5 to 4.5
5. Treatment by amylolytic and proteolytic enzymes followed by inactivation.
6. Clarification using activated carbon followed by separation of coal particles.
7. Precipitation of pectin fractions by ethyl alcohol and separation of the residue.
8. Drying the residue.

Industrial production of pectin is a very time consuming and resource-intensive process. It requires serious thermal power and a large excess of ethanol to precipitate. This alcohol is regenerated after use, which also obliges to high energy costs.

Scientific and engineering studies aimed at improving the efficiency of pectin production have been carried out continuously in recent decades and affect various areas of the technological process [15, 16]. This is an increase in the output of the finished product, as well as chemical [17], and design solutions.

Structurally, pectin in plant tissues is part of complexes containing protein and hemicellulose, which, in turn, are associated with cellulose, forming a fairly strong and plastic frame. To obtain pectin from this complex system, various techniques are used: chemical, biochemical, and physical.

The present studies are aimed at usage of ultrasonic waves for partial disintegration of this system and separation of pectin fractions from other components. The effectiveness of ultrasound for such purposes was confirmed by scientific studies [18, 19].

High-frequency elastic oscillations in medium cause appearance of areas with high and low pressure. In areas with low pressure, microscopic vacuum cavities are systematically formed, which collapse releasing mechanical energy into nearby areas, which distorts/breaks plant tissue. It is difficult to predict specific break points of the cell cages, but the fact of mechanical destruction can be observed instrumentally, for example, by accumulation of substance in the medium (in our case, pectin).

An ultrasonic wave generator constructed on the basis of piezoelectric elements was used in our work. The electric field generated by the device causes oscillations of the source of ultrasonic waves, which is transmitted through the walls of the device to the material under study. The frequency of ultrasonic vibrations was 42 KHz, and the power of ultrasonic waves was 30 and 50W. As a smoothly varying variable we selected the sample exposure time. The research results are presented in Figure 1.

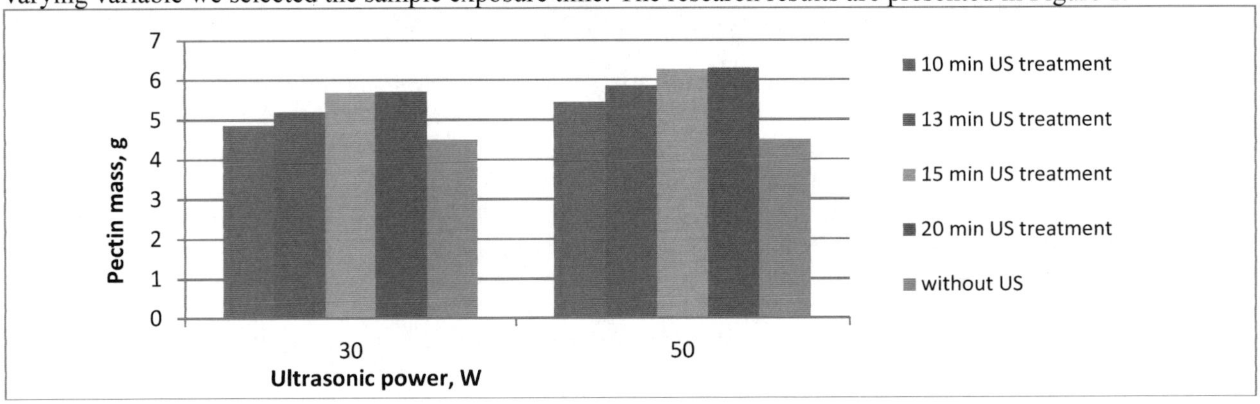

Figure 1. Pectin output depending on the ultrasound treatment mode

As is seen from Fig.1, the output of pectin depends on the time of sample treatment and while time is passing, this growth decreases and eventually stops, which may indicate that a certain amount of destruction has been reached at a given power. The maximum increase in the pectic substances output is 26%. Power also affects the output of pectic substances, with a 60% increase in power, the amount of pectin released increases, on average, by 10%.

Further we conducted studies aimed at identifying the temperature optimum for extraction of pectic substances with water from apple pomace. For this purpose, extraction was performed at various temperatures, and then the pectin precipitated with alcohol was isolated from medium in two ways: by filtration and separation. The obtained data are shown in Figure 2. From the data obtained, it is seen that the maximum output of pectic substances is in the range between 60 and 70 °C.

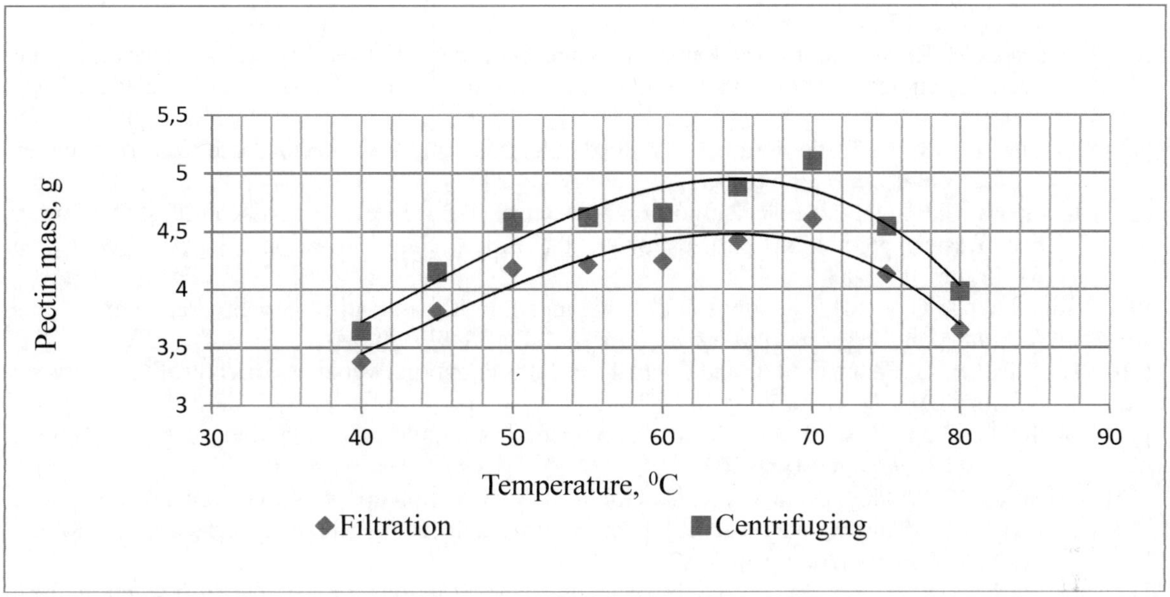

Figure 2. Influence of temperature on pectin substances output

The optimum range of temperature extraction can be explained by the plant nature of the raw materials used. Apparently, apples grades and conditions for pomace obtaining can affect the structure of cell walls and, consequently, the pectin output at various temperatures.

Usage of mixture separation is very promising. As our research shown, in comparison with traditional filtration, separation makes it possible to extract soluble fractions of pectin more fully, on average, by 10%. In addition, the process of separation itself is significantly accelerated. Filtering under vacuum takes 120 minutes or more, and using separation the whole process, including loading and unloading the apparatus, takes no more than 50 minutes. Thus, the operation of pectin separation can be accelerated more than twice, with a noticeable increase in its output.

3. Conclusions

Summarizing the obtained results we can draw the following conclusions:

1. Ultrasonic processing of apple pomace allows one to increase the efficiency of pectin extraction by more than 20%, and the power of acoustic treatment also has a positive impact on the process efficiency.

2. The use of separation at the stage of solution partitioning from pectin allows one to speed up the process more than twice, which, as applied to production scales, will result in significant time savings, which today is a decisive economic factor.

References

[1] Prade R A, Zhan D, Ayoubi P and Mort A J 1999 Pectins, pectinases and plant-microbe interactions *Biotechnol Genet Eng Rev* **16** pp 361-91

[2] Donchenko L V 2000 *Technology of pectin and pectin products* (Moscow: DeLi)

[3] Buchweitz M, Speth M, Kammerer D R and Carle R 2013 Impsct of pectin type on the storage stability of black currant (Ribes nigrum L.) anthocyanins in pectic model solutions *Food Chem.* **139**(1-4) pp 1168-78.

[4] Jaramillo-Flores M E, Gonzalez-Cruz L, Cornejo-Mazon M, Dorantes-Alvarez L, Gutierrez-Lopez G and Hermandez-Sanchez H 2003 Effect of termal treatment of the antioxidant activity and content of carotenoids and phenolic compounds of cactus pear cladodes (Opuntia ficus-indica) *Food science and technology international* **9** p 273

[5] Aymeson A 2012 *Food thickeners, stabilizers, gelling agents* (SPb: Professia)
[6] Gonzalez M, Rivas C, Caride B, Lamas M A and Taboada M C 1998 Effects of orange and apple pectin on cholesterol concentration in serum, liver and faeces *J Physiol Biochem* **54**(2) pp 99-104
[7] Tuinier R, Rolin C, de Kruif C G 2002 Electrosorption of pectin onto casein micelles *Biomacromolecules* **3** pp 632-8
[8] Gomez B, Gullón B, Yáñez R, Parajó J C and Alonso J L 2013 Pectic oligosacharides from lemon peel wastes: production, purification, and chemical characterization *J Agric Food Chem* **61**(42) pp 10043-53
[9] Solopenkova O V 2013 Stuffings and its composition: the impact of carbohydrate components *Konditerskoe i hlebopekarnoe proizvodstvo* **3-4** (138-139) pp 24-5
[10] Liu L, Chen G, Fishman M L and Hicks K B 2005 Pectin gel vehicles for controlled fragrance delivery *Drug Deliv* **12**(3) pp 149-57.
[11] Shalini R, Gupta D K 2010 Utilization of pomace from apple processing industries: a review *J. Food Sci Technol* **47**(4) pp 365-71.
[12] Priyangini F, Walde S G and Chidambaram R 2018 Extraction optimization of pectin from cocoa pod husks (Theobroma cacao L.) with ascorbic acid using response surface methodology *Carbohydr Polym* **202** pp 497-503.
[13] Chaharbaghi E, Khodaiyan F and Hosseini S S 2017 Optimization of pectin extraction from pistachio green hull as a new source *Carbohydr Polym* **173** pp 107-13.
[14] Jafari F, Khodaiyan F, Kiani H and Hosseini S S Pectin from carrot pomace: Optimization of extraction and physicochemical properties *Carbohydr Polym* **157** pp 1315-22.
[15] Thirugnanasambandham K, Sivakumar V 2015 Application of D-optimal design to extract the pectin from lime bagasse using microwave green irradiation *Int J Biol Macromol* **72** pp 1351-7.
[16] Naghshineh M, Olsen K and Georgiou C A 2013 Sustainable production of pectin from lime peel by high hydrostatic pressure treatment *Food Chem* **136**(2) pp 472-8.
[17] Gimaev I N, Romanova N K and Reshetnik O A 2004 Influence of parameters of hydrolysis-extraction process on the output and quality of pectin from fruit and berry raw materials *Vestnik kazanskogo tekhnologicheskogo universiteta* **1** pp 214-8.
[18] Smotraeva I V, Balanov P E and Tretyakov N A 2014 The use of ultrasound in the processing of plant materials *Izvestiya Sankt-Peterburgskogo gosudarstvennogo agrarnogo universiteta* **37** pp 264-7.
[19] Balanov P E and Smotraeva I V 2016 The impact of ultrasound and microwave radiation on the output of juice from wild berries *Izvestiya Sankt-Peterburgskogo gosudarstvennogo agrarnogo universiteta* **42** pp 52-6.

Solving the problem of using dairy waste in Russia

T Pilipenko[1], E Belokurova[1], I Pankina[1] and A Shlykova[1]

[1]Peter the Great St. Petersburg Polytechnic University, 29, Polytechnicheskaya str., St. Petersburg, 195251, Russia

Corresponding authors: pilipenko_tv@spbstu.ru, belokurova_es@spbstu.ru, pankina_ia@spbstu.ru,.shantonina@inbox.ru

Abstract. The present work is devoted to the problem of whey utilization, which is a by-product in the production of cheese and curd. We carried out analysis of modern Russian and foreign inventions aimed at waste-free milk treatment. We propose a modern waste-free method for curd production, which ensures the full use of the milk components. We carried out modeling of curd production process. The curd production process was presented as a combination of the most important technological operations – ultrafiltration whey treatment, mixing, thermocoagulation and clot treatment. We developed a structure of information flows, which totally characterizes waste-free technology of curd production. Alongside with that we proposed a structure of information flows accepted for mathematical modeling of the process. We propose models for formation of mathematical expectation of protein content in curd depending on the process factors, for which we used the regression equations. In this case, it turned out to be expedient to use linear dependencies. This can be explained by the fact that theory, research methods and calculation of linear models are developed deeper than others and require less computation. The question of applicability of such a model was resolved on the basis of checking its adequacy. The task of calculating the parameters of the model was solved on the basis of experimental data processing, which are the results of a passive experiment on study the waste-free curd production. Mathematical processing of experimental data allowed us to derive a regression equation and establish a correlation between the factors of curd production process and its protein content.

1. Introduction

During production of most types of dairy products (curd, cheese, casein) whey is formed, which can account for up to 75-85% of the raw material - milk. Currently, in Russia, most of the whey, unfortunately, is either not used at all and is discharged into sewage, or is not used rationally - as a raw material for production of feed for farm animals [1, 2].

It should be noted that a significant amount of useful substances of milk passes into the whey. The degree of transition of proteins into whey is 24.3%. Milk sugar (96.0%) is almost completely transferred to whey, as well as 22.5% of casein and 95% of whey proteins. Whey is saturated with lacto albumins and lactoglobulins - from 1.0 to 1.1%, which is important for adequate nutrition of a man of today [3, 4].

A scientific team from the Department of Food Sciences, School of Food Industry, University of Campinas (Brazil) found that whey protein is a potential and versatile ingredient at the development of a new and natural component of food. It can work as a texture modifier, thickener, carrier, gelling agent, surfactant component and foaming agent among other related functions and bioactivity. Nanocomposites containing whey proteins, for example, have been used by them as effective systems for encapsulating active food and drug components, increasing their solubility, transport, dispersibility, and bioavailability [5].

In the countries of the European Union, the discharge of whey into sewage is strictly prohibited, since it is known that when producing 10 tons of cheese or curd per day, the discharge of organic matter is equal to discharge of organic matter from 3000 people. In such countries where the technical base of cheese production is highly developed (France, Holland, USA), more than 90% of whey is processed. In the European Union, industrial processing of cheese whey for food products is 60%, and for feed and technical equipment it is 40%. In the USA, cheese whey processing is carried out into a wide range of products: 40% for dry whey; 30% for lactose; 10% for whey protein concentrate and dry demineralized whey; dry whey with hydrolyzed lactose; feed is only 3% and other technical products are 2%. Thus, in the USA for food purposes 95% of whey obtained in the production of cheese is used [6-8].

In Russia in 2011, 2,040 thousand tons of cheese whey were obtained, and only 779.2 thousand tons were processed by industry, which amounted to only 38.2%. Over the next 7 years, the percentage of whey processing decreased. The volumes of non-returnable waste (not processed whey) are constantly growing, as according to Art. 5 of the Tax Code of the Russian Federation, they are not taxed [9].

At the same time, Russian scientists are developing scientific foundations, methods and techniques for using whey in the production of various food products, including that for healthy nutrition [10, 11].

Membrane technology has revolutionized the production of dairy products. With the development of new technologies in membrane processes, it becomes possible to increase the use of whey in food production. Thanks to the introduction of membrane technologies, the full use of all milk substances occurs and the environmental friendliness of cheese and curd production is improved, preventing environmental pollution [12].

Nelson B.K, Barbano D.M developed a multi-stage MF process that removes a high percentage of whey proteins from skimmed milk, while creating a retentate with low concentrations of soluble minerals, non-protein nitrogen (NPN) and lactose [13].

The main developments both in Russia and abroad are aimed at increasing the efficiency of using dairy raw materials through introduction of new technologies based on development of membrane technology. Thus, the use of whey in production of food products, fodder and technical products will completely dispose such waste from the production of cheese and curd, which causes significant harm to the environment, polluting it with biological waste. However, it should be noted that in our country the volume of production of new types of milk protein products with WPC-UV (whey protein concentrate, obtained by ultrafiltration) is very insignificant [14-16].

2. Material and methods

In this work a waste-free method for curd production is proposed, which ensures the full use of the milk constituents. Herewith the organoleptic characteristics of the end product are improved, and the production process time is reduced. Coagulation is carried out by introducing of 10% ultrafiltration whey concentrate (WPC-UV) with an acidity of 250 °T and solids content of 18-20% in pasteurized milk with a temperature of 90-95 °C [17-19].

WPC-UV is obtained as follows: the whey from the previous batch of curd is pasteurized and subjected to ultrafiltration. The protein concentrate is pasteurized at a temperature of 72-74 °C with 20 s exposure time. For a more complete coagulation and compaction of the clot, the mixture is kept for 35-40 minutes. The prepared curd is of soft, spreadable consistence. The moisture content in the end product is 64%, the fat content is 10%, the acidity is 68 °T. Ultrafiltrate after pasteurization and introduction of flavoring components and aromatizing agents can be used for the production of refreshing drinks.

For a more reliable assessment of the process of curd waste-free production, we carried out a simulation of the technological process of curd production. As a basis for it we used an already existing model of a similar chemical-technological process. For the model being developed the curd production process was presented as a set of technological operators: ultrafiltration processing of whey -1, mixing - 2, thermocoagulation and processing of clot - 3. (Figure 1)

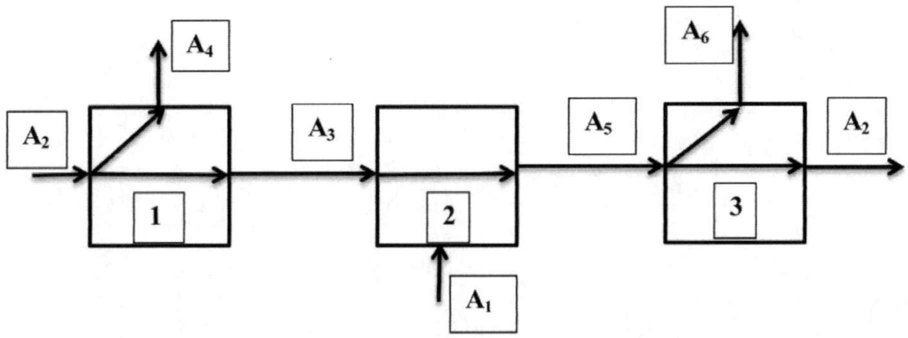

Figure 1. Operator diagram of waste-free process of curd production

The purposeful functioning of the thermocoagulation tank is ensured by both interconnection of technological operators with each other through technological flows of curd clot and whey masses, and by interaction with external technological operators through technological flows of milk (A_1), whey (A_2), WPC-UV (A_3), filtrate A_4 and end product (A_6).

During the operation process, each technological operator converts physical parameters of the input process flows into the physical parameters of the output process flows in accordance with a given direction of technological connections of the curd production process.

The properties of each process flow can be characterized by a certain set of information variables. The process flow of milk A_1 supplied to the container for thermocoagulation (VDP) is characterized by flow rate F_M, fat content Zh_M, protein content B_M, acidity K_M, temperature t_M. The process flow WPC-UV (A_3), also supplied to VDP, is characterized by consumption F_K, fat content Zh_K, acidity K_K, protein content B_K, and temperature t_K.

The intermediate flow of clot and whey (A_5) is characterized by the physicochemical composition and properties of clot and whey (Q_{clot}, Q_W). The flow of end product (A_6) is characterized by flow rate F_T, fat content Zh_T, moisture content V_T, protein content B_T, physicochemical and organoleptic properties Q_T and temperature t_T.

Output process flows of intermediate product (whey A_2) are characterized by flow rate F_S, fat content Zh_S, protein content B_S, temperature t_S. To maintain the temperature mode of the process, cooling water flows A_9, A_{10}, characterized by the flow rates F_9, F_{10} and temperature t_v are fed to the process operators.

3. Results and discussion

The obtained conclusions concerning relations between various stages of crud production process as well as complexity and multifactor nature of simultaneously occurring phenomena, justify the possibility of representing such objects as a single link of directional action for their mathematical description. The structure of information flows, which fully characterizes the waste-free method of curd production according to its protein content, is presented in Figure 2 (a).

We present each information variable that corresponds to a certain process which affects the protein content in the curd as an external input information flow. Then one can consider the waste-free crud production process as an information operator which transforms the input information flows into the output information flow - the protein content in crud.

When optimizing the waste-free curd production process, one should take into account only technological factors which provide better performance. In addition, when selecting process factors, one should consider only those that have a clear metrological meaning and meet the requirements for factors and its combination. All this allows excluding from further consideration some technological and other factors [20, 21]. The structure of information flows adopted for mathematical modeling of the process is shown in Figure 2 (b).

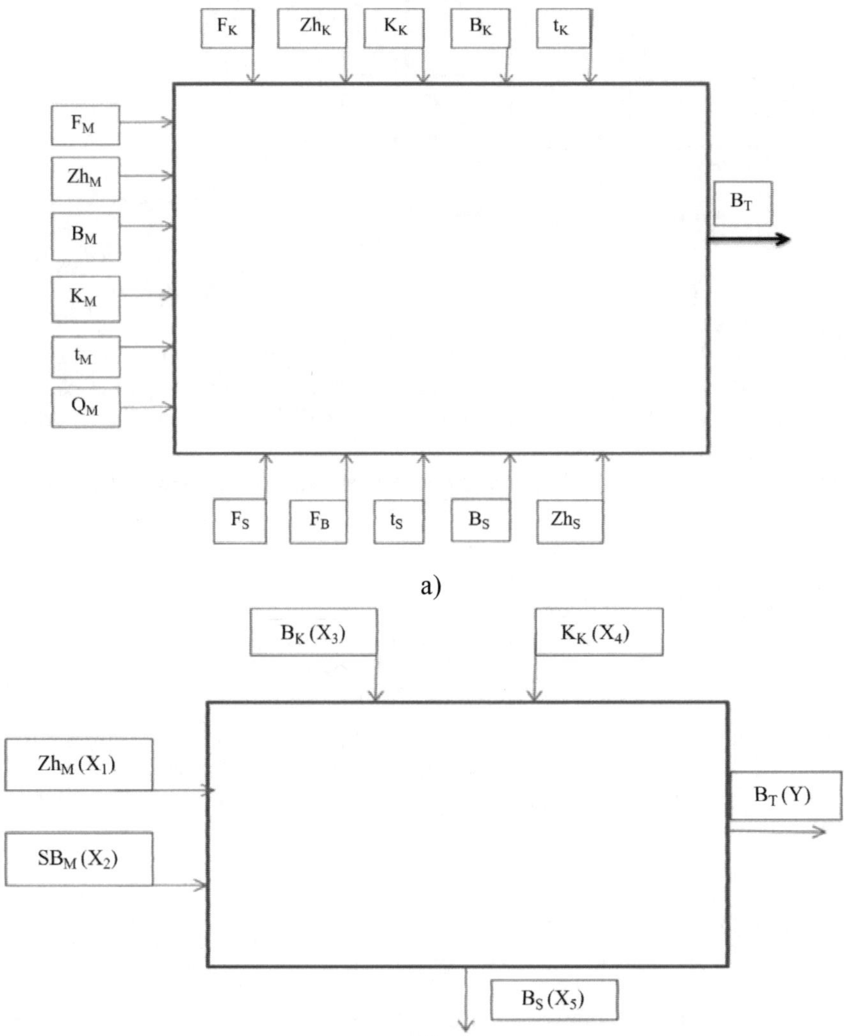

Figure 2. Parametric diagrams of crud production process:

a) Fully characterizing the crud production process,

b) Adopted for mathematical modeling of the process

Due to the complexity and multifactorial nature of crud production for a healthy nutrition, it is not possible to obtain an analytical relationship between the process factors and protein content in crud. Therefore, to simulate this dependence, it is advisable to use experimental statistical methods. The validity of this approach is justified in a number of works devoted to modeling of technological processes [22, 23].

In this case, after selection of main group of factors, which determine the formation of protein content in the curd during the process stages, it is necessary to choose the form of relations between factors and the protein content. For this relationship, the main process factors are non-random values. The influence of unaccounted factors may cause discrepancies between the calculated values of the protein content in curd and the actual ones.

Therefore, the protein content should be considered as a random variable. In this case it is advisable to model the formation of mathematical expectation of the protein content in crud, depending on the process factors, for which the regression equations are used. In this case, it is advisable to turn to a linear relationship. This is due to the fact that the theory, methods of research and calculation of linear models are developed deeper than others and require less computation. The question of the applicability of such a model is decided on the basis of checking its adequacy.

The empirical model describes the process only in the intervals of change of factors which will be considered during the experiment. The levels of factors characterizing the raw materials were in the following ranges: milk fat content from 2.7% to 3.0%, milk acidity from 14 to 19 °T, protein content from 2.8% to 3.0%, dry matter content from 10.0% to 11.7%; protein content WPC-UV from 16.4% to 20.4%, acidity WPC-UV from 247.9 to 325.4 °T. During the processing cycle of a homogeneous batch of raw materials, the levels of these factors can be considered constant, while the factors ZH_M, SB_M, B_K, during the continuous technological process of curd production are uncontrollable. The level of factor B_K has an impact on the crud production. With its increase, the quality of the end product deteriorates, and organoleptic indicators (consistency) also deteriorate. In this regard, the factor F_K is taken constant and is 10% by weight of milk. This value was obtained experimentally. To stabilize the protein content in the end product, it is most convenient to use the change in acidity level of the UV concentrate (Kk). In view of the aforesaid, the equation for this process of production of cheese will be written in the form:

$$B_T=B_0+B_1Zh_M+B_2SB_{M3}+B_3B_K+B_4K_K+B_5B_S \quad (1)$$

For convenience of subsequent mathematical processing the factors Zh_M, SB_M, B_K, K_K, Bs were assigned code marks X_1, X_2, X_3, X_4, X_5, respectively. Then equation 1 can be rewritten in the following form:

$$Y= Bo+B_1X_1+B_2X_2+B_3X_3+B_4X_4+B_5X_5 \quad (2)$$

The task of calculating the model parameters was solved on the basis of processing the experimental data, which are the results of a passive experiment to study the waste-free crud production process. In an experimental study, the actually possible ranges of factor changes included in equation 2 were considered. This fact makes it possible to use the results of modeling in real production conditions.

As a result, a linear model was obtained in the following form:

$$Y= 27.19-3.71X_1 +0.54X_2-0.29X_3 +0.01X_4-2.62X_5 \quad (3)$$

Adequacy testing is an integral part of modeling, since the possibility of applying the obtained mathematical models in practice is determined by the required accuracy of describing the dependencies between protein content and process factors.

As a criterion of proximity of protein content in curd calculated from equation 3 and its experimental (real) values, i.e. the model adequacy, the coefficient of multiple correlation P and Student's criterion t can be used.

The following values were obtained: P=0.773; t=5.15 (at Pmin=0.75; tmin=3.1). The given data testifies adequacy of the obtained mathematical model of a waste-free crud production process. In addition, checking the model adequacy was also carried out using the F-test. The hypothesis of the model adequacy is accepted if the calculated value of the F criterion is less than the Ptabl value. In our case, Rcrit=1.484, and Ftabl=3.33, with a significance level q=5% and the number of freedom degrees equal to 5. Thus, the obtained regression equation is adequate at a five percent level of significance.

4. Conclusions

The obtained results allow us to draw the following conclusions:

1. Basing on theoretical studies, we proposed modeling of the formation of the mathematical expectation of the protein content in curd, produced using waste-free technology.

2. We experimentally established technological factors which provide obtaining a product with high protein content. Mathematical processing of experimental data allowed us to derive a regression equation and establish a correlation between the factors of crud production process and its protein content.

References

[1] Pronina O V and Polyansky K K 2015 Current state of whey processing *Innovacii v nauke* **47** pp 59-63.

[2] Khramtsov A G 2017 Realization of the phenomenon of whey in the technological platform of the food industry *Industriya pitaniya* **3** (4) pp 23-9.
[3] Khramtsov A G and Vasilisin S V *Handbook of dairy production technologist. Technology and recipes. V. 5. Products from skimmed milk, buttermilk and whey* (SPb.: GIORD) pp 48-59.
[4] Sizenko E I, Gudkov S A and Serebryakova T G 2008 The main problems of scientific research in dairy industry *Pischevaya promyshlennost* **3** pp 14-6.
[5] Empson J 1989 Annual review of the international dairy situation *Dairy Industries Int.* **54** (2) pp 16-7.
[6] 2017 Review. Whey protein as a key component in food systems: Physicochemical properties, production technologies and applications *Food Structure* **14** pp 17-29.
[7] Smolnikova F Kh 2013 Use of whey in food production. *Torgovo-ehkonomicheskie problemy regional'nogo biznes prostranstva* **1** pp 346-8.
[8] Ozhgikhina N N and Volkova T A 2012 Cheese whey processing as a factor for increasing the efficiency and safety of cheese production *Int. scientific-practical conference dedicated to the memory of Vasily Matveevich Gorbatov* **2**(2) pp 15-22.
[9] Tax Code of the Russian Federation. Order No. 146-FZ of July 31, 1998
[10] Kadieva T A, Margieva F T, Vanieva B B and Kokoeva A T 2016 The use of whey in the production of fermented milk drinks *In the collection: Achievements of science to agriculture. Materials of the regional scientific-practical conference* pp 125-9.
[11] Florinskaya E E 2016 Functional dairy products using whey protein concentrate *Voprosy pitaniya* **85** S2 p 217.
[12] Kumar P, Sharma N, Ranjan R and Kumar S 2013 Perspective: A Review *Asian-Australas J Anim Sci* **26** (9) pp 1347–58.
[13] Nelson B K and Barbano D M 2005 Removal of serum proteins from skim milk before cheese making *J Dairy Sci* **88** (5) pp 1891-900.
[14] Sizenko E I, Gudkov S A and Serebryakova T G 2008 The main problems of scientific research in the dairy industry *Pischevaya promyshlennost* **3** pp 14-6.
[15] Goroschenko L G 2018 Russian production of cheese and cottage cheese in 2017 *Syrodelie i maslodelie* **3** pp 4-6.
[16] Khomuk I V 2018 Production of crud by ultrafiltration. In the collection: Innovative technologies in science and education *collection of articles of the VII International Scientific and Practical Conference: in 2 parts* pp 119-21.
[17] Pilipenko T V and Florinskaya E E 2012 The use of high-tech industries for the implementation of innovative technologies in the field of food *In the book: Innovative technologies in the field of food products and catering products for functional and specialized purposes* (St. Petersburg) pp 66-75.
[18] Etuev T M and Pilipenko T V 2012 Innovations in the production of safe curd products with functional properties *In the book: Innovations in the field of technology products for functional and specialized purposes Collective monograph* (St. Petersburg) pp 109-23.
[19] Florinskaya E E and Pilipenko T V 2016 Competitiveness of dairy products based on membrane technologies *International scientific-practical conference dedicated to the memory of Vasily Matveevich Gorbatov* **1** pp 323-5.
[20] Savitskaya T V and Belkov V P 2005 Mathematical models of typical operations and one-step periodic processes: the text of lectures (Moscow: RHTU named after D.I. Mendeleev)
[21] Dvoretsky S I, Egorov A F and Dvoretsky D S 2003 *Computer modeling and optimization of technological processes and equipment. Textbook* (Tambov: Publishing House Tamb. State Tech. University)
[22] Klinov A V and Mukhametzyanova A G 2009 *Mathematical modeling of chemical-technological processes: textbook* (Kazan: Kazan State Technological University)

Characteristics of the fatty acid composition of naked oats of Russian selection

G A Batalova[1], V N Krasilnikov[1], V S Popov[1*], E E Safonova[1]

[1] Peter the Great St. Petersburg Polytechnic University, Polytechnicheskaya 29, St. Petersburg, Russian Federation, Russian Federation

* Corresponding author: popovitaly@yandex.ru

Abstract. The fatty acid composition of acyl lipids of 7 naked oat cultivars was studied. Lipids are source of energy and plastic material; promote metabolic processes in the body. In the grain of studied lines the content of raw lipids varied from 5.91 to 7.87 %. The main fatty acids of the investigated lines of naked oats are linoleic (36.2-38.7 %), oleic (33.5-36.7 %), palmitic (15.3-17.8 %). According to the content of oleic and linoleic fatty acids and their ratio (1: 1), lipids of naked oats belong to the oleic-linoleic group of vegetable oils.

1. Introduction

It is known that the basis of the world food system is made up of cereals. Annually, the production of cereals is approximately 3 times higher than the total production of other crops, including leguminous plants.

Today, all nutritionalists advise to keep cereals as the central part of our diet because of their natural high content of carbohydrate complex, characterized, in particular, by their large prebiotic and bifidogenic ability.

Epidemiological studies have shown that consumption of whole-grain products, obtained from such traditional crops for northern regions of Russia as rye and oats, is associated with a decrease in the incidence of a number of chronic diseases, in particular diabetes, cardiovascular diseases.

At present, extensive work is carried out on the selection of naked oats, effective for growing in various soil and climatic zones and resistant to pathogenic organisms and abiotic factors [1].

Benchmarking of the production and processing of naked oats has shown that the strong side of the naked oat is a high nutritional value [2].

The presence of lipids distinguishes oat grain from other cereals [3,4,5]. Oat grain lipids mainly consist of unsaturated oleic (18:1) and linoleic (18:2), and saturated palmitic (16:0) acids and low levels of stearic (18:0) and linoleic (18:3) acids [6,7].

Currently, in different oat cultivars, the lipid content is in the range of 5-8 % [7,8]. It makes the oat a potentially oilseed crop. The oil content in the seeds of such cereal crops as wheat, barley, rye and triticale is only 1-3 % [8,9].

Acyl (saponifiable) lipids are sources of acyl radicals ω-9, ω-6 and ω-3 fatty acids, of which ω-6 and ω-3 are considered as essential nutrition factors. In general, speaking about the functionality of saponifiable lipids, as sources of essential fatty acids, we can state:

- monounsaturated fatty acids (ω-9) are equivalent in their effectiveness to polyunsaturated fatty acids in terms of lowering the level of cholesterol in the blood;

- (ω-3) fatty acids have a healing effect and are considered as indispensable nutritional factors, especially in childhood;
- there is a relationship between specific fatty acids (γ- and α-linolenic acid) and physiological response in some diseases [10].

The purpose of this study was to study the fatty acid composition of acyl lipids of naked oats of domestic selection.

2. Objects and methods of research.

For the study were selected samples of cultivated and new cultivars of naked oats. As the morphological characteristics of the grain were used the parameters of the grain volume weight and 1000 seeds weight (Table 1).

Samples of grain were provided by the Zonal Scientific Research Institute of Agriculture of the Northeast named after N.V. Rudnitsky (city of Kirov).

Table 1. Morphological characteristics of naked oat cultivars of domestic selection

No.	Cultivar	Origin (parental forms)	1000 seeds weight, g	Grain volume weight, g/L
1	1h07	K-2108 Pushkinsky Golozerny	25.9	698
2	766h05	Borys x Mozart	26.0	701
3	41h04	Faust x Mernime	24.6	701
4	74h12	86h04 x BAI 3	27.8	703
5	8h12	1h07 x 543h02	24.5	710
6	857h05	OA-503-1 x Tyumensky Golozerny	26.1	710
7	37h12o	Sallust x Vyatsky	25.8	723

The content of lipids in the selection oat cultivars was determined in accordance with GOST 10857-64 "Oilseeds. Method for determination of oil content".

The fatty acid composition of oat oil was determined by gas-liquid chromatography according to GOST R ISO 5508:2010 "Animals and vegetable fats and oils. Determination of methyl esters of fatty acids (FAME) by gas chromatography". Preparations of methyl esters were carried out in accordance with GOST R ISO 5509:2000 "Animal and vegetable fats and oils: Preparation of methyl esters of fatty acids".

3. Results and discussion.

The content of raw lipids varies from 5.91 to 7.87 % in the grain of the investigated naked oat cultivars (Table 2). The average lipid content is 6.9 % ± 0.98 %.

The noted limits of variation of this measurement are characteristic for grain of cultural types of oats, both hulled and naked. The content of lipids above 7-9 % is very rare for them [11]

Table 2. Fat content in selection naked oat cultivars

No	Cultivar	Humidity, % on dry matter	Fat content, % on dry matter
1	1h07	9.13	7.71
2	766h05	9.43	5.91
3	41h04	9.18	7.87
4	74h12	8.99	6.59
5	8h12	8.78	7.58
6	857h05	9.44	6.10
7	37h12o	8.88	6.32

The lipid complex of investigated naked oat cultivars was characterized by fatty acid composition (Table 3).

Table 3. Fatty acid composition of lipids of naked oats

Fatty acids	Design ation	Content, % of the amount						
		1h07	**766h05**	**857h05**	**41h04**	**37h12o**	**74h12**	**8h12**
Saturated acids								
Hexanoic (caproic)	C6:0	0.17	0.07	0.14	0.19	0.23	0.21	0.08
Tetradecanoic (myristic)	C14:0	0.36	0.38	0.37	0.31	0.35	0.28	0.27
Hexadecanoic (palmitic)	C16:0	16.63	17.41	17.8	16.60	16.58	15.38	15.34
Octadecanoic (stearic)	C18:0	1.70	1.05	1.25	1.87	1.38	1.23	1.24
Monounsaturated acids								
cis-9-hexadecenoic (palmitoleic)	C16:1	0.19	0.37	0.29	0.17	0.36	0.33	0.17
cis-11-octadecenoic (vaccenic)	C18:1	0.89	1.03	1.00	0.89	0.66	0.87	0.74
cis-9-octadecenoic (oleic)	C18:1	35.74	33.74	33.49	36.42	36.18	36.34	36.73
cis-11- eicosenoic (gondoic)	C20:1	0.82	0.88	0.73	0.85	0.79	0.85	0.73
cis-11- docosenoic (cetoleic)	C22:1	0.70	0.24	0.24	-	-	-	0.24
cis-13-docosenoic (erucic)	C22:1	1.11	0.68	0.89	0.88	0.76	0.67	0.91
Diunsaturated acids								
all cis-9,12-octadecadienoic (linoleic)	C18:2	36.25	38.68	38.37	35.89	37.12	37.61	37.83
all cis-9,12-tetracosanoic	C24:2	1.13	1.31	1.00	1.06	1.29	1.26	1.48
Triunsaturated acids								
all cis-9,12,15-octadecatetraenic (α-linolenic)	C18:3	1.46	1.5	1.61	1.12	1.43	1.69	1.25
No data	-	2.85	2.67	2.83	3.75	2.86	3.29	2.99

Triglycerides of oat lipids of the studied genotypes are represented by three types of fatty acids – saturated (palmitic acid), monounsaturated (oleic acid), dinasaturated (linoleic acid). The proportion of these acids in the lipids of the studied lines accounted for an average 89.4 %. The ratio of linoleic acid to oleic acid is close to 1. It allows to attribute oat oil to the oleic-linoleic group of oils (Table 4). The noted features of fatty-acid composition of lipids of naked oats are characteristic for tetraploid species, such as A.barbata, A.magna, A.vaviloviana [12].

It is actual to study lipid changes during grain storage so the oats is marked by a significant variety of lipases and lipoxidases [18]. Human body does not have effective chemical systems for the transformation of the compounds of one family of essential fatty acids into compounds of another family, therefore in assessing the nutritional properties of oat lipids metabolic pathways of linoleic acid, the family of n(ω)-6 essential fatty acids, are considered (Figure 1).

n–6 Fatty acids	
linoleic	18:2
↓	Δ 6-desaturase
gamma-linolenic	18:3
↓	elongase
Dihomo-gamma linolenic	20:3
↓	Δ 5-desaturase
arachidonic	20:4
↓	elongase
tetracosatetraene	24:4
↓	Δ 6-desaturase
tetracosapentaene	24:5
↓	β-oxidase
docosapentaenoic	22:5

Figure 1. Metabolism of linoleic acid in mammals [16].

Essential fatty acids, including linoleic (ω-6), are important as precursors of hormonal compounds. So linoleic acid is the basis of the formation of hormone-like compounds called eicosanoids, such as prostaglandins and leukotrienes, which play an important role in maintaining the body's homeostasis. There is a positive effect of linoleic acid on the lipid profile of blood serum of patients with diabetes mellitus. On the contrary, a deficiency of linoleic acid contributes to the development of atherosclerosis. Optimum consumption of linoleic acid is 5-8 % of the total caloric value [10,13].

Fatty acid composition of selected lipids allows attribute oat oil to the oleic-linoleic group, which includes such vegetable oils as sesame, peanut (Table 4).

Table 4. Comparative characteristics of the fatty acid composition of vegetable oils

Vegetable oils	Fatty acids content, %			
	Saturated	Monounsaturated	Polyunsaturated	
			Linolic	Linolenic
Oat	18.4±1.1	38.5±1.4	37.4±1.4	1.5±0.2
Peanut *	19.5	37.0	42.0	1.5
Pumpkin *	19.2	28.0	52.0	0.8
Corn *	14.5	32.5	52.0	1.0
Sesame *	13.5	42.0	44.0	0.5
Wheatgerm *	16.0	22.0	57.0	5.0
Sunflower *	12.5	24.0	63.0	0.5

* borrowed from [14]

The optimal ratio in the daily diet of ω-6 to ω-3 fatty acids should be 5-10 : 1 [15].

The lines of the naked oats 41h04 and 857h05 were transferred to the State Test, respectively, as Bekas and Baget cultivars for the production of functional and gluten-free foods – the content of oleic acid is 36.42 and 33.49 %, linoleic 35.89 and 38.37 % respectively, gluten is less than 0,2 mg/g of product [16,17].

4. Conclusions.

1. Lipids of naked oats by fatty acid composition belong to the oleic-linoleic group of vegetable oils according to the content of oleic and linoleic acids and their ratio (1:1);

2. The content of crude lipids varied from 5.91 % to 7.87 % and averaged 6.9 % ± 0.98 % in the grain of the studied lines;

3. Linoleic (36.2-38.7 %), oleic (33.5-36.7 %), and palmitic (15.3-17.8 %) acids are the main fatty acids of lipids of naked oats of the selection studied lines;

4. When designing the food matrix of functional and specialized products based on or with the addition of oatmeal, it is necessary to take into account the metabolic aspects of the main nutrients taking into account their bioavailability.

References

[1] Batalova G A 2014 Perspectives and results of naked oats breeding *Legumes and groat crops* **2** (10) pp 64-9.

[2] Peltonen-Sainio P, Kirkkari A-M and Jauhiainen L 2004 Charactering strengths, weaknesses, opportunities and threats in producing naked oat as a novel crop for northern growing conditions *J. Agricultural and Food Science* **13** (1-2) pp 212-28.

[3] Banas A, Debski H and Banas W 2007 Lipids in grain tissues of oat (Avena sativa): differences in content, time of deposition, and fatty acid composition *J. of Experimental Botany* **58**(10) pp 2463-70.

[4] Koehler P and Wieser H 2013 Chemistry of Cereal Grains *Handbook on Sourdough Biotechnology* (New York: Springer Science end Business Media) pp 11-45.

[5] Heneen W K, Karlsson G and Brismar K 2008 Fusion of oil bodies in endosperm of oat grains *Planta* **228**(4) pp 589-99.

[6] Zhou M X, Robards K, Glennie-Holmes M and Helliwell S 1999 Oat lipids *J. of the American Oil Chemists Society* **76**(2) pp 159-69.

[7] Leonova S, Shelenga T, Hamberg M, Konarev A V, Loskutov I and Carlsson A S 2008 Analysis of oil composition in cultivars and wild species of oat (Avena sp.) *J. of Agricultural and Food Chemistry* b(17) pp 7983-91.

[8] Liu K 2011 Comparison of lipid content and fatty acid composition and their distribution within seeds of 5 small grain species *J. of Food Science* **76**(2) pp 334-42.

[9] Ragaee S, Abdel-Aal E M and Noaman M 2006 Antioxidant activity and nutrient composition of selected cereals for food use *Food Chemistry* **98**(1) pp 32-8.

[10] Ipatova L G, Kochetkova A A, Nechaev A P and Tutelyan V A 2009 *Fat products for a healthy diet. A modern look.* (Moscow: DeLiPrint) pp 14-121.

[11] Sichkar N M and Lishkevich M I 1958 Biochemistry of oats *In the book "Biochemistry of cultivated plants", Vol. 1, Bread and cereals cultures* (Moscow-Leningrad: State publishing of agricultural literature) pp 339-40.

[12] Loskutov I G 2007 *Oats (Ovena L.) Spread, taxonomy, evolution and selection value* (St. Petersburg, SRC of the VIR)

[13] McDonald B E and Fitzpatrick K 1998 Designer vegetable oils *In the book «Functional Foods. Biochemical and Processing Aspects»* (Lancaster-Basel: Technomic Publishing Co.inc) pp 274-6.

[14] Anemueller H *1993 Lebensmittelkunde und Lebensmittelqualitätin der Ernährungsberatung* (Stuttgart: HippokratesVerl)

[15] 2009 *Norms of physiological needs for energy and nutrients for different population groups of the Russian Federation. Methodical recommendations* (M.: Federal Center for Hygiene and Epidemiology of Russian Agency for Health and Consumer Rights)

[16] Chekina M S, Meledina T V and Batalova G A 2016 Prospects for the use of oats in the production of special purpose products *Vestnik of the St. Petersburg State Agrarian University* **43** pp 20-5.

[17] Batalova G A, Andreev N R, Nosovskaya L P, Adikaeva L V, Goldstein V G and Shevchenko S N Evaluation of technological properties of some naked oat cultivars as raw material for the production of starch *Leguminous and cereal crops* **1** (17) pp 83-9.

[18] Young V L 1986 Oat lipids and lipid-related enzymes *In: Oats chemistry and technology* (Minnesota, USA) pp 205-26.

IOP Conf. Series: Earth and Environmental Science **337** (2019) 012040 doi:10.1088/1755-1315/337/1/012040

Mathematical modeling of pneumatic installations for transportation of solid domestic waste

R Yusupov[1] and S Andreev[1*]

[1]Russian State Agrarian University-Moscow Agricultural Academy named after K.A. Timiryazev Institute of Mechanics and Energy named after V.P. Goryachkin, 127550, Moscow, St. Timiryazevskaya, 49, Russia

* Corresponding author: jusupow@mail.ru

Abstract. The article is devoted to the consideration of the method of mathematical modeling of a pneumatic transport installation. The peculiarity of the approach is that it is aimed at studying low-frequency processes in a simulated object, i.e. those processes that affect its energy. This will allow you to get a mathematical model of small dimension. For this purpose, the model is decomposed into subsystems separated from each other by weak elastic relations. A method for estimating the numerical value of the dynamic compliance of these bonds is proposed.

1. Introduction

In a number of industrially developed countries of Europe, people faced an exacerbation of the problem of shortage of space for municipal waste in large cities much earlier than Russia. Practically the only way to solve it radically now is the thermal neutralization of municipal solid waste with the generation of thermal and / or electrical energy [1-6].

The process of incineration of solid waste not only provides sanitary and hygienic treatment of waste, but also reduces the amount of disposal of residues from the processing of solid waste to 10% of the original. Waste incineration is a thermal oxidation process to reduce waste, extract valuable materials, ash, or generate energy from it. The lowest calorific value per working mass for Moscow wastes is, on average, 7500–8000 kJ / kg, humidity is about 30–40%, ash content is 25–30% [1,2].

Due to changes in the morphological composition (increasing the share of packaging, plastic, paper, reducing food waste), there is a tendency to increase the heat of combustion of MSW (in Western European countries, this value reaches 10,500–12,500 kJ / kg). Using solid waste as a fuel for generating heat and electricity as opposed to dumping them saves traditional fuels (coal, gas, fuel oil).

The calorific value of household waste roughly corresponds to brown coal. On average, the calorific value of household waste ranges from 1000 to 3000 kcal / kg. It was also revealed that the calorific value of 10,500 kg of municipal solid waste is equivalent to 1000 kg of oil; caloric content of household waste is only 2 times lower than that of coal; about 5000 kg of garbage emit as much heat when burning as 2000 kg of coal or 1 ton of liquid fuel.

Currently, the level of household waste incineration varies from one country to another. From the total volumes of household garbage, the share of burning varies in such countries as Austria, Italy, France, Germany, from 20 to 40%; Belgium, Sweden - 48-50%; Japan - 70%; Denmark, Switzerland 80%; England and the United States - 10% [7-9]. In our country, only about 2% of household garbage is incinerated so far, and in Moscow - about 10% [1,2,9].

According to foreign data, it is advisable to use waste incineration in cities with a population of at least about 15 thousand. About 300-400 kWh of electricity can be generated from each ton of waste. Each city resident gives out, approximately, from 500 to 800 kg of waste per year.

The experience of industrial operation of TPPs operating on solid waste in Moscow showed that the installations operate in compliance with the environmental requirements adopted in Russia and the EU for thc process of thermal disposal of solid waste. The equipment of new power plants on alternative fuel meets modern requirements and allows to solve both the problems of sanitary cleaning of solid waste and the problem of energy supply to the adjacent areas of the city.

One of the important conditions for improving the efficiency of equipment and solving environmental problems in waste recycling enterprises is the use of pneumatic transport in production processes.

Pneumatic transport has a number of significant advantages over mechanical. Therefore, during the construction of new and reconstruction of existing enterprises, it becomes necessary to design and implement pneumatic conveying units (PTU). When designing pneumatic installations, it should be taken into account that air is used in installations other than transportation and for technological purposes, which allows reducing the number of pieces of process equipment (for cleaning, surface treatment), as well as ventilation costs [10-14].

Pneumatic transport installations are a complex of devices that ensure the movement of bulk materials (powdered, granular, crushed, etc.) or special vehicles (capsules, containers with raw materials, finished products, etc.) using compressed air or discharged gas.

Pneumatic transport is one of the progressive ways to mechanize and automate the movement of goods. This type of transport has been used in almost all sectors of the economy.

Pneumatic transport is widely used to move materials due to their significant performance and large radius of operation in the most cramped working conditions, i.e., using areas unsuitable for other transportation methods, saving production space, complete absence of residues and losses of the product being moved in lines sanitary and hygienic conditions of its transportation; with the exception of violations of technological and hygienic modes of air in industrial premises due to the absence of dust; ease of installation, reduction of staff and simplified maintenance; flexibility in operation and the possibility of full automation control. To move solid household waste (MSW) through pipelines, it is the most advisable to use a method of pneumatic transportation in suspension. At the same time the speed of the transporting air in the pipeline must be quite high - 25 - 35 m / s.

There are disadvantages of pneumatic transport: the relatively high specific energy consumption per unit mass of the transported product, the complexity of manufacturing and operating equipment for cleaning the transporting and exhaust air, significant wear. However, the correct choice of the method and equipment for pneumatic conveying of this product allows to eliminate them partially or completely. Of interest to specialists is the question of mathematical modeling of the elements of a pneumatic transport installation when solving the problem of synthesizing an automatic control system (ACS) for the process of transporting solid waste. (Figure 1).

IOP Conf. Series: Earth and Environmental Science **337** (2019) 012040 doi:10.1088/1755-1315/337/1/012040

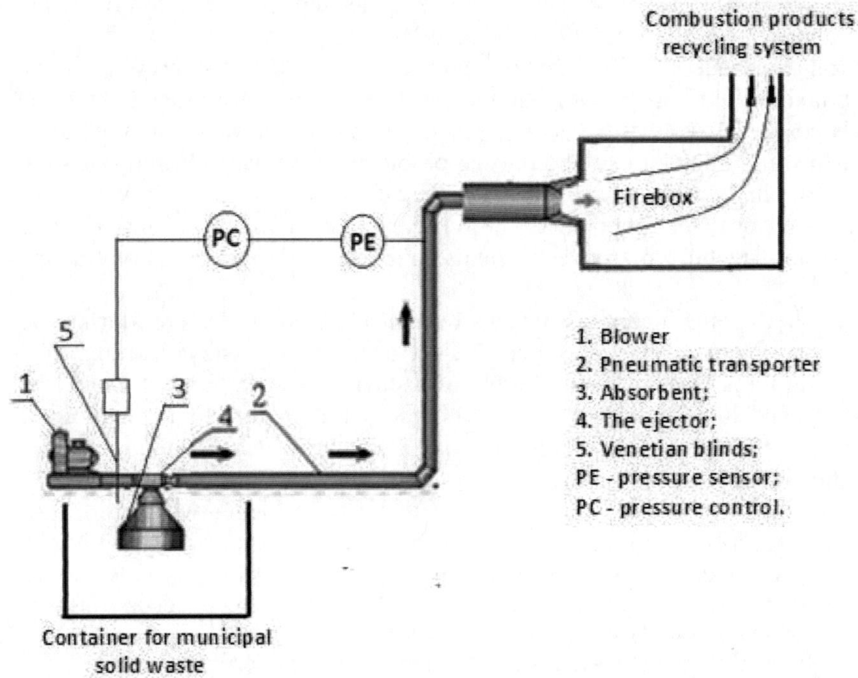

Figure 1. Functional-technological scheme of automation of the system of pneumatic transport of municipal solid waste

2. Method

Associated with the use of an equivalent circuit for replacing elements of a pneumatic conveying installation method is proposed. The essence of the method lies in the fact that the dynamic system of a real installation can be modeled as a definitely connected set of parameters of the following elements: [15,16].

- An element that has the property to accumulate kinetic energy. In a mechanical system, it is an inertial element (mass), in an electrical one, inductance.
- An element with the property of accumulating potential energy. In a mechanical system, it is a spring; in an electrical system, a capacitor (capacitance).
- An element with the property of irreversibly dissipating energy. In a mechanical system, it is a mechanical resistance, it characterizes irreversible energy losses, friction losses, in an electric system, electrical resistance, and it characterizes irreversible energy losses due to heating.

Here one can also consider such a concept as mechanical conductivity, which characterizes irreversible energy losses due to slipping. If we consider the electrical system, then it is characterized by electrical conductivity, which characterizes the loss of energy leakage.

Figure 2 shows the replacement pattern of the dynamic system of a pneumatic conveying installation for solid waste transportation.

The tasks of designing complex mechanical systems, predicting their dynamic and energy characteristics require the creation of mathematical models. Special software systems have been developed in our country and in some countries abroad to automate the process of forming mathematical models of complex objects. They are based on the use of methods of electromechanical analogy, matrix-topological and tabular methods. In turn, the task of saving machine time when performing calculations of complex dynamic systems with a large scatter of time constants, ensuring clarity and simplicity of formalizing mathematical models determines the search for ways to reduce the dimensions of the latter. This is evident because there is a class of tasks that do not require drawing the high-frequency components of dynamic processes. This is especially true for mechanical systems of traction and

IOP Conf. Series: Earth and Environmental Science 337 (2019) 012040 doi:10.1088/1755-1315/337/1/012040

transport vehicles, the efficiency of which is largely determined by the influence of low-frequency components of the oscillations [15, 16].

However, despite the progress achieved in this direction, models often have too much detail, and therefore a higher dimension. A way of reducing the dimension of a mathematical model is proposed by introducing into it an integral parameter reflecting the action of a set of elements and functional units of the system. Analysis of the dynamic systems of real mechanisms and machines shows that in the low-frequency range of oscillations, this parameter can be the compliance of quasi-elastic bonds through media of different physical nature: electric, hydraulic, pneumatic, etc. Thus, in an internal combustion engine, such a connection implies a fuel mixture in a piston system, in hydraulic machines - a hydraulic connection, in electric machines - a connection through an electromagnetic field, in a pneumatic conveying installation - airflow in the pipeline [15, 16].

In the process of modeling a complex dynamic system are based on the proposed approach, it is important to identify quasi-elastic constraints. In some cases, this task can be solved heuristically with a focus on the experience of the researcher. In addition, theoretical methods described in [17–22] can be used here.

Much attention is paid to the methods of studying dynamic systems that are based on equivalent replacement schemes by foreign authors. In works [23-25] some terms are clarified, sufficient and necessary conditions for the similarity of mechanical and electrical systems are identified. Some methods for the formation of equivalent schemes are specified. For example, the concepts "loop method", the node method, by analogy with the methods known in electrical engineering, the method of loop currents, the method of node potentials, are introduced.

A common drawback of all the theoretical works discussed above is that they do not reveal at all or weakly reveal the issues of studying the dynamic properties of systems composed of components of different physical nature (electrical, hydraulic, mechanical).

In general, according to the review of analytical methods for studying the dynamic properties of power transmissions, it can be concluded that the scope of application of existing methods is limited to one specific physical nature (mechanical, hydraulic, electrical).

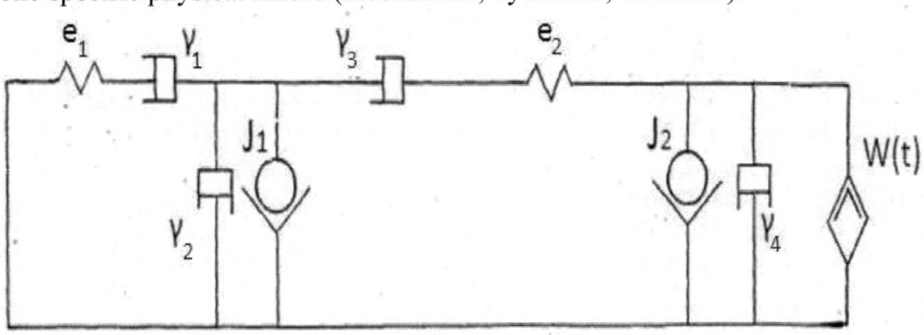

Figure 2. The equivalent circuit of the dynamic system of pneumatic transport installation

e_1 – compliance of a weak elastic electromagnetic coupling in a dynamic system of an asynchronous electric motor of a blower, $(N\ m)^{-1}$;

γ_1 – the mechanical conductivity, which characterizes the losses in terms of the velocity factor ibid, $(N\ m)^{-1}$;

γ_2 – mechanical conductivity, the reciprocal of the mechanical resistance β_2, characterizes the power factor losses in the same place, $(N\ m\ s)^{-1}$;

J_1 – the moment of inertia of the rotor of an electric motor, $kg\ m^2$;

γ_3 – mechanical conductivity, characterizes leakage losses in a dynamic system in a pneumatic conveying installation, $(\mathrm{N\ m\ s})^{-1}$;

e_2 – compliance of weak elastic coupling in the dynamic system of the pneumatic conveying installation, $(N\ m)^{-1}$

J_2 – the mass of the transported medium in the pneumatic conduit, reduced to rotor speed, $kg\ m^2$

e_4 - mechanical conductivity, the reciprocal of the mechanical resistance β_4, characterizes the friction losses in the dynamic system of pneumatic conveying installation, $N\,m\,s^{-1}$;

$W(t)$ – the source of angular velocity, s^{-1}.

3. Results

Let us consider the transition process occurring in the subsystem (Figure 3). Suppose it is oscillatory in nature. In this case, there is a periodic conversion of the potential energy of the elastic element into the kinetic energy of the inertial element. At the same time, some of the energy is irreversibly lost in dissipative elements (friction loss and leakage). Moreover, friction losses cause a decrease in torque, which means potential energy, and leakage losses cause a decrease in speed and, consequently, kinetic energy. Hence, the energy balance equation for the part of the transition process corresponding to a certain time $\Delta t / 2$

$$\Delta W_p = \Delta W_K + \Delta W_{\Omega 1} + \Delta W_{M1},$$

where: ΔWp – the potential energy stored in the elastic element at the initial moment of the considered time interval;

ΔW_k – the increment of the kinetic energy of the inertial element over time $\Delta t / 2$;

$\Delta W_{\Omega 1}$ – the increment of dissipation losses by the speed factor for the same time;

ΔW_{m1} – the increment of dissipation losses by the garden factor for the same time.

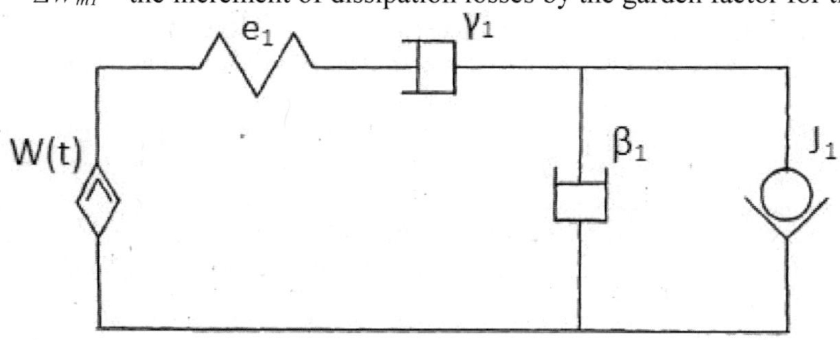

Figure 3. The equivalent circuit of the dynamic link system

Further, we note that $\Delta W\Omega 1$ affects the kinetic energy of the inertial element, and ΔW_{m1} affects the potential energy of the elastic element. Thus, $\Delta W_{\Omega 1}$ and ΔW_{m1} can be expressed as ratios:

$$\Delta W_{\Omega 1} = n\,\Delta W_K, \quad \Delta W_{M1} = m\,\Delta W_p, \tag{1}$$

where: n and m are coefficients. Then equation (1) can be rewritten:

$$\Delta W_p = \Delta W_K + n\,\Delta W_K + m\,\Delta W_p.$$

Hence, $\Delta W_p = \Delta W_K \frac{1+n}{1-m}$.

Since $\Delta W_p = e\Delta M_1^2/2$, $\Delta W_K = J\Delta\Omega_1^2/2$,

where: $\Delta\Omega_1/\Delta M_1$ – the steepness of the mechanical characteristics of the subsystem.

Thus, the calculation of compliance is possible if the slope of the characteristic $M = f(\Omega)$ and the coefficients n and m are known.

To determine the latter, we first find their connection with the decrement of the oscillations in the subsystem. The ratio is known

$$d = exp(\alpha T), \tag{2}$$

where: α is the attenuation coefficient of free oscillations;

T – the period of free oscillations.

Consider the equivalent circuit of the subsystem (Figure 3). The analysis of the scheme suggests that, under the assumption of linearity of its parameters, the transition process in it can be represented as the superposition of two transient processes: one in speed, the other in torque. Moreover, the transient process in the speed factor is implemented on the circuit section with the series-connected elements γ_1, J_1, e_1, and on the force factor - on the section with the parallel-included elements J_1, β_1, e_1. Greater visibility is achieved if the replacement scheme of the subsystem is brought into conformity '; with the preceding arguments and present it in the form of a set of two simplified schemes. The first one is constructed from parallel-connected 2β, e, J, and the other is built from the successively inserted 2γ, e, J (Figure 4).

Let us verify the validity of the statement that the real transient process in the mechanical circuit of a subsystem that can be represented as the result of the superposition of two transients. We use the operator equations describing the circuit in Figure 4. For the circuit in Figure 46, we can write

$$peM(p) + 2\gamma M(p) + M(p) / (pJ) = \Omega(p)$$

The parameters of the transient process (rate of attenuation αΩ and the frequency of free oscillations ωΩ) are determined from the characteristic equation.

$$p^2 + 2\gamma p/e + 1/(eJ) = 0$$

$$\alpha_\Omega = -\gamma/e,$$

$$\omega_\Omega = \sqrt{\frac{1}{eJ} - \left(\frac{\gamma}{e}\right)^2}\,. \tag{3}$$

Similarly for the circuit in Figure 4a

$$PJ\Omega(p) + 2\beta\Omega(p) + \Omega(p)/(pJ) = M(p)$$

$$p^2 + 2\beta p/J + 1/(eJ) = 0,$$

$$\alpha_M = -\beta/J,$$

$$\omega_M = \sqrt{\frac{1}{eJ} - \left(\frac{\beta}{J}\right)^2} \tag{4}$$

Finally, for the equivalent circuit of a dynamic link system (Figure 3)

$$peM_1(p) + \gamma_1 M_1(p) + \gamma_2 M_1(p) - \gamma_2 M_2(p) = \Omega(p)$$

$$-\gamma_2 M_1(p) + \gamma_2 M_2(p) + M_2(p)/(pJ) = 0$$

$$p^2 + p(J\gamma_1\gamma_2 + e)/(Je\gamma_2) + (\gamma_1 + \gamma_2)/(Je\gamma 2) = 0$$

$$\alpha'' = (J\gamma_1\gamma_2 + e)/(2Je\gamma_2),$$

$$\omega'' = \sqrt{[(\gamma_1 + \gamma_2)/(eJ\gamma_2)] - (J\gamma_1\gamma_2 + e)^2/(2Je\gamma_2)^2}\,. \tag{5}$$

Let's compare the characteristic equation (5) with the equation obtained after adding (3) and (4)

$$p^2 + p(J\gamma_1\gamma_2 + e)/(Je\gamma_2) + 1/(eJ) = 0 \tag{6}$$

For systems with weak dissipation, it is characteristic that γ2 >> γ1, then it can be assumed that equations (5) and (6) are identical. From this, it follows that the transient process in the subsystem is the result of the imposition of two transient processes, conventionally observed in mechanical circuits, the equivalent circuits of which are shown in Figure 4a, b. It is interesting to note that the attenuation coefficient α " and the frequency of free transition of the transient process ω " in the subsystem (Figure 3) are arithmetic means of the corresponding parameters of transition processes in simplified mechanical circuits (Figure 4a, b), i.e.

$$\alpha'' = (\alpha_\Omega + \alpha_M)/2, \qquad \omega'' = (\omega_\Omega + \omega_M)/2.$$

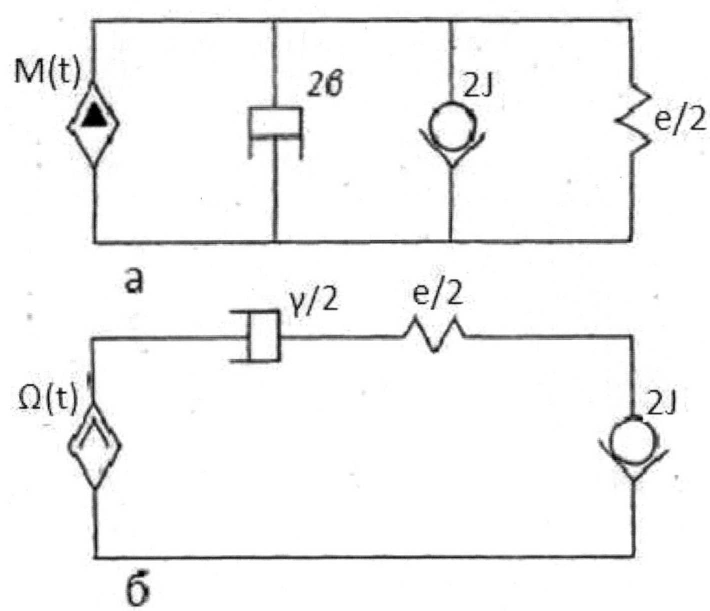

Figure 4. Substitution schemes: a - by force factor; b - by speed factor

If the first relation is obvious, then the second is proved as follows:

$$(\omega_\Omega+\omega_M)^2/4=\left(\sqrt{\frac{1}{eJ}-\left(\frac{\gamma}{e}\right)^2}+\sqrt{\frac{1}{eJ}-\left(\frac{\beta}{J}\right)^2}\right)^2/4=\left[2/(eJ)-(\gamma_1/e)^2-(\beta/J)^2+2\sqrt{\frac{1}{eJ}-\left(\frac{\gamma}{e}\right)^2}\cdot\sqrt{\frac{1}{eJ}-\left(\frac{\beta}{J}\right)^2}\right]/4. \quad (7)$$

Taking into account the fact that for dynamic systems with weak dissipation, the proximity of the frequencies ω_Ω and ωM is characterized, we can assume that the arithmetic mean and geometric mean of these frequencies are equal;

$$\omega_{s.a}\approx\omega_{s.r} \quad or \quad (\omega_\Omega+\omega_M)/2\approx\sqrt{\omega_\Omega\,\omega_m}.$$

Then in expression (7)

$$\sqrt{\frac{1}{eJ}-\left(\frac{\gamma}{e}\right)^2}\cdot\sqrt{\frac{1}{eJ}-\left(\frac{\beta}{J}\right)^2}=\omega_{s.2}^2\approx\omega_{s.a}^2.$$

After substitution of the obtained relation in (7), the transformations take the following form:

$$2\omega_{s.a}^2\approx 2/(eJ)-(\gamma_1/e)^2-(\beta/J)^2, \textit{ since } \beta=1/\gamma_2, \textit{ then}$$

$$2\omega_{s.a}^2\approx(2eJ\gamma_2^2-\gamma_1^2\gamma_2^2J^2-e^2)/(eJ\gamma_2)^2=$$

$$=[2eJ\gamma_2(\gamma_1+\gamma_2)-(J\gamma_1\gamma_2+e)^2]/(eJ\gamma_2)^2=$$

$$=2(\gamma_1+\gamma_2)/(eJ\gamma_2)-(J\gamma_1\gamma_2+e)^2/(eJ\gamma_2)^2,$$

$$\textit{hence} \quad \omega_{sr}^2\approx\sqrt{(\gamma_1+\gamma_2)/(eJ\gamma_2)-(J\gamma_1\gamma_2+e)^2/[2(eJ\gamma_2)^2]} \quad (8)$$

As we see, relations (8) and (5) practically are coincided. This is confirmed by concrete calculations. The insignificant discrepancy between the calculated data and the expressions (8) and (5) themselves is explained by the accepted assumption about the equality of $\omega_{sr.a}$ and $\omega_{sr.g}$. We note that the result obtained agrees well with the well-known propositions from the theory of oscillations on the superposition of oscillations with similar frequencies [26].Another feature that confirms the connectedness of transients in the studied circuits is interesting. To do this, we define relations for calculating the resonant frequencies in terms of speed and moment in the subsystem (Figure 3) by transforming the expressions for the complex mechanical conductivities Y and the resistances Z.

Expression for mechanical conductivity Y:

$$Y=\gamma_1+\gamma_2/(1+\gamma_2^2\omega^2J^2)+j\omega\,[e-\gamma_2^2J/(1+\gamma_2^2\omega^2J^2)].$$

Equating to zero, the coefficient at the imaginary part of the complex mechanical conductivity will find the resonant frequency of the moment [27]

$$\Omega_{rezm} = = \sqrt{1/(eJ) - (\beta/J)^2} \quad (9)$$

As we can see, the obtained relation coincides with expression (4) for calculating the natural oscillation frequency in the simplified subsystem (Figure 4a).

Similarly, we write the expression for the mechanical resistance Z:

$$Z = 1/\gamma_2 + \gamma_1/(\gamma_1^2 + \omega^2 e^2) + j\omega\, [J - e/(\gamma_1^2 + \omega^2 e^2)].$$

Hence the resonant frequency in speed:

$$\Omega_{rez.\Omega} = \sqrt{1/(eJ) - (\gamma_1/e)^2} \quad (10)$$

From a comparison of the obtained relation with (3), it follows that they coincide.

Thus, in accordance with the known theoretical principles [26], the resonant frequencies are $\omega_{rez.M}$ and $\omega_{rez.\Omega}$ can be considered the partial frequencies of the subsystem (Figure 3), and the simplified subsystems (Figure 4a, b) - as partial systems.

Now let us return to the further presentation of the procedure for deriving a formula for calculating the dynamic compliance of a quasi-elastic connection in a subsystem.

The speed transition observed in the scheme in Figure 4b will be characterized by the decrement of the oscillation d_Ω, the attenuation coefficient $\alpha\Omega$, the free oscillation period T_Ω. Accordingly, the torque transient process is d_M, α_M, T_M. A graphical representation of transients is presented in Figure 5.

For further reasoning, it is advisable to use the decrement of vibration to understand the ratio of the magnitudes of factors (torque, speed) changing in the transient process at times $t_1 + \Delta t$, not $t_1 + T$ [27].

Then:

$$d_M = exp\,(\alpha_M \Delta t),\ d_\Omega = exp\,(\alpha_\Omega \Delta t). \quad (11)$$

Taking into account the decrement d_Ω, we find the dissipative losses in speed over the segment Δt as the difference of the increments of the kinetic energies in the regions (t_1, t_0) and (t_0, t_2):

$$W''_\Omega = \Delta W_{k1} - \Delta W_{k2} = J\Delta\Omega_1^2/2 - J\Delta\Omega_2^2/2 = J\Delta\Omega_1^2/2 - J\,(\Delta\Omega_1 d_\Omega)^2/2 = (1 - d_\Omega^2)\, J\Delta\Omega_1^2/2, \quad (12)$$

where: $\Delta Wk1$ – the increment of kinetic energy in the area (t_1, t_0),

ΔWk_2 – the same at the site (t_0, t_2),

$\Delta\Omega_1$ –the increment of the velocity factor in the segment (t_1, t_0).

$\Delta\Omega_2$ – the same on the segment (t_0, t_2).

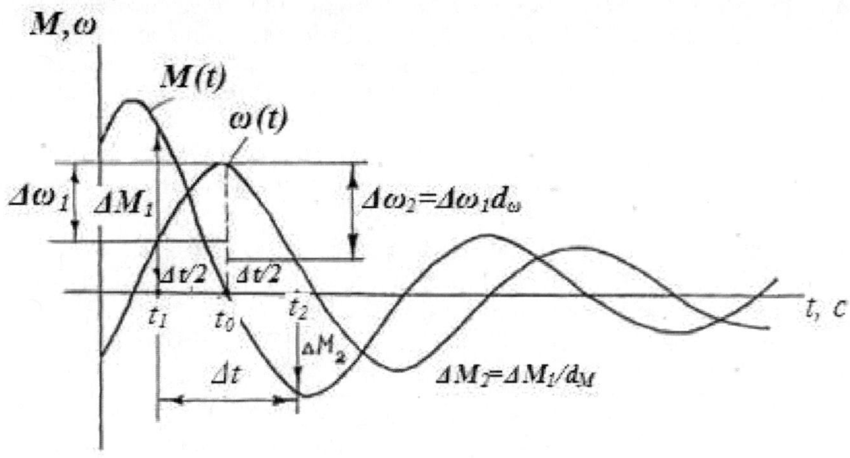

Figure 5. To the definition of the energy balance in the area of the oscillatory transition process in subsystem

Similarly, we define the dissipative losses by the force factor as the difference between the increments of potential energies in the (t_1, t_0) and (t_0, t_2) sections:

$$W''_M = \Delta W_{P1} - \Delta W_{P2} = J\Delta M_1^2/2 - J\Delta M_2^2/2 = e\Delta M_1^2/2 - e\,(\Delta M_1)^2/(2d_M^2) = (1 - 1/d_M^2)\, e\Delta M_1^2/2, \quad (13)$$

where: ΔWp_1 - the increment of kinetic energy in the area (t_1, t_0),

ΔWp_2 - the same at the site (t_0, t_2),

ΔM_1 is the increment of the speed factor at the site (t_1, t_0).

ΔM_2 - the same at the site (t_0, t_2).

Denote by $\Delta W_{\Omega 1}$ - dissipative energy loss by the speed factor in the area (t_1, t_0), $\Delta W_{\Omega 2}$ – in the area (t_0, t_2), then

$$W''_\Omega = \Delta W_{\Omega 1} - \Delta W_{\Omega 2}\,. \quad (14)$$

Similarly, for dissipative losses by the force factor:

$$W''_M = \Delta W_{M1} - \Delta W_{M2} \quad (15)$$

Express the ratio of dissipative losses in the areas (t_1, t_0) and (t_0, t_2) through decrements:

$$W_{\Omega 1}/\Delta W_{\Omega 2} = (\Delta\Omega_{s.1}^2)/\gamma/\,(\Delta\Omega_{s2}^2)/\,\gamma = d_\Omega^2, \quad (16)$$

$$W_{M1}/\Delta W_{M2} = (\Delta M_{s.1}^2)/\beta/\,(\Delta M_{s.2}^2)/\,\beta = d_M^2, \quad (17)$$

where: $\Delta\Omega_{s.1}^2$, $\Delta\Omega_{s.2}^2$, $\Delta M_{s.1}^2$, $\Delta M_{s.2}^2$ - average values of increments of speed and power factors in the areas (t_1, t_0) and (t_0, t_2).

Taking into account relations (14), (15), we get:

$$\Delta W_{\Omega 1} = W''_\Omega/\,(1+d_\Omega^2),\ \Delta W_{M1} = W''_M d_M^2/\,(1+d_M^2). \quad (18)$$

After substitution in (18) of expressions (12), (13):

$$\Delta W_{\Omega 1} = (1 - d_\Omega^2)\, J\Delta\Omega_1^2/\,[2(1+d_\Omega^2)], \quad (19)$$

$$\Delta W_{M1} = (d_M^2 - 1)\, e\Delta M_1^2/\,[2(d_M^2+1)], \quad (20)$$

Comparing (19) and (20) with (1), we find

$$n = (1 - d_\Omega^2)/\,(1+d_\Omega^2),\ m = (d_M^2 - 1)/\,(d_M^2+1) \quad (21)$$

Taking into account relation (2), we obtain the final expression for calculating the dynamic compliance of the elastic connection in the pneumatic duct:

$$e = J\,(1+d_M^2)/\,(1+d_\Omega^2)\,(\Delta\Omega_1/\Delta M_1)^2. \quad (22)$$

In practical calculations, compliance is calculated by successive approximation. The first approximate value is found from the assumption that the oscillatory process is continuous, i.e. decrements $d_\Omega = d_M = 1$. At the same time

$$e = J\,(\Delta\Omega_1/\Delta M_1) \quad (23)$$

Then, the approximate values of the attenuation coefficients α_Ω, α_M from (3) and (4) are calculated. Based on them, for a given period Δt, the refined values of decrements and compliance are determined by (11). The cycle is repeated until the value of duration is not satisfied with the desired accuracy.

Dynamic elasticity constraint. Above is a method for calculating compliance in the subsystem of a pneumatic conveying installation (PTU), based on the assumption that the steepness of the static and dynamic characteristics $\Omega = f(M)$ coincide.

Here we propose a technique that allows to take into account the deformation of the steepness of the characteristic depending on the oscillation frequency of the variable component of the external load. An expression characterizing the relationship between the slope and frequency can be obtained from the equation describing the replacement circuit of the mechanical chain of the subsystem (Figure 3). As is known, the expression of the total mechanical conductivity of the equivalent circuit of a mechanical chain simultaneously corresponds to the steepness of the dynamic characteristic:

$$(j\omega) = d\omega/dM = \gamma_1 + j\omega e_1 + \gamma_2/\,(1+j\omega J_1\gamma_2) = (\gamma_1+\gamma_2)(1 - \omega^2 T_1T_2 + j\omega T_{12} + j\omega T_1)/\,(1+j\omega T_2). \quad (24)$$

For the module of complex mechanical conductivity, we have:

$$Y(\omega) = (\gamma_1+\gamma_2)\sqrt{(1-\omega^2T_1T_2)^2 + (\omega T_{12} + \omega T_1)^2/(1+\omega^2T_2^2)}. \quad (25)$$

This expression corresponds to the unregulated subsystem. The influence of the regulator on the formation of the slope of the characteristic $f_2s = f(M)$ is taken into account by the coefficient K [192].

$$Y(\omega)_{reg.} = Y(\omega) = K\,(\gamma_1+\gamma_2)\sqrt{(1-\omega^2T_1T_2)^2 + (\omega T_{12} + \omega T_1)^2/(1+\omega^2T_2^2)}. \quad (26)$$

Here: γ_1, γ_2- coefficients (mechanical conductivities), characterizing dissipative losses;

$T_1 = e_1/\,(\gamma_1+\gamma_2)$, $T_2 = J_1\gamma_2$, $T_{12} = J_1\gamma_1\gamma_2/\,(\gamma_1+\gamma_2)$ are time constants;

e_l – elastic bond compliance in the subsystem;
J_l – the moment of inertia of translationally and rotationally moving masses;
ω – the angular frequency of harmonic oscillations in the subsystem.
As a result, the formula for calculating the compliance takes the form:

$$e = J_1 K^2 (\gamma_1+\gamma_2)^2 \frac{(1+d_M^2)[(1-\omega^2 T_1 T_2)^2+(\omega T_{12}+\omega T_1)^2]}{(1+d_\Omega^2)(1+\omega^2 T_2^2)}. \quad (27)$$

Under conditions of low-frequency vibrations from 12.0 rad / s, automatic regulators of both the internal combustion engine and transmission can be attributed to proportional-type regulators, i.e. in this case, it is permissible to assume that the coefficient K is at fixed points of the static characteristic; remains constant.

The correctness of the adopted approach is evidenced by the data given in [28]. Analyzing the amplitude-frequency characteristic of the automatic controller of an internal-combustion engine, it is possible to establish that the regulator is a static link in the range of vibration frequencies $\omega \leq 40$ rad / s, which is three times greater than the frequency range within which dynamic and energy processes in the present work.

From this it can also be concluded that in the range of oscillation frequencies $\omega \leq 12$ rad / s, the automatic controller of the pneumatic transport unit is a speedless link and the proportionality factor is a constant value in a given neighborhood of the static characteristic $M = f(\Omega)$ of the pneumatic transport unit. Considering the fact that in (27)

$$J_l K^2 (\gamma_l+\gamma_2)^2 (1 + d_M^2)/(1 + d_\Omega^2) = e_{ART.}$$

Corresponds to the compliance calculated on the basis of the static characteristic $M = f(\Omega)$, we can write that the dynamic elasticity of the elastic connection in the pneumatic conveying installation

$$e_d = e_s [(1- \omega^2 T_1 T_2)^2+(\omega T_{12} + \omega T_1)^2 / (1+\omega^2 T_2^2),$$

where: d_m and d_Ω are the decrements of the oscillations of the force and velocity factors.
Taking into account $T_l = e_d/(\gamma_1+\gamma_2)$, we get

$$e_d = \frac{(\gamma 1+\gamma 2)^2(1+\omega^2 T_2^2)-2(\gamma 1+\gamma 2)e_{CT}\omega^2(T_{12}-T_2)}{2e_{CT}\omega^2(1+\omega^2 T_2^2)} - \sqrt{\frac{(\gamma 1+\gamma 2)^2(1+\omega^2 T_2^2)-2(\gamma 1+\gamma 2)}{4e_{CT}^2\omega^4(1+\omega^2 T_2^2)^2}} - \frac{(\gamma 1+\gamma 2)^2(1+\omega^2 T_{12})}{\omega^2(1+\omega^2 T_2^2)} \quad (28)$$

4. Conclusions

1. Practically the only way to solve the problem of the shortage of space for municipal waste in large cities is currently the thermal neutralization of solid household waste with the generation of heat and / or electric energy.

2. The equipment of new power plants on alternative fuel meets modern requirements and allows to solve both the problems of sanitary cleaning of solid waste and the problem of energy supply to the adjacent areas of the city.

3. One of the important conditions for improving the efficiency of the equipment and solving environmental problems in waste recycling enterprises is the use of pneumatic transport in production processes. In pneumatic installations, air is used, besides transportation, and for technological purposes, which reduces the number of pieces of process equipment, as well as the cost of ventilation.

4. It is of interest to specialists the question of mathematical modeling of the elements of a pneumatic transport installation when solving the problem of synthesizing an automatic control system (ACS) in the process of transporting solid waste.

5. Analysis of published analytical methods for studying the dynamic properties of complex systems shows that the scope of existing methods is limited to one specific physical nature: mechanical, hydraulic, electrical.

6. A universal method is proposed associated with the use of an equivalent circuit for replacing elements of a pneumatic conveying installation. A feature of the method is that in order to reduce the dimension of the model, the ability to assess the energy performance of the system, an integral parameter is introduced into the mathematical model - the compliance of a weak elastic connection.

7. An energy method has been developed for determining the numerical value of the dynamic compliance of a weak elastic connection in inhomogeneous elements of a complex system.

IOP Conf. Series: Earth and Environmental Science **337** (2019) 012040 doi:10.1088/1755-1315/337/1/012040

References

[1] Dreyer A A, Sachkov A N, Nikolsky K S, Marinin Yu I and Mironov A V 2002 *Solid industrial and household wastes, their properties and processing* (Moscow)

[2] Gumerov I R, Tuhvatulina D R and Titov A V 2017 Existing problems of processing and utilization of solid domestic waste in the population *Scientific electronic j. Science Alley* **16**

[3] U.S. Pat. No.5669317 Plant for Thermal Waste Disposal and Process for Operating Such a Plant. 1977.

[4] Tchobanoglous G, Theisen H and Vigil S 1993 *Integrated Solid Waste Management Engineering Principles and Management Issues* (New York: McGraw-Hill)

[5] U.S. Environmental Protection Agency. Municipal Solid Waste in the United States: 2005 Facts and Figures

[6] U.S. Environmental Protection Agency. Municipal Solid Waste in the United States: 2005 Data Tables.

[7] Palprasert Ch 1996 *Organic Waste Recycling: Technology and Management. Second Edition* (New York: John Wiley & Son) pp 115-65.

[8] U.S. Environmental Protection Agency. Municipal Solid Waste. Frequently Asked Questions about Recycling and Waste Management.

[9] Peshkov M M 2015 The economic side of waste processing in Russia *Scientific notes of young researchers* **3** pp 20-3

[10] Kalinushkin M P, Koppel M A, Seryakov V S and Shapunov M M 1986 *Pneumatic conveying equipment: A Handbook* (Leningrad: Mechanical Engineering)

[11] Artykov N A 1984 *Pneumatic transport of easily damaged materials* (Tashkent)

[12] Weldshof G 1964 *Pneumatic transport at high concentrations of relocatable material* (Moscow)

[13] Evstifeev V N 1989 *Pipeline transport of plastic and bulk materials in construction* (Moscow)

[14] Malis A Y and Castor M G 1985 *Pneumatic transport for bulk materials* (Moscow: Agropromizdat)

[15] Yusupov R H 1991 *The interaction of the elements of the engine-transmission system of the tractor* (Krasnoyarsk: Publishing house of Krasnoyarsk State Agrarian University)

[16] Yusupov R H 1993 *Improving the efficiency of the operation of machine-tractor units by improving the static and dynamic characteristics of its energy part* diss. on the competition uch. step. Dr. tech. Sciences (St. Petersburg)

[17] Banach L Y 1976 Simplification of the design schemes of dynamic systems *Oscillations and strength of machine elements* (Moscow: Science) pp 12–6.

[18] Banakh L Ya and Perminov M D 1972 The study of complex dynamic systems using weak links between subsystems *Machine science* **4** pp 3–8.

[19] Berdnikov V V 1980 *Applied theory of hydraulic circuits* (Moscow: Mashinostroenie)

[20] Vantusov Yu A 1984 *Dynamics of mechanical chains of agricultural units* (Saratov: Publishing house of Saratov State Agrarian University) pp 20-4.

[21] Yusupov R H 1997 Features of modeling of electromechanical circuits in the low frequency range *Electricity* **8** pp 72–5.

[22] Antonov A S 1975 *Power transmission of wheeled and tracked vehicles: theory and calculation* (Leneingrad: mechanical engineering).

[23] Lenk A 1978 *Electromechanical circuits: Systems with lumped parameters* (Moscow: Mir).

[24] Temesh G and Mitra S 1977 *The modern theory of filters and their design* (Moscow: Mir)

[25] Wasserrab T 1980 Uber electromechanische System mit Konzentrierten Parametern *Electrotechnic und Mashinenbau* **97** 10 pp 422–5.

[26] Migulin V V, Medvedev V I, Mustel E R and Parygin V N 1988 *Fundamentals of the theory of oscillations* (Moscow: Science).

[27] Neumann L P 1981 *Theoretical foundations of electrical engineering* (Leningrad: Energoizdat).

[28] Kutkov G M 1980 *Traction dynamics of tractors* (Moscow: Mashinostroenie).

IOP Conf. Series: Earth and Environmental Science **337** (2019) 012041 doi:10.1088/1755-1315/337/1/012041

Technique of determining the effective temperature conductivity coefficient of organic raw material porous layer

V Bukhmirov[1], O Kolibaba[1], R Gabitov[1] and D Dolinin[1]

[1]Ivanovo State Power Engineering University, 153003 Ivanovo, Russia

Corresponding authors: buhmirov@tot.ispu.ru, koli-baba@mail.ru, Ramilgab@mail.ru, tevp@tvp.ispu.ru

Abstract. It is necessary to know for calculation of solid household wastes thermal processing thermophysical characteristics. Existing methods consider the solid household waste layer as a multicomponent mix which thermophysical characteristics are defined taking into account volume fractions and properties of components, layer porosity thus not. A technique is proposed for determining the effective coefficient of thermal diffusivity of a multicomponent porous layer, constructed on the basis of the solution of the inverse heat conduction problem. The base of problem-solving procedure is the method of the discrete satisfaction of boundary conditions (DSBC). According to the proposed method, the dependence of the effective coefficient of thermal diffusivity of solid communal waste of the average morphological composition on the temperature and porosity of the layer was determined. The proposed technique can be used to determine the effective coefficient of thermal diffusivity of bodies with different porosity structures.

1. Introduction

The municipal solid wastes (MSW) are the waste generated in the population, in commercial enterprises, institutions, municipal services. MSW are a heterogeneous mixture of complex morphological structure, including: paper, textiles, plastics, food and vegetable ingredients, stones, bones, leather, rubber, wood, ferrous and non-ferrous metals, glass.

Widespread waste disposal technologies are thermal methods, which include pyrolysis and gasification. Thermal processing of raw materials not only allows you to get a flammable gas, but also to solve the environmental problems associated with urban pollution and land reclamation.

Municipal solid wastes are subjected to thermal treatment in special shaft furnaces - thermal reactors [1, 2]. The reactor layer of solid waste passes successively processes heating, drying, pyrolysis and gasification of solid carbonaceous residue. The resulting processing combustible gas is used as fuel. Frequently, a pre-treatment of MSW takes place which consists in sorting and drying.

To calculate the heating process layer of solid waste on the basis of the differential equation, the heat conductivity is necessary to know the thermophysical properties (TPP).

In engineering practice, the thermophysical properties of the MSW are determined by the properties of the mixture components known from the literature and the value of their volume fractions [3]. At the same time, the porosity of the layer and its influence on TPP are, as a rule, not taken into account, which leads to an increase in the error in calculating the process parameters of the process reactor (temperature field and consumption of combustible gas).

2. Statement of the Problem

Consider a layer of solid waste as a conventional isotropic porous body, endowed with effective thermal properties: density ρ_{eff}, thermal conductivity λ_{eff} and heat capacity c_{eff}, bound together the effective thermal diffusivity a_{eff}.

$$a_{eff} = \frac{\lambda_{eff}}{c_{eff}\rho_{eff}} \tag{1}$$

One of the promising approaches to the determination of TPP of porous bodies is the conduct of thermophysical studies based on the processing of experimental data by inverse thermal conductivity problem (ITCP) methods, when, based on the results of measuring the temperature or heat flux at a given boundary and the temperature at internal points of the sample, effective thermophysical characteristics of this body [4, 5].

The technique of ITCP solution for the MSW layer is proposed, which does not require knowledge of the parameters of external heat exchange [6].The essence of the technique is that the sample is heated by the random monotonically changing heat flow. On the base of the experimentally measured temperature surface T (1, Fo) and thermal center T (0, Fo) layer and the approximation of a given value of the effective thermal diffusivity is calculated the sample temperature field T (X, Fo). The calculated values of the temperature at the center point are compared with the experimental data at the same point while minimizing the standard deviation calculation and experiment.

The method of discrete satisfy the boundary conditions (DSBC) is put on the basis of the proposed solution of the ITCP [6]. According to the DSBC method, the thermal conductivity differential equation satisfied continuously, while the boundary conditions - in discrete time points, g, evenly on the viewing of the selected time interval Fo. Time moments for which the boundary conditions are formulated, defined by the expression $\frac{Fo_k}{g}i$ where $i \in \{1, \overline{g}\}$. The higher the value, the more accurate approximate solution, but the solution of the problem more difficult. If $g \to \infty$, an approximate solution tends to the exact. Authors recommend, in engineering practice for the calculation of the thermally thin bodies temperature fields is enough to take $g = 3 \div 5$, and for the calculation of the heat thermally massive bodies $g = 3 \div 5$. Where in the calculation error does not exceed 3% [6]. The initial conditions in the DSBC method are also catered discretely at m points evenly distributed over the heated body thickness. When solving applications with an error not exceeding 3%, you can take $m = 2 \div 3$ [6].

The method is based on the heat conduction problem exact analytical solution for the simplest forms bodies with the type 1 boundary conditions. For an infinite plate this solution has the form:

$$T(X,Fo) = \phi(Fo) - 2\sum_{l=1}^{\infty}\frac{(-1)^{l+1}}{\delta_l}\cos(\delta_l \cdot X)\cdot\exp(-\delta_l^2\cdot Fo)\int_0^{Fo}\exp(\delta_l^2\cdot Fo)\frac{\partial\phi(Fo)}{\partial Fo}dFo + \\ 2\sum_{l=1}^{\infty}\cos(\delta_l\cdot X)\cdot\exp(-\delta_l^2\cdot Fo)\int_0^1\left[T(X,0)-T(1,0)\right]\cdot\cos(\delta_l\cdot X)dX, \tag{2}$$

where $\phi(Fo) = T(1, Fo)$ - given boundary conditions type I, $T(X,0)$ - initial condition, δ_l - characteristic numbers, $\delta_l = \frac{2l-1}{2}\pi$, $X = \frac{x}{R}$ - dimensionless coordinate; $Fo = \frac{a\tau}{R^2}$ -dimensionless time; R -calculated plate size, m; a - coefficient of thermal diffusivity, m^2/s; τ - time, s.

The first two terms on the right hand side of equation (2) - a component of the temperature field of action of the boundary conditions. The third term - a component of the temperature field of action of the initial conditions.

In accordance with the DSBC method solution (2) is represented as

$$T(X,Fo) = T(1,Fo) - \sum_{n=1}^{g}A_n\cdot\Phi_{X,n}(Fo) + \sum_{n=1}^{m}a_n\cdot f_{X,n}(Fo), \tag{3}$$

where $\Phi_{X,n}(Fo)$ - auxiliary functions of the temperature field component from the boundary conditions; $f_{X,n}(Fo)$ - auxiliary functions of the temperature field component from the initial conditions;

$$\Phi_{X,n}(Fo) = 2\sum_{l=1}^{\infty}\frac{(-1)^{l+1}}{\delta_l}\cdot\cos(\delta_l\cdot X)\cdot\Phi_{o,n}(l,Fo)\ , \tag{4}$$

where $\Phi_{o,n}(l,Fo) = \frac{n}{2}\exp(-\delta_l^2\cdot Fo)\int_0^{Fo}\exp(\delta_l^2\cdot Fo)\cdot Fo^{\frac{n}{2}-1}dFo$.

$$f_{X,n}(Fo) = 2\sum_{l=1}^{\infty}\cos(\delta_l\cdot X)\cdot f_{o,n}(l,Fo)\ , \tag{5}$$

where $f_{o,n}(l,Fo) = \exp(-\delta_l^2\cdot Fo)\int_0^1 (X^n-1)\cos(\delta_l\cdot X)dX$.

In the DSBC method the heated surface temperature of the body, as a function of time, is approximated by a polynomial

$$T(1,Fo) = T_o + \sum_{n=1}^{g} A_n Fo^{\frac{n}{2}}\ , \tag{6}$$

and the initial temperature field of the body – by a polynomial

$$T(X,0) = T(0,0) + \sum_{n=1}^{m} a_n\cdot X^n \tag{7}$$

Here A_n - the coefficients defined by discrete satisfaction of the boundary conditions; a_n - the coefficients defined by discrete satisfaction of the initial conditions.

From the solution of (3) the temperature of the unexposed surface of the plate

$$T(0,Fo) = T(1,Fo) - \sum_{n=1}^{g} A_n\cdot\Phi_{o,n}(Fo) + \sum_{n=1}^{m} a_n\cdot f_{o,n}(Fo) \tag{8}$$

Pre-calculated values of the functions $\Phi_{X,n}(Fo)$, $\Phi_{1,n}(Fo)$, $\Phi_{0,n}(Fo)$, $\Phi_{0\div1,n}(Fo)$, $f_{X,n}(Fo)$, $f_{1,n}(Fo)$, $f_{0,n}(Fo)$, $f_{0\div1,n}(Fo)$ are presented in [7].

The coefficients a_n are determined by a discrete satisfaction of given initial temperature distribution $T(X,0)$ in m coordinates, evenly spaced over the plate thickness.

$$T(X_j,0) = T(0,0) + \sum_{n=1}^{m} a_n\cdot X_j;\quad j\in\{1,\overline{m}\}\ , \tag{9}$$

where $T(X_j,0)$ initial temperature set points.

3. Solution algorithm

The algorithm for finding the effective temperature coefficient is based on the use of equation (6) and reduces to solving systems of algebraic equations (3) and (8). The values of *T (1, Fo)* and *T (0, Fo)* are given from the experimental data.

Solving the system (3) we obtain the values of the coefficients A_n as a function of yet unknown values of the effective diffusivity a_{eff}. Substituting $A_n(a_{eff})$ in (8), we obtain g equations with the unknown a_{eff}. The system is solved using one of the iterative methods. The proposed algorithm for calculating the effective coefficient of thermal diffusivity is implemented in the software package MATHCAD.

4. Results and Discussion

The developed technique was used to determine the effective coefficient of thermal diffusivity of a sample of the MSW layer of the average morphological composition [8, 9]. For this purpose, a series of laboratory experiments was performed to heat MSW samples in an installation, the scheme of which is shown in Figure 1. The experimental setup includes a closed container, insulated from the ends and from below with asbestos. A heating element was used as the container cover.

The experimental procedure was as follows. The test material (a sample of a MSW layer of an average morphological composition 160 mm in height with a previously determined porosity of 0.5 and zero moisture) was placed in a container. On the height of the layer along the axis, as shown in Fig. 1, two thermocouples of the type TXA (k) were installed. The signal from the thermocouples came to the analog module MBA-8 and then processed on the computer. The thermocouple readings were recorded with an interval of one second.

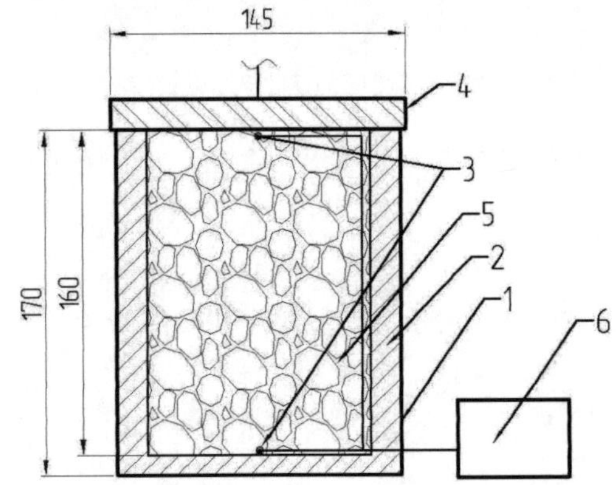

Figure 1. The scheme of the experimental setup:
1 - container,
2 - asbestos,
3 - thermo-couples,
4 - heating element,
5 - MSW,
6 - input module.

To increase the accuracy of temperature measurements in the dried sample, thermocouple openings were indicated with a thin steel needle with a movable limiter. The stopper allows you to place the thermocouples precisely at a given distance from the surface. To avoid heat transfer through the electrodes, thermocouples were installed from the end of the container and paved in isothermal surfaces. To minimize the experimental error, the temperature measurement points were chosen on the body axis, in which the temperature gradients along the width of the container were zero.

During the heating of the MSW sample, the temperature in the calculated cross sections along the height of the layer was measured (Figure 1). To obtain more reliable results and to avoid the effect of random errors, the same experiment was repeated at least 5 times. As initial data for further calculations, the temperatures averaged over the number of runs for the given heating regime were taken (Fig. 2). The experimental results are presented graphically in Figure 2.

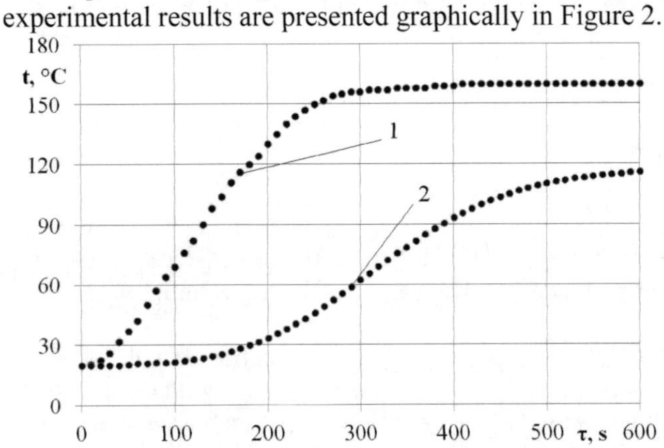

Figure 2. The experimental temperature values of the MSW sample with a porosity of 0.5:
1 - temperature of the heated surface T (1, Fo);
2 - temperature of the unheated surface T (0, Fo).

To assess the effect of porosity on the value of the effective coefficient of thermal diffusivity, similar experiments were performed for MSW samples with a porosity of 0.62 and 0.7. Using the values of *T(1,Fo)* and *T(0,Fo)* obtained from the experiment, by accessing the software implementation of the ITCP solution algorithm, the effective temperature-conductivity coefficient values were found fulfilling the condition of providing a mean-square deviation of the calculated and experimental data not more than 3 %.

The dependence of the effective coefficient of thermal conductivity on temperature for different values of the porosity of the MSW layer is shown in Figure 3.

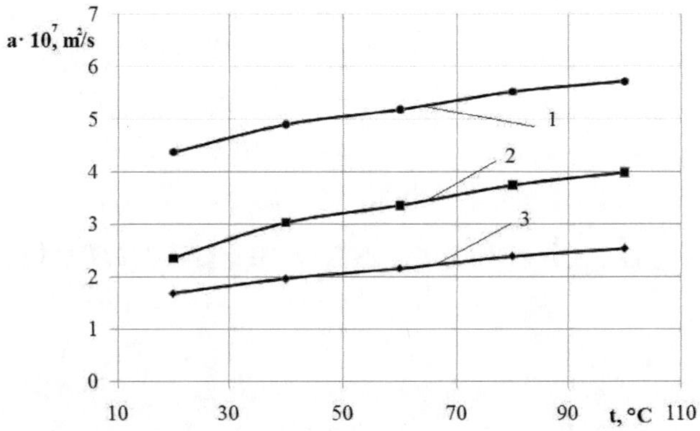

Figure 3. Effective coefficient of thermal diffusivity when the MSW layer is porous: 1 - 0.7; 2 - 0.62; 3 - 0.5

From the analysis of Figure 3 it can be seen that the porosity of the layer exerts a significant influence on the value of the effective coefficient of thermal diffusivity. With increasing porosity, the numerical value of the effective coefficient of thermal diffusivity increases. With increasing temperature, the effective coefficient of thermal diffusivity increases insignificantly for small values of the porosity of the layer (up to 0.5) and is more significant for large values of porosity (more than 0.7).

5. Conclusions

1. A technique based on the method of discrete satisfaction of boundary conditions, is proposed for determining the effective coefficient of thermal diffusivity of a porous layer.

2. The dependence of the effective coefficient of temperature conductivity on tem-perature for a layer of previously dried solid municipal waste with a porosity of 0.5 to 0.7 in the temperature range from 20 to 100 ° C was recommended, which is recommended for calculation of heat exchange in technological reactors for the processing of MSW.

References

[1] Basu P 2010 *Biomass gasification and pyrolysis. Practical design and Theory* (Burlington: Elsivier)

[2] Ryzhkov A F, Bogatova T F, Val'Tsev N V, Gordeev S I, Khudyakova G I, Osipov P V, Abaimov N A, Chernyavskii N V and Shul'Man V L 2013 Development of low-temperature thermochemical conversion reactors for coal power engineering *J.Thermal Engineering* **60** (12) pp 895–903

[3] Bukhmirov V V, Kolibaba O B and Gabitov R N 2016 Experimental determination of thermal characteristics of municipal solid waste *J. MATEC Web Conf.* **72** p 01047 DOI: 10.1051/matecconf/20167201047

[4] Beck J V, Blackwell B, St Clair C R Jr 1095 *Inverse Heat Conduction. Ill-posed problems* (New York: A Wiley – Interscience Publication)

[5] Alifanov O M 1994 *Inverse Heat Transfer Problems* (Heidelberg: Springer-Verlag) DOI 10.1007/978-3-642-76436-3

[6] Krylova L S 2001 *Design and operation of forging heat technology installations of machine building plants* (Ivanovo)

[7] Sokolov A K 1975 *Study guide for using the tables of auxiliary functions to calculate the unlimited plate warming* (Ivanovo)

[8] Ilinykh G V and Sangadzhieva T N 2012 Importance of municipal solid waste generation rate, composition and properties researches in the waste management system development *J. Bulletin of Perm National Research Polytechnic University* **3** pp 39–48

[9] Sotnezov A, Zaytsev V and Tarasova N P 2015 Morphological composition of solid municipal waste *J. Safety in Technosphere* **4** (4) pp 10–5 DOI: 10.12737/14429

Modelling the static and dynamic characteristics of pneumatic muscle

A Donskoj[1], L Kotkas[1,*], T Salova[1], V Barilovich[1], V Girshov[1], K Mertens[1], G Cherkesov[1], A Donskoy[1], G Fokin[1], I Akhmetova[2]

[1]Peter the Great St. Petersburg Polytechnic University, 195251, St. Petersburg, Russia
[2]Kazan State Power Engineering University, Kazan, Russian Federation

*Corresponding athor: kotkas334@gmail.com

Abstract. Pneumatic muscle is a new type of actuator that has a number of advantages over pneumatic cylinders, such as higher power-to-weight ratio, smooth speed adjustment, and longer operating life. Pneumatic muscle-based industrial devices can be used in combustible, explosive, and contaminated environments in different industry fields. This article is dedicated to the modeling of pneumatic muscles manufactured by FESTO, a global leader in the production of this type of pneumatic actuators. The existing pneumatic muscle models have several significant drawbacks: they can only be applied to specific types of pneumatic muscles and are inappropriate for FESTO pneumatic muscles, their accuracy is low, and they contain too many adjusting factors. The model presented in this article makes it possible to obtain static characteristics with up to 10% accuracy. The aim of this study is to assess the feasibility of application of the developed expression for modeling the dynamic characteristics of pneumatic muscle and to develop a comprehensive mathematical model of pneumatic muscle for describing the processes of raising and lowering a load.

Pneumatic muscle is a linear single acting actuator. It consists of a rubber rube reinforced by braided mesh and with fitting caps on both sides. When filled with compressed gas, the pneumatic muscle expands in the radial direction and contracts in the axial direction, generating a significant force (for FESTO MAS 40, the maximum force is 5,700 N [1]). Compared to a pneumatic cylinder, the pneumatic muscle has a higher power-to-weight ratio, smooth speed adjustment [2], and longer operating life (50 million cycles for FESTO pneumatic muscles [1]). Pneumatic muscles are also explosion-fireproof and can operate in hazardous and dusty environments.

Nowadays, there are many studies dedicated to the development of industrial manipulators using pneumatic muscles [3-6]. At the same time, different ways of pneumatic muscle positioning are being developed [7-9], and some progress has already been made in this regard (for example, study [8] has achieved ± 0.1° accuracy for rotation movement and ± 0.02 mm [9] for linear movement).

By now, many authors have attempted to model the static characteristics of the pneumatic muscle. The basic expressions are presented in studies [10, 11]. In order to improve convergence with experimental data, various authors make allowance for the thickness of the pneumatic muscle bladder [10], stretching properties of mesh threads [12], diameters and conical shape of the ends of pneumatic muscle [13]. Study [3] represents the force of pneumatic muscle as a function of relative contraction of the pneumatic muscle ε and applied pressure p, $F(\varepsilon, p)$, and adds an empirical function $k(p)$ to the expression. The expressions in the above-mentioned studies show good convergence with the

IOP Conf. Series: Earth and Environmental Science **337** (2019) 012042 doi:10.1088/1755-1315/337/1/012042

experiment (8-15% error). However, they are true only for specific types of pneumatic muscles and cannot be applied to FESTO pneumatic muscles. Expressions [14, 15] have been obtained specifically for the purposes of modeling of FESTO pneumatic muscles, but they are empirical and contain at least five adjusting factors.

The dynamic characteristics of muscle-based manipulators with FESTO pneumatic muscles are presented in studies [16-18]. Their authors use empirical expressions for the force generated by the pneumatic muscle, which require a large number of adjusting factors. The aim of this study is to examine the dynamic characteristics of pneumatic muscle with a new refined model to assess its efficiency with regard to different loading weights and to assess the applicability of the model in the further development of a muscle-based manipulator.

A refined geometric model has been developed for examining the static characteristics of the pneumatic muscle [19, 27-30]. In order to establish a correlation between the force generated by the pneumatic muscle, applied pressure, and geometric parameters of the pneumatic muscle, the proposed model examines a cell of the bladder with length l, width r, and mesh angle 2α (Figure 1, a).

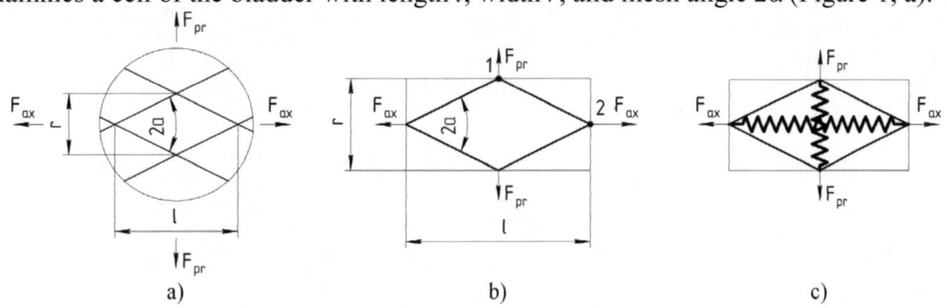

Figure 1. Pneumatic muscle calculation model: a) fragment with crossing cord threads, b) cell of the pneumatic muscle bladder with cord mesh, c) cell elasticity calculation model.

After pressurization, forces start to affect the cell, changing the muscle's physical dimensions. The cell is affected by pressure force F_{pr}, which is a result of the impact of pressurized gas on the bladder of the pneumatic muscle, and by axial force F_{ax}, which results from its contraction in the axial direction and impedes this contraction. The cell is also affected by restoring forces F_1 and F_2 that causes elastic bladder to return to its initial parameters.

In order to determine the correlation between the resulting forces, it is necessary to examine the equilibrium at points 1 and 2 (Figure 1, b). The expression for force F_{ax} generated by the pneumatic muscle per one cell is as follows:

$$F_{ax} = \frac{F_{pr} - F_1 - F_2 \text{tg}\alpha}{\text{tg}\alpha} \quad (1)$$

where α is the mesh angle after pressurization.

The equation of forces F_{pr} obtained in [19] look as follows:

$$F_{pr} = \frac{(p_S - p_A)Dl}{2} \quad (2)$$

where p_S is the air pressure in the supply line, p_A is the atmospheric pressure, D is diameter of the tube after pressurization, L is the length of the tube after pressurization.

In order to determine restoring forces F_1 and F_2, the cell is represented as two springs positioned in the radial and axial direction (Figure 1, c). Forces F_1 and F_2 will then be equal to:

$$F_1 = c_r \Delta y \quad (3)$$

$$F_2 = c_l \Delta x \quad (4)$$

where Δx is the change of cell size in the axial direction, Δy is the change of cell size in the radial direction, c_r is the cell rigidity along the perimeter, c_l is the cell rigidity along the length. The rigidities indicated below are expressed through the specific rigidity of the material c_0 and aspect ratio of the cell:

$$c_r = \frac{c_0 l_0}{r_0} \tag{5}$$

$$c_l = \frac{c_0 r_0}{l_0} \tag{6}$$

where r_0 is the initial altitude of cell, l_0 is the initial length of cell.

In order to find the applied forces, it is first necessary to determine the geometric relationship between the change of cell size in the radial and axial direction. Figure 2 shows the parameters of a triangle – a quarter of a rhombus of the cord inside the cell before and after pressurization: $r_0/2$, $l_0/2$, b, α_0 are the initial altitude, length, hypotenuse, and angle of the triangle; $\Delta x/2$ and $\Delta y/2$ are changes in the length and altitude after pressurization, $r/2$, $l/2$, b, α are the altitude, length, hypotenuse, and angle after pressurization.

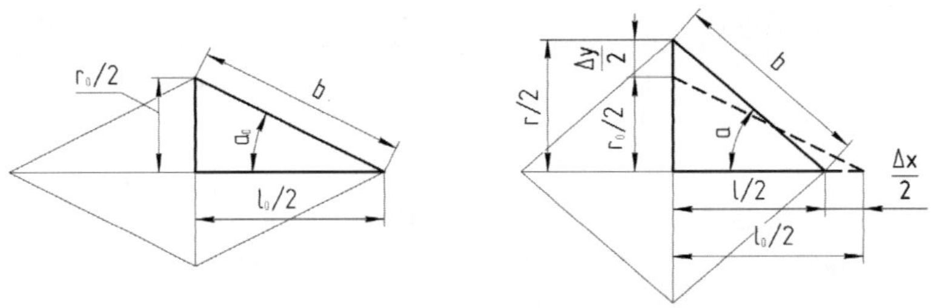

Figure 2. Calculation model for pneumatic muscle mesh rhombus deformation: a) mesh size before pressurization, b) mesh size after pressurization.

According to Figure 2, the following ratios for length l, angle α, altitude $r_0/2$, and altitude change $\Delta y/2$ are found [19]:

$$\frac{l}{2} = \frac{l_0}{2} - \frac{\Delta x}{2} \tag{7}$$

$$\frac{\Delta y}{2} = \frac{(l_0 - \Delta x)\mathrm{tg}\alpha}{2} - \frac{r_0}{2} \tag{8}$$

$$\frac{r_0}{2} = \frac{l_0}{2}\mathrm{tg}\alpha_0 \tag{9}$$

By substituting in (1) the above ratios (2)-(9) an expression for the force generated by one cell (1) is obtained:

$$F_{\mathrm{ax}} = \frac{\frac{(p_S - p_A)D(l_0 - \Delta x)}{2} - \frac{c_0}{\mathrm{tg}\alpha_0}[(l_0 - \Delta x)\mathrm{tg}\alpha - l_0\mathrm{tg}\alpha_0] - c_0\mathrm{tg}\alpha_0\mathrm{tg}\alpha\Delta x}{\mathrm{tg}\alpha} \tag{10}$$

Thus, the expression for the force generated by the pneumatic muscle with allowance for the pressure on the fitting caps looks as follows [19]:

$$T = \pi D_0 \frac{\frac{(p_S - p_A)D\left(1 - \frac{x}{L_0}\right)}{2} - \frac{c_0\left[\left(1 - \frac{x}{L_0}\right)\mathrm{tg}\alpha - \mathrm{tg}\alpha_0\right]}{\mathrm{tg}\alpha_0} - c_0\mathrm{tg}\alpha_0\mathrm{tg}\alpha\frac{x}{L_0}}{\mathrm{tg}\alpha_0\mathrm{tg}\alpha} - \frac{\pi D_0^2(p_S - p_A)}{4} \tag{11}$$

where x is the position coordinate of pneumatic muscle.

The following expression for the change of mesh angle is obtained [19]:

$$\alpha = arccos\left[\left(1 - \frac{x}{L_0}\right)cos\,\alpha_0\right] \tag{12}$$

Next, it is necessary to consider the change of the pneumatic muscle diameter as the bladder is filled with compressed air [21]:

$$D = \frac{D_0\left(1-\frac{x}{L_0}\right)\text{tg}\alpha}{\text{tg}\alpha_0} \tag{13}$$

For the purpose of modeling the dynamic characteristics, a mathematical model of the pneumatic muscle has been developed, which comprises a system of differential equations, including: equation of pneumatic muscle motion; equation of pressure change in the bladder of the pneumatic muscle; equation of changes in the pneumatic muscle geometry during movement.

Figure 3 shows the principle of operation of the pneumatic muscle when lifting a load with mass *m*.

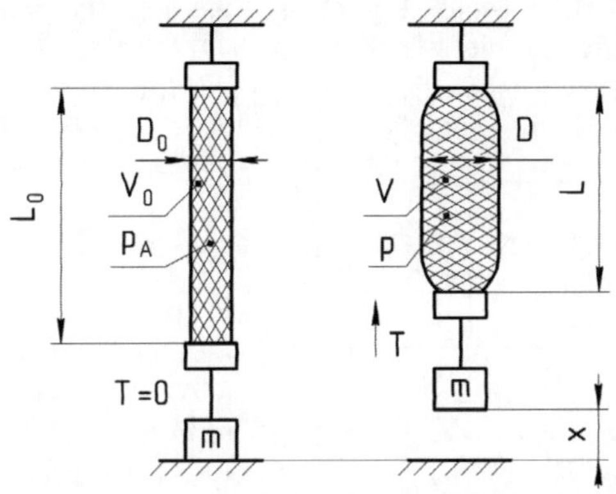

Figure 3. Principle of operation of the pneumatic muscle.

According to Figure 3, in the initial state, pressure in the bladder of the pneumatic muscle is equal to atmospheric pressure p_A, force T is zero, diameter equals D_0, mesh angle is α_0. When compressed air is delivered from the supply line with pressure p_S, pressure in the bladder becomes equal to p, the pneumatic muscle generates force T, thereby lifting the load to coordinate x. At the same time, the diameter of the pneumatic muscle becomes equal to D ($D > D_0$), length becomes equal to L ($L < L_0$), volume becomes equal to V ($V > V_0$), mesh angle becomes equal to α.

When the pneumatic muscle is emptied and the load with mass m is lowered, all parameters of the pneumatic muscle become equal to their initial values ($D = D_0$, $L = L_0$, $V = V_0$, $\alpha = \alpha_0$, $T = 0$).

The equation of pneumatic muscle motion will then look as follows:

$$mx'' = T - P - hx' \tag{14}$$

where m is the load mass, T is the force generated by the bladder of the pneumatic muscle, P is the weight of the moving parts, h is the damping coefficient.

The equation of pressure change in the bladder of the pneumatic muscle when it is filled/emptied can be written as [20]:

$$p' = a\left[\frac{kf_S\sqrt{Rt_S}}{V\sqrt{\zeta_S}}\sqrt{p_S^2 - p^2} - \frac{kp}{V}V'\right] + +(1-a)\left[-\frac{kf_E\sqrt{Rt_S}\left(\frac{p}{p_S}\right)^{\frac{k-1}{2k}}}{V\sqrt{\zeta_E}}\sqrt{p^2 - p_A^2} - \frac{kp}{V}V'\right] \tag{15}$$

where t_S is the air temperature in the supply line; k is the adiabatic coefficient for air ($k = 1.4$); R is the gas constant ($R = 287$ J/(kg·K); f_S, ζ_S is the area and resistance factor of the supply line; f_E, ζ_E is the area and resistance factor of the exhaust line; V is the volume of the pneumatic muscle bladder; a is the logical factor that determines the nature of the dynamic pneumatic process in the pneumatic muscle: when $a = 1$, the pneumatic muscle is filled, when $a = 0$ it is emptied.

The equations that describe the changes in pneumatic muscle geometry during movement look as follows [21-26]:

$$V = \frac{\pi D^2}{4}(L - x) \tag{16}$$

$$V' = \frac{\pi D}{2}(L - x)D' - \frac{\pi D^2}{4}x' \tag{17}$$

$$D' = \left(1 - \frac{x}{L_0}\right)\frac{D_0}{\text{tg}\alpha_0 \cos^2\alpha}\alpha' - \frac{D_0}{L_0\text{tg}\alpha_0}x' \tag{18}$$

$$\alpha' = \frac{\cos\alpha_0}{L\sqrt{1-\left(1-\frac{x}{L_0}\right)\cos\alpha_0}}x' \tag{19}$$

The numerical modelling of Eq. (11)-(13), (14)-(19) allows to obtain the static and dynamic characteristics for the pneumatic muscle. Figure 4 illustrates the relationship between the static characteristics of the pneumatic muscle MAS-20-600 (D_0 = 20 mm, L_0 = 600 mm) obtained during modelling with experimental curves taken from catalog [1]. It can be stated that the calculated results are close to the experimental ones (divergence in modes p = 0.7; 0.6; 0.5; 0.4; 0.3 MPa lies within 10 %).

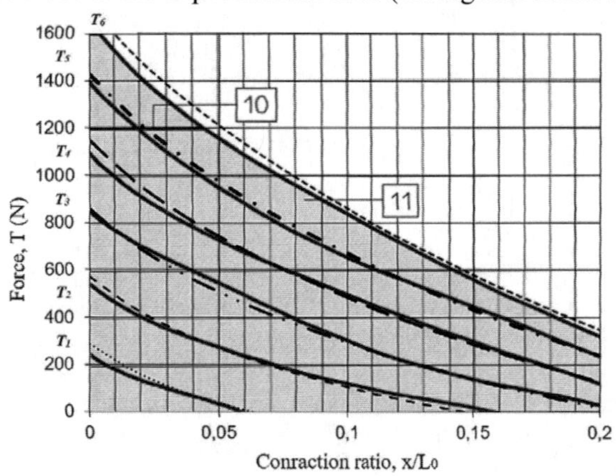

Figure 4. Diagram of correlation between the force generated by pneumatic muscle MAS 20 and relative contraction: solid lines represent experimental curves from the FESTO catalog, dashed lines stand for the results of calculation according to model; T1 with pS = 0.2 MPa; T2 with pS = 0.3 MPa; T3 with pS = 0.4 MPa; T4 with pS = 0.5 MPa; T5 with pS = 0.6 MPa; T6 with pS = 0.7 MPa.

Figure 5 shows the dynamic characteristics during the lifting and lowering of a 40 kg and 60 kg load by the pneumatic muscle with pressure p_S = 0.4 MPa. The diagrams prove that the developed expression for the force generated by the pneumatic muscle adequately describes the behavior of the pneumatic muscle when lifting/lowering a load as well as its ability to position the load.

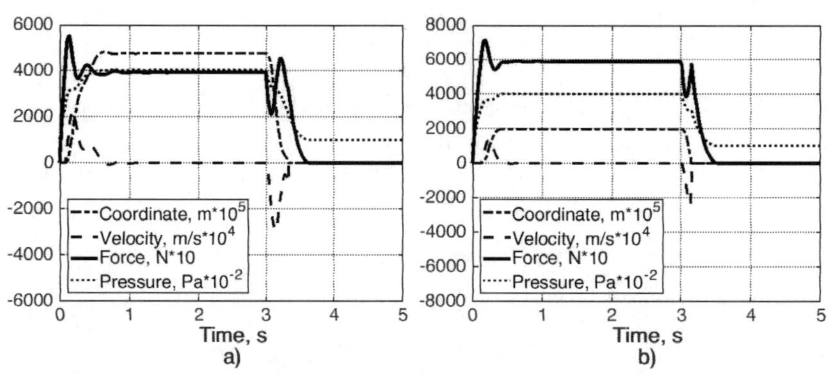

Figure 5. Dynamic characteristics of the MAS-20-600 pneumatic muscle when lifting/lowering a load with the following mass: a) 40 kg, b) 60 kg.

This article is dedicated to the modeling of the static and dynamic characteristics of the pneumatic muscle. A refined equation for the force generated by the pneumatic muscle with a minimum number of parameters and good convergence with the experiment was proposed in this study. Further the applicability of this expression for the evaluation of the dynamic characteristics was assessed. For this purpose, a comprehensive mathematical model of the pneumatic muscle comprising differential equations of motion and equations of pressure change in the bladder of the pneumatic muscle were developed. It shows good results, which correspond to experimental data and can be used in subsequent development of pneumatic muscle manipulators.

References

[1] Fluidic *muscle MAS-...* [online] Available at: http://www.festo.com/cat/ru_ru/data/doc_ru/PDF/RU/MAS_RU.PDF
[2] Jouppila V, Gadsden S A and Elman A 2014 *Syst. Meas. and Con.* **136** (4)
[3] Tondu B and Lopez P 2000 *IEEE Con. Syst. Mag* **20** (2) pp 15-38 (2000)
[4] Nazarov F M, Forental V I and Forental M V 2015 *Izvestiya Samarskogo Nauchnogo Centra Rossijskoy Akademii Nauk* **17** pp 851-8.
[5] Deaconescu T and Deaconescu A 2017 *MATEC Web Conf.* **112** (2017)
[6] Caldwell D G, Tsagarakis N, Medrano-Cerda G A, Schofield J and Brown S 2001 *Con. En. Pract.* **9** pp 23-36
[7] Caldwell D G, Medrano-Cerda G A and Goowin M 1995 *IEEE Contr. Syst. Mag.* pp 40-8.
[8] Ahn K K, Tu D C T 2004 *KSME Int. J.* **18** (8) pp 1388-400.
[9] Sarosi, Gyeviki J, Veha A and Toman P 2009 *7th Int. Symp. Intell. Syst. Inform. (SISY 09)* pp 301-5
[10] Schulte H F 1961 *App. Ext. Pow. Prosthet. Orthotics.*
[11] Chou C P and Hannaford B 1996 *IEEE T. Robotic Autom* **6** (1) pp 90-102
[12] Davis S, Tsagarakis N, Canderle J and Caldwell D G 2003 *Int. J. Robotic. Res.* **22** (3) pp 213-27
[13] Tsagarakis N and Caldwell D G 2000 *IEEE Int. Conf. Robotic. Autom. (ICRA'00)* pp 3641-6
[14] Sarosi J and Fabulya Z 2012 *Int. J. En., Ann. F. En. Hunedoara* **10** (2) pp 105-10
[15] Sarosi J 2012 *7th IEEE Int. Symp. Appl. Comput. Intell. Inform. (SACI 2012)* pp 229-33
[16] Pitel J and Tothova M 2016 *MATEC Web Conf* **76**
[17] Hosovsky A and Havran M 2012 *Technicki vjesnik* **19** (3) pp 673-81
[18] Boržíková J, Piteľ J, Tóthová M and Šulc B 2011 *12th Int. Contr. Carpath. Conf. (ICCC 2011)*
[19] Donskoy A S and Kotkas L A 2018 *Gidravlicheskie mashiny`, gidropnevmoprivody` i gidropnevmoavtomatika. Sovremennoe sostoyanie i perspektivy` razvitiya* pp 297-307
[20] Donskoy A S 2009 *Mathematical modeling of processes in pneumatic actuators*
[21] Donskoy A S and Kotkas L A 2018 *Gidravlicheskie mashiny`, gidropnevmoprivody` i gidropnevmoavtomatika. Sovremennoe sostoyanie i perspektivy` razvitiya* pp 345-54
[22] Kolodyazhnyi D Y, Nagornyi V S, Sipatov A M and Modorskii V Y 2017 *Russian Aeronautics* **60** pp 97–102 doi:10.3103/S1068799817010147
[23] Danilishin A M, Kartashov S V, Kozhukhov Y V and Kozin E G 2017 *IOP Conference Series: Materials Science and Engineering* **232** doi:10.1088/1757-899X/232/1/012069
[24] Zhemelev G 2018 *CEUR Workshop Proceedings* **2135** pp 6–12
[25] Nagorny V S, Smirnovsky A A, Chernyshev A S and Kolodyazhny D Y 2015 *Technical Physics Letters* **41** pp 859–62 doi:10.1134/S1063785015090072
[26] Karazin V I, Kozlikin D P, Sukhanov A A and Khlebosolov I O 2016 *Lecture Notes in Mechanical Engineering* pp 75–85 doi:10.1007/978-3-319-29579-4_8
[27] Panchenko I A, Volkov A V and Bolotin A E 2014 *Teoriya i Praktika Fizicheskoy Kultury* 32–4
[28] Petritchenko M R, Subbotina S A, Khairutdinova E F, Reich E V, Nemova D V, Olshevskiy V Ya and Sergeev V V 2017 *Magazine of Civil Engineering* **73** (5)
[29] Borovkov V M, Zysin L V and Sergeev V V 2002 *Izvestiya Akademii Nauk. Energetika* **6**

[30] Petritchenko M R, Kotov E V, Nemova D V, Tarasova D S and Sergeev V V 2018 *Magazine of Civil Engineering* **77** (1)

IOP Conf. Series: Earth and Environmental Science **337** (2019) 012043 doi:10.1088/1755-1315/337/1/012043

Influence of material and thickness of slot insulation of the electric machine on the bearing voltage ratio

K Vostrov[1], V Frolov[1], V Shestakov[1], E Popkov[1], A D Mitrofanov[2]

[1]Peter the Great St. Petersburg Polytechnic University, 195251 St. Petersburg, Russia
[2]Kazan State Power Engineering University, Kazan, Russian Federation

Abstract. The Modern electrical drive systems which able to operate in variable-speed mode became a conventional setup since power electronics-based frequency converters became wide spread. Nevertheless, specific parasitic phenomena of bearing currents always accompanying drive systems fed by frequency converters. In electrical machines, especially in small size motors, major role in parasitic current patch plays capacitance between stator winding and rotor surface. In this paper it will be considered influence of winding insulation type on the bearing voltage ratio and corresponding winding-to-rotor capacitance.

1. Introduction

One of the reasons for the failure of modern variable speed electrical drive systems is the premature failure of the bearing units of the drive motor. When the motor is powered by a frequency converter, the accelerated wear of the bearings may occur. The bearing failure usually happens due to the parasitic currents flowing through them because of undesired voltage induced on the drive shaft due to the operation of the inverter. The mechanism of bearing currents occurrence and its variety have been repeatedly described by various authors [1-8].

The origins of bearing currents flowing can be generally separated into two groups.

The first and convenient origin of bearing currents is related to the machine's inaccurate geometry. The asymmetry of the magnetic system of the drive motor can arise because of reasons such as: different values of the resistances and inductances of the stator and rotor windings; clearances in the joints between the packages of the stator and rotor lamination stacks; asymmetrical arrangement of poles; uneven air gap between the rotor and the stator. The mentioned factors of electromagnetic imbalance are a consequence of the imperfect assembly of the drive motor and possible inaccurately in production technology or inappropriate operating conditions. The asymmetry of the electromagnetic system of the machine leads to the appearance of undesired potential on the shaft ends and other massive parts of the electric motor.

Another reason for the appearance of bearing currents caused by mechanical reasons may come from misalignment of the machine's rotor. Due to mechanical imbalance, the rotor of the drive motor starts to make elliptical movements. Such a motion of the rotor causes electromagnetic imbalance of the drive motor, since the interaction of the magnetic fields of the stator and the rotor becomes pulsating. This leads to the inducing of potentials on the shaft ends of the electric motor, which causes bearing currents flowing.

It should be noted that such potential can be induced, even if the stator feeding voltage waveform is absolutely sinusoidal. The resulting voltage causes the appearance of bearing currents. Thus, the mechanical asymmetry and imbalance of the drive motor can cause bearing currents either in direct on-line or frequency converter fed drive systems.

The second source of bearing currents is related to modern electrical drive systems and became a major one since frequency converters began to be widely introduced. The main reason for the appearance of bearing currents flowing in modern electric drives is the common-mode voltage. The three-phase voltage at the inverter output operating in the pulse-width modulation mode has the form of a high-frequency pulse sequence. Being generated with a shift, such sequences forming a 3-phase output and supply the motor terminals. Wherein the instant sum of the phase voltages is not equal to zero and forms a high frequency voltage between the motor's neutral point and grounded parts of the motor. From the bearing currents arisen by such a mechanism only the drives fed by the frequency converter suffer.

Depending on the bearing currents flowing path, a different kinds of them are introduced:

1. Circulating bearing current, caused mainly by the machine inaccurate design, flows along the shaft inside of the machine and closes through the motor housing. Being passed through both bearings, circulating current causes its accelerated degradation.
2. Non-circulating bearing currents, which induced by common-mode voltage, flows from the motor's stator windings to the grounded parts of the machine through the capacitance couplings and bearings. Depending on grounding conditions, such a current may damage either motor's or driven load's bearings.
3. Stator ground current appears due to common-mode voltage as well, but flows to ground point through stray capacitance between winding and stator and not endangers bearings. But, if motor grounding is malfunctioning, the stator ground current may flow through the load's grounding connection, and in such a case, and in this case the bearings will be harmed, both the engine's and the load's.
4. Shaft current may occur in case of bad motor protection earth connection when either non-circulating or stator ground current flow through the shaft and further through the driven load to the load's ground potential.

A schematic explanation of different kinds of bearing currents patch is presented in Figure [9].

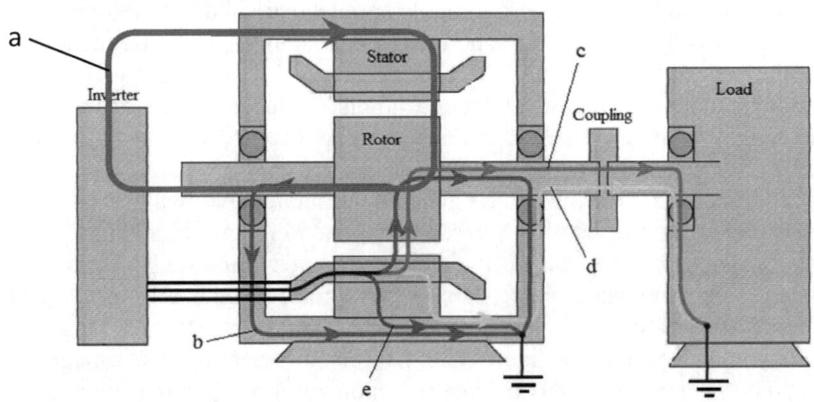

Figure 1. Schematic representation of the electric drive and possible ways of leakage of the bearing currents through parasitic capacitances, where a) – circulating bearing current, b) – non-circulating bearing current, c) and d) – shaft current caused by winding to rotor and winding to stator capacitances respectively, e) – stator ground current.

In low power scale electrical machines, the predominant type of bearing currents are non-circulating currents caused by the flow of a high-frequency common-mode induced current through parasitic capacitances between the machine rotating elements and grounded parts.

It was reported about different techniques of bearing currents mitigation [1, 5, 10-15]

One of the methods to reduce non-circulating bearing currents is to increase the impedance of the elements that form its path. Taking into account that this current is mainly capacitive in nature, an effective and promising method is to reduce the parasitic capacitances of an electrical machine. The largest contribution to the transmission of non-circulating bearing current belongs to the capacitance between the stator winding of the motor and the rotor surface [3,4]. There are a number of publications aimed at studying and reducing this capacity [13 – 15, 26-29]. To assess the machine's risk of non-circulating bearing currents, the bearing voltage ratio (*BVR*) is commonly used. The ratio shows the share of the winding-to-rotor capacitance from to the sum of all parasitic capacitances and usually is calculated by the equation [3]:

$$BVR = \frac{C_{wr}}{C_{wr}+C_{sr}+2C_b} \quad (1)$$

where C_{wr} – capacitance between winding and rotor, C_{sr} – capacitance between stator and rotor, C_b – capacitance between inner and outer bearing races.

Bearing voltage ratio will be used further in this paper to evaluate and compare results. Current study will be focused on the one certain possible winding-to-rotor capacitance reduction by applying different insulating materials with different thickness as a winding insulation at slot opening.

2. Objective and methods

The purpose of this work is to investigate the influence of the use of various insulating materials on the bearing voltage ratio, and, as a consequence, on the amplitude of the induced bearing current.

The method of carrying out the research implies the examination of the one slot of the electric motor with a power of 30 kW. Modeling is done by the finite elements method investigation of the slot with the use of different types and sizes of slot insulation. Figure 2 shows the stator slot of the electric machine being modeled, on which the main elements have been modeled are marked.

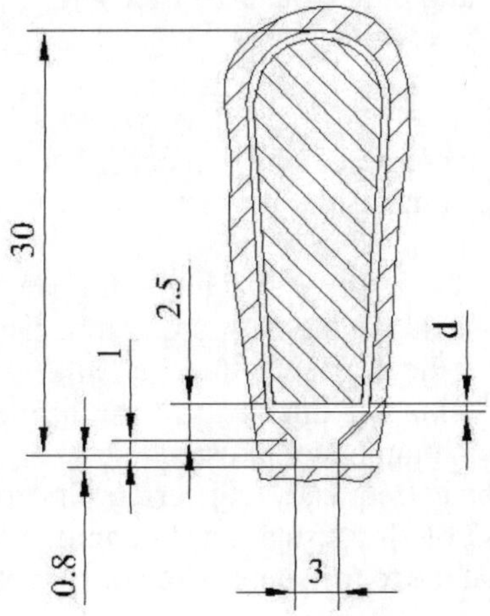

Figure 2. Geometry of the investigated slot (all dimensions in mm). "d" is the thickness of the insulation at the slot opening. Different values of "d' were investigated.

As slot insulation, the materials listed in Table 1were reviewed. [16-18].

Table 1. List of reviewed materials.

Material	Dielectric permeability
Mikanit	2
Polyethylene terephthalate	3,1
Fiberglass	5,5
Plastics with inorganic filler	7.7

Also in the research the influence of the thickness "d" of the materials used was investigated. The results of the studies are shown in Figure 3. When calculating the bearing voltage ratio, the capacitance of one bearing was assumed to be at average level of 0.5 nF [19], while the capacitance between the rotor and the stator assumed as 1.3 e-11 F [20-25].

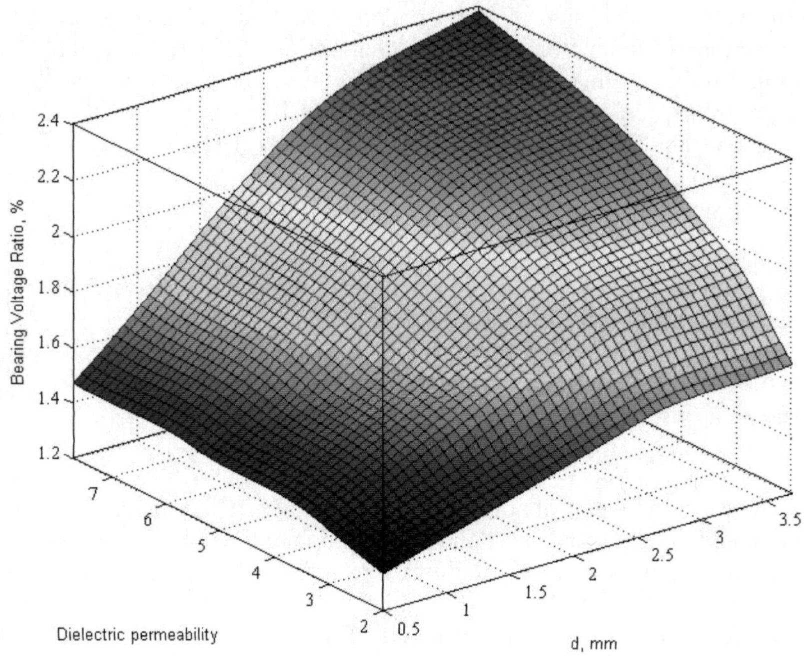

Figure 3. Dependence of the bearing voltage ratio on the material used and the thickness of the insulation layer d.

3. Summary

The analysis of the obtained data shows that for the maximum possible reduction of the bearing currents of the drives it is recommended to use the most possible thin insulating materials with a lowered coefficient of dielectric permittivity.

Since non-circulating bearing currents are most affected by low-power machines, the use of ultrathin micaceous plates as insulation for slots is most justified in such machines. Also Figure 3 shows that when using other materials with higher dielectric permittivity, the bearing voltage ratio may still remain above 1.5% if thin films of such materials are used. Increasing the thickness of the insulation layer adversely affects the electrical machine regardless of the insulation material used.

Thus, when designing electrical machines powered by a frequency converter, first of all it is necessary to strive to utilize the smallest possible thickness of slot insulation. In addition, the use of materials with a lower dielectric permeability will also contribute to the suppression of parasitic bearing currents.

References

[1] Vasilev B Yu and Kozyaruk A E 2016 Bulleten of the South Ural State University. Ser. Power Engineering **16(3)**, 93-100 (2016)

[2] 2011 ABB drives. Bearing currents in modern AC drive systems *Technical guide* **5** [Online]. Available: - https://library.e.abb.com/public/8c253c2417ed0238c125788f003cca8e/ABB_Technical_guide_No5_RevC.pdf

[3] Muetze A 2004 *Bearing Currents in Inverter-Fed AC-Motors* – PhD dissertation (Darmstadt: Technischen Universitaet Darmstadt)

[4] Muetze A and Binder A 2006 IEEE Industry Applications Magazine **12**(4) pp 22-31

[5] Boldea I and Nasar S A 2006 *Electric Drives*, 2 nd edition, (Boca Raton: CRC Press) p 13

[6] Dobrodeev E and Rozhkov L 1979 *Electrical equipment of thermal power plants*, 2 nd edition, (Moscow: Energia)

[7] Romanenko A, Muetze A and Ahola J 2017 *IEEE Transactions on Energy Conversion* **32**(1) pp 99-110

[8] Romanenko A, Ahola J and Muetze A 2015 *IEEE Energy Conversion Congress and Exposition (ECCE)* pp 4851-2

[9] Gonski P M 2012 *Diagnosing & Understanding Motor Bearing Currents* (Frazer: Keystone Engineering Group, Inc.) [Online]. Available: https://www.ecmweb.com/motors/diagnosing-understanding-motor-bearing-currents

[10] Busse D, Erdman J, Kerkman R J, Schlegel D and Skibinski G 1996 *Conference Record of the 1996 IEEE Industry Applications Conference Thirty-First IAS Annual Meeting* **1** pp 610-7

[11] Heino T 2014 *Bearing Currents and Their Mitigation in Frequency Converter-driven Induction Motors* – Bachelor's thesis (Vaasa: Electrical Engineering)

[12] Muetze A and Oh H W 2008 *IEEE Transactions on Industry Applications* **44**(1) pp 135-43

[13] Hyypio D B 1998 *Dynamoelectric machines with shaft voltage prevention method and structure* US patent 5 821 652 A, Oct. 13, 1998.

[14] Muetze A and Binder A 2003 *Electric Machines and Drives Conference 2003. IEMDC'03. IEEE International* **3** pp 1859-65

[15] Yaskawa Electric America, Inc. 2008 *Motor Bearing Current Phenomenon, Application Note*

[16] *New chemical technologies. Analytical portal of the chemical industry* [Electronic resource]. Available: http://www.newchemistry.ru (in Russian)

[17] Vakser N 1985 *Insulation of electrical machines*. Tutorial, L., ed. (SPb: St. Petersburg Polytechnic University)

[18] Likhachev V 2005 *Handbook of the winding constructor of asynchronous electric motors*

[19] Romanenko A, Ahola J, Muetze A and Niskanen V 2014 *2014 16th European Conference on Power Electronics and Applications* pp 1-10

[20] Magdun O, Gemeinder Y and Binder A 2010 *IECON 2010 - 36th Annual Conference on IEEE Industrial Electronics Society* pp 962-7

[21] Smolovik S V, Liamov A S and Chudny V S 2017 *Proceedings of the 2017 IEEE Russia Section Young Researchers in Electrical and Electronic Engineering Conference, ElConRus 2017* pp 1581–4 doi:10.1109/EIConRus.2017.7910875

[22] Sakhno L I and Smorygin N A 2018 *Proceedings of the 2018 IEEE Conference of Russian Young Researchers in Electrical and Electronic Engineering, ElConRus 2018* pp 227–30 doi:10.1109/EIConRus.2018.8317071

[23] Bahrami M R 2018 *MATEC Web of Conferences* **224** doi:10.1051/matecconf/201822402021

[24] Rudenko G V and Bolotin A E 2015 *Teoriya i Praktika Fizicheskoy Kultury* 97–9

[25] Lyamov A, Makarova M and Smolovik S 2015 *2015 IEEE Eindhoven PowerTech, PowerTech 2015* doi:10.1109/PTC.2015.7232543

[26] Anikina I D, Sergeyev V V, Amosov N T and Luchko M G 2017 *Int. J. of Hydrogen Energy* **42** (1)

[27] Borovkov V M, Zysin L V and Sergeev V V 2002 *Izvestiya Akademii Nauk. Energetika* (6)

[28] Kiseleva M A, Kondratyeva N V, Kuznetsov N V, Leonov G A and Solovyeva E P 2014 *IFAC Proceedings Volumes (IFAC-PapersOnline)* **19** pp 5872–7

[29] Kaikko J, Mankonen A, Vakkilainen E and Sergeev V 2017 *Energy Procedia* **120**

IOP Conf. Series: Earth and Environmental Science **337** (2019) 012044 doi:10.1088/1755-1315/337/1/012044

Micro turbine installations classification principles based on a systematic approach

V Barskov[1], S Besedin[1], V Rassokhin[1], A Smetankin[1,*], O Nikiforova[1], A Denisov[1], V Shashikhin[1], B Sidel'nikov[1], B Korotkov[1], I Akhmetova[2]

[1]Peter the Great St.Petersburg Polytechnic University, 195251 Polytechnicheskaya, 29, St.Petersburg, Russian Federation
[2]Kazan State Power Engineering University, Kazan, Russian Federation

Corresponding author: smetanckin.anatoly@yandex.ru

Abstract. This article reviews a novel approach toward classification of micro turbine installations based on a systematic approach. Classification principles are described, and general requirements to classification system are formulated. Novel classification system for micro turbine installations is proposed.

1 Introduction

In recent years, there has been a steady trend in the global energy sector for the advanced development of distributed energy resources. Micro turbine installation (MTI) is one of the most promising energy sources for distributed energy generation [1-10], but in order to develop commercially efficient durable energy sources based on MTI, a number of scientific and technology issues have to be resolved. Developing modern technology devices employs development and testing of mathematical models, called mathematical modeling [11-17]. The foundation for mathematical modeling approach is the system of classification, which is based on classification criteria [21-27].

2 Methods

This paper presents a novel approach towards the classification of MTI which is based on a systematic approach. The aim of classification is to limit the choice of system representation approaches, to match chosen classes with certain methods and rules of system analysis, and to recommend selection of method for the corresponding system class. Classification is always relative. In theory, any system in general can be characterized by a number of attributes and assigned to different classifications, which can be useful to selection of modeling method. Classification system may have a specific problem focus: design and manufacturing, operation, maintenance, disposal. Classification implies is creation of a metric space to be used in analysis and synthesis of systems. The unified classification system approach is hampered by the lack of a logical vector. In this case categorization of classifications (metric spaces) must be conducted at the first place. The flow chart of the MTI system classification method is presented on figure 1.

IOP Conf. Series: Earth and Environmental Science **337** (2019) 012044 doi:10.1088/1755-1315/337/1/012044

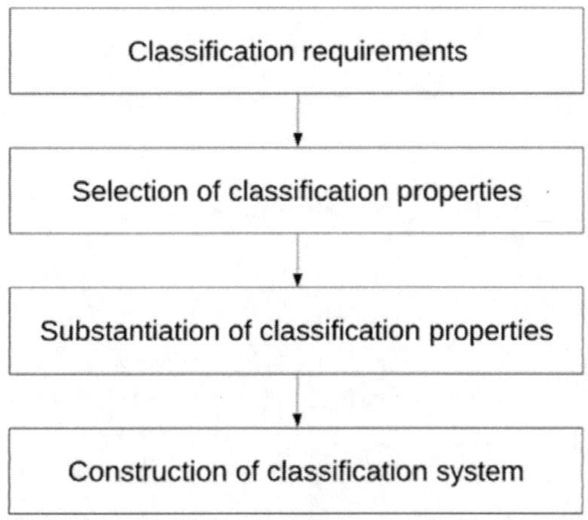

Figure 1. Flow chart of the MTI system classification method.

Various classification properties are used to form system classes. Systems are divided into real and abstract ones by nature of their elements. Depending on their origin, systems can be natural or artificial. Systems can be permanent or temporary, depending on their life span. Depending on variability of their properties, systems are divided into static and dynamic. Depending on the complexity of the system, they are divided into simple and complex. Depending on interactions with an ambient environment, systems can be isolated, closed, open- balanced and open- dissipative. Isolated systems do not exchange energy and matter with the environment. The entropy of an isolated systems moves to its maximum value. Closed systems do not exchange matter with ambient environment, but can exchange energy. Open system exchange matter and energy with an ambient environment. The change in entropy of an open system ds is defined by algebraic sum of entropy produced inside the system $d_p s$ and entropy coming from or leaving to an ambient environment, $d_c s$. In a state of lasting equilibrium (steady state) ds=0. Open in the equilibrium state are such systems that return to their steady state exponentially (without oscillations), when they deviate from the steady state. For open systems in equilibrium, the rule of entropy economy is observed according to the theory of I. Prigogine. Open dissipative systems are observed as a result of cooperative processes. Their behavior is nonlinear. The mechanism of dissipative structure formation is the following: subsystems fluctuate, sometimes reaching the point of bifurcation, after which the order of higher level may occur. Transitions to states of dynamic ordering, coherence, self-oscillations, autocatalytic reactions are similar to phase transition (as a result of increased fluctuations).

There are large numbers of classification properties for elements and systems. Common feature is the amount of information which the model carries. The distinctness of information determines the boundaries at which the model and the object preserve the homomorphism i.e. may be considered adequate in the sense of certain proximity criteria. The definition of hierarchy of characteristics of the classification system (Figure 2):

Area C1 – attributes that determine properties of the object as a system in general, regardless of its structure – systematic;

Area C2 – attributes that determine properties of the object, based on its structure (morphology) – structural;

Area C3 – attributes that determine properties of the object, based on its structural elements – elemental.

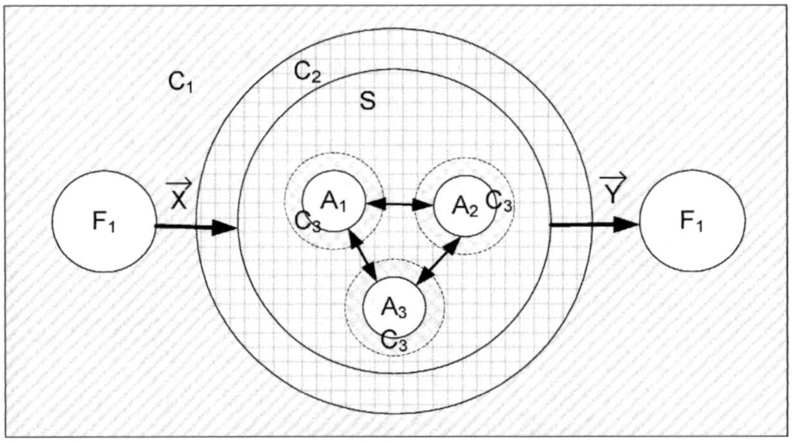

Figure 2. Graphical representation of hierarchy of attributes of system classification.

According to the general set theory, the inclusion of elements of a set being classified into a certain class, in respect to equivalence of an element attribute to a certain class attribute, is determined by the axioms of reflexivity, symmetry and transitivity. Then classification is determined as a partition of a set (of objects) into subsets (classes) by similarities or distinctions in accordance with methods (rules) chosen. Classification system is an aggregate of classification methods and rules, and their results.

The hierarchical classification method is a consistent division of a set of objects into subordinate classification groups. The faceted classification method is a parallel division of a set of objects into independent classification groups. Level of classification is the stage of classification in the hierarchical classification method, as a result of which a set of classification groups is obtained. The depth of the classification is the number of levels in classification. An attribute of classification is a property or characteristic of an object by which the classification is carried out. An aspect, property, characteristic of an object can be used as attributes for identifying a given object. In the faceted (parallel) classification, the set of data is divided into independent groups according to properties, aspects, characteristics. Each facet is formed according to essential attribute that corresponds to the analyzed one.

The number of facets can be increased without affecting the existing facets. Classification inside facets is mainly based on a hierarchical principle. Hierarchical sequential classification principle establishes a subordination relation between the classification groups based on the following logical division laws:

1. Only one principle of division should be applied at one particular level of classification. Therefore, the simplest division is the enumeration, or listing of classification groups. Division of elements by ranking of attributes is also common;
2. The number of levels of classification (depth of classification) is equal to the number of attributes selected for classification as the basis for division;
3. Any element of the divisible set may be present only in one classification group;
4. The sum of the elements of division should equal the original set.

This classification structure of information is a type of the systematic approach, which is called "tree-like" hierarchy. "Tree-like" hierarchy is the type of information system structure that allows, if necessary, to supplement it with the additional data (information).

3 Results and Discussion

Summing up what has been mentioned above and having described the purpose of the classification system and its functions, we can formulate the following basic requirements for the classification system of the TS MTI:

1. All the information should be divided into separate classification groups according to one attribute or a certain combination (set) of attributes;

IOP Conf. Series: Earth and Environmental Science **337** (2019) 012044 doi:10.1088/1755-1315/337/1/012044

2. All objects should be divided among classification groups according to a common attribute that is typical for similar objects, and attributes that can distinguish objects from one another;
3. Classification groups should be independent of each other;
4. Objects and features must be uniquely determined for each corresponding classification groups;
5. The classification system should be general and should provide for problem execution;
6. The system should be flexible: inclusion of new classification groups and objects or exclusion of existing ones should be made without deconstruction of the system;
7. Properties that determine the quality of the group of objects or a particular object should be taken as a characterizing attribute.

Modern MTI constitute complex technical system (TS) consisting of a large number of interconnected elements. These systems interact with ambient environment and other systems [19-20]. To describe the structure of the MTI system let us define the “structural element” of MTI. Structural element is a technical object (part of the MTI system), that has functional independence in relation to the whole system. The indivisibility of an element will be considered as criteria for disregarding internal structure of such element in regard to the model. Structural elements of MTI, for example, are: turbine, compressor, generator, bearings, control block. To select the class of the system, the principle of the hierarchy of characteristics of the classification system is used. Three levels of attributes are determined. Upper level (1st level) systematic attributes are such attributes that determine the properties of an object as a system in general, regardless of its structure. Middle (2nd) level is structural attributes that determine properties of an object, based on its structure (morphology). Bottom (3rd) level is elemental attribute that determines properties of an object based on its structural elements. Within one level, we apply a parallel (faceted) structure of classification attributes. Therefore, for 1st level of systematic attributes facet features are determined:

Purpose – by fields of application, mobility and types of application;

Functional – by the type of general function carried out;

Technical – by technical properties (power, efficiency, etc.)

The result of a valid selection of classification attributes of MTI was the classification system shown in figure 3.

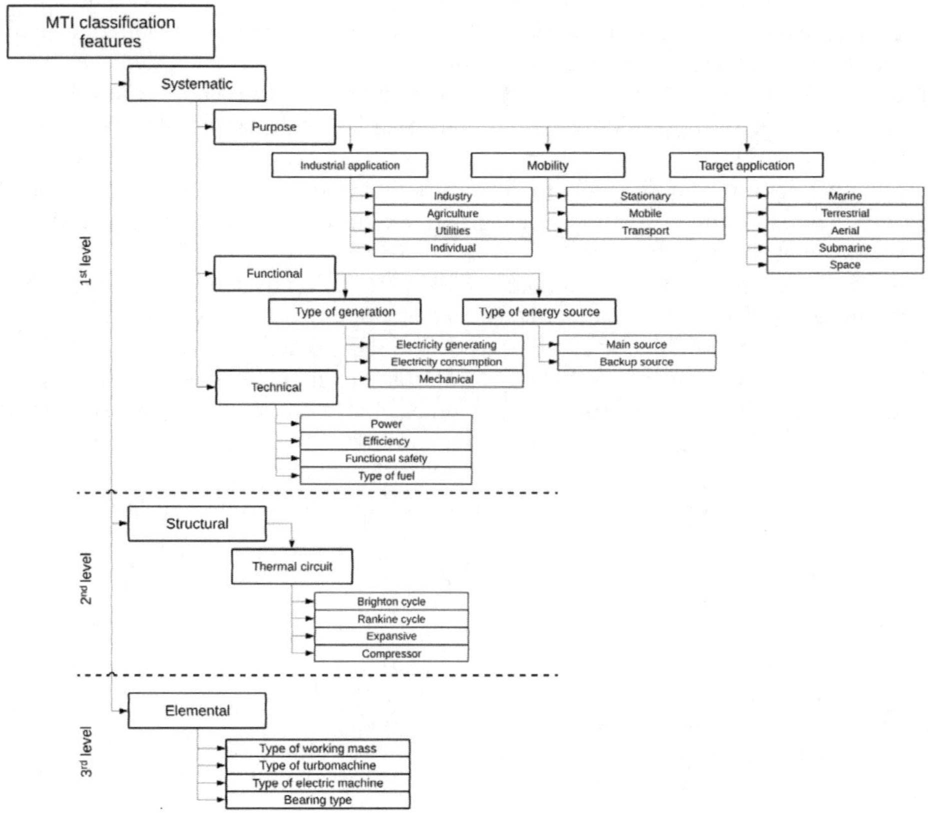

Figure 3. Flow chart of MTI classification.

4 Conclusions

The validity of the selected set of classification attributes is supported by its relevance to all requirements for classifying technical systems. This approach to the classification of MTI allows to perform R&D based on analysis and synthesis of MTI as a complex technical system.

References

[1] Barskov V, Besedin S, Besedina K, Rassokhin V, Smetankin A and Nikiforova O 2018 *J. of Advanced Research in Dynamical and Control Systems* **10** pp 384-92

[2] Asgari H, Chen X, Menhaj M and Sainudiin R 2013 *J. of Engineering for Gas Turbines and Power* **135** p 092601

[3] Asgari H, Chen X and Sainudiin R 2013 *Int. J. of Modelling, Identification and Control* **20** pp 253-70

[4] Mohammadi E and Montazeri-Gh M 2015 *J. of Engineering for Gas Turbines and Power* **137** p 071202

[5] Duan J, Sun L, Wang G and Wu F 2015 *Energy* **91** pp 168-75

[6] Duan J, Fan S, An Q, Sun L and Wang G 2017 *Energy* **134** pp 400-11

[7] Saha A, Chowdhury S, Chowdhury S and Crossley P 2009 *IEEE Transactions on Energy Conversion* **24** pp 529-38

[8] Sun Y, Zhang C, Chen H, Lin P and Xu X 2018 *IEEE Access* **6** p 8409454

[9] Cagnano A and De Tuglie E 2018 *Electric Power Systems Research* **157** pp 145-56

[10] Peirs J, Reynaerts D and Verplaetsen F 2004 Sensors and Actuators *A: Physical* **113** pp 86-93

[11] El-Khattam W and Salama M 2004 *Electric Power Systems Research* **71** pp 119-28
[12] Barskov V, Besedin S, Besedina K, Zabelin N, Matveev Y, Rassokhin V, Lavrov N and Fokin G 2017 *Int. j. of advanced biotechnology and research* **8** pp 1708-15
[13] Breńkacz L, Żywica G and Bogulicz M 2019 *Mechanisms and Machine Science* **60** pp 223-35
[14] Fu L, Feng Z and Li G 2018 *Microsystem Technologies* **24** pp 2333-47
[15] Fu L, Feng Z and Li G 2018 *Microsystem Technologies* **24** pp 1433-42
[16] Peirs J, Reynaerts D and Verplaetsen F 2003 *J. of Micromechanics and Microengineering* **13** pp 190-5
[17] Fu L, Shi Y, Deng Q and Feng Z 2009 *Hsi-An Chiao Tung Ta Hsueh/Journal of Xi'an Jiaotong University* **43** pp 15-9
[18] Epstein A, Senturia S, Al-Midani O and Anathasuresh G 1997 *28th Fluid Dynamics Conference* **12**
[19] Arav B L, Besedin S N and Yaichnikov M Yu 2012 *Young scientist* **4** pp 57-62
[20] Gaides M A 2005 *M. GLOBE-PRESS* p 201
[21] Mamrayeva D, Stybaeyeva A and Tashenova L 2018 *Economic Annals-XXI* **167** pp 4–7 doi:10.21003/ea.V167-01
[22] Iben U, Makhnov A and Schmidt A 2018 *J. of Physics: Conference Series* **1038** doi:10.1088/1742-6596/1038/1/012128
[23] Vdovin R A, Smelov V G, Sufiiarov V S and Borisov E V 2018 *IOP Conference Series: Materials Science and Engineering* **441** doi:10.1088/1757-899X/441/1/012058
[24] Vatin N, Gorshkov A, Nemova D and Tarasova D 2014 *Applied Mechanics and Materials* pp **633** p 634
[25] Konakov V G, Kurapova O Y, Lomakin I V, Novik N N, Sergeev S N, Solovyeva E N, Zhilyaev A P, Archakov I Y and Ovid'Ko I A 2017 *Reviews on Advanced Materials Science* **50** pp 1-12
[26] Yaichnikov I K and Suschenko V P 2014 *Teoriya i Praktika Fizicheskoy Kultury* pp 30–32
[27] Barsi D, Perrone A, Qu Y, Ratto L, Ricci G, Sergeev V and Zunino P 2018 *J. of Thermal Science* **27** pp 259–69 doi:10.1007/s11630-018-1007-2

IOP Conf. Series: Earth and Environmental Science **337** (2019) 012045 doi:10.1088/1755-1315/337/1/012045

The fuel burning intensification in the swirl burner using air ionization on the power-generating heat-and-power plant

S Maspanov[1,*], I Bogov[1], V Suhanov[1], V Shashikhin[1], A Denisov[1], Y Smirnov[1], Y Lypar[1], B Sidel'nikov[1], B Korotkov[1], T Akhmetov[2]

[1]Peter the Great St.Petersburg Polytechnic University, 195251, Polytechnicheskaya, 29, St. Petersburg, Russia.
[2]Kazan State Power Engineering University, Kazan, Russian Federation

*Corresponding athor: s.maspanov.turbo@mail.ru

Abstract. According to experts, the power-generating complex in the entire chain from fuel production to the thermal energy application efficiency of the use of energy resources is only 44 %. That necessitates the researching of ways to enhance efficiency of organic fuel burning. One of the way to solve the problem is the burning fuel in the swirl burner with using of ionizated air. Domestic and international researches have shown that atomic oxygen increases the completeness combustion and reduces the number emitted into the atmosphere of pollutant emissions. At present, the reduction of air pollution of toxic substances allocated by industrial enterprises, is one of the most important problems facing humanity. Air pollution has a unhealthy exposure for humans and the environment. In this paper estimated effectiveness of existing chemical kinetic mechanisms of natural gas burning for the swirl burner and mechanisms unhealthy emissions produced based on the results of the evaluation. The purpose of work is to increase the effectiveness of fuel combustion and technical and economic, as well as environmental indicators in the swirl burner on the heat-and-power plant.

1. Introduction

Existing kinetic mechanisms of natural gas combustion in the swirl burner in terms of atomic oxygen ionization are not validated because the lack of validation experimental data. Development of heat and power engineering in current days is connected with the necessity of the most important task of the combustion chamber (CC) and firebox furnaces (FF) toxicity reducing, used in gas, steam turbines and in combined-cycle plants [1 — 8]. Various methods are used for the toxic emissions from CC and FF suppression [1, 4, 5, 9 — 13], the most promising of which to the method of air ionization, supplied to the fuel combustion zone [14 — 22]. So, this work dedicated to experimental investigation of fuel burning intensification in swirl burner using air ionization on the heat-and-power engineering plant, which the most important. By the object of research was chosen swirl burner because it proved its high characteristics of ignition, durability, technological effectiveness reliability and as well the possibility of considering the swirl burner as physical model of fuel combustion in CC and FF. The purpose of paper is to increase of fuel combustion efficiency as well, as environmental indicators in the swirl burner on the heat-and-power object. For achieve this target we developed testing stand, consisting of: swirl burner, ionizing device releasing atomic oxygen and the measuring equipment for carrying out experiments and obtain results [18]. Simultaneously following research works were solved: have been carried out experimental studies, information processing and analysis of experimental studies [29-34].

2. Methods

For carrying out scientific experimental researches the physical simulation method was used as the most optimal. In accordance with compatible schematic diagram of testing stand in published works [18, 20], special testing stand was developed, where new experimental data were obtained. The photograph of this stand is shown on Fig. 1.

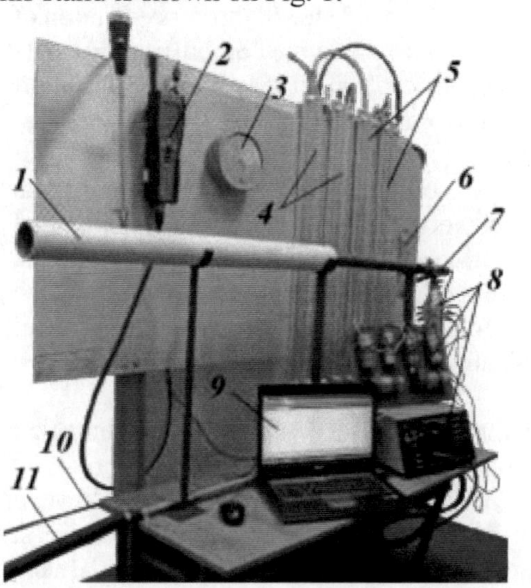

Figure 1. Purpose-designed polyfunctional testing stand:
1 — exhaust pipe;
2 — flue gas analyzer;
3 — aneroid barometer;
4 — pressure gauges (flammable gas);
5 — pressure gauges (air);
6 — thermometer;
7 — swirl burner;
8 — ionizing device;
9 — laptop pc;
10 — soothing pipe for flammable gas;
11 — soothing pipe for air.

The main requirements for modern gas turbine, steam turbine and combined-cycle plants are low production costs and compliance with permissible emission levels NO_x and CO. Oxidation of fuel is a finite rate process and a minimum residence time is required for high combustion efficiency and low emissions [2, 11, 15]. As a flammable gas, a propane-butane mixture of the following composition was used: propane (C_3H_8) — 80 %; butane (C_4H_{10}) — 13 %; ethane (C_2H_5) — 3,5 %; ethylene (C_2H_4) — 3,5 %. Preliminarily flammable gas and air were fed into the swirl burner (7) from sources. At the same time, the following indicators were measured: air temperature — $T_{air} =$ 300,15 °*K*; gas temperature — $T_{gas} =$ 300,15 °*K*; static air pressure — $P_{static\ air} =$ 3236 *Pa*; static gas pressure — $P_{static\ gas} =$ 4423 *Pa*; total air pressure — $P^*_{air} =$ 3481 *Pa*; total gas pressure — $P^*_{gas} =$ 4444 *Pa*. Further, according to the received data, the following were calculated: air speed — $c_{air} =$ 20,2*m/s*; gas speed — $c_{gas} =$ 4,7*m/s*; air flow rate — $G_{air} =$ 0,005 kg/*s*; gas flow rate — $G_{gas} =$ 0,0003 kg/*s*. In accordance with the results of the calculations, the reaction mass were fed into the swirl burner (7) through soothing pipes (10) and (11). Before the air was supplied to the combustion, it was passed through the ionizing device (8). That is air passed through an inhomogeneous stationary electric field. This produces the formation of atomic oxygen, which is the most powerful oxidant. More detailed description of the work ionizing device is contained in a published work [19]. In the swirl burner (7) simulated the processes occurring in CC. Combustion products were discharged through the exhaust pipe (1) and certain chemical compounds were determined by means of a flue gas analyzer (2).

3. Results and discussion

Based on the results of the experimental data processing, the dependence of the concentration of carbon monoxide in the combustion products of gaseous fuel when it is burned in a swirl burner from the temperature of the exhaust gases. This dependence is shown on Fig. 2. In this case, the experimental values of the concentration CO, obtained without of ionization air, are indicated — by points 1 and with of ionization air — by points 2. The set of points 1 was approximated — by line 3, and points 2 — by line 4.

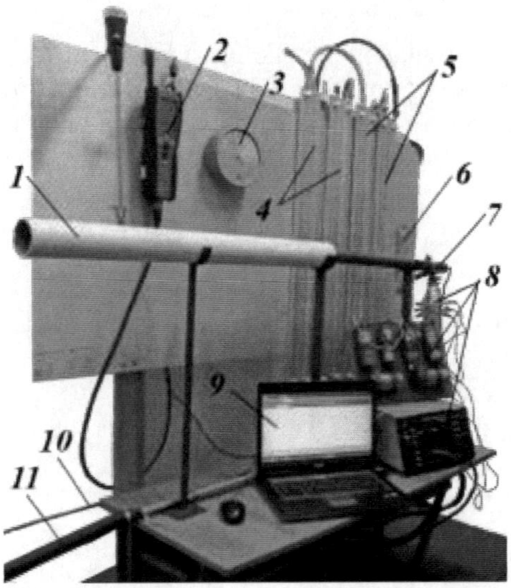

Figure 2. Dependence of the concentration of carbon monoxide in the combustion products of gaseous fuel when it is burned in a swirl burner from the temperature of the exhaust gases:
1 — experimental points corresponding to carbon monoxide concentrations without of ionization air before the combustion process;
2 — experimental points corresponding to the concentrations of carbon monoxide with of ionization air before the combustion process;
3, 4 — approximating lines.

From the analysis of the graphs shown on Fig. 2. it follows that with an increase in the temperature of the combustion zone, corresponding to an increase in the temperature of the exhaust gases from 470 to 600°C, emission of carbon monoxide increases both in the ionization of air and without it. However, with increasing temperature in combustion zone, in lack of ionization air, the CO emission increases from 0 *ppm* to $0{,}23\cdot10^4$ *ppm* and in availability of air ionization from 0 *ppm* to $0{,}14\cdot10^4$ *ppm*. Thus, in availability of ionization air at exhaust gases temperature of $t = 600$°C, emission CO is reduced by $0{,}09\cdot10^4$ *ppm*, id est by 39,1 %. For the rational use of air ionization at the entrance to the swirl burner, it is necessary to create scientifically grounded methods for their design and operation. For the development of such methods requires a large amount of experimental research. However, the number of published results of such studies is still small. Of particular interest are the papers [11, 17, 23 — 28].At the same time, it should be noted that the published reports do not contain sufficient information on the effect of air ionization before the entrance to the swirl burner on fuel combustion in CC [29-35].

4. Conclusion

The performed analysis of experimental studies allows us to draw the following conclusions:

1. The air ionization air before entering the swirl burner in the research range of temperature changes in the combustion zone leads to a reduction in emissions CO by 39,1 %;
2. Scientific novelty lies in the fact that we are acting not on fuel, with the air agent in the combustion zone, we activate atomic oxygen in a way that the activation energy without temperature increase will be sufficient for efficient combustion, and this economical effectiveness;
3. The performed experimental studies have shown the efficiency of purpuse-designed polyfunctional testing stand and the expediency of its further use for expanding the range of experimental studies.

References

[1] Bogov I A, Kulandin A A, Suhanov V A, Tolmachev V V, Bezuhov A P, Smirnov A A, Bodrov A I and Korihin N V 2007 *Osnovi tehnicheskoi termodinamiki, termohimii, teorii goreniya i ciklov gazoturbinnih i parogazovih ustanovok* (SPb.: IA «Energomashinostroenie»)

[2] Mitrofanov V A 2000 *Zakonomernosti obrazovaniya vrednih veschestv i povishenie ekologichnosti GTD*: avtoref. dis. … kand. teh. Nauk (SPb)

[3] Mitrofanov V and Rudakov O 2003 Mathematical simulation and calculation of flame ignition and quenching in gas-turbine engine *Combustion and atmospheric pollution* (Moscow: TORUS PRESS Ltd) pp 273-6
[4] Mitrofanov V A 2004 *Kameri sgoraniya gazoturbinnih dvigatelei: matematicheskoe modelirovanie, metodologiya rascheta koncepciya optimalnogo proektirovaniya*: avtoref. dis. … dokt. teh. Nauk (SPb)
[5] Mitrofanov V A, Rudakov O A and Sigalov Y V 2005 *Thermal Engineering* **52**(11) pp 874-8
[6] Bulysova L A, Vasil'ev V D, Berne A L and Gutnik M M 2018 *Thermal Engineering* **65**(6) pp 362-70
[7] Roslyakov P V, Khokhlov D A and Proskurin Y V 2017 *Thermal Engineering* **64**(8) pp 574-84 (2017)
[8] Roslyakov P V and Zakirov I A 2001 *Nestehiometricheskoe sjiganie prirodnogo gaza i mazuta na teplovih elektrostanciyah* (Moscow: Izdatelstvo MEI)
[9] Maspanov S N, Bogov I A, Tolmachev V V and Rassohin V A 2015 *XLIV Nedelya nauki SPbPU: materiali nauchnogo foruma s mejdunarodnim uchastiem. Institut energetiki i transportnih sistem*. 30 noyabrya — 05 dekabrya 2015 g. Chast I. (SPb.: Izd-vo Politehn. Un-ta) pp 49-51
[10] Rizk N K and Mongia H C 1993 Gas Turbines Power **115** (3) pp 612-9
[11] Coogan S, Gao X, McClung A and Sun W 2016 *ASME Turbo Expo 2016: Turbomachinery Technical Conference and Exposition Volume 4A: Combustion, Fuels and Emissions*
[12] Tumanovskii A G, Gutnik M N and Artemenko A A 2000 *Thermal engineering* **47**(10) pp 879-82
[13] Vedeshkin G K and Sverdlov E D 2005 *Thermal Engineering* **52**(11) pp 829-32
[14] Wilk M and Magdziarz A 2010 *Polish J. of Environ. Stud.* **19** (6) pp 1331-6
[15] Dragin V A 2000 *Razrabotka i issledovanie elektroozonatora dlya povisheniya effektivnosti sjiganiya pechnogo topliva v kotelnih APK*: avtoref. dis. … kand. teh. Nauk (Krasnodar)
[16] Indoor Air Chemistry: Cleaning Agents, Ozone and Toxic Air Contaminants URL: https://www.arb.ca.gov/research/apr/past/01-336.pdf (Date of access: 26.09.2018)
[17] Lantz A, Larfeldt J, Ehn A, Zhu J, Subash A A, Nilsson E J K , Li Z and Aldén M 2016 *ASME Turbo Expo 2016: Turbomachinery Technical Conference and Exposition Volume 4A: Combustion, Fuels and Emissions*
[18] Maspanov S N, Bogov I A, Tolmachev V V and Suhanov V A 2017 *Stend dlya issledovaniya vliyaniya ozona na processi v kamere sgoraniya gazoturbinnoi ustanovki* Patent na poleznuyu model RUS 175593 (10.03.2017)
[19] Maspanov S N, Bogov I A and Suhanov V A 2017 *XLVI Nedelya nauki SPbPU: materiali nauchnogo foruma s mejdunarodnim uchastiem. Institut energetiki i transportnih sistem. 13 noyabrya — 19 noyabrya 2017 g.* Chast I. (SPb.: Izd-vo Politehn. Un-ta) pp 114-6
[20] Maspanov S N, Bogov I A and Suhanov V A 2017 *XLVI Nedelya nauki SPbPU: materiali nauchnogo foruma s mejdunarodnim uchastiem. Institut energetiki i transportnih sistem. 13 noyabrya — 19 noyabrya 2017 g.* Chast I. (SPb.: Izd-vo Politehn. Un-ta) pp 116-9
[21] Andreev S A and Petrova E A Eksperimentalnoe issledovanie goreniya v ozono-vozdushnoi srede, http://elibrary.ru/download/elibrary_21691449_76600167.pdf (Date of access: 11.10.2017)
[22] Andreev S A and Petrova E A Ocenka energozatrat na ozonirovanie topochnogo prostranstva vodogreinih kotlov, URL: http://elibrary.ru/download/elibrary_23681676_16510445.pdf (Date of access: 16.03.2018)
[23] Hung W S Y 1993 *Gas Turbines Power* **115** (3) pp 588-93
[24] Vasil'ev A Y, Sviridenkov A A, Go'tsev V F, Tret'yakov V V, Furletov V I, Maiorova A I and Yagodkin V I 2005 *Thermal Engineering* **52**(11) pp 838-48
[25] Tolmachev V V and Bogov I A 2012 *Nauchno-tehnicheskie vedomosti SPbPU. Estestvennie i injenernie nauki* **147-1** pp 85-94

[26] Tolmachev V V, Bogov I A and Vohmyanin S M 2012 *Nauchno-tehnicheskie vedomosti SPbPU. Estestvennie i injenernie nauki* **147-2** pp 176-80
[27] Tolmachev V V 2012 *Modelirovanie i issledovanie vihrevogo zapalno-stabiliziruyuschego modulya kamer sgoraniya gazoturbinnih ustanovok*: avtoref. dis. … kand. teh. Nauk (SPb)
[28] Schennikov K A, Bogov I A and Tolmachev V V 2000 *Sovremennoe turbostroenie: Sb. nauchnih trudov* **3** pp 68-85
[29] Sokolov N P, Polishchuk V G, Andreev K D, Rassokhin V A and Zabelin N A 2015 *Thermal Engineering (English translation of Teploenergetika)* **62** pp 423–33 doi:10.1134/S0040601515060105
[30] Mishin M V, Protopopova V S and Alexandrov S E 2015 *Russian J. of General Chemistry* **85** pp 1209–21 doi:10.1134/S1070363215050394
[31] Emelyanov O, Kliuiko P and Shemet M 2016 *Proceedings of the 2016 IEEE International Conference on Dielectrics, ICD 2016* **2** pp 1171–4 doi:10.1109/ICD.2016.7547826
[32] Schegolev V A and Novoseltsev O V 2015 *Teoriya i Praktika Fizicheskoy Kultury* pp 15–7
[33] Kuptsov V D, Katelevsky V Y and Valyukhov V P 2015 *Proceedings of SPIE - The International Society for Optical Engineering* **9506** doi:10.1117/12.2178594
[34] Shevkunov S V 2014 *Colloid J.* **76** pp 490–508 doi:10.1134/S1061933X14040139
[35] Shevkunov S V 2014 *Russian J. of Electrochemistry* **50** pp 1127–36 doi:10.1134/S102319351412009X

Problems of Petroleum industry in Iraq

H Alzuwaini[1,*], D Vasil'kov[1], N Kirillov[1], A Khitrov[1], V Tolmachev[1], R Okorokov[1], O Gatsenko[1], D Zaripova[2]

[1]Peter the Great St.Petersburg Polytechnic University, 195251, Polytechnicheskaya, 29, St. Petersburg, Russia
[2]Kazan State Power Engineering University, Kazan, Russian Federation

*Corresponding author: halzwiny@yahoo.com

Abstract. Iraq has been faced a lot of challenges from internal and external conflicts' security situation is stable and war with terrorism, international transactions are not in favour of Iraq. Impact on investors, oil companies in the oil field.

1. Introduction

Despite security and financial troubles, Iraq's crude oil production has been grown substantially over the last 5 years. Moving into 2017, what challenges remain for Iraq's oil sector? Since 2010, Iraq oil production has grown by two million b/d and reached a record high of 4.4 million b/d in the third quarter of 2015.

The growth is in part due to the push by lawmakers to boost oil production to generate badly needed income, as the government continues to struggle financially to contain Islamic State militants in the northern and eastern provinces of Iraq.

Looking ahead to 2016, Iraq's oil production is contingent on several political risk challenges that Iraqi leaders must address to maintain strong production levels.

Iraqi oil reserves are considered the world's fifth largest ones, with 140 billion barrels, and Iraq is OPEC's second largest producer behind Saudi Arabia.

However, a decade of war, domestic political disputes, insufficient investments, and low oil prices has been placed a strain on Iraq's oil sector.

2. The impact of war on oil imports and exploitation & government policy

Defense spending has been risen significantly since the U.S. invasion in 2003. The takeover of Mosul by Islamic State militants in June 2014 fueled defense spending over the last year and a half, pinching the national budget so much that investments towards oil production are often insufficient to operate at max capacity.

Iraq's defense spending is likely to continue to place a strain on the national budget in 2016 until ISIS is effectively pushed out of Ramadi and Mosul. While security forces have made recent gains against ISIS in Ramadi, capturing Mosul could take months.

Low oil prices and rising imports have been also placing strain on the Central Bank of Iraq and oil revenues. Foreign currency reserves have reportedly shrunk from $78 billion at the end of 2013 to $59 billion in mid-2015. The Iraqi dinar also has been fallen sharply across markets this year.

IOP Conf. Series: Earth and Environmental Science **337** (2019) 012046 doi:10.1088/1755-1315/337/1/012046

The low prices of oil is also undervaluing current exports and inhibiting the government from making larger gains. Figures show that a year ago, the government pulled in about $300 million per day in revenue. Now it's down to roughly $240 million a day.

Low oil prices are likely to negatively impact Iraq's oil revenues in 2016, especially as Iran gears up to begin exporting its oil to western nations, further flooding the supply market.

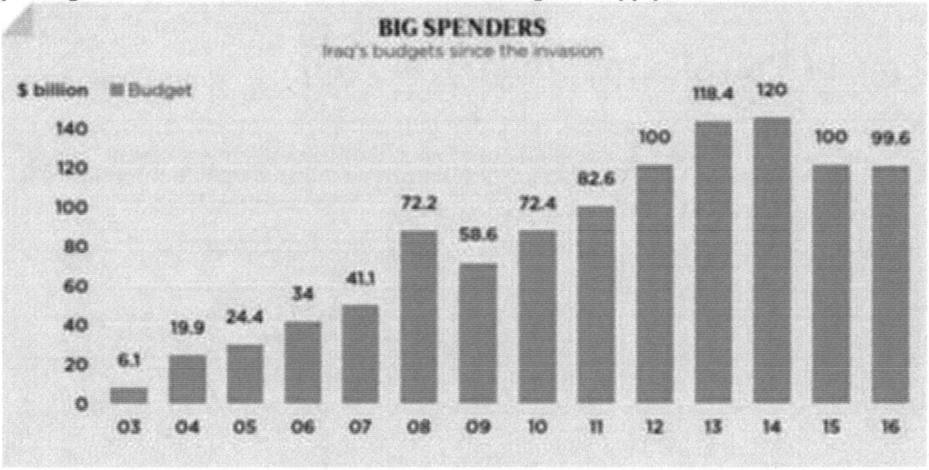

Figure 1. Source: Brookings.

Internal political challenges: Domestic political disputes over oil production between the central government in Baghdad and Kurdish officials in Erbil is further impacting the rate of oil production [1].

As of mid-November, Kurdistan is exporting over 500 000 b/d of oil, every seventh barrel of OPEC's second largest exporter. The exports, however, are controversial as Bagdad accuses Erbil of bypassing existing legislation on oil exports and failing to transfer agreed volumes of oil to the government.

In December 2014, Kurdistan and Bagdad signed a deal in which Kurdistan is required to transfer some average 550 000 bpd to Iraqi state oil firm SOMO over the course of 2015 while receiving 17% of Iraqi budget or over $1,1 billion a month.

However, the deal quickly unraveled as Kurdish officials say the volume of oil transfer is too high with current «insufficient» funds from the government.

Both Erbil and Baghdad have a need to fund their own security forces to counter the threat of ISIS within Iraq, and revenue from oil plays a key role in supporting security operations.

According the consideration, it is unlikely a deal will be reached between both governments unless Baghdad is able to accept a lower threshold of oil transfers from Kurdistan. Meanwhile, current political challenges could scupper production output by 10%.

3. Humanitarian crisis

In addition to security and political problems, investment for oil is being hindered by the mass exodus of refugees and internally displaced people.

The influx of million of refugees and displaced people have placed a stain on the government, especially with respect to provide adequate aid and security to displaced people.

Likewise, a worsening security situation may cause skilled laborers such as oil field works and technicians that are critical to maintaining Iraqi oil exports to leave Iraq.

Remaining steps forward: In order for Iraq to effectively reach full operating capacity, Iraq must ultimately improve the vastly deteriorating security situation and strengthen counterterrorism efforts to stop the spread of violence and boost investor confidence.

Luckily, Iraq's largest oil refineries in the southern areas have largely been unaffected by recent violence. The recapture of the Baiji oil field from ISIL militants can also be seen as a strategic success [2].

IOP Conf. Series: Earth and Environmental Science **337** (2019) 012046 doi:10.1088/1755-1315/337/1/012046

Additionally, Iraqi lawmakers must place serious commitment and investment towards oil infrastructure modernization. Commercial and technical difficulties continue to plague Iraq oil production, though the recent surge in production output is a positive sign for Iraq's future production potential.

Low oil prices will be remained a challenge for Iraq's oil sector, especially as Iran is expected to enter the market in 2016. In the long-run, Iraq must work to diversify its economy and energy exports to soften the blow for future price drops.

4. Iraqi oil & OPEC

Iraq is the one of the major reasons why OPEC was able to increase oil production over the last year, even as oil prices were dropped to six-year lows.

The war-torn country averaged 3,2 million barrels per day (mb/d) of production in 2014. But despite the onslaught from ISIS and the collapse in oil prices, Iraq succeeded in achieving steady gains in output, surpassing 4,1 mbd in September. Along with Saudi Arabia's increase of around 600,000 barrels per day since 2014, Iraq accounted for a majority of OPEC's production gains over the last year, allowing the cartel to produce more than 31,5 mb/d - well in excess of its stated production target of 30 mb/d [3].

5. KRG policy and its impact on oil production

However, although Iraq has succeeded in defying gravity thus far, the cracks in the country's oil success story are started to show.

Genel Energy, a London-based oil exploration company that is concentrated in the Kurdistan region of Iraq, trimmed its production forecast for the year, due to the late payments from the Kurdish Regional Government (KRG). The Iraqi central government in Baghdad and the KRG have been to odds over oil sales. In December 2014, the two sides reached an accord that would see the KRG exporting oil under the purview of the Iraqi state, in exchange for access to its portion of national revenues.

The collapse in oil prices has sapped the state of resources, however. That has held up payments to the KRG, which in turn has halted the transfer of funds to private oil companies operating within Kurdistan.

As a result, oil companies in Iraq are hit with a double whammy - not only are prices for their oil low, but they are not even receiving payments on time. Genel Energy said that the payment delays mean the company will only be able to produce 85 000 to 90 000 barrels per day this year, down from its previous forecast of 90 000 to 100 000 barrels per day. The company's share price fell by 6,34 percent on October 20 after the announcement.

«Given the payment situation has been irregular, we stopped investing in the drilling of the fields», Genel's CFO Ben Monaghan told Bloomberg. «Inevitably with conventional oil fields, if you stop investing, the production will begin to decline». Genel cut its capital expenditure guidance to \$150-\$175 million for the year, down from \$150-\$200 million.

For the company, there are some reasons for hope. The KRG decided to skirt its agreement with Baghdad over the summer and begin exporting oil on its own terms. That has allowed for partial payments to oil companies, including Genel Energy, but also Gulf Keystone Petroleum and DNO ASA, a Norwegian E&P company.

Genel received a \$16,5 million payment from the KRG on October 20 for its oil exports, the second month in a row that it was paid. In its October 20 statement, the company said that «The receipt of predictable export payments…would allow Genel to deliver the investment required to increase production». The second payment offers some cautious optimism, and the company is striking a positive tone.

The KRG also submitted its second consecutive payment to Gulf Keystone, another important producer in Kurdistan. Collectively, companies operating in Kurdistan say they are owed more than \$1 billion [4-8].

Kurdistan is dependent on oil for its economy, so the KRG is doing as much as it can to increase oil exports and square up with private companies. Kurdistan exported 620 000 barrels of oil per day in

IOP Conf. Series: Earth and Environmental Science **337** (2019) 012046 doi:10.1088/1755-1315/337/1/012046

September through a pipeline that runs to the Turkish port of Ceyhan. That is a significant increase over the 470 000 barrels per day that the semi-autonomous region exported in August.

6. International transactions and their impact on Iraq's production

For now, this appearance is a short-term problem. But if low oil prices, security concerns, or political instability (or a combination of all three) deter investment today, Iraq could fall short of its production goals over the long-term.

Actually, Iraq is the oil market's biggest problem. There's a bigger threat to oil prices than Iran.

Last week, Iran struck a nuclear deal with several world powers that, among other things, will end an oil embargo and allow the country to increase production for export.

Analysts have begun to worry that the extra supply - estimated by Goldman Sachs at up to 400 000 barrels a day - could worsen the already flooded market.

Iran's agreement with the P5+1 solidifies the case of increased production in Q1. However, it is arguably additional supplies from Iraq that are pressuring world oil prices more than Iran at this juncture. Iraqi production has been risen consistently over recent quarters, but surged ahead in 2Q 2015 prompting us to conclude that Iraq will likely overtake Saudi Arabia as the biggest contributor to OPEC growth this year [5].

The full potential of Iranian oil exports is not expected within the next year, because the production infrastructure is still needed to be brought back to life. And so Iraq's surging production is a more immediate threat to oil prices, which last week fell for a third straight week.

Iraqi excess production at 920 kb/d above its inferred Iran Agreement quotas is second only to Saudi Arabia at 1,670 kb/d above quota.

Here's JPMorgan's forecast of Iraqi oil production:

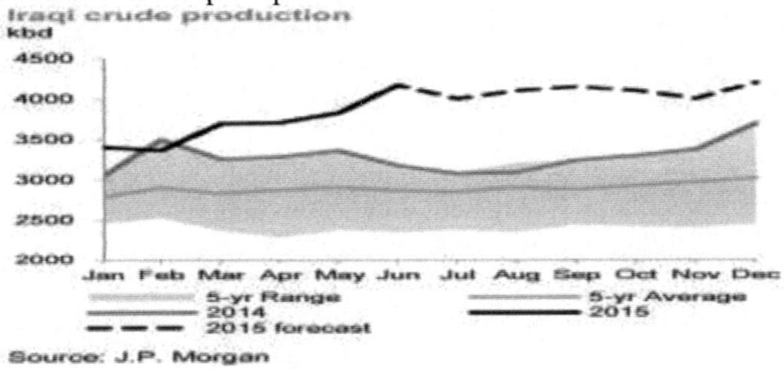

Figure 2. JPMorgan's forecast of Iraqi oil production.

7. The basic requirements are needed to increase production

The introduction of directional drilling technologies has helped Iraq dramatically increase its production, and these technologies had respect for the promise of even greater yields in the future.

As measured by its ever-increasing volume of oil exports, the country's oil industry has been fully recovered from the damage and mismanagement that it suffered during Saddam Hussein's reign.

Notwithstanding Iraq's already-known reserves, it seems likely that recent new exploration efforts by oil companies will lead to an even greater tally of Iraq's oil wealth.

It's also important to note that Iraq isn't just blessed with large reserves, it also happens that the geologic formations holding Iraq's oil are much more accessible to drilling than they are in most countries, which means the costs of extracting that oil are much lower.

Given the world's insatiable appetite for oil and gas, Iraq's dominant supply role is likely to become even stronger going forward. Although it's difficult to predict the future with certainty, it seems likely that sometime in the near future Iraq will grow into a regional and international energy super-power based on its rich reserves.

It won't be easy to get oil out of Iraq. After a slow start in 2015, Iraq's south is expected to export more oil over the next few months. But there are obstacles standing in the way.

In recent years, it's become apparent that Iraq's crude oil will not age with grace. Consequently, some oil production has been shut down in order to preserve the quality. And in order to get around that, a new blend has been introduced, which «will unlock volumes». However, ultimately, Iraq's southern export infrastructure will determine the potential boost from this new blend.

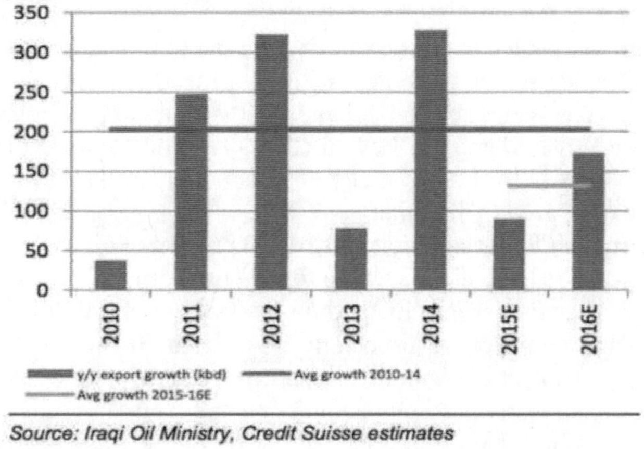

Figure 3. Export growth from southern fields – base case.

Generally, 3 links could hamper exports from the south, namely (a) the network of pipelines, (b) the lack of storage facilities and (c) the pumping station that provides a connection between the fields and the main export depot at Fao [peninsula in the southeast of Iraq]. Furthermore, the available pumping and storage capacity at Fao itself, an important link between the fields and the offshore laoding facilities, could be an additional blockage.

In any case, there's some potential for more exports, but it may not be executed well.

8. The security situation and its negative impact on the investing companies

Instability threatens Iraq's oil production: oil companies operating in the Basra region could potentially become targets for the region's well-armed militant groups that are increasingly acting with impunity [Reuters].

While for much of 2015 oil traders were fixated on the United States shale oil producers, Chinese energy consumption and the actions of Saudi Arabian oil minister Ali al-Naimi, this summer they are also likely to be keeping a close eye on OPEC's second-largest producer, Iraq, where instability could curtail the nation's oil production and move global crude markets.

Over the past 18 months the south of Iraq has seen a surge in crime, including murders, kidnapping, armed robberies, extortion and drug trafficking.

High-profile incidents included the shooting of an Italian oil worker in August 2015 in Basra, the murder of two oilfield engineers who were working on the Az-Zubair field in November, and the kidnapping of 26 Qatari hunters in Muthanna province in December.

Tribal violence has also increased as state control has receded and tribes have filled the vacuum, playing a greater part in dealing out justice and resolving regional disputes.

Since the beginning of 2015 tribal conflicts have left dozens dead, including women and children. These deaths have sparked mass protests demanding better state-provided security and prompted Iraq's top Shia cleric, Grand Ayatollah Ali al-Sistani, to call for an end to the clashes.

In early 2016 Baghdad moved to try to clamp down on criminal and tribal violence in the south, sending federal forces armed with heavy weapons and tanks into Basra to try to confiscate weapons from militant groups in the region.

On January 15, the government announced that more than 30 people had been detained in the operation and weapons including rocket-propelled grenades and mortars had been seized during raids on houses.

The failure of the central government to disarm renegade militants and sooth tribal tensions at the beginning of the year has underlined the limits of Baghdad's influence in the region.

However, the operation's success was short-lived and what was originally intended as a show of force quickly turned into an embarrassing demonstration of the central government's declining influence in Iraq's south.

Stability in Iraq's south is expected to further deteriorate over 2016 as budget problems in Baghdad limit the government's ability to provide basic necessities to citizens.

In 2015 Baghdad has seen its budget deficit expand rapidly owing to extra spending on the ongoing war against the Islamic State in Iraq and the Levant (ISIL) and the collapse in oil prices, which have fallen by two thirds since June 2014 due to increased production in the US, OPEC's decision not to reduce production, and slowing growth in China.

In October, Iraq's government forecast deficit worth 11,9 percent of gross domestic product for 2016. That figure is now expected to be even higher owing to further declines in the price of oil.

Baghdad's financial crisis has frozen activity on thousands of important infrastructure projects across the country - something that could prove especially significant in the south, which regularly sees disruptive protests during the summer months because of electricity blackouts, high pollution levels, and inadequate supplies of fresh water.

In the past year, almost nothing has been done to improve power and water infrastructure around Basra - and the budget crisis has created new problems, including job losses and delayed salary payments.

Owing to the government's failure to address the grievances of communities in Basra, Iraqi officials are expecting the summer of 2016 to be the most disruptive season of protests on record, and some fear it could potentially affect the region's oil exports.

Over the past three years Iraqi oil exports have shown incredible resilience, hitting new highs despite ISIL taking control of large swaths of the country in June 2014.

One of the key factors behind Iraq's success in increasing oil output amid political turmoil is geographical.

More than three quarters of the country's oil is produced and exported from Basra and neighbouring provinces in the south, which lie more than 500km from the fighting with ISIL.

In January Iraq exported an average of 3,88 million barrels a day, including exports from Iraqi Kurdistan,

This is 660 000 barrels a day more than it was exporting before ISIL's advance across the country in the summer of 2014, which saw the group secure around a third of the country's territory.

Just how long this impressive export growth can be sustained remains to be seen, and 2016 could well prove to be a key test for oil companies active in Basra.

Though the region's oilfields and related facilities are well-fortified and international oil companies can rely on several layers of security, including government-provided oilfield protection forces and private contractors, the environment in which oil companies are operating is almost certainly going to get more difficult.

The failure of the central government to disarm renegade militants and soothe tribal tensions at the beginning of the year has underlined the limits of Baghdad's influence in the region. Additionally, failure to provide adequate power and water infrastructure is likely to fuel public anger and separatist sentiment, which could potentially be channelled by political groups looking to further undermine the central government.

9. Conclusion

All this has ramped up risk for oil companies operating around Basra, which could potentially become targets for the region's well-armed militant groups that are increasingly acting with impunity. As Basra's oil exports are the biggest single source of income for Iraq's government, any kind of sustained period of disruption to operations would be devastating for the country's already fragile economy and would significantly affect its efforts to fight ISIL in the north.

Additionally, an unexpected cut in Iraqi oil production would deliver a shock beyond the country's borders, applying upward pressure to global crude prices and making Iraq a key market to observe for oil traders in 2016 along with the US, Saudi Arabia, and Iran.

References

[1] Rt.com *Iraq oil output hits record high ahead of production freeze talks* https://www.rt.com/business/339163-iraq-oil-production-prices/ (retrieved 16 April 2016)

[2] Klare M T 2009 Will Iraq Be a Global Gas Pump? The (Re)Making of a Petro-State *The Huffington Post* (14 July 2009)

[3] Eia.gov *International Energy Statistics* http://www.eia.gov/cfapps/ipdbproject/IEDIndex3.cfm?tid=5&pid=57&aid=6 (Retrieved 11 January 2015)

[4] iraqoilreport.com

[5] OPEC *OPEC Share of World Crude Oil Reserves* http://www.opec.org/opec_web/en/data_graphs/330.htm (Retrieved 11 January 2015)

[6] Bakayev V 2015 *J. of Physical Education and Sport* **15** pp 615–8 doi:10.7752/jpes.2015.04093

[7] Zhukov S V and Kopytin I A 2017 *World Economy and International Relations* **61** pp 36–42 doi:10.20542/0131-2227-2017-61-12-36-42

[8] Elvidge C D, Bazilian M D, Zhizhin M, Ghosh T, Baugh K and Hsu F-C 2018 *Energy Strategy Reviews* **20** pp 156–62 doi:10.1016/j.esr.2017.12.012

IOP Conf. Series: Earth and Environmental Science **337** (2019) 012047 doi:10.1088/1755-1315/337/1/012047

Multifunctional electronic devices protected and automatic of modern electricity system

M G Popov[1,*], Yu N Bocharov[2], E I Gurevich[3], F H Halilov[2],V V Popov[3], T R Akhmetov[4]

[1]Department Power Plants & Power Automation of Peter the Great Saint-Petersburg Polytechnic University (SPbPU), Russia, 195251, Saint-Petersburg, Polytechnicheskaya st., 29
[2]Department High Voltage Engineering, Electrical Insulation and Cable Technolo-gy of Peter the Great Saint-Petersburg Polytechnic University (SPbPU), Russia, 195251, Saint-Petersburg, Polytechnicheskaya st., 29
[3]Department Theoretical Foundations of Electrical Engineering and Electrome-chanics of Peter the Great Saint-Petersburg Polytechnic University (SPbPU), Russia, 195251, Saint-Petersburg, Polytechnicheskaya st., 29
[4]Kazan State Power Engineering University, Kazan, Russian Federation

*Corresponding author: PopovMG@eef.spbstu.ru

Abstract. The report reflects the long-term solutions for the modernization of main means of emergency control and protection of integrated power systems in the implementation of adaptive control algorithms.

1. The relevance of research area and the degree of its development.

The existing financial and economic, and as well operational and technical interrelations be-tween the electric power engineering entities are characterized by a reduction in costs for depre-ciation, ageing of the resource and poorly justified prolongation of the operating lifetime of the main equipment of power plants and substations. In these complicated conditions, ensuring the reliability and stability of the power supply in the integrated power system (IPS) can be partially solved by introducing into the operation practice a new electrical power equipment for regula-tion and control of the interchange power (FACTS), as well as the by creating of ring circuits of electric networks [1-8]. However, this approach significantly complicates the problem of fast and high-quality supervisory control of transient operation modes of power systems [9-11]. The appearance of new measuring and information systems, monitoring means (WAMS) of transient modes and systems of centralized (system) control allowed to some extent to improve the effi-ciency of dispatch and emergency control of the integrated power systems modes [12,13]. An important role in this has the information and technical perfection of communication devices.

Many researches and scientific publications have been devoted to the issues of stability and reliability growth of electrical power equipment operation of power systems. By means of these studies, advanced control means of the excitation of synchronous machines have been created [2, 4, 11],

[14-17], the technical possibilities and conditions for the transmission of electric power by alter-nating current in extended power pools have been scientifically justified. Despite significant ad-vances in research on sustainability and the principles of emergency control of the integrated power systems,

IOP Conf. Series: Earth and Environmental Science **337** (2019) 012047 doi:10.1088/1755-1315/337/1/012047

there is much concern about the issue of ensuring the speed performance, selec-tivity and operating reliability of the protection and automation means. It must be emphasized that practically all (with rare exception) scientific research and publications of authors are based on the use of an idealized mathematical description and linear models of power equipment [1, 2, 4, 14, 16-19], and of a robust description of the filter synthesis problem [15]. And by extension, such an approach is characterized not only by the presence of a method error, but also by errors in practical and quantitative assessments of the limit conditions and, as a consequence, in the choice of parameters of technical protection means and other means of emergency control that do not satisfy basic regulatory requirements.

The elimination of this substantial defect is possible only with the development of the theory of nonlinear filtration [3, 20-23], the application of numerical methods and rigorous mathemati-cal models in the study of quasi-stationary asynchronous and oscillatory synchronous modes of the integrated power systems. Such approach becomes possible now due to growth of computing capacities and functional capabilities of industrial microcontrollers, and as well due to the in-crease in the automation level of modern power facilities. Obviously, the validity of the new obtained research results should be confirmed based on carrying out and improving physical, full-scale experiments and experimental-industrial trials.

All these current problems, the issues of qualitative and reliable control of modern integrated power systems, predetermined the subject of complex research, reflected in this report on the author's research.

2. New generalized stability criteria and the fundamentals of adaptive highly sensitive means of emergency instability prevention automation (APNU) of intersystem (interstate) power transmission lines of extended power pools

The modern development of energy systems is characterized by the inclusion of maneuverable and powerful blocks for parallel operation with the previously introduced aggregates of power plants and large power pools [1-4, 7, 8]. Within the integrated power systems, individual plants and even parts of the power systems can be located at a considerable "electrical" removal from each other. In systems of a simple structure, asynchronous regimes as a rule are caused by loss of the synchronism of one of the generators. Though, due to the fact of the creation of integrated power systems with an advanced electrical network circuit, the conditions for resynchronization after the appearance of an asynchronous regime turned out to be rather complicated and negative. Moreover, with asynchronous modes in complex systems, phenomena and processes that are unusual to simple structure systems have been revealed. Their main feature is the possibility of an asynchronous mode occurrence of many generated sources, due to additional loss of stability. To reveal the area boundary of oscillatory stability, the author proposes an algorithm for controlling the generalized stability criteria (Jacobian) [1, 3, 8], whose coefficients are calculated by differentiating the expressions for the active and reactive components of the nodal and exchange power.

Research of the stability is made to structural diagrams, and chain ring of power (figures 1, 2) when the variation of stiffness and the installed capacity equivalent power systems [1, 3, 19]. The figures show the maximum conditions of convergence modes, the parameters of the elements of the circuit and ring structures are specified for each voltage class in the reduced units with respect to the base power S_{bas} = 10000 MVA. The value and direction of the exchange power were set by reducing the volume of generation in the receiving part of the power system with a balanced load of intermediate power systems.

Synthesis of the measuring body of emergency automation to prevent violations of the stability of intersystem transmission lines is feasible on the example of the design scheme figure 1. In this case we assume that the interconnection 500 kV double circuit is made by overhead lines. Taking into account stated earlier, means of emergency automation in this scheme are placed on substations of 500 kV (buses of knots 501, 502 and 503, figure 1) of intersystem transit 503-501.

IOP Conf. Series: Earth and Environmental Science **337** (2019) 012047 doi:10.1088/1755-1315/337/1/012047

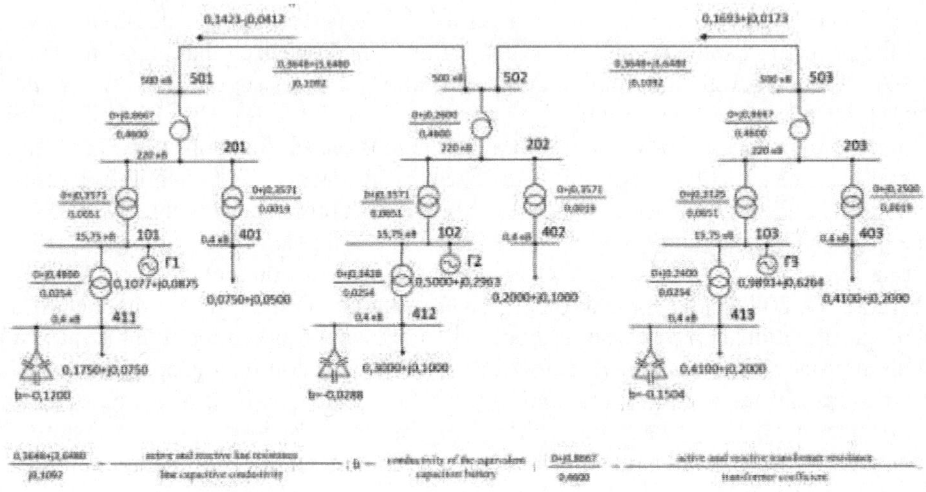

Figure 1. The simplified scheme is equivalent scheme of the power pools chain structure.

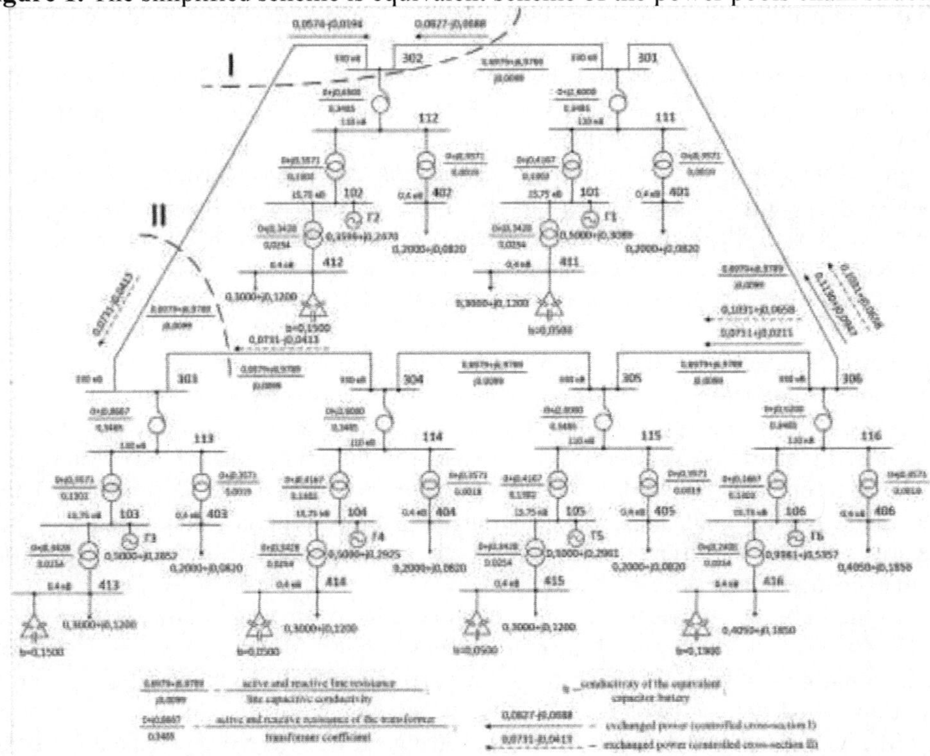

Figure 2. The simplified scheme is equivalent scheme of the power pools ring structure

Thus, the assessment of stability of intersystem transit of 500 kV of the three-machine settlement scheme of power connection (figure 1) is made as a result of calculation of the Jacobian (determinant) of the matrix of coefficients formed according to (1).

$$J=\begin{bmatrix} \frac{\partial P_{501}}{\partial U_{m\,501}} & \frac{\partial P_{501}}{\partial \gamma_{u_{501}}} & \frac{\partial P_{501/502}}{\partial U_{m\,502}} & \frac{\partial P_{501/502}}{\partial \gamma_{u_{502}}} & 0 & 0 \\ \frac{\partial Q_{501}}{\partial U_{m\,501}} & \frac{\partial Q_{501}}{\partial \gamma_{u_{501}}} & \frac{\partial Q_{501/502}}{\partial U_{m\,502}} & \frac{\partial Q_{501/502}}{\partial \gamma_{u_{502}}} & 0 & 0 \\ \frac{\partial P_{502/501}}{\partial U_{m\,501}} & \frac{\partial P_{502/501}}{\partial \gamma_{u_{501}}} & \frac{\partial P_{502}}{\partial U_{m\,502}} & \frac{\partial P_{502}}{\partial \gamma_{u_{502}}} & \frac{\partial P_{502/503}}{\partial U_{m\,503}} & \frac{\partial P_{502/503}}{\partial \gamma_{u_{503}}} \\ \frac{\partial Q_{502/501}}{\partial U_{m\,501}} & \frac{\partial Q_{502/501}}{\partial \gamma_{u_{501}}} & \frac{\partial Q_{502}}{\partial U_{m\,502}} & \frac{\partial Q_{502}}{\partial \gamma_{u_{502}}} & \frac{\partial Q_{502/503}}{\partial U_{m\,503}} & \frac{\partial Q_{502/503}}{\partial \gamma_{u_{503}}} \\ 0 & 0 & \frac{\partial P_{503/502}}{\partial U_{m\,502}} & \frac{\partial P_{503/502}}{\partial \gamma_{u_{502}}} & \frac{\partial P_{503}}{\partial U_{m\,503}} & \frac{\partial P_{503}}{\partial \gamma_{u_{503}}} \\ 0 & 0 & \frac{\partial Q_{503/502}}{\partial U_{m\,502}} & \frac{\partial Q_{503/502}}{\partial \gamma_{u_{502}}} & \frac{\partial Q_{503}}{\partial U_{m\,503}} & \frac{\partial Q_{503}}{\partial \gamma_{u_{503}}} \end{bmatrix} \quad (1)$$

where P_{501}, Q_{501} - respectively, the active and reactive components of the analytical signal power node 501, measured in [W] and [var]; U_{m501}, $\gamma_{u_{501}}$ - respectively, the instantaneous amplitude and the instantaneous angle of the voltage controlled on the buses of the node 501, measured in [V] and [degree];$P_{501/502}$, $Q_{501/502}$ - active and reactive components of the analytical power exchange signal in section 501-502, controlled by node 501 and measured in [W] and [var].

Explanations for the values of the system of equations (1) are given for one of the combinations (501 and 502) of the indices of the nodes of the calculation scheme in figure 1. The physical meaning of the other node and the exchange capacity is completely analogous.

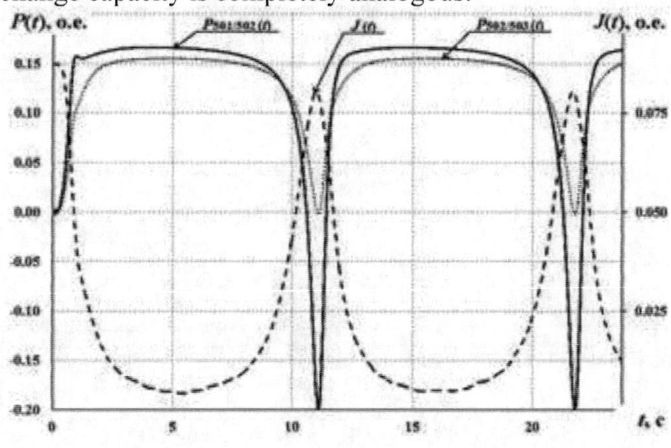

Figure 3. Changing the criteria J (dashed line) and active capacities $P_{501/502}$ (solid line), $P_{502/503}$ (dashed line) in intersystem sections 501-502, 502-503

Figures 3 and 4 (using the example of a chain circuit with controlled system interconnections 501-502 and 502-503) show the results of studies of dynamic regimes of IPS and the revealed regular patterns of the change in the generalized criteria - the dynamic determinant J(t). As a result of research it was established that the boundary of the oscillatory stability area of extended power pools corresponds to the minimum of the dynamic determinant J(t), which is always in the range of the angles of conditional stability (Figure 4, dashed line).

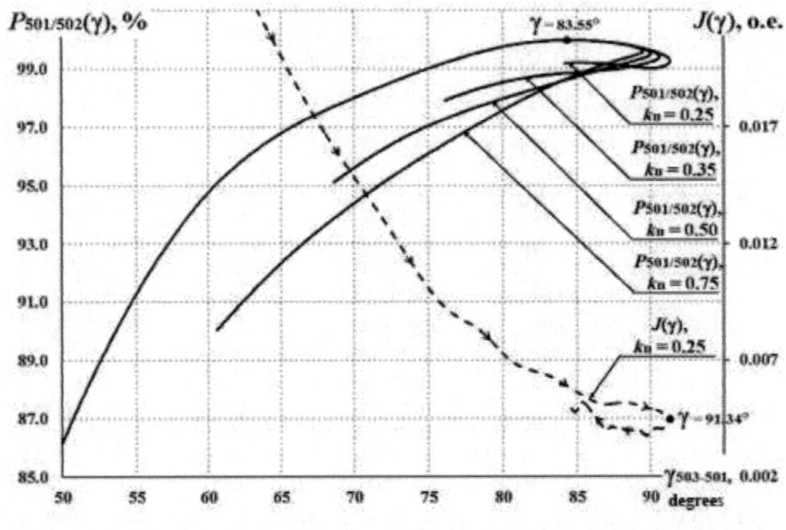

Figure 4. The Jacobian variation J (dashed line, k_p = 0.25 p.u.) and active power $P_{501/502}$ (solid lines) in coefficient k_p variation of the emergency response on unloading in the receiving part of IPS

The maximum of the generalized criterion characteristic $J(t)$ (Figure 3, dashed line) is located in the range of angles γ = 270-280° and corresponds to the limiting negative active power in the controlled intersystem section 501-502. The dependences of the criteria J and the angular characteristics of the controlled section power for the volume variation (k_p = var) of its emergency shutdown are shown in Figures 4, 5.

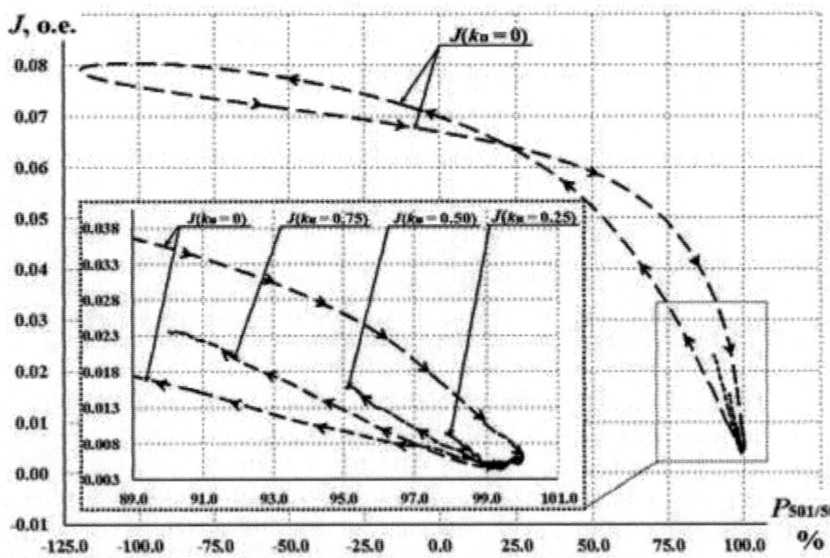

Figure 5. Characteristics of the Jacobian variation J (dashed lines) in coefficient kp variation of the emergency response on unloading in the receiving part of IPS.

Taking into account the generalized criteria for the violation of stability, the equality to zero of the of the determinant ($dJ(t)$ = 0), the thesis is proposed and, as a result of analytical and numerical studies, the law of emergency control of the exchange capacity (dP) of power systems with the control of the change in the generalized stability criterion $J(t)$:

$$dP = \frac{J}{1 + pT_S} \cdot \left(k_p + \frac{k_{\text{int}}}{pT_{\text{int}}} \cdot \frac{p}{1 + pT_{\text{diff}}}\right) \tag{2}$$

where T_S - time constant, taking into account the complete delay of the introduction of emergency response, sec; k_p, k_{int} - respectively, the coefficients of proportional and integral channels of emergency power control, p.u.; T_{int} - is the integration time constant taken as 0.2 c; T_{diff} - is the time constant of the differential link equal to 5 msec.

As a result of the equation roots research (2), it is shown that in the absence of changing the sign of the criterion characteristic (*J*), the emergency control system will be stable for positive values of the coefficients k_p, k_{int} of the amplification of the proportional and integral component of the control signal:

$$C_{1,2} = sign(J) \cdot \frac{-\left(\frac{1}{T_{\text{diff}}} + \frac{1}{T_S}\right) \pm \sqrt{\left(\frac{1}{T_{\text{diff}}} + \frac{1}{T_S}\right)^2 - 4 \cdot \left(1 + \frac{k_{\text{int}}}{k_p \cdot T_{\text{int}}}\right) \cdot \frac{1}{T_{\text{diff}} \cdot T_S}}}{2} \tag{3}$$

Dropping of delay ($T_S = 0$) in the structure of the emergency control does not introduce qualitative changes into the stability assessment of the pro-posed law of emergency control of extended power pools.

The development and implementation of the above-described APNU de-vices of extended power pools will allow to manage the congestion of intersystem power transmission lines (with reserves of up to 2-5%) effectively and safely, while ensuring their stability.

3. Fundamentals of adaptive self-adjusting means of protection and emergency control of electrical power equipment of electrical systems.

According to modern estimates, the degree of reactive power compensation in backbone networks is on average no more than 50% for 500-1500 kV lines [1, 3, 25-29]. The acuteness of this problem has increased even more recently due to a certain decrease in power consumption and, as a result, a decrease in active power flows with a significant undercompensation of the reactive power of extended electrical networks. Due to this situation, most of the large integrated power systems (IPS) have a very difficult situation with reliability and sensitivity of the main (differential) protections of electrical power equipment of the backbone network of 500-1500 kV - the minimum operating current is about 0.7-1.0 p.u. In addition, the problem of reliability of modern microprocessor differential protection systems is exacerbated by the large nomenclature and variety of used devices (half-sets), as well as the insufficient experience (lifetime) in operation of digital protections and the absence of a scientifically grounded approach to the selection (calculation) of their response parameters.

In connection with the abovementioned negative factors, cases of non-selective protection operation, accompanied by overloading of the lower voltage class lines entering the controlled sections, have become more frequent, which, eventually can cause a breakdown in the stability of the parallel operation of power pools.

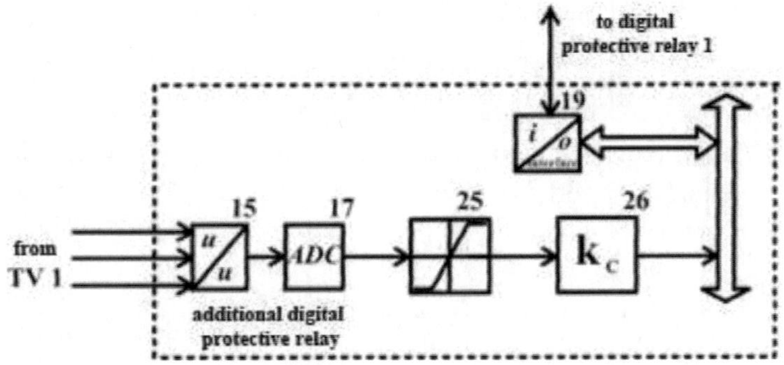

Figure 6. Generalized diagram of correction signals formation.

To ensure the selectivity, increase the speed performance and sensitivity of the improved current lengthway differential protection of power equipment inter-system (interstate) power transmission lines, the adaptive correction of the response characteristic must be performed (Figure 6). Approbation and evaluation of the efficiency of correction of the response characteristic of the current lengthway differential protection is performed using a database of digital oscillograms of full-scale experiments on the study of non-stationary modes of electrical power equipment of extended power pools.

Figure 7 shows the conditions of normative sensitivity and selectivity, which corresponds to the space of the protection operation parameters for which the white surface (the vector function of the protection operation parameters) prevails over the shaded for the damaged phases (phase A, figure 7, a). The selective operation of the measuring elements of undamaged phases (phase C, figure 7, b) is characterized by opposite conditions for the arrangement of surfaces.

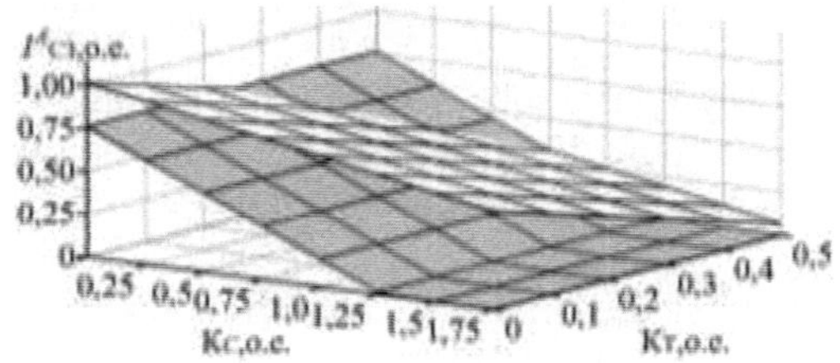

a) protection tripping current of phase A

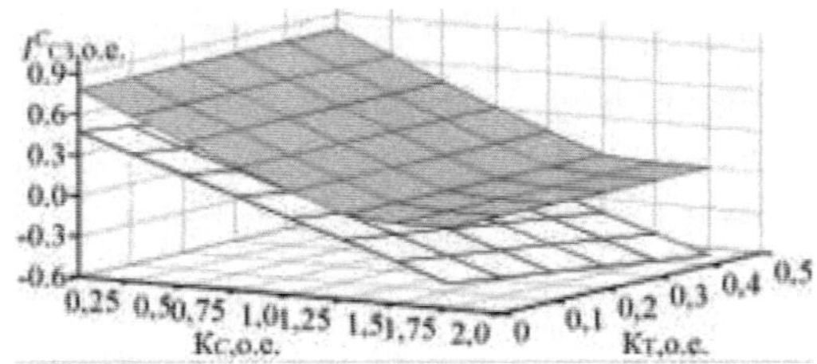

b) protection tripping current of phase C

Figure 7. Protection tripping currents of phases A (a) and C (b) in the space of parameters KT and KC when the line is switched to idle (shaded surface) and in short-circuit mode of phases A and B (white surface).

Taking into account the complexity of the perception of new principles for calculating the triggering parameters of current differential protections, the author, in the concluding part of the fourth chapter, presents a mathematical description of the method for selecting the operating values ($I_{prot\ min}$, K_T, K_C) of advanced current lengthway differential protections with dynamic correction of their operating and brake signals. Then, the vector functions of the selectivity of protection in idle mode $W_{I_{\text{prot select idle}}}$ and short circuit mode $W_{I_{\text{prot select. fault}}}$ (Figure 7) are described by expressions (with explanations only for phase A):

$$W_{I_{\text{prot select idle}}} = I_{\text{prot min}} - max \left\| \begin{matrix} I^A_{\text{prot idle max}}(K_C) \\ I^B_{\text{prot idle max}}(K_C) \\ I^C_{\text{prot idle max}}(K_C) \end{matrix} \right\| \geq 0, \quad (4)$$

$$W_{I_{\text{prot select fault}}} = min \left\| \begin{matrix} I^A_{\text{prot fault min}}(K_T, K_C) - I^A_{\text{prot idle max}}(K_C) \\ I^B_{\text{prot fault min}}(K_T, K_C) - I^B_{\text{prot idle max}}(K_C) \\ I^C_{\text{prot fault min}}(K_T, K_C) - I^C_{\text{prot idle max}}(K_C) \end{matrix} \right\| \geq 0, \quad (5)$$

where $I^A_{\text{prot idle max}}$, $I^A_{\text{prot fault min}}$ - the maximum and minimum values of the operating current of phase A protection in the idle and short circuit modes, p.u., respectively.

The requirements for protection according to the conditions of a given sensitivity ($K_{sense\ set}$) relative to the target set point ($I_{prot\ min}$) are described by a vector function:

$$W_{I_{\text{prot sense fault}}} = \frac{I^{min}_{\text{prot fault min}}(K_T, K_C)}{I_{\text{prot min}}} - K_{\text{sense set}} \leq 0, \quad (6)$$

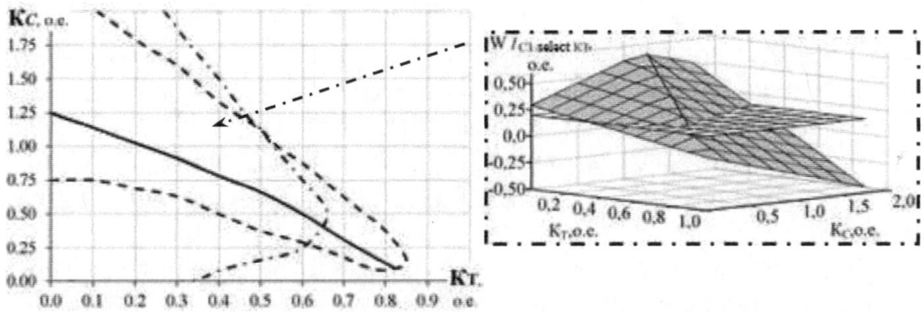

Figure 8. The selectivity area (dot-dash line), the sensitivity area (dashed line) and the characteristic (solid line) of the maximum selectivity and sensitivity of protection in the space of the braking parameters (KT) and correction (KC).

Absolute selectivity of the protection, characterized by zero value of the minimum operating current (Iprot min = 0), is achieved with 100-120% correction of operating signals (Figure 8, small area K_T = 0-0.2 p.u.). From the analysis of figure 8 it follows that the normative sensitivity is achieved with the braking coefficient $K_T \leq 0.2$ p.u. and the correction coefficient K_C, equal to 0.70-1.25 p.u. The increase of the braking coefficient more than K_T = 0.35-0.40 p.u. leads to a narrowing of the selectivity area (Figure 8).

4. Conclusion

The research and development carried are aimed at improving the basic technical high-response means of emergency control of power systems that ensure the continuity of their operation, the quality of electrical energy, economy, technical, information security and automation of its production.

1. The development and implementation of the above-described APNU devices of extended power pools will effectively and safely manage the congestion (with reserves of up to 2-5%) of intersystem power transmission lines, while ensuring their stability. The developed means of APNU should be classified as adaptive, since the measuring element is synthesized on the basis of the control of the generalized stability criteria, which reflects the dynamic properties of the power pools of a free structure.
2. The basis of adaptive means of differential protection of power equipment of power systems is developed. They are based on the use of adaptive correction of operating and brake protection signals. Self-tuning of protection means is achieved as a result of applying gradient parametric numerical methods for identifying the parameters of power equipment that affect the value of the minimum operating protection current. Absolute selectivity of protection, characterized by zero value of the minimum operating current (IC3 min = 0), is achieved only by correction of operating signals in the volume of 100-120%.

The research was conducted with the financial support under contract Peter the Great Saint-Petersburg Polytechnic University (SPbPU) No. 13.8886.2017/warhead "Improvement of the theory of working processes of production, conversion and transport of energy and the development of scientific bases of design of highly efficient thermal power plants, energy and transport machines, systems and complexes".

References

[1] Zakharova E 2016 *Kriteriy ekspress-otsenki staticheskoy ustoychivosti obyedinennykh energosistem : avtoref.dis....kand.tekhn.nauk* (SPb: Sankt-Peterburgskiy politekhnicheskiy universitet Petra Velikogo)

[2] Lyubarskiy D 2006 *Povysheniye ustoychivosti funktsionirovaniya ustroystv releynoy zashchity i protivoavariynoy avtomatiki v usloviyakh elektromagnitnykh i elektromekhanicheskikh perekhodnykh protsessov*: dissertatsiya ...doktora tekhnicheskikh nauk (Moskva)

[3] Popov M 2018 *Sovremennye sredsva protivoavariynogo upravleniya obyedineniykh energosistem*: dis.... d-ra.tekhn.nauk (SPb)

[4] Ragozin A 1998 *Obobshchennyy analiz dinamicheskikh svoystv energoobyedineniy na osnove strukturnogo podkhoda*: dis.... d-ra tekhn.nauk (SPb)

[5] Akke M 1997 *Some Control Applications in Electric Power Systems.* (Lund: Doctoral Dissertation. Department of Industrial Electrical Engineering and Automation, Lund University, Sweden)

[6] Martinez E 2002 *Power Systems and Communication Systems Infrastructures for the Future, Beijing, People's Republic of China*

[7] Ragozin A 2002 *Power of Russia* **2**

[8] Ragozin A 2002 *Power of Russia* **10**

[9] Ellithy K, Said S and Kahlout O 2014 *Int. J. of Electrical Power and Energy Systems* **63** pp 933–9

[10] Ilango G S, Nagamani C, Sai A V S S R and Aravindan D 2009 *Electric Power Systems Research* **79** pp 595–605

[11] Mahmud M A, Hossain M J and Pota H R 2014 *Int. J. of Electrical Power and Energy Systems* **58** pp 57–63

[12] Perez S 2006 *Modeling relay for power system protection studies*. Doctoral Dissertation. Department of Electrical Engineering (Saskatoon: University of Saskatchewan Saskatoon, Canada)

[13] Phadke A G and Kasztenny B 2009 *IEEE Transactions on Power Delivery* **24** pp 89–95

[14] *Metodicheskiye ukazaniya po ustoychivosti energosistem, utverzhdennyye prikazom Minenergo Rossii* 30.06.2003 № 277

[15] Mokeyev A 2011 *Metody i algoritmy obrabotki signalov v intellektualnykh elektronnykh ustroystvakh energosistem razlichnogo funktsionalnogo naznacheniya*: dis.... d ra tekhn.nauk (SPb)

[16] *Obshchiye trebovaniya k sistemam protivoavariynoy i rezhimnoy avtomatiki.releynoy zashchity i avtomatiki.telemetricheskoy informatsii.tekhnologicheskoy svyazi v EES Rossii*.Prilozheniye 1 k prikazu №57 OAO «RAO EES Rossii» ot 11.02.08

[17] Standart OAO RAO «EES Rossii» ot 26.07.05 *Pravila predotvrashcheniya razvitiya i likvidatsii narusheniy normalnogo rezhima elektricheskoy chasti energosistem*

[18] Bessolitsyn A 2010 *Polevyye metody opredeleniya elektricheskikh parametrov vozdushnykh liniy elektroperedachi* avtoref.dis.... kand.tekhn.nauk (SPb: SPbGPU)

[19] Popov M, Zakharova E, Sinyanskiy I and Goryachevskiy K 2015 *Elektricheskiye stantsii* **11**

[20] Vanin V, Popov M, Popov S and Ambrosovskaya T 2015 Elektricheskiye stantsii **11**

[21] Mokeev A et al 2011 *Digital Filters, InTech*

[22] Vanin V K, Ambrosovskaya T D, Popov M G and Popov S O 2016 *Power Technology and Engineering* **50** pp 87–92

[23] Jia Q Q, Dou C X, Wang C, Wang N, Tian J and Bo Z Q 2010 *Int. J. of Electrical Power and Energy Systems* **32** pp 416–20

[24] Castellanos B R, Calderón G J G, Olguin S D, Sarmiento U H and Messina A R 2006 *Electric Power Systems Research* **76** pp 169–79

[25] Popov M 2018 *Nauchno-tekhnicheskiye vedomosti SPbPU. Estestvennyye i inzhenernyye nauki* **24.1**

[26] Kholodnykh P V 2017 *Proceedings of the 2017 IEEE Russia Section Young Researchers in Electrical and Electronic Engineering Conference, ElConRus 2017* pp 896–900 doi:10.1109/EIConRus.2017.7910699

[27] Bolotin A E, Bakayev V V and Vazhenin S A 2015 *J. of Physical Education and Sport* **15** pp 724–9 doi:10.7752/jpes.2015.04110

[28] Belyaev N, Egorov A, Korovkin N and Chudny V 2018 *E3S Web of Conferences* **58** doi:10.1051/e3sconf/20185801010

[29] Vavilov D, Melezhik A and Platonov I 2015 *IEEE International Conference on Consumer Electronics - Berlin, ICCE-Berlin* pp 3–6 doi:10.1109/ICCE-Berlin.2014.7034262

IOP Conf. Series: Earth and Environmental Science **337** (2019) 012048 doi:10.1088/1755-1315/337/1/012048

Serious Games as innovative tools in HR policy

D S Bylieva[1,*], V V Lobatyuk[1] and T A Nam[1]

[1]Peter the Great St. Petersburg Polytechnic University, Polytechnicheskaya st., 29, Saint-Petersburg, 195251, Russia

*Corresponding author: bylieva_ds@spbstu.ru

Abstract. In the life of modern society information and communication technology occupy a significant place, being an active principle of social changes. Practically all processes, one way or another, go along with technologies. The total penetration of information and communication technologies into all spheres of life initiates the transformation of established social relations and norms. Particular attention we pay to the study of social norms in virtual reality. We observe the main forms of deviant behavior in the network, and the mechanism of social control, aimed at their regulation. At the same time, we prove that the areas of regulation are quite mobile and do not have a clear framework. Although we try to define three main areas of this type: area of law regulation, area of moral regulation, unregulated field. All the presented areas describe the visible phenomena. The essence of the virtualization process, however, turns out to be undetected. But after all, society itself can set a socially acceptable framework for the functioning of information and communication technologies, which can vary depending on religion, national circumstances or level of economic development. As a result, it can be noted that in the modern world the social landscape is changing dramatically.

1. Introduction

The increasing role of Information and Communication technology changes people's life in all areas of human activity dramatically [1]. Having become an important part of the global Internet space, game universes enter into daily existence of people [2]. Currently there are 2.3 billions of active gamers all over the world, i.e. people who are considered as regularly playing computer or videogames for about thirteen hours per week on average [3]. But there are gamers who are not satisfied by having been playing for several hours a day, for instance there's more than six million of people in China that spend at least twenty two hours a week. In Great Britain, France and Germany above ten million of people spend minimum twenty hours a week for this activity and more than five million of "extreme" gamers play for forty five hours weekly on average in the USA [4].

The appearance of a phrase "serious games" that initially sounded as a paradoxical collocation means that games are playing more significant role for society, but not limited to the sphere of entertainment. A game framework facilitated achieving serious goals faster and with pleasure. The first definition of the term "serious game" belongs to Abt who claimed that these games have an explicit and carefully thought-out educational purpose and are not intended to be played rimarily for amusement [5].

Traditionally, it's supposed that the US army was the first who used games for serious purposes during World War II in order to improve its image among the people. Nevertheless, the early stage of the games with serious intentions was dominated by educational ones. Sometime later the scope of serious games applications has expanded tremendously. In 2018 Çiftci conducted a study of scientific

IOP Conf. Series: Earth and Environmental Science **337** (2019) 012048 doi:10.1088/1755-1315/337/1/012048

publications devoted to serious games (dated from 2007 and later in) and found some researches in such fields as psychology, health care sciences, environmental sciences, ecology, public environmental occupational health, rehabilitation, business economics and psychiatry and other disciplines [6]. Alvarez and Michaud identified seven serious Games markets: Defense, Training & Education Games, Advertising, Information & Communication, Health, Culture, Activism [7]. Djaouti, Alvarez and Jessel expand the list: State & Government, Military & Defense, Healthcare, Education, Corporate, Religious, Culture & Art, Ecology, Politics, Humanitarian, Advertising, Scientific Research [8]. Laamarti, Mohamad and Saddik distinguish Education and Training, Well-Being, Advertisement, Cultural Heritage, Interpersonal Communication, Bio-medical and Health Care [9]. However in the sphere of HR policy serious games have been spread not so long ago.

2. Methodology

A thorough analysis of hundreds of games allowed us to define goals, applications and peculiarities of construction of games in the field of personnel policy. Research methods are represented by a set of different approaches of theoretical and applied scientific knowledge, providing the implementation of analytical tools. In particular, methods of system, functional, complex and process comparative analysis and synthesis, grouping of actual data.

In this study an interdisciplinary approach was used, which allowed to consider innovative tools in HR policy as a complex social phenomenon in the perspective of the theory of organization based on the achievements of Economics, sociology and psychology. The analysis of several hundred serious games allowed to define the purposes of application and peculiarities of game construction and to make their categorization in the field of HR policy.

We have summarized various approaches that reveal the features of attracting potential employees, adaptation of newcomers, engagement and motivation, personnel selection, training and education of employees with the help of serious games in the conditions of modern reality. We have also used such methods of scientific research as the construction of logic frameworks, graphical interpretation of theoretical information and empirical data.

3. Results and Discussion

In terms of areas of use games could be divided into five categories in HR management: 1) staff education and training; 2) attracting prospective manpower; 3) employee selection; 4) adaptation of newcomers; 5) engagement and motivation. The figure 1 shows games' characteristics that are defined as key features according to analyses of corresponding spheres of implementation in HR policy.

3.1. Games for staff education and training.

This is the most extensive area of games used for staff education and training. The history of this kind of game development in HR stemmed from 1991 when recruits were taught how to drive the tank with the help of the game "The Bradley Trainer" in US armed forces.

In the category of educational games there are many specific goals: hard and soft skills training, teambuilding, management, diversity-related conflict management, leadership, ensuring information security of the company, the rules of behavior in the workplace, including good habits in terms of ecology.

Depending on employer's requirements of staff acquisition of special knowledge and skills the educational games could get various forms. The simplest principle of learning is based on the gamification of the educational/controlling process. For instance, correct answers to questions can be visualized as climbing a mountain ("Guru" is used when launching the complex digital marketing tool "Mobile Connect" at "Exact Target") or as a good monster hit (in the game "Coaching Champion" of the company "Designing digitally" for coaching skills development of managers).

The majority of games aimed at developing specific professional skills. For example, in the company Proplex the game helps to work out the skill of ongoing dialogue with clients, in the chain of American stores "Giant Eagle" cashiers work out routine activities thanks to the games. And some more facts, the game "True Office" in the field of financial services provides compliance training, Bankers Lab specializes in building award-winning "flight simulators" for bankers.

It should be noted that a special and wide category of staff training games are business games, which are part of the managers' training practice, and can be used in the staff training process in different enterprises. Such kind of games place people into a virtual economic environment and help them "to acquire some economic and financial skills" [10]. Most commonly in these games traditional business procedures could be practiced: procurement, controlling, production, sales and marketing (e.g. Business Model Gaming, Up Start). Some of the games emphasize logistics (e.g. Beer game). J.k.h. Riedel, J. B. Hauge underscore the need of use of multidisciplinary and multi-person games at the enterprise for hard and soft skills development [11]. But games with the aim to develop soft skills appeared much later than the classic business games and they can often contain a large amount of entertainment content. For example, communication skills can be practiced in the game "Merchants", where the player learns to negotiate and resolve conflicts while playing the role of the Venetian merchant of the 15th century. And in the company Slalom Consulting gamification includes a special game to remember names, faces, and technical expertise of coworkers on the floor.

A new trend is the development of games related to ethics and environmental behavior. The game Simulated Ethics Serious 2D Game from Designing Digitally demonstrates situational harassment topics in 2D from the point of view of the offender and the victim, helping to master the norms of behavior.

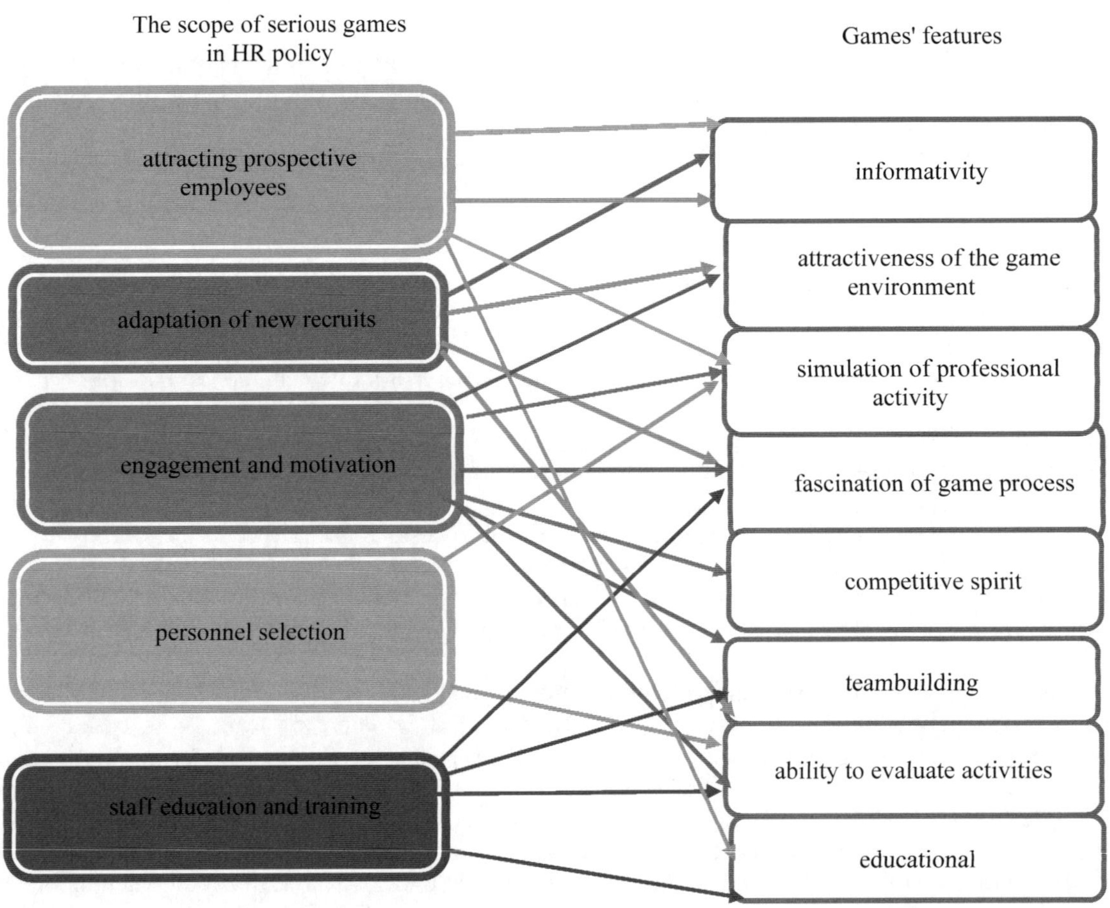

Figure 1. Key features of serious games in different spheres of HR policy.

3.2 Games for attracting prospective manpower.
American Army's first-person shooter game was created in 2002. It includes training and military missions and it can be recognized as one of the first games designed to attract potential employees. The game was distributed for free and was named "the first successful and well executed serious game that gained total public awareness" [12].

Games can be a great way to create an image of an organization. however, educational games can sometimes serve the same purpose [13].

In this context, it is interesting that some games could combine several goals at the same time, which is certainly beneficial for the company and allows to create a high-level game. The clear example is My Marriott Hotel (the game in Facebook), where you can try to work in the hotel, and this can both attract potential employees and give them an idea about the features of the work, and also it can be used training process of new recruits. The game Plantville from Siemens is used for the same purposes due to training such activities as plant management, financial and production processes management, purchase and installation of Siemens equipment and others.

Games that serve to attract and select candidates often allow the player to get feedback.

So playing Supply Chain, Finance, Sales and Marketing, you can get an estimate of your success in each of the sections.

3.3 Games for the employee selection
The most difficult thing is the specification of criteria for assessing gaming activities in the process of games development for staff selection. And the higher the required qualification, the more difficult this task becomes. For instance, the game Wasabi Waiter (California) perfectly copes with the assessment of potential waiters on the criteria of speed, quality and empathy to virtual customers. The game The Secret Service (the United Kingdom) offers ooperations officer to work in a situation of lack of time to assess such qualities as judgment and decision-making. And for the selection of qualified personnel the company Connect Cubed (USA) initially tests the existing employees of the company, which are the "standard" of the necessary qualities, and then selects the candidates which are most appropriate to this profile.

Wibi Data uses another creative approach in seeking for people who share the ideals of the company. The company deliberately added new levels in the game "Portal 2", where the company presents its virtual office. In order to send resume, it is necessary to do all the game tasks and get a password.

3.4. Games for adaptation of newcomers (new recruits)
We have singled out the category of games which associated with the adaptation of newcomers to the enterprise, although this function can often be considered as optional when attracting or training personnel. But some big companies are developing purposefully adaptive games. For example, a large Russian company "Gazprom Neft" began gamification of work with personnel with the development of the game - a virtual tour, quests for beginners, allowing to get acquainted them with the various divisions of the holding (Corporate center, oil field, refinery, petrol station and so on) and understand the business peculiarities).

3.5. Games for an engagement and motivation
Motivational games set goals that are most difficult to quantify. In this area it is impossible to do separate game elements. Most commonly, a whole stimulating communicative game system is developed, which allows to link on-line and off-line activities, interact and create ratings of employees. For instance, the company Enter uses a motivational game "OlimpiaDa" (which is connected with intranet) that allows to transfer not only the success of employees, but also participation in various social activities into a unified game rating in the field of immediate results (e.g. the recommendations of experts in the company, conducting/ participating in training workshops, a help to newcomers, adopted innovative ideas).

Moreover, the game is quite flexible, its rules are modified in accordance with the current goals of the company: the most demanded activities are encouraged by large points. The most popular universal online system of non-financial incentives is "Gingerbread" in the field of motivation in Russia. The virtual "Gingerbread" is a virtual thanks, the way of to thank colleagues (necessarily reasoned), as well as a virtual currency that can be exchanged in the store for the benefits offered by the company. In addition, the system provides for the distribution of badges, which can be assigned automatically according to the results or manually by managers, as well as work with the rating associated with the performance of tasks [21-25].

If the work itself takes place in a computer environment, the motivational system can be tied to the direct execution of tasks. There's an interesting experience when turning work with bugs in the Russian social network Odnoklassniki in the game: the so called "box bugs" were compiled accompanied by photos of complaining users, each of this was supported by a description of pumped skills in the process of elimination of these bugs. Accordingly, the task was the maximum "pumping" skills.

However, experiments to transfer responsibility for the control of employees' activities to artificial intelligence (for example, in the Russian office of Phobos) show that it is impossible to automate the assessment of all aspects of activity even working in the field of computer technology.

One of the issues related to the game motivation system is the question of the final rewards of employees with the highest ratings, which is carried out off-line in the majority of the considered games. For example, in the game "Alfa-city", that is implemented by the Russian Bank "Alfa Bank", a player can spend the points received for the successful completion of working tasks on the construction of buildings.

4. Conclusion

The technologies developed in the field of personnel motivation demonstrate the direction of future development in the field of gamification in HR.

Interactive formation of a stimulating communication system that demonstrates success in work, training, social activities and other encouraged actions allows employees not only to clearly see the possible goals and prospects of their own development, but also the general logic of the organization. As a consequence there is a virtual gamified model of the organization. Merging such a system with a game training will bring personnel management to a new level, which allows to track the weaknesses of the workforce and eliminate them in a proper time.

References

[1] Shipunova O D, Berezovskaya I P, Mureyko L M, Evseev V V and Evseeva L 2018 *ESPACIOS* **39**(40)

[2] Shipunova O D, Berezovskaya I P, Mureyko L M, Kolomeyzev I V Serkova V V 2017 Cultural Code in Controlling Stereotypes of Mass *Consciousness European Research Studies J.* **20**(4)

[3] Shipunova O D, Berezovskaya I P and Gashkova E M 2017 *St. Petersburg State Polytechnical University J. Humanities and Social Sciences* **8**

[4] Silkina G Yu 2017 *Proceedings of the 29th International Business Information Management Association Conference - Education Excellence and Innovation Management through Vision 2020: From Regional Development Sustainability to Global Economic Growth* pp 1165-76

[5] Didenko N I, Skripnuk D F and Mirolyubova O V 2017 *Proceedings of 2017 10th International Conference Management of Large-Scale System Development* p 8109611

[6] Babkin A V, Karlina E P and Epifanova N Sh 2016 *Proceedings of the 28th International Business Information Management Association Conference - Vision 2020: Innovation Management, Development Sustainability, and Competitive Economic Growth* pp 11-17

[7] Lichy J, Kachour M and Khvatova T 2017 *J. of Marketing Management* **33**(9)

[8] Korneychuk B 2018 *Digital Transformation and Global Society. DTGS 2018. Communications in Computer and Information Science* **858**

IOP Conf. Series: Earth and Environmental Science **337** (2019) 012048 doi:10.1088/1755-1315/337/1/012048

[9] Markeeva A V 2015 *Leadership and Management* **2**(3)
[10] Wijman T 2018 *Mobile Revenues Account for More Than 50% of the Global Games Market as It Reaches $137.9 Billion in 2018.* Available at: URL https://newzoo.com/insights/articles/global-games-market-reaches-137-9-billion-in-2018-mobile-games-take-half (accessed 4 November 2018)
[11] McGonigal J 2011 (NewYork: Penguin)
[12] Abt C C 1970 *Serious Games* (NewYork: TheVikingPress)
[13] Çiftci S 2018 *J. of Education and Training Studies* **6**(2)
[14] Alvarez J and Michaud L 2008 *Serious Games: Advergaming, edugaming, training and more* (France: IDATE)
[15] Djaouti D, Alvarez J and Jessel J P 2011 *Handbook of Research on Improving Learning and Motivation through Educational Games: Multidisciplinary Approaches* (Hershey, PA: IGI Global)
[16] Laamarti F, Mohamad E and Saddik A E 2014 *J. of Computer Games Tech.*
[17] Dufourny S and Dhaenens C 2016 *IFAC-PapersOnLine* **49**(12)
[18] Riedel J and Hauge J B 2011 *17th International Conference on Concurrent Enterprising* (Aachen:IEEE)
[19] Djaouti D, Alvarez J, Jessel J-P and Rampnoux O 2011 *Origins of serious games, Serious Games and Edutainment Applications* (Berlin: Springer)
[20] Fersman N G, Zemlinskaya T Ye and Novak-Kalyayeva L 2017 *Proceedings of the 30th International Business Information Management Association Conference, IBIMA 2017 - Vision 2020: Sustainable Economic development, Innovation Management, and Global Growth*
[21] Anfinogentova A A, Dudin M N, Lyasnikov N V and Protsenko O D 2018 Economy of Region **14** (2)
[22] Bonnechère B, Sholukha V, Omelina L, Jan S V S and Jansen B 2018 *Sensors (Switzerland)* **18** (7)
[23] Bonnechère B, Sholukha V, Omelina L, Van Vooren M, Jansen B and Jan S V S 2017 *Gait and Posture* **57**
[24] Bonnechère B, Sholukha V, Omelina L, Jansen B and Jan S V S 2016 *ACM International Conference Proceeding Series* pp 45-48
[25] Mavridis A, Leftheris I, Tsiatsos T and Kudryavtseva N 2012 *15th International Conference on Interactive Collaborative Learning, ICL 2012* p 6402169

IOP Conf. Series: Earth and Environmental Science **337** (2019) 012049 doi:10.1088/1755-1315/337/1/012049

Social Assessment of Innovations and Professional Responsibility of Future Engineers

E G Pozdeeva[1,*], O D Shipunova[1] and L I Evseeva[1]

[1]Peter the Great St. Petersburg Polytechnic University, Polytechnicheskaya st., 29, Saint-Petersburg, 195251, Russia

*Corresponding author: elepozd@mail.ru

Abstract. This article focuses on the social aspects of technological development in the digital age. The assessment of feasibility and effectiveness of implementing new engineering developments implies an assessment of the sociological context of their future use. The position of the authors arises from the general concept of socially responsible innovations. It is emphasized that the global nature of implementing digital technology requires an ethical approach in analyzing the impact of implemented engineering developments on economic, political and social relations. The formation of professional culture of the future professional implies the availability of skills of social assessment of innovations and the readiness to be held responsible for innovative activities.

1. Introduction

In a digital society, engineering activity features a high degree of system integration [1], it acquires a global nature and becomes the backbone of any activity. The specifics of the network culture of the information society changes the requirements to charging the intellectual potential of engineering activity [2, 3, 4]. Aspiring engineers, that constituted the foundations of the creative class of modern scientific and technological progress, should possess not only knowledge in their professional field, but also a high level of responsibility in relation to the prospect of implementing their developments. In this regard, the training of technical specialists should include a body of knowledge on applied ethics and ethics review [5]. The development of the discourse of ethics review based on technical and technological innovations under the control of competent specialists becomes the basis for sustainable development of the society [6, 7].

This article aims to conduct a sociological analysis of students' readiness for responsible engineering activities. The authors associate the responsibility with the integrative quality of the individual, which shows the ability of a person to be held responsible for its actions before oneself, society and the state.

The concept of "Responsible Research and Innovation" (RRI) has been updated in the context of the digital society [8]. In Russia, the research of this field is still in its infancy. The concept of “responsible research” implies a focus on proactive management, active participation of representatives of social groups and various agents, which determines the socio-humanitarian parameters of techno-scientific development [9].

From a perspective of responsible research, the concept of “Social Technology Assessment” (TA) has become widespread. Under social technology assessment it is understood an epistemological claim for systematic and multidimensional research and early recognition of the possible consequences of scientific and technological development on the basis of all available knowledge [10]. The goal of social assessment is to develop and implement innovations in a more acceptable, socially desirable and sustainable way.

Social technology assessment (TA) is interpreted as a theory of assessing the consequences of scientific and technological development based on all available knowledge of the natural, engineering and human sciences [11]. The essence of TA is captured in the Western European program "Responsible Research and Innovation" (RRI), where a three-element communication model has been formed, which includes politicians, experts and civil society [12].

The development of social assessment agenda led to the decision that "responsible innovations" are one of the conditions for the implementation of the European Union's innovation program "Horizon 2020" (Horizon 2020). In this program, responsible innovation refers to an approach that allows you to "anticipate and assess the possible impact and social expectations pertaining to scientific research and innovations in order to facilitate the development of inclusive and sustainable research and innovations" [13,14]. At the same time, an emphasis is made on the integral approach that combines a socio-humanitarian assessment of innovative projects, a techno-scientific paradigm, and applied ethics [15, P.10].

Sociological studies emphasize that students recognize the need to begin work while still studying, having made a choice in favor of entrepreneurship. However, the availability of intentions is of consequence in 30% of cases of further involvement of students in real activities [16]. In addition, an analysis of students' opinions shows the absence of such an important element as the responsibility of the individual towards oneself in the course of activities) [17].

2. Research Methods

The methodological background for studying students' readiness for responsibility and social assessment is the theory of planned behavior, according to which purposeful actions of an individual are to be preceded by the formation of intentions [18].

In this study we used the following empirical research methods: questionnaire survey, web interviewing (sample of 110 students), several sociological surveys conducted among the students of Peter the Great St. Petersburg Polytechnic.

The empirical base of the study is a sample among engineering professions amounted to 4 thousand respondents. Among the respondents to the survey, 68% of respondents have a clue of their future job. Moreover, only 64% of the surveyed students are going to be involved in a job connected with the received training specialty, 28% were undecided [19-25].

3. Results

The results of this exploratory study (web interviewing, sample of 110 students) allowed to analyze how the students understand the concept of professional responsibility and responsible behavior. Among the respondents, 51% agree that the responsibility of an individual for its actions increases. Students rated their capability to take risks as follows: almost half of the respondents (54%) are ready to take risks; 16% are not ready; 30% found it difficult to answer.

Based on the answers, the following groups have been identified:

1. Most of the students (69%) believe that professional responsibility is first and foremost, discipline, organized nature and ability to meet the deadlines.
2. Running second (about 50%) are the parameters as follows: independence, ability to use the available resources; yielding the desired result; ability to assess the consequences of their actions and behavior.
3. 44% of students consider high quality of work to be the essence of professional responsibility.
4. Getting achieved the objectives set by superior officers gained 31% of supporters among students.
5. Readiness to take risks as the essence of professional responsibility was noted only by 13% of respondents.

While analyzing the opinions of students, it is possible to distinguish specific strategies of responsible behavior:

- 46% of students tend to take all available opportunities and attract those who can help in achieving results;
- 21% of respondents always take up the matter on their own, get it done and bear responsibility for this;
- 18% of students choose a job according to a previously known algorithm;
- 15% of students learn, at first, the opinion of experts, weigh up everything, and then get down to work.

4. Conclusion

As can be seen from the above, students are characterized by a fairly clear strategy of responsible behavior, which meaning and model they form on their own. Among the main elements of professional responsibility, organizational and managerial skills, optimization approach and a trusting relationship prevailed. The ability to bear responsibility while getting things done was in the last place.

Modern engineering education should be aimed at the development of a complex of skills where an important place belongs to responsible attitude and the skills of social assessment of innovations. In order to ensure the readiness of future engineers for responsible research and social assessment of innovative developments it is essential to establish an interdisciplinary professional and educational environment.

References

[1] Gorokhov V G and Grunwald A A 2011 *Higher education in Russia* **5**
[2] Byl'eva D S 2016 *St. Petersburg State Polytechnical University. Humanities and Social Sciences* **2** (244)
[3] Bylieva D, Lobatyuk V and Rubtsova A 2018 *SHS Web of Conferences* **44** p 00021
[4] Shipunova O S, Berezovskaya I P, Mureyko L M, Evseev V V and Evseeva L I 2018 *ESPACIOS* **39**(40)
[5] Pozdeeva E G, Trostinskaya I R and Evseeva L I 2018 *St. Petersburg State Polytechnical University. Humanities and Social Sciences* **9**(3)
[6] Steghofer J 2017 *IEEE Technol. and Society Magazine* **36**(3)
[7] Lintern G and Kugler P N 2017 *Systems Engineering* **20**(4) pp 307-17
[8] Dobryakova M and Kotelnikova Z 2015 *Foresight Russia* **9**(1)
[9] Grebenshchikova E G 2015 *Hum. Scient. Research* **12**
[10] Seredkina E V and Chernikova I V 2015 *Bulletin of PNRPU. Culture Story. Philosophy. Right* **2**
[11] Makarova E N and Kolesova O V 2016 *Bulletin of PNRPU. Culture Story. Philosophy. Right* **2**
[12] Seredkina E V 2017 *Bulletin of PNRPU.Culture Story. Philosophy. Right* **2**
[13] Shilina M G 2017 *Mediascope* **2**
[14] Shipunova O D and Berezovskay I P 2017 *Civil Engineering and Technology* **9**(10)
[15] Grunwald A 2011 *Enterprise and Work Innovation Studies* **7**
[16] Bogatyreva K and Shirokova G 2017 *Foresight and STI Gov.* **11**(3)
[17] Emelyanova T P 2013 *Inf. Hum. Por. Knowledge. Understanding. Skill* **2**
[18] Ajzen I 1991 *Organizational Behavior and Human Decision Processes* **50**(2)
[19] 2017 *Polytech eyes of students: Education, science, politics*. Almanac of sociological research Ed by Pazinkina E
[20] Razinkina E 2018 *E3S Web of Conferences* **03043**
[21] Kuzmin A 2018 *Engineering for Rural Development* **17**
[22] Bogach N, Boitsova E, Lezhenin Y and Zhuikov A 2018 *Engineering for Rural Development* **17**
[23] Bolotin A E and Selyukin D B 2015 *Teoriya i Praktika Fizicheskoy Kultury* **32–34**
[24] Kasyanik P M, Gulk E B, Olennikova M V, Zakharov K P and Kruglikov V N 2017 *Advances in Intelligent Systems and Computing* **544**

[25] Kabanova N and Kogan M 2017 Lecture Notes in Computer Science (including subseries Lecture Notes in Artificial Intelligence and Lecture Notes in Bioinformatics) p 10295

IOP Conf. Series: Earth and Environmental Science **337** (2019) 012050 doi:10.1088/1755-1315/337/1/012050

Modelling of the developed system for wood waste gasification

A A Aizyzan[1], A V Fedyukhin[1], K V Strogonov[1], E L Shaburov[2*], S A Kravchenko[2], N B Kirillov[2], N V Nikitkov[2], Yu M Pankratov[2] and T R Akhmetov[3]

[1]National Research University "Moscow Power Engineering Institute", Krasnokazarmennaya 14, Moscow, 111250 Russia.
[2]Peter the Great St. Petersburg Polytechnic University, St. Petersburg, Russian Federation
[3]Kazan State Power Engineering University, Kazan, Russian Federation

* Corresponding author: shel7439245@mail.ru

Abstract. The aim of this work is mathematical modelling of the developed technological scheme of the combined-cycle gas generator unit. An efficient technological scheme for gasification of wood waste with electricity generation is proposed in the paper. Mathematical modelling of the developed biomass gasification scheme was carried out. The payback period of the proposed combined-cycle plant with gas generator is calculated.
Keywords: combined-cycle gas generator, mathematical modelling.

1. Introduction

During gasification, biomass usually includes wood materials, waste from logging and woodworking industries, agricultural organic waste from plant growing and animal husbandry, and agricultural processing industries [1-6]. Plant biomass is considered one of the most "noble" types of fuel and in many countries is considered as a promising source of energy for the near future. The annual reproducible potential of biomass is estimated to be 10 times higher than world mineral production. However, the availability and economic feasibility of using different types of biomass is different [7-11].

One of the main problems in regions with sufficient forest resource base is the inefficient use of wood waste. A huge amount of waste, representing valuable raw materials, remains unused. For their disposal large financial costs are consumed.

Recycling of wood waste shows a high energy potential for supply of industry and utilities, as well as economic potential.

2. The developed system for biomass gasification

Figure 1 shows the basic scheme of a combined-cycle plant based on biomass gasification [12]. Dried biomass and vapor-air mixture under pressure of 20 atm are fed into the fluidized bed gas generator. Gasification produces generating gas. Before being supplied to the gas turbine unit, the generator gas is subjected to cleaning from dust and ash in the filter. Exhaust gases after the gas turbine are sent to the waste-heat boiler, in which superheated steam is produced.

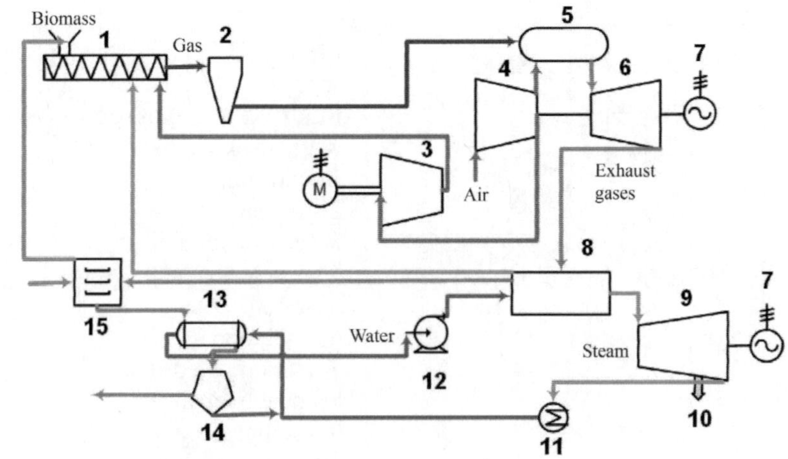

Figure 1. Gas-steam plant based on biomass gasification (basic scheme):
1 - gasifier,
2 - cyclone,
3 - booster compressor,
4 - air compressor,
5 - combustion chamber of gas turbine,
6 - gas turbine,
7 - generator,
8 - waste heat boiler,
9 - steam turbine,
10 - heat extraction,
11 - condenser,
12 - feed pump,
13 - condensation heat exchanger, 14 - separator,
15 - drying drum

The paper proposes a technological scheme for biomass gasification with a combined-cycle plant (Figure 2). The principle of operation of the scheme is similar to the basic scheme. Advantages of the developed scheme are the following:

- An increase in the energy efficiency of the circuit due to a decrease in energy consumption for own needs (absence of drying unit and additional air compressor for gas generator);
- Heating of compressed air in a new gas generator 4;
- Obtaining a certain pressure in the developed gas generator using an ejector 6.

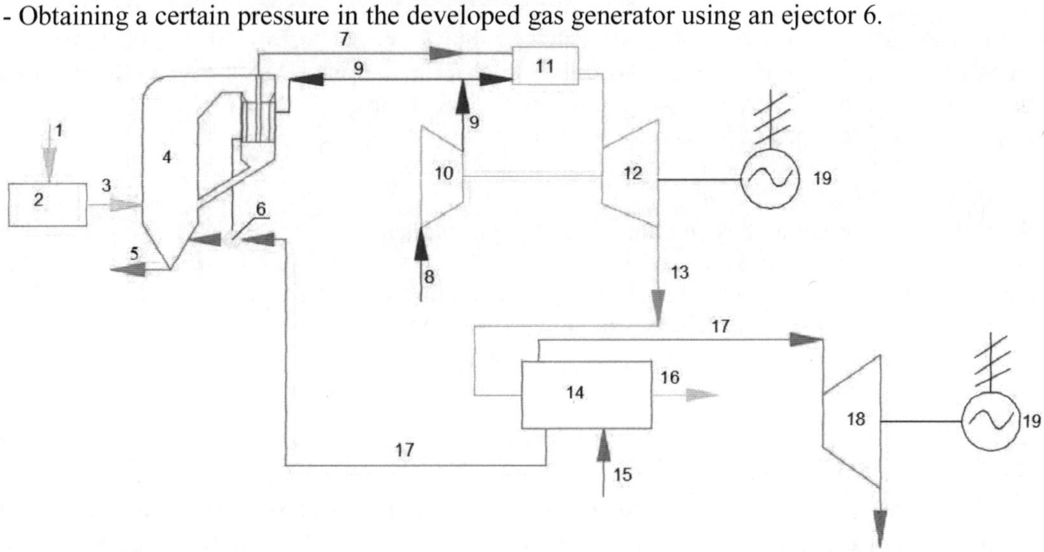

Figure 2. The developed technological scheme of the combined-cycle plant with a gas generator:
1 - biomass; 2 - crusher; 3 - crushed biomass; 4 - circulating fluidized bed gas generator with a heat exchanger; 5 - ash; 6 - ejector; 7 - generating gas; 8 - cold air; 9 - compressed air; 10 - compressor; 11 - combustion chamber; 12 - gas turbine; 13 - waste gases; 14 - waste heat boiler; 15 - feedwater; 16 - flue gases; 17 - superheated steam; 18 - steam turbine; 19 - generator.

IOP Conf. Series: Earth and Environmental Science **337** (2019) 012050 doi:10.1088/1755-1315/337/1/012050

An efficient circulating fluidized bed gas generator with a heat exchanger (Figure 3) has been developed.

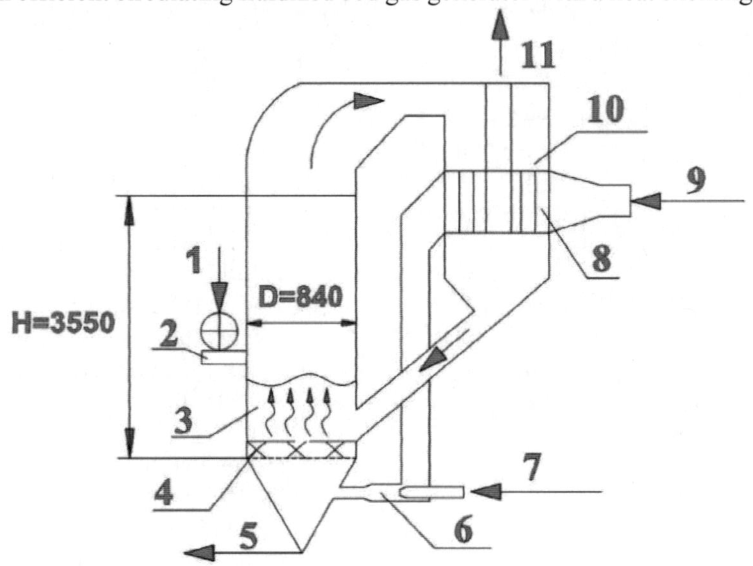

Figure 3. The developed scheme of the circulating fluidized bed gas generator with a heat exchanger:
1 - biomass;
2 - feeder;
3 - gasification zone;
4 - a layer of fuel;
5 - ash;
6 - ejector;
7 - superheated steam;
8 - heat exchanger;
9 - compressed air;
10 - cyclone dust collector;
11 - generator gas.

The features of the proposed gas generator are the following:

- The use of physical heat of the generator gas 11 in the heat exchanger 8;
- Increase of pressure of compressed air 9 in the ejector 6 using superheated steam 7;
- Cleaning of the generator gas from dust and ash in a vertical cyclone dust collector 10;
- Reduction of carbon loss in the fluidized bed due to recycling of captured particles of dust and ash.

3. Mathematical modeling of the developed system for biomass gasification

To assess the energy potential of the products of thermal conversion of wood waste, one can apply a mathematical model, which with sufficient accuracy allows one to perform the calculation without additional experimental research. The main results of such a simulation will be the composition, volume, and calorific value of the combustible gas at the outlet of the reactor.

To calculate the developed gas generator we took:

- Pressure in the gas generator $p = 27$ at,
- Air flow coefficient $\alpha = 0.3$;
- The amount of superheated steam mn = 0.2 kg/kg of biomass.

Assumption: all carbon enters the gasification process due to recirculation in the gas generator, ash leaves the gas generator with the temperature of the generating gas. The calculation was carried out in the Mathcad 15 software environment according to the method [13].

Table 1 presents the composition of wood waste per dry mass.

Table 1. Elementary composition of wood waste [12].

Sample	C_c %	H_c, %	O_c, %	N_c, %	S_c, %	A_c, %	W, %	Q_h^p, MJ/kg
Wood waste	49.75	6	43	0.5	0.05	0.7	10	16.3

Knowing the composition of biomass and the main thermal parameters of gasification, a calculation was made on the basis of the equation of the material and heat balance of the gas generator, as a result of which we obtained:

- Heat of combustion of generator gas $Q_{hc} = 5016.4$ kJ/m^3;
- Temperature of generator gas $t = 843$°C;
- Volume output of generator gas $V_{gg} = 2.7$ m^3/kg;
- Composition of generator gas (Table 2):

IOP Conf. Series: Earth and Environmental Science **337** (2019) 012050 doi:10.1088/1755-1315/337/1/012050

Table 2. Composition of gasification products of wood waste.

Components	CH_4	CO	H_2	CO_2	H_2O	N_2	SO_2
%	1.79	17.74	19.78	11.21	13.35	36.12	0.01

- Gasification efficiency: chemical is 83.81%, thermal is 97.65%.

As a result of thermal calculation of the gas turbine plant, we obtain: the electric power is 16 MW; efficiency is 36%.

The electric power of steam turbine is 8 MW, and the total electric power of the entire installation is 24 MW, while the efficiency of the combined-cycle plant is 53%.

In order to compare and verify the obtained results, mathematical modeling of the developed technological scheme for biomass gasification with a combined-cycle gas installation was carried in the Aspen Plus 8 software environment out according to the procedure [14-15] (Figure 4).

At the beginning simulation of the developed gas generator takes place, which includes:

- 5 blocks: DECOMP, MIXER, GASIF, SPLIT and TO;
- 10 material flows: BIO, STEAM, COMPAIR, IN-GASIF, AIRSTEAM, COMPAIR1, PRODUCTS, GENGAS, ASH, GEANGAS1.

At the initial stage, biomass BIO is subjected to pyrolysis in the DECOMP unit. Then, the IN-GASIF pyrolysis products are sent to the GASIF reactor, which produces the generator gas using the AIRSTEAM vapor-air mixture. Oxidizer obtained from the mixer (in Figure 2 ejector) MIXER. After thermal conversion in the GASIF reactor, the resulting generating gas is cleared of dust and ash in the SPLIT unit. At the end, purified gas GENGAS heats the compressed air COMPAIR. The heated air COMPAIR1 enters the mixer MIXER together with the superheated steam STEAM.

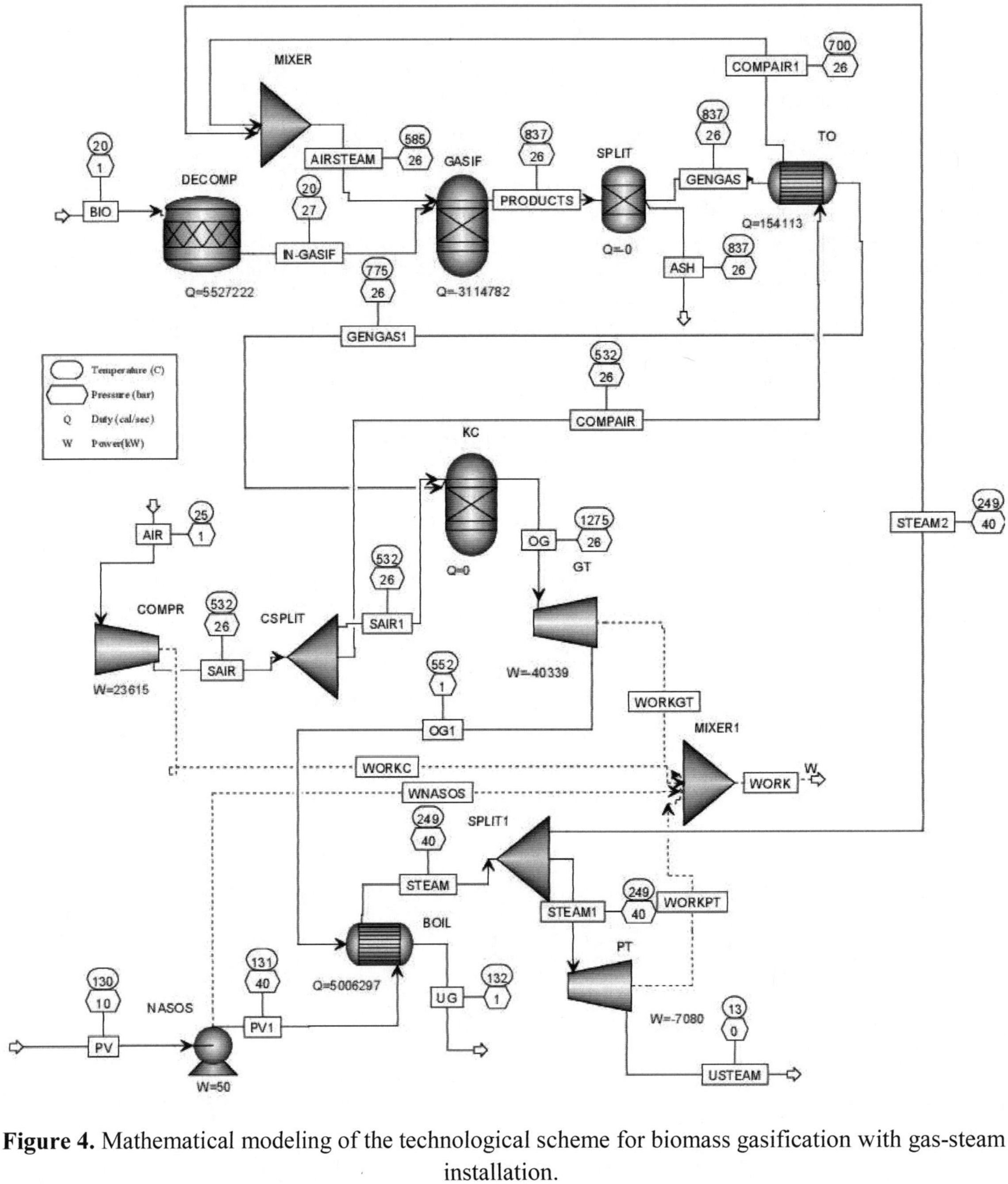

Figure 4. Mathematical modeling of the technological scheme for biomass gasification with gas-steam installation.

After the gas generator there is a combined-cycle plant, having the following parts in the gas turbine plant:

1) Units:
   - Compressor COMPR;
   - Combustion chamber KC;
   - Gas turbine GT;

2) Material flows:
   - Atmospheric air AIR;

IOP Conf. Series: Earth and Environmental Science **337** (2019) 012050 doi:10.1088/1755-1315/337/1/012050

- Compressed air SAIR;
- Some part of the pressed air to the combustion chamber SAIR1;
- Waste gases from the combustion chamber OG;

And in the steam turbine plant:

1) Units:
   - Pump NASOS;
   - Waste-heat boiler BOILER;
   - Steam turbine PT;
2) Material flows:
   - Feed water PV;
   - Overheated stream STEAM;
   - Spent steam STEAM1;
   - Steam for gasification STEAM2;
   - Exhaust gases UG.

Units CSPLIT and SPLIT1 allow one to separate streams. The power of compressor, gas and steam turbines is provided by the flows WORKC, WORKGT and WORKPT.

In the simulated scheme, the parameters in each flow are shown above the names: above – temperature, and below – pressure. This allows one to monitor the changes in the parameters in each stream as the source data changes.

This program allowed us to simulate the developed technological scheme.

As a result of modeling the developed scheme, we obtained the following characteristics:

- Molar consumption of generator gas components (Table 3):

Table 3. Molar consumption of gasification products.

Components	CH_4	CO	H_2	CO_2	H_2O	N_2	SO_2
kmol/h	31.74	233.84	213.74	106.41	101.55	421.23	0

- Molar gas flow rate of 1108.64 kmol/h or 25807.4 kg/h;
- The electrical power of the gas turbine plant was 16.72 MW, the electrical power of the steam turbine unit was 7.08 MW and the total electrical power of the entire plant was 23.75 MW.

To assess the economic feasibility of the developed scheme of combined-cycle plant with a gas generator, we performed the calculation of the payback period and comparison with the basic scheme (table. 4).

Table 4. Comparison between the capital expenses, mln. rubles.

	Basic scheme	Proposed scheme
1. Gas generator	450	450
2. Steam-gas installation	3000	1201
3. Construction and installation work	700	700
4. Other facilities for construction	300	5
Total amount	4450	2356
Simple payback period	7	5

Table 4 presents a comparison of projects construction costs. Complex equipment under the basic scheme is 4.45 billion rubles [11]. According to the proposed scheme, the cost of the gas generator is 11.9 thousand rubles/kW [16], the cost of combined-cycle plant is $ 0.8 thousand/kW [17], and other facilities for construction include the cost of crusher [18].

As a result of the economic calculation, it turns out that the proposed scheme pays off in 5 years. At the same time, the cost of generated electricity is equal to 1.88 rubles/(kW·h).

Economic calculations have shown that the payback period of the investment of a combined-cycle gas turbine unit on biomass is about 5 years, which is less than that for the project carried out according to the basic scheme, with a payback period of 7 years.

4. Conclusions

The technological scheme of the combined-cycle plant based on biomass gasification has been developed. The advantages of the proposed scheme are the following:

- An increase in energy efficiency of the circuit due to a decrease in energy consumption for own needs (absence of a drying unit and an additional air compressor for a gas generator).
- Heating of compressed air in the new gas generator 4.
- Obtaining a certain pressure in the developed gas generator using an ejector 6.

An efficient circulating fluidized-bed gas generator with a heat exchanger has been developed, in which the efficiency of gasification of biomass is increased when using the physical heat of the generator gas in the heat exchanger to heat the air.

The efficiency of the proposed scheme is more than 50% when the electric power of the installation is 24 MW.

The software environment Aspen Plus 8 allowed us to simulate the developed system of gasification of wood waste, in which the difference between the results of calculation and modeling is less than 5%.

References

[1] Huang N, Zhao P, Ghosh S, Fedyukhin A 2019 *Applied Energy* **240** pp 882-92

[2] Smyatskaya Y, Politaeva N, Toumi A, Olshanskaya L 2018 *MATEC Web of Conferences* **245** 8004

[3] Mankonen A, Kaikko J, Vakkilainen E, Sergeev V 2018 *MATEC Web of Conferences* **245** 07007

[4] Maksimova Y G, Vasil'ev D M, Zorina A S, Ovechkina G V, Maksimov A Y *Applied Biochemistry and Microbiology* **54**(2) pp 173-8

[5] Namsaraev Z B, Gotovtsev P M, Komova A V, Vasilov R G 2018 *Renewable and Sustainable Energy Reviews* **81** pp 625-34

[6] Sultanguzin I A, Fedyukhin A V, Kurzanov S Y, Gyulmaliev A M, Stepanova T A, Tumanovsky V A, Titova D P 2015 *Thermal Engineering (English translation of Teploenergetika)* **62**(5) A008 pp 359-64

[7] Kirpichnikov I M and Fil N S 2012 Wood waste as a renewable source for the generation of heat and electric energy *Bulletin of the South Ural State University, Series: Energy* **16** pp 18–21

[8] Politaeva N A, Kuznetsova T A, Smyatskaya Y A, Trukhina E V and Atamanyuk I 2018 Energy Production from Chlorella Algae Biomass Under St. Petersburg Climatic Conditions *Chem. Pet. Eng.* **53** pp 11–12

[9] Zhao Y, Wang L, Huang L, Maximov M Y, Jin M, Zhang Y, Wang X and Zhou G 2017 Biomass-derived oxygen and nitrogen co-doped porous carbon with hierarchical architecture as sulfur hosts for high-performance lithium/sulfur batteries *Nanomaterials* **7**

[10] Fedyukhin A, Sultanguzin I, Gyul'Maliev A and Sergeev V 2017 Biomass pyrolysis and gasification comprehensive modeling for effective power generation at combined cycle power plant *Eurasian Chem. J.* **19** pp 245–53

[11] Fedyukhin A V 2014 *Development of systems for the combined generation of heat and electrical energy based on the study of the processes of pyrolysis and gasification of biomass* dis. Cand. tech. Sciences (Ivanovo: ISEU)

[12] Moscow Timiryazev Agricultural Academy Zoo Engineering Faculty [Electronic resource] URL: http://www.activestudy.info/ximicheskij-sostav-drevesiny-i-kory-xarakteristika-organicheskix-veshhestv/

[13] Popov S K and Ippolitov V A 2016 *Processes and installations for gasification of fuel: textbook* (Moscow: Publishing House MEI)

[14] Popov S K, Svistunov I N and Ippolitov V A 2017 *Simulation of high-temperature processes and installations in the environment Aspen Plus: textbook* (Moscow: Publishing House MEI)

[15] Fedyukhin A V, Sultangusin I A and Kurzanov S Yu 2016 *Application of applied software for solving problems of industrial heat and power engineering: a tutorial on the courses Applied Software in Heat Power Engineering, Mathematical Modeling and Optimization of Heat*

Power Engineering Systems, Numerical Simulation Methods, etc. in the direction of Heat Power Engineering and Heat Engineering (Moscow: Publishing House MEI)
[16] INTERREMASH URL: http://www.adaptika.ru/
[17] The center for online auctions URL: https://www.b2b-center.ru/info/experts.html?id=4102
[18] Tiu.ru URL: https://moskva.tiu.ru/Drobilka-drevesiny.html

IOP Conf. Series: Earth and Environmental Science **337** (2019) 012051 doi:10.1088/1755-1315/337/1/012051

Cybersport within non-classical ergonomics of immersive and interactive environments

S Sergeev[1,*], I Burmistrov[2]

[1]Peter the Great St. Petersburg Polytechnic University, 29, Polytechnicheskaya str., St. Petersburg, 195251, Russia
[2]Founder and CEO interUX Usability Engineering Studio OÜ Pärnu mnt 139E/11 Tallinn 11317 Estonia

*Corresponding author: sergeev_sf@spbstu.ru

Abstract. The article is devoted to non-classical ergonomic of eSports. The theoretical foundations of non-classical and post-non-classical ergonomics of immersive and interactive environments are considered as applied to the activities of cybersportsmen. It is shown that the use of tools and the conceptual basis of classical ergonomics and engineering psychology are not enough to consider the self-organizing processes of mutual orientation of the gameplay participants in an artificial dynamic environment of activity. The new conceptual basis includes philosophical and natural science positions of radical and epistemological constructivism, which allows considering the recursive processes of autopoietic self-organization that take place in the subjective and communication sphere of gamers in the process of achieving a game result from a unified systemic position. The issues of professional training of gamers, the inclusion in the gaming environment, immersiveness and interactivity of artificial environments are considered. The problem of presence in the game world as criteria for the incarnation of the gamer's "I" in the "I" of an effective game character is analyzed.

1. Introduction

Development of technologies for information processing and presentation led to the creation of computer games that unite gamers in a single gaming environment using network technologies, which was the beginning of the massive development of eSports [1]. In artificial gaming virtual worlds, an illusion of the presence and complete immersion of a gamer in the plot and environment of the game is created, which starts the work of his psychophysiological and cognitive mechanisms in the direction required by the script. An e-sport (cybersport) industry appeared. Its development required the involvement of knowledge and technology from various areas of technical and humanitarian knowledge in order to improve the means of imitation and enhance the emotional impact on the audience. In the first place, this concerned the creation of computer games scenarios generating the achievement motivation, which required an unconditional knowledge of the foundations of psychology. In addition, a significant role was played by experts in the field of design and technical aesthetics, engaged in the problems of ensuring the interactivity and aesthetic qualities of the gaming environment. Note that the gaming environment does not necessarily have to exactly repeat a certain physical world similar to the world of human activity, but it should present at least minimal opportunities to conduct competitive activities, realizing the motive to achieve some goal understood and accepted by gamers.

The spontaneous nature of appearance and organization of gaming community at the initial stages of its development did not suggest any advantages to its participants due to the possession of specially designed interfaces and a workplace organized in accordance with the requirements of ergonomics. There was a natural selection of the best gamers based on the games results, each of which used its own tactics, strategy and means of achieving goals. They became carriers of gaming experience, which was transmitted in the process of gaming communication and communication between gamers. However, over time, due to increased competition and the scale of gaming activities, appearance of eSports, it became clear that ergonomically improved controls and information display allow gamers to have significant advantages over opponents. This led to the emergence of specialized gaming workstations that use powerful computers with advanced graphic capabilities, with a more advanced workstation, controls and manipulation. All this showed the emerging need for ergonomic knowledge related to the design of the working environment of the gamer. However, it should be recognized that the artificial world of the game should be ergonomic, since arbitrary projects that do not take into account the specifics of including a gamer in the game can have a significant destructive effect on his body and psyche [2]. There arose not only issues of designing and organizing a gamer's workplace (physical ergonomics), but also more serious issues of creating a dynamic environment of activity and optimization of interface systems connecting the gamer to the game world. There were problems of training gamers, including them in the gaming environment, ensuring gaming interactivity, communication and motivation, translating into an effective game character, etc.

These issues cannot in principle be solved by the methods of design, classical ergonomic design and usability, since the methodological basis of these disciplines is mainly engineering, cybernetic and informational representations and models. This is completely insufficient for the effective design of complex systems that immerse the user in artificial gaming environments of activity [3].

2. Problems and phenomenology of gaming

The main task of the designer of the artificial world of a computer game is to create an interactive environment including (immersing) the gamer in dynamic content, perceived and interpreted by the subject as a form of a certain reality in which the participants of the game involved in the joint activity are present and interact interactively. The interaction is carried out according to certain rules with the help of real governments that have their virtual counterparts and tools in an artificial environment, through which the observed effect on the virtual participants (avatars) and the objective world of the game is provided. In contrast to the environment of the natural world of our life, the environment in the world of computer games is a complex multi-level system, perceived by the participant as something to be studied and used for personal gain to achieve the game result. In essence, the game can be considered as an unproductive form of human activity in the artificial world, aimed at the effective use of its resources in the interests of achieving the goals of the game and the personal growth of the gamer. In the process of the game, an individual history of the participants is realized, who acquire a social status and position in the artificial world, realizing the motives of achievement. The most popular gaming platforms used in eSports are: Dota 2 (tournaments ESL One Birmingham, MDL Changsha Major, Epicenter XL, WESG 2017, DAC and The Bucharest Major), Counter-Strike: Global Offensive (World Electronic Sports Games tournaments, support of Electronic Sports League, SL i-League StarSeries, ELEAGUE Major, etc.), League of Legends (League of Legends World Championship tournaments), Overwatch (tournaments Major League Gaming, DreamHack, and Intel Extreme Masters). All of them have widely branched game worlds and historical lines of plots, create equal conditions for all participants in the game.

Note that the immersion into the game world does not pass without a trace for the gamer, since gaming effects occur, reflecting the processes of personality modification in the face of the physical and social factors affecting the game. First of all, it should be noted that these phenomena are often associated with the form of psychological gambling addiction [4]. The following phenomena are observed:

- Well-being or euphoria at the computer;

- Inability to stop;
- Increased amount of time spent at the computer;
- Neglect of family and friends;
- Feelings of emptiness, depression, irritation not at the computer;
- Lying to employers or family members about their activities;
- Problems with work or study [5].

At the same time, these signs, in our opinion, can be interpreted as signs of a person's transition to another artificial sphere of life activity, in which a gamer has a new sense of presence, activity and readiness for action. It is these indicators that reflect the subject's involvement in the game world. The engaging properties of the gaming environment can be so strong that it becomes for the gamer a complete substitute for reality and leads to pathological processes of social self-isolation. It is necessary to counteract the negative influence of gaming through the organization of the gaming environment and communication, the formation of the professional community of cybersportsmen.

Classic ergonomics works well when creating games with low levels of immersiveness and involving effects, but it is not effective when developing highly immersive environments that trigger mechanisms of environmental and personal self-organization. To solve the problems arising in highly immersive gaming environments, which include the MOBA (Multigamer Online Battle Arena) and RTS (Real-time Strategy) genres, new theoretical schemes are required that take into account the complex, recursive, self-organizing nature of the communicative processes arising in gaming environment.

3. Fundamentals of the theory of immersive media

The way out of this situation may be referring to methods of non-classical and post-non-classical ergonomics of complex systems, reflecting the views of man and communication as self-organizing and developing systems of autopoietic type, creating and realizing the individual history of the subject [6]. Further we consider the main provisions of these concepts and the possibility of their application in the creation of e-sports technologies.

Post-non-classical ergonomics of immersive environments is based on the ideas of radical and epistemological constructivism, synergetics, second-order cybernetics and autopoietic self-organization and philosophy of complex systems (V.I. Arshinov, J. Gibson, E.N. Knyazeva, S.P. Kurdyumov, V.A. Lektorsky, S.A. Tsokolov, P. Watzlawick, E. Von Glasersfeld, K.J. Gergen, F. Varela, N. Luhmann, H. Maturana, I. Prigogine, H. Haken, H. von Foerster, C.H. Waddington, M. Eigen) [7].

They develop philosophical and natural science concepts reflecting the processes of self-organization in man and society, determining the type and properties of mental reflection and social self-organization. They determine the activity of a person. It is fundamentally important to define the concept of "environment", which becomes the result of human constructive activity and is associated with the processes of its vital activity [8], and is not simply a physical structure, as it is considered in classical ergonomics.

The following interpretations of the term "environment" are used:

- The environment of the ergatic system is a product of constructive activity of the psyche of the human operator and cannot be considered outside its mental content;
- The environment reflects the phenomenon of dynamic integrity of cyclically forming chains of a person's relationship with physical and social realities in the process of ensuring his life activity, speaking to the subject simultaneously as a subjective reality and as an external objective, objective structure of the world in which the subject acts;
- The environment of activity in the content plan always arises as a dynamic process of forming networks of relations in the subject. Various elements of the external and/or internal environment are selectively involved in this process in order to ensure autopoiesis of the organism, stability of the personality, and continuity of its history.

The environment is a theater of activity, on the scene of which events are occurring which are accessible to an external observer describing the subject's behavior. We note a number of important properties of self-organizing media:

- Redundancy and observability;
- Accessibility to cognitive experience (constructibility);
- Out-of-subject spatial localization;
- Autonomy of existence;
- Media synchronization;
- Integrity of the environment;
- Environmental mobility.

Redundancy, as a property of the environment, allows the autopoietic system to reproduce itself and preserve the organization with a variable structure and elements, a plurality of relationships with it, the result of which is the construction of a certain picture of reality. Redundancy is one of the properties that allows us to consider an entity as an environment. The surrounding physical environment is fundamentally multivalued with respect to its objectivity, its laws and meanings generated by the acting brain. Observability is a consequence manifested in the process of constructing reality. That which is not observable is not the environment, although it may be present in physical reality. As an observer, you can act as a subject, as well as other subjects with similar mechanisms for generating reality, making distinctions. At the same time, the observed phenomenon is always refracted through the prism of a person's individual experience, and has a personality coloring. The accessibility of the environment to the cognitive experience of the subject reflects its fundamental constructibility and, as a result, causes a representation in the form of reality in the subjective world of the subject. Inaccessible (not constructed) to cognitive experience elements of reality are not perceived by the subject in the form of an environment. Accessibility to cognitive experience implies that the subject has tools to influence the environment. These tools exist in both ideal and physical form, and reflect the property of interactivity.

Extrasubject spatial localization is manifested in brain constructs separating the properties belonging to the environment from the properties of the subject. In brain constructs, the environment is always located outside the subject.

The autonomy of the existence of the environment is manifested in the presence of its own history not directly related to the experience of the subject, which appears only in the process of interaction with the environment.

The synchronization of the environment determines the presence of individual temporal qualities inherent in all elements of the environment and associated with the processes of modification of the gamer's experience.

The integrity of the environment characterizes the unity and interconnectedness of its multi-quality content of social and physical nature with the properties of the subject. It manifests itself in the constancy of the perception of a person who constructs and constitutes the environment as the world of his activity.

Motivation ability of environment characterizes the possibilities and mechanisms of influence of the environment on the motivational sphere of the subject, modulating its activity, forcing to accept the rules of behavior in the environment, forcing and encouraging, absorbing and immersing it in activity. It presents the form of the dictates of the environments. It is provided mainly by the content of the medium, the storyline and the gamer's role-playing behavior.

Cognitive processes in the psyche of a gamer are associated with the acquisition and use of knowledge in the gaming environment. However, the concept of "knowledge" in post-non-classical ergonomics has a different meaning from the concepts of "knowledge" adopted in traditional instrumental theories:

- Knowledge, unlike information, cannot be extracted from the person in which it exists in an implicit form;
- It cannot be transferred directly from person to person. It can be built only by man, grown in it;
- Knowledge arises and develops with a person, improves in the process of life, acquires properties that take into account the experience of the subject;
- Knowledge does not possess a material form, operations similar to operations with physical, material objects are not applicable to it;

- Knowledge is related to the work of the mechanism of understanding;
- Knowledge bears the features of a social construct that reflects interpretations generated and shared by community members;
- Language acts as a means of constructing knowledge [9].

In the works of Sergeev S.F. a theory of activity in immersive environments was developed [7], which proved to be effective in creating simulators and aircraft systems with increased automation [10].

4. The problem of similarity in the game world

The accuracy of simulating an artificial game environment within the framework of a medium-oriented approach reflects its similarity to the environment of the physical world and is a characteristic of medium redundancy [11]. The richer the environment, the easier it is for the subjects of the game to find individually determined forms of behavior leading to the solution of the game problem. The saturation of the environment is manifested in the properties of the world of tasks that arise before the actor. The larger the class of possible interactions with the gamer is provided by the medium, the wider is its potential. However, it makes no sense to seek to saturate the environment only with analogues of the elements of real activity. It is advisable to enter into it objects with properties that differ from the properties of objects of the game activity. Higher accuracy of imitation is inexpedient for technical and economic reasons.

You can apply a three-component definition of simulation accuracy in the game: the fidelity of the equipment (workplace), the fidelity of the environment, and the psychological fidelity. Equipment fidelity refers to the degree to which the game interface duplicates the appearance and dynamic properties of a certain equipment environment and artifacts used. Here, the accuracy of modeling should be maximal (indistinguishable by the senses). Environment fidelity refers to the extent to which imitation is perceived as the real environment. Here, as we have already noted, the simulated reality must correspond to the dynamic qualities of the real environment, but its objective content can be saturated with objects of various degrees of dynamic and physical similarity, which is related to the goals of the future game activity. And, finally, the psychological similarity refers to how much the gamer perceives the artificial world and its subject filling as a copy of the real world. Psychological similarity is the concept which is driven by previous experience of the gamer. Its high accuracy at the initial stages of learning the game does not matter much. The general principle of choosing the degree of similarity of the gaming environment is that the maximum dynamic similarity is required to ensure the gamer's motor and sensorimotor coordination in the form of artifact management skills, and the minimum is to solve problems involving the higher mental functions of the operator (in the form of knowledge and skills). At the same time, in all cases, it is useful to change the modeled parameters within the limits providing the functional reserve.

It is important to note that in the gaming environment the results of influences on her by the learner are saved, which can be interpreted in terms of behavior, solving the game problem, working out modes of operation, etc. The game control system records the results of each gamer's activities and ranks them according to their achievements of the game goals. However, the full automation of meaningful interpretations of these processes is to a certain extent limited. The structure and capabilities of the artificial world modeled in the information-dynamic model and intersubjective communication gives rise to gaming motivation to learn. Thus, in the game, an interaction of the autopoietic systems of the gamers and game communication with the objects and subjects of the game world occurs. It is important that the games for e-sports should not include a random factor in achieving the result, which should depend only on the knowledge and skills of the gamer.

At the present stage of development of the gaming industry emergent gameplay (emergent game play) consists of arising in the game, not provided by the developers of forms of game behavior, non-linear narration and gamer-generated content, which allows one to generate new content directly in the game based on the gamer's activity [12].

5. Immersiveness, presence and interactivity in the game world

IOP Conf. Series: Earth and Environmental Science **337** (2019) 012051 doi:10.1088/1755-1315/337/1/012051

The main interface properties of the gaming environment are expressed in terms of immersiveness, presence, interactivity [13]. Immersiveness is a property of the gaming environment, showing its ability to involve the subject in the current system of relations, determined by the content of the environment. Immersiveness can be defined as the properties of technological part of the gaming environment, providing the psychological state of a person in which his "I" perceives himself shrouded, turned on, and interacting with a certain environment, providing him with a continuous stream of incentives and experience. This is a dialog experience and the gamer, in turn, can also be included in the gaming environment, using internal mechanisms for immersion. A game can be considered as an immersion in dialog experience produced by an artificial (or natural) gaming environment. The ability to play is connected with the gamer's ability to immerse himself in the game plot, the game environment and act in it, having rebuilt from the surrounding distracting and interfering influences.

Immersiveness is associated with the depth and breadth of the content of the environment. The depth depends on the amount of data encoded and transmitted in a separate communication channel formed in the medium, and the "width of information" determines the number of sensory measurements presented simultaneously. The breadth of information is achieved through polymodal forms of information. A high degree of immersiveness of the game can be achieved in various ways, including: the use of virtual reality systems, immersion into the world of the game scenario, instructions and algorithms of behavior in the game [14]. As a result of immersion in the environment, the subject of the gaming experience has a sense of presence.

Presence expresses the feeling of being in a certain environment, including one that is different from the environment of direct sensory experience. The difference between immersiveness and presence is that the first concept is more connected with the technological, physical and narrative characteristics of the gaming environment, and the second defines the subjective components of the environmental experience. The experience of presence can occur not only in physical environments, but also, for example, in the process of reading a literary work, but the author of the work must be provided with the immersiveness of the environment in which the plot unfolds. Books are still an effective means of creating artificial worlds, allowing readers to experience what is happening in them, like what is happening in the real world.

Presence, despite its external terminological simplicity, is a complex concept associated with various aspects of a person's relationship with the environment(s). In the works of C. Heeter, three different types of presence were distinguished: ecological, social, and personal [15]. Ecological presence shows the degree to which the environment itself seems to recognize your existence and respond to you. Many environments formed by virtual reality systems are indifferent to users. Social presence means that if a lot of people are immersed in a virtual environment, the presence of others provides additional evidence that the environment "exists", and thus each participant experiences higher levels of presence. Most gamers of online computer games and chats are strongly attracted to social interaction in these environments. Personal presence is a measure of the degree to which and for what reasons everyone feels to be "in" a virtual environment. Six types of presence are identified in studies of M. Lombard and T. Ditton:

- Presence as social wealth. This is the degree to which the environment is perceived as helping to communicate when interacting with other people, warm, sensitive to the personnel using it;
- Realism is the extent to which the environment may seem perceptual and/or socially realistic to the observer;
- Transportation is your sense of "being there", "here" and/or "together";
- Immersiveness is the degree to which a person's sense organs are occupied by his environment;
- Social actor within the environment is the extent to which the user responds socially to the person's representation through the environment;
- Environment as a social actor is the extent to which the environment itself is perceived as a social actor [16].

The presence associated with virtual reality is most often considered in accordance with the concept of "presence as transportation". At the same time, people are usually considered to be in an immersive virtual reality in the case when they communicate about their feelings of being in the virtual world ("You are there"). The term presence in the company of people or "social presence" shows the feeling of being together in the virtual world ("we are together").

D.W. Schloerb distinguishes two types of presence:

- Subjective presence is associated with the person's judgment that he is physically located in a real, remote, or virtual environment;
- Objective presence, probability of successful completion of the task [17].

Schloerb's definitions of subjective and objective presence are fully empirical. Questioned is the value of subjective presence. Objective presence, ability to work, should be the most important criteria for virtual reality.

An important distinction between "presence" and "immersiveness" is suggested by Slater M. & Wilbur S.:

- Immersiveness is an objective description of aspects of the system, type of representation field and visualization technologies;
- Presence is a subjective phenomenon such as the feeling of being in a virtual environment [18].

Schubert et al. propose an embodied structure of knowledge as a means to explain presence [19]. The mental representation of the environment is made in terms of patterns of possible actions based on perception and memory: "Presence is experienced when these actions involve a perceived ability to conduct and move one's own body in virtual reality (VR)." In predicting the outcome of actions, people have the ability to suppress the contributions of the current environment, which explains why we can experience a presence in VR, despite the sense of the conflicting characteristics of the real environment. Presence in the game world can be considered as one of the criteria for the incarnation of the gamer's "I" in the "I" of an effective game character.

Opportunities for gamer activity in the environment are provided by its interactivity showing up to which extent users can participate in changing and shaping the content of an established environment in real time. Interactivity is the user's power to manage changes to a given environment. In this case, the virtual world must respond to the actions of the gamer. Interactivity requires dynamic modeling and is determined by the technological structure of the gaming environment, the properties of its interface. Interactivity reflects the compliance of the form of the medium and its content to the gamer's actions.

The degree of interactivity depends on many factors, which include:

- The speed with which the system handles normally and the speed of assimilation of input effects in the environment;
- Range is the number of opportunities for action at any given time;
- Mapping is the ability of a system to control changes in an artificial environment in a natural and predictable form [20].

Examples of the interface through which interactivity is realized in computer gaming environments are the keyboard, mouse, gloves, tablets, speech recognition systems, gaze directions, virtual reality helmets with induced media, etc.

6. Conclusion

Creating efficient gaming environments which provide competitive cybersports requires a serious attitude to the processes of self-organization in the gamer's psyche embodied in a game character immersed in the game world and interacting with the game environment. Integration of all aspects of the game is possible with the use of methods and ideas of non-classical ergonomics of immersive environments, allowing one to take into account the self-organizing nature of the psyche of the gamer involved in game communication.

Acknowledgements

The authors are grateful to colleagues for their help, for the special work of the technical staff on translation and editing the article. The work was performed within the implementation of the national task of the Ministry of Education and Science of Russia No. 25.8444.2017/BCh under the project

"Engineering and psychological design of simulators and training systems based on symbiotic polymodal man-made environments".

References

[1] Ivanova N A 2017 *Uchenye zapiski universiteta im. P.F. Lesgaft* **11**(153) pp 321–5

[2] Volkova E N and Grishina A V 2015 *Psychological Journal* **2** pp 20–31

[3] Sergeev S F 2009 *Educational and professional immersive environments*. (Moscow: Public Education)

[4] Nikitina L N 2016 *World of Science* **4**(4) pp 1–7

[5] Fatkhutdinova A A 2017 *Vestnik VEGU* **2**(88) pp 139–44

[6] Sergeev S F 2009 *School technologies* **2** pp 9-11

[7] Sergeev S F 2010 *Ergonomics of immersive environments: methodology, theory, practice:* dis. ... Dr. Psychol. Sciences (St. Petersburg)

[8] Sergeev S F 2010 *Theoretical and applied problems of acmeology*. Collection of scientific. articles. Issue 14. (SPb: SPbIPiA) pp 20-52

[9] Sergeev S F 2008 *Vestnik St. Petersburg University.* **12**(3) pp 418-27

[10] Sergeev S F 2007 *Mechatronics, Automation, Control* **11** pp 6-11

[11] Sergeev S F, Zakharevich M A and Sokolov V N 2011 *Russian scientific j.* **6**(25) pp 102–9

[12] Kazakova N Yu 2016 *Design. Materials. Technology* **4**(44) pp 83-8

[13] Sergeev S F 2006 Immersiveness, presence and interactivity in learning environments *School technologies* **6** pp 36–42

[14] Ryan M-L 1994 Immersion vs. Interactivity: Virtual Reality and Literary Theory *Semiotics*

[15] Heeter C 1992 *Presence: Teleoperators and Virtual Environments* **1** (2) pp 262–71

[16] Lombard M and Ditton T 1997 *J. of Computer-mediated communication* **3**(2)

[17] Schloerb W D 1995 *Presence* **4** pp 64–80

[18] Slater M and Presence S 1997 Wilbur **6** pp 603–16

[19] Schubert T W, Friedman F and Regenbrecht H T 1999 Embodied presence in virtual environments In: *Visual representations and interpretations* (London: Springer-Verlag) pp 268–78

[20] Steuer J S 1992 *J. of Communication* **4**(24) pp 73–93

[21] Bakayev V 2015 *J. of Physical Education and Sport* **15** pp 615–8 doi:10.7752/jpes.2015.04093

[22] Bakaev V V, Bolotin A E and Aganov S S 2016 *Teoriya i Praktika Fizicheskoy Kultury 2016* pp 6–8

[23] Bolotin A and Bakayev V 2017 *J. of Physical Therapy Science* **29** pp 1092–4 doi:10.1589/jpts.29.1092

Immersive media in simulators of complex ergatic systems

S Sergeev[1,*], I Burmistrov[2]

[1]Peter the Great St. Petersburg Polytechnic University, 29, Polytechnicheskaya str., St. Petersburg, 195251, Russia
[2]Founder and CEO interUX Usability Engineering Studio OÜ Pärnu mnt 139E/11 Tallinn 11317 Estonia

*Corresponding author: sergeev_sf@spbstu.ru

Abstract. This paper shows a change in the paradigm of medium-oriented training of complex systems operators on simulators using high-precision models of real control systems and their application environments. Evolution of the technogenic environment leads to an increase in the complexity of professional activity, which hinders the implementation of simple algorithmic learning technologies. Forecasts to reduce the role of man in ergatic systems are not justified. New ergatic systems using artificial intelligence technologies require from operators qualities that exceed human capabilities, which requires an examination of projects and an assessment of user control complexity. The reasons for the non-classical "complex learning" methodology, which takes into account the student's self-organization and orientation in a complex dynamic context, are shown. Promising methodological approaches to the development of simulators that implement symbiotic medium-oriented learning are considered.

1. Introduction

Historically, the emergence of simulators was a reaction of technology developers to the need to use when working with new machines and mechanisms unusual forms of human experience, which arose from systems of mediated control and information presentation. Natural adaptation mechanisms and habitual sensorimotor forms of response for real-time control of artificial dynamic systems turned out to be insufficient, and the task was to form special experience, structures of professionally important qualities, knowledge, skills and abilities [1]. This problem was solved on simulators by actively immersing future operators in simulated conditions psychologically and perceptually similar to the conditions of real activity [2]. A training direction has been formed associated with repeated training of typical control algorithms in teaching tasks and conditions which showed the main features of the professional environment of work [3].

In different years, well-known Russian scientists, mainly representing the Moscow and Leningrad scientific schools of psychology, dealt with the problems of psychological support of professional activity. Their works revealed the classical content of professional activity as a subject of psychology, its features were investigated (K.A. Abulkhanova-Slavskaya, V.A. Bodrov, A.K. Gastev, S.G. Gellerstein, M.A. Kotik, N.D. Levitov, V.N. Myasishchev, A.P. Nechaev, G.S. Nikiforov, O.G. Noskova, K.K. Platonov, V.F. Rubakhin, I.M. Sechenov, V.A. Tolochek, I.N. Shpielrein). The methodology and theory of professional activity were developed (B.G. Ananiev, V.A. Bodrov, A.V. Brushlinsky, A.I. Galaktionov, L.G. Dikaya, E.M. Ivanova, A.V. Karpov, B.F. Lomov, V.I. Medvedev, G.V. Sukhodolsky, V.D. Shadrikov), the development directions of the professional character were investigated, the problem of professional suitability was developed (L.I. Antsyferova, M.A. Dmitrieva, A.L. Zhuravlev, D.N. Zavalishina, E.F. Zeer, A.M. Zimichev, E.A. Klimov, G.L. Koroteev, A.A. Krylov, A.I. Naftulev, A.A. Oboznov, Yu.P. Povarenkov, N.S. Pryazhnikov, A.A. Rean, I.L. Solomin, Y.I. Filimonenko), methods and technologies of engineering-psychological design and operator activity were created (V.F. Venda, Yu.Ya. Golikov, G.M. Zarakovsky, V.P. Zinchenko, A.N. Kostin, B.F. Lomov, V.M. Munipov, G.V. Sukhodolsky, Y.K. Strelkov, D.A. Oshanin), the problems of professional selection and training were solved using technical means of training and simulators (V.A. Bodrov, V.N. Druzhinin, V.N. Zankovsky, G.L. Koroteev, V.N. Sokolov, A.A. Frumkin, V.E. Shukshunov).

As a result, by the eightieth years of the last century, a stable opinion was formed in the engineering environment about the need for accurate modeling of the conditions of real activity in order to achieve the required levels of professional readiness [4]. A whole branch of simulator engineering appeared, using the latest achievements of science and technology of modeling and presenting information for simulating on the simulator all aspects of the perceptual experience of operators practically indistinguishable for a subject from the experience he gets in real activity [5].

Training on simulators began to be reduced to the repetition of algorithms of real activity in standard conditions and formation of forms of operator behavior appropriate to typical situations [6]. To imitate the conditions of activity, computer technologies are now widely used that implement models of variable realism, covering the Milgrem continuum from complete artificial virtual and augmented reality to the physical in-situ conditions of real activity [7]. At the same time, the modeling of immersive environments that provide the formation of a sense of presence and a full multimodal immersion of the student in the learning environment [8, 9, 10] has become technologically available.

However, all the achievements in this area of simulator development end there, since the methodological, psychological and pedagogical, side of simulator training is still based on a simple repetition of training tasks that do not take into account the specifics of individual training and self-study of a person. The world of problems solved in modern ergatic systems, in its complexity has gone far from the implementation of primitive sensorimotor coordination and algorithmic operations, the development of which focused the first simulators [11, 12].

2. The challenges of complex learning in training environments

Despite the fact that all levels of human psyche organization are included in the operator's activity, the methods of simulator training are still based on the didactic formula "Repetiljo est inaier studiorum" (Latin) known to us as "Repetition is the mother of learning". As a result of repetition, according to the adherents of this trend, all professional skills and abilities arise and improve [13].

Revolutionary changes in the design of controlled machines associated with the evolution of man-made environment and means of network electronic communications radically change the formulation and solution of the problem of taking into account the human factor in training [14]. The simple repetition of elements of professional tasks is not enough to effectively include the operator in solving problems related to the symbiotic relationship of a person with the technogenic world, requiring the development of complex forms of intellectual activity [15, 16].

Over the last 20 years, processes of intensive introduction of automation technologies when executing control algorithms for complex technical systems are observed. This created among designers an illusion that a person would soon be removed from the control circuit of the machine. In their opinion, it became possible to replace the operator with systems and means of automation and robotization [8]. A rapid decline in the significance of the human operator was expected when working with complex systems, up to its complete exclusion in connection with the transition to automatic control. This was facilitated by the intensive development of electronic communications systems and artificial intelligence methods that promised to completely replace a person at the point of "technological singularity" [17, 18, 19]. The methodological basis of the design also included the philosophy of materialism prevalent in the engineering environment, which simplified man to the level of a complex machine.

However, these predictions for reducing the role of man in ergatic systems have not yet come true. Moreover, the inclusion of a person in various forms of orienting and controlling communication circulating in control systems is more intense than before. New tasks appeared, such as the formation of missions for robots, copying control in virtual environments, the control of automatic devices [20, 21, 22].

In the field of professional training, for the first time we faced with the paradoxical fact that developers of technology presented requirements for the professional qualities of operators, exceeding the capabilities of their psychophysiological system for processing and interpreting the information necessary for making adequate decisions. This means that no operator, in principle, under any circumstances, will be able to work effectively with such a system. However, simulators for these systems are produced with the hope that methodologists and instructors will still be able to teach the operator the desired behavior. This does not happen, and we are faced with a potentially dangerous, in terms of the human factor, systems with currently unsuitable operators. At the same time, the latter may not be aware of their professional inadequacy, focusing on false conceptual models of the reflexive

IOP Conf. Series: Earth and Environmental Science **337** (2019) 012052 doi:10.1088/1755-1315/337/1/012052

consciousness. The current situation requires developers to estimate the limit level of operator activity in the future system. It is necessary to determine the potential for mastering its algorithmic and cognitive components. One requires to perform works on adaptation the equipment to the operator serving it, which differs from the classical human adaptation to the equipment.

In contrast to the classical task of professional selection, which implies the absolute presence in the general population of the required volume of potentially eligible applicants, in this case there may not be any of them at all. Only controlling the level of complexity of an operator's activity in an ergatic system can create conditions for error-free control and decision-making. This is a new task of ergonomic and technical design requiring the creation of activity models with variable functional load on the operator.

The intellectualization of the technogenic management environment is accompanied by the emergence of technologies for modeling almost all aspects of perceived professional and educational experience, which removes the problem of similarity in simulators for a long time considered basic when creating systems for professional training of operators of complex technical systems. Instead of the question "How to model?" the questions "What to model?" and "Why model?" were raised. Engineers do not yet have an answer to these questions, as they lie in the plane of pedagogy and psychology of education.

Education in the traditional concepts of classical pedagogy is considered as an active interaction of the two systems - the teacher and the student through the educational content. The teacher is the source of knowledge (structured information), and the student is its consumer, to a certain extent a passive element. By virtue of this, all teaching methods are aimed at presenting information and training tasks to the student with the subsequent implementation of activities in modeling environments containing essential elements and algorithms of real activity. Such a scheme of the educational process underlies all e-learning methods, but it is not effective in learning how to work with modern complex control systems. In them, the operator should set the task of management himself, and not be a passive conductor of standard ideas. With the increasing complexity of the technogenic environment of human civilization, it became necessary to change the paradigm of professional training, the transition to complex learning, by which we understand the use of representations of non-classical and post-non-classical psychology and pedagogy, considering the processes of directional orientation and coordination of the student's self-organizing cognitive system in the environment of his experience [23, 24].

3. Pedagogical prolegomena of complex learning

In complex learning, a controlled technological environment must provide the user with the capabilities available to him, allowing him to solve control problems and not limit his field of action. There are questions of distribution of operator functions in the management environment of the ergatic system, which are considered in the context of self-organization of a particular subject in the process of influencing him on self-organizing communication. The appearance of a managed order in the final system is a special topic of complex learning, distinguishing it from the classical concepts associated with the use of the category of knowledge.

Classical models of teaching in the form of a didactic triangle, including a student, a teacher and educational content, reflect a number of didactic principles widely used in pedagogy, which are considered as the main active element of the teacher's learning environment which implements teaching methods. The pupil is to a certain extent passive and is the object and subject of pedagogical influence. It is the attitude of the teacher and the student that determines the quality of the pedagogical process. Consequently, in the opinion of designers of the methodical component of electronic learning systems and simulators, in order to create an effective e-learning system, it is sufficient to simulate the working environment, functions and logic of the teacher's actions using technology.

This largely controversial point of view of classical psychology and pedagogical psychology is widely replicated in the engineering and pedagogical environment and is the basis of e-didactics, which is a complex discipline about teaching methods in the new pedagogical reality of the century of technology. The main problem of this approach is that it reflects, within the framework of classical rationality, the engineering understanding of learning as an externally controlled information process associated with transfer of knowledge considered as portions of structured information. This contradicts modern scientific data from the field of human learning, according to which the learning process is of communication nature, orienting the student in the area of educational content [25, 26]. At the same time, the categories "knowledge" and "learning" reflect the processes of self-organization of the human cognitive system in the learning environment [27].

In complex learning, communication plays a special role in professional and learning environments. The concept of orienting communication is proposed, which allows explaining the learning processes in self-organizing systems [25].

The main provisions of the concept are the following:

- The brain is a physical system of autopoietic type which is capable, together with human sensorimotor systems, to create and maintain the integrity of the mental content generated in it in the form of subjective world with a subject acting in it, receiving and using conscious experience in the process of creating chains of orienting relationships in recursive cycles of self-reproduction and communication.
- A person, through orienting communication, continuously assimilates into the structures of his constructive experience, which he assesses as the positive aspects of interactive contacts of perceptual systems with the world, giving the subject potential for self-preservation and the continuation of biological and social evolution.
- In the process of learning, interconnected information processing cycles and formation cycles of information processing tools occur. The process of searching and creating efficient cognitive tools that allows one to explore the world in the course of creating a personal history of a person is continuous at all time levels.
- Learning is intervening in the cycles of experience and cognitive tools through the communication orientation of a subject and the introduction of relevant information.
- Self-learning of a person is built on the basis of reflexive self-orientation, proceeding in the form of the process of internal communication of the subject with himself and his inner world. Consciousness at the same time is an instrument of social and ego communication, involving a person in the autopoiesis processes of one's self.
- Knowledge as a result of the pedagogical process is formed on the psychological and neurobiological levels and is a systematic, inseparable from a person property of his psychobiological organization embodied in it.
- The subject works with his subjective world through the exchange and interpretation of the information circulating in it, and the brain works with the physical world by fixing the changes that occur at the inputs of perceptual systems [25].

In accordance with the concept of orienting cooperation, learning systems are systems that organize and maintain the information environment in the form of a dynamic system in which communication processes orient the cognitive and personal mechanisms of a student in the area of educational content aimed at generating (designing) a learning result.

In the process of pedagogical communication, a self-organizing social system arises, producing in the course of its evolution directional changes in the cognitive and personality systems of a person. Note that such self-organizing metasystems arise during the course of any communication. The specificity of the learning communication metasystem is its focus on obtaining a pedagogical result. The role of the teacher is to maintain the vector of educational communication in the direction of ensuring the learning effect. At the same time, we note that the teacher creates the conditions for the emergence of educational communication, and in this he is independent. However, in the implementation of communication, his freedom and activity are limited by the properties and mechanisms of self-organization that arise and operate in a particular learning situation.

4. Non-classical medium-oriented education

The development of modeling technologies has led in the last decade to the widespread use of technology and methodology of medium-oriented education, its classical version postulating the special role of the similarity of the real activity model. Practically, the simulators have turned into high-precision models of real activity. Learning in the environment is a variant of the trial and error methodology transferred to complex activities. However, the classical approach exaggerates the value of external means of activity, leaving in the shadow the internal determinants of a professional formation. As a result, the quality of education suffers, internally inefficient structures and knowledge systems are formed. The effect of an amateur, sharply narrowing the range of effective professional activity appears. It can be said that classical medium-oriented education engenders middle-level professionals en masse.

Designing of training systems for training operators of complex technical systems required new post-non-classical ergonomics of simulator engineering [28, 29], which considers the learning process as a process of autopoietic self-organization and orientation of social communication and the organism [30, 31].

The main theses of the hypothetical scheme of this direction are the following:

- The concepts of "teaching environment" and "learning environment" are introduced, which separate the formal model of activity and the technology of its creation (learning environment) from the learner's cognitive-mental and personal structures, included in the teaching communication (teaching environment);
- The environment of the ergatic training system is a product of the constructing activity of the human operator psyche and cannot be considered outside its mental content;
- Environment reflects the phenomenon of dynamic integrity of cyclically formed chains of a person's relationship with physical and social reality in the process of ensuring his life activity. The environment appears before the subject at the same time in the form of subjective reality and as an external objective, objective structure of the world in which the subject operates;
- The teaching environment in terms of content always arises as a dynamic process of forming a network of relationships in the subject of training, in which he personally (not always consciously) selectively engages the most diverse elements of the external and/or internal environment in order to ensure: autopoiesis of the organism, consciousness, personality stability and the continuity of its history.

In the environment agents (actors) of various physical nature can act. They can be a machine or a person, or both people or both machines. Elements of network can be of different nature. All education actors accumulate or use knowledge to generate new knowledge or distribute existing knowledge among carriers with different levels of knowledge. This is the main function of education - the distribution of knowledge in the teaching environment. Note that education does not create new knowledge, although in the process of distributing knowledge, elements of new may arise [32]. Using the experience of successfully trained for the creation of training procedures can be an effective direction for improving simulators.

5. Symbiotic environment-oriented learning

Symbiotic learning is the use of artificial intelligence technologies to enhance the capabilities of the operator, which becomes an element of the communication association of a person with an artificial or distributed intelligent controlled system in the form of an intelligent symbionte [15]. Intellect manifests itself only in an organized environment and is associated with the search, creation and use of tools that harmonize the system's relationship with the structured environment. Intellectual symbiontes include various types of integration of the intellectual and cognitive systems of a person with various kinds of organized physical entities, which make it possible to increase the efficiency of the emerging system in solving intellectual problems.

When a student is included in interaction with complex systems (simulator, control environment), various forms of inter-system associations arise in which active forms of transformation and structuring of the learning environment manifest themselves, which can be useful in real activity. One can speak about the formation and development of symbiotic conceptual models of activity, which, unlike classical mental conceptual models, are not fixed, by largely an algorithmic product of the human psyche stored in his memory, but is the result of the joint work of the operator and the artificial intelligence of the machine.

The problem of adaptability is central while creation of technical means for training operators of complex ergatic systems of all types. It shows the general property of the living to adaptation by developing new flexible cognitive and executive structures and underlies all teaching methods. Practically all parameters of a learning task can be used to include and improve those or other professionally important functions, knowledge and skills of a person.

The recursive nature of complex learning is associated with cyclic processes of self-reproduction, operating in the human body, which makes it possible to single out recursive repetition as one of the main methods and methodological principles of learning in artificial environments.

The artificial learning environment in the simulator, presented to the student in the form of the world of his activity, should include all types of training leading to the emergence of effective professional behavior.

When creating a learning environment in the simulator, it is necessary to take into account the effects of self-organization arising in the system "learning environment - student". It is important to ensure the activity of the student in the process of incorporation into the learning environment, which ensures further activity in the professional environment. The main problems arising in the synthesis of simulators as systems of adaptive and symbiotic information interaction are the following:

- Search for patterns of stochastic determination of human behavior by a controlled object and environment, by means of machine learning;
- Creation of a data logging network for qualitative and quantitative analysis of factors and criteria for the complexity of solving professional tasks in the course of the life cycle of a controlled object and simulators.
- Selection of optimal models, levels, methods and means of adapting the structure and means of interaction in normal and extreme conditions for various functions;
- Development of models of adaptive communication interaction of the operator with artificial intelligence systems;
- Study of the psychological patterns of perception of thought, mnemic and perceptual processes in the conditions of cyclic self-organization of the structures of professional experience;
- Using the history of the operator's interactions with the simulator and real technology for its optimization and improvement [32].

6. The technogenic world, its educational and modifying properties

Observations on the speed of civilization evolution show that it is continuously accelerated, it acquires a global character and obeys a simple logarithmic law. The point at which this function goes to infinity is the point of singularity. At this point, qualitative changes in the content and forms of human development should occur [33, 34, 17]. According to the author of the term "technological singularity" Vernor Vinge by 2030, humanity will have technological means to create superhuman intelligence, namely:

- - Large computer networks (and their combined users) will be able to "realize themselves" as superhuman intelligent entities;
- - The human-machine interface will become so close that the user's intelligence can reasonably be considered superhuman;
- - Biology will provide us with the means to improve the natural human intelligence [18].

Belief in scientific and technological progress is supported by other researchers. Thus, according to A.P. Nazaretyan "Among the conditions for the optimal development of civilization in the coming decades is the inevitable "denaturalization" of the human body and mind, the accelerated formation of man-machine complexes and the artificial restructuring of the material substrate of mind" [33, p. 276]. The world is becoming more organized and predictable [35]. The formation of the techno-biode has begun which is a single self-organizing planetary unity that unites man and technology [36, 37].

The evolution of the technogenic world, and appearance of a global intellectual environment, lead to a change in the fundamental human values which are the basis of the stability of the human person. The technogenic environment involves almost the whole of humanity in all spheres and forms of its life activity in the logic of its evolution. A parallel digital virtual world of total connectedness of mankind is being formed. It can be assumed that the technogenic environment will have the properties of a global distributed learning system that forms all the elements necessary for self-reproduction. This will be the solution to the problem of simulator training.

Acknowledgements

The authors are grateful to colleagues for their help, for the special work of the technical staff on translation and editing the article. The work was performed within the implementation of the national task of the Ministry of Education and Science of Russia No. 25.8444.2017/BCh under the project "Engineering and psychological design of simulators and training systems based on symbiotic polymodal man-made environments".

References

[1] Bodrov V A 2006 *Psychology of professional activity. Theoretical and applied problems.* (Moscow: Publishing House "Institute of Psychology of the Russian Academy of Sciences)

[2] Liskin V M, Sergeev S F and Meshkov A N 1987 *Trudy VNIITE. Ergonomics* **33** pp 29–37

[3] Sergeev S F 2006 *School technologies* **3** pp 58–65

[4] Sergeev S F, Zakharevich M A and Sokolov V N 2011 *Russian scientific journal* **6**(25) pp 102–9

[5] Sergeev S F 2010 *Biotechnosfera* **2**(8) pp 15–20

[6] Sergeev S F 2011 *Scientific and Technical Bulletin of St. Petersburg State University of Information Technologies, Mechanics and Optics* **6**(76) pp 118–23

[7] Milgram P and Kishino A F 1944 *IEICE Transactions on Information Systems and Systems* pp 1321-9

[8] Sergeev S F 2006 *School technologies* **6** pp 36–42

[9] Sergeev S F 2010 *Theoretical and applied problems of acmeology. Collection of scientific articles. Issue 14.* (SPb: SPbIPAA) pp 20-52

[10] Sergeev S F 2011 *Presence and immersiveness in learning environments.* (SPb: Polytechnic University Publishing House)

[11] Sergeev S F 2014 *Man in the Techno Medium: Convergent Technologies, Global Networks, Internet of Things. Collection of scientific articles. Issue 1.* (Vologda: VoGU) pp 146-50

[12] Sergeev S F 2016 *Automation. Computer science* **2**(39) pp 39-44

[13] Sergeev S F 2011 *Scientific and Technical Journal of St. Petersburg State University of Information Technologies, Mechanics and Optics* **1**(71) pp 109-14

[14] Sergeev S F 2018 *Psychology of Technointegration and Technomodification of Man: Theoretical and Methodological Basis. Actual Problems of Labor Psychology, Engineering Psychology and Ergonomics. Issue 8*. (Moscow: Publishing House "Institute of Psychology of the Russian Academy of Sciences) pp 30-49

[15] Sergeev S F 2013 *Scientific and Technical Journal of Information Technologies, Mechanics and Optics* **2**(84) pp 149–54

[16] Sergeev S F 2014 *Psychological J.* **35**(5) pp 88–98

[17] Sergeeva A S and Sergeev S F 2017 *Psychological J.* **3**(38) pp 124-8

[18] Vinge V 1993 *The Coming Technological Singularity: Survive in the Post-Human Era*

[19] Kurzwell R 2005 *The Singularity is Near: When Humans Transcend Biology* (New York: Viking Penguin)

[20] Sergeev S F 2013 *Science and technology of the XXI century. Communications and NBICS-convergence Global Future 2045. Convergent Technologies (NBICS) and Transhumanist Evolution*. (Moscow: Publisher LLC MBA) pp 158-68

[21] Sergeev S F 2013 *Engineering-psychological and pedagogical problems and prospects of simulator engineering Actual problems of labor psychology, engineering psychology and ergonomics Issue 5*. (Moscow: Publishing House of the Institute of Psychology of the Russian Academy of Sciences) pp 13-38

[22] Sergeev S F and Sergeeva A S 2014 *Robotics and Technical Cybernetics* **2**(3) pp 23–8

[23] Sergeev S F 2017 *Methodological problems of education in self-organizing technogenic environments Year of Ecology in Russia: pedagogy and psychology in the interests of sustainable development: a collection of articles of the scientific-practical conference* (December 4–5, 2017, Moscow) (Moscow: Publishing House "Pero") pp 438-43

[24] Sergeev S F 2017 *Paradigms of engineering and vocational education in the evolving technological world Fundamental and applied research of modern psychology: results and development prospects* (Moscow: Publishing House of the Institute of Psychology of the Russian Academy of Sciences) pp 2384-92

[25] Sergeev S F and Sergeeva A S 2014 *Open Education* **5**(106) pp 41–8

[26] Sergeev S F 2014 *Public education* **8** pp 137-45

[27] Sergeev S F 2008 *Philosophy of Education* **1**(22) pp 286-94

[28] Sergeev S F 2009 *Engineering and psychological design of complex ergotechnical environments: methodology and technology Actual problems of labor psychology, engineering psychology and ergonomics Issue 1*. (Moscow: Publishing House "Institute of Psychology of the Russian Academy of Sciences") pp 429-49

[29] Sergeev S F 2012 *Open Education* **1**(90) pp 90-100

[30] Luhmann N 2007 *Introduction to systems theory* (Moscow: Publishing house "Logos")

[31] Varela F J, Maturana H and Uribe R 1974 *Biosystems* 5 pp 187-96

[32] Sergeev S F 2010 *Ergonomics of immersive environments: methodology, theory, practice*: dis. ... Dr. Psychol. Sciences (St. Petersburg)

[33] Nazaretyan A P 2015 *Bulletin of the Russian Academy of Sciences* **84**(3) pp 275-7

[34] Panov A D 2013 *Metaphysics* **3**(9) pp 141-88

[35] Lepsky V E 2012 *Self-developing innovation environments in the context of the formation of the VII socio-humanitarian technological order Organization of self-developing innovation environments* (Moscow: "Kogito-Center") pp 5-25

[36] Chekletsov V V 2010 *Philosophical Sciences* **6** pp 36-53

[37] Sergeev S F 2014 *European J. of Psychological Studies* **3**(3) pp 93-101

[38] Bakayev V 2015 *J. of Physical Education and Sport* **15** pp 615–8

[39] Bakaev V V, Bolotin A E and Aganov S S 2016 *Teoriya i Praktika Fizicheskoy Kultury 2016* pp 6–8

[40] Anfinogentova A A, Dudin M N, Lyasnikov N V and Protsenko O D 2018 *Economy of Region* **14** pp 638–50

IOP Conf. Series: Earth and Environmental Science **337** (2019) 012053 doi:10.1088/1755-1315/337/1/012053

3D Printed Plastic Tool for Al Thin-Sheet Forming

L B Aksenov[1,*] and I Y Kononov[1]

[1]Peter the Great Saint-Petersburg Polytechnic University, 194064, Polytechnicheskaya 29, Russia

*Corresponding author: l_axenov@mail.spbstu.ru

Abstract. Shown prospects of 3D printed plastic tooling in the processing of metal forming. The study of the process of thin sheet forming of aluminum 3003 ANSI established that in this process of destruction and plastic deformation of the plastic tool does not occur. Lubrication of plastic matrices and punches were not required, as they have anti-friction properties. Computer simulation of this process in the software system "Simufact.forming" accurately reflects the force parameters, stress-state of plastic tool and the actual course of the process of forming the blank, including the spring-back of the part after forming. Technology of sheet forming with plastic tool can be recommended for single and small batch production.

1. Introduction

3D printing is common technology and it is applied for rapid manufacturing of finished parts or prototypes of the materials supported by 3D-printers. The positive experience was obtained in various industries: medicine, manufacture of household appliances, furniture, clothing, in the construction of buildings and structures and [1]. Technologies of 3D printing have a high-performance, high-speed and cost-effective for small-lot and piece production. The main purpose of the use of products manufactured using 3D printing is reducing the cost of manufactured products, accelerations of prototyping and product development, customizing products for a specific customer.

Achievements of 3D printing in metal forming are not so wide, but one of the first positive applications in the dies manufacturing was done more than 20 years ago [2]. The role of 3D printers is also significant and promising part of the design and implementation of production with technologies of metal forming [3]. As for to sheet forming, tool manufactured by techniques of 3D printing in some cases is able replace metal tools. There is evidence that the use of 3D printing is capable of 80% reduction in time spent on development and manufacture of tooling and to reduce the cost of tooling by 70% [4-5].

Among the many types of 3D printers, printers for the manufacture of plastic products are much cheaper compared with printers which are designed to operate with metal or other materials [6]. Therefore, the possibility of using the plastic tool is the most economically advantageous in cases when it is possible and, first of all, it is the field of small-batch and unit production [7].

Great attention to this process pays the automotive industry, seeing the future in the production of spare parts and individual tuning of motor cars [8-9]. Interesting can be called the use of plastic tool for sheet hydroforming of parts of the aviation industry [10]. However, the area of possible application of the plastic tool in metal forming is just beginning to be defined.

The use of 3D printing to improve design processes die forging can be successfully carried out already at the present stage of development of the technology of 3D printing. So, it is possible to print models of dies and to use them for physical modeling of processes with a certain ratio of geometric similarity for the purpose of determining the deformation of metal during stamping, and also to check the adequacy of the methods of computer modeling [11]. In this case the cost and time of conducting the physical simulation is significantly reduced compared to the use of metal tool which is made in traditional ways.

In the present paper shown the possibility of application of 3D printed plastic dies for cold forming of thin sheet aluminum.

2. Used equipment

For the manufacture of plastic dies on the basis of information from open sources [12,13], was designed and assembled special 3D-printer (Fig.1), which belongs to the group of Fused Deposition Modeling (FDM) 3D-printers.

Figure 1. 3D-printer to manufacture plastic dies.

The following print settings of the printer were used:

- Type of plastic: polyethyleneterephthalate.
- Print resolution on the X and Y axes, depending on the diameter of the extruder nozzle - 0.4 mm.
- Print resolution on the vertical Z-axis (layer thickness) - 0.2 mm.
- Print speed - 30 mm/s.
- The temperature of the extruder - 245 °C.
- Temperature of heating table - 110 °C.

3. Thin sheet forming with plastic tooling

A study of forming by plastic tooling of part for micro heat exchanger from thin sheet aluminum ANSI-3003 (Fig. 2) was done (Fig. 2). The parameters of used blanks as follows: aluminum sheet material of ANSI 3003; thickness: 0.1 mm; length and width are 30 mm. In the forming process were used three sets of dies. The working surface of the dies had periodic profile with the tooth height h=2, 3 and 4 mm with the number of bends 12, 9 and 5, respectively.

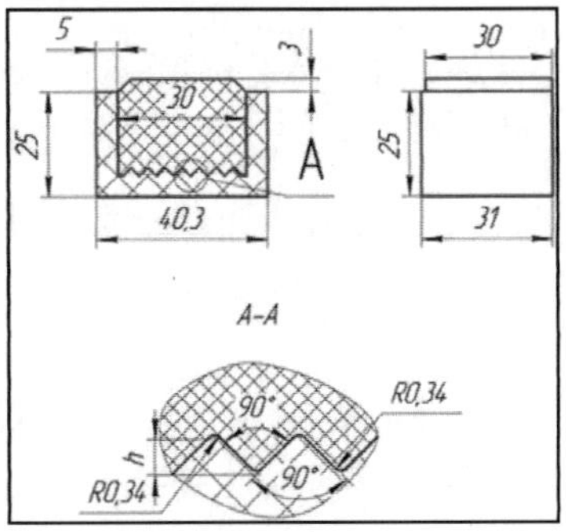

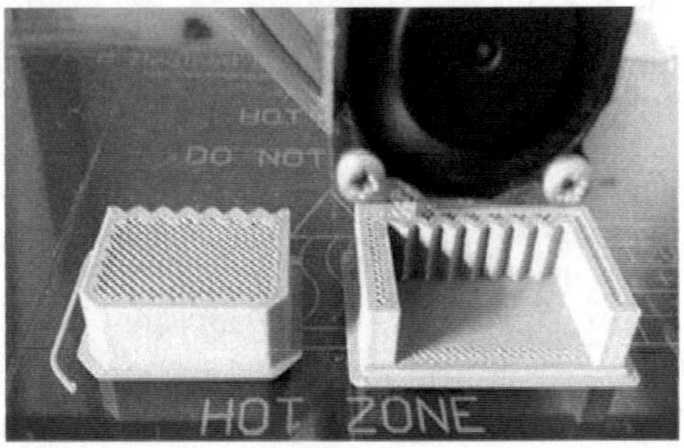

a. b.

Figure 2. Sketch of plastic dies with the relief height h=2, 3, 4 mm (a) and the process of their manufacture on a 3D printer with a view of the internal structure (b).

In Fig. 3a presented the main stages of process of forming of sheet blanks in the plastic dies, and in Fig. 3b examples of elements of micro heat exchangers manufactured using this technology.

The maximum of forming force amounted to 1350 N for the profile with the tooth height of 2 mm, 700 N for the profile with the tooth height of 3 mm and 500 N for the profile with the tooth height 4 mm. At equal size of the workpiece the forming force increases with decrease of the height of the formed profile. This reflects the dependence of the force from the number of formed teeth, i.e. the number of bends in the workpiece, which increases with number of teeth.

The final height of formed tooth profile is in the range of 45-65% of the height of the tooth on the die. It is connected with the elastic deformation (spring-back) of the sheet workpiece after forming. There is also uneven distribution of the height of the tooth along width of the workpiece.

The smallest tooth height is observed in the central parts and the height gradually increases from the center to the edges of the part. In addition to the elastic spring-back there is a phenomenon associated with the presence of friction and resistance to kinking that occurs due to "tightening" of the material to the center of the die during the forming

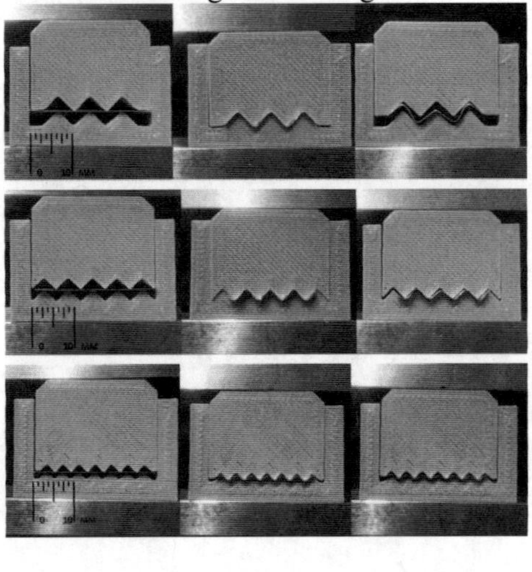

a. b.

IOP Conf. Series: Earth and Environmental Science **337** (2019) 012053 doi:10.1088/1755-1315/337/1/012053

Figure 3. The process of sheet forming in plastic dies with a teeth height h= 4, 3, 2 mm (a) and elements of micro heat exchangers made by forming in plastic dies (b).

In addition, as a result of the described factors, the formed parts received have some deviation from the plane. For these details, it doesn't affect their performance characteristics, as do not change parameters of heat exchange, and the deviation from the plane is compensated when connecting heat exchanger with other parts of the unit.

4. Computer simulation of sheet forming with plastic tooling

Computer simulation of sheet forming process was performed in the software package „Simufact.forming" [14]. For the calculation were set to: characteristics of the workpiece material; the geometry of the dies and the workpiece; the type of process -cold sheet forming; the movement of the punch, providing at the final moment of forming the gap between the dies equal to the thickness of formed material. In the calculations it was used the elastic-plastic model material of the blank DB.AlMn from the database „Simufact.material" taking into account the deformation hardening of metal corresponding to the ANSI 3003 aluminum [15].

In the simulation was used a mesh with rectangular finite elements (Fig. 4a). Nine elements were located at the thickness of the workpiece. Totally in the model were used for about 30 thousand elements. One calculation takes approximately 60-90 minutes.

The distribution of values of effective stresses during forming of the workpiece (Fig. 4a) shows that the highest values of stress are observed in the outer layers of the bent workpiece, where summation takes place the stresses of the same sign from bending and stretching of the workpiece.

During the forming the maximum force of deformation reached 1350 N. On the area of the workpiece 900 mm2 average pressure on the tool was 15MPa. At this pressure, fracture, plastic deformation or scratches on the working surface of the plastic tool was not observed.

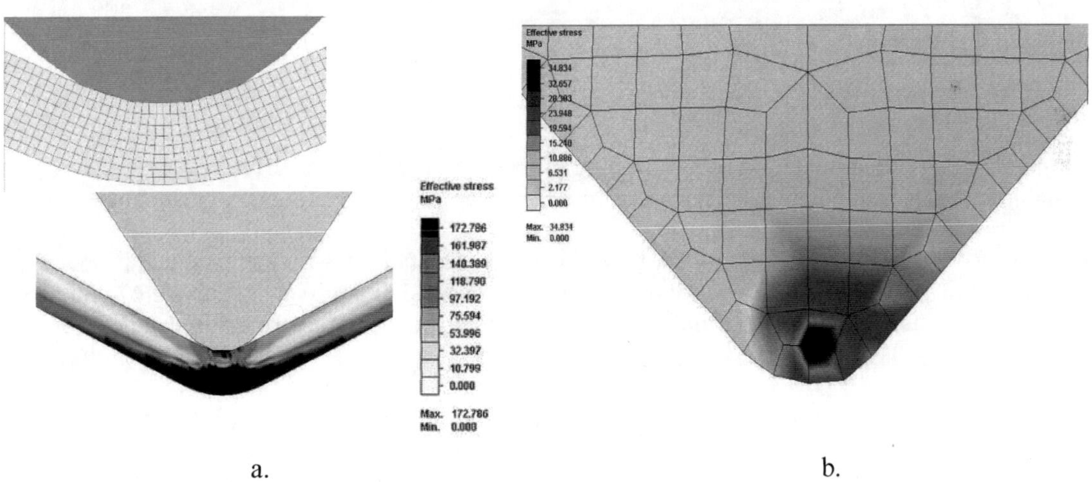

a. b.

Figure 4. Simulation of thin sheet forming process: a - finite element mesh and distribution of effective stress in a sheet blank; b - the effective stresses in the top plastic die.

Lubrication of the dies was not held because they have anti-friction properties. Similar results are obtained by computer simulation of stress state of plastic dies (Fig. 4b). As a model for material of dies was chosen an elastic body with a modulus of elasticity equal to 2400 MPa, obtained according to test method ASTM D790 [16]. The simulation showed that the maximum value of the effective stress in the dies is corresponding to the end of the forming, about 35 MPA, which is less than the value of yield stress, which is 52 MPa. Thus, simulations and experiments show that plastic dies in the studied process are in the elastic state.

After forming a few dozens of parts noticeable wear of the plastic dies was not detected. Steps of simulation of forming the workpiece is shown in Fig. 5. The model of formed material of the workpiece has elastic properties, which allows reproducing the change of geometry (elastic spring-back) of parts after forming.

The simulated dimensions of the parts differ from those obtained when forming not more than 0.2 mm in height molded profile, which is of the order ±15%. The calculation of the required forming force coincides to the experimental data with an accuracy of ±10%. Thus, computer simulation of the forming process by plastic dies accurately reproduces the geometrical and force parameters of the real process.

Studied process of sheet metal forming has a main advantages of 3D printing: economic efficiency, as the forming by plastic dies are much cheaper then forming by metal dies or with the use of the mix of polyurethane and metal dies because of the low cost of the tool and a significant acceleration of production.The conducted research allow to recommend the technology of thin sheet forming with the use of plastic tools, for single and small batch production.

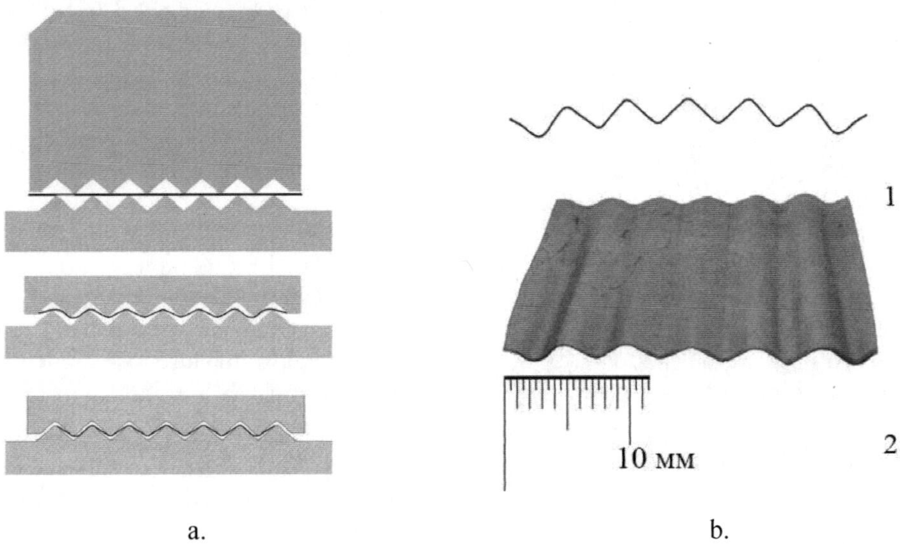

Figure 5. Stages of simulation of the forming process (a) and the shape of the product (b) obtained by simulation in the software package „Simufact.forming" (1) and experimentally (2).

The future development of plastic tool for metal forming technologies is seen in the improvement of technology of manufacturing plastic tool (their structure, method of applying a layers, etc.) and their application to a greater number of technological processes and to manufacture not only the forming elements but also the die blocks in assembly.

5. Conclusion

- On the example of forming process of thin sheet forming of aluminum it is shown that plastic tool, manufactured using 3D printing, can be successfully used for metal forming if the effective stresses in the tool do not exceed the elastic limit for plastic.
- When forming plastic tool is not destroyed, plastically not deformed and possess sufficient wear resistance. Lubrication of plastic dies is not required, as they have anti-friction properties. Technology of sheet forming with the use of a plastic tool can be recommended for single and small batch production.
- Computer simulation in the software package "Simufact.forming" accurately reflects the required force for thin sheet forming with plastic tool, distribution of stresses in the tool and also actually course of forming of the blank, including the shape changing of the part after the forming as a result of the elastic spring-back.

Acknowledgments

The authors express their gratitude to the Government of St. Petersburg and the Charity Foundation of Mr.V. Potanin for the financial support of this work.

References

[1] Henshaw S *What is 3D printing?* Retrieved: http://3dprinting.com/what-is-3d-printing/ (Accessed 03.03.2017)

[2] Gupta P and Lee J 1993 *The Society of manufacturing Engineers Conference*.pp 11-3

[3] Rosochowski A and Matuszak A 2000 *J. of Materials Processing Technology* **106** (1/3) pp 191-8

[4] Du Z H, Chua C K, Chua Y S, Loh-Lee K G and Lim S T 2002 *Int. J. of Advanced Manufacturing Technology* **19**(1) pp 411-7

[5] Cheah C M, Chua C K, Lee C W, Lim S T, Eu K H and Lin L T 2002 *Advanced Manufacturing Technology* **19**(7) pp 510-5

[6] Collins P C et al 2014 *JOM* **66** pp 1299–309

[7] Liewald M and de Souza J H C 2008 *Production Engineering Research and Development* **2**(1) pp 63-72

[8] Durgun I 2015 Rapid Prototyping J. **21/4** pp 412–22

[9] Leacock G, Cowan G, Cosby M, Volk G, McCracken D and Brown D 2015 *Key Engineering Materials* **639** pp 325-32

[10] Stratasys Ltd *Metal Hydroforming 3D Printing for Forming Tools* Retrieved: http://www.stratasys.com/solutions/additive-manufacturing/tooling/metal-hydroforming (Accessed 02.03.2016).

[11] Aksenov L B and Kononov I Y 2016 *Blanking production in machine building* **9** pp 22-4

[12] Prusa J *Prusa i3 Documentation* Retrieved: http://reprap.org/wiki/Prusa_i3 (Accessed 02.02.2017).

[13] Excell J and Nathan S 2010 *Engineer* doi:10.3201/eid1905.121071.

[14] *Simufact Simulation of Manufacturing: Simufactforming*. https://www.simufact.com/module-hot-forging.html . (Accessed 05.10.2017).

[15] Kaufman J G 1999 *Properties of Aluminum Alloys: Tensile, Creep, and Fatigue Data at High and low temperatures* (Washington, DC: ASM International)

[16] REC® LLC *Acrylonitrile butadiene styrene properties* Retrieved: http://www.rec3d.ru (Accessed 27.10.2017).

Information and communication technologies as an active principle of social change

D S Bylieva[1,*], V V Lobatyuk[1] and A V Rubtsova[1]

[1]Peter the Great Saint-Petersburg Polytechnic University, 195251, Polytechnicheskaya 29, Russia

*Corresponding author: bylieva_ds@spbstu.ru

Abstract. In the life of modern society information and communication technology occupy a significant place, being an active principle of social changes. Practically all processes, one way or another, go along with technologies. The total penetration of information and communication technologies into all spheres of life initiates the transformation of established social relations and norms. Particular attention we pay to the study of social norms in virtual reality. We observe the main forms of deviant behavior in the network, and the mechanism of social control, aimed at their regulation. At the same time, we prove that the areas of regulation are quite mobile and do not have a clear framework. Although we try to define three main areas of this type: area of law regulation, area of moral regulation, unregulated field. All the presented areas describe the visible phenomena. The essence of the virtualization process, however, turns out to be undetected. But after all, society itself can set a socially acceptable framework for the functioning of information and communication technologies, which can vary depending on religion, national circumstances or level of economic development. As a result, it can be noted that in the modern world the social landscape is changing dramatically.

1. Introduction

Information and communication technologies change the life of a modern person. Technical means become an integral part that accompanies a person almost constantly. Lack of access to the Internet is becoming a problem that makes it impossible to fully exist. Changes in people's lives due to information and communication technologies (ICT) are so global that they cause serious concern among researchers, who note that the digital technological environment programs new forms of interaction, communication and expression, new roles and forms of social institutions [1].

Moreover, the perfect essence of man exist in the the network mediated by the technical means. The man and his or her digital twin are not identical. It is doubtless that several people can stand behind a virtual person, and on the contrary one person can act for an unlimited number of network entities. Even one network identity shows the world a complex combination of desired perceptions and manifested personality traits. C. Diamant asserts that virtual person ‘is a representation that is neither fact nor fiction, interacting with humans yet on different evolutionary and ontological levels, thus virtually disembodied’ [2].

The significance of human life in the network, in which more and more areas from education to health care take part, requires comprehension of behavioral norms on the Internet. B. Peter shows the interdependence of the development of society and technology network [3]. Shipunova, J., & Kuznetsov, D. note that traditional ‘social order associated with the administrative and legal framework of the

IOP Conf. Series: Earth and Environmental Science **337** (2019) 012054 doi:10.1088/1755-1315/337/1/012054

government is being replaced by the most open communication in form of free associations of individuals' [4].

2. Methodology

The activity-communication approach is used in the study, which defines the essence and role of ICT in modern society. At the same time, a general scientific group of formal and logical methods is used at both the empirical and theoretical levels. These methods are analysis, synthesis, induction, deduction, abstraction and the systems approach. The systems approach allowed to reveal the integrity of the ICT phenomenon, to identify the relationships between technologies and social changes occurring under their influence. In addition, the principles of the systems approach were used in the analysis of forums, multiplayer games, cyber crimes and legislation in this area. We also used such methods of scientific research as dynamic and structural analysis, the construction of logic schemes, a graphical interpretation of theoretical information and empirical data. In this study we also use the interdisciplinary approach, which allowed to consider ICT as a complex social phenomenon from the perspective of social and philosophical problems based on the achievements of economics, sociology, and law.

3. Results and Discussion

Some researchers believe that virtual reality is free from social aspects [5].However, the existence and interaction in the network of an increasing number of individuals naturally forms the norms of behavior. Social norms on the Internet appear, disappear and transform, and have varying degrees of rigidity. In figure 1 we show the main forms of deviant behavior in the network, and the mechanism of social control, aimed at their regulation. The most rigid are the norms where the virtual world is most closely associated with the physical. There are legal rules regulating primarily the process of computer equipment, software, confidential information and money protection, which largely moved into the virtual sphere. A huge number of ways of electronic money thievery have been developed from phishing to carding [6]. Virtual reality does not allow to cause any physical harm. Nevertheless, the harm done to various individuals, organizations and social institutions can be very significant. It is easier to influence digital objects in the virtual world, as well as destroy traces and remain anonymous. Researchers point to a high degree of latency of Internet crimes, which is from several tens to several thousand percent [7].

However, in addition to the economic gain obtained by one means or another, ideology, passion, and even revenge can be the motive for destructive activity on the web [8]. And if, in the case of vested interests, deviance of behavior is recognized by the perpetrator, in the case of ideological, religious or political motives, people can act within their own subcultural norms. Western researchers point out that by separating themselves from violators of norms, they call themselves white hat versus black hat hackers which acknowledges the use of malicious or ethical applications of hacking [9].

Illegal assignment of objects in the virtual world looks completely different than in the physical world. Transfer from one owner to another in one case, and free copying of what one would like to restrict in access. Property as possession is significantly different from property as the desire to limit the range of users. In this area, falling under the regulation of copyright, the rules are more fluid and unclear. The ban on the dissemination of digital goods can sometimes be an infringement of the Internet freedom, and free use of articles, films, books, programs without the consent of the copyright holder for many people is a norm. Barbrook noted that the design of the Internet involves the easy creation and distribution of copies throughout the network, which allows us to talk about a real-life form of anarcho-communism [10]. Romele, A., Severo M. consider a specific type of network relationships associated with non-economic and the social distribution of goods, through the prism of digital gifts [11].

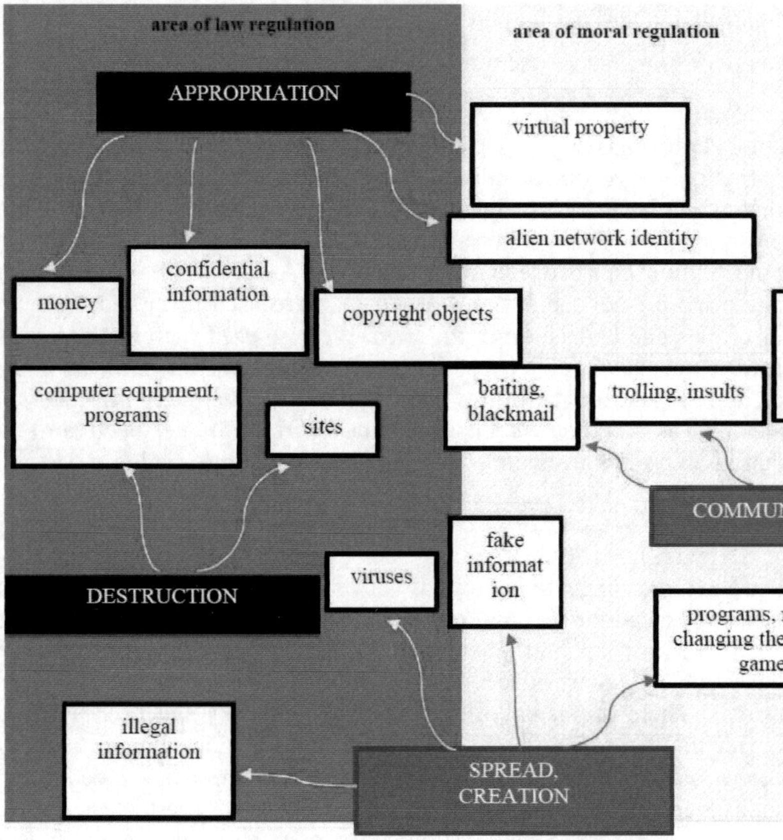

Figure 1. Social control in virtual reality.

The question of restrictions on the dissemination of a particular information is debatable. Many people recognize the value of the Internet as a space of freedom that is best suited for the open exchange of information by stakeholders. On the other hand, there is information that can cause harm. For example, an online crime is creating, distributing, and in some countries viewing information related to child pornography. Also, information that calls for suicide, related to drug trafficking and other socially dangerous topics, may be prohibited by law. At the same time, officially banned sites with similar topics can live in the "illegal" part of the network.

The area of social and communicative interaction in the network is governed by local rules of communication, prescribed by a communicative platform (general type: forums, dating sites, etc., and private type, for example, on pages of popular bloggers). Standards on the network communication sites are formed as a response of administrators to undesirable behavior. Therefore, these rules vary at different sites depending on the views of administrators, although they usually have much in common with the avoidance of conflict situations. It should be noted that in the absence of the possibility to impose penalties on a person behind one subject or another in the network, the range of sanctions for violating the norms in the network is quite limited: from deleting a disagreeable comment to blocking an account, that is, completely prohibiting the activity of a given virtual person in this "place". However, the maximum possible punishment is the eternal “ban” as the destruction of a virtual person in a certain “point” of the Internet space. Nonetheless it is not irreparable for a person, since he can create a new virtual person, an account and continue his activities.

Opposite to the legally regulated part of the Internet it is the pole of diverse virtual worlds, where the concept of norms has nothing to do with the world of off-line. The ability to create illusions of different nature, to accept any personage significantly expands the perception of what is permissible compared to what is accepted in the physical world. In the most illusory part of the network (in the game universes), the norms of behavior are determined only by the technical and gaming capabilities provided by the developer. Virtual theft, murder, kidnapping, destruction, vandalism in games most often do not appear to be a violation of moral standards, since such actions are possible, on the contrary, they can be the goal of the game. In the "least dense" area of virtual reality, the violation will be precisely the use of undocumented features, the desire to play the game not in the way its demiurge planned. Šisler proposes to determine the game code as a social norm that governs the behavior of players [12]. For example, using programs and macros (in particular, cheat codes, trainers, etc.) to overcome difficult or unpleasant episodes (using, for example, passing through walls, invulnerability, restoring "health", unlimited arsenal of weapons, etc.) will be the violation of the norms. In single-player games, such programs and macros appeared to be an area of personal characteristics and moral preferences. Sanctions could be made only by the mockery of developers (for example, in subsequent episodes of the game, the use of code caused sarcastic messages on the screen or a decrease in the useful game characteristics of the character to a minimum). Chen & Ong, exploring deception (cheating) in online games reveal a different understanding of what is cheat. For example, some players consider a violation of the rules only of the external program, which serve to change the game itself, but not intentionally designed by the creators of the game. Even more interesting phenomenon is when a cheat usually becomes a norm if it has been used in a game for a long time [13].

In modern multiplayer games, where players interact with each other, cheat codes are already a serious violation. To prevent them, special protective programs are developed for finding and applying sanctions. In professional e-sports there is a whole system of penalties for using of unauthorized codes: disqualification, penalty, prohibition of participation in professional leagues. In a number of Eastern countries (for example, South Korea) there is a penalty for developers of cheat software in the form of fines and imprisonment, but for users, the punishment remains usual - an eternal ban.

4. Conclusion

The question of the reality of the virtual goes beyond the philosophical consideration, the legalization of virtuality becomes a legal problem. In most countries (including Russia), the current regulation of game actions remains within the framework of user agreements that each player must accept when entering the virtual world. However, in eastern countries, there has already been a tendency to recognize virtual gaming money and things as the equivalents to objects in the physical world. Fearing responsibility, today the creators of multiplayer games indicate in the user agreement that all game objects remain their property, and the player gets only the opportunity to use a certain account and objects.

The behavior of a virtual person on the network is not limited by physical laws, and social norms can vary greatly depending on the "location" of the network. Areas of the Internet, tightly connected with the world outside the network, are regulated most strictly and are bound by legal norms. However, in this area hackers who do not pursue economic benefits do not consider their behavior deviant, but act according with the norms of their group. Features of virtual reality change attitudes towards norms, for example, attitudes toward property. Digital objects are recognized as common and copyright is rejected. It is difficult to link the ideal being of a person in a network with a physical figure. It makes difficult to apply sanctions for violating the norms. In the virtual environment, the maximum punishment is the complete cessation of the existence of this identity (ban), which does not prevent the creation of new incarnations and come back again. In the most non-physical reality of the virtual worlds, social norms are created by their demiurge and are not similar to those adopted outside. A violation will be the intervention in the "world order", a variety of "cheat codes" help to achieve goals by easier ways. Today the norms of behavior in the network continue to be built thanks to lawmaking, the sanctions of administrators, moderators of games, communication platforms and interaction of subjects in the network.

References

[1] Shipunova O D, Berezovskaya I P, Mureyko L M, Evseev V V and Evseeva L I 2018 *ESPACIOS* **39(40)** p 1015

[2] Shipunova O D, Berezovskaya I P, Mureyko L M, Kolomeyzev I V and Serkova V 2017 *Consciousness European Research Studies J.* **20**(4) pp 694-705

[3] Shipunova O D, Berezovskaya I P and Gashkova E M 2017 *St. Petersburg State Polytechnical University Journal. Humanities and Social Sciences* **8** pp 57-64

[4] Silkina G Y 2017 *in Proceedings of the 29th International Business Information Management Association Conference - Education Excellence and Innovation Management through Vision 2020: From Regional Development Sustainability to Global Economic Growth* pp 1165-76

[5] Didenko N I, Skripnuk D F and Mirolyubova O V 2017 *Proceedings of 2017 10th International Conference Management of Large-Scale System Development* p 8109611

[6] Diamant C 2018 *Caietele Echinox* **34** pp 218-28

[7] Nikiforova N 2017 *Technology and Culture* **58**(1) pp 300-1

[8] Shipunova O D and Kuznetsov D I 2015 *J. of Social Sciences MCSER Publishing* **6**(3) pp 251-60

[9] Taratuta E E 2007 *Philosophy of virtual* (St. Petersburg: St. Petersburg state University)

[10] Bataev A V 2017 *Proceedings of the 29th International Business Information Management Association Conference - Education Excellence and Innovation Management through Vision 2020: From Regional Development Sustainability to Global Economic Growth* pp 25-35

[11] Bataev A V 2017 *Proceedings of the 30th International Business Information Management Association Conference, IBIMA 2017 - Vision 2020: Sustainable Economic development, Innovation Management, and Global Growth* pp 89-99

[12] Lyukevich I and Nadezhina O S 2017 *Proceedings of the 30th International Business Information Management Association Conference, IBIMA 2017 - Vision 2020: Sustainable Economic development, Innovation Management, and Global Growth* pp 1-10

[13] Sklyarov S N and Evdokimov K N 2016 *J. of Criminology of the Baikal State University of Economics and Law* **10**(2) pp 322-30

[14] Leukfeldt R, Kleemans E and Wouter S 2017 *American Behavioral Scientist* **61**(11) pp 1387–402

[15] Furnell S 2002 *Cybercrime: Vandalizing the Information Society* (London: Addison-Wesley)

[16] Holt T J 2010 *Social Science Computer Review* **28** pp 466-81

[17] Barbrook R 1998 *First Monday* **3**(12)

[18] URL:http://firstmonday.org/ojs/index.php/fm/article/view/631/552 (accessed 4 December 2018).

[19] Romele A and Severo M 2016 *Culture & Society* **33**(5) pp 43–63

[20] Sisler V 2009 *Middle East J. of Culture & Communication* **2** pp 275–92

[21] Chen V and Ong J 2018 *Information, Communication & Society* **21**(2) pp 273-87

Innovative activity of enterprises in the digital economy

L Goncharenko[1,*], S Sybachin[1], S Rakhimova[1] and G Podbiralina[1]

[1]Plekhanov Russian University of Economics, Moscow, Russia

*Corresponding author: inn.invest@mail.ru

Abstract. This article describes core innovative activity vectors of enterprises in the digital economy environment. The main threats, associated with the innovative activity of enterprises, and obstacles to the digital economy development in Russia have been identified.

1. Introduction

Entrepreneurial activity is carried out in the context of a constant search for more efficient business conduct methods, which will help to gain competitive advantages, expedite the implementation of new products, and form the customer experience. In addition, the market demand is also moving towards more technologically-efficient and eco-friendly products. Often, new process solutions, successfully implemented in the market, may be a subject to the intellectual property rights (in particular, protected by patents), which allows an organization to receive unique competition protection rights for a certain period of time, which is a good opportunity to maximize economic benefits.

In the Federal Law "On Science and the State Science and Technology Policy" [1], the innovation activity is defined as an activity aimed to implement innovative projects as well as to develop and support operability of an innovation infrastructure. Innovations in a general sense are characterized as a new or significantly improved product (goods, service) or process, a new sales method or a new organizational method in business practices, workplace organization or in external relations, which has been brought into use [1]. However, there are many other definitions that may differ from the definition stated in the law. Nevertheless, despite the difference in definitions, scientific and technological innovations are changing business models in seemingly well-established spheres. A typical example is the Netflix company, which has successfully implemented an artificial intelligence technology in the entertainment industry.

Another defining aspect of this paper is to describe the "digital economy" term. Yet a decade ago, there was no practical difference between the digital and IT words. However, for the time being, the difference between these terms is becoming more obvious. IT is a narrower term that defines technology as a work tool. In its turn, digital means qualitative structural changes as a result of the implementation of IT technologies, which modify the business - end user interaction models and expand the market. The possibility of implementing innovative solutions directly at manufacturing plants is of specific interest.

2. Enterprise Innovative Activity Management

Supply and demand studies in the IT solution market show that business processes, related to the development and implementation of advanced solutions by production organizations, have changed in this sector. The total number of companies, which have own in-house departments for studying advanced technologies, has decreased over the past ten years. Despite the availability of resources (e.g., business incubators, start-up platforms, crowd sourcing platforms, manufacturing companies, open source software environments, university laboratories), most enterprises still rely on technology providers and services provided by professional consultants, in particular, industry analysts& developers as well as on competitive intelligence data. Establishing project teams and using outsourcing services (including those outside the country where the organization runs its core activities) have become more popular. Investments in key technologies by the industry with associated benefits from investments are showed in the table 1.

Table 1. Investments in key technologies by the industry with associated benefits from investments[a].

	Internet of things	Artificia l intellig ence	Robotics	Augment ed reality	3D printers	Dron es	Virtual reality	Blockc hain
Today:	64%	50%	45%	7%	8%	10%	4%	1%
In 3 years:	50%	54%	46%	12%	14%	24%	6%	5%

[a]study data of PwCDigital IQ

3. Internet of Things

The concept of the Internet of Things first appeared in 1999. However, at that time such decisions were not yet realizable. At the moment, the Internet of Things technology finds its application in all spheres of the economic activity, and the industrial sector is no exception. In a general sense, the idea of the Internet of Things is that enterprises integrate into a common system based on a digital platform. As a result, major part of typical processes is controlled without human intervention, and information from any point of the chain goes to a common data processing center. At the same time, global network connectivity creates more vulnerabilities for cyber threats. That's why the Internet of Things based investments in process automation also involve the development of security systems. The Internet implementation at enterprises is already a rather actively used service provided by such companies as Cisco, SAP, etc.

4. Artificial Intelligence

Artificial intelligence (AI) is characterized as: "designing and construction of intelligent agents, which perceive environmental objects and take actions affecting the environment" [13]. The main difference between the artificial intelligence and software is the ability to make independent decisions and "take action". Usually, the development of artificial intelligence is advisable in the presence of large data arrays (big data).

The number of start-ups, which render services for the development and implementation of technologies based on artificial intelligence in various kinds of enterprises, as well as venture investments in these startups, have grown significantly. AI developing start-ups (pcs) are showed in figure 1.

IOP Conf. Series: Earth and Environmental Science **337** (2019) 012055 doi:10.1088/1755-1315/337/1/012055

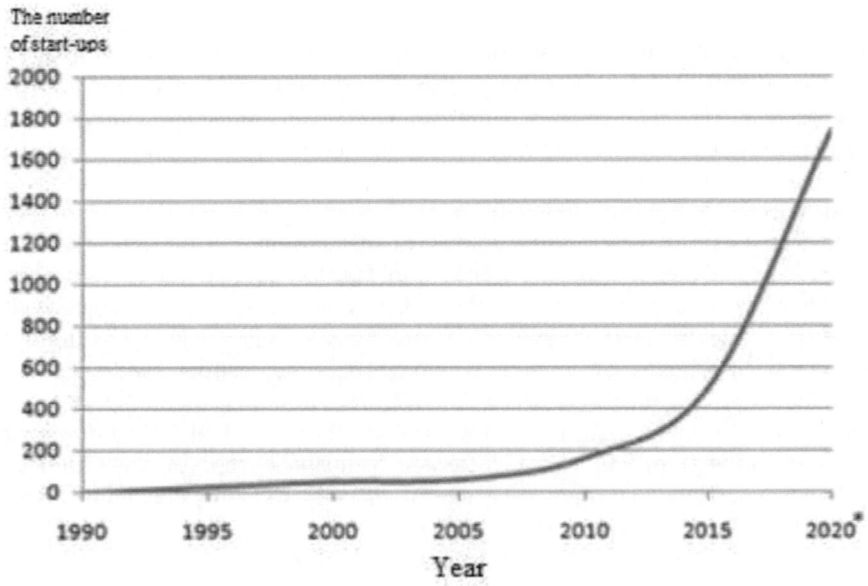

Figure 1. AI developing start-ups(pcs) base on data presented by Standford AI lab, AI Index.

*- predicted value

Caterpillar, an American manufacturer, is among the companies, which successfully implemented the artificial intelligence technology. This company applies an assisting system to analyze and predict the demand for new and already sold products. The company uses a video analysis system to evaluate the efficiency of products and their interactions during the working cycle [19].

5. Innovative activity on non-digital technology implementation

Innovation activities are often associated exclusively with IT-technologies. However, this is not entirely true. The industrial sector of the economy faces problems, whose solution can be found in such branches of science as biotechnology, neuroscience and renewables. Using power purchase agreements (PPA) is a good example of application of these solutions. In a standard model of these agreement, the seller establishes a power (solar or wind) generation system and provides its maintenance on the buyer's territory. Similar agreements are actively used by IT companies to service data centers. Amazon, Google and Microsoft are the leaders in the use of alternative energy sources [18]. The use of power purchase agreements are showed in figure 2.

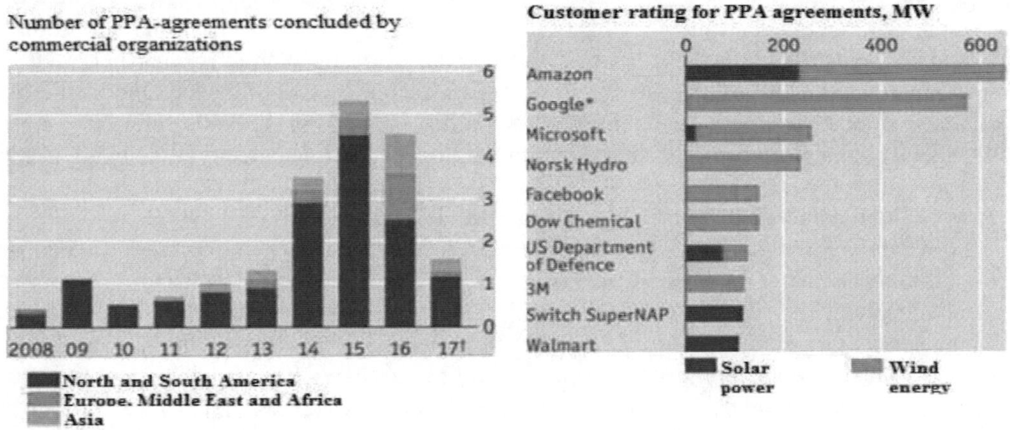

Figure 2. The use of power purchase agreements.

IOP Conf. Series: Earth and Environmental Science **337** (2019) 012055 doi:10.1088/1755-1315/337/1/012055

The technological solution under review has already managed to go from the category of innovative technologies to the category of actively used practices. The renewable energy sector is one of the most developing ones. Despite the US withdrawal from the Paris Agreement, regulating greenhouse gas emissions, many organizations (Caterpillar, Amazon, Google, etc.) have decided not to abandon sustainable development policies and continue to implement and develop renewable energy systems.

6. Innovative activities in the Russian Federation

The Federal State Statistics Service (Rosstat) publishes official data on the innovation activities of Russian enterprises. It should be noted that only a negligible partof organizations (about 9% on average) uses and implements innovative approaches to doing business. However, it should be taken into account that innovation costs include, but are not identical to, R&D expenditures, which account to a relatively small percentage. According to Rosstat, only 1326 commercial organizations conducted R&D in 2016. The total costs for the same year amounted to RUB 8,737,778.7 million [20]. At the same time, a group of European Union member countries (28 countries) spends about 1.3% of GDP on innovations, with its value of EUR 15,326,468.0 million in 2017. Russian companies spend on innovation a little more than 0.5% of the country's GDP [21]. Innovative activity of enterprises are showed in figure 3.

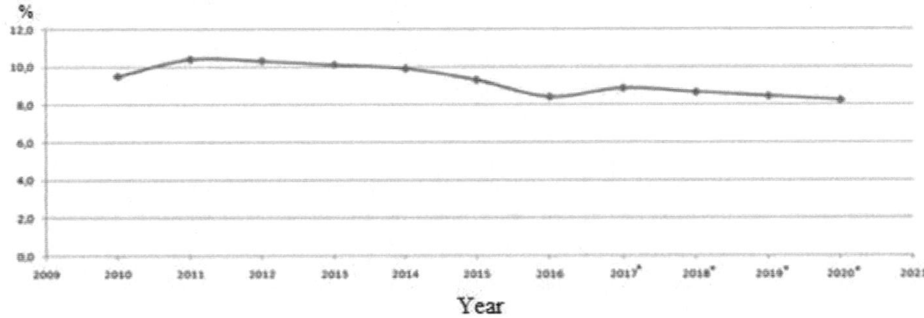

Figure 3. Innovative activity of Russian enterprises.

7. Threats related to innovation activities

Recent studies show that, in spite of the improvement, systematization and rationalization of business processes, which may become a result of a competent implementation of innovations, organizations may suffer from a loss of confidence from their clients, especially in case of implementation of IT technologies. According to PwC, 67% of managers of the world's largest companies expects a drop in stakeholder confidence, specifically, due to the implementation of artificial intelligence and block chain technologies [11].

Furthermore, it's obvious that these innovations require an organization to step up to a completely different level of protection against cyber risks. The epidemics of the WannaCry and Petya viruses in 2017 showed that Russia was among the countries most affected by these attacks. One of the reasons behind this is the unpreparedness of enterprises for digital space threats. This applies to both cyber attacks and social engineering methods, encouraging employees to violate safety rules by their actions.

8. Obstacles to the development of the digital economy in Russia

It becomes obvious that the market is already oriented towards the digital economy and its consistent propagation is only a matter of time. However, there are some factors, which can slow down the development of enterprises. The first and most obvious are the complications between suppliers of products and services, which are capable to transfer business processes to the digital level, and organizations. Unfortunately, foreign developers, the interaction with which is complicated by sanctions, are at present major suppliers of technologies. Not all companies are capable to finance their own developments, and the use of free software is also associated with a fairly high level of implementation of IT technologies at enterprises. Restrictions, imposed by the sanctions, also reduce the potential financial sources in foreign markets, which may be used for funding digital conversion of Russian companies. Another factor is the legal risks, caused by a legislative requirement to acquire, store and process information, as well as the lack of unambiguous position and regulation on some issues directly related to the innovation implementation prospects (cryp to currency, intellectual property protection, etc.).

In conclusion, it should be noted that innovations at present are not only a method of gaining competitive advantages, but they also move into the category of factors, vital for sustainable development of enterprises. Business conduct methods become obsolete so quickly that business models need updates almost immediately after their implementation. Innovations enable more efficient cost management, which is necessary to maintain competitive positions in the market. The digital economy opens up new potential for growth and development not only for financial and IT companies, but also for organizations in the manufacturing sector.

Acknowledgments
This article was prepared as part of the project section of the government contract as requested by the Ministry of Education and Science of the Russian Federation on the subject formulated as 'Development of Methodological Principles and Organizational Economic Mechanism of Strategic Management of Economic Security in Russia' (Assignment No 26.3913.2017/PCh)

References

[1] Federal Law No. 127-FZ "On Science and State Science and Technology Policy" dated 23 August 1996 (as amended as of 23.05.2016) (entered into force on 01 January 2017)

[2] Alsaad A, Mohamad R, Taamneh A and Ismail N A 2018 *Technology Analysis and Strategic Management* **30**(8) pp 980-92

[3] Bondarenko T G, Orekhov S A, Sokolnikova I V, Soltakhanov A U and Khmelev I B 2018 *Espacios* **39**(36) p 12

[4] Deng Z, Jean R, Sinkovics J B and Rapid R R 2018 *J. of Int. Business Studies* **49**(8) pp 1010-32

[5] Eckhardt J T, Ciuchta M P and Carpenter M 2018 *Strategic Entrepreneurship J.* **12**(3) pp 369-91

[6] Frolova I V, Panfilova E A, Matytsyna T V, Lebedeva N Y and Likhatskaya E A 2016 *Social Sciences and Interdisciplinary Behavior - Proceedings of the 4th International Congress on Interdisciplinary Behavior and Social Science ICIBSOS* pp 317-24

[7] Geraskina I N, Zatonskiy A V and Petrov A A 2017 *Int. J. of Civil Engineering and Technology* **8**(10) pp 1432-47

[8] Gryzunova N V, Zaharova G S and Ordov K V 2018 *Espacios* **39**(6) p 32

[9] Kazakova N A, Bolvachev A I, Gendon A L and Golubeva G F 2016 *Studies on Russian Economic Development* **27**(6) pp 638-48

[10] Kuznetsov Y V, Kapustina N V and Maslova E V 2017 *Managing Service, Education and Knowledge Management in the Knowledge Economic Era - Proceedings of the Annual International Conf., on Management and Technology in Knowledge, Service, Tourism and Hospitality* pp 19-22

[11] Moritz E B 2017 *20 Years Inside The Mind Of A CEO...What's Next?* 20th Survey of CEOs of the World's Largest Companies, PwC, (CEOSurvey PwC) - https://www.pwc.com/jg/en/

[12] Romanchenko O, Shemetkova O, Piatanova V and Kornienko D 2019 *Advances in Intelligent Systems and Computing* **850** pp 245-53

[13] Russell S and Norvig P 2009 *Artificial Intelligence: A Modern Approach* (Upper Saddle River, NJ: Prentice Hall)

[14] Sedova N V, Gagiev N N and Melnikova D M 2017 *European Research Studies J.* **20**(2) pp 251-60

IOP Conf. Series: Earth and Environmental Science **337** (2019) 012055 doi:10.1088/1755-1315/337/1/012055

[15] Tauscher K and Laudien S M 2018 *European Management J.* **36**(3) pp 319-29
[16] Trachuk A and Linder N 2018 *J. of Innovation and Technology Management* **15**(3) p 1850027
[17] Zimin A, Otto V, Filimonova N, Fedosova R and Kuznetsov Y 2016 Advanced Science Letters **22**(8) pp 2002-6
[18] Big Business Sees The Promise Of Clean Energy The Economist https://www.economist.com/business/2017/06/10/big-business-sees-the-promise-of-clean-energy Retrieved 2017-06-25
[19] Official website of Caterpillar - https://www.caterpillar.com/en/
[20] Official website of the Federal State Statistics Service - http://www.gks.ru/
[21] Official websiteof the Organization for Economic Co-operation and Development (OECD) - http://stats.oecd.org/

Possible use of processed ferromanganese concretions for production sewage purification

M A Sulimova[1,*]

[1]Mining Museum, Saint-Petersburg Mining University

*Corresponding author: kafmet@spmi.ru

Abstract. This article describes a possible scenario for ferromanganese concretions usage as a sorbent for purification of wastewater obtained from mining enterprises.

1 Introduction

Sewage is produced by all mining enterprises. Most of the enterprises dump sewage in nearby reservoirs that will eventually lead to environmental disaster. In this regard, penalties are increased and measures for sewage dumping become tougher [1*]. However, according to researches [2], pollution of water reservoirs of the Russian Federation increases.

Huge variety of the existing chemical compositions of water systems, conditions of their formation and the necessity of individual researches for each case makes waste water purification from phenols, cyanides, oil products and surfactant almost impracticable task [3-8]. The majority of efficient ways of deep cleaning are associated with economic and resource expenses, usage of scarce reagents with its subsequent regeneration, utilization or waste disposal. Therefore search for new efficient ways of sewage purification is still relevant [9-14]. Its implementation will allow one to reduce negative impact on environment and to decrease investment and operational expenditure for environmental actions.

A great number of enterprises have water treating facilities [15-17] which, however, are not capable to cope with the amount of pollutants in drains. To choose the optimal sewage disposal method one should take into account sewage characteristics and the possibility of pollutants extraction.

There are a lot of ways for sewage disposal. Destructive methods of sewage disposal comprise oxidizing methods, thermooxidizing, electrochemical oxidation and hydrolysis [18-22]. Copper, manganese and their compounds, as well as metals of variable valency mainly of the VIII group, their oxides and salts are used as catalysts. This leads to repeated pollution of drains. One of the most widespread and inexpensive purification methods is sorption. Any sorbent has to correspond to the following characteristics: high sorption capacities, high reagent resistance to acids and alkalis. The sorbent has to be low-toxic or non-toxic, it has to possess low abradability, high mechanical strength, and low prime cost [23, 24].

Table 1. FMC sample parameters: specific surface area, porosity, etc.

Characteristic	Quantity
Specific area capacity	3116 mm^3/g
Number of closed pores	46564
Specific volume of closed pores	3.2 mm^3/g
Specific surface area of closed pores	345.9 mm^2/g
Occluded porosity	0.64 %
Total specific volume of occluded porosity	2620.5 mm^3/g
Open porosity	84.11 %
Total specific volume of pore space	2623.7 mm^3/g
Common porosity	84.21 %

Ferromanganese concretions (FMC) are one of the most perspective sorbents, due to its low prime cost in comparison with other sorbents [25]. FMC distribution territory is quite large: from the Arctic tundra to tropics [3]. FMC chemical and phase composition is quite diversified, and is able to change depending on the deposit and depth of its burial, as well as on the time of their formation.

Table 2. Chemical composition of ferromanganese concretions of various deposites [25, 26].

Macrocomponents		Microcomponents	
Element	Mass fraction %	Element	Mass fraction 10-4 %
Fe	19.97	Ni	228
Mn	12.4	Co	120
P	0.94	Cu	52
Si	11.0	Zn	166
Al	2.0	Mo	254
Ca	1.35	Pb	36
Mg	0.83	Ba	2500
K	1.36	V	114
Na	1.0	W	36
CO2	1.65	Ge	1.5
Copr	1.25	Ga	11
S	0.08	Cr	50
Ti	0.22	Zr	93
Mn/Fe	0.62	Hf	2.2
Ta	$0.5*10^{-4}$	Sr	410
As	$327*10^{-4}$	Rb	62
Th	$5.2*10^{-4}$	Cs	2.5
Ra	$15.4*10^{-4}$	Sc	6

The surface morphology of FMC ore is heterogeneous (Figure 1), some part of the surface has dense texture formed from globules, having pores of micron size between globules. Sample morphology is more bulky at the FMC granule cleavage

IOP Conf. Series: Earth and Environmental Science **337** (2019) 012056 doi:10.1088/1755-1315/337/1/012056

than at its surface. Also Figure 1 presents tomographic images of FMC sample, which show specific surface and porosity of structure.

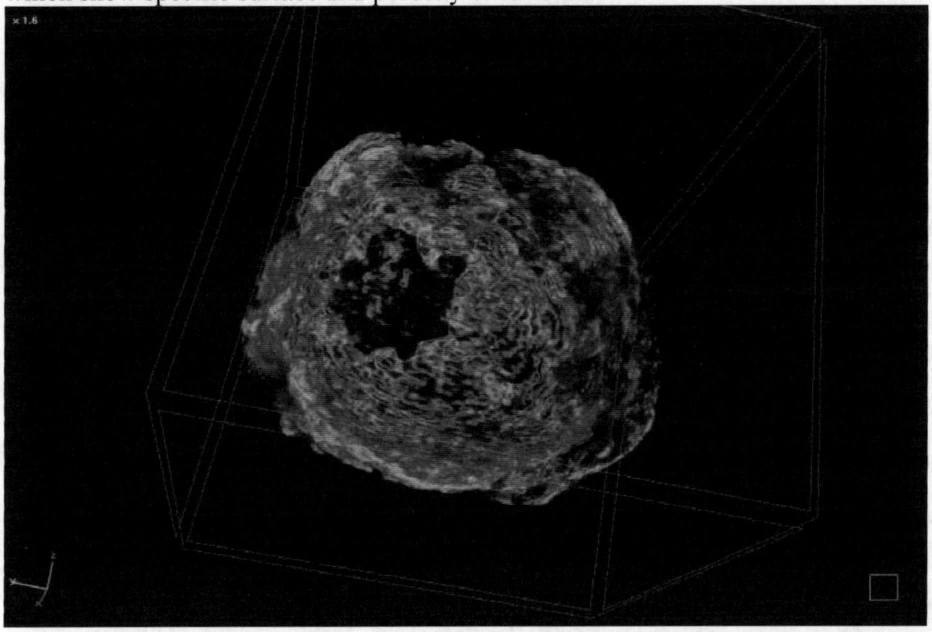

Figure 1. Tomography of FMC sample.

The concretions phase composition was defined by x-ray diffraction method using Shimadzu XRD-7000. The graphs presented in Figures 2-3 show that the main crystal phase of concretions is manganese oxide. By comparing this result with data from Table 1, one can conclude that iron-containing component of concretions is in X-ray amorphous state. Usually X-ray amorphous hydroxides have a developed surface and, as a result, high sorption capacity.

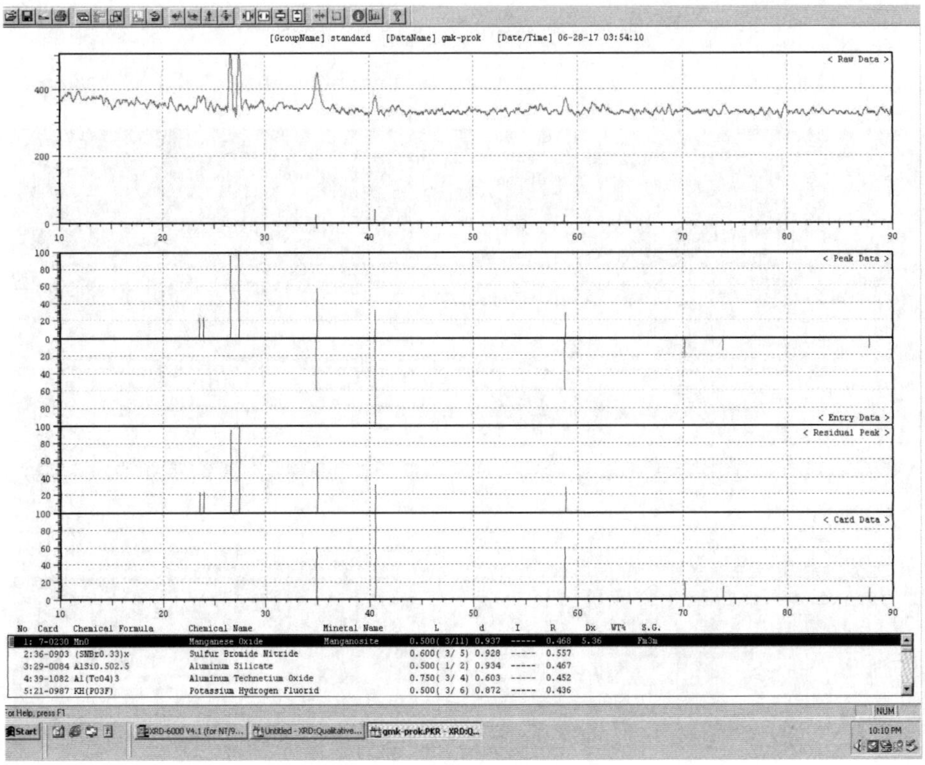

Figure 2. X-ray diffractometry of the FMC sample.

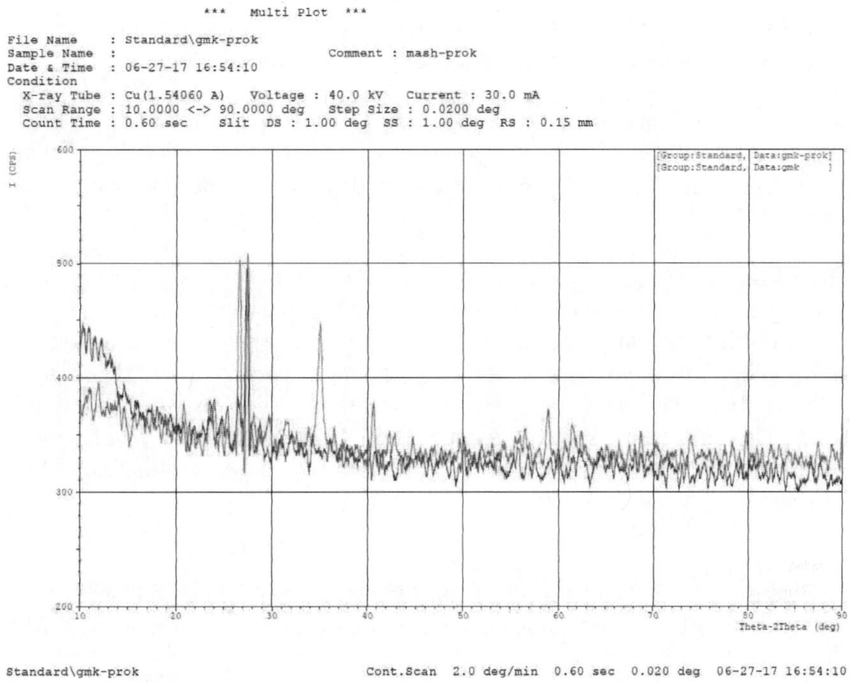

Figure 3. X-ray diffractometry of the FMC sample.

According to the results of X-ray diffractometry, the ferromanganese concretion samples are completely amorphous.

Ferromanganese concretions are extracted in order to obtain manganese concentrates and iron oxides. The samples less than 0.01 mm are dumped. This fraction can be used as a sorbent. Thus usage of ferromanganese concretions as a sorbent for sewage disposal is perspective and low-cost [26].

Table 3. Results of FMC usage as a sorbent for the reference pollutants of sewage from metallurgical enterprises [27].

Pollutant	MPC for water for housekeeping needs	Concentration of pollutants before interreacting with FMC	Concentration of pollutants after interreacting with FMC
Phenol	0.001 mg/L	1-1.2 g/L	0.001 mg/L
Nitrile	0.1 mg/L	1.1 g/L	0.07 mg/L
Dissolved oil product into water	0.1 mg/L	0.5 g/L	0.2 mg/L
Anionic surface active agents	0.5 mg/L	0.3 g/L	0.4 mg/L
Cu^{2+}	0.01 mg/L	5.1 g/L	3.9 g/L
Pb^{2+}	0.006 mg/L	3.933 g/L	2.8 g/L
Co^{2+}	0.01 mg/L	7.4 g/L	7.04 g/L
Hg^{2+}	0 mg/L	5 g/L	2.6 g/L

IOP Conf. Series: Earth and Environmental Science **337** (2019) 012056 doi:10.1088/1755-1315/337/1/012056

The method is based on oxidizing adsorption using ferromanganese concretions of the Finnish Gulf, which possess an oxidizing function and catalytic properties [24]. The sewage purified using FMC might contain suspended matters, heavy metals and oil products. These pollutants do not influence the FMC oxidizing ability. During sorption of cations of heavy and non-ferrous metals on FMC, there is no decrease in sorbent capacity, as the process is driven by ion-exchange mechanism. Thus, the presence of heavy metals cations in sewage does not have an impact on oxidizing ability of FMC.

2 Conclusion

Ferromanganese concretions have high specific surface area, due to the structure amortization due to a large number of pores and roughnesses of structure, which are able to act as active sites. The FMC chemical composition is composed of amorphous forms of manganese and iron oxides, which are a catalyst and an oxidizer. This increases sorption capacity of FMC and, as a result, increases sorption capacity of a sorbent based on it. Therefore FMC is reasonable to be used as a sorbent for sewage treatment.

References

[1] Bazhin V Yu 2017 *Notes of the Mining Institute* **227**(1) pp 547-53

[2] State report *On the state and use of water resources of the Russian Federation in 2016*

[3] Aleksandrov P A, Anikeeva L I, Andreev S I and Roosters S I 2009 *Thalassochemistry of World Ocean Ore Generation* (SPb.: Federal State Unitary Enterprise VNII Okeangeologiya)

[4] Barashkova P S, Molodkina L M and Korovina M D 2017 *Magazine of Civil Engineering* **71**(3) pp 68-75

[5] Baturova L P, Demidov A I, Maslov V V and Ferapontova O E 2015 *Russian J. of Applied Chemistry* **88**(1) pp 59-64

[6] Politaeva N, Bazarnova Y, Smyatskaya Y, Slugin V and Prokhorov V 2017 *J. of Industrial Pollution Control* **33**(2) pp 1617-21

[7] Chechevichkin A V, Vatin N I, Samonin V V and Grekov M A 2017 *Magazine of Civil Engineering* **76**(8) pp 201-13

[8] Politaeva N A, Slugin V V, Taranovskaya E A, Alferov I N, Soloviev M A and Zakharevich A M 2017 *Izvestiya Vysshikh Uchebnykh Zavedenii, Seriya Khimiya i Khimicheskaya Tekhnologiya* **60**(7) pp 85-90

[9] Barashkova P S and Molodkina L M 2018 *Magazine of Civil Engineering* **77**(1) pp 112-20

[10] Arakcheev E N, Brunman V E, Brunman M V, Brunman M V, Konyashin A V, Dyachenko V A, Petkova A P and Nekrasov R E 2017 *Gigiena i Sanitariya* **96**(3)

[11] Andrianova M and Bondarenko E 2016 *MATEC Web of Conferences* **53** p 01037

[12] Politayeva N A, Smyatskaya Y A and Slugin V V 2018 *Comptes Rendus de L'Academie Bulgare des Sciences* **71**(6) pp 766-71

[13] Ulyanova V V and Sobgayda N A 2016 *Ecology and Industry of Russia* **20**(7) pp 4-9

[14] Ul'yanova V V and Sobgaida N A 2015 *Chemical and Petroleum Engineering* **51**(1-2) pp 133-7
[15] Aksenov V I, Ladygichev M G, Nichkova I I, Nikulin V A, Klein S E and Aksenov E V 2005 *Water industry industrial enterprises*
[16] Aksenov V I 1983 Closed water management systems of metallurgical enterprises, Metallurgy (1983)
[17] Matveyeva A N, Pakhomov N A and Murzin D Y 2016 *Industrial and Engineering Chemistry Research* **55**(34) pp 9101-8
[18] Qadeer R and Rehan A H 2002 *Aq. J. Chem.* **26** p 357
[19] Palma M, Paiva J L, Zilic M and Converti A 2007 *Chem. Eng. Process.* **4** p 7646
[20] Xu A, Yang M, Du H and Sun C 2009 *Appl. Clay Sci.* **43** p 435
[21] Pan B C, Xiong Y, Su Q, Li A M, Chen J L and Zhang Q X 2003 *Chemosphere* **51** p 953
[22] Pavlov R D 2010 *Mining Notes* pp 194-9
[23] Shaderman F I 1998 *Laboratory and technological research of mineral raw materials* (Moscow: IMGRE) pp 25-6
[24] Bochkarev G R and Pushkareva G I 1998 On a new natural sorbent for the extraction of metals from aqueous media *FTPRPI* pp 46-51
[25] Sulimova M A, Sizyakov V M, Litvinova T E and Vasilyev V V 2016 *Black metals* **8** pp 43-9
[26] Nefedov Yu A 2010 *Geology and minerals of the world ocean* pp 40-7
[27] Sulimova M A and Litvinova T E 2016 *Geoconference SGEM 16* **5**(2) pp 569-76

IOP Conf. Series: Earth and Environmental Science **337** (2019) 012057 doi:10.1088/1755-1315/337/1/012057

Evaluation of the efficiency of rectangular separators to collect the particles from the gas flows

Oksana S Dmitrieva[1*], Vu L Nguen[2], Nikolay D Yakimov[2] and Evgeniy G Sheshukov[2]

[1]Kazan National Research Technological University, 420015, Kazan, Karl Marx str., 68, Russia
[2]Kazan State Power Engineering University, 420066, Kazan, Krasnoselskaya str., 51, Russia

* Corresponding author: ieremiada@gmail.com

Abstract. Analysis of the separation devices` operation and improvement of efficient gas mixture purification processes is relevant issue and is a great interest for the specialists of different industrial brunches. Modernization of internal devices in the cases of existing separators is the main goal in the field of increasing the efficiency of fine dispersed particles` collection process. Design of a rectangular separator with a number of I-beams has been developed. Operation principle of the device is described. When the multiphase flow moves between the device elements, the centrifugal force appears, that ensures the coagulation of finely dispersed liquid drops, facilitating their uniform settling on the entire surface of I-beams. The advantage of these separators is noted. High values of centrifugal force are achieved at relatively low gas flow rates that allows to carry out the separation processes at low energy costs. Numerical study of the gas mixture flow through the rectangular separator is shown at different values of the flow rate and different diameter of particles. The results of separation efficiency are shown for different designs of I-beams. Increase in the rate leads to efficiency increase. Ranges of efficient operation of the separator have been determined. Studies showed the advisability of using the rectangular separators to purify the gases from finely dispersed particles.

1. Introduction

The main share of electricity in the Russian Federation is generated by thermal power plants. Combustion of a large amount of fossil fuels leads to pollution of atmosphere with toxic substances. The amount of emissions of harmful substances is usually determined by the type of the fuel used and the combustion process efficiency [1–5].

Devices for purification of gases from solid and liquid impurities play an important role when equipping the power, chemical, petrochemical plants with process devices. Nowadays, liquid impurities are collected by means of different devices at the enterprises, equipped with gas treatment facilities. As a result, the gas purification process becomes complicated and cost intensive, while over 60% of the costs is accounted for desulphurization. In recent years, a number of countries (Japan, USA, Germany, Russia, Poland, etc.) are developing the new methods and devices, designed for simultaneous purification of flue gases from several pollutants. New technologies are more compact and less cost intensive [6–11]. A number of scientific papers have been published regarding gas purification [12–19], as well as some experience in solving the complex issues to increase the energy& resource process efficiency has been accumulated. At the same time, the issue of designing the new gas purification devices and improving their efficiency is still relevant.

2. Methods of studies

Purpose of this paper is to study the process of purification of flue gases from liquid impurities by means of the developed separator [20]. Since there is a significant release of combustion products into the environment as a result of intensive operation process of power plants, it can be said that the current measures are not so effective, but cost-intensive. We propose to emit the flue gases evenly through the separation columns, for complex purification of them. In order to calculate the rate of flow and collecting the particles in rectangular separators, the finite-element method in ANSYS Fluent software program was applied.

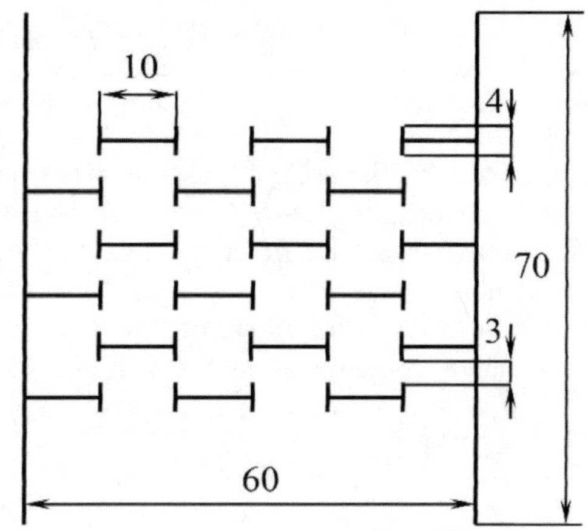

Figure 1. Analytical model of device.

The separator consists of several rows of I-beams, arranged in a rectangular case (Figure 1). In order to ensure a minimum deflection, the beams are connected with each other by means of two transverse plates, which are height wise fixed to the case of device. The operation principle of device is the following: when a multiphase flow moves between the device elements, the centrifugal force appears, which provides coagulation of particles, contributing to their uniform settling on the entire surface of I-beams. Availability of several rows of I-beams increases the efficiency of collecting the fine dispersed particles due to a more structured gas flow. In order to ensure the inform flow in the device, and hence the minimum hydraulic resistance, the subsequent row of I-beams shall be located from the previous one at a distance equal to 52–60% of the length of I-beam [21].

In order to carry out the numerical studies, 3D model of device with a width of 60 mm, a length of 70 mm and a height of 20 mm was created. Width of elements was b = 10 mm, height of elements – 4 mm, width of gaps – 3 mm (Figure 1). The boundary conditions were the following: rate at the inlet of device, Win and pressure at the outlet of device, the medium was – air at a temperature of 20°C. Diameter of drops varied in the range of d = 1–10 μm, the initial rate of the drops was 0 m/s. Drops were located at the inlet of device. Since the diameter of drops was small enough, in order to simplify the calculations and reduce the time for carrying out of them, it was decided to ignore the gravity. When the drop came into contact with any device surface, it was considered as settled.

Operation efficiency of device was determined by the following formula:

$$E = 1 - n_l/n_0, \quad (1)$$

where n_0 – number of particles at the inlet of device, n_l – number of particles at the outlet of device.

3. Results of studies and discussion of them

As a result of studies, it was found that an increase in the flow rate leads to an increase in purification efficiency. This is due to the growth of centrifugal force. Slight impact of the gas flow rate on the particles with a size of 1 μm is probably explained by low inertia of them (Figure 2).

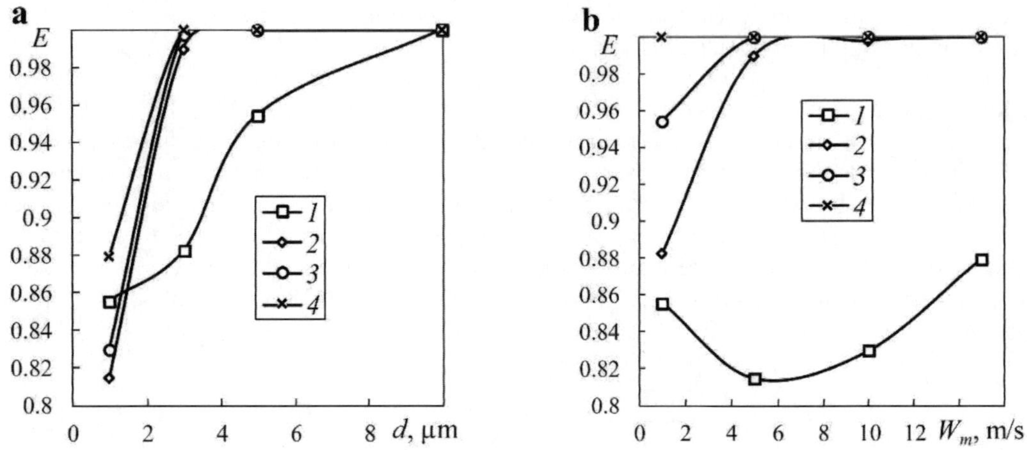

Figure 2. Dependency of efficiency of E on diameter of particles d (a) Wm, m/s: 1 – 1, 2 – 5, 3 – 10, 4 – 15) and average gas flow rate Wm (b) d, μm: 1 – 1, 2 – 3, 3 – 5, 4 – 10), b = 10 mm.

Width of elements has insignificant impact on efficiency of device. As the width is increased, the efficiency slightly decreases. For example, when particles with a diameter of 3 μm settle on the elements with a width of 10 mm, the efficiency varies in the range of 0.88 to 1; as well as on elements with a width of 20 mm – range of 0.97 to 0.99; on elements with a width of 40 mm – range of 0.85 to 0.96 (Figure 3).

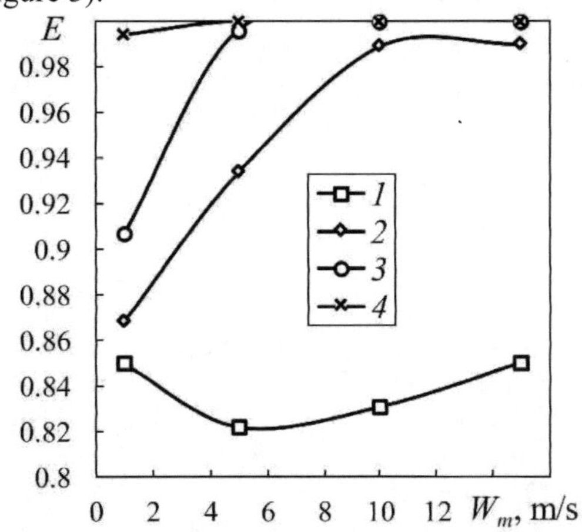

Figure 3. Dependency of efficiency of *E* on average gas flow rate W_m, b = 20 mm, d, μm: *1* – 1, *2* – 3, *3* – 5, *4* – 10.

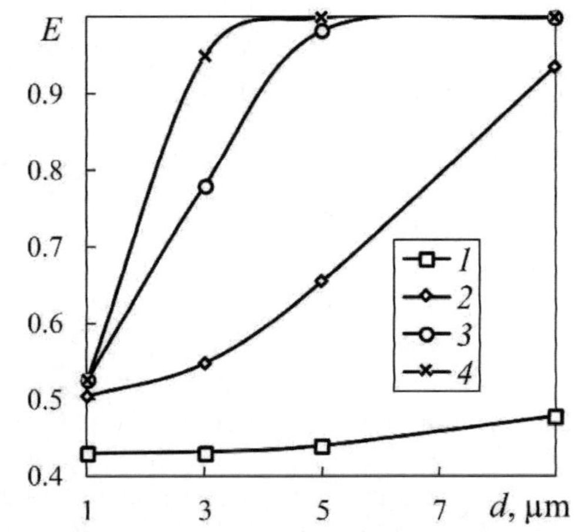

Figure 4. Dependency of efficiency of *E* of one row of device on diameter of particles *d*. Settled particles – polyurethane, b = 10 mm, W_m, m/s: *1* – 0.1, *2* – 1, *3* – 5, *4* – 10.

It was found that particles with a diameter of 1 μm settle with an efficiency of about 42%. Rate has quite insignificant impact. Particles with a diameter of 9 μm begin to settle on the surface of elements with an efficiency close to 100% at rate of 2–3 m/s (Figure 4).

4. Conclusion

Thus, the advantages of the developed device are simple manufacturing design, low metal consumption, while providing the required stiffness of the load-bearing elements. High values of centrifugal forces near I-beams in the proposed device for the fine dust/gas purification lead to an increase in the overall efficiency of collecting the solid dispersed particles.

Higher values of centrifugal forces at relatively low rates of heterophase medium are created in a rectangular separator. This is due to the fact that at the same width of separators, the average radius of curvature of the circulating currents of I-beams, having concavity on both sides, is less almost by 2 times, that leads to a significant change in the gas flow structure and an increase in the centrifugal force. This means that settling efficiency of medium- and fine dispersed particles by the separators, designed in the form of I-beams, is higher when other operation conditions are equal.

As a result of studies, the following conclusion can be drawn: the devices are operating with sufficient efficiency and can be used for pre-collecting the fine dispersed particles prior to the fine purification filters.

Acknowledgments
The study was performed with funding from the RF President's grant project No. MK-4522.2018.8.

References

[1] Prabhansu M Kr, Karmakar P C and Chatterjee P Kr 2015 A review on the fuel gas cleaning technologies in gasification process *J. Environ. Chem. Eng.* **3**(2) 689–702 DOI: 10.1016/j.jece.2015.02.011

[2] Merentsov N A, Bokhan S A, Lebedev V N, Persidskiy A V and Balashov V A 2018 System for Centralised Collection, Recycling and Removal of Waste Pickling and Galvanic Solutions and Sludge *Mater. Sci. Forum* **927** 183–9 DOI: 10.4028/www.scientific.net/MSF.927.183

[3] Wang C, Liu Y, Zheng S and Jiang A 2018 Optimizing combustion of coal fired boilers for reducing NOx emission using Gaussian Process *Energy* **153** 149–58 DOI: 10.1016/j.energy.2018.01.003

[4] Arkhipov A M, Kanunnikov A A, Lipov Yu M, Tretyakov Yu M, Roor A V and Solovev N I 2010 Optimizing the technology for firing Kuznetsk coal in a K-50-14-250 steam boiler at a heating boiler house *Therm. Eng.* **57**(12) 1058–64 DOI: 10.1134/S0040601510120104

[5] Zhelev T K and Semkov K A 2014 Cleaner flue gas and energy recovery through pinch analysis *J. Cleaner Prod.* **12(**2) 165–70 DOI: 10.1016/S0959-6526(02)00192-0

[6] Delitsyn L M, Vlasov A S, Borodina T I, Ezhova N N and Sudareva S V 2013 Innovative technologies for full utilization of ash generated at coal-fired thermal power stations for producing alumina and construction materials *Therm. Eng.* **60**(4) 231–8 DOI: 10.1134/S0040601513040034

[7] Su X, Zhang L, Xiao Y, Sun M, Gao X and Su J 2015 Evaluation of a flue gas cleaning system of a circulating fluidized bed incineration power plant by the analysis of pollutant emissions *Powder Technol.* **286** 9–15 DOI: 10.1016/j.powtec.2015.07.038

[8] Guan G 2017 Clean coal technologies in Japan: A review *Chin. J. Chem. Eng.* **25**(6) 689–97 DOI: 10.1016/j.cjche.2016.12.008

[9] Dmitriev A V, Madyshev I N and Dmitrieva O S 2018 Cleaning of Industrial Gases from Aerosol Particles in Apparatus with Jet-Film Interaction of Phases *Ecology and Industry of Russia* **22**(6) 10–4 DOI: 10.18412/1816-0395-2018-6-10-14

[10] Li X, Wei T, Wang D, Hu H, Kong L and Xiang W 2016 Study of gas–liquid two-phase flow patterns of self-excited dust scrubbers *Chem. Eng. Sci.* **151** 79–92 DOI: 10.1016/j.ces.2016.05.010

[11] Lee B K, Jung K R and Park S H 2008 Development and application of a novel swirl cyclone scrubber: Experimental *J. Aerosol Sci.* **39**(12) 1079–88 DOI: 10.1016/j.jaerosci.2008.07.010

[12] Heck U 2016 Customized solvers optimize filtration device design *Filtr. Sep.* **53**(5) 14–6 DOI: 10.1016/S0015-1882(16)30205-1

[13] Golovanchikov A B, Balashov V A and Merentsov N A 2017 The filtration equation for packing material *Chem. Pet. Eng.* **53** 10–3 DOI: 10.1007/s10556-017-0285-y

[14] Laptev A G and Lapteva E A 2015 Model of gas purification from the fine-dispersed phase in the bubbling layer based on the concept of active input section *Theor. Found. Chem. Eng.* **49**(2)

157–62 DOI: 10.1134/S0040579515020074

[15] Dmitrieva O S, Dmitriev A V and Kruglov L V 2016 Calculation of the Average Velocity of the Liquid in the Stream-Film Contact Devices *Procedia Eng.* **150** 753–60 DOI: 10.1016/j.proeng.2016.07.101

[16] Shvab A V and Evseev N S 2015 Studying the separation of particles in a turbulent vortex flow *Theor. Found. Chem. Eng.* **49**(2) 191–9 DOI: 10.1134/S0040579515020128

[17] Wu J, Zhang Y, Wang H 2014 Numerical study on tangential velocity indicator of free vortex in the cyclone *Sep. Purif. Technol.* **132** 541–51 DOI: 10.1016/j.seppur.2014.06.007

[18] Shvydky V S, Yaroshenko Y G, Spirin N A and Lavrov V V 2016 Gazification of coal dust particles in the blast furnace tuyere apparatus *IOP Conf. Ser.: Mater. Sci. Eng.* **150** 012021 DOI: 10.1088/1757-899X/150/1/012021

[19] Moskalev L N and Ponikarov S I 2016 Application of a contact vortex condenser to absorb methanol from a contact gas *Theor. Found. Chem. Eng.* **50**(2) 194–200 DOI: 10.1134/S004057951602007X

[20] Dmitriev A V, Zinurov V E and Dmitrieva O S 2018 Intensification of gas flow purification from finely dispersed particles by means of rectangular separator *IOP Conf. Ser.: Mater. Sci. Eng.* **451** 012211 DOI: 10.1088/1757-899X/451/1/012211

[21] Dmitriev A V, Zinurov V E and Dmitrieva O S 2018 Influence of elements thickness of separation devices on the finely dispersed particles collection efficiency *MATEC Web Conf.* **224** 02073 DOI: https://doi.org/10.1051/matecconf/201822402073

The use of heat pump installations as part of waste energy convertion complexes in the joint generation of electrical and thermal energy

Lyudmila Plotnikova[1], Elena Kostyleva[1*], Irina Chilikova[1] and Sergey Sitnikov[1]

[1]Kazan State Power Engineering University, 420066 Kazan, Krasnoselskaya, 51, Russia

* Corresponding author: dunysha@bk.ru

Abstract. The possibility of using heat pump installations at thermal power plants is considered. An estimation of the sources of low-potential heat for the operation of the heat pump in the joint generation of electrical and thermal energy was carried out. The results of a comparative analysis of the schemes for the inclusion of a heat pump in the cooling system of recycled water are presented using the example of the Kirov thermal power plant.

1. Introduction

The priority direction in the policy of energy saving is the development of energy-saving technologies aimed at the utilization of heat of secondary energy resources – thermal waste of power plants, industrial production, municipal, domestic, residential facilities.

As follows from the second law of thermodynamics, it is impossible to turn all the heat supplied to the working body in the heat-power cycle into work, since part of the heat supplied will need to be taken to a cold source. The task of energy-saving technology is that before giving the environment this waste energy, it is necessary to take the maximum amount of remaining exergy and use it with due efficiency [1].

The maximum use of the exergy of low-potential secondary energy sources will allow the heat pump installation. The experience of introducing heat pump plants into heat and technological schemes of chemical, oil-refining, food, pulp and paper industries is well known [2-5]. With the removal of heat from low-potential sources and the expenditure of mechanical energy, they receive heat with a temperature at which it can be used for heating, hot water and for technology. The most favorable conditions for the use of heat pumps for the simultaneous reception of heat and cold, where the ratio of demand for them is close to the ratio of heat output of the heat pump cycle and cooling capacity.

The main sources of low-potential heat in the joint generation of electrical and thermal energy are:

- heat of cooling systems with water of the equipment of thermal power plant (TPP), state district power station;
- condensate heat from schemes of continuous blowdown expanders;
- secondary waste heat of the reverse network water.

The combined production of electric and thermal energy at steam turbine TPPs, which allowed to save substantially fuel and energy resources of the country during the Soviet period, is currently not effective enough. One of the main reasons for the reduction in the efficiency of the TPP is a significant reduction in the generation of electricity by thermal consumption [6].

IOP Conf. Series: Earth and Environmental Science **337** (2019) 012058 doi:10.1088/1755-1315/337/1/012058

A hidden reserve of improving the energy efficiency of the stations are the cooling systems of steam turbine condensers of TPP. Thermal power plants consume a significant amount of water to condense steam in condensers of steam turbines, provided by technical water supply. Two schemes of organization of technical water supply for cooling steam turbine condensers are most common: co-current water taken from a river passes through turbine condensers and then merges back downstream and circulating with repeated use of water after cooling it in cooling towers or splashing pool. However, both schemes of cooling the condensers of steam turbines are associated with large losses of thermal energy to the environment, which leads to a decrease in the efficiency of the TPP, an increase in the degree of negative impact on the environment.

Improvement of the cooling system is possible due to the introduction of a heat pump unit, which will allow taking heat from the cooling water turbine heated in the condenser, increasing its potential and using it for the own needs of the TPP [7-8]. It's also possible in heat pump systems to use the heat of the so-called "dirty" condensate from the scheme of the continuous blowdown of boiler units, which due to the salt content is unsuitable for return to the cycle of the station is useless and is discharged into the hydro-, ash removal system.

2. Object of research

The object of research is one of the thermal power plant of the city of Kirov with a total thermal capacity of 1600 MW with installed electric power of 320 MW [9]. Six turbines are installed at the plant: one ПТ-60/75-130/13, one ПТ-60-130/13 and four T-50-130. To provide cooling of condensers of turbines on the territory of the power plant, there are 4 typical counterflow cooling towers. Cooling water is used at the power plant according to the circulating scheme. From the collector by circulating electric pumps it is supplied respectively to the condensers of turbogenerators. At the first stage, turbogenerators use pumps with a capacity of 6300 m3/h and 5000 m3/h. After the condensers, the heated water enters the cooling tower, from where it returns to the circulation collector. In the summer, three cooling towers operate, the cooling water flow rate in the circulation circuit reaches 28000-30000 m3/h, in the autumn-winter period there is one cooling tower in operation with a flow rate close to the nominal flow of 12000-14000 m3/h.

As a result of the analysis of the characteristics of the main equipment of the TPP, it was revealed that:

- a significant amount of heat in the TPP is lost - discharged into the atmosphere in the cooling towers. So, for example, for a single turbine T-50-130, even when operating in the most efficient combined mode with a fully closed diaphragm of a low-pressure cylinder, the standard venting discharge of steam into the condenser is 10 t/h or 6,88 MW;
- a large amount of thermal energy is spent on heating the complex of industrial and administrative buildings of the enterprise. The share of own heating needs in relation to the release of thermal energy during the heating period 2016-2017 was 2,1 %;
- to ensure the cost of these coolants expends electrical energy. The share of the expense for own needs in relation to the expenditure on leave during the heating period 2016-2017 was 8,2 %.

As can be seen from the presented data, the thermal power plant under consideration has a significant potential for the use of low-potential heat of the cooling water of the circulation circuit. Thus, as a source of low-potential heat at the considered TPP selected water flow with low temperature – the flow of recycled water. The received heat energy is proposed to be used for the heating system of industrial buildings of the same CHP, namely for heating the boiler compartment of the main building.

The proposed scheme for using a heat pump (HP) is shown in Figure 1. Here, the condensate after the turbine condensers 23, previously directed to the cooling tower 36, is proposed to be partially redirected to the evaporator of the heat pump. In the evaporator (Figure 1), the condensate is cooled to a temperature corresponding to the temperature of the cooled water in the cooling tower and is then directed to the return line of the circulating water supply.

The thermal potential of the condensate in the heat pump is increased by the compression of the refrigerant vapor, which received heat from the condensate flow. The increased heat potential of the flow is transferred to the heat pump condenser where the heat transfer to the consumer is due to the refrigerant condensation.

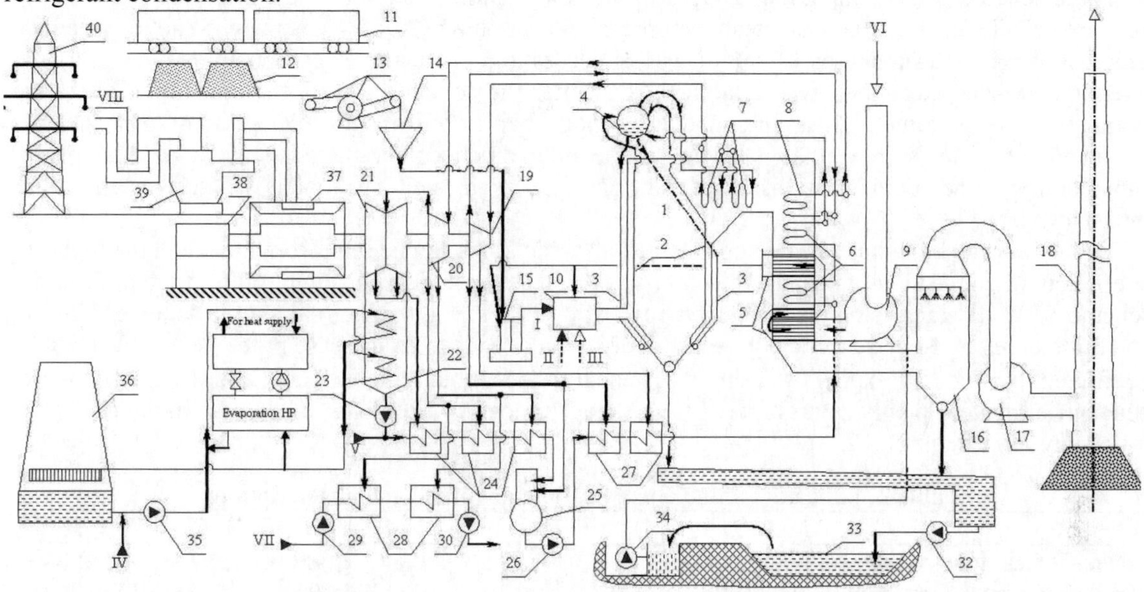

Figure 1. The scheme of the TPP with the inclusion of a heat pump: 1 – boiler unit, 2 - furnace shields, 3 - lowering pipes, 4 - drum of the boiler, 5 - air heater, 6 - water economizer, 7 - I and II stages of supersteamheater, 8 - intermediate superheater, 9 - 10 - burner, 11 - train with coal, 12 - coal storage, 13 - coal transportation and primary crushing system, 14 - raw bunker, 15 - mill, 16 - ash collector, 17 - smoke exhaust, 18 - chimney , 19-electric generator, 20-turbo installation, 21-cylinders of high, medium and low 22 - condensate pump, 24.27 - regenerative heaters of low and high pressure, 25 - deaerator, 26 - feed pump, 28 - water heaters I and II stages, 29.30 - water supply pumps I and II stages, 31 - hydraulic ash removal channel, 32 - shoal pump, 33 - ash dump, 34 - clarified water pump, 35 - circulating water pump, 36 - cooling tower, 37 - power generator, 38 - generator exciter, 39 - boosting transformer.

3. Results

The calculation of the heat pump system for heating the 1-st stage of the boiler section of the main building of the Kirov TPP plant in the winter period as a basic source of heat has been made. The source of low-potential heat is a circulation system with a cooling water flow rate up to 21000 m3/h and an average annual temperature of 28 °C.

The following were taken as input data (Table 1).

Table 1. Initial data for the calculation of the heat pump installation.

№	Parameter	Value	Unit
1	The thermal load Q_{TH} for the boiler room building as a heated heat pump unit	1731	KW
2	The temperature of the low-potential heat carrier of the direct circulating water after the condenser to the cooling tower at the entrance to the heat pump t_{H1}	28	°C
3	The temperature of the low-potential flow of return circulating water for cooling into the turbine condenser after the heat pump t_{H2}	22	°C
4	The temperature of the high-potential flow at the inlet to the heat pump of the heating system t_{B1}	42	°C
5	The temperature of the hot water after the heat pump – the direct line of the heating system according to the temperature schedule t_{B2}	55	°C
6	Average ambient temperature during the heating season t_O	5,4	°C

7	Temperature of superheating of steam in the intermediate heat exchanger Δt	20	°C

Three schemes of switching the heat pump into the cooling system of the circulating water of the TPP are calculated: 1 - a classical steam compression heat pump [10]; 2 - scheme with an intermediate heat exchanger (supercooling of liquid and superheating of steam is combined in an additional intermediate heat exchanger, where the hot freon after the condenser heats the cold freon after the evaporator); 3 - scheme with an intermediate heat exchanger and a subcooler (if a hot heat carrier is produced for water supply, that is, it enters the inlet much colder than it leaves the heat pump, supercooling of the liquid is possible in an additional heat exchanger - subcooler, which is installed after the condenser).

Energy efficiency indicators are summarized in Table 2. As can be seen from the table, the highest coefficient of conversion of heat and electricity, as well as the lowest consumption of primary fuel - in scheme No. 3. By increasing the consumed power by 13.85 kJ/kg, the thermal load in schemes No. 2 and 3 increases by 33 and 57.5 kJ/kg, respectively, which leads to an increase in the heat conversion coefficient μ. Since the exergy of electricity is equal to the amount of electricity, and the exergy of the heat carrier makes up only a part of its heat, this leads to loss of exergy and a decrease in the exergy efficiency for scheme No. 2.

Table 2. Indicators of the energy efficiency of the calculated options.

№ scheme	**1**	**2**	**3**
Specific heat load of the heat pump, kJ/kg	264,9	297,9	322,4
Specific energy consumed by the electric motor, kJ / kg	68,95	82,8	
Degree of compression in the compressor		3,48	
Conversion factor of heat	5,05	4,74	5,12
Coefficient of electricity conversion	3,84	3,6	3,89
Specific consumption of primary energy	0,68	0,73	0,68
Exergy efficiency	0,389	0,371	0,506

Based on the results presented in Table 2, Scheme No. 3 was chosen. The cycle of the heat pump system for the selected option is presented in the form of the *p, h*-diagram in Figure 2.

The cost-effectiveness of introducing HP is shown in Table 3.

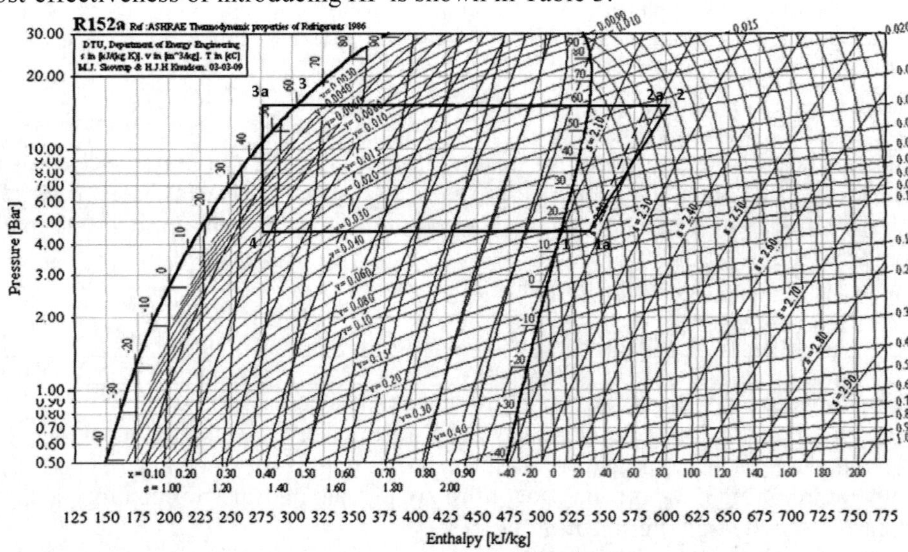

Figure 2. Cycle of heat pump system with intermediate heat exchanger and subcooler

Selected heat pump unit HT-3000 with heat recovery and supercooler, designed and selected the main equipment for the heat pump installation – capacitor brand 1000КНГ-1,6.

To assess the economic efficiency, a feasibility study of the introduction of a heat pump installation was carried out, the payback period of the main equipment was determined, which amounted to about 5 years.

Table 3. Results of technical and economic calculation.

Economic parameter	Year of operation						
	1	2	3	4	5	6	7
Income tax savings due to depreciation, th. ₽.	93,4	93,4	93,4	93,4	93,4	93,4	93,4
Savings in heat supply from HP, th. ₽.	5336,45	5336,45	5336,45	5336,45	5336,45	5336,45	5336,45
Operating costs, th. ₽.	3220	3220	3220	3220	3220	3220	3220
Net cash flow, th. ₽.	2209,85	2209,85	2209,85	2209,85	2209,85	2209,85	2209,85
Net cash flow cumulative total, th. ₽.	-11684,00	-9474,15	-7264,30	-5054,45	-2844,60	-634,75	1575,10

References

[1] Malyshev A A and Kireev V S 2017 *Scient. J. of NIU ITMO. Series: Refr. eq. and air con.* **4** 22
[2] Plotnikova L V, Chilikova I I, Sitnikov S Y and Efremov G I 2016 *Int. J. of Ph. and Tech.* **8**(4) 26727
[3] Fetisova E A, Zherdev A A and Smorodin A I 2012 *Engi. J.: Sci. and Inn.* **5**(5) 7
[4] Nosov G A, Popov D A, Belskaya V I, Zhiltsov V S, Yakovlev D S and Kairalieva T G 2015 *F. chem.l tech.* **10**(6) 64
[5] Shelginsky A Ya and Malenkov A S 2014 *Ind.l en.* **7** 32
[6] Zamaleev M M, Sharapov V I, Salikhov A A and Yaparov I V 2013 *Izv. Vyssh. Uch. Zav. Probl. of en.* **11-12** 45
[7] Burov V D, Dudolin A A, Ilyina I P, Oleynikova E N, Sedlov A S and Tymoshenko N I 2013 *Her. of the Mosc. P. Eng. Ins.* **3** 044
[8] Anikina I D and Sergeev V V 2013 *Sci. and tech. bull. of the St. P. St. Pol. Uni.* **178** 56
[9] Sokovnin O M, Zagoskina N V and Zagoskin S N 2007 *Thermal power plants of the Kirov branch of JSC "TGC-5"*(Kirov: Kirov) p 378
[10] Krasavina E O and Plotnikova LV 2016 *Bull. of Kaz. St. P.r Eng. Uni.* **4**(32) 95

IOP Conf. Series: Earth and Environmental Science **337** (2019) 012059 doi:10.1088/1755-1315/337/1/012059

Investigation of the multi-layer open cell foam filter model using numerical simulation and experimental studies

Olga V Soloveva[1,*], Sergei A Solovev[1], Emil V Shamsutdinov[1] and Evgeny G Sheshukov[1]

[1]Kazan State Power Engineering University, 420066, Kazan, Krasnoselskaya str., 51, Russia

* Corresponding author: solovyeva.ov@kgeu.ru

Abstract. In this work, we studied the influence of the internal features of an open cell foam material structure on the efficiency of particle deposition and the pressure drop value of the medium. Constructing of the open cell foam filter models is based on the Distinct Element Method, which allows creating geometry with a random arrangement of pores in space. Calculations of the flow field and particle trajectories were performed using the ANSYS Fluent software package (v. 19.0). Experimental studies were conducted using samples created on the basis of the inverse matrix of 3D printed computational models. Also, models of various geometries with a fixed porosity of the medium are considered. Conducted numerical and experimental studies show that the filter model with the lowest porosity provides the highest deposition efficiency of particles. For a given fixed porosity of the medium, a filter with a change in the cells in diameter across the layers gives a lower value of the pressure drop, which is a significant advantage. The results can serve as the basis for creating aerosol filters with improved characteristics.

1. Introduction

In many energy companies, there is a problem of the environment pollutant emission. The applied filters are often ineffective, as they quickly become clogged or do not catch small particles. The high porosity of cellular materials leads to a decrease in pressure drop and an increase in mechanical strength when used as filters. In addition, open cell foam materials (OCFM) can act as heat accumulators, if they are made of ceramics, or serve as heat transfer enhancers in the case of using aluminum [1].

Open cell foam materials are actively used to enhance heat transfer. In work [2], studies of heat transfer in porous material were carried out taking into account forced convection, radiation, and boiling. In [3], it was shown that the use of OCFM in regenerative heat exchangers increases their efficiency. The authors of [4] demonstrated that the heat transfer in devices with low thermal conductivity can be increased by inserting an open cell foam aluminum metal matrix, the cooling rate with a metal matrix and without it was analyzed. The use of OCFM increased the cooling rate, that demonstrates the feasibility of using these materials as heat transfer enhancers. Despite the fact that many studies have obtained good results using porous media as cooling elements, there are still a few papers on the accumulation of energy in materials at high temperatures.

For use as aerosol filters, the OCFMs are made from polyurethane for low temperatures of the gas to be filtered or from ceramics for high gas temperatures [5]. In addition to being used as filters, open cell foam materials can be used in samplers to determine the concentration and type of pollutants [6-8]. The

IOP Conf. Series: Earth and Environmental Science **337** (2019) 012059 doi:10.1088/1755-1315/337/1/012059

semi-empirical dependencies for the coefficient of particle penetration through a porous medium were proposed by the authors of [9-10], expressions are based on taking into account such physical mechanisms of precipitation as gravity, diffusion, and entanglement. In the case of polyurethane porous material, the experimental data are in good agreement with the numerical simulation data. The three-dimensional model of the porous medium can be obtained either by computer tomography or generated on the basis of an algorithm that includes the parameters of the internal structure [11-12]. The use of filters with predetermined geometric characteristics, taking into account the physical properties of the emissions (for example, particle size), can reduce the pressure drop across the porous medium and increase the efficiency of particle deposition [14-17].

2. Problem formulation

The geometry of the computational domain is created in layers with three cell diameters, each layer is 2 cm. The cells in the first layer have a diameter of $d_c = 4$ mm, in the second – $d_c = 5$ mm, in the third – $d_c = 6$ mm. The total length of the porous insert is 6 cm. The length of the inlet and outlet nozzles is 4 cm each. The entrance is located closer to the layer containing the largest cell size of the porous medium. This geometry is a model of a multilayer filter and is aimed at reducing the overall filter resistance, while larger particles should be deposited in the first layer, and smaller particles – in subsequent layers. Figure 1b shows tubes, printed on a 3D printer used for experimental studies. These tubes are created on the basis of the inverse matrix of the geometry of a porous medium used in numerical simulation.

To calculate the flow of a gas suspension in a porous medium, the internal volume of the geometry is used. The number of grid elements averaged 20 million, which ensured sufficient accuracy of the calculation.

The following boundary conditions were set at the boundaries of the computational domain: the mass flow rate at the inlet and the atmospheric pressure at the outlet, the remaining zones were taken as walls by default.

Hydrodynamic calculations were carried out on the basis of solving the Navier-Stokes equations in the CFD package ANSYS Fluent using the finite volume method. Direct numerical simulation was performed in the approximation of a laminar viscous incompressible gas.

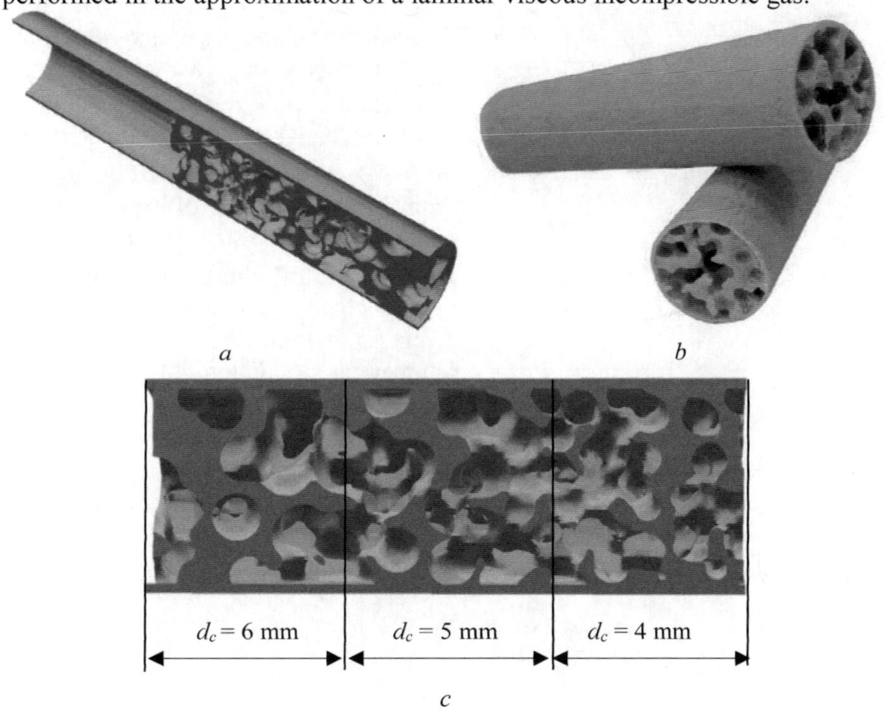

Figure 1. The geometry of a porous medium: *a* is a model of a porous medium used to create experimental samples with different cell diameters in layers, *b* is a picture of experimental samples with porosity of the medium $\varepsilon = 0{,}6$, lower tube has cell diameter $d_c = 4$ mm and upper tube has $d_c = 5$ mm, *c* – is the central section of the model with cell sizes varying in layers.

Equations of motion of the carrier medium:

$$\nabla \cdot \vec{U} = 0, \tag{1}$$

$$\left(\vec{U} \cdot \nabla\right)\vec{U} = -\frac{1}{\rho}\nabla P + \nu\Delta\vec{U}, \tag{2}$$

where $\vec{U}$ – is the gas velocity vector, ρ – is the gas density, P – is the pressure, ν – is the kinematic coefficient of gas viscosity.

3. Results of numerical simulation

3.1. Calculation of the pressure drop value

The complex geometry and porosity of the medium provide a nonlinear change in pressure drop depending on the average filtration rate (Fig. 2). A detailed simulation of the gas flow in a porous medium was carried out taking into account the surface morphology. To verify the accuracy of the results of the numerical simulation, experimental studies were carried out using tubes, printed on a 3D printer containing an insert made of porous material (Fig. 1, b). The air in the tube was pumped by a compressor. The magnitude of the pressure drop was measured using the "testo 510" manometer at a short distance from the porous material to directly evaluate the resistance of the medium. The mean value of the flow velocity was measured using the "testo 450" hot-wire anemometer.

It was assumed that the creation of a porous medium with cells that decrease in diameter over the layer thickness will reduce the value of the pressure drop while maintaining the efficiency of the deposition of particles. Figure 2 demonstrates the agreement of the results of experimental studies with numerical simulation data. The use of a medium with a change in the diameter of the cells (a total of three layers of 2 cm each) gives less aerodynamic drag compared to a medium with a fixed diameter of the cells, which, if the deposition efficiency of the particles is maintained, provides a significant advantage in assessing the quality of the filter.

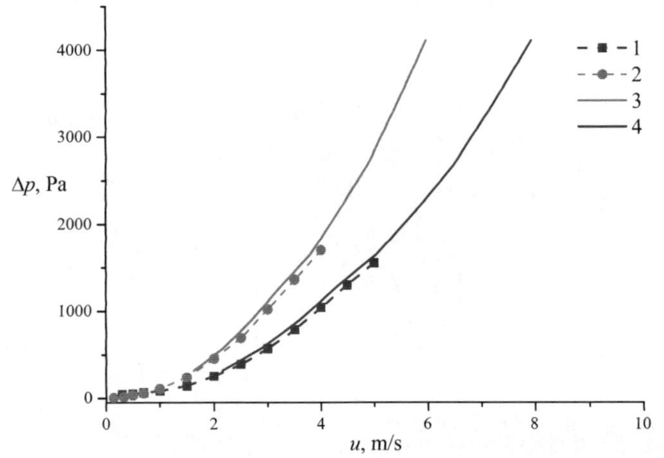

Figure 2. The dependence of the pressure drop on the average filtration rate in a tube with porosity $\varepsilon = 0{,}6$: 1 – the results of experimental studies with a change in pore diameter in layers ($d_c = 4$ mm, $d_c = 5$ mm, $d_c = 6$ mm), 2 – the results of experimental studies with pore sizes $d_c = 4$ mm, 3 – the results of numerical simulation with a change pore diameter by layers (mm, mm, mm), 4 – the results of numerical simulation with pore sizes $d_c = 4$ mm.

3.2. Calculation of the particle deposition efficiency

The efficiency of particle deposition was determined in the gas velocity field found under the assumption of a small concentration of the second phase.

The Lagrange equations of particle motion in disregard by all forces, except the force of aerodynamic drag will be written in the form:

$$\frac{d\vec{V}}{dT} = \frac{(\vec{U} - \vec{V})}{\tau_p} \tag{3}$$

$$\frac{d\vec{R}}{dT} = \vec{V}, \tag{4}$$

where m_p is the mass of the particle, m_f is the mass of the gas with a volume equal to the volume of the particle, d_p is the particle diameter, ρ is the gas density, μ is the dynamic viscosity of the gas, $\vec{V}$ and $\vec{U}$ is the particle and gas velocities respectively, $\tau_p = m_p / 3\pi\mu d_p = \rho d_p^2 / 18\mu$ is the particle relaxation time (the parameter that determines the particle inertia), $\vec{R}$ is the radius vector position of the particle.

The found particle trajectories determine the deposition efficiency, as the ratio of the number of settled particles to the number of total started particles:

$$E = \frac{N}{N_0}, \tag{5}$$

Figure 3 shows the curves of changes in efficiency for the four cases of cell diameter at a fixed porosity of the medium. It is interesting to note that the geometry made as the layers with a change in the diameter of the cells while maintaining the porosity of the medium provides the maximum value of the deposition efficiency with small particle diameters, which can be explained by the hydrodynamics feature of the air flow with a reduced resistance on the first two layers compared to the constant diameter of the cells along the entire length of the porous medium. With an increasing particle diameter, the deposition efficiency curve with cell diameters varying in layers coincides with the efficiency curve for $d_c = 5$ mm. Thus, it can be concluded that the use of a filter with a variable diameter of the cells is advisable only in the case of trapping small particles, since the efficiency in the case of inertial particles is determined by the average pore diameter of the material.

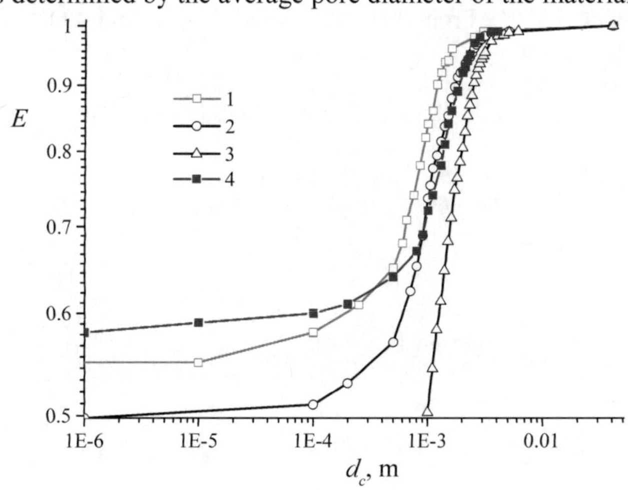

Figure 3. Particle deposition efficiency for four cases: 1 – the geometry with porosity $\varepsilon = 0{,}6$ and cell diameter $d_c = 4$ mm, 2 – the geometry with porosity $\varepsilon = 0{,}7$ and cell diameter $d_c = 4$ mm, 3 – the geometry with porosity $\varepsilon = 0{,}8$ and cell diameter $d_c = 4$ mm, 4 – the geometry with porosity $\varepsilon = 0{,}6$ and cell diameters $d_c = 6$ mm, $d_c = 5$ mm, $d_c = 4$ mm, created by layers.

4. Conclusion

Experimental studies and numerical modeling of gas flow in tubes containing a porous medium were carried out. Four variants of a porous medium with cell diameters of 4, 5 and 6 mm, as well as with cells varying in diameter across the layers, were considered. In two cases, the porosity of the medium remained constant $\varepsilon = 0{,}6$. The dependence of the pressure drop on the average filtration rate is nonlinear, indicating a complex internal structure of the medium containing numerous channels and turns (pore system), the results of numerical calculations for pressure drop agree well with the experimental data. Geometry with a variable cell diameter in layers creates a lower aerodynamic drag compared to a geometry with a cell diameter of 4 mm, which is an advantage of its use in aerosol filter design. At the same time, the efficiency of particle deposition is determined by the average pore diameter for inertial particles and turns out to be significantly greater than the efficiency of particle deposition compared with the option of the smallest cell diameter. Studies show that creating a filter with a variable pore size in layers is advisable only in cases where even a small change in pressure drop is essential for the process in the case of capturing inertial particles. For small particles, a variable pore filter is preferred.

Acknowledgments
The reported study was funded by RFBR according to the research project № 19-07-01188. Also, the reported research was funded by Russian Foundation for Basic Research and the government of the Republic of Tatarstan of the Russian Federation, grant № 18-41-160005.

References

[1] Kim S and Lee C-W 2014 *Proc. Mater. Sci.* **4** 290
[2] Kumar P and Topin F 2017 *Int. J. Mult. Flow.* **94** 123
[3] Regulski W, Szumbarski J, Gumowski K, Skibiński J, Wichrowski M and Wejrzanowski T 2015 *Chem. Eng. Sci.* **137** 320
[4] Hellmann A, Pitz M, Schmidt K, Haller F and Ripperger S. 2015 *Aerosol Sci. Tech.* **49** 16
[5] Horneber T, Rauh C, Delgado A 2014 *Chem. Eng. Sci.* **117** 229
[6] Clark P, Koehler K A and Volckens J 2009 *J. Aerosol Sci.* **40**(7) 563
[7] Yang X, Li Y, Zhang L, Jin L, Hu W and Lu T J 2018 *J. Heat Transf.* **140**(1) 014502
[8] Schmidt K and Becker J. 2013 *Adv. Eng. Mater.* **15**(1-2) 40
[9] Saw L H, Ye Y, Yew M C, Chong W T, Yew M K and Ng T C 2017 *App. Energy.* **204** 1489
[10] Hu X, Wan H and Patnaik S S 2015 *Int. J. Heat Mass Transf.* **88** 617
[11] Arbak A, Dukhan N, Bağcı O and Özdemir M 2017 *Exp. Therm. Fluid Sci.* **86** 180
[12] Bianchi E, Groppi G, Schwieger W, Tronconi E and Freund H 2015 *Chem. Eng. J.* **264**, 268
[13] Zafari M, Panjepour M, Emami M D and Meratian M 2015 *App. Therm. Eng.* **80** 347
[14] Solovev S A, Soloveva O V and Popkova O S 2018 *Rus. J. Phys. Chem. A.* **92**(3) 603
[15] Soloveva O V, Solovev S A, Khusainov R R, Popkova O S and Panenko D O 2018 *J. Phys: Conf. Ser.* **944**(1) 012113
[16] Dmitriev A, Madyshev I and Dmitrieva O 2018 *Ecol. Ind. Rus.* **22**(6) 10
[17] Dmitriev A V, Zinurov V E and Dmitrieva O S 2018 *MATEC Web Conf.* **224** 02073

Enrichment of ash and slag waste generated by burning of fuels with additives

Elvira R Zvereva[1*], Olga S Zueva[1], Kamil K Gilfanov[1] and Valeria P Tutubalina[1]

[1]Kazan State Power Engineering University, 420066, Kazan, Krasnoselskaya str., 51, Russia

* Corresponding author: 6elvira6@list.ru

Abstract. The existence of opportunities and prospects for scarce metals extraction from ash and slag waste of high-sulphur fuel oil burning was noted. The volatile ash produced during fuel combustion consisting mainly of metal oxides contains highly toxic substances. Their binding in a zone of burning of fuel is important not only for the purpose of extraction and further use in industry, but also to prevent their harmful environmental impact. It was shown that in the case of petroleum fuel oil an effective way to improve the performance of the fuel while maximizing the adverse environmental consequences of its combustion is to develop new generation effective fuel additives and improve filtration systems for flue gases. It has been established that the use of additives based on carbonate sludge makes it possible to achieve greater binding of metal oxides to non-volatile compounds in the burning zone of the fuel and their reduction in the composition of the emission gases. The mutual using of carbonate additives with the admixture of carbon nanotubes the effect may be even greater. Rational ash and slag waste utilization through the use of their beneficial properties will allow to free occupied areas and reduce environmental damage to the environment.

1. Introduction

The activities of energy industry leads to the emergence of a large amount of mineral ash and slag waste generated during the combustion of liquid and especially solid fuels. Combustion of fuel leads to formation of volatile substances and solid non-combustible waste containing various, including valuable substances. However, toxic compounds are formed in ash and slag waste also, therefore rational waste utilization will make it possible to use the components of ash and slag waste and improve the ecological situation. In Russia a huge amount of ash and slag wastes was formed, requiring close attention in order to process them to improve the ecological situation in the country and to solve production problems related to the fabrication of new materials and products [1].

In Russia, no more than 10% of the total amount of generated waste annually is used as secondary resources for further processing in various industries. In developed countries, 70-95% of ash and slag waste is recycled, and 100% in some Scandinavian countries, for example, in the Netherlands and Denmark. This work is devoted to study of some aspects of controlled change in the composition of ash and slag waste generated during the combustion of fuels with additives with a view to their further use.

2. Features of ash and slag waste

Ash and slag waste from thermal power plants is a technogenic raw material that can be used to extract valuable metals and to manufacture composite materials and products for various purposes. The value of ash is due to its complex chemical composition. In ash waste a large amount of calcium, iron, aluminium, chromium, nickel, manganese, rare and trace elements compounds are concentrated: vanadium, germanium, gallium. In connection with the natural resources depletion the processing of ash and slag waste in order to extract valuable and noble metals and the further use of waste from their processing could help in implementing a resource-saving policy [2, 3].

Rational multi-stage disposal of ash and slag waste should include measures to maximize the binding of highly toxic substances formed during fuel combustion with solid ash residues suitable for further processing; removal of harmful and valuable components from waste and the subsequent use of neutralized residues in various industries. As valuable features of ash and slag waste it should be noted that they contain noble metals (gold, platinum), rare and trace elements and there are a possibilities of extracting iron-containing magnetic concentrate and secondary coal. Currently there are effective technological solutions that allow high-quality ash and slag waste processing with the aim of extracting from them for example by leaching, gold, titanium, lithium, beryllium, vanadium, tungsten, copper, manganese, zinc, rare earth elements, etc. [4-8].

Measures for ash and slag waste utilization should take into account the inconstancy of the ash composition and its dependence on the type and quality of extracted fuel. Thus the composition and the properties of ash and slag waste as a secondary resource are determined by [9]:

- the type and the features of employed fuel formation (natural factors);
- the features of the fuel combustion process, methods for ash and dust removing (technological factors);
- the storage characteristics of ash and slag waste (environmental factors).

Technological factors may be most manipulated. The employed fuels are natural sorbents containing impurities of many valuable elements including rare metals. When incinerated their content in ash increases many times (100-1000 times) and may be of industrial interest. However some fraction of the valuable elements form volatile compounds and escape into the atmosphere with flue gases. This fraction can and should be minimized. Moreover the metal oxides formed during fuel combustion and emitted through a chimney into environment are highly toxic substances that are harmful to the environment. Therefore the capture of metal oxides is important not only for the purpose of their further use in industry, but also to prevent their harmful environmental impact.

There are several ways to solve the problem of ash and slag waste enrichment while improving the quality of flue gases by trapping metal oxides. Some of them are associated with using of highly efficient methods of burning fuel, improving the design features of boiler furnaces and developing filters that delay some of the harmful emissions. Another simple and cheap way to improve the performance properties of oil and alternative fuels and to perfect the environmental safety of the fuel and energy complex is the special additives introduction into it. Often this method is not only the most affordable, but the only possible.

The use of petroleum fuel oil remains environmentally inappropriate. Therefore the focus should be on improving of performance fuel characteristics while maximizing elimination the adverse environmental consequences of its combustion. Accordingly solutions to the problem should be sought in the development of effective new generation fuel additives and in improvement of flue gas filtration systems. In this work it was shown that use of new generation additives containing nanoparticles allows to achieve greater bonding of metal oxides with nonvolatile compounds in a zone of fuel burning and reduction volatile toxic compounds in the emission gases composition.

3. Materials

Samples of high- sulphur fuel oil brand M100 produced by the Nizhnekamsk oil refinery were taken as boiler fuel. This type of fuel is used in thermal power plants and in boiler rooms as emergency and backup fuel. To improve the physicochemical properties of organic fuels, to increase the efficiency and environmental friendliness of combustion, and to bind volatile compounds, dehydrated carbonate sludge was used. Carbonate sludge is a waste of TPPs chemical water treatment process. Carbonate sludge as a product obtained by chemical precipitation has a complex of specific physicochemical properties, among which should be noted the diverse chemical composition (calcium carbonates, magnesium and iron hydroxides, aluminium compounds, etc.), as well as its high dispersion and surface activity due to the high porosity arising in the process of water evaporation [10]. As nanoscale additives, multi-walled carbon nanotubes of the Taunit carbon nanomaterial (http://www.nanotc.ru) dispersed in diproxamine (liquid non-ionic surfactant well soluble in oils) were used.

4. Results and discussion

When oil processing organometallic compounds of vanadium, nickel, rare and trace elements are concentrated in the heavy fraction, i.e. in fuel oil and go into ash when at burning [1]. Fuel oil obtained from high-sulphur oil varieties contain the highest concentrations of organometallic compounds [4-7]. Solid fuel oil combustion products are the most affordable raw material for the production of valuable metals after iron ores. On the other hand burning of high-sulphur fuel oils is also characterized by appearance of the greatest amount of harmful emission gases. In order to study the possibilities of binding harmful volatile substances based on organometallic compounds, samples of ash residues of high-sulphur brand M100 fuel oil used at thermal power plants and boilers as emergency and backup fuel were studied. Chemical analysis of ash and slag waste showed that most of them are composed of compounds of iron, vanadium, nickel and sulphur. The chemical composition of ash besieged on the regenerative surfaces of boilers from fuel oil burning is given in Table 1.

Table 1. The chemical composition of ash resulting from the combustion of brand M100 fuel oil.

Element	S	V	Ni	Ca	Zn	Mo	Cr	Mn	Mg
Content,%	11.5	18.4	5.9	0.7	0.2	0.2	0.1	0.1	0.9

Solid and liquid additives are added to highly viscous and sulphurous boiler fuels to improve various properties of fuels. The additives currently used are mainly intended to combat low temperature corrosion. However, environmental degradation requires use of additives that reduce the content of harmful emissions in flue gases.

In matters of additives creation it should be noted that the composition of fuel oil has its own features. It is characterized by the presence of a large number of sulphur compounds and a significant amount of vanadium compounds. Therefore we are not talking here only about the completeness of fuel combustion, i.e. about converting residual hydrocarbons (CH_x), nitrogen oxides (NO_x) and carbon monoxide (CO) into harmless nitrogen gas (N_2), water (H_2O) and carbon dioxide (CO_2), but also about the binding of sulphur and vanadium in the combustion fuel zone with the purpose of preventing them from entering into atmosphere. Sulphur and vanadium which are present in fuel oil in the form of various compounds are oxidized during combustion with the formation of oxides which lead to corrosion damage in furnaces and to acid rain in the atmosphere. In addition a high level of sulfur in fuel is one of the factors behind appearance of so-called anthropogenic carbon nanomaterials in the emission gases incoming the atmosphere.

One of the ways to reduce emissions of sulphur compounds can be achieved by adding substances that bind sulfur and vanadium into non-volatile compounds in the combustion zone of fuel. When choosing additives we were guided by the well-known fact that various solid mixtures based on calcium hydroxide, magnesium and some other metals (zinc, copper, etc.) can be effective for binding sulfur in the combustion fuel zone for preventing equipment corrosion. As an additive to boiler fuel we took a finely dispersed fraction of dehydrated water treatment carbonate sludge [10–12]. The development of effective technologies for the use of carbonate sludge has another positive side. To a large extent it would also solve the problem of recycling waste water treatment processing.

Laboratory and industrial tests have shown that the additive in the form of dehydrated carbonate sludge water treatment makes it possible to chemically bind the sulfur contained in fuel oil, which is usually released together with vanadium compounds into the atmosphere during fuel combustion. Due to this, emissions of sulfur oxides and vanadium compounds into the atmosphere are reduced. Industrial tests have shown that the addition of 0.1 wt.% leads to a decrease in emitted sulfur oxides by 36.5 wt. % (analysis of test results of the boiler unit TGM-84B station No. 7, Naberezhnye Chelny). The use of

IOP Conf. Series: Earth and Environmental Science **337** (2019) 012060 doi:10.1088/1755-1315/337/1/012060

additives also contributes to the formation of a looser structure of deposits in the zone of high-temperature heating surfaces which reduces corrosion of these surfaces [13].

Thus based on dehydrated carbonate sludge additives developed by us [11, 12] significantly improve the viscosity characteristics of fuel oil and at the same time reduce the content of sulphur oxides in flue gases. Their combined use with additives based on pre-dispersed carbon nanotubes [14, 15] made it possible to create composite fuel with improved properties containing small additives of these substances [16-21]. The use of boiler fuel with developed additives will allow for a more complete combustion of the fuel and more efficient binding of highly toxic substances with solid ash residues suitable for further processing.

5. Conclusion

The existence of opportunities and prospects for scarce metals extraction from ash and slag waste of high-sulphur fuel oil burning was noted in the work. Since a part of the metal oxides formed during fuel combustion are volatile highly, toxic, and harmful to the environment substances, their binding in the fuel combustion zone is important not only for their further use in industry, but also to prevent their harmful environmental impact. It was shown that in the case of petroleum fuel oil an effective way to improve the performance of the fuel while maximizing the adverse environmental consequences of its combustion is to develop new generation effective fuel additives and improve filtration systems for flue gases. It has been established that the use of additives based on carbonate sludge makes it possible to achieve greater binding of metal oxides to non-volatile compounds in the burning zone of the fuel and their reduction in the composition of the emission gases. The mutual using of carbonate additives with the admixture of carbon nanotubes the effect may be even greater. Rational ash and slag waste utilization through the use of their beneficial properties will allow to free occupied areas and reduce environmental damage to the environment.

References

[1] Tsygankova M V, Bukin V I, Lysakova E I, Smirnova A G and Reznik A M 2011 *Russ. J. Non-Ferr. Met.* **52** 19

[2] Malchik A G, Litovkin S V, Rodionov P V, Kozik V V and Gaydamak M A 2016 *IOP Conf. Ser.: Mater. Sci. Eng.* **127** 012024

[3] Lihach S A, Kulesh R N, Nikolaeva V I, Orlova K Y and Ilyasova A S 2017 *MATEC Web Conf.* **92** 01051

[4] Al-Malack M H, Bukhari A A, Al-Amoudi O S, Al-Muhanna H H and Zaidi T H 2013 *Int. J. Environ. Res.* **7** 455

[5] Al-Ghouti M A, Al-Degs Y S, Ghrair A, Khoury H and Ziedan M 2016 *Chem. Eng. J.* **173** 191

[6] Rasoulnia P and Mousavi S M 2016 *RSC Adv.* **6** 9139

[7] Nourbaghaee H, Hamidi A G and Pourabdoli M 2018 *Waste Manage.* **74** 297

[8] Rastegar S O, Mousavi S M and Shojaosadati S A 2015 *RSC Adv.* **5** 41088

[9] Shabarov A N, Aleksandrova T N and Nikolaeva N V *To the question of complex use of ashes and slag waste CHP plant.* In: Litvinenko V. (eds) XVIII International Coal Preparation Congress. Springer, Cham. 603 (2016)

[10] Abdullazyanov E Y et al, *New technologies, materials and equipment in power engineering* (Kazan, KSPEU, Vol. 3, 267, 2018).

[11] Zvereva E R, Mingaleeva G R, Khabibullina R V and Akhmetvalieva G R 2016 *Petroleum Chemistry* **56** 65

[12] Zvereva E R, Zueva O S and Khabibullina R V 2016 *Mater. Sci. Forum* **870** 666

[13] Zvereva E R, Dmitriev AV, Shageev M F and Akhmetvalieva G R 2017 *Thermal Eng.* **8** 50

[14] Zueva O S, Makshakova O N, Idiyatullin B Z, Faizullin D A, Benevolenskaya N N, Borovskaya A O, Sharipova E A, Osin Yu N, Salnikov V V and Zuev Yu F 2016 *Russ. Chem. Bull.* **65** 1208

[15] Zueva O S, Mongush Y K and Makarova A O 2018 *Solid State Phenomena* **284** 713

[16] Zvereva E R, Khabibullina R V, Akhmetvalieva G R, Makarova A O and Zueva O S 2017 *Advances in Engineering Research* **133** 914
[17] Zvereva E R, Khabibullina R V and Zueva O S 2017 *Solid State Phenomena* **265** 374
[18] Zvereva E R, Makarova A O and Zueva O S 2018 *Solid State Phenomena* **284** 767
[19] Zvereva E R, Zueva O S, Khabibullina R V, Mingaleeva G R, Akhmetvalieva G R, Salikhzyanova D R and Khatmullina Z F 2016 *Chem. Technol. Fuels Oils* **597** 15
[20] Zvereva E R, Zueva O S, Khabibullina R V and Makarova A O 2017 *JEAS* **11** 2950
[21] Zvereva E R, Khabibullina R V, Makarova A O, Akhmetvalieva G R, Burganova F I, Ermolaev D V and Zueva O S 2019 *Petroleum Chemistry* **59** 89

IOP Conf. Series: Earth and Environmental Science **337** (2019) 012061 doi:10.1088/1755-1315/337/1/012061

The effect of fine particle influence on numerical simulation of bidisperse fluidized bed

Olga V Soloveva[1,*], Sergei A Solovev[1] and Emil V Shamsutdinov[1]

[1]Kazan State Power Engineering University, 420066, Kazan, Krasnoselskaya str., 51, Russia

* Corresponding author: solovyeva.ov@kgeu.ru

Abstract. In this paper, we consider methods for choosing the effective diameter of solid particles for fluidized bed modeling using the Euler-Euler approach to describing a multiphase medium. Calculations are carried out for the modal average diameter D_{mod} and average diameter of Sauter D_{32}. Studies have also been conducted for the bidisperse composition of particles, with separation into large and small fractions. The results of numerical simulation were compared with experimental data.

1. Introduction

Fluidized bed (gas-solid particles) devices are widely used in the petrochemical and oil refining industry and the gasification of coal and biomass [1-3]. Reactors are widespread due to the high efficiency of heat and mass transfer between components. Solid particles are a mixture of approximately spherical granules of different diameters, thus, constitutes a polydisperse discrete phase.

The particle size distribution plays an important role in the behavior of the fluidized bed. In some cases, the addition of small particles leads to the expansion of the layer and an increase in the porosity of the dense phase [4, 5]. However, in other cases, the addition of these particles leads to a decrease in solid-phase porosity [6, 7]. The authors [8] showed that the addition of small particles just slightly reduces the effective viscosity. Experimental results showing a decrease in bubble size using fast X-ray tomography were obtained in [9]. The effect of fine particles on fluidization was investigated in [10] by numerical simulation. It was shown that the addition of a large number of small particles suppresses the bubbles, and the device can achieve a stable quasistationary mode of operation.

One of the possible ways to obtain data is numerical simulation. Numerical calculations of the fluidization process are usually based on the Euler-Lagrange and Euler-Euler approaches. The solid phase is considered as discrete particles in the Euler-Lagrange approach (for example, [11, 12]). In this approximation, each particle is considered separately. This approach requires significant computational resources and laborious for industrial devices that work with hundreds of tons of particles. In the Eulerian-Eulerian approach, the carrier (gas, liquid) and the discrete (solid) phase are considered continuous. To take into account the interaction of particles in the fluidized bed, by analogy with the kinetic theory of gases, an equation was added describing the change in the kinetic energy of the particles due to their collisions (for example, [13]). This approach is less demanding in terms of computational resources and requires discretization of particle characteristics. Discretization can be carried out using a multi-group approach, the essence of which is to divide the domain of definition into groups of diameters and assign them to correspond some value of the cumulative distribution function that characterizes the fraction of the group in the particle layer, and the transport equations are solved for each group separately. A multi-group approach to simulate a fluidized bed was used in [14, 15]. In [16], it was shown that the method of taking into account polydispersity can affect the hydrodynamic characteristics of the flow obtained by numerical simulation.

Thus, a universal method for the discretization of the characteristics of polydisperse solids is required. The method should provide the optimal choice of the smallest number of particle fractions for the numerical calculation and reflect the effects associated with polydispersity.

2. Problem formulation

2.1. *Effective diameter model*

Let us consider several ways of choosing the number and size of discrete phases for numerical simulation using the Eulerian-Eulerian approach. The calculation results are verified with experimental data obtained at the laboratory facility. The experimental setup is a glass tube with an internal diameter of 2.2 cm. A porous material is at the bottom of the tube to ensure a uniform airflow. The process of fluidization was controlled on a high-speed camera, the bed height was estimated by frames.

In this paper, we consider discretization methods based on the choice of the effective diameter and the particle terminal velocity. The method essence is to obtain two effective diameters, the first for coarse particles forming a dense phase, and the second for fine particles carried away by the flow. As effective diameters, we consider two models: D_{32} is Sauter mean diameter, D_{mod} is modal diameter.

$$D_{32} = \frac{\int_0^\infty f(D)D^3 dD}{\int_0^\infty f(D)D^2 dD}, \quad D_{\text{mod}} = \arg\max_D \left(f(D) \right). \tag{1}$$

Terminal velocity

$$u_t = \left(\frac{4gD(\rho_p - \rho_f)}{3\rho_f C_D} \right)^{0.5},$$
$$C_D = \frac{24}{\text{Re}_p} + \frac{6}{1+\sqrt{\text{Re}_p}} + 0.4, \quad \text{Re}_p = \frac{\rho_f \left| u_p - u_g \right| D}{\mu_f} < 200000. \tag{2}$$

Equating the known gas velocity u_g to the terminal velocity (1), it is necessary to obtain the diameter, relative to which the initial mixture of particles can be divided into the two groups described above. D_s is the root of equation (3).

$$u_g = \left(\frac{4gD(\rho_p - \rho_f)}{3\rho_f C_D} \right)^{0.5}. \tag{3}$$

When under the action of the forces of gravity and resistance the particle of diameter D_s is in equilibrium ($u_p = 0$), the particles with a diameter less than D_s will be carried away by the stream, and particles with a large diameter tend down. This fact allows us to determine the effective diameters (4).

$$D^1_{32} = \frac{\int_0^{D_s} f(D)D^3 dD}{\int_0^{D_s} f(D)D^2 dD}, \quad D^2_{32} = \frac{\int_{D_s}^\infty f(D)D^3 dD}{\int_{D_s}^\infty f(D)D^2 dD},$$
$$D^1_{\text{mod}} = \underset{D\in[0,\, D_s]}{\text{argmax}} \left(f(D) \right), \quad D^2_{\text{mod}} = \underset{D\in(D_s,\, \infty)}{\text{argmax}} \left(f(D) \right). \tag{4}$$

The effective diameters and volume fractions of the experimental data are presented in Table 2.

Table 1. The effective diameters and initial volume fractions.

Velocity, m/s	Fine particle diameter, μm		Coarse particle diameter, μm		Fine particle volume fraction	Coarse particle volume fraction
	D^1_{32}	D^1_{mod}	D^2_{32}	D^2_{mod}	α^1_0	α^2_0
0.0716	30.5	32.4	66.9	56.4	0.005	0.411
0.0892	33.4	35.6	67.2	56.4	0.014	0.402
0.1088	36.4	39.6	67.7	56.4	0.033	0.383
0.1213	38.2	42.0	68.3	56.4	0.052	0.364

All tests were carried out for 30 g of particles, the initial volume fraction of solid particles in a fixed bed was $\alpha_0 = 0.41639$.

2.2. Fluidized bed model

In the numerical simulation of the fluidized bed, a continuous Eulerian-Eulerian multiphase model was used, supplemented by the kinetic theory of gases to take into account collisions of solid particles. For each phase, the laws of conservation of mass, momentum are satisfied. In the study of this problem, the following equations were solved:

Conservation of mass

$$\frac{\partial \alpha_i \rho_i}{\partial t} + \nabla \cdot (\alpha_i \rho_i \vec{v}_i) = 0, \tag{5}$$

where α_i is volume fraction for *i*-th phase, ρ_i is density, $\vec{v}_i$ is velocity.

Conservation of momentum

$$\frac{\partial \alpha_i \rho_i \vec{v}_i}{\partial t} + \nabla \cdot (\alpha_i \rho_i \vec{v}_i \vec{v}_i) = -\alpha_i \nabla p + \nabla \cdot \overline{\overline{\tau}}_i + \alpha_i \rho_i \vec{g} + \sum_j \vec{R}_{ij}, \tag{6}$$

where $\overline{\overline{\tau}}_i = \alpha_i \mu_i \left(\nabla \vec{v}_i + \nabla \vec{v}_i^T\right) + \alpha_i \left(\lambda_i - (2/3)\mu_i\right) \nabla \cdot \vec{v}_i \overline{\overline{I}}$ is stress tensor, p is pressure, μ_i, λ_i are bulk and shear viscosities, $\overline{\overline{I}}$ is unit tensor, $\vec{R}_{ij} = K_{ij}\left(\vec{v}_i - \vec{v}_j\right)$ is interfacial interaction force.

The equation for the solid phase granule temperature is represented as [17]

$$\frac{3}{2}\left[\frac{\partial}{\partial t}(\alpha_s \rho_s \Theta_s) + \nabla \cdot (\alpha_s \rho_s \vec{v}_s \Theta_s)\right] = \left(-p_s \overline{\overline{I}} + \overline{\overline{\tau}}_s\right) : \nabla \vec{v}_s + \nabla \cdot \left(k_{\Theta_s} \nabla \Theta_s\right) - \gamma_{\Theta_s} + \phi_{sg}, \tag{7}$$

where $k_{\Theta s}$ is granule energy diffusion coefficient, Θ_s is solid phase granule temperature, $\gamma_{\Theta s}$ is energy dissipation from particle collisions, ϕ_{sg} is energy exchange between solid and gas phases.

To close the equation system, semi-imperial models of interphase interaction were used. For the liquid (gaseous) phase (*i*-th) and solid granular phase (*j*-th), the interaction model was taken from [13]; in the case of two solid granular phases, the model [18] was used. The energy dissipation due to particle collisions was described by the model obtained in [19].

3. Results

The results of numerical simulation using the methods described for selecting an effective diameter for bidisperse particles were compared with experimental data, as well as with the results of calculations performed with one effective diameter for monodisperse particles. The distribution of the volume fraction of the catalyst at the height of the tube, obtained in the calculations, is shown in Figure 1.

It can be seen that in all calculations the height of the dense layer of particles is within the limits of the observed minima and maxima. From Figure 1 *a* and *b*, it can be seen that at low speeds the addition of fines practically does not affect the average heights of the dense layer obtained in the calculations, this is because the volume fraction of particles with a diameter less than D_s is small.

The time-averaged layer heights obtained experimentally and numerically are shown in Figure 2. The average heights of the dense layer, obtained in calculations with two effective diameters D^1_{32} and

IOP Conf. Series: Earth and Environmental Science **337** (2019) 012061 doi:10.1088/1755-1315/337/1/012061

D_{32}^2 (for small and coarse fractions, respectively), showed better agreement with experiment than the heights obtained from calculations with one effective diameter D_{32}. In the case when the effective diameters were chosen as a mode, it turned out that the heights obtained in the calculations with two effective diameters D_{mod}^1 and D_{mod}^2 turned out to be greater than the experimental average, moreover, the heights in the calculations with one effective diameter D_{mod}.

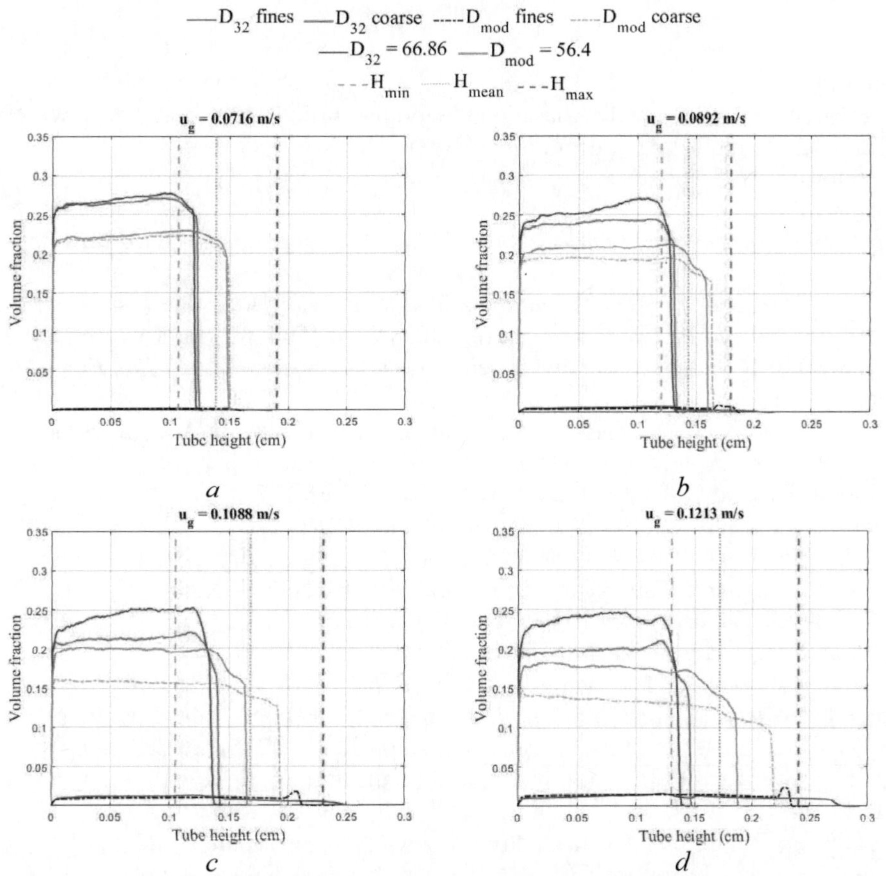

Figure 1. The distribution of the volume fraction of particles at the height of the tube for four values of flow velocity.

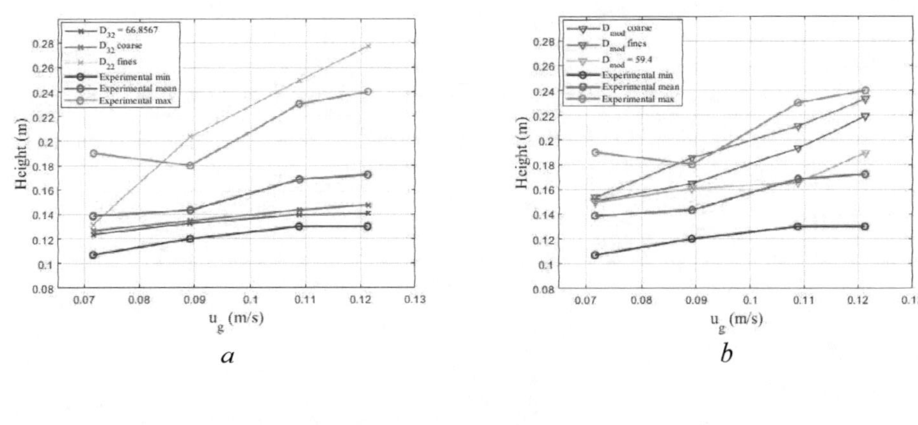

Figure 2. Time-averaged layer height as a function of flow velocity.

Based on this fact, for polydisperse particles and with a similar distribution function, we can recommend the following conditions for choosing effective diameters

$$D^1_{32} \le D^1_{eff} \le D^1_{\mathrm{mod}},\ D^2_{32} \le D^2_{eff} \le D^2_{\mathrm{mod}} \qquad (8)$$

Acknowledgments

The reported research was funded by Russian Foundation for Basic Research and the government of the Republic of Tatarstan of the Russian Federation, grant № 18-48-160006.

References

[1] Yang W C 2003 *Handbook of Fluidization and Fluid-Particle System* (Oxford: Taylor and Francis Group)
[2] Basu P 2006 *Combustion and Gasification in Fluidized Beds* (Boca Raton: CRC press)
[3] Sadeghbeigi R 2012 *Fluid Catalytic Cracking Handbook* (Oxford: Butterworth-Heinemann)
[4] Van Ommen J R, Nijenhuis J, van den Bleek C M and Coppens M 2007 *Ind. Eng. Chem. Res.* **46** 4236
[5] Soloveva O V, Solovev S A, Egorova S R, Lamberov A A, Antipin A V and Shamsutdinov E V 2018 *Chem. Eng. Res. Des.* **138** 212
[6] Khoe G K, Ip T L and Grace J R 1991 *Powder Technol.* **66** 127
[7] Saayman J, Ellis N and Nicol W 2013 *Powder Technol.* **245** 48
[8] Beetstra R, Nijenhuis J, Ellis N and van Ommen J R 2009 *AIChE J.* **55** 2013
[9] Brouwer G C, Wagner E C, van Ommen J R and Mudde R F 2012 *Chem. Eng. J.* **207-208** 711
[10] Gu Y, Ozel A and Sundaresan S 2016 *AIChE J.* **62**(7) 2271
[11] Chen X and Wang J *Powder Technol.* 2014 **254** 94
[12] Zhou Q, Wang C, Wang H and Wang J 2016 *Int. J. Miner. Process.* **151** 40
[13] Gidaspow D 1994 *Multiphase Flow and Fluidization* (San Diego: Academic Press)
[14] Shah S, Myöhänen K, Kallio S, Ritvanen J and Hyppänen T 2015 *Powder Technol.* **274** 239
[15] Solovev S A, Soloveva O V, Antipin A V and Shamsutdinov E V 2018 *J. Phys.: Conf. Series.* **944** 012114
[16] Loth E, O'Brien T, Syamlal M and Cantero M 2004 *Powder Technol.* **142** 209
[17] Ding J and Gidaspow D 1990 *AIChE J.* **36** 523
[18] Syamlal M 1987 *The Particle-Particle Drag Term in a Multiparticle Model of Fluidization* (Morgantown WV: EG & G Washington Analytical Service Center)
[19] Lun C K K, Savage S B and Jeffrey D J 1984 *J. FluidMech.* **140** 223

IOP Conf. Series: Earth and Environmental Science **337** (2019) 012062 doi:10.1088/1755-1315/337/1/012062

Simulation of a reactor for thermal processing of coal-water fuel to produce syngas of specified composition

Guzel Mingaleeva[1,*], Aygul Galkeeva[1], Emil Shamsutdinov[1] and Olga Afanaseva[1]

[1]Kazan State Power Engineering University, 420066, Kazan, Krasnoselskaya str., 51, Russia

* Corresponding author: mingaleeva-gr@mail.ru

Abstract. The main physical and chemical regularities for development of a flow gasifier for thermal processing of coal-water fuel (CWF) on condition of the set structure received syngas are presented in this article. Based on the calculations made on the residence time of a CWF drop in the reactor and the time of gasification reactions the gasifier was designed with vertical heating surfaces. Calculation of the main sizes of a gas generator for gas productivity on synthesis gas 500, 1000 and 1500 m^3/h is carried out. The results of a constructive calculation show that in order to ensure the required residence time of a CWF particle in the volume of the gasifier it is necessary to ensure a certain length of the reactor, regardless of the gas throughput.

1. The theoretical basis for the development of a gasifier

The using of coal-water fuel (CWF) for the production of energy and chemical products is a promising direction due to the huge reserves of coal which are exceed reserves oil and natural gas the modern industry uses a huge amount of fossil fuels [1]. The thermal processing of CWF is a continuous process that can be implemented in a column type gasifier. Currently, standard designs and methods for calculating these devices are absent, despite the fact that entrained flow gasifier are widespread in the world [2-10]. The largest number of such plants is located in China, where methanol, ammonia, acetic acid, fuel gas, hydrogen, carbon monoxide and other products are produced on their basis.

The scientific works in the field of gasification are mainly aimed at determining the composition of the generating gas, depending on the composition of the initial components and operating parameters of the process [2, 4, 11-15]. But in this work a different approach is used: the composition of the gas is set for the subsequent release of a certain product. For example, for the production of methanol, which is a large-tonnage product, the world production of which is 160-1275 thousand tons per year, it is necessary to produce syngas with a molar ratio of CO / H_2 in the range of 0.4-0.6. The best conditions for the production of such gas are created precisely during the gasification of CWF [11].

1.1. Process mode parameters

The authors earlier determined the temperature at which CWF obtained from coal of various grades decompose with the formation of gas with a molar ratio of CO / H_2 equal to 0.4-0.6 and the minimum content of other components. In the process of oxygen gasification, the reactions of interaction of the components of the fuel with oxygen are exothermic and flow intensively, while the resulting gas contains a significant amount of unreacted vapor. In oxygen-free gasification, steam acts as an oxidizing agent and its amount in the gas decreases. In this case, the reactions of the interaction of the components of coal with steam are endothermic and heat is required for their flow. The optimum mode of the gasification process was revealed, at which the time to reach equilibrium concentrations during thermal conversion of CWF to syngas at 1150 K is about 3 sec. [16].

1.2. The main calculated dependencies

The main calculated dependences are presented for the process of oxygen-free gasification of coal-water fuel taking into account the processes of heat and mass transfer between a drop of CWF and the flow of hot gas.

In the physical formulation of the problem a CWF drop after spraying by a mechanical injector enters the environment of hot synthesis gas, which is heated due to heat transfer from the beam of flame tubes, inside which the products of combustion of a portion of the produced syngas are circulated. Conventionally, a drop passes through 3 zones of the reactor: evaporation of moisture, extraction of volatile components and reaction of the components of the mixture (Figure 1).

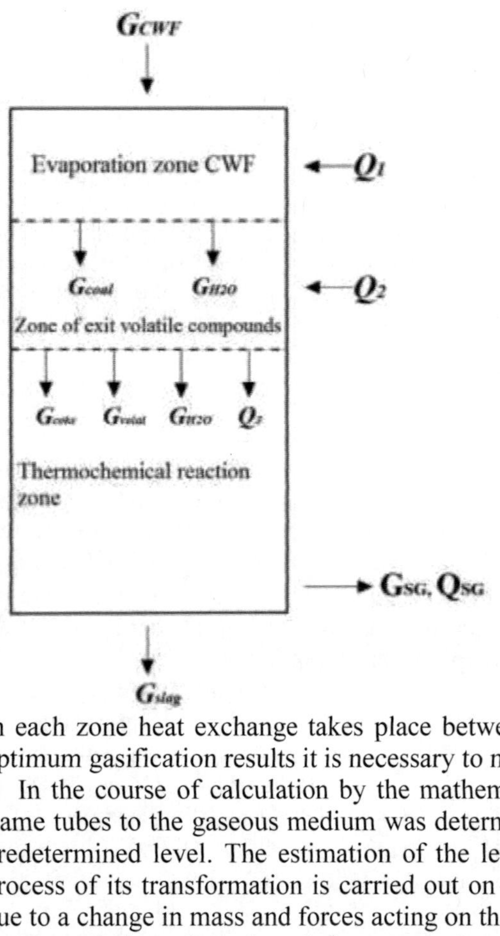

Figure 1. The scheme of gasification of coal-water fuel.

In each zone heat exchange takes place between the drop of CWF and the environment. To ensure optimum gasification results it is necessary to maintain the temperature in the volume at 1150 K.

In the course of calculation by the mathematical model, the amount of heat transferred from the flame tubes to the gaseous medium was determined in order to maintain the operating conditions at a predetermined level. The estimation of the length of a path that a drop of CWF overcomes in the process of its transformation is carried out on the basis of determining the loss of velocity of a drop due to a change in mass and forces acting on the drop.

The solution of the heat balance equation for the moisture evaporation zone is to determine the amount of heat required to evaporate moisture, warm the fuel particle to the evaporation temperature and heat the steam to the environment temperature:

$$Q_{reql} = rm_{H_2O} + c_{d0}m_{d0}(T_{evap} - T_0) + c_{H_2O}m_{H_2O}(T_{SG} - T_{evap}),$$

where m_{d0}, m_{H_2O} – is the weight of drop CWF and steam, kg; c_{d0} c_{H_2O},– is the specific heat capacity of a drop CWF and steam; r_{H_2O} – is the heat of water vaporization, J / kg. T_{evap} – is evaporation temperature, K; T_0 – is environment temperature, K; T_{SG} – is syngas temperature, K.

For the exit zone of volatile components, the amount of heat needed to evaporate them and heat the particle to ambient temperature is determined:

$$Q_{req2} = r_{H2O}m_{H2Ovolat} + r_{CH4}m_{CH4volat} + c_{coke}m_{coke}(T_{SG} - T_{volat}) + c_{volat}m_{volat}(T_{SG} - T_{volat})$$

where $m_{H2Ovolat}$, $m_{CH4volat}$, m_{coke}, m_{volat} – is weight of volatile steam, methane, coke particle and volatile compounds, kg; c_{coke}, c_{volat}, – is specific heat of coke particle, and volatile substances, J / (kg · K); r_{H2O} r_{CH4} – heat of water and methane vaporescence, J/kg; T_{volat} – is temperature of volatile compounds, K.

The equations of heat and mass transfer are solved using the theory of similarity according to known dependencies, taking into account the mechanism for the transformation of a CWF drop.

For the zone of thermochemical reaction of the components of the mixture on the basis of heat balance, the temperature of syngas at the outlet from the gasifier is determined:

$$T_{SG3} = \frac{c_{SG2} \cdot m_{SG2} \cdot T_{SG} + c_{volat2} \cdot m_{volat2} \cdot T_{SG} + c_{coke} \cdot m_{coke} \cdot T_{SG} + Q_{c.r.1d}}{c_{react} \cdot m_{react}},$$

where c_{SG2}, c_{react} - is specific heat of syngas and reaction mixture, J / (kg · K); m_{react}, m_{SG2} – is weight of reaction mixture and syngas, kg; $Q_{c.r.1d}$ - the total thermal effect of gasification reactions on 1 CWF drop, J / kg.

The calculation of the velocity of a drop of variable mass moving in a stream of hot gas which based on the determination of the forces acting on the drop with Meshchersky equation [17]:

$$m_d \cdot \frac{du_d}{d\tau} = \sum_{i=1}^{l} F_i + u_d \cdot \frac{dm_d}{d\tau},$$

where $\sum_{i=1}^{l} F_i$ – is as et of forces acting on the drop, N; m_d – is weight of CWF drop, kg; u_d – is velocity of a CWF drop, m/s; τ – time, s.

As a result of the calculation using the mathematical model it was determined that the length of all reaction zones, and, accordingly, the length of the entire gas generator can be affected by various factors - temperature, size of drops, speed at the nozzle exit. The size of a CWF drop sprayed by a nozzle has a significant effect on the length of the drying zone and the yield of volatile components. This is due with necessity to increase the time for the processes occurring with a moving drop in the respective zones. At a given gasification temperature, it is necessary to ensure a certain time for a CWF drop in the gas generator volume. Therefore in the process of mathematical modeling, the drop rate at the nozzle outlet was determined, at which the required CWF drop time in the gasifier will be provided with acceptable overall dimensions of the gasification reactor.

2. Gasifier design

2.1. Principle of operation

Based on the calculations made on the residence time of a CWF drop in the reactor and the time of gasification reactions the gasifier was designed with vertical heating surfaces. There are made in the form of flue tubes with hot combustion products circulating inside them, which are formed when a certain fraction of syngas is burned obtained by gasification of CWF (Figure 2).

Between the vertical flue pipes and the gaseous medium in the volume of the gasifier the heat transfer occurs which is necessary to ensure the heat of all stages of oxygen-free CWF gasification occurs.

The gasifier works as follows. CWF is supplied through a pipeline to the ring collector 8 and sprayed by nozzles 7 into the environment of synthesis gas at a temperature of 1150 K. Part of the synthesis gas is discharged from the pipe 3, cooled, cleaned and with the air sent through the pipe 10 to the combustion in the flame tube 6. By means of burner 9 a stable supply of the gas-air mixture and its combustion in the flame tube 6 is carried out. The resulting flue gases rise up through the flue tubes 5 and are withdrawn through pipe 11. The syngas which fills the gas generator is heated by means the heat transfer from the hot flue gases passing through the flue tubes 5. As a result the temperature in the gasifier becomes sufficient for gasification of CWF, where the moisture contained in CWF is used as an oxidizer. Formed during the gasification process syngas is discharged through the pipe 3, and the slag through the outlet 4 enters the system of ash and slag removal.

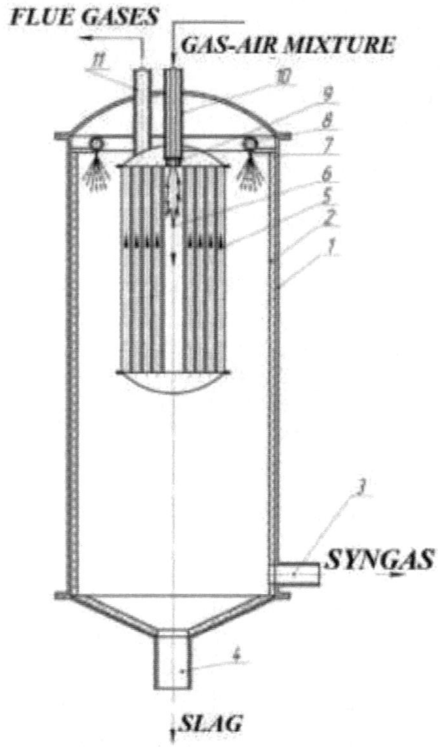

Figure 2. Entrained flow gasifier : 1 - the case; 2 - lining; 3 - pipe for removal of syngas; 4 - pipe for slag removal; 5 - flue tubes; 6 - flame tube; 7 - CWF nozzles; 8 - ring collector; 9 - burner; 10 - supply of gas- air mixture; 11 - flue gas discharge.

2.2. The determination of basic dimensions

The results of a constructive calculation show that in order to ensure the required residence time of a CWF particle in the volume of the gasifier it is necessary to ensure a certain length of the reactor, regardless of the gas throughput. Table 1 presents the results of a constructive calculation of the gasifier with different gas productivity. Increased productivity is achieved by increasing the number of flame tubes, and, accordingly, the volume of the reactor. At the same time its length remains unchanged.

Table 1. The results of a constructive calculation for different gas productivity

Gas productivity, m^3 / h	Volume gasifier, m^3	Amount of tubes, pcs	Diameter gasifier, m
500	54.91	365	2.28
1000	112.83	750	3.27
1500	171.5	1140	4.03

In this way the calculation of the gasification process at the design stage of the design parameters of the gas generator makes it possible to achieve unification of the type of fuel when regulating the operating parameters.

Acknowledgments
The reported study was funded by RFBR according to the research project No.17-08-00295 «A».

References
[1] S&P Global Platts 2017 *BP Statistical Review of World Energy 2017* https://www.bp.com/en/global/corporate/energy-economics/statistical-review-of-world-energy.html
[2] Wang M, Liu G and Hui C W 2018 *Clean Techn. Environ. Policy* **20** 259–69
[3] Zhang J, Ma L, Zhou Z and Li Z 2013 *Energy* **51** 137–45
[4] Higman C and Burgt M 2003 *Gasification* (Houston: Elsevierscience)
[5] Huang, N., Zhao, P., Ghosh, S., Fedyukhin, A. Co-hydrothermal carbonization of polyvinyl chloride and moist biomass to remove chlorine and inorganics for clean fuel production, 2019, Applied Energy, 240, p. 882-892
[6] Solovev, S.A., Soloveva, O.V., Antipin, A.V., Shamsutdinov, E.V. Investigation of internal elements impaction on particles circulation in a fluidized bed reactor, 2018, Journal of Physics: Conference Series, 944(1),012114
[7] Soloveva, O.V., Solovev, S.A., Egorova, S.R., Lamberov, A.A., Antipin, A.V., Shamsutdinov, E.V. CFD modeling a fluidized bed large scale reactor with various internal elements near the heated particles feeder, 2018, Chemical Engineering Research and Design, 138, p. 212-228
[8] Fedyukhin, A.V., Sultanguzin, I.A., Stepanova, T.A., Voloshenko, E.V., Kurzanov, S.Y., Isaev, M.V. Improving solid-fuel pyrolysis and gasification for effective cogeneration 2013 Coke and Chemistry, 56(8), p. 302-306
[9] Saari, J., Sermyagina, E., Kaikko, J., Vakkilainen, E., Sergeev, V. Integration of hydrothermal carbonization and a CHP plant: Part 2 –operational and economic analysis, 2016, Energy, 113, p. 574-585
[10] The Values of Specific Consumption of Energy in the Assessment of the Level of Future Demand Aleshina, A., Vladimirov, I., Kozhukar, E. 2019 2018 International Multi-Conference on Industrial Engineering and Modern Technologies, FarEastCon 2018, 8602833
[11] Zaitsev A V, Ryzhkov A F, Silin V E et all 2010 *Gazogeneratornye tehnology v energetike* ed Ryzhkov A F (Ekaterinburg: Sokrat)
[12] Fedyukhin, A., Sultanguzin, I., Gyul'Maliev, A., Sergeev, V. Biomass pyrolysis and gasification comprehensive modeling for effective power generation at combined cycle power plant. 2017, Eurasian Chemico-Technological Journal, 19(3), p. 245-253
[13] Sultanguzin, I.A., Fedyukhin, A.V., Kurzanov, S.Y., Gyulmaliev, A.M., Stepanova, T.A., Tumanovsky, V.A., Titova, D.P. Prospects for the development of independent power supply systems on the basis of solid fuel thermal conversion technology 2015, Thermal Engineering (English translation of Teploenergetika), 62(5),A008, p. 359-364
[14] Sergeev, V.V., Aleshina, A.S. A gas-generator combined-cycle plant equipped with a high-head heat-recovery boiler. 2011, Thermal Engineering, 58(3), p. 268-270
[15] Shestakov, V., Novikova, O., Melnichenko, A., Luchnikova, A Methodical approaches to energy supply with usage of renewable energy sources on objects of transport infrastructure of federal importance, 2018 MATEC Web of Conferences, 245,15005
[16] Mingaleeva G R, Ermolaev D V and Galkeeva A A 2016 *Clean Techn. Environ. Policy* **18** 297–304
[17] Pomerantcev V V, Arefyev K M, Ahmedov D B et all 1986 *Osnovy prakticheskoy teoriy goreniy* (Leningrad: Energoatomizdat)

IOP Conf. Series: Earth and Environmental Science **337** (2019) 012063 doi:10.1088/1755-1315/337/1/012063

Practical applicability of the method for measuring pressure of controlled medium on the example of a pulsating combustion boiler

Anton Sinitsyn[1*] and Liliya Mukhametova[2]

[1]Vologda State University, Heat Gas Supply and Ventilation, 160000 Lenin st., 15, Russia
[2]Kazan State Power Engineering University, Kazan, Russian Federation

* Corresponding author: sinitsyn.science@mail.ru

Abstract. This paper proposes a new way to study the nature of vibration component of pressure change over time for an energy device. This method makes it possible to record and process measurement results, compare frequency characteristics at various points of the device working space, and evaluate the presence of defects in its work. The authors have identified the main problems of developing a measuring complex to study the nature of vibration component of the pressure change over time for an energy device, and numerical and field tests have been carried out.

1. Introduction

The urgency of the problem of increasing the efficiency of heat sources is an indisputable argument while developing new heat-generating devices and modernization of existing ones [1-7]. Traditional methods, such as increasing the power of burner, the use of fuel with increased heat release during combustion, an increase in the surface of heat transfer, etc. are not always feasible and are often costly. A simple mechanical transfer of accumulated experience in the design, construction and operation of gas consumption facilities is absolutely unacceptable in modern conditions. Therefore, it is necessary to develop heating equipment for new and more efficient ways of technical development and environmental safety [8-11]. In this direction, the implementation of pulsating combustion processes in heat and power plants seems very promising. This mode of combustion allows providing maximum completeness of heat generation of fuel, significantly intensifies heat and mass transfer processes and increases the thermal stress of the combustion chamber. Under such conditions, a decrease in the metal content of the structure, reducing the cost of installation and maintenance of thermal power plants is obvious. In addition, the products of combustion meet the most stringent environmental requirements.

Nowadays, in many countries intensive research is being conducted in the field of design and implementation of heat and power plants based on pulsating combustion in technological processes. The absence of a burner as such and significant flue gas pressure at the outlet that does not require a high chimney, maintaining performance even at an overpressure close to zero in the supply gas pipeline are additional advantages of the pulsating combustion boiler, ensuring its attractiveness in the heat and power sector of the market. However, the widespread introduction of pulsating combustion devices into technological processes is hampered by the absence of a reliable theory of the working

IOP Conf. Series: Earth and Environmental Science **337** (2019) 012063 doi:10.1088/1755-1315/337/1/012063

process for calculating design parameters during their design, as well as calibration calculations for determining the effectiveness of their work. In previous works [8, 12], the mathematical apparatus and the method for calculating energy devices based on a non-equilibrium method of burning fuel are described in detail. However, the main problem with the implementation of such a technique is determination of the actual dependence of the change in pressure of the medium over time, the accuracy of which strongly influences the results of the calibration calculation of the energy device. The method should be implemented through a measuring complex, which can be applied to configure the operating modes of the device under study.

Therefore, an urgent task is simulation of vibration processes and creation of a universal measuring complex for determining the nature of the vibration component of the pressure over time for an energy device.

2. A review on pressure measurement techniques

The developed method relates to the measuring technique and active non-destructive testing and can be used to measure the pressure of the controlled medium. The method includes measuring the nature of vibration component of the pressure change over time for the energy device, allows one to work out the start-stop mode, start-up mode, long-term operation and shutdown of the energy device, as well as transients at changing the main (resonance frequency, fuel and air composition and flow) and additional (fuel pressure, temperature of the operating medium) operating parameters of the energy device, makes it possible to record and process measurement results, to compare frequency characteristics in various points of the working device space, and to evaluate the presence of defects in their work.

The analysis of methods for measuring pressure showed the following. Nowadays, a method for measuring the pressure of the medium is known [8]. Atmospheric pressure Patm acts on one end of a U-shaped tube, partially filled with working fluid. The other end of the tube is connected to the area of the measured pressure Pchan by means of various types of supply devices. At Pchan> Patm, the liquid located in the part of the connected measured pressure will be forced out to the part connected to the atmosphere. As a result, a liquid column is formed between the levels of liquids located in different parts of the U-shaped tube, the height of which is the measured excess pressure. However, this method is acceptable for determining the amplitude of fluctuations of the medium; it is impossible to determine the nature of the change in pressure of the medium using it.

In addition, there is a method of pressure measuring, which consists in placing a pressure sensor in a test medium, placing a temperature sensor on a pressure sensor, recording the output signals of the pressure sensor and the temperature sensor. These signals determine the pressure of medium, form mechanical oscillations with a frequency greater than the possible frequency of oscillations of the working pressure of the medium in the medium under study, extract the variable signal from the output signal of the sensor with the frequency of specified mechanical oscillations. Diagnostic functions are determined from this signal and the output signals of the sensor and the temperature sensor, by the deviation of which from the nominal value, the error of pressure measurement is judged.

The disadvantages of the existing method are the difficulty of calibrating and assigning units of measurement by a sensor to units of pressure, while the sensor is not acoustically isolated from the environment, which makes the measurement dependent on any noise.

The closest to the proposed method is a method of measuring blood pressure, when the compressor pumps air into the cuff to achieve a pressure value that is obviously greater than the maximum pressure of the patient, after which the cuff pressure decreases through the decompression valve, during which the decision block enters cuff pressure fluctuation signals from a pressure sensor. The microprocessor receives these signals through an analog-to-digital converter. All of these signals, in addition, are recorded by a multi-channel recorder.

The disadvantage of such a system is low noise immunity of measurements, especially under conditions of uneven and intense fluctuations in the pressure of medium, that is, precisely when pressure measurements are of great practical interest. When starting the boiler, internal combustion engine or their transient modes, the form of pressure measurement over time is distorted, which can lead to incorrect measurement and technical conclusion about the performance of the power facility. An additional disadvantage is also the functional failure of the method, consisting in a limited range of the studied objects.

Thus, the development is relevant and can find direct application in various sectors of energy and utilities.

3. Description of the proposed method

The purpose of the study was to improve the accuracy and informativeness of pressure measurement methods, to create the ability to measure and predict system operation in non-stationary conditions of vibration combustion, to assess the presence of defects in the operation of energy devices.

The advantages of the proposed solution in comparison with analogues are the following:

1. The ability to measure the parameters of the working process at high temperatures of the environment in the heat-stressed zone of the energy device (over 700 °C).

2. Acoustic isolation of the mechanical signal-to-analog conversion system, which improves the measurement accuracy at the lowest pressure amplitudes and reduces the measurement time.

3. When testing the energy device for analysis, not only the maximum, average and minimum pressure of the working medium are displayed, but also the dynamics of pressure change over a given period of time. This allows evaluating the accuracy of measurement in the presence of any errors and, in cases where interference caused erroneous measurement, do not take into account the values obtained; to reconfigure the amplitude of oscillations for the obtained curve of pressure change, without changing the overall picture of the process (in the case of adjusting the amplitude to the desired value); to assess the relationship of oscillations at various points of the working path of the energy device; to evaluate the influence of secondary parameters of the workflow on the pressure in the combustion chamber of the device (for example, the influence of the configuration of the working space).

This goal is achieved by the fact that the pressure of medium in at least one measuring point on the impulse tube is perceived by a mechanical-electrical converter placed in a heat-sound-insulated acoustic capacitance, in which the mechanical oscillation is converted into an electrical signal by an electronic circuit consisting of EMF source and resistance. The signal is transmitted to an analog-to-digital conversion device, where a digital signal is formed in dimensionless units, the conversion to the pressure dimension of which carried out using two U-shaped manometers, configured so that one of them measures the maximum pressure, and the second – the minimum one, while the non-return valve in the case of measuring the maximum pressure passes a drop in fluid levels towards the atmosphere and blocks towards the measured medium. In case of measuring the minimum pressure it passes a drop in fluid levels in the direction of the measured liquid and blocks in the direction of the atmosphere. Using the software of the computing unit, the digital signal of the sound change is converted into pressure in dimensionless units, and the pressure is converted from dimensionless units to pressure dimensions in order to measure and predict the operation of the energy system under non-stationary conditions of vibration burning, to evaluate the presence of defects in the energy devices.

For clarity, figure 1 shows the scheme of pressure measurement, and figure 2 shows the results of the study using the proposed method and a comparison with the known method of measuring pressure.

Figure 1 schematically shows a measuring device consisting of a heat and sound insulated body 1 in which acoustic the following parts are placed: acoustic capacitance 2, a pulse tube 3 connected to acoustic capacity at one end, and heat and sound insulation 4, which fills the free space of enclosure 1. In acoustic capacitance an electric transducer, in particular, a carbon microphone 5, included in an electrical circuit consisting of an EMF source 6 and resistance 7. The signal output is connected by an analog-to-digital conversion and registration device 8.

IOP Conf. Series: Earth and Environmental Science **337** (2019) 012063 doi:10.1088/1755-1315/337/1/012063

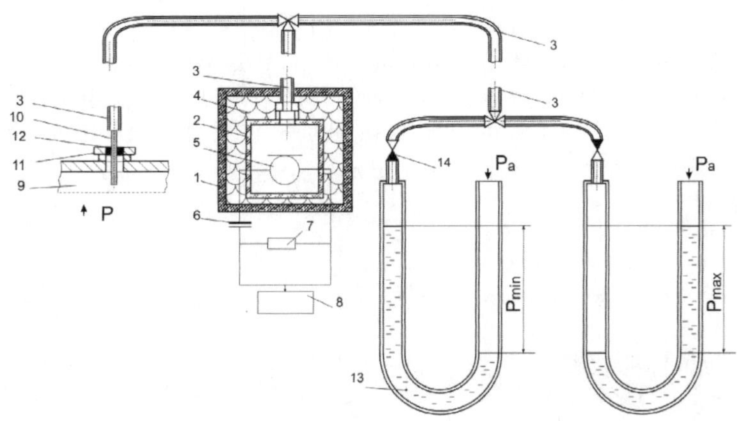

Figure 1. Principal scheme of the measuring device.

The scheme for measuring the nature of the pressure oscillation consists of an impulse tube 3 attached to the object under measurement 9, a contact tube 10 installed in the sleeve 11 and fixed in it by means of heat and sound insulating material 12.

The oscillation amplitude measurement scheme consists of two U-shaped pressure gauges 13 for measuring the pressure of the medium, connected by a pulse tube 3 with a check valve 14, installed so that for measuring the minimum pressure Pmin it is located in the direction of the pressure gauge 13, and for measuring the maximum pressure Pmax it is placed in the opposite direction. Here P is the measured pressure of the medium, Pa is the atmospheric pressure.

The device for implementing the proposed method works as follows. Depending on the test task, at least one measurement point is selected at which the pressure of medium will be measured. In the measurement object 9, a sleeve 11 is preliminarily arranged, in which the contact tube 10 is fixed. To avoid thermal and noise effects, a heat and sound insulating material 12 is used on the measurement system. The contact tube 10 is connected to the pulse tube 3 from one side, and from another one is placed into the object under test 9 in such a way to interact with the environment. The system for measuring the amplitude of pressure fluctuations is calibrated using standard methods and instrumentation (conventionally not shown).

When the object under study 9 is operating, the medium pressure on the pulse tube 3 is perceived by the mechanical-electrical converter 5, while the external environmental noise is extinguished by heat and sound insulation 4. In the converter 5, the received sound signal is converted by an electrical circuit consisting of an EMF source 6 and resistance 7 to an analog signal transmitted to the analog-to-digital conversion device 8, in particular, the PC sound card, where a digital signal is formed, which is, in particular, a periodic process of sound change in dimensionless units in time.

Measurement of the amplitude of pressure fluctuations using two U-shaped pressure gauges 13, configured so that one of them is set to measure the maximum pressure, and the second one to measure the minimum pressure. Through the impulse tube 3, the pressure of the medium is transferred to the working fluid of the manometer 13, which is exerted by the atmospheric pressure on the open side of the pressure gauge, causing the fluid level to change depending on whether the measured pressure is higher or the atmospheric one. The valve 14 in case of measuring the maximum pressure passes the differential levels of fluid in the direction of the atmosphere and blocks in the direction of the measured medium; in the case of measuring the minimum pressure – on the contrary: it passes in the direction of the measured liquid and blocks in the direction of the atmosphere.

Using the software of the computing unit (conventionally not shown), the digital signal of the sound change is converted into pressure in dimensionless units, and also converted from dimensionless units into pressure dimension.

By processing the data of the analog-digital conversion device 8 and pressure gauges 13 on a computer, it is possible to obtain the frequency-pulse nature of the propagation of thermo-acoustic oscillations in the coordinates P-τ.

IOP Conf. Series: Earth and Environmental Science **337** (2019) 012063 doi:10.1088/1755-1315/337/1/012063

4. Description of mathematical methods

An example of the implementation of the proposed method is given for study of frequency-pulse nature of the propagation of thermo-acoustic oscillations in the operating boiler of pulsating combustion of the PV type based on the Helmholtz resonator under the following conditions:

– Pressure in gas pipe is 102 kPa;
– Pressure in air pipe is 100 kPa (atmospheric pressure);
– Fuel consumption is 36 m3/h;
– Water temperature at the boiler inlet is 44 ºC;
– Water temperature at the boiler outlet is 50 ºC;
– Excess air ratio is 1.25;
– Resonance frequency is 33 Hz;
– Adiabat index of flue gas γ=1.4.

– The studies were conducted in three stages: obtaining an analog signal with its conversion to digital using an audio computer editor; mathematical processing of the audio signal in a periodic process and determining the values of the amplitudes of oscillations.

– At first stage, a sound signal was received from a 5-angle microphone signal converter. Mathematical processing of the sound file was carried out using the MathCAD program and the additional Signal Processing package designed to process audio signals. The signal was read from the sound file, obtaining complete information about it and building the original signal graph. Next, to go from the voltage dimensions to the pressure dimensions, the signal was calibrated using a U-shaped manometer.

5. Experimental results

For comparison, the data on the pressure amplitude are correlated with the results of similar experiments by Korean researchers [13] for the Helmholtz chamber with an aerodynamic valve – PCS (Pulsating Combustion System). Figure 2 presents the obtained frequency characteristics using the method [13] and the proposed method for a period of time in ms.

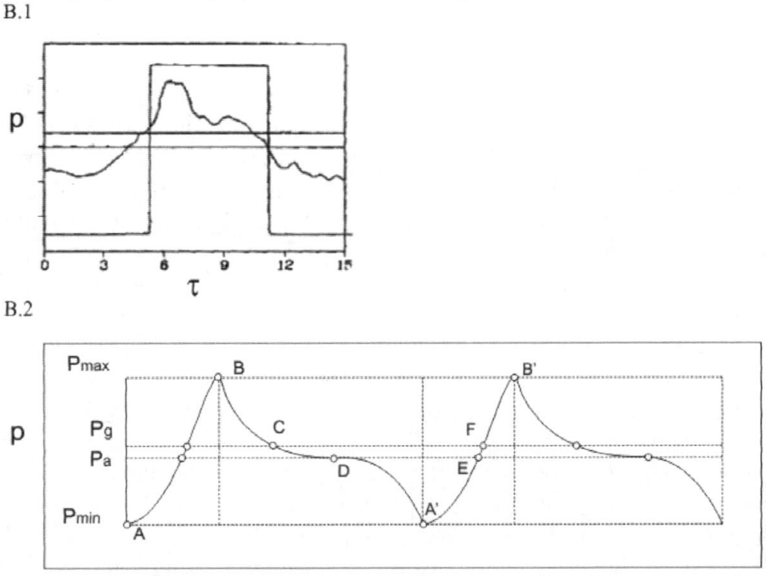

Figure 2. Comparison of the cycle of pulse-frequency diagrams: B.1 - PCS [13]; B.2 - PV-400.

Based on the results of an experimental study using this method, it became possible to present the process of pressure change in the path of pulsating combustion boiler in the form of a diagram describing the processes occurring in this process. Here: A-B is pressure increase in the process of combustion of the fuel-air mixture; B-A' is the flue gas cooling process; C-F is the process of natural gas intake through the gas-pulsating valve; D-E is the process of air flow through the air-pulsating valve; A-A' is time of one cycle (determined by the acoustic properties of the resonator), Pg is pressure in the gas fuel pipe, Pa is atmospheric pressure (pressure in the air pipe).

6. Conclusions

A new method for studying the nature of vibration component of pressure change over time for an energy device is proposed. It allows one to work out the start-stop mode, start mode, long-term operation and stop of the energy device, transient modes when the main (resonance frequency, consumption and flow of fuel and air) and additional (fuel pressure, temperature of the working environment) parameters of the energy device change. Also, it allows one to record and process measurement results, compare frequency characteristics in various points of the working space devices to evaluate the presence of defects in their work. Literary and patent review has shown that this method is not currently used, and devices that implement this method are not available in the market for measuring instruments. The use of measuring complex based on the proposed method will allow testing the engineering methodology of design and calibration calculations of devices based on vibratory, pulsating or detonation fuel combustion, and apply energy-efficient devices of the energy complex to the further development of the industry. The authors have identified the main problems of developing a measuring complex to study the nature of vibration component of the pressure change over time for an energy device, conducted numerical and field tests. A prototype of a measuring complex has been developed that implements all the possibilities of such a method. Tests were carried out on the operating boiler of pulsating combustion PV-400 installed in a water-heating boiler house of one of the educational buildings of Vologda State Technical University. Based on the simulation results, recommendations have been developed for improving the design of pulsating combustion boilers. The results are implemented at the enterprise producing such boilers.

References

[1] Anikina I D, Sergeyev V V, Amosov N T and Luchko M G 2017 Use of heat pumps in turbogenerator hydrogen cooling systems at thermal power plant *Int. J. of Hydrogen Energy* **42**(1) pp 636–42

[2] Akhmetova I G and Chichirova N D 2017 Estimation of the effective heating systems radius as a method of the reliability improving and energy efficiency *J. of Physics: Conference Series* **891**(1) p 012169

[3] Sebelev A, Kirillov A, Porshnev G, Lapshin K and Laskin A 2018 Thermodynamic analysis of design and part-load operation of a novel waste heat recovery unit *MATEC Web of Conferences* **245** p 04010

[4] Saari J, Sermyagina E, Kaikko J, Vakkilainen E and Sergeev V 2016 Integration of hydrothermal carbonization and a CHP plant: Part 2 –operational and economic analysis *Energy* **113** pp 574–85

[5] Paramonov A P, Kadyrov M R and Trinchenko A A 2017 Research on Influence of the Furnace Chamber Aerodynamics on Ecological Indicators of Boiler Plants (Part 1: Model of a Low-temperature Swirl Furnace) *Procedia Engineering* **206** pp 546–51

[6] Trinchenko A A and Paramonov A P 2017 Introduction of low-temperature swirl technology of burning as a way of increase in ecological of low power boilers *IOP Conference Series: Earth and Environmental Science* **90**(1) p 012094

[7] Kharlamova T and Osipova K 2018 The state of modern heat power engineering and increasing the economic efficiency of heat supply *MATEC Web of Conferences* **245** p 05002

[8] Sinitsyn 2012 *Science, Technology and Higher Education* Westwood, Canada

[9] Ovchinnikov P, Borodiņecs A and Millers R 2017 Utilization potential of low temperature hydronic space heating systems in Russia *J. of Building Engineering* **13** pp 1–10

[10] Anikina I and Suslov V 2018 Influence of heat pumps inclusion in deaeration scheme of heating network make-up water on the operating modes of the TPP *MATEC Web of Conferences* **245**

p 15004
[11] Akhmetova I and Akhmetov T 2018 Analysis of additional factors for determining heat network pipelines failure rates *E3S Web of Conferences* **58** p 01011
[12] Shepherd J 2005 *Technical Report FM2005.002 GALCIT* California
[13] Keel S and Shin H D 1991 *Institute of Energy* **64**

IOP Conf. Series: Earth and Environmental Science **337** (2019) 012064 doi:10.1088/1755-1315/337/1/012064

Development of a mobile power plant for thawing frozen soils

Anton Sinitsyn[1*], Anton Mihin[2] and Daniya Zaripova[3]

[1]Vologda State University, Heat Gas Supply and Ventilation, 160000 Lenin st., 15, Russia
[2]Peter the Great St. Petersburg Polytechnic University, St. Petersburg, Russian Federation
[3]Kazan State Power Engineering University, Kazan, Russian Federation

* Corresponding author: sinitsyn.science@mail.ru

Abstract. The article proposes a new device for radiation-convective heating of working fluids, which can be used in particular for emergency and repair and construction works for heating of frozen soil, concrete, road surface and other building materials. The studies were carried out as part of implementation of the Innovation Assistance Fund for Energy Sector Enterprises.

1. Relevance of research

The essence of thawing and defrosting of frozen soil is to change the structural state of one of the components - water (transition from solid to liquid) when additional energy is introduced from outside without disturbing the structural state of the soil itself. When using a number of methods for preparing frozen soils for exploitation by introducing additional energy, the same thermophysical processes occur in soil, which in the literature are called methods of artificial thawing of soil. Defrosting of frozen soil is a complex thermodynamic process occurring in an inhomogeneous capillary-porous medium, complicated by the presence of ground moisture phase transitions.

The search for new technological solutions and soil preparation systems for repair, construction and emergency work is an important task in northern latitudes due to their harsh natural and climatic conditions [1-6]. The increased mechanical strength of frozen soils impedes their exploitation by conventional technical means, which necessitates the preliminary preparation of such soils. The following basic methods of soil preparation in winter conditions are used: thawing followed by exploitation of thawed soil, preliminary mechanical loosening and exploitation of frozen soil. Mechanized methods are currently the most common, but usage of special means of loosening or cutting into separate pieces in a complex urban underground economy is difficult and sometimes impossible due to technical requirements and safety considerations. In urban conditions, the problem of exploitation of frozen soils arises, first of all, in front of emergency, municipal and energy supply services and organizations. A large number of emergency situations in winter is associated with ruptures of pipelines for various purposes. The recovery time of site depends on the speed of the repair team of the damaged pipeline. Another important aspect of the problem raised above is preparation of soil before construction - a task that arises before construction organizations conducting work in the winter period. The use of special means of loosening or cutting the soil does not allow the mixture to be used without a sufficient amount of thawed soil for erecting embankments, filling trenches and sinuses of pits due to the impossibility of good compaction, as it leads to subsidence of soil after

IOP Conf. Series: Earth and Environmental Science **337** (2019) 012064 doi:10.1088/1755-1315/337/1/012064

thawing. In such conditions, usage of alternative methods of soil preparation during repair and construction and emergency work expands the technical capabilities of the organization and allows one to ensure the efficiency of the work being done. These tasks can be solved with the help of organization of soil thawing [7].

2. Review of existing devices and analysis of its efficiency

Currently, there are various studies in the field of frozen soils, including [8, 9], as well as methods and devices for heating frozen soils [10], among which there are devices of convective, radiation, and chemical types. Among the most used are thermomats of various modifications and manufacturers. Despite their wide distribution, their use in real conditions is accompanied by a very long warm-up period. A device is known (Pat. 2208741RU F24D5/08), comprising a housing, the inner side of which is made with a cavity, to specify the channel in which the furnace tube is installed. The heating unit has a burner associated with the first end of pipe, and the first fan device connected with pipe, to move the gaseous products of combustion through the pipe from the first end to the second end. The housing has a hole on its upper surface, the hole is connected to the second fan device for directing air into the channel inside the housing, and a heat transfer plate installed between the pipe and the hole, while the heat transfer plate is shaped so that it surrounds the upper surface of pipe in order to absorb radiation from it and to exclude the penetration of air from the hole directly to the pipe, as a result of which the air is heated by the heat exchanger plate before passing down through the lower end of the channel.

The drawbacks of the device are that the heat flux generated by movement of heated air due to its heating from the heat exchange plate will always be less than the heat flux created by movement of hot flue gases (as proposed in the claimed solution), so the device heat output will be low and it will not provide warming of the working fluid (for example, thawing the mass of frozen soil in the winter). In addition, the device does not use the heat of exhaust gases, which significantly reduces the efficiency of installation.

Also, there exists device (Pat. 2161672RU E01C23/14), comprising a housing in which there is a chamber having an inlet for receiving hot gas from the burner and a radiating surface with many holes. The holes in the radiating surface are of such dimensions that the hot gas heats the radiating surface to create radiation heat transfer to the asphalt surface and pass through the holes to create the conventional heat transfer to the asphalt surface.

The drawbacks of the device are that the heat flux is generated only by radiation from a special surface heated by flue gases; such device has a lower efficiency as compared with the stated solution due to only one heat transfer component – radiant. If convective flow appears here, it will be insignificant.

The device closest to the claimed device is that (Application No. 20100117129 published on November 10, 2011), containing a housing with a smoke tube located in it, at one end of which a gas burner is installed with gas and air channels, at the other end is fan with air heater with reflective screen, while the pipe is U-shaped or П-shaped; the air heater is connected by one channel to the air supply pipe to the gas burner, and the other channel is connected via gas exhaust pipe to inlet fan and a fixed or rotating outlet divider, which, in turn, is placed between the fire tube and the reflective screen, and is made or curved expanding channels.

The main disadvantage of this device is rapid overheating and failure of fan, which creates the convective movement of hot gases, and placing equipment in a single package increases the weight of device and complicates its operation. In addition, such a device must be connected to the mains, which is impossible under the conditions of use.

Thus, our development is relevant and can find direct application in various sectors of energy and utilities.

3. Description of the installation construction

The purpose of the device being developed is to improve the quality, reliability and efficiency of devices used to heat working fluids, improve the methods of surface heating and warming of frozen

IOP Conf. Series: Earth and Environmental Science **337** (2019) 012064 doi:10.1088/1755-1315/337/1/012064

soil, concrete, road pavement by radiation-convective method, with minimum energy and time requirements, taking into account the environmental safety requirements with emissions of exhaust gases into the air, the detection and study of defects in the working bodies.

This goal is achieved by the fact that device for heating the working bodies includes a housing with a smoke tube located therein, at one end of which a gas burner is installed with channels for supplying gas and air, at the other end is a fan, an air heater, and a reflective screen. The pipe is U-shaped or П-shaped; the air heater is connected by one channel to the air supply pipe to the gas burner, and the other channel is connected via a gas exhaust pipe with an inlet fan and a fixed or rotating outlet divider, which, in turn, is placed between the fire tube and the reflective screen, and is made or curved expanding channels.

The main equipment of the invention includes a burner device using which the organization of fuel and air mixing with formation of fuel-air mixture and maintaining its sustainable combustion is carried out, a fire tube in which the torch develops and flue gases move, as well as the pipe is heated to create radiation -convective heat transfer from the pipe to the environment, a fan (smoke exhauster), which induces the movement of flue gases from the burner unit along the entire length of the smoke tube to its outlet on the air heater. In air heater, heat is exchanged between the outgoing flue gases and cold air, as a result of which the source air is heated to a temperature of 100..250 °C, which ensures high efficiency (up to 85-90%) and increases the combustion efficiency of the mixture. In the upper part of the working space of apparatus, above the smoke tube and fan, there is a reflective screen that includes a layer of thermal insulation. To create radiation-convective heat transfer the divider is placed in the apparatus, which carries out movement of flue gases in working space of body, bounded by a reflective screen and heated by working fluid. The divider can be made in the form of straight or curved expanding channels, combined into a single system of cylindrical shape, which can be both stationary and rotating under the action of a moving stream of flue gases.

The temperature range during operation of device is flexible. It is determined by tasks and depends on the properties of working fluids, parameters and operating modes, as well as climatic conditions of operation.

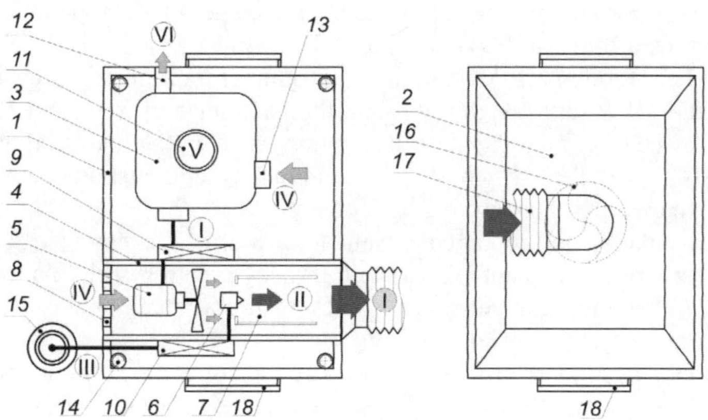

Figure 1. Schematic diagram of the installation.

Figure 1 shows a schematic diagram of the installation, which consist of base part 1 and cap 2. The base part includes electric generator 3 and heat generator, which includes heat-insulating body 4, fan unit 5, burner 6, combustion chamber 7, air grill 8, power control system 9 and gas fuel supply 10. Electric generator 3 is equipped with a cap for filling by liquid fuel 11, exhaust gas pipe 12 and air supply nozzle 13. Base part 1 is installed on an uneven ground surface using telescopic legs 14. A separate cylinder 15 is provided for gas fuel supply. The cap 2 includes a swirler 16. The base part and cap are connected by a corrugated tube 17. To move the base part and the cap 2, the handles 18 are arranged.

To create a hot coolant I, consisting of flue gases II from the combustion of gas fuel III and air IV in generator 3, combustion of liquid fuel V with exhaust gases VI is provided.

IOP Conf. Series: Earth and Environmental Science **337** (2019) 012064 doi:10.1088/1755-1315/337/1/012064

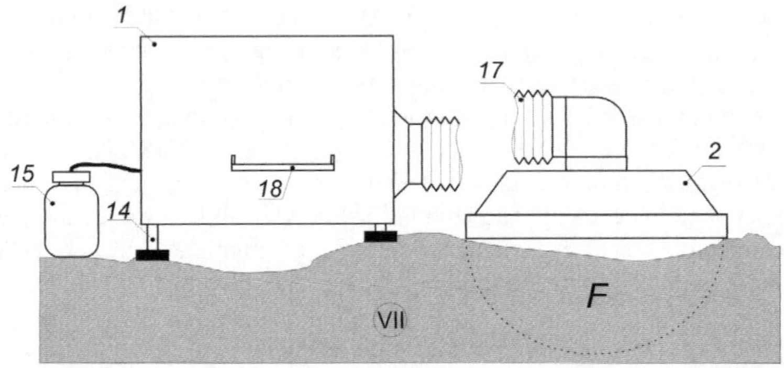

Figure 2. The layout of the installation on the ground surface.

Figure 2 shows the layout of the installation on the ground VII for heating a limited area F.

4. Features of the installation operation

The device works as follows. The base part 1 is transferred using handles 18 to the ground VII, set horizontally using telescopic legs 14. The cap 2 is placed using handles 18 on the surface of the working fluid in the zone of its warming up F. The base part 1 and cap 2 are connected using corrugated tube 17. The cylinder 15 is placed at a distance of at least 1 m from the base part and connected to it with a gas rubber or rubber-polymer hose (conventionally not shown), after which the gas reducer of the cylinder is opened (conventionally not shown).

Refueling of electric generator 3 is carried out through a cover for filling with liquid fuel 11. Starting is performed using a control unit (conventionally not shown). Due to the combustion of the fuel-air mixture consisting of liquid fuel V and air IV, electrical energy is generated, which will ensure the smooth operation of the heat generator. The resulting products of combustion are removed through the exhaust pipe 12.

After starting the generator and exiting it to a stationary mode, the heat generator is turned on and the heating mode is selected through the power control system 9 and the gas fuel supply system 10. The fan unit 5 is turned on, forcing air IV from the environment through the air grill 8 to the burner 6. At the same time gas fuel III is supplied to it through the gas fuel supply system 10 from the cylinder 15, where it mixes with air, forming a gas-air mixture that burns in the combustion chamber 7. At the same time, flue gases II are exhaled, which are guided along the combustion chamber to the output of the heat source, are mixed with air IV, forming hot coolant I.

Through corrugated pipe 17 hot coolant I is sent to cap 2, at the entrance to which swirl 16 is located, in which the flow of hot coolant I is twisted, thereby creating a circulation movement of the medium under cap 2. After the heat generator comes to a stationary mode, the plant capacity is controlled using a regulation system power 9.

When using the device as a source of thermal energy during technical and energy inspections of various construction elements, structures, buildings, convective heat transfer is made to the material under examination, and its thermal characteristics and latent defects are determined by the nature of heat flow distribution. The survey is carried out in accordance with the method of determining the density of heat fluxes (GOST 25380-82) passing through single-layer and multilayer enclosing structures of residential, public, industrial and agricultural buildings and structures during experimental research and under their operating conditions, as well as by non-destructive testing of the thermal type (GOST 18353-79), used in the study of thermal processes in products.

The advantage of the proposed device is that, compared with other means of heating, heating, heating working fluids, for example, frozen soil, concrete, pavement, the proposed device operates autonomously without being connected to power supply systems, does not depend on the quality of the installation surface (up to 30% slope of the surface) and its condition (ice, water, sand). At this, fast speed and great depth of heating of the working fluid is achieved.

In addition, depending on the heating area of working fluid, various cap configurations can be selected: according to the surface shape (round, square, etc.), according to the surface size (from a zone of 0.25 m^2 to 2 m^2 and more). Usage of this device makes it possible to carry out research on identification of internal defects in working fluids by non-destructive testing in the most inaccessible places in any cap position (horizontally, vertically).

The proposed device is mobile and can be serviced by a brigade of two people.

Thus, the proposed device allows one to solve the problem of heating, and warming of working fluids (frozen soil, concrete, pavement) by radiation-convection method during emergency and repair and construction works, to improve the quality, reliability and efficiency of devices of this class used to heat any working bodies, conduct energy and technical surveys of various construction elements, buildings, and structures, ant to carry out flaw detection of working bodies.

5. Experimental data

As shown by computational and experimental studies of the device in [11] for convective thawing of frozen sandy soil, conducted by the Department of Heat and Gas Supply and Ventilation of Vologda State University, the depth of thawing increases according to dependencies similar to those determined by various well-known mathematical models and experimental data. In this case, the convective method of heating allows defrosting the working fluids several times faster than the other known methods.

So, for example, when 15 kW thermal energy is supplied to a sandy soil from combustible reduced natural gas for 3 hours, the depth of heating of frozen soil of the 40%-moisture with an initial temperature of -10 °C at outside air temperature -2 °C with periodic removal of the thawed part of the pit was 1 m, which is approximately 30 times faster than using devices based on heat transfer due to thermal conductivity (thermomats) [12,13].

For such indicators as specific power of heating per 1 m^2 of soil, the time of thawing to a depth of 0.5 m, the cost of thawing, expressed in monetary terms ($RUB/h \cdot m^2$), as well as heating temperature, the proposed installation is more efficient. Considering such indicators as size and weight of the installation, the thermomat is more attractive from the point of view of movement and transportation, however, taking into account its low efficiency during thawing (about 30 hours), these advantages are relegated to the background. In view of the fact that the second analogue – a device for heating the soil and concrete WACKER NEUSON E700M is designed for large areas of warming up (up to 400 m^2), for such conditions, such device is inefficient. Similar results confirm viability of the proposed technical solution and are the basis for implementation of the installation in power supply and utility companies.

6. Conclusions

The authors have proposed a new method for warming frozen soils for municipal power industry using a mobile power plant operating on organic fuel and providing a warm layer of frozen soil in the cramped conditions of the urban environment. The installation is multifunctional for various fields of activity, including heat and power engineering and construction, and can be used as a source of thermal energy for heating any working fluids of various orientations in space, detecting internal defects in working fluids by thermal non-destructive testing method, during emergency and repair construction works for heating, warming frozen soil, concrete, road pavement and other building materials.

Acknowledgements

The presented work was performed within the framework of financing under the federal target program of the Ministry of Science and Education of the Russian Federation, as well as within the framework of research funded by the Innovation Support Fund. This work was supported by a grant named after the outstanding organizers of the oil and gas industry of the International Fuel and Energy Association (IFPA). The research results received international recognition at the 2nd International

IOP Conf. Series: Earth and Environmental Science **337** (2019) 012064 doi:10.1088/1755-1315/337/1/012064

Exhibition of Inventions and Inventors of Mediterranean Countries INVENTECO EXPO in Catania (gold medal), as well as at the youth competition Russia Power and HydroVision Russia 2011 (PennWell, United Kingdom).

References

[1] Ivanova A, Pavlov S, Akimov L and Zakharova L 2018 Zoning of the territory with snow removal using snow melting plants *MATEC Web of Conferences* **170** p 02023

[2] Nikolskiy S and Pertseva O 2017 Substantiation of Reference Method for Determining Concrete's Freeze-Thaw Resistance *J. of Physics: Conference Series* **790**(1) p 012023

[3] Vishnyakov I E, Borchsenius S N, Kayumov A R and Rivkina E M 2016 Mycoplasma Diversity in Arctic Permafrost *BioNanoScience* **6**(4) pp 516–9

[4] Nikolskiy S and Pertseva O 2016 *Accuracy evaluation of rapid method for concrete's frost resistance determination* Advances and Trends in Engineering Sciences and Technologies – Proc. of the Int. Conf. on Engineering Sciences and Technologies, ESaT 2015 pp 335–40

[5] Lessovaia S N, Dultz S, Plötze M, Andreeva N D, Polekhovsky Y, Filimonov A and Momotova O 2016 Soil development on basic and ultrabasic rocks in cold environments of Russia traced by mineralogical composition and pore space characteristics *Catena* **137** pp 596-604

[6] Leksin V N and Profiryev B N 2017 Socio-economic priorities for the sustainable development of Russian arctic macro-region *Economy of Region* **4** pp 985–1004

[7] McRoberts E C and Morgenstern N R 1974 The stability of thawing slopes Can. Geotech. J. 11(4) pp 447–69

[8] Tukada K and Ogawa S 1997 Failure criterion of unsaturated soils subjected to freezing and thawing *Ground freezing* pp 327–32

[9] Nixon J F and Morgenstern N R 1973 Practical to a theory of consolidation for thawing soil *Permafrost. 2nd Int. Conf. On Permafrost, Washington, D.C.*

[10] Taylor G S and Luthin J N 1978 A model for coupled heat and moisture transfer during soil fruzing *Canad. Geotech. Journal* **15**

[11] Sinitsyn A A 2013 The calculation of the characteristics and of the geometrical sizes of power devices *European Applied Sciences* **4**

[12] Sinitsyn A A 2013 Simulation of fire engineering processes in energy devices aimed at their optimization and improvement of reliability *Life Science Journal, the Acta Zhengzhou University Oversea Version. Marsland Press: New York* **10** pp 442–7

[13] Mukhametshina R M, Petrov A V 2018 Technics and Technology of Transport **1** (6) p 13 URL: http://transport-kgasu.ru/files/N6-13EB118.pdf

Simulation of heat transfer processes in cooling systems with heat recovery

Anton Sinitsyn[1*], Vyacheslav Rakov[1], A P Eperin[2], Yu A Rundygin[2], R V Rusinov[2], R A Izmailov[2], A M Simonov[2], V K Yun[2], Anton Mihin[2] and Irina Akhmetova[3]

[1]Vologda State University, Heat Gas Supply and Ventilation, 160000 Lenin st., 15, Russia
[2]Peter the Great St. Petersburg Polytechnic University, St. Petersburg, Russian Federation
[3]Kazan State Power Engineering University, Kazan, Russian Federation

* Corresponding author: sinitsyn.science@mail.ru

Abstract. The article considers modeling of modern heat exchange systems which use thermal energy for its beneficial use. The proposed mathematical model makes it possible to investigate systems with utilization of thermal energy. The article presents the results of simulation of heat exchange processes using heat energy recovery for internal combustion engines.

1. Relevance of the investigation

Today, energy conservation is a worldwide strategic task of national importance. Meanwhile, many enterprises have significant energy losses due to insufficient use of heat generated in technological processes. In particular, the heat of gas heated during the process of any production is either not used efficiently or not used at all and the heated gas is released into the atmosphere. This leads to enormous energy losses in the volume of the enterprise, country, world, and also determines various environmental problems [1-6]. This is especially typical for high-temperature productions with coolant temperatures from 1000 °C and more, where energy losses are the greatest, as well as when using gases containing a large amount of impurities and aggressive substances.

In the metallurgical industry, increasing the energy efficiency of heat exchange systems is a very urgent task. In ferrous and non-ferrous metallurgy, as well as in the metalworking process, a great amount of heat is expended, much of which is discharged into the environment. In such systems, reducing heat energy losses is most relevant, for example, through the use of heat energy recuperators. A typical heat exchanger is a surface-type heat exchanger which uses waste gas heat, in which heat exchange between the heating and heated medium is carried out through a separation wall. Typical designs of recuperators of metallurgical furnaces are ceramic, metal loop, block, etc. Such devices allow using a part of waste heat to improve the process, or for other needs, such as heating. The authors are working on development and implementation of gas-liquid heat exchangers in production, transport, metallurgy and other industries. Several experimental samples were made, original ways of increasing the efficiency were found, described in [7-9].

2. Description of the mathematical model

IOP Conf. Series: Earth and Environmental Science **337** (2019) 012065 doi:10.1088/1755-1315/337/1/012065

Evaluation of efficiency of heat recovery device introduction into heat exchange system at the design stage is associated with problems of modeling the heat exchange system. The authors propose to use mathematical modeling to compare operation of the heat exchange system before and after installation of heat energy recuperator.

To simulate the heat exchange process, a block model was used, consisting of 3 main blocks: heating unit (HU), recuperator of thermal energy (RTE) and a cooling circuit (CC).

The system carries out the process of receiving thermal energy from HU Q_{HU} and from RTE $Q_{\mathrm{RTE}}^{\mathrm{out}}$ during heating. Thermal energy Q_{HU} and $Q_{\mathrm{RTE}}^{\mathrm{out}}$ depend primarily on the mode of operation of the HU. The use of heat in the system is carried out through the cooling circuit Q_{rem}^{CC}, the exhaust system Q_{EG}^{HU}, the radiation of thermal energy into the environment through the heated surfaces of the heating unit $Q_{\mathrm{rem}}^{\mathrm{HU}}$.

In RTE, heat transfer occurs, which is characterized by supply of a portion of thermal energy $Q_{\mathrm{RTE}}^{\mathrm{in}}$ from exhaust gases and removal of the utilized thermal energy into the coolant $Q_{\mathrm{RTE}}^{\mathrm{out}}$ (J).

In the cooling unit, the process of removal of thermal energy from the coolant for the needs of the process is carried out. The intensity of heat transfer in the cooling unit depends on the ambient temperature, the speed of coolant in the heat exchanger, coolant temperature in the heat exchanger and the speed of its circulation. All blocks are connected by a single circulation circuit.

The output parameter in this case is time to reach the required temperature of liquid coolant in the cooling unit.

The ratio between temperature changes will be as follows:

$$M_1'c_{p1}(T_{1,out} - T_{1,\mathrm{in}}) - M_2'c_{p2}(T_{2,out} - T_{1,\mathrm{in}}) = 0 \quad (1)$$

where M_1', M_2' is mass flow of gas and liquid coolant (kg/s); c_{p1}, c_{p2} is average specific heat capacity of coolant at constant pressure for the given temperature ranges (J/(kg·K)); $T_{1,\mathrm{out}}, T_{2,out}$ is gas and liquid temperature at the output of heat exchanger; $T_{1,\mathrm{in.}}, T_{2,in.}$ is gas and liquid temperature at the input of heat exchanger (°C).

$$C_1\Delta T_1 - C_2\Delta T_2 = 0 \text{ or } Q_1' - Q_2' = 0 \quad (2)$$

where C_1, C_2 is the product of specific heat capacity and the mass flows, Q_1', Q_2' are heat fluxes (W).

Taking into account that heat exchanger is not ideal, some of heat energy will be spent on hydrodynamic losses, heat transfer to the atmosphere through the outer walls.

During heating, the heat energy of the coolant will also be spent on heating the walls of RTE, reducing its efficiency [10].

Further we consider an example of a heat exchange system with an air-heating unit, the temperature of the liquid coolant in which is equal to T^0, energy Q at this time moment is equal to zero. After starting the heating unit, at each moment of time a certain amount of thermal energy $Q_i$enters the system, it depends on the amount of combustible fuel and is equal to:

$$\int_{\mathrm{ti\text{-}1}}^{\mathrm{ti}} Q_{\mathrm{ci}}\mathrm{dt} = q_h \times \rho \times H_U \times 0.3/3600 = 3 \times 0.7 \times 43500 \times 0.3/3600 = 7.61$$

where q is hourly heat consumption, l/h;

ρ is fuel density, kg/m3;

HU is specific heat of combustion, kJ/kg.

The intensity of the system heating, depending on the incoming additional thermal energy, is determined experimentally and is described by a logarithmic or polynomial law. And its confidence level can be estimated by the trend determination coefficient R2

$T = ALn(t) - B$, for R2=0.986,

$T = At^3 - Bt^2 + Ct - D$, for R2=0.998,

where A, B, C, D are coefficients taking into account environmental parameters and uneven heating of the heat source.

Received thermal energy in the heating unit will be equal to:

$$Q_{HU} = \int_{\mathrm{ti\text{-}1}}^{ti}(Q_{\mathrm{hi\text{-}1}} + Q_{\mathrm{HUi}} - Q_{\mathrm{remi}}^{\mathrm{HU}} - Q_{\mathrm{CC}}^{\mathrm{out.}})\mathrm{dt} \quad (3)$$

where $Q_{acci\text{-}1}$is accumulated heat energy; $Q_{HUi}$is heat energy obtained from fuel combustion; $Q_{remi}^{HU}$is heat energy removed to the atmosphere (in this case it is unknown); $Q_{CC}^{out}$is heat energy removed to the cooling circuit.

The energy removed by the cooling circuit per unit time is equal to the difference between input $Q_{\text{in}}^{\text{CC}}$ and output $Q_{\text{out}}^{\text{CC}}$ heat energy for the given time period:

$$Q_{\text{rem}}^{\text{CC}} = \int_{ti-1}^{ti}(Q_{\text{in}}^{\text{CC}} - Q_{\text{out}}^{\text{CC}})\text{dt} \quad (4)$$

The heat balance of the entire system can be expressed by a general integral equation:

$$Q_{\text{sys}} = \int_0^t(Q_{HU} + Q_{\text{RTE}} - Q^{CC})\text{dt} \quad (5)$$

The temperature of liquid coolant in the system will depend on balance of receipt of thermal energy and its removal. Based on the presented dependencies, the heating process of HU is modeled by theoretical means, which allows one to estimate the effect of RTE and various factors on the rate of its heating.

Experimental and theoretical studies on efficiency of usage the heat exchanger allow one to a priori estimate the process of the entire system after launch. However, such studies involve a number of complex tasks. The first is the need to find the dependence of the coolant temperature on time $T_C = f(t)$. The second is to find the amount of thermal energy released in the system as a function of time $Q_{\text{HU}}^{C} = (t)$. The third is to find the amount of thermal energy, removed by the exhaust gas system as a function of time $Q_{\text{HU}}^{\text{EG}} = (t)$. The fourth is to find CC capacity and the amount of thermal energy removed by it as a function of time $Q_{\text{rem}}^{\text{CC}} = f(t)$. The fifth is to find RTE capacity as a function of the coolant temperature $Q_{\text{RTE}} = f(t)$. The sixth is to find HU thermal characteristics.

3. Description of mathematical methods

When modeling the heat exchange system, the methods of mathematical analysis and statistical calculations described in [11, 12] were used.

3.1 Change of temperature of the heat carrier of the heating unit.

Figure1 shows theoretical and experimental dependences of the heating. Immediately after switching on HU, the coolant temperature is equal to the ambient temperature -27 ° C, after which the temperature rises rapidly, then the temperature rise decreases and at 22 minutes the cooling circuit opens and the temperature stabilizes at 93 ° C.

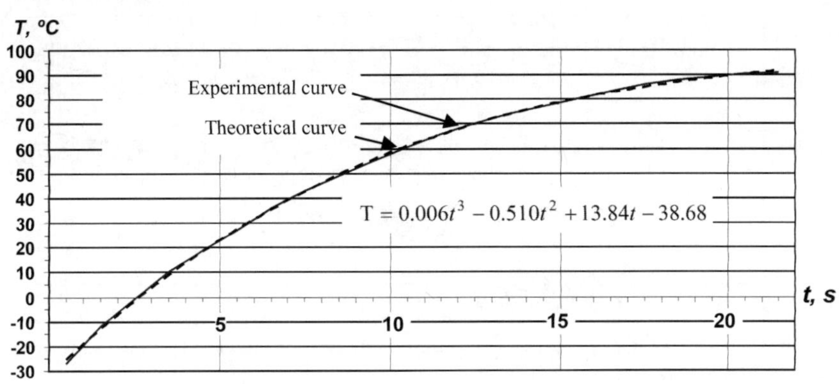

Figure 1. The intensity of temperature change of the coolant system without RTE.

3.2 The amount of heat energy released into the system.

It is known that part of the energy of combustion of the air-fuel mixture is discharged into the heat exchange circuit [8]. Thus, knowing the amount of combustible fuel, the amount of thermal energy entering the heat exchange circuit is established. The heat capacity of fuel combustion is HU = 46000 kJ/kg. The density of fuel is 0.75 g/cm3. At low coolant temperature (Figure2), the amount of combustible fuel is greater, while the amount of energy discharged into the heat exchange system decreases and stabilizes after the operating temperature is reached.

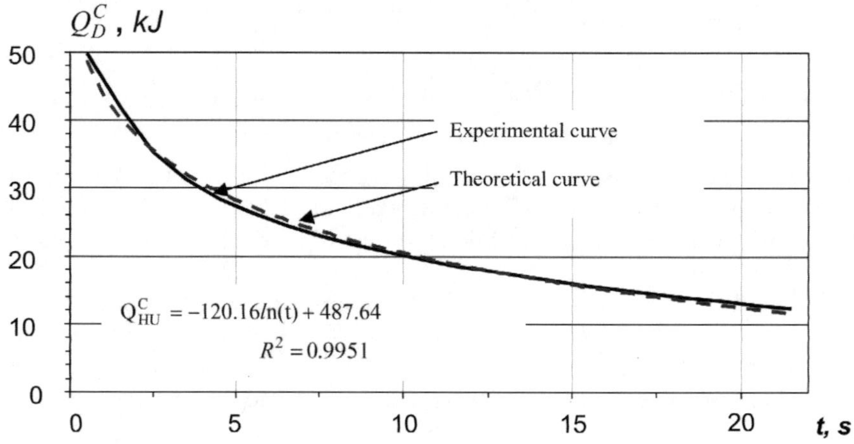

Figure 2. Relationship between the thermal energy removed to CC and time.

After the heating unit is turned on, the heat from exhaust gases is consumed mostly for heating elements of the exhaust system; therefore, during this period gas temperature is not significant. Intensive temperature risc bcgins after about 40 s, and after 2 min exhaust temperature is already more than 200 °C, which is equivalent to 1.5 kW of heat output. After warming up to the operating temperature, 40% of the thermal energy from the combustible air-fuel mixture is already discharged into the exhaust system.

The results of calculating the characteristics of HU heating in different modes of operation of the cooling circuit (Figure 3) show that when its capacity is increased to 100%, the coolant does not warm above 55 ° C, and the maximum thermal energy entering the cooling circuit will not exceed 1 kW, which indicates inefficiency of the cooling unit.

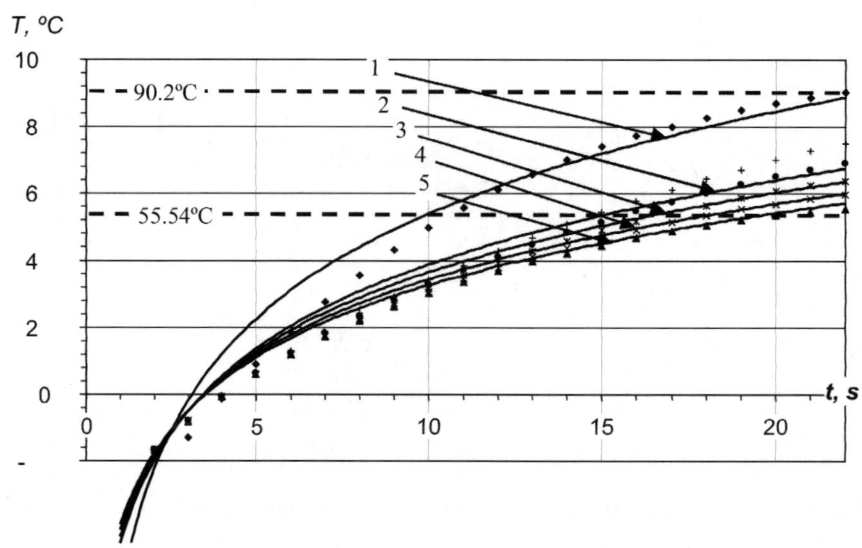

Figure 3. Coolant temperature raise intensity for cooling circuit power: 1 – 0%; 2 – 20%; 3 – 40%; 4 – 60%; 5 – 80%; 6 – 100%.

IOP Conf. Series: Earth and Environmental Science **337** (2019) 012065 doi:10.1088/1755-1315/337/1/012065

When the cooling circuit is turned off, the coolant warms up to the operating temperature only at the 21st minute. When turning on the cooling circuit: at 0.5 kW, the temperature at the 21st minute will reach 88 °C; when switched on at full capacity, the coolant at the specified time will only warm up to 55 °C. The cooling circuit will receive only 1.4 kW of thermal power. Further, the temperature of the coolant will not rise at all.

The RTE power depends on the exhaust gas temperature. From previous studies, it is known that after starting the heating unit, the heat energy of the exhaust gas is spent mostly on heating the HU parts and the exhaust system. As shown by tests, the efficiency of conversion of heat energy of the exhaust gas during heating of the coolant can reach 50%, and the energy return begins only after the 2nd minute of the heating unit.

The results of theoretical studies show that the use of a device for heat energy recovery makes it possible to reduce the warm-up time to working temperature from 22 to 10 minutes, and when the cooling circuit is turned on, an additional 2.5 kW of heat energy can be obtained already at the 12-minute after switching on, while without using RTE, the heating unit cannot provide this amount of energy for up to 22nd minute.

4. Conclusions

Summing up the research on the creation of a new type of heat exchangers, we can note the following:

1. A method for study of heat exchange systems with a heat recuperator is proposed.
2. The results of the study of the heat exchange system in various modes of operation before and after application of heat recuperator are presented.
3. The adequacy of the adopted mathematical model and calculation program is confirmed by the analysis of laboratory studies and operating experience of heat exchangers.

References

[1] Zlotin V E 2011 Efficient heat recovery heat exchangers *Novosti teplosnabzheniya* **1**

[2] Potekhin V V, Pantyukhov D N and Mikheev D V 2017 Intelligent control algorithms in power industry *EAI Endorsed Transactions on Energy Web* **17**(11)

[3] Grasmanis Dz, Sovetnikov D O and Baranova D V 2017 Energy performance of domestic hot water systems *Magazine of Civil Engineering* **76**(8) pp 140–55

[4] Szkarowski A, Janta-Lipińska S and Gawin R 2016 Reducing emissions of nitrogen oxides from dkvr boilers [Obniżenie emisji tlenków azotu z kotłów dkvr] *Rocznik Ochrona Srodowiska* **18**(1) pp 565–78

[5] Zubkova M, Stroganov A, Chusov A and Molodtsov D 2017 Hydrogenous fuel as an energy material for efficient operation of tandem system based on fuel cells *Key Engineering Materials* **723** pp 616–21

[6] Sebelev A, Kirillov A, Porshnev G, Lapshin K and Laskin A 2018 Thermodynamic analysis of design and part-load operation of a novel waste heat recovery unit *MATEC Web of Conferences* **245** p 04010

[7] Sinitsyn A 2012 *Science, Technology and Higher Education* Westwood, Canada

[8] Rakov V A 2012 Using exhaust heat to reduce engine warm-up time and vehicle cabin heating *Autotransportnoe predpriyatiye* **8**

[9] Petukhov B S and Shikov V K 1987 *Energoatomizdat* **1**

[10] Kruglov G A 2010 *Thermal Engineering*

[11] Migay V K 1987 *Simulation of heat exchange energy equipment*

IOP Conf. Series: Earth and Environmental Science **337** (2019) 012066 doi:10.1088/1755-1315/337/1/012066

Method for determining the basic energy characteristics of elements of a hybrid car engine

Vyacheslav Rakov[1*], V N Ivanov[2], Yu G Karpov[2], V F Melehin[2], R A Izmailov[2], V K Yun[2] and D A Zaripova[3]

[1]Vologda State University, Department of automobiles and automobile industry, 160000 Lenin st., 15, Russia
[2]Peter the Great St. Petersburg Polytechnic University, St. Petersburg, Russian Federation
[3]Kazan State Power Engineering University, Kazan, Russian Federation

* Corresponding author: vyacheslav.rakov@mail.ru

Abstract. A method for determining the basic energy characteristics of a hybrid car engine is presented. The technique takes into account parameters of the vehicle and parameters of the driving mode during operation. The basis of the technique is traction power and energy calculation. According to the source data, the flow of energy passing through mechanical, electrical chains and through the storage of electrical energy is determined. The technique is implemented in the form of algorithms for the calculation program and allows one to simultaneously receive the main technical characteristics of the vehicle when it is equipped with a hybrid engine of a sequential, parallel and mixed type. The presented method allows one to obtain the following values: the required power of the internal combustion engine, the required power of the motor generator, the minimum capacity of the electrical energy storage device, the estimated fuel consumption. This article presents an example of calculating the basic parameters of a combined power plant for a car weighing 1500 kg when driving it along the standard European driving test cycle. Similar calculations can also be carried out for any other cycle and parameters of the vehicle, including the bus, truck, and tractor.

1. Key points

1.1 A method for determining the basic characteristics of a hybrid power plant of car by the specified parameters is proposed.

1.2 A program has been developed for calculating the basic energy parameters of a hybrid power plant of car.

1.3 The calculation results for basic energy characteristics for a passenger car with a series, parallel and mixed scheme of elements at the power plant are presented.

2. Introduction

Internal combustion engines (ICE) and electric or hybrid power plants (HPP) are used as power plants at modern vehicles operating in periodic mode. The latter are used in cars, buses; tractors [1-2].

The method of assessment the ICE efficiency in unsteady operation was described in [3-4]. However, defining the basic energy characteristics of power plant elements is sometimes difficult and is the subject of an exact energy analysis and calculation [6-11]. First of all, it is because characteristics of power plant elements depend on the vehicle parameters and conditions of its movement during operation.

To determine the optimal power of ICE, working in combination with an electric motor, a calculation method is needed, which takes into account the vehicle parameters (mass, drag area of car), specified dynamic driving conditions and characteristics of the power plant elements (transmission, electric motor, energy storage, power converter, generator).

The method proposed in this paper includes: traction calculation; energy calculation of various types of schemes of hybrid power plants. All sections are connected in such a way to make a comparative analysis of characteristics of elements and to choose the optimal drive scheme for the vehicle, depending on the specified route. For convenience, the technique is implemented in the form of algorithms of a mathematical model and a calculation program, which is called GSU-AVTO [3], which allows theoretical studying and obtaining various dependencies.

3. Materials and methods

The basis of the calculation is driving cycle, which is a plot showing the relation between the vehicle speed, the longitudinal slope of road plane and time (V=f(t), α=f(t)). These dependencies can be obtained in a practical way by recording the change in parameters when moving along a given route, or can be set theoretically. In this case, we used the standard urban cycle according to the test protocol for passenger cars (EU directive 80/1268 2004), (Figure 1).

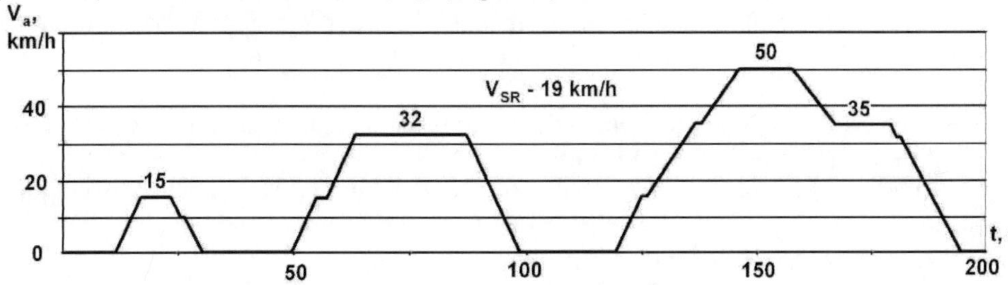

Figure 1. The standard European driving cycle for cars testing for toxicity.

Using well-known formulas of traction calculation, the resistance forces acting on the vehicle in each i-th point were found. N_{Ki} is rolling resistance force; $N_{Wi}$is air resistance force; N_{Ii} is inertia resistance force; $N_{Pi}$is lifting resistance force.

The total resistance force acting on the vehicle at each measured point is the sum of all forces

$$N_{Ti} = N_{Ki} + N_{Wi} + N_{Ii} + N_{Pi} \tag{1}$$

The graph (Figure 2) shows the curve for average power change when tested on a standard driving cycle.

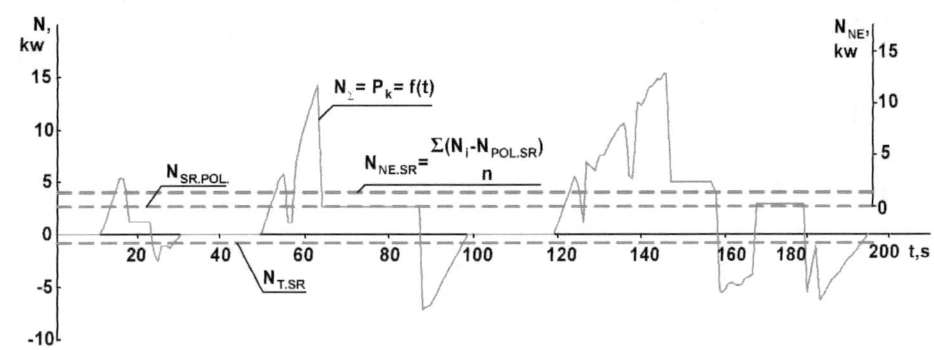

Figure 2. The chart for total resistance power (traction force) on driving wheels: $N_{Ti} = f(t)$; $Nsr.pol.$ is average positive power at a stage; $N_{T.SR}$ is average negative power (average braking power); $N_{NE.SR}$ is average peak load power (average power storage power).

IOP Conf. Series: Earth and Environmental Science **337** (2019) 012066 doi:10.1088/1755-1315/337/1/012066

In a sequential scheme, the mechanical energy produced by the engine (ICE) is converted into electrical energy using a generator (M/G), then part of it is used to charge the traction energy storage device (BAT), and part of it is transmitted to the wheels by an electric motor (W). The block diagram of the common chain of losses in the serial drive is shown in Figure 3, a [5]. ICE in it regulates the speed of movement, and works in an optimal stationary mode, constantly giving all the generated power to TES charge and movement.

If there is no load on the wheels and the battery is fully charged, the power cannot be realized, in this case, ICE is turned off. Based on this, one can make a conclusion that all power generated by ICE will be spent on movement and loss in drive. It can also be affirmed that some spent energy will return due to braking recovery.

Some energy will be transferred from ICE to wheels in a straight chain (Figure 3.b), and some other through the energy storage device (c). The losses will be different for these cases. On this basis, each chain of losses is considered separately.

The average drive power at the stage transmitted directly to the car wheels will be equal to the entire average power NSR.POL. minus the one that is transmitted through the energy storage NNE.SR.

A part of braking energy transmitted from wheels will return to BAT, after which it will be reused on movement, passing twice through the chain of losses (Figure 3, d).

The power that is additionally obtained due to recovery at the (N_3) stage is based on the average braking power on wheels, given by recovery efficiency and efficiency of the drive loss chain (Figure 3 g).

While calculating NICE we take into account the most economically efficient ICE load mode ($K \approx 0.75$), and energy transfer losses in the drive chain.

From a theoretical experiment, it was found that for a car weighing 1500 kg when driving in regards to the standard European urban cycle, according to the calculation results, the required ICE power of a hybrid power plant will be 4.75 kW (Table 1).

The parallel scheme of hybrid drive has fundamental differences. As in a conventional car, it has a manual transmission and automatic transmission. The electric motor is installed parallel to the transmission and is activated only during acceleration, thus unloading the ICE. The excess engine energy when moving is consumed by the generator to charge the drive.

The electric motor in parallel circuit is reversible, i.e. can work in as a generator, but the charge is possible only with medium loads and in the mode of engine braking. Peaks of load are smoothed by an electric motor, which allows one to reduce ICE power, but inefficient modes of operation and idling cannot be eliminated, which causes ICE energy efficiency ratio(K) [5] to decrease.

According to the peculiarities of parallel circuit, the algorithm for calculating the average required power $N_{SR.POTR.}$of the hybrid traction drive on the route has been included only in the time periods where ICE is involved (NVIH. > 0).

The ICE power, taking into account the losses in energy transmission, will be a sum of the power transmitted in a straight line, taking into account the losses in mechanical transmission, and power transmitted through electric motor (parallel circuit). It is also necessary to take into account the losses in the drive chain when transferring energy from ICE to drive, losses in the drive itself and losses in transferring energy to car wheels.

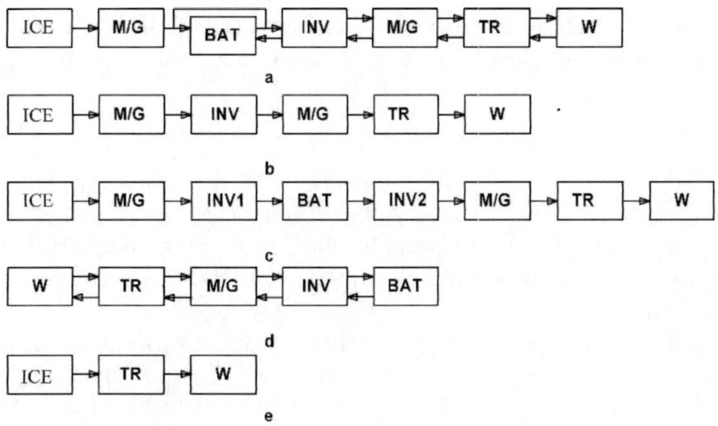

Figure 3. Chains of energy losses in drive: a) total chain of losses; b) in mechanical transmission; c) in electrical transmission; d) when realizing the power returned during braking; e) chain of losses in mechanical transmission (ICE - internal combustion engine; M/G - motor-generator; INV - inverter; BAT - energy storage; TR - transmission; W - wheel)

The block diagrams of loss chain in mechanical and electrical gears are shown in Figure 3b.

In the chain of electric drive, the flow of electrical energy during charging passes from the generator to the traction energy storage and back to the generator operating in the electric motor mode. Electric energy in a parallel circuit passes twice through an energy converter and only after that is transferred to an electric motor and transmission.

The recovered energy is transferred to the accumulator, after which, at the time of peak loads, it returns back, passing twice through the transmission, generator-motor, energy converter (Figure 3d).

The mixed scheme of HPP includes the advantages of sequential and parallel circuits. The energy flows in it are calculated in the same way as in the mixed scheme, however, it is assumed that ICE can be turned off at low energy consumption, and movement can be carried out from an electric motor powered by an energy storage device.

4. Results

The main energy characteristics are calculated for 3 typical drive circuits: serial, parallel and mixed. Table 1 presents the calculation results.

Table 1. Calculation data on main energy characteristics of a hybrid engine

Characteristics	Gear type		
	Serial	Parallel	Mixed
Initial energy in the storage, kJ	60	62	133
Max. energy store in the storage, kJ	219	122	162
Max. ICE power, kW	4.75	8.99	4.45
Max. Drive power, kW	10.8	9.0	11.5
Consumption, l/100 km	4.55	4.4	3.68

According to the results of study of the HPP power characteristics, when driving a car along the standard European driving cycle, the lowest fuel consumption of 3.68 l/100 km is achieved with a mixed HPP. As compared to other schemes, it is lower by about 20%.

The smallest energy storage is required for a parallel circuit and is 122 kJ, for a mixed circuit it will be slightly larger. The drive with the largest capacity is required for a sequential drive circuit - 219 kW.

The highest ICE power (8.99 kW) should be in parallel to the drive circuit. The ICE power of sequential and mixed circuit will be approximately the same.

5. Conclusions

Thus, this method allows one to accurately calculate the basic energy characteristics of a hybrid power plant with a series, parallel or mixed scheme for each specific vehicle and a given driving cycle.

IOP Conf. Series: Earth and Environmental Science **337** (2019) 012066 doi:10.1088/1755-1315/337/1/012066

Also, this method is used to calculate the optimal characteristics of the electric energy storage device of the hybrid power plant and can be used in assessing the technical condition of a hybrid car according to the method developed by the authors.

References

[1] Pistoia G 2010 *Electric and Hybrid Vehicles. Power Sources, Models, Sustainability, Infrastructure and the Market* (Oxford: The Netherlands Linacre House)

[2] Aleksandrov I K and Rakov V A 2012 Analysis of prospects for the development of electrified road transport [Analiz perspektiv razvitiya ehlektrificirovannogo avtomobil'nogo transporta] *Transport of the Russian Federation* **2** pp 50–2

[3] Kok D B 2001 Design optimisation of a flywheel hybrid vehicle (Eindhoven: Technische Universiteit Eindhoven)

[4] Campeau S 2001 Design and analysis of a parallel hybrid electric vehicle (Canada: Ecole de Technologie Superieure)

[5] Kapustin A and Rakov V A 2016 *Hybrid cars [Gibridnye avtomobili]* (Vologda: Vologda State University)

[6] Koshcheev L, Popkov E, Seit R 2018 *MATEC Web of Conferences* **245** 07011

[7] Sebelev A, Kirillov A, Porshnev G, Lapshin K, Laskin A 2018 *MATEC Web of Conferences* **245** 04010

[8] Kasatkin I, Egorov M, Kotov E, Zakhlebaev E 2018 *MATEC Web of Conferences* **245** 15003

[9] Shamsutdinov E V, Vachagina E K, Nazmeev Yu G 1999 *Teploenergetika* **46**(5) pp. 72-4

[10] Sultanguzin I A, Fedyukhin A V, Kurzanov S Y, Gyulmaliev A M, Stepanova T A, Tumanovsky, V A, Titova D P 2015 *Thermal Engineering (English translation of Teploenergetika)* **62**(5) A008 pp. 359-64

[11] Mudrov A G, Andreev A V 2018 *Technics and Technology of Transport* **1**(6) p 1. URL: http://transport-kgasu.ru/files/N6-01IT118.pdf

IOP Conf. Series: Earth and Environmental Science **337** (2019) 012067 doi:10.1088/1755-1315/337/1/012067

Thermal and aerodynamic efficiency of a stationary switching regenerative heat exchanger

Nikolay Monarkin[1*], Sergey Lukin[1], Yu M Anurov[2], B A Tihomirov[2], G A Agasiants[2], S M Galileev[2] and T R Akhmetov[3]

[1]Vologda State University, Institute of civil engineering, 160000 Vologda Lenina st. 15, Russia
[2]Peter the Great St. Petersburg Polytechnic University, St. Petersburg, Russian Federation
[3]Kazan State Power Engineering University, Kazan, Russian Federation

* Corresponding author: nikolay-monarkin@yandex.ru

Abstract. The article is devoted to comparison of thermal and aerodynamic efficiency of two configurations of the regenerative nozzle used in the regenerative heat exchanger for room ventilation. The impact of nozzle channel diameter, nozzle length and stage time on the thermal efficiency of the nozzle, its aerodynamic resistance and the required fan power is considered.

1. Introduction

In civilian buildings (public, administrative, residential, etc.) there may be problems concerning the sustainable air exchange [1-9]. This is caused, for example, by the exception of natural influx when using modern hermetic windows, low functionality of existing ventilation systems due to its wear and clogging. One of solutions of such problems is usage of stationary switching regenerative heat exchangers (SSRHE). SSRHE is a compact decentralized ventilation device that allows utilizing the heat of exhaust air in the cold season [10-13]. For effective ventilation of one or several adjacent rooms, SSRHE is installed in a pair in antiphase mode. A reversing fan is used to induce air movement in the heat exchanger. The heat exchange element is a regenerative nozzle, which is a cylindrical beam of channels of small cross-sectional dimensions. The principle of operation of the SSRHE consists in alternately changing modes of external air inflow and exhaust of the internal one after a certain period of time. In the exhaust mode, the internal warm air of the room passes through the nozzle and heats it in order to transfer some thermal energy to the outside air after switching over to the inflow. Thus, the device alternates the stages of accumulation and regeneration of thermal energy.

The purpose of this study is to solve the optimization problem. For a given cross-sectional area (a given nozzle diameter) and a given air flow rate V, it is required to determine the diameter of a single channel d, the nozzle length L, the duration of stages for regeneration and accumulation of thermal energy at which the nozzle will work efficiently enough and with the least aerodynamic resistance.

2. Methods

The paper [14] shows the scheme of the SSRHE regenerative nozzle (Figures 1 and 2) and proposes a mathematical model (1)-(3) of heat exchange processes in a single channel of nozzle. The solution of mathematical model (1)-(3) allows one to find the temperature fields of nozzle and air depending on the coordinate z and time τ: $T_a(z,\tau)$, $T_n(z,\tau)$. The proposed mathematical model is implemented using the Matlab software package.

$$c_n \cdot \rho_n \cdot S_n \cdot \frac{\partial T_n(z,\tau)}{\partial \tau} = P \cdot \alpha \cdot \left(T_a(z,\tau) - T_n(z,\tau)\right), \quad 0 < \tau < n \cdot \tau_0, \quad 0 < z < L. \tag{1}$$

$$G \cdot c_a \cdot \frac{\partial T_a(z,\tau)}{\partial z} + P \cdot \alpha \cdot \left(T_a(z,\tau) - T_n(z,\tau)\right) + c_a \cdot \rho_a \cdot S_a \cdot \frac{\partial T_a(z,\tau)}{\partial \tau} = 0, \ 0 < \tau < n \cdot \tau_0, \ 0 < z < L. \tag{2}$$

$$T_a(0,\tau), \quad T_a(z,0), \quad T_n(0,\tau), \quad T_n(z,0). \tag{3}$$

Equations (1) and (2) describe the temperature fields of the nozzle and air, respectively. Expressions (3) are the boundary conditions of the problem that are specified.

In model (1)-(3): G is mass flow rate of air in the channel, kg/s; c_n, c_a are mass heat capacity of nozzle and air, respectively, J/(kg·K); ρ_n, ρ_a is density of nozzle and air, respectively, kg/m^3; P is channel perimeter, m; S_n, S_a is cross-sectional area of nozzle and air channel, respectively, m^2; α is heat transfer coefficient, W/(m^2·K); τ0 = τacc+τreg is duration of the regenerator cycle, s; τacc is duration of the accumulation period; τreg is duration of the regeneration period.

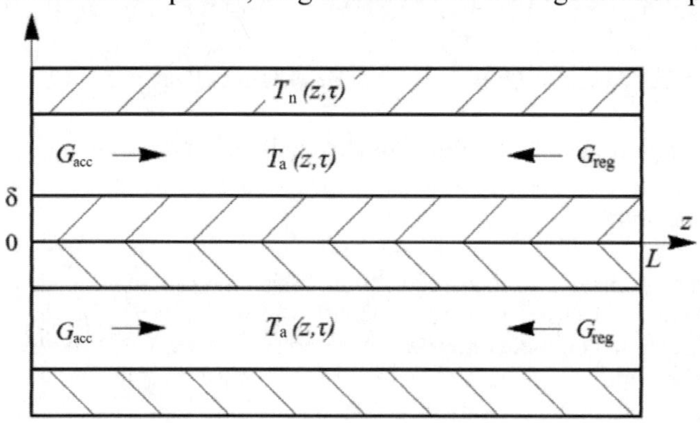

Figure 1. The scheme of adjacent channels of the SSRHE regenerative packing: δ is wall thickness of the channel, G_{acc} and G_{reg} are the mass air flow at the stages of heat accumulation and regeneration.

To assess the efficiency of the nozzle, one can use the following factors for the stages of accumulation and regeneration:

$$E_{acc} = \frac{\delta T_1}{\Delta T_{max\ reg} \frac{\delta T_2}{\Delta T_{max}}}, \tag{4}$$

where $\Delta T_{max} = (T_{in} - T_{out})$ is maximum available nozzle air cooling or heating from the internal air temperature Tin to the external one T_{out}, °C;

δT1 is cooling the internal air in the nozzle at the end of the accumulation stage, °C;

δT2 is heating the outside air in the nozzle at the end of the regeneration stage, °C.

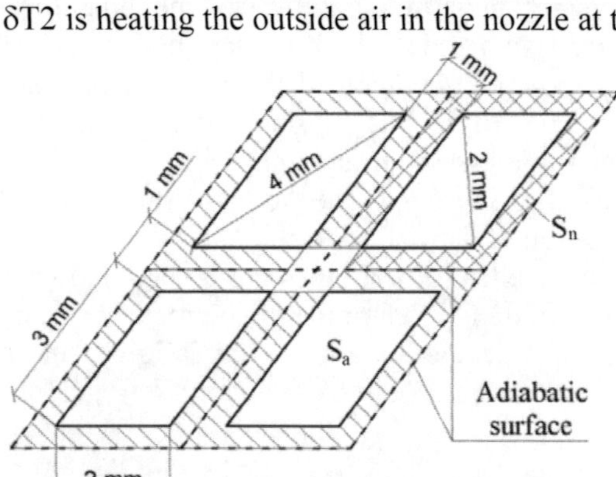

Figure 2. Cross-section of 4 adjacent channels of the nozzle: Sn and Sa is cross-sectional area of the nozzle and for air passage.

Changing of parameters such as diameter of the channels d, the length of the nozzle L, the duration of regeneration and accumulation stages will entail a change in δT1 and δT2.

The coefficients E_{acc} and E_{reg} cannot be greater than 1, but, for example, when the air flow is large, or the heat exchange area is small, or the duration of accumulation and regeneration stages τacc, τreg are large, these coefficients can be significantly less than one. For example, if E_{acc}=0.5, this means that at the end of the accumulation stage, the internal air gives the regenerator only half of its heat from the maximum possible. The operation of the regenerator can be considered sufficiently effective if E_{acc}≥0.9, and E_{reg}≥0.9, that is, more than 90% of the heat of the internal air is utilized in the regenerator during the whole process.

The diameter of the regenerator nozzle is limited, based on design considerations. However, the length of the available prototype has a margin to increase, if we start from the average thickness of the walls of buildings in Russia. Keeping the nozzle diameter constant will cause a decrease in the number of channels in the nozzle as the diameter of a single channel increases. As the channel diameter d increases, the total heat exchange area in the regenerator nozzle decreases, and the coefficients E_{acc} and E_{reg} for the same duration τacc, τreg will begin to decrease at some point in time. To prevent this, one can simultaneously reduce the duration of τacc, τreg. Then, each channel diameter d will correspond to its time τacc=τreg, which preserves the efficiency of the regenerator.

3. Results

Table 1 shows a comparison of two nozzle configurations that differ in size of a single channel d, while the wall thickness between the channels is kept constant. The comparison was made at changing the nozzle length and the time of one stage of accumulation (regeneration) of thermal energy.

Table 1. Comparison of efficiency factors for different channel lengths, channel diameters and stage time.

Nozzle length, m	Stage time, s	Efficiency factors for channel diameter, m			
		d1=0.0016		d2=0.004	
		Eacc	Ereg	Eacc	Ereg
L1=0.2	20	0.8777	0.9502	0.7937	0.8457
L1=0.2	40	0.8804	0.9350	0.7916	0.8290
L1=0.2	60	0.8845	0.9200	0.7904	0.8120
L2=0.64	20	0.8935	0.9840	0.8692	0.9486
L2=0.64	40	0.8940	0.9788	0.8698	0.9415
L2=0.64	60	0.8948	0.9730	0.8705	0.9347

Using the data from Table 1, the graphs were plotted, shown in Figure 3.

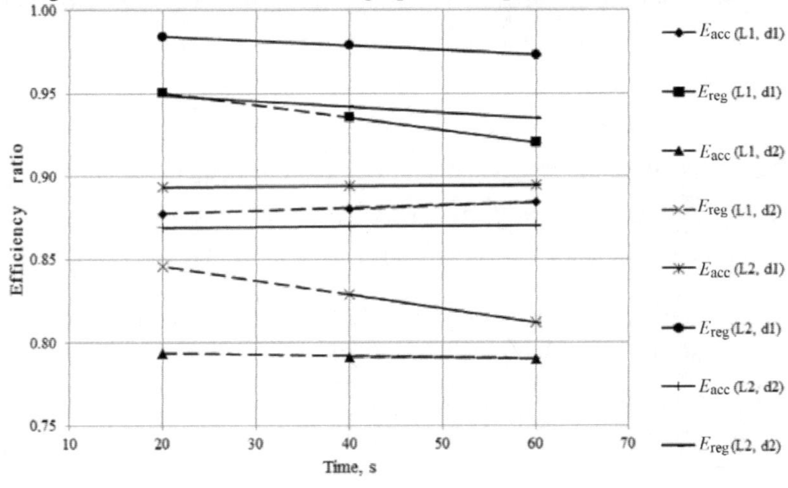

Figure 3. Comparison of efficiency factors for different channel lengths, channel diameters and stage times.

According to table 1 and figure 3 we can draw the following conclusions:

1. With an increase in the length of channel (nozzle) and maintaining the diameter of channel, the efficiency factors Eacc and Ereg increase. At the same time, it is seen from Figure 1 that the graphs corresponding to a larger channel length are located predominantly in the upper part of the coordinate plane, which allows one to conclude that an increase in the channel length has a favorable effect on the thermal efficiency of the nozzle.
2. An increase in channel diameter leads to a decrease in efficiency factors Eacc and Ereg regardless of the stage time and channel length.
3. An increase in time of accumulation (regeneration) stage causes a decrease in regenerative capacity of nozzle (Ereg) for both diameters and channel lengths. In this case, accumulation capacity (Eacc) of nozzle with an increase in stage time increases in all cases, moreover, when length is L = 0.2 m, diameter d2 = 0.004 m.

It should be noted that in the future it is more convenient to operate with coefficients Eacc and Ereg per cycle, and they will be higher than the coefficients determined at the end of the accumulation and regeneration stages. In general, the average Eacc and Ereg should be equal to each other, and this will simply be coefficient of thermal efficiency of regenerator E. Then the amount of heat recovered by the regenerator can be determined as follows:

$$Q = E \cdot c \cdot V(t_{\text{in}} - t_{\text{out}}), \text{W} \tag{5}$$

where V is volume air consumption, m^3/s;
c is air capacity, kJ/(kg·°C);
t_{in} and t_{out} are indoor and outdoor air temperatures, °C.

For the regenerator, an important indicator is power consumption of fan, which can be estimated by expression:

$$N = \frac{V \cdot \Delta p_{\text{sum}}}{\eta}, \text{W}, \tag{6}$$

where η is fan efficiency; Δp_{sum} is total pressure loss in regenerator, Pa.

The main element that creates resistance to air passage in regenerator is nozzle. Since the regenerator is of switching type, the air inlet into the channels of the nozzle is carried out from different sides, depending on the stage. Therefore, we can assume that the total pressure loss in the nozzle (Δp_{sum}) consists of loss at the entrance to channel (Δp_{in}) and the friction loss in channel (Δp_{fr}):

$$\Delta p_{\text{sum}} = \Delta p_{\text{in}} + \Delta p_{\text{fr}}, \text{Pa}. \tag{7}$$

Since the nozzle channels are identical and parallel to each other, to determine the total pressure loss in the nozzle, it is sufficient to determine the loss in a single channel.

Loss of pressure on the friction in the channel can be estimated by the expression:

$$\Delta p_{\text{fr}} = \xi \cdot \frac{L}{d} \cdot \frac{\rho \cdot w^2}{2}, \text{Pa}, \tag{8}$$

Where for laminar flow mode the coefficient ξ is determined by the formula:

$$\xi = \frac{64}{Re \frac{\frac{64}{w \cdot d}}{\left(\frac{\mu}{\rho}\right)} \frac{64 \cdot \mu}{w \cdot d \cdot \rho}} \tag{9}$$

If we substitute (9) into the expression for Δp, we get:

$$\Delta p_{\text{fr}} = \frac{64 \cdot \mu}{w \cdot d \cdot \rho} \cdot \frac{L}{d} \cdot \frac{\rho \cdot w^2}{2} = \frac{32 \cdot L \cdot \mu \cdot w}{d^2}, \text{Pa}, \tag{10}$$

where the air velocity in the channel is determined for the average air density at a temperature $T_{av} = (T_{in} + T_{out})/2$.

We assume that the speed of movement of air in a room is sufficiently small, so the pressure loss when air enters the channel will be equal to one speed (dynamic) pressure. That is, the coefficient of local resistance of the input will be equal to 1 [15]. Then

$$\Delta p_{\text{sum}} = \Delta p_{\text{in}} + \Delta p_{\text{fr}} = \frac{\rho \cdot w^2}{2} + \frac{32 \cdot L \cdot \mu \cdot w}{d^2}, \text{Pa}. \tag{11}$$

Table 2. Comparison of pressure loss and fan power for various lengths and channel diameters.

Nozzle length, m	Pressure losses Δp_{sum}, Pa		Fan power N, W	
	For the channel diameter, m			
	d1=0.0016	d2=0.004	d1=0.0016	d2=0.004
0.2	27.2	3.0	0.261	0.028
0.64	86.6	9.3	0.829	0.089

The calculation of pressure losses and fan power was performed for air consumption V = 0.00861 m³/s; air speeds in a single channel were: w1= 0.62; w2=0.41.

From table 2 it follows:

1) A 2.5 times increase in diameter reduces the pressure loss and the required fan power by about 9.5 times.
2) Increasing the channel length from 0.2 to 0.64 m (for 3.2 times) with a constant diameter increases the pressure loss and the required fan power by about 3.2 times.

Table 3. Pressure losses in a regenerative nozzle

Nozzle length, m	Pressure losses, Pa					
	At the inlet Δp_{in}		Per friction Δp_{fr}		Total Δp_{sum}	
	For the channel diameter, m					
	d1=0.0016	d2=0.004	d1=0.0016	d2=0.004	d1=0.0016	d2=0.004
0.2	0.242	0.107	27.0	2.9	27.2	3.0
0.64	0.242	0.107	86.4	9.2	86.6	9.3

According to Table 3, it can be concluded that the pressure loss at the air inlet to the channels of the nozzle is extremely small compared to the friction loss. Therefore, they can be neglected in the future.

In this case, from formula (10) it can be seen that $\Delta p_{fr} \sim \frac{1}{d^2}$, whereas the heat transfer coefficient $\alpha \sim \frac{1}{d}$. That is, as the diameter of the channel decreases, pressure losses grow much faster than the heat transfer coefficient increases.

4. Conclusions

As a result, we can draw the following conclusion. An increase in diameter of a single channel leads to a decrease in the heat transfer coefficient, to a decrease in air velocity (at the same flow rate) and to a sharp decrease in pressure loss. Increasing the length of the regenerator nozzle increases the pressure loss almost in direct proportion, but significantly increases the thermal efficiency coefficients Eacc and Ereg. Therefore, to achieve maximum thermal efficiency of the regenerator, it is advisable to increase the nozzle length, while increasing the channel diameter can be used as a way to reduce the aerodynamic resistance of the nozzle and, as a result, reduce the required fan power.

In general, optimization of regenerator parameters should be carried out according to economic or energy criteria. The economic criterion is saving money on heating costs in the presence of a regenerator minus the cost of electricity consumed by fan. Energy criterion is saving heat energy for heating less energy costs for the production and transportation of electricity consumed by fan. The optimal parameters of the regenerator must meet the maximum energy or economic criterion.

References

[1] Petritchenko M R, Kotov E V, Nemova D V, Tarasova D S and Sergeev V V 2018 Numerical simulation of ventilated facades under extreme climate conditions *Magazine of Civil Engineering* **77**(1) pp 130–40

[2] Borodinecs A, Zemitist J, Millers R, Tuanova K, Geikins A and Nefedova A 2016 Specifics of multi-apartment building deep complex retrofitting *CESB 2016 - Central Europe Towards Sustainable Building 2016: Innovations for Sustainable Future* pp 49–55

[3] Zubkov D, Chusov A and Molodtsov D 2016 Measuring the thermal resistance of typical multi-layer building walls using the accelerated measurement method *MATEC Web of Conferences* **53** p 01020
[4] Petrichenko M R, Nemova D V, Kotov E V, Tarasova D S and Sergeev V V 2018 Ventilated facade integrated with the HVAC system for cold climate *Magazine of Civil Engineering* **77**(1) pp 47–58
[5] Statsenko E, Ostrovaia A, Musorina T and Sergievskaya N 2018 Thermal Properties of the Building with Low Energy Consumption (LEB) *Advances in Intelligent Systems and Computing* **692** pp 417–31
[6] Staritcyna A, Pollock E, Sokolova E and Martynenko E 2016 Energy efficiency in multi-story buildings *MATEC Web of Conferences* **73** p 02009
[7] Sinitsyn A A 2013 *Life Science Journal* **10**(4) pp 1421-26
[8] Gamaunova O, Musorina T 2018 *MATEC Web of Conferences* **251** 05041
[9] Mokhireva A, Loginova P, Glebova E, Bespalov V 2018 *MATEC Web of Conferences* **245** 16001
[10] Sinitsyn A A, Monarkin N N and Rogulina T V 2014 Automation and energy saving of machine-building and metallurgical production, technology and reliability of machines, devices and equipment Materialy devyatoj mezhdunarodnoj nauchno-tekhnicheskoj konferencii pp 202–5
[11] Sinitsyn A, Filippova E, Derevianko O 2019 *2018 International Multi-Conference on Industrial Engineering and Modern Technologies, FarEastCon 2018*, 8602553
[12] Petritchenko M R, Subbotina S A, Khairutdinova F F, Olshevskiy V Ya, Sergeev V V 2017 *Magazine of Civil Engineering* **73**(5) pp 40-8
[13] Meshalkina M, Sushnikov V, Kryzhova N 2018 *MATEC Web of Conferences* **245** 03003
[14] Monarkin N N, Lukin S V and Kochkin A A 2018 Privolzhskij nauchnyj zhurnal 3 (47) pp 57–62
[15] Taliev V N 1979 Aerodynamics of ventilation (Moscow: Stroiizdat)

IOP Conf. Series: Earth and Environmental Science **337** (2019) 012068 doi:10.1088/1755-1315/337/1/012068

Mathematical modeling of heating temperature mode for a heat exchange system of the type "pipe in pipe"

Alijon Naimov[1], Anton Sinitsyn[2*], Farit Gazizov[3], A P Eperin[4], Yu A Rundygin[4], G A Agasiants[4], S M Galileev[4] and T R Akhmetov[3]

[1]Vologda State University, Department of Informatics and Computer Technology, 160000 Lenin st., 15, Russia
[2]Vologda State University, Department of Heat Gas Supply and Ventilation, 160000 Lenin st., 15, Russia
[3]Kazan State Power Engineering University, Kazan, Russian Federation
[4]Peter the Great St. Petersburg Polytechnic University, St. Petersburg, Russian Federation

* Corresponding author: sinitsyn.science@mail.ru

Abstract. A mathematical model of heating of inhomogeneous medium "TEH-sand-air" was constructed and investigated as an initial-boundary value problem for the heat equation in polar coordinate system with boundary conditions that take into account the peculiarities of thermal processes at the boundaries of the inhomogeneous medium. A solution is given to the initial-boundary problem and an algorithm is proposed for calculating the thermal characteristics of the heating process and regular temperature conditions. The study is relevant for development of a computer virtual model of the heating process of an inhomogeneous medium in order to clearly demonstrate the heating process over a short period of time, as well as to calculate the corresponding thermal characteristics based on real experimental data.

1. Introduction

At present, the theory of modeling the thermal state of bodies of various shapes describing real objects for various technological processes is widely developed and used [1-12]. Quite a large number of variants of formulation of the heat conduction problem for bodies of various shapes are given in the works of A.V. Lykov, which considers the solution of the unsteady heat conduction equation (semi-infinite body, unbounded plate, solid cylinder, ball, hollow cylinder) by several methods (separation of variables, operational, Fourier and Hankel integral transforms). However, the mathematical description of thermal state of body for specific production conditions has significant differences in determining the shape of the body and the boundary conditions.

This article is devoted to study of temperature during the process of heating an inhomogeneous medium consisting of a thermoelectric heater (TEH) placed in a steel cylindrical pipe, sand which fills the space between the inner steel pipe and the outer copper pipe, and the air surrounding the outer pipe. This study is relevant for the studies described in articles [13-14] on development of a computer virtual model of heating process of a inhomogeneous medium in order to clearly demonstrate the heating process over a short period of time, as well as to calculate the corresponding thermal characteristics based on real experimental data.

In order to calculate the temperature mode in the process of heating the inhomogeneous medium "TEH-sand-air", three experiments were carried out with different TEH electric capacities, and temperature was measured using a thermocouple at four points (on the surface of the outer and inner pipes and at two internal points) at successive points in time (every 5 milliseconds) from the beginning of the process until the temperature stabilizes. The thermocouple readings were recorded by an analog-to-digital converter and processed by the computer program Table Curve 2D.

This raises the following questions:

1. To find relationships between coefficients of the obtained formulas and the thermal characteristics of process: coefficients of thermal conductivity, heat capacity, and heat transfer.
2. Whether it is possible to find heat transfer coefficients from TEH to sand and heat transfer of sand to air.
3. What is the temperature dynamics in space and in time.

To answer these questions, the methodology of work [15] is used, consisting of the following steps:

I. Construction of a mathematical model of heating process in the form of an initial-boundary problem for the heat equation with boundary conditions that take into account the inhomogeneity of the medium.
II. The solution of the boundary problem by the method of separation of variables, where it is necessary to construct a basis of eigenfunctions that satisfy the boundary conditions.
III. Calculation of thermal characteristics of the process based on experimental data using the formulas of the stationary and regular modes.

The "TEH-sand-air" heating process under study differs qualitatively from the process considered in [13]. As a consequence, we obtain another initial-boundary value problem, where the basis of eigenfunctions cannot be found explicitly. The existence and completeness of the eigenfunctions of the corresponding Sturm-Liouville problem was proved in [16]. With its help, a formula for solving an initial-boundary value problem, and, in particular, a regular-mode formula, is derived. The regular mode determines the "acceleration curve" which is the temperature change curve from the beginning of the heating process until the moment of stabilization.

2. Construction of mathematical model

To build a mathematical model of the heating process of an inhomogeneous medium "TEH-sand-air" we give the following assumptions and notations.

1. Temperature $U(t,x)$ is a function of variables t and x, where t is time $t \geq 0$, x is distance from the center of the inner pipe to the point where the temperature is measured $r \leq x \leq R, r > 0$, r is the radius of the inner pipe, R is the radius of the outer pipe; in the conducted experiments r = 10 mm, R = 50 mm. Therefore, the temperature depends only on the distance to the center of the inner pipe and it is constant along any cylindrical surface concentrically located relative to the inner pipe.
2. We denote the temperature on the surface of the inner pipe at time t as $U(t,r)$, $U(t,r+0)$ is the temperature of the sand at the boundary with the inner pipe, $U(t,x)$ is the temperature of the sand at the inner point x, $r < x < R$, $U(t,R-0)$ is the temperature of sand at the border with the outer pipe, $U(t,R)$ is the air temperature near the outer pipe. In this case, the temperature of the copper pipe is neglected due to the large value of copper coefficient of thermal conductivity.
3. An inhomogeneous medium is characterized by the following constants: c_r is specific heat capacity of the inner (steel) pipe, c_R specific heat capacity of the outer (copper) pipe, Q is amount of heat released from a unit area of the inner pipe in a unit period of time, h_r is heat transfer coefficient from the inner pipe to sand, h_R is heat transfer coefficient of sand to air, α_R is air heat loss coefficient near the outer pipe, k, ρ, c is heat conductivity, density and specific heat of sand, $a^2 = \frac{k}{c\rho}$ thermal diffusivity of sand.

IOP Conf. Series: Earth and Environmental Science **337** (2019) 012068 doi:10.1088/1755-1315/337/1/012068

The mathematical model of the heating process will consist of the balance equations in the inner space of sand, on the inner surface of pipe, at the boundaries of "TEH-sand", "sand-air", in the air near the external pipe.

The equation of heat balance in the internal space of sand, according to the Fourier law, [13], has the form:

$$\frac{1}{a^2}\frac{\partial U}{\partial t}(t,x) = \left[\frac{\partial^2 U}{\partial x^2}(t,x) + \frac{1}{x}\frac{\partial U}{\partial x}(t,x)\right], t > 0,\ r < x < R, \tag{1}$$

where the right-hand side is the Laplace operator in the polar coordinate system with the radius-vector x.

On the surface of the inner pipe, we set the heat balance as

$$c_r\frac{\partial U}{\partial t}(t,r) - k\frac{\partial U}{\partial x}(t,r+0) = Q, t > 0, \tag{2}$$

where $c_r\frac{\partial U}{\partial t}(t,r)$ is the amount of heat for inner pipe warming, $-k\frac{\partial U}{\partial x}(t,r+0)$ is the amount of heat given to sand.

At the "TEH-sand" boundary, the heat balance condition is:

$$-k\frac{\partial U}{\partial x}(t,r+0) = h_r[U(t,r) - U(t,r+0)], t > 0 \tag{3}$$

At the "sand-air" boundary we have:

$$-k\frac{\partial U}{\partial x}(t,R-0) = h_R[U(t,R-0) - U(t,R)], t > 0 \tag{4}$$

The heat balance in the air near the outer pipe is set by condition

$$-k\frac{\partial U}{\partial x}(t,R-0) = c_R\frac{\partial U}{\partial t}(t,R) + \alpha_R U(t,R), \tag{5}$$

where $c_R\frac{\partial U}{\partial t}(t,R)$ is the amount of heat for warming of the outer pipe, $\alpha_R U(t,R)$ is the amount of heat, lost by air near the outer pipe.

At the initial moment, the inhomogeneous medium has an ambient temperature:

$$U(0,x) = u_0,\ r \le x \le R \tag{6}$$

Thus, the mathematical model of heating process is an initial-boundary problem for the differential heat conduction equation (1) in polar coordinate system with boundary conditions (2), (3) on the inner boundary, (4), (5) on the outer boundary, and the initial condition (6). By solution $U(t,x)$, $t \ge 0, r \le x \le R$ of the initial-boundary problem (1) - (6) a temperature mode is defined during the heating of the inhomogeneous medium "TEH-sand-air". The function $U(t,x)$ at each $t > 0$ have discontinuities in points $x = r$ and $x = R$.

2.1 Stationary mode

Stationary mode, as in [15], is called stabilization of temperature during heating:

$$\bar{U}(x) = \lim_{t\to+\infty} U(t,x),\ r \le x \le R,$$

where $\bar{U}(x)$ is temperature at distance x from the center of TEH when stabilized. In (1) - (5), passing to the limit at $t \to +\infty$, we derive that the function $\bar{U}(x)$ should meet the following conditions:

$$\frac{d^2\bar{U}(x)}{dx^2} + \frac{1}{x}\frac{d\bar{U}(x)}{dx} = 0, r < x < R,$$

IOP Conf. Series: Earth and Environmental Science **337** (2019) 012068 doi:10.1088/1755-1315/337/1/012068

$$-k\frac{d\bar{U}}{dx}(r+0)=Q,\ -k\frac{\partial\bar{U}}{\partial x}(r+0)=h_r[\bar{U}(r)-\bar{U}(r+0)],$$

$$-k\frac{\partial\bar{U}}{\partial x}(R-0)=\alpha_R\bar{U}(R),\ -k\frac{\partial\bar{U}}{\partial x}(R-0)=h_R[\bar{U}(R-0)-\bar{U}(R)].$$

From these expressions we obtain $\bar{U}(x)$:

$$\bar{U}(x)=\begin{cases}\bar{U}(r+0)+\frac{Q}{h_r}, & x=r,\\ \left(1+\frac{\alpha_R}{h_R}\right)\frac{rQ}{R\alpha_R}+\frac{rQ}{k}ln\frac{R}{x}, & r<x<R,\\ \frac{rQ}{R\alpha_R}, & x=R.\end{cases} \tag{7}$$

Using the formula (7), basing on the experimental data, one can obtain the values for thermal constants α_R, k, h_r, h_R. Namely, we assume that the given values are temperatures in four measurement points at stabilization: $\widetilde{U}(r), \widetilde{U}(r_2), \widetilde{U}(r_3), \widetilde{U}(R)$, where $r<r_2<r_3<R$. By assuming $\bar{U}(r)=\widetilde{U}(r), \bar{U}(r_2)=\widetilde{U}(r_2), \bar{U}(r_3)=\widetilde{U}(r_3), \bar{U}(R)=\widetilde{U}(R)$, in relation to the unknown values α_R, k, h_r, h_R we obtain the following system of linear equations:

$$\begin{cases}\left(1+\frac{\alpha_R}{h_R}\right)\frac{rQ}{R\alpha_R}+\frac{rQ}{k}ln\frac{R}{r}+\frac{Q}{h_r}=\widetilde{U}(r)\\ \left(1+\frac{\alpha_R}{h_R}\right)\frac{rQ}{R\alpha_R}+\frac{rQ}{k}ln\frac{R}{r_2}=\widetilde{U}(r_2)\\ \left(1+\frac{\alpha_R}{h_R}\right)\frac{rQ}{R\alpha_R}+\frac{rQ}{k}ln\frac{R}{r_3}=\widetilde{U}(r_3)\\ \frac{rQ}{R\alpha_R}=\widetilde{U}(R).\end{cases}$$

Further we find the unknown values:

$$\begin{cases}\alpha_R=\frac{rQ}{R\widetilde{U}(R)},\quad k=\frac{rQ\,ln\frac{r_3}{r_2}}{\widetilde{U}(r_2)-\widetilde{U}(r_3)},\\ h_R=\left(\frac{R}{rQ}\cdot\frac{\widetilde{U}(r_2)\,ln\frac{R}{r_3}-\widetilde{U}(r_3)\,ln\frac{R}{r_2}}{ln\frac{r_2}{r_3}}-\frac{1}{\alpha_R}\right)^{-1},\\ h_r=\left(\frac{\widetilde{U}(r)}{Q}-\frac{r}{R}\left(\frac{1}{\alpha_R}+\frac{1}{h_R}\right)-\frac{r}{k}ln\frac{R}{r}\right)^{-1}.\end{cases} \tag{8}$$

The solution of the problem (1) - (6) we find in the following form

$$U(t,x)=\bar{U}(x)+\sum_{i=1}^{\infty}A_iV_i(x)e^{-\lambda_i a^2 t}, \tag{9}$$

where $0<\lambda_1<\lambda_2<\ldots<\lambda_n<\ldots$ and functions are defined as eigenvalues and eigenfunctions of the Sturm-Liouville problem in the form of Kneser [13]:

$$\begin{cases}\frac{d^2V}{dx^2}(x)+\frac{1}{x}\frac{dV}{dx}(x)+\lambda V(x),\ r<x<R,\\ \lambda c_r a^2V(r)+k\frac{dV}{dx}(r+0)=0,\quad -k\frac{dV}{dx}(r+0)=h_r\big(V(r)-V(r+0)\big),\\ (-\lambda c_r a^2+\alpha_R)V(R)=-k\frac{dV}{dx}(R-0),\quad -k\frac{dV}{dx}(R-0)=h_R\big(V(R-0)-V(R)\big).\end{cases} \tag{10}$$

The authors of the work [16] proved the existence and completeness of the eigenfunctions of the problem (10) in the space of functions

$$H = \left\{ f(x): \int_r^R f^2(x)dx < +\infty,\ |f(r)| + |f(R)| < \infty \right\}$$

with dot product

$$\langle f, g\rangle = \int_r^R x f(x) g(x) dx + \frac{ra^2 c_r}{k} f(r) g(r) + \frac{Ra^2 c_R}{k} f(R) g(R) .$$

In other words, any function $f(x) \in H$ is uniquely representable as a series in eigenfunctions of the problem (10):

$$f(x) = \sum_{i=1}^{\infty} B_i V_i(x) ,$$

where $B_i = \langle f, V_i\rangle$, $\langle V_i, V_i\rangle = 1$, $i = 1,2,\ldots$ and $\langle V_i, V_j\rangle = 0$ at $i \neq j$. At the same time, for the first eigenvalue λ_1 there exists an inequality:

$$\frac{\alpha}{c_R} \leq \lambda_1 \leq \frac{4\alpha}{c_R} .$$

By assuming $f(x) = u_0 - \bar{U}(x)$, we find the coefficients A_i, $i = 1,2, \ldots$ of the representation (9):

$$A_i = \langle u_0 - \bar{U}, V_i\rangle,\ i = 1,2, \ldots \tag{11}$$

In this case the series in the right part of the formula (9) at each $t > 0$ converges in the common sense.

2.2 Regular mode

According to the terminology of [17], the regular temperature regime is determined by the approximate formula

$$U(t, x) \approx \bar{U}(x) + A_1 V_1(x) e^{-\lambda_1 a^2 t},$$

where $\lambda_1$is the first eigenvalue, and $V_1(x)$ is the corresponding normalized eigenfunction of the Sturm-Liouville problem (10); they can be found using approximately numerical-analytical methods. The coefficient A_1 is determined by the formula $A_1 = \langle u_0 - \bar{U}, V_1\rangle$.

Thus, you can set the following algorithm for calculating the regular temperature mode duting the process of heating the inhomogeneous medium "TEH-sand-air":

1. From experimental data $\widetilde{U}(r)$, $\widetilde{U}(r_2)$, $\widetilde{U}(r_3)$, $\widetilde{U}(R)$, obtained during temperature measurements at four points during stabilization, we find the approximate values of thermal constants using formulas (8).
2. Using numerical-analytical methods, we find the approximate values of the first eigenvalue λ_1 of the Sturm-Liouville problem (10) and the corresponding normalized eigenfunction $V_1(x)$.
3. We find the stationary mode $\bar{U}(x)$ from the formula (7).
4. We calculate the coefficient A_1 from the formula

$$A_1 = \int_r^R x\big(u_0 - \bar{U}(x)\big) V_1(x) dx + \frac{ra^2 c_r}{k}\big(u_0 - \bar{U}(r)\big) V_1(r) + \frac{Ra^2 c_R}{k}\big(u_0 - \bar{U}(R)\big) V_1(R) .$$

5. The regular mode is calculated from the formula

$$U(t, x) \approx \bar{U}(x) + A_1 V_1(x) e^{-\lambda_1 a^2 t},\ r \leq x \leq R,\ t \geq 0 \tag{12}$$

3. Conclusions

Thus, we constructed and investigated a mathematical model of the process of heating an inhomogeneous medium "TEH-sand-air" in the form of an initial-boundary value problem for the heat equation in a polar coordinate system with boundary conditions which take into account the peculiarities of thermal processes at the boundaries of an inhomogeneous medium was. We present a solution to the initial-boundary problem and propose an algorithm for calculating the thermal characteristics of the heating process and a regular temperature regime using numerical-analytical methods for solving the Sturm-Liouville problem in the form of Kneser.

This study is relevant for the development of a computerized virtual model of the heating process of an inhomogeneous medium in order to clearly demonstrate the heating process in a short period of time, as well as to calculate the corresponding thermal characteristics based on actual experimental data.

The work was carried out on the basis of the results of research conducted with the support of the Ministry of Education and Science of the Russian Federation as part of the fulfillment of the state task to higher educational institutions for project No. 1816.

References

[1] Mityakov A, Mityakov V, Sapozhnikov S, Gusakov A, Bashkatov A, Seroshtanov V, Zainullina E and Babich A 2017 Hydrodynamics and heat transfer of yawed circular cylinder *Int. J. of Heat and Mass Transfer* **115** pp 333–9

[2] Vilchevskaya E and Sevostianov I 2016 Overall thermal conductivity of a fiber reinforced composite with partially debonded inhomogeneities *Int. J. of Engineering Science* **98** pp 99–109

[3] Zaitsev D K, Pozhilov A A, Smirnov E M and Smirnovsky A A 2016 Numerical simulation of conjugate heat and mass transfer in a loop heat pipe evaporator *CEUR Workshop Proceedings* **1576** pp 512–20

[4] Vitokhin E Y and Ivanova E A 2017 Dispersion relations for the hyperbolic thermal conductivity, thermoelasticity and thermoviscoelasticity *Continuum Mechanics and Thermodynamics* **29**(6) pp 1219–40

[5] Mityakov V, Gusakov A, Seroshtanov V and Grekov M 2018 Investigation of flow and heat transfer at the circular fins *MATEC Web of Conferences* **245** p 06001

[6] Krivtsov A M, Sokolov A A, Müller W H and Freidin A B 2018 One-dimensional heat conduction and entropy production *Advanced Structured Materials* **87** pp 197–213

[7] Karyakin Y E, Kuzin V M, Pletnev A A and Fedorovich E D 2018 Numerical Modeling of the Thermal State of a Metal–Concrete Container with Spent Nuclear Fuel During Its Transportation *J. of Engineering Physics and Thermophysics* **91**(4) pp 991–8

[8] Mankonen A, Kaikko J, Vakkilainen E, Sergeev V 2018 *MATEC Web of Conferences* **245** 07007

[9] Akhmetova I, Akhmetov T 2018 *E3S Web of Conferences* **58** 01011

[10] Soloveva O V, Solovev S A, Egorova S R, Lamberov A A, Antipin A V, Shamsutdinov E V 2018 *Chemical Engineering Research and Design* **138** pp 212-28

[11] Agafonova N D, Egorov M Y, Sergeev V V, Gotovskii M A, Kruglikov P A, Lebedev M E, Sudakov A V, Fedorovich E D, Fokin B S 2018 *Atomic Energy* **123**(3) pp 154-1

[12] Akhmetova I G, Chichirova N D 2017 *Journal of Physics: Conference Series* **891**(1) 012168

[13] Sinitsyn A A and Timoshenko P O 2009 *University science to the Region* **1**

[14] Telkov M G and Naimov A N 2012 *University science to the Region*

[15] Samarskii A A 2003 *Selected works*

[16] Samarskii A A and Mikhailov A P 2002 *Mathematical modeling*

[17] Timoshenko P O and Sinitsyn A A 2009 *Young researchers to the Region* **1**

Computer modelling of heat energy regeneration process in a reversible heat exchanger

Alijon Naimov[1], Anton Sinitsyn[2*], M M Vetyukov[3], V A Palmov[3], V F Kosmach[3], Yu F Titovec[3] and I G Akhmetova[4]

[1]Vologda State University, Department of Informatics and Computer Technology, 160000 Lenin st., 15, Russia
[2]Vologda State University, Department of Heat Gas Supply and Ventilation, 160000 Lenin st., 15, Russia
[3]Peter the Great St. Petersburg Polytechnic University, St. Petersburg, Russian Federation
[4]Kazan State Power Engineering University, Kazan, Russian Federation

* Corresponding author: sinitsyn.science@mail.ru

Abstract. The article considers approbation of the developed mathematical model of the process of thermal energy accumulation in a stationary switching regenerative heat exchanger using the Matlab computer system. The model was constructed and numerically investigated using a difference scheme. This method is necessary for development of algorithms for calculating the thermal characteristics of heat and mass transfer processes under conditions of regeneration and phase transitions, as well as for analyzing the criteria for optimal control of the operation of this apparatus under nominal and critical environmental conditions.

1. Introduction

While construction of modern buildings and structures, various heat exchangers are used in their systems of heat supply and ventilation, in which complex processes of heat and mass transfer take place [1-5]. At the design stage of these devices, it is necessary to repeatedly calculate its characteristics in order to optimize the operating parameters [6].

To evaluate their efficiency, either complex mathematical models or empirical formulas of limited scope are used. The use of existing methods of calculation is often unacceptable either in terms of counting time or accuracy when going beyond the scope of application [7].

The most difficult problem in analyzing the efficiency of such devices is calculation of the heat and mass transfer processes occurring in them, taking into account the regenerative nature and phase transitions. For such processes, empirical relationships are used, as a rule which are valid for a narrow region of regime and design parameters [8].

At the same time, creation of simple engineering methods for calculating heat and mass transfer processes in various gas-dynamic devices, as, for example, in [9-15], will make it possible to identify their modes of operation, optimize their modes, and promptly evaluate the efficiency of various design schemes during the development of new devices or technologies.

The Matlab system was chosen as a universal tool for solving the problem. The choice of the system is due to the fact that among the analogues Matlab is distinguished by a simple programming language that makes it easy to create one's own algorithms. The system is specially designed for research on the development of new equipment, including engineering research. Also, Matlab has powerful visualization tools, which is important when determining the optimal parameters of the designed equipment [16].

The objects of simulation are the processes of heat and mass transfer in a stationary switching regenerative heat exchanger (SSRHE). The existing mathematical and computer models have been developed either for recuperative and mixing heat exchangers, or for regenerative heat exchangers operating at high coolant temperature [17], as well as for regenerative heat exchangers with a rotary nozzle. The most adequate mathematical model developed for the SSRHE is given in [16]. This model was constructed and numerically investigated using a difference scheme. But such issues as the existence and uniqueness of a solution, as well as analytical methods for finding a solution have not yet been investigated. The analytical method of finding a solution to the problem is relevant from the point of view of developing algorithms for calculating the thermal characteristics of the process and optimal control of the operation of the specified apparatus.

2. Development of a mathematical model

We introduce the following notations: $T_a(\tau, z)$ is air temperature at the time moment τ in the point z, where $\tau \geq 0$, $0 \leq z \leq L$, °C; $T_n(\tau, z)$ is nozzle temperature at the time moment τ in the point z, where $\tau \geq 0$, $0 \leq z \leq L$, °C; G_{ac} is air consumption at the accumulation stage, m^3/h; c_a is air heat capacity, kJ/kg·°C; ρ_a is air density, kg/ m^3; S_a is the cross-section area of channel along which air goes, m^2; P is the cross-section perimeter of channel along which air goes, m; α is heat transfer coefficient for air, W/m^2·°C; T_{in} temperature of the room, from which the air is supplied, °C; T_{out} is ambient temperature where some part of heat enters, °C; λ_n is thermal conductivity of the nozzle material, W/m·°C.

We construct a mathematical model of the process of heating the air in the passageway. To do this, we note that $-G_{ac}c_a\frac{\partial T_a}{\partial z}(\tau,z)$ - the amount of heat entering the flow area at point z at time τ; $c_a\rho_aS_a\frac{\partial T_a}{\partial \tau}(\tau,z)$ is the amount of heat spent on the heating of the flow section at the point z at the moment of time τ; $P\alpha(T_a(\tau,z)-T_n(\tau,z))$ is the amount of heat transferred from the air nozzle in the flow area at the point z at time τ.

The balance equation of thermal energy in the flow section at the point z at the moment of time τ:

$$-G_{ac}c_a\frac{\partial T_a}{\partial z}(\tau,z) = c_a\rho_aS_a\frac{\partial T_a}{\partial \tau}(\tau,z) + P\alpha(T_a(\tau,z)-T_n(\tau,z)), \tau > 0, 0 < z < L. \quad (1)$$

Equation (1) is a mathematical model of the process of heating air in the passage channel. It is necessary to take into account the initial condition

$$T_a(0,z) = T_{in} - \frac{T_{in}-T_{out}}{L}\cdot z, \quad 0 \leq z \leq L. \quad (2)$$

i.e. at the initial moment of time, the temperature in the flow channel is linearly distributed between Tin and Tout, as well as the boundary (edge) condition:

$$T_a(\tau,0) = T_{in}, \quad \tau \geq 0. \quad (3)$$

i.e. at the left end of the passage, the temperature is constant all the time and is equal to room temperature T_{in}.

Thus, we obtain the initial-boundary (mixed) problem (1), (2), (3) for the unknown function $T_a(\tau, z)$.

Now we build a mathematical model of the process of heating the nozzle. In the process of heating the nozzle $P\alpha(T_a(\tau,z)-Tn(\tau,z))$ is the amount of thermal energy supplied from the air nozzle in the flow area at the point z at the moment of time τ; $c_n\rho_nS_n\frac{\partial T_n}{\partial \tau}(\tau,z)$ is the amount of thermal energy spent on heating the nozzle at point z at time τ;$-\lambda_nS_n\frac{\partial^2 T_n}{\partial z^2}(\tau,z)$ is the amount of thermal energy transferred inside the nozzle according to the Fourier law.

The balance equation of thermal energy in the cross section of the nozzle at the point z at the moment

IOP Conf. Series: Earth and Environmental Science **337** (2019) 012069 doi:10.1088/1755-1315/337/1/012069

of time τ:

$$P\alpha(T_a - T_n) = c_n\rho_n S_n \frac{\partial T_n}{\partial \tau} - \lambda_n S_n \frac{\partial^2 T_n}{\partial z^2}, \quad \tau > 0, \quad 0 < z < L. \tag{4}$$

Equation (4) is a mathematical model of the process of heating the nozzle. It is necessary to consider the initial condition:

$$T_n(0, z) = T_{in} - \frac{T_{in} - T_{out}}{L} \cdot z, \quad 0 \le z \le L. \tag{5}$$

i.e. at the initial moment of time the temperature in nozzle is distributed linearly between Tin and Tout, and the boundary (edge) conditions:

$$\frac{\partial T_n}{\partial z}(\tau, 0) = 0, \frac{\partial T_n}{\partial z}(\tau, L) = 0, \quad \tau > 0. \tag{6}$$

i.e. nozzle ends are insulated all the time. Thus, we obtain the initial-boundary (mixed) problem (4), (5), (6) for the unknown function $T_n(\tau, z)$.

The mixed problem (1) - (6) is a mathematical model of the process of accumulation of thermal energy in the thermodynamic air-nozzle system

3. Approximate solution

The mixed problem (1) - (6) is rewritten as follows:

$$\frac{\partial Ta}{\partial \tau}(\tau, z) + a\frac{\partial T_a}{\partial z}(\tau, z) + bT_a(\tau, z) = bT_n(\tau, z)), \tau > 0, \quad 0 < z < L, \tag{7}$$

$$Ta(0, z) = T_{in} - \frac{T_{in} - T_{out}}{L} \cdot z, \quad 0 \le z \le L, T_a(\tau, 0) = T_{in}, \quad \tau \ge 0, \tag{8}$$

$$\frac{\partial T_n}{\partial \tau}(\tau, z) - c\frac{\partial^2 T_n}{\partial z^2}(\tau, z) + dT_n(\tau, z) = dT_a(\tau, z)), \tau > 0, \quad 0 < z < L, \tag{9}$$

$$T_n(0, z) = T_{in} - \frac{T_{in} - T_{out}}{L} \cdot z, \quad 0 \le z \le L, \frac{\partial T_n}{\partial z}(\tau, 0) = 0, \frac{\partial T_n}{\partial z}(\tau, L) = 0, \quad \tau > 0. \tag{10}$$

where

$$a = \frac{G_{ac}}{\rho_a S_a}, b = \frac{P\alpha}{c_a\rho_a S_a}, c = \frac{\lambda n}{c_n\rho_n}, d = \frac{P\alpha}{c_n\rho_n S_n}.$$

Taking into consideration (7) and (8) the function $T_a(\tau, z)$ can be expressed using $T_n(\tau, z)$:

$$T_a(\tau, z) = \Phi\ _a T_a(\tau, z), \tag{11}$$

where

$$\Phi\ _a T_n(\tau, z) = \begin{cases} e^{-b\tau}\left(T_{in} - \dfrac{T_{in} - T_{out}}{L}(z - a\tau)\right) + b\displaystyle\int_0^\tau e^{b(s-\tau)} T_n(s, z + a(s-\tau))ds, & z \ge a\tau, \\ e^{-\frac{b}{a}z} T_{in} + b\displaystyle\int_{\tau - \frac{z}{a}}^{\tau} e^{b(s-\tau)} T_n(s, z + a(s-\tau))ds, & z < a\tau. \end{cases}$$

In view of (9) and (10) the function $T_n(\tau, z)$ can be expressed using $T_a(\tau, z)$:

$$T_n(\tau, z) = \Phi\ _n T_a(\tau, z), \tag{12}$$

where

IOP Conf. Series: Earth and Environmental Science **337** (2019) 012069 doi:10.1088/1755-1315/337/1/012069

$$\Phi\ {}_nT_a(\tau,z) = \int_0^L G(\tau,z,\eta)\left(T_{in} - \frac{T_{in}-T_{out}}{L}\eta\right)d\eta + d\int_0^\tau\int_0^L G(\tau-s,z,\eta)T_a(s,\eta)d\eta ds,$$

$$G(\tau,z,\eta) = \frac{1}{L}\left(e^{-\beta_0\tau} + 2\sum_{k=1}^{\infty} e^{-\beta_k\tau}\cos\frac{\pi k}{L}z\cos\frac{\pi k}{L}\eta\right), \beta_k = d + c\left(\frac{\pi k}{L}\right)^2, k = 0,1,2, \ldots .$$

Integral representations (11) and (12) are derived from well-known formulas for solving mixed problems of the form (7), (8) and (9), (10).

Thus, the mixed problem (7) - (10) is equivalent to the system of integral equations (11), (12). System (11), (12) is solved by the method of successive approximations. Namely, we construct the following sequences of functions:

$$T_n^{(0)}(\tau,z) = T_{in}, T_a^{(n)}(\tau,z) = \Phi_a T_n^{(n-1)}(\tau,z), T_n^{(n)}(\tau,z) = \Phi n T_a^{(n)}(\tau,z) \qquad (13)$$

The sequence of functions n = 1,2,... converges to the solution ($T_n(\tau, z)$, $T_a(\tau, z)$) of the mixed problem (7) - (10): $(T_n^{(n)}(\tau,z), T_a^{(n)}(\tau,z))$

$$\lim_{n\to\infty} T_n^{(n)}(\tau,z) = T_n(\tau,z), \lim_{n\to\infty} T_a^{(n)}(\tau,z) = T_a(\tau,z).$$

Therefore, for large numbers n a couple of functions $(T_n^{(n)}(\tau,z), T_a^{(n)}(\tau,z))$ can be taken as an approximate solution ($T_n(\tau, z)$, $T_a(\tau, z)$) of the mixed problem (7) - (10):

$$T_n(\tau,z) \approx T_n^{(n)}(\tau,z), \quad T_a(\tau,z) \approx T_a^{(n)}(\tau,z).$$

For large numbers n and fixed points in time τ, the function graphs $(T_n^{(n)}(\tau,z), T_a^{(n)}(\tau,z))$ are shown in Figures 1, 2.

4. Numerical solution

We find the approximate values of solution of the mixed problem at the grid points $\Omega(N_\tau, N_z) = \{ (\tau_i, z_j): \ i = \overline{1, N_\tau + 1}, \ \ j = \overline{1, N_z + 1} \}$, where $\tau_i = (i-1)h_\tau$, $z_j = (j-1)h_z$, $h_\tau = \frac{h_z}{a}, h_z = \frac{L}{N_z}$, N_τ, N_z are set natural numbers. The values of the required functions $T_a(\tau,z)$, $T_n(\tau,z)$ in each point (τ_i, z_j) of grid we denote as follows: $TA_{i\ j} = T_a(\tau_i, z_j)$, $TN_{i\ j} = T_n(\tau_i, z_j)$. At the given grid $\Omega(N_\tau, N_z)$ the two matrices are unknown $TA = \left(TA_{i\ j}\right)_{i=\overline{1,N_\tau+1}}^{j=\overline{1,N_z+1}}$, $TN = \left(TN_{i\ j}\right)_{i=\overline{1,N_\tau+1}}^{j=\overline{1,N_z+1}}$.

For a mixed problem, we will find out what conditions the required TA and TN matrices should satisfy. According to the initial conditions, the first rows of these matrices are known: $TA_{1\ j} = T_a^0(z_j)$, $TN_{1\ j} = T_n^0(z_i)$, $j = \overline{1, N_z + 1}$. For the *TA* matrix, in view of boundary condition we know the first column:

$$TA_{i\ 1} = T_a^1(\tau_i), \quad i = \overline{1, N_\tau + 1}. \qquad (14)$$

Now we find out what the system of integral equations with respect to the grid $\Omega(N_\tau, N_z)$ turns into. For the first integral equation we have:

$$T_a(\tau_i, z_j) = \begin{cases} e^{-b\tau_i}T_a^0(z_j - a\tau_i) + b\displaystyle\int_0^{\tau_i} e^{b(s-\tau_i)}T_n(s, z_j + a(s-\tau_i))ds, \ if \ z_j \geq a\tau_i, \\ e^{-\frac{b}{a}z_j}T_a^1\left(\tau_i - \frac{z_j}{a}\right) + b\displaystyle\int_{\tau_i - \frac{z_j}{a}}^{\tau_i} e^{b(s-\tau_i)}T_n(s, z_j + a(s-\tau_i))ds, \ if \ z_j < a\tau_i, \end{cases}$$

where $i \geq 2$, $j \geq 2$. We should note that for the introduced notations:

$z_j \geq a\tau_i \Leftrightarrow (j-1)ah_\tau \geq a(i-1)h_\tau \Leftrightarrow j \geq i, \quad z_j - a\tau_i = (j-1)h_z - a(i-1)h_\tau = (j-i)h_z = z_{j-i+1},$

$$\tau_i - \frac{z_j}{a} = (i-1)h_\tau - (j-1)\frac{h_z}{a} = (i-j)h_\tau = \tau_{i-j+1}, T_a^0(z_j - a\tau_i) = T_a^0(z_{j-i+1}) = TA_{1,\ j-i+1},$$

$$T_a^1(\tau_i - \frac{z_j}{a}) = T_a^1(\tau_{i-j+1}) = TA_{i-j+1,\ 1}.$$

Next, we replace each integral with a finite sum, applying the following rectangle formula:

$$\int_\alpha^\beta f(s)ds \approx h\sum_{k=1}^{N} f(s_k), \text{ where } h = \frac{\beta-\alpha}{N}, \quad s_k = \alpha + h\cdot(k-1).$$

In case of the first integral we assume $\alpha = 0$, $\beta = \tau_i$, $N = i - 1$. So

$$h = \frac{\beta-\alpha}{N} = \frac{\tau_i - 0}{i-1} = \frac{(i-1)h_\tau}{i-1} = h_\tau,\ s_k = \alpha + h(k-1) = h_\tau(k-1) = \tau_k,$$

$$z_j + a(\tau_k - \tau_i) = (j-1)h_z + (k-i)h_z = z_{k+j-i}, T_n(\tau_k,\ z_j + a(\tau_k - \tau_i)) = Tn(\tau_k, z_{k+j-i}) = TN_{k,\ k+j-i}.$$

$$\int_0^{\tau_i} e^{b(s-\tau_i)} T_n(s, z_j + a(s-\tau_i))ds \approx h_\tau \sum_{k=1}^{i-1} e^{b(s_k-\tau_i)}\, T_n(s_k,\ z_j + a(s_k - \tau_i)) =$$
$$= h_\tau \sum_{k=1}^{i-1} e^{b(\tau_k-\tau_i)}\, T_n(\tau_k,\ z_j + a(\tau_k - \tau_i)) = h_\tau \sum_{k=1}^{i-1} e^{b(\tau_k-\tau_i)}\, TN_{k,\ k+j-i},$$

In case of the second integral by assuming $\alpha = \tau_i - \frac{z_i}{a}$, $\beta = \tau_i$, $N = j - 1$, we get:

$$h = \frac{\beta - \alpha}{N} = \frac{z_j}{a(j-1)} = \frac{a(j-1)h_\tau}{a(j-1)} = h_\tau,$$

$$s_k = \alpha + h(k-1) = \tau_i - \frac{z_j}{a} + (k-1)h_\tau = (k+i-j-1)h_\tau = \tau_{k+i-j},$$

$$z_j + a(\tau_{k+i-j} - \tau_i) = (j-1)h_z + (k-j)h_z = (k-1)h_z = z_k, T_n(\tau_{k+i-j}, z_k) = TN_{k+i-j,\ k}.$$

$$\int_{\tau_i - \frac{z_j}{a}}^{\tau_i} e^{b(s-\tau_i)} T_n(s, z_j + a(s_k - \tau_i))ds \approx$$
$$\approx h_\tau \sum_{k=1}^{j-1} e^{b(\tau_{k+i-j}-\tau_i)}\, T_n(\tau_{k+i-j}, z_j + a(\tau_{k+i-j} - \tau_i)) = h_\tau \sum_{k=1}^{j-1} e^{b(\tau_{k+i-j}-\tau_i)}\, TN_{k+i-j,\ k},$$

Thus, the first integral equation in the grid $\Omega(N_\tau, N_z)$ takes the following form:

$$TA_{i\ j} = \begin{cases} e^{-b\tau_i} TA_{1,\ j-i+1} + bh_\tau \sum_{k=1}^{i-1} e^{b(\tau_k-\tau_i)}\, TN_{k,\ k+j-i}, & if\ j \geq i, \\ e^{-\frac{b}{a}z_j} TA_{i-j+1,\ 1} + bh_\tau \sum_{k=1}^{j-1} e^{b(\tau_{k+i-j}-\tau_i)}\, TN_{k+i-j,\ k}, & if\ j < i, \end{cases} \qquad (15)$$

where $i = 2,\ \ldots,\ N_\tau + 1$, $\quad j = 2,\ \ldots,\ N_z + 1$.

For the second integral equation we have:

$$Tn(\tau_i, z_j) = \int_0^L G_m(\tau_i, z_j, \eta) T_n^0(\eta) d\eta + d \int_0^{\tau_i} \int_0^L G_m(\tau_i - s, z_j, \eta) T_a(s, \eta) d\eta ds,$$

where $i \geq 2, j \geq 1$,

$$G_m(\tau, z, \eta) = \frac{1}{L}\left(e^{-\beta_0 \tau} + 2\sum_{k=1}^{m} e^{-\beta_k \tau} cos\frac{\pi k}{L} z \cdot cos\frac{\pi k}{L}\eta\right),$$

$$\beta_k = d + c\left(\frac{\pi k}{L}\right)^2, \; k = 0,1,2, \; \dots .$$

The first integral is replaced by a finite sum using the formula of rectangles:

$$\int_0^L G_m(\tau_i, z_j, \eta) T_n^0(\eta) d\eta \approx h \sum_{l=1}^{N_z} G_m(\tau_i, z_j, \eta_l)\, T_n^0(\eta_l),$$

where

$$h = \frac{L-0}{N_z} = h_z, \quad \eta_l = h_z(l-1) = z_l, \quad T_n^0(\eta_l) = T_n^0(z_l) = TN_{1,\, l}.$$

Consequently,

$$\int_0^L G_m(\tau_i, z_j, \eta) T_n^0(\eta) d\eta \approx h_z \cdot \sum_{l=1}^{N_z} G_m(\tau_i, z_j, z_l)\, TN_{1,\, l}.$$

Now we replace the second integral with the final sum:

$$\int_0^{\tau_i} \int_0^L G_m(\tau_i - s, z_j, \eta) T_a(s, \eta) d\eta ds \approx h_\tau h_z \sum_{k=1}^{i-1} \sum_{l=1}^{N_z} G_m(\tau_i - \tau_k, z_j, z_l)\, T_a\,(\tau_k, z_l) =$$

$$= h_\tau h_z \sum_{k=1}^{i-1} \sum_{l=1}^{N_z} G_m(\tau_i - \tau_k, z_j, z_l)\, TA_{k,\, l}.$$

The first integral equation in the grid $\Omega(N_\tau, N_z)$ takes the following form:

$$TN_{i\ j} = h_z \sum_{l=1}^{N_z}\left[G_m(\tau_i, z_j, z_l) TN_{1,\, l} + d h_\tau \sum_{k=1}^{i-1} G_m(\tau_i - \tau_k, z_j, z_l) TA_{k,\, l}\right], \tag{16}$$

where $i = 2, \dots, N_\tau + 1, \quad j = 1, \dots, N_z + 1$.

Thus, the required matrices TA and TN must satisfy conditions (14), (15), (16). These conditions together represent a mixed problem with respect to the grid $\Omega(N_\tau, N_z)$. In other words, (14) - (16) are a discrete analogue (discretization) of the mixed problem.

Discrete problem (14) - (16) is solved by the method of successive approximations. To do this, we set the number of iterations (approximations) - the positive integer n. Next, we do the following:

1. For the TN matrix, we define a zero approximation, by assuming

$$TN_{i\ j}^{(0)} = T_n^0(z_j), \quad i = \overline{1, N_\tau + 1}, \quad j = \overline{1, N_z + 1}.$$

2. We consistently find the matrices $TA^{(1)}$, $TN^{(1)}$, ... , $TA^{(n)}$, $TN^{(n)}$ using the following formulas:

$$TA_{1\ j}^{(iter)} = T_a^0(z_j), \quad TA_{i\ 1}^{(iter)} = T_a^1(\tau_i), \quad j = \overline{1, N_z + 1}, \quad i = \overline{1, N_\tau + 1}, \tag{17}$$

$$TA_{i\ j}^{(iter)} = \begin{cases} e^{-b\tau_i} TA_{1,\, j-i+1} + b h_\tau \sum_{k=1}^{i-1} e^{b(\tau_k - \tau_i)}\, TN_{k,\, k+j-i}^{(iter-\ )}, & if\ j \geq i, \\ e^{-\frac{b}{a} z_j} TA_{i-j+1,\, 1} + b h_\tau \sum_{k=1}^{j-1} e^{b(\tau_{k+i-j} - \tau_i)}\, TN_{k+i-j,\, k}^{(iter-1)}, & if\ j < i, \end{cases} \tag{18}$$

$$i = \overline{2,\ N_\tau + 1}, \quad j = \overline{2,\ N_z + 1},$$

$$TN_{1\ j}^{(iter)} = T_n^0(z_j), \quad j = \overline{1, N_z + 1}, \tag{19}$$

$$TN_{ij}^{(iter)} = h_z \sum_{l=1}^{N_z} \left[G_m(\tau_i, z_j, z_l) TN_{1l} + dh_\tau \sum_{k=1}^{i-1} G_m(\tau_i - \tau_k, z_j, z_l) TA_{kl}^{(iter)} \right], \quad (20)$$

$i = \overline{2,\ N_\tau + 1}, \quad j = \overline{1,\ N_z + 1}, \ iter = 1,\ \dots\ ,\ n.$

3. We take $TA \approx TA^{(n)}, TN \approx TN^{(n)}$ is approximate numerical solution of the discrete problem (14) - (16).

The question of the error of an approximate numerical solution, i.e. at every point (τ_i, z_j) of the grid $\Omega(N_\tau, N_z)$ the values $TA_{ij}^{(n)}$, $TN_{ij}^{(n)}$ are slightly different from $TA(\tau_i, z_j)$, $TN(\tau_i, z_j)$ exact solution $(TN(\tau, z), TA(\tau, z))$ of the mixed problem. For this, it is necessary to estimate the difference moduli $\left|TA_{ij}^{(n)} - TA(\tau_i, z_j)\right|$, $\left|TN_{ij}^{(n)} - TN(\tau_i, z_j)\right|$ in each point (τ_i, z_j) of the grid $\Omega(N_\tau, N_z)$.

5. Features of computerization of the mathematical model

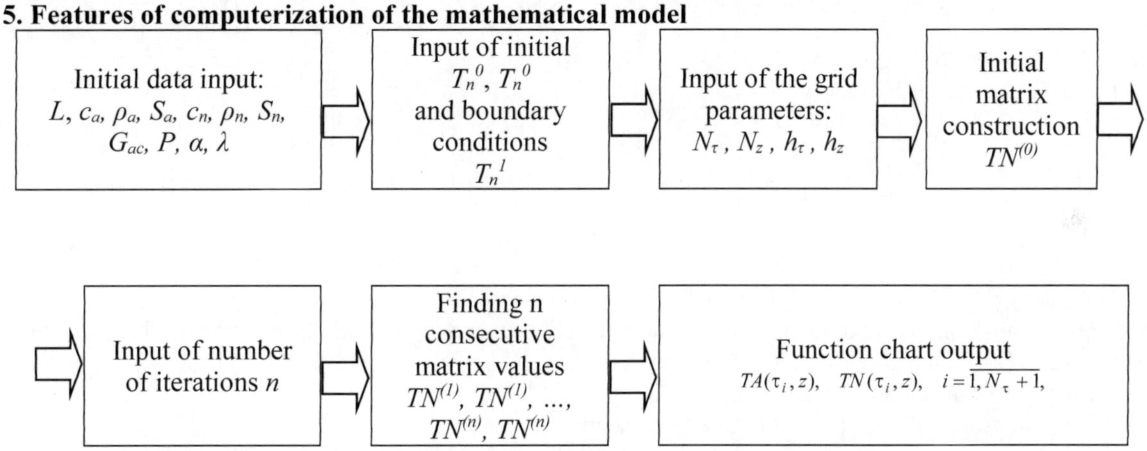

Figure 1. The block diagram of the algorithm for calculating the mathematical model in the Matlab system.

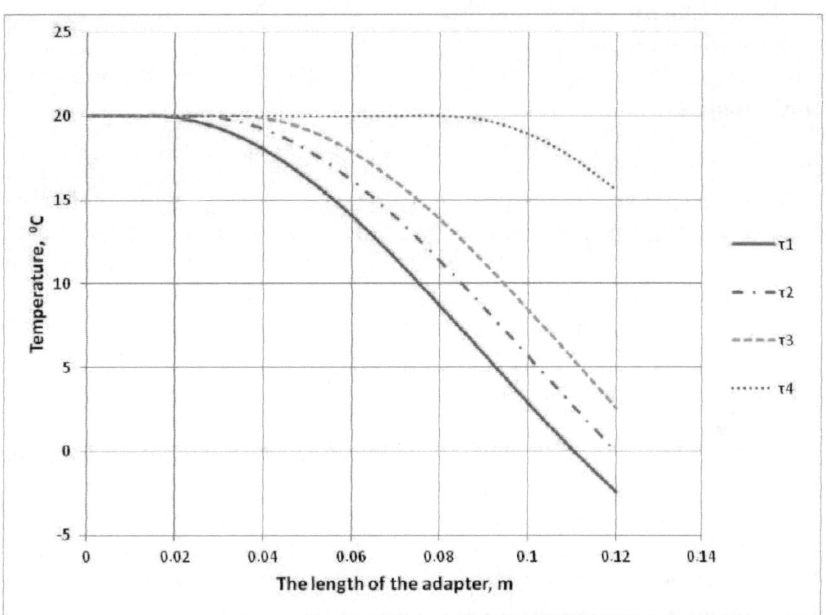

Figure 2. Plotted function TA(τ, z).

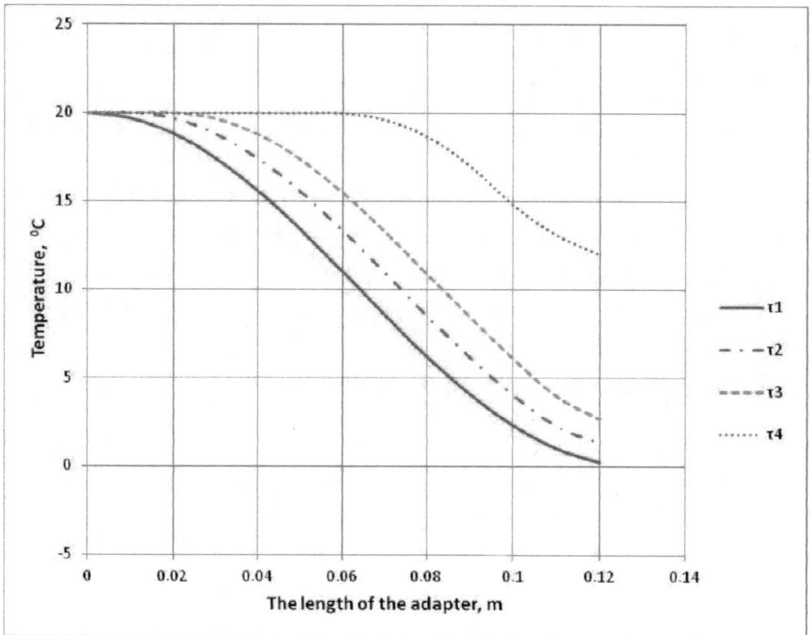

Figure 3. Plotted function TN(τ, z).

Figures 2 and 3 show air temperature and nozzle temperature change along the nozzle, respectively. Moreover, each graph shows the temperature change at the initial (τ1) and final (τ4) time points, as well as for a pair of intermediate moments with an interval of 4 s.

From the graphs in Figure 2 and 3 it can be seen:

1) When visualizing the proposed mathematical model in Matlab environment, the graphical results to a certain extent correspond to the actual heat exchange processes in the SSRHE;
2) The density of the nozzle affects the uniformity of heating of the nozzle, i.e. the greater the density of the nozzle, the more uneven it is heated;
3) Heat transfer coefficient α depends on the air flow G_{ac};
4) With an increase in the value of the heat transfer coefficient α, the uneven distribution of temperature along the nozzle increases.

6. Conclusions

As a result of the work, a computer simulation of the process of heat energy regeneration in a reversible heat exchanger was carried out. The proposed method can be used to develop algorithms for calculating the thermal characteristics of heat and mass transfer processes under conditions of regeneration and phase transitions, as well as for analyzing the criteria for optimal control of the operation of this apparatus under nominal and critical environmental conditions. Compiled and proposed algorithm allows one to:

1) Change the values of initial data, initial and boundary conditions, so it becomes possible to determine the optimal parameters and conditions for real experiments;
2) Change the grid parameters and the number of iterations, thereby changing the accuracy of the calculations;
3) Adjust and add individual steps for calculating the mathematical model.

The work was carried out on the basis of the results of research conducted with the support of the Ministry of Education and Science of the Russian Federation as part of the fulfillment of the state task to higher educational institutions for project No. 1816.

References

[1] Kostenko V A, Gafiyatullina N M, Semchuk A A and Kukolev M I 2016 Geothermal heat pump

in the passive house concept *Magazine of Civil Engineering* **68**(8) pp 18–25
[2] Egorov M Y 2018 Methods of Heat-Exchange Intensification in NPP Equipment *Atomic Energy* **124**(6) pp 403–7
[3] Ivanov N, Ris V, Tschur N and Zasimova M 2016 Numerical simulation of conjugate heat transfer in a tube bank of a subsea cooler based on buoyancy effects *J. of Physics: Conference Series* **745**(3) p 032058
[4] Mityakov V, Gusakov A, Seroshtanov V and Grekov M 2018 Investigation of flow and heat transfer at the circular fins *MATEC Web of Conferences* **245** p 06001
[5] Antonova O V, Boldyrev Y Y, Borovkov A I and Voinov I B 2017 On the Development of a Design Procedure of the Hydrodynamic Characteristics of Hydrogenerator Thrust Bearings Taking into Account the Heat Exchange *J. of Machinery Manufacture and Reliability* **46**(6) p 572–8
[6] Sinitsyn A A 2013 *Life Science Journal* **10**
[7] Sinitsyn A A and Belyanskiy D A 2014 *Measurement Techniques* **56**
[8] Sinitsyn A A, Belyanskiy D A and Sukhanov I A 2013 *Life Science Journal* **10**
[9] Vasil'ev V A 2010 Methods for calculating thermal processes in a stationary switching regenerative heat exchanger (SPb)
[10] Danilishin A M, Kozhukhov Y V, Neverov V V, Malev K G and Mironov Y R 2017 The task of validation of gas-dynamic characteristics of a multistage centrifugal compressor for a natural gas booster compressor station *AIP Conference Proceedings* **1876** p 020046
[11] Aksenov A A, Danilishin A M, Kozhukhov Y V and Simonov A M 2018 Numerical simulation of gas-dynamic characteristics of the semi-open 3D impellers of the two-element centrifugal compressors stages *AIP Conference Proceedings* **2007** p 030025
[12] Galerkin Y, Soldatova K and Drozdov A 2018 The application of mathematical models for industrial centrifugal compressor optimal design *ACM International Conference Proceeding Series* pp 187–91
[13] Lebedev A, Gileva L, Danilishin A and Sokolov M 2018 Surge protection system development in centrifugal compressor with an indicative method using numerical simulation of unsteady processes and analysis of pressure fluctuation signals *MATEC Web of Conferences* **245** p 09010
[14] Aksenov A, Kozhukhov Y, Sokolov M and Simonov A 2018 Analysis and modernization of real gas thermodynamic calculation for turbocompressors and detander units *MATEC Web of Conferences* **245** p 09005
[15] Rekstin A F, Semenovskiy V B, Galerkin Y B and Sokolov A A 2018 The analysis of design and measured gas-dynamic characteristics of the centrifugal compressor within turboexpander aggregate *AIP Conference Proceedings* **2007** p 030028
[16] Sinitsyn A A and Nikiforov O Y 2013 *World Applied Sciences J.* **27**
[17] Klein H and Eigenberger G 2001 *Int. J. of heat and mass transfer* **44**

IOP Conf. Series: Earth and Environmental Science **337** (2019) 012070 doi:10.1088/1755-1315/337/1/012070

Improving methods for identifying electric motor parameters in case of stator winding damage

A N Alyunov[1*], O S Vyatkina[1], V N Ivanov[2], Yu G Karpov[2], V F Melehin[2], A M Andreev[2] and L R Mukhametova[3]

[1]Vologda State University, 160000 Lenina st., 15, Russia
[2]Peter the Great St. Petersburg Polytechnic University, St. Petersburg, Russian Federation
[3]Kazan State Power Engineering University, Kazan, Russian Federation

* Corresponding author: alyunov@mail.ru

Abstract. A method is proposed for estimating the parameters of electric motors. It has new properties that increase the sensitivity and efficiency in terms of detecting internal short circuits in the windings of controlled electric motors in non-stationary conditions.

1. Introduction
Asynchronous electric motors (AD) are the most widespread receivers of electrical energy [1-2]. To ensure reliable and failure-free AD operation as a part of technological processes, it is necessary to implement completely new methods for their protection with the possibility of preventive action, carrying out continuous monitoring of motor parameters. In contrast to the systems of classical protection, which are triggered by faults, the system being developed makes it possible to avoid severe emergencies by predicting and detecting damages at an early stage [3-7]

The peculiarity of the classical methods of protection is the delayed reaction to the occurrence of damage due to the finite time of measurement and evaluation of parameters. Therefore, localization of damage due to lack of time is of an emergency nature, which inevitably leads to undesirable disturbances for the power system. For the same reason, a detailed diagnosis of damage in this case, as a rule, is impossible.

2. Methods
The implementation of the preventive action protection system, which ensures the early detection of possible damage to the motor and the prediction of their evolution, is based on continuous monitoring with extrapolation and parameter estimation. This approach allows one to win some time for detailed diagnostics and development of response measures that impede the development of emergency situations and, consequently, sudden disturbances of the power system.

The method is based on the use of mathematical model of the device being protected, the input of which receives the same disturbances as the object itself [8]. The reaction of the protected object is compared with the reaction of the mathematical model. In accordance with the selected criterion for the proximity of the model and the object, the model parameters are set. The model parameters obtained in this way and available for measurement are used to assess the state of the protected object (Figure 1).

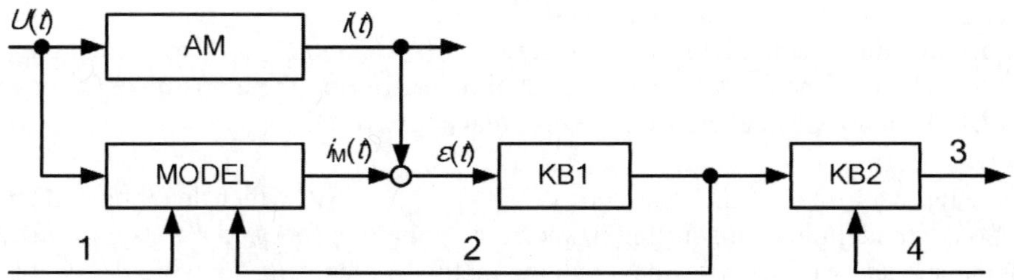

Figure 1. The scheme for parameters identification.

The following notations are used in Figure1:

AM is the object being protected (asynchronous electric motor);

MODEL is mathematical model of the object being protected;

KB1 is block of the parameters correction;

KB2 – is block of the parameters control;

1 is setting of initial parameters;

2 is correction;

3 is alert signal (turning off);

4 is setting acceptable values.

The input action is the supply voltage signal $U(t)$, which is supplied to the AM object and at the same time to its model: a microprocessor device or a computer program. Output signals (currents of object $i(t)$ and model $i_M(t)$) are compared with each other and the difference in their values ε(t) goes to the parameter correction unit (KB1). KB1 determines the parameters of the model L_M and R_M according to the observed input and output signals of the object. As the parameters of mathematical model approach the parameters of the object, $i_M(t)$ tends to $i(t)$, so $\varepsilon(t)$ tends to zero. The task of KB1 is to set the parameters of the model in such a way that the signals $i(t)$ and $i_M(t)$ differ in the least possible way according to some criterion. Only when this condition is fulfilled we can we speak about the compliance of mathematical model with the object under study.

AD equivalent circuit values corresponding to the nominal mode of operation are taken as the initial tuning parameters of the model. Then, in the working mode, the current parameters of the model corresponding to the object are calculated and compared with the allowable values. In case of output of the obtained parameters beyond the permissible area, which is a prerequisite for the emergency mode, the unit for controlling the parameters generates a warning signal to the maintenance personnel or a signal to turn off the protected object.

The definition of the initial tuning parameters of the model of an asynchronous motor is carried out on the basis of a standard Γ-shaped equivalent circuit [9], which can be converted into an equivalent circuit using successive transformations (Figure2).

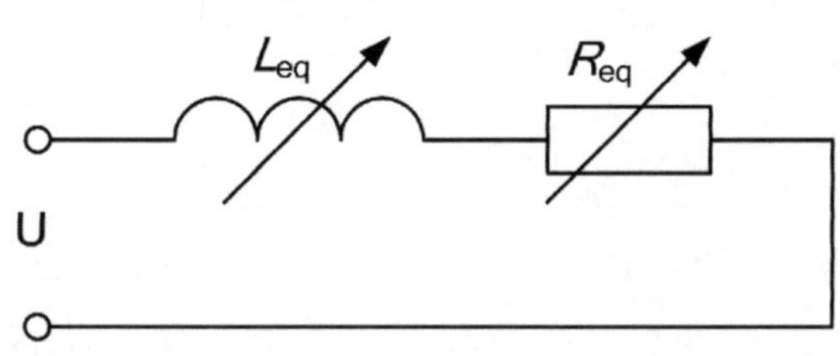

Figure 2. Equivalent scheme for the asynchronous electric motor.

So, R_{eq} and L_{eq} can serve as initial parameters of the model.

A mathematical model of an asynchronous motor can be described by the equation:

$$L_M(t)\frac{di(t)}{dt} + R_M(t)i(t) = u_M, \tag{1}$$

where $L_M(t)$, $R_M(t)$ are adjustable model settings;

u_M is the output voltage of the model;

i(t) is the instantaneous value of the motor current.

It is necessary to adjust the model so that the parameters L_M, and R_M tend to the engine parameters. The degree of adequacy of the model to the protected object during the adjustment is set according to the functional *E*, which directly depends on the instantaneous error

$$\varepsilon = i(t) - i_M(t). \tag{2}$$

As L_M, and R_M approach the engine parameters, $i_M(t) \to i(t)$, $\varepsilon \to 0$, which leads to a decrease in the functional E. So the model must be adjusted in such a way that the functionality is minimized.

To adjust the model, we will use the gradient method, according to which each of the model parameters is adjusted in proportion to the rate of change of the functional according to the same parameter:

$$\frac{dX_j}{dt} = -\rho_j \cdot \mathrm{grad}_x E, \tag{3}$$

where X_j is adjustable model parameters (j=0…n);

E is identification criterion;

ρ_j is positive amplification coefficient.

If we select the functional $E=\varepsilon^2/2$ as the criterion being minimized, then the equations of the tuning contours taking into account equations (1) and (3) will take the form [10]:

$$\begin{cases} \frac{dR_M}{dt} = -\rho_R \cdot \varepsilon \cdot i(t); \\ \frac{dL_M}{dt} = -\rho_L \cdot \varepsilon \cdot \frac{di(t)}{dt}; \\ \varepsilon = u(t) - L_M(t)\frac{di(t)}{dt} - R_M(t) \cdot i(t), \end{cases} \tag{4}$$

where ρ_R, ρ_L are amplification coefficient.

To implement the described method and to obtain the instantaneous values of currents and voltages necessary to solve the resulting system of equations, a pilot plant was created, with the help of which both the normal operating conditions of the engine and various circuits are fixed.

The scheme contains the object under study – the AIR100S4U3 asynchronous motor of the brand (P_n = 3 kW). A distinctive feature of this electric motor is the presence of taps from phase coils with different numbers of turns, which allows one to simulate interturn circuits in different operating modes. The ends of the motor windings are brought out and connected as a star.

3. Results

Using this installation, instantaneous voltages and currents were obtained in the mode of starting the engine, as well as when different numbers of turns in the stator winding were closed. For instantaneous values, taking into account the initial conditions L_{eq} and R_{eq}, the system of equations (4) is solved by the Runge – Kutta method using the MATLAB software package. The results of the calculation of equivalent parameters in different modes are presented in Figure 3.

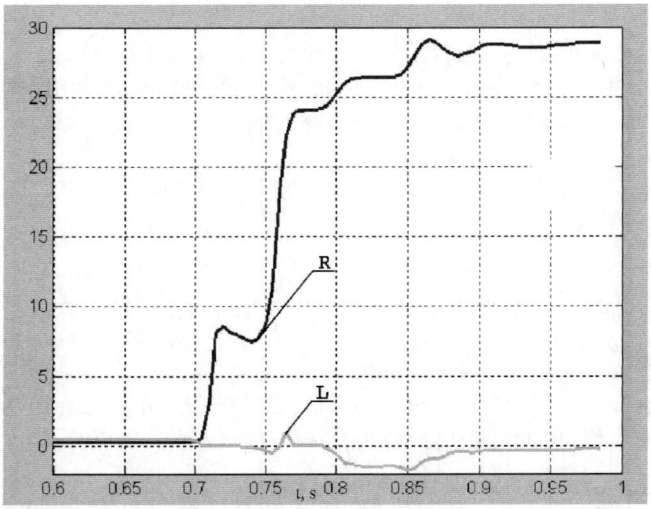

a)

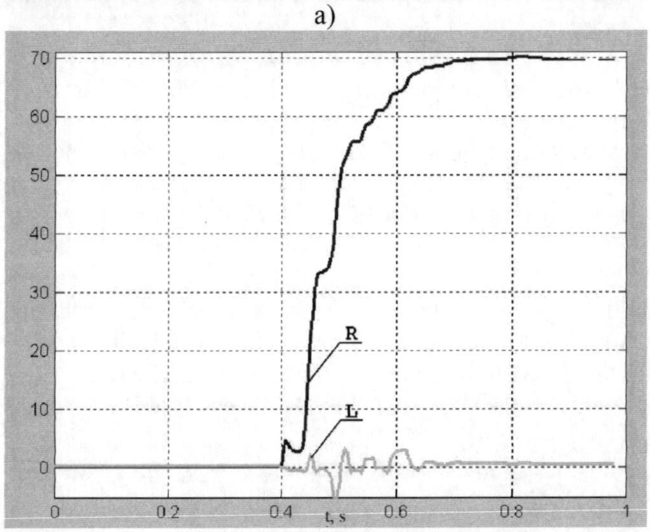

b)

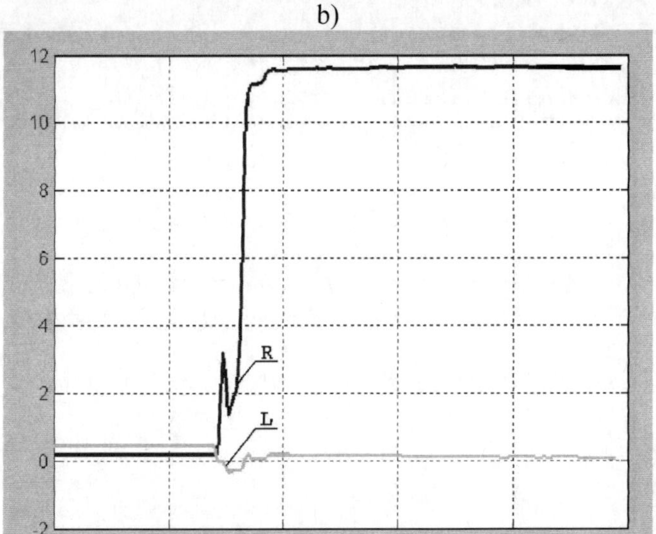

c)

Figure 3. R(t) and L(t) Charts for different modes: a) when the engine is started; b) when closing the first tap for the beginning; c) when closing the third tap for the beginning.

IOP Conf. Series: Earth and Environmental Science **337** (2019) 012070 doi:10.1088/1755-1315/337/1/012070

4. Conclusions

The desired quality of the transition process has a great influence on the choice of model parameters. From the analysis of differential equations (4), it follows that it is primarily determined by the correct choice of ρ_R and ρ_L. To increase the protection performance, i.e. to reduce the time of transition process, you need to increase the ρ_R and ρ_L.

Thus, on the basis of the proposed method, technical means and systems for continuous monitoring of AD parameters can be developed without shutting it down.

References

[1] Boguslawsky I, Korovkin N and Hayakawa M 2016 *Large A.C. Machines: Theory and Investigation Methods of Currents and Losses in Stator and Rotor Meshes Including Operation with Nonlinear Loads* (Tokyo: Springer) pp 1–550

[2] Smolovik S V, Liamov A S and Chudny V S 2017 *Proceedings of the 2017 IEEE Russia Section Young Researchers in Electrical and Electronic Engineering Conference, ElConRus 2017 7910875* pp 1581–4

[3] Silin N, Korovkin N and Minevich T 2017 *2017 International Conference on Industrial Engineering, Applications and Manufacturing, ICIEAM 2017 - Proceedings 8076412*

[4] Frolov V Y, Ivanov D V and Belsky R A 2017 *Proceedings of the 2017 IEEE Russia Section Young Researchers in Electrical and Electronic Engineering Conference, ElConRus 2017 7910861* pp 1520–3

[5] Vanin V K, Ambrosovskaya T D, Popov M G and Popov S O 2016 *Power Technology and Engineering* **50**(1) pp 87–92

[6] Stogov A Y and Belyaev A N 2016 *Proceedings of the 2016 IEEE North West Russia Section Young Researchers in Electrical and Electronic Engineering Conference, EIConRusNW 2016 7448278* pp 694–8

[7] Baboraik A M, Usachev A E, Ildarkhanov R G and Kubarev A Y 2018 NEW METHOD OF DETERMINING SIZE AND PROVISION OF DEFECTS IN THE INSULATION OF WINDINGS OF STATOR OF TURBOGENERATORS AT MEASUREMENT OF PARTIAL DISCHARGES IN THEM «Izvestiya Vyss. uchebnykh Zaved. Probl. Energ. 20 41–53

[8] Kaziev V M 2013 Introduction to the analysis, synthesis and modeling of systems (Moscow: Internet University of Information Technologies)

[9] Kopylov I P 2017 *Electric Machines* (Moscow: Yurait)

[10] Bahmann Yu, Vanin V K, Ginkover A M, Pavlov G M and Pechkovsky A V 1987 *Electricity* **4**

[11] Silin N, Korovkin N and Minevich T 2017 *2017 International Conference on Industrial Engineering, Applications and Manufacturing, ICIEAM 2017 - Proceedings 8076412*

[12] Frolov V Y, Ivanov D V and Belsky R A 2017 *Proceedings of the 2017 IEEE Russia Section Young Researchers in Electrical and Electronic Engineering Conference, ElConRus 2017 7910861* pp 1520–3

[13] Vanin V K, Ambrosovskaya T D, Popov M G and Popov S O 2016 *Power Technology and Engineering* **50**(1) pp 87–92

[14] Stogov A Y and Belyaev A N 2016 *Proceedings of the 2016 IEEE North West Russia Section Young Researchers in Electrical and Electronic Engineering Conference, EIConRusNW 2016 7448278* pp 694–8

[15] Sakhapov R L, Zamilev M A 2018 *Technics and Technology of Transport* **3**(8) p 11 URL: http://transportkgasu.ru/files/N8-11ITS318.pdf

IOP Conf. Series: Earth and Environmental Science **337** (2019) 012071 doi:10.1088/1755-1315/337/1/012071

Innovative technology for dismantling the windings of electric motors using ultrasonic radiation

A E Nemirovskiy[1], A I Kashin[2*], V F Kosmach[3], Yu F Titovec[3], I N Toptygin[3] and D A Zaripova[4]

[1]Vologda State University, Lenina Street 15, Vologda, 160000 Russia
[2]Murmansk State Technical University, Sportivnaya Street, 13, Murmansk, 183010 Russia
[3]Peter the Great St. Petersburg Polytechnic University, St. Petersburg, Russian Federation
[4]Kazan State Power Engineering University, Kazan, Russian Federation

* Corresponding author: kashin-tema@mail.ru

Abstract. The use of ultrasonic radiation showed an advantage over the existing methods of dismantling the windings of electric motors (EM). The method is based on the cavitation impact of ultrasonic vibrations and the capillary effect, which helps the caustic soda solution (NaOH) to quickly penetrate into the thickness of the lacquer impregnation and contribute to the destruction of the insulation from the inside. Analysis of dismantling of EM windings on the basis of ultrasound was carried out by the method of active experiment planning. As a result, a mathematical model of the process of removing the "burnt" EM winding was obtained, and the optimal parameters of power and duration of ultrasound, the temperature and the concentration of the NaOH solution were determined. Asynchronous electric motors (AEM) are the most widespread receivers of electrical energy. To ensure reliable and trouble-free operation of the AEM as a part of technological processes, it is necessary to implement completely new methods for their protection with the possibility of preventive action, carrying out continuous monitoring of motor parameters. In contrast to the systems of classical protection, which are triggered by faults, the system being developed makes it possible to avoid severe emergencies by predicting and detecting damages at an early stage.

1. Introduction

Currently, electric motors are the largest consumers of electrical energy [1-5]. According to many sources, they consume over 80% of all electricity produced in the country. In industry and agricultural production, EMs are operated under very harsh conditions. High humidity, dustiness, aggressive environment, low quality of the supply voltage annually lead to failure of up to 20% EM [6]. EM failure results in great damage to enterprises of the agro-industrial complex and other industries. The specifics of agricultural enterprises are associated with the process of life support of biological objects, smelting of metal in metallurgical plants [7-10]. Consequently, the failure of technological operations due to the failure of EM leads to a decrease in productivity. Failure of EM leads to process downtime and product damage. In addition to the main damage to the enterprise, there may be additional losses - reducing the levels of electrical and fire safety associated with possible short circuits in the windings of the stator or rotor of the damaged EM. In the technology of their repair, the dismantling of the burned-down stator

IOP Conf. Series: Earth and Environmental Science **337** (2019) 012071 doi:10.1088/1755-1315/337/1/012071

winding is important. Repaired EM must meet the requirements of reliability and efficiency in further work.

2. Methods

The existing methods of winding dismantling have a number of disadvantages: increased energy consumption, deterioration of steel electrical characteristics, mechanical damage to the EM stator package, negative impact on the environment, etc. [11]. The basis of the developed method is cavitation impact of ultrasound. The use of ultrasound increases the capillary effect, as a result of which the working solution of NaOH penetrates faster into the thickness of winding. As a result, the process of dissolving the impregnating composition and releasing the burnt winding is accelerated. In contrast to the known methods, the proposed one is the most efficient since it does not have a destructive effect on the useful elements of EM construction, is environmentally friendly and less energy consuming.

To find the optimal parameters of factors which act on the EM winding in order to dismantle it, a multifactorial active experiment was conducted. The temperature of NaOH alkaline solution, the solution concentration, the power of ultrasonic radiation and the duration of exposure were chosen as influencing factors. A number of exploratory experiments were conducted in advance to select the intervals of factors variation [12]. For this, it was necessary to develop the hardware complex and the design of the ultrasonic bath [11].

Figure 1 shows the block diagram of the hardware complex [13].

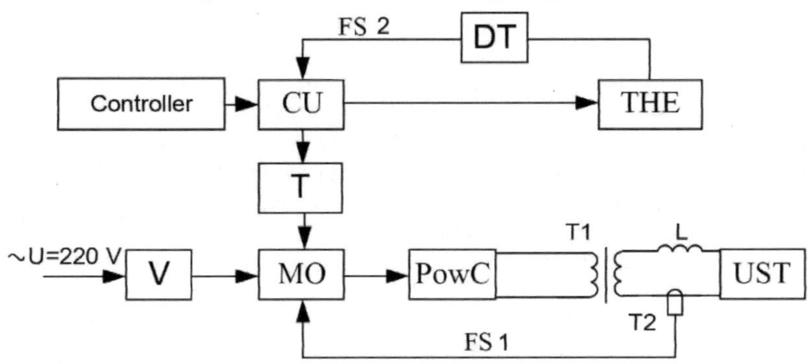

Figure 1. Block diagram of the hardware complex.

An alternating voltage of 220 V is supplied to the input of the hardware complex. The power converter PowC is designed to convert an alternating voltage of one frequency into an alternating voltage of another frequency. Rectifier V converts AC voltage to DC voltage. The master oscillator MO converts the DC voltage into a rectangular AC of a given frequency. The main function of the CU control unit is to turn on or turn off the generator through predetermined temporal characteristics using timer T. CU, using a thyristor voltage regulator, changes the voltage level applied to tubular heating element (THE), thereby maintaining the desired bath temperature. The THE heating temperature is adjusted through feedback (FS 2) using DT. The feedback signal (FS) is removed from DT and fed to CU. It is the signal level of FS which determines the THE temperature. The transformer T1 provides galvanic isolation of the power converter (PowC) from the ultrasonic transducer (UST). The inductance L is chosen in such a way that, coupled with MO and UST, it creates a voltage resonance and bring the UST to a working state. The current transformer T2 is necessary for galvanic isolation from the mains voltage and for setting a specific mode of the piezoelectric ultrasound transducer: manual, semi-automatic and automatic. Feedback FS 1 adjusts the frequency of MO with temperature changes in the characteristics of ultrasound. The controller module is included in CU and is designed to control the process: a change of THE temperature and UMO power.

The exploratory studies using the hardware complex and the ultrasonic bath were carried out using the specially made EM winding layouts, known as statorets (Figure 2). The statorets were made according to GOST-IEK 60851-3–2011 and were models of EM slot and interphase insulation.

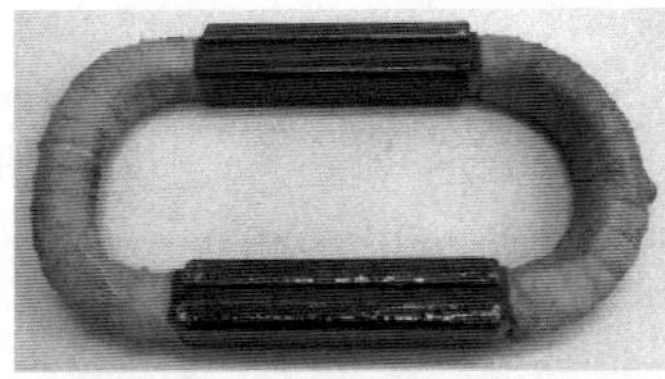

Figure 2. The stator model.

The model is a section of a square shaped section pipe, inside of which there is a single-layer winding made of PETV-2 copper wire with of 1 mm diameter with a number of turns equal to 120. The slot insulation is made of a synthetic cardboard. To give strength to the frontal parts of the winding, a band of cotton electrical tape was applied. The stator model was impregnated with varnish ML-92.

3. Results

The results of exploratory research are shown in Figure 3. As it can be seen, for 2% concentration of NaOH and ultrasound exposure (curve 3"), we get almost the same result as for 10% concentration of the solution without ultrasound (curve 1). So, ultrasound enhances the effect of lacquer impregnation destruction for about 5 times, which is quite significant.

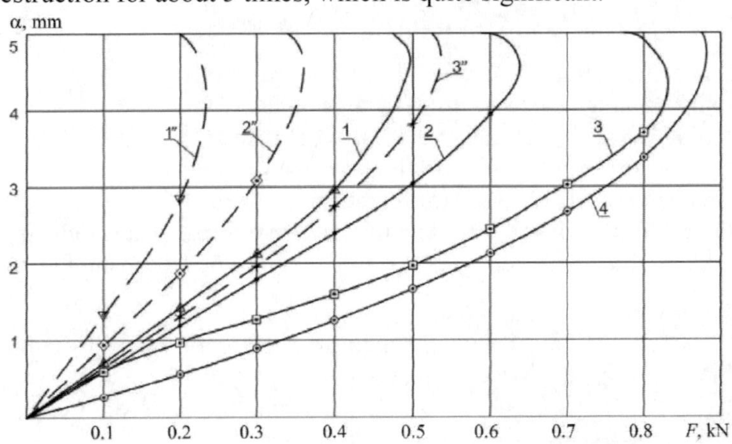

Figure 3. Relationship between the bending of EM winding frontal parts and the force of exposure and NaOH concentration:
1 - for k=10%;
2 - for k=5%;
3 - for k=2%;
4 - or k=0%;
1"- for k=10% and ultrasound treatment;
2"- for k=5% and ultrasound treatment;
3"- for k=2% and ultrasound treatment.

These experiments allowed us to determine the factor variation levels (Table 1).

Table 1. The factor variation levels.

Independent parameters	Designation	Variation levels			Variation level
		−1	0	+1	
Treatment duration, min	X1	30	100	170	70
Temperature, °C	X2	20	35	50	15
Ultrasonic radiation power, W	X3	40	90	140	50
Solution concentration, %	X4	2	4.5	7	2.5

For the correct formulation of the experiment and for obtaining reliable results, we used a second-order rotatable plan with star points 3^4. The advantage of such plans is a small number of experiments. So, with k = 4, the total number of experiments is 25. When using the rotatable plan 3^4, it is necessary to add the matrix of the plan 2^4 to the matrix of the second order plan. 16 points are used as the "core" of the matrix, then 8 "star" points and one zero point are added. The value of the star shoulder was 1.414.

The mechanical characteristics were chosen as the response function: residual strength of the winding and adhesion of the windings between each other [13]. To remove the mechanical characteristics we developed a stand (Figure 6) [14]. It allows one to explore the impact of destructive factors on the impregnation of the EM winding with different levels of exposure, presented in table 1. The levels of influence of factors vary in each experiment and in different ways destroy the solidity of the lacquer impregnation of the EM winding.

The stand (Figure 4) consists of a bed 1, on which fork 2 of arm mounting 5 is mounted, a platform for interchangeable nozzles 16, and supporting table 4 for a strain gauge 9 of an electronic dynamometer. Supporting table 4 has a height adjustment and is fixed with a bolt. The platform for replaceable nozzles consists of welded corners on the sides of the profile of the frame and the stop. To install the nozzle, simply insert it into the grooves and push it in until it stops. The arm 5 is attached to the fork 2. The shoulders of arm 5 are perpendicular to the axis of its rotation.

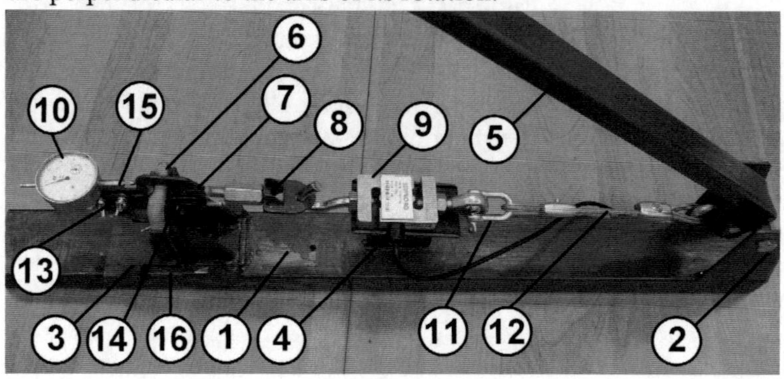

Figure 4. The stand construction: 1 - bed; 2 - fork; 3 - replaceable nozzle; 4 - table for strain gauge; 5 - arm; 6 - stator model; 7 - indenter; 8 - tightening device; 9 - strain gauge; 10 - dial indicator; 11 - ligging loop; 12 - wire rope; 13 -supporting arm; 14 - fixing clamp; 15 - dial indicator stock 16 - area for replaceable nozzles.

The stator models are fixed on interchangeable nozzles (Figure 5) with the help of metal plates tightened with a wing nut. The arrows indicate the forces applied to the stator model when determining the response functions.

Figure 5. Replaceable nozzle of the stand:
1 - to determine the residual strength of the frontal part;
2 - to determine the residual adhesion of the windings between each other;
3 - to determine the effort of pulling out the winding.

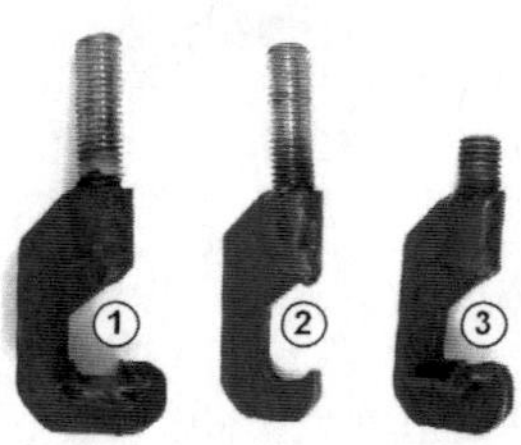

Figure 6. Indenters:
1 - to determine the residual strength of the frontal part;
2 - to determine the residual adherence of the windings between each other;
3 - to determine the effort of pulling out the winding.

For each nozzle, an indenter is provided, without which it is impossible to test the mechanical characteristics of the EM winding (Figure 6).

The stand (see Figure 4) makes it possible to test all the response functions for constructing mathematical models of the process of dismantling the "burnt" EM windings.

As a result of the experiment, mathematical models (1) and (2) of destruction of a burnt EM winding were obtained with the aim of its further "easy" dismantling. The significance of the regression coefficients was tested according to the Student's criterion, the adequacy of the model was determined by the Fisher criterion, the reproducibility of the experiments was measured by the Cochren criterion.

Regression equation for the function of residual strength:

$$\begin{gathered} Y(X_1,X_2,X_3,X_4) = 0.381 - 0.097X_1 - 0.073X_2 - 0.051X_3 - 0.018X_4 - \\ -0.013X_1X_2 + 0.037X_2X_3 - 0.046X_2X_4 - 0.03X_3X_4 - \\ -0.049X_1X_2X_4 + 0.023X_1X_3X_4 - 0.116X_2X_3X_4 + \\ +0.134X_1X_2X_3X_4 + 0.04X_1^2 - 0.023X_2^2 + 0.065X_4^2. \end{gathered} \quad (1)$$

Regression equation for the function of residual adherence:

$$\begin{gathered} Y(X_1,X_2,X_3,X_4) = 0.529 - 0.077X_1 - 0.057X_2 - 0.059X_3 - 0.03X_4 - \\ -0.029X_1X_2 + 0.035X_2X_3 - 0.018X_1X_3 - 0.04X_2X_4 - 0.023X_3X_4 - \\ -0.028X_1X_2X_4 - 0.097X_2X_3X_4 + 0.112X_1X_2X_3X_4 + \\ +0.055X_1^2 - 0.015X_2^2 - 0.026X_3^2 + 0.083X_4^2. \end{gathered} \quad (2)$$

From equations (1) and (2) it can be seen that duration of factors X_2, X_3, X_4 exposure has the greatest contribution to the destruction of the monolithic varnish impregnation of the EM winding. The value of the factor X_1 can be reduced several times compared with the value of this factor in the existing methods of dismantling and thereby reduce the time for dismantling the EM winding. The factors of NaOH solution temperature X_2 and ultrasound power X_3 are also very significant and make almost the same contribution to softening of EM winding. The NaOH concentration X_4, i.e. concentration of the chemically active environment, has the least impact on the process of dismantling the EM winding, which is very important because it provides an opportunity to significantly reduce environmental damage and preserve the health of staff. In our case, it was possible to reduce the concentration of NaOH by almost 2 times.

To obtain the optimal parameters of the influencing factors on the process of destruction of the winding of a burnt EM, we find the partial derivatives of each factor and solve the resulting system of equations for each response function:

$$\frac{\partial Y(X_1,X_2,X_3,X_4)}{\partial X_1} = 0; \quad \frac{\partial Y(X_1,X_2,X_3,X_4)}{\partial X_3} = 0;$$
$$\frac{\partial Y(X_1,X_2,X_3,X_4)}{\partial X_2} = 0; \quad \frac{\partial Y(X_1,X_2,X_3,X_4)}{\partial X_4} = 0. \quad (3)$$

Having solved the system of equations (3), we obtain in relative units: for residual strength: X_1=1.024; X_2=1.03; X_3=0.525; X_4=0.738, for adhesion of the windings: X_1=0.83; X_2=1; X_3=0.596; X_4=0.659.

In actual values:

$$X_1 = \frac{X_1^a - X_1^0}{\Delta X_1}; \quad (4) \qquad X_2 = \frac{X_2^a - X_2^0}{\Delta X_2}; \quad (6)$$
$$X_3 = \frac{X_3^a - X_3^0}{\Delta X_3}; \quad (5) \qquad X_4 = \frac{X_4^a - X_4^0}{\Delta X_4}; \quad (7)$$

Where X_1^a, X_2^a, X_3^a, X_4^a are actual values of the factors; X_1^0, X_2^0, X_3^0, X_4^0 are the zero-level variation of factors; ΔX_1, ΔX_2, ΔX_3, ΔX_4 are the intervals of factors variation.

Taking into account expressions (4) - (7), the optimal values X_1, X_2, X_3, X_4 are obtained in actual units for the process of the most efficient dismantling of EM windings using ultrasonic treatment. Table 2 shows the optimal values of factors for which the residual strength and adhesion of the EM winding are minimal. At the same time, the optimal solution temperature for NaOH is 30% lower than the temperature which is characteristic for the leaching method [6]. This makes it possible not only to reduce energy consumption with the developed method of disassembling the EM windings, but also to maintain the insulation between the plates, as well as electrophysical characteristics of the stator steel by reducing the penetration of NaOH molecules into the space between the magnetic core plates.

Table 2. The values of optimal parameters.

Factors	Minimal values of residual strength	Minimal values of residual adhesion
X_1^a, min	171	159

X_2^a, °C	50.45	50
X_3^a, W	116.25	120
X_4^a, %	6.3	6.1

From equations (1) and (2), partial regression equations were obtained, which were used to construct the surfaces of response functions for residual strength and residual adhesion of the EM windings using the Mathcad mathematical package, shown in Figure 7. At $X_1 = 171$ min. residual strength $X_4 = 6.3\%$, other parameters: $X_2 = 51$ °C and $X_3 = 116$ W. At $X_1 = 159$ min. Adhesion $X_4 = 6.1\%$, other parameters: $X_2 = 50$ °C and $X_3 = 120$ W.

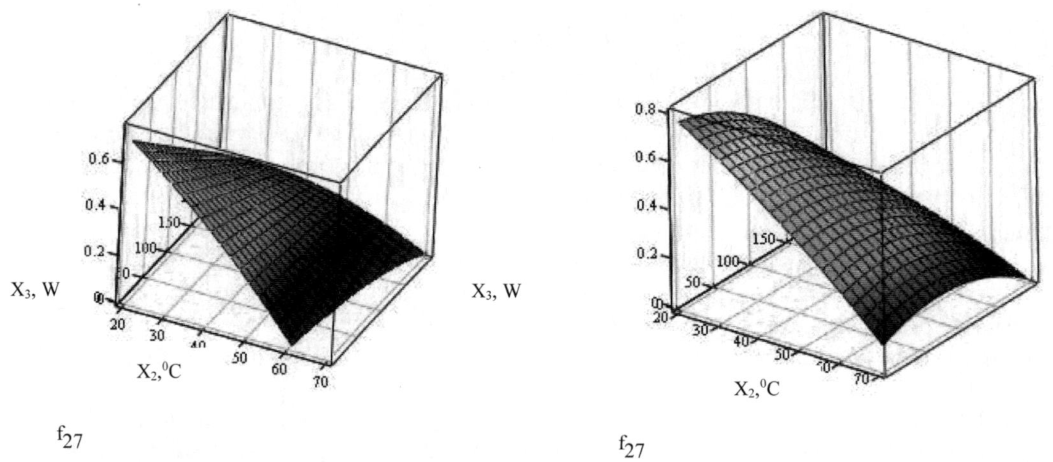

a)

$$y=0.5535-7.7\cdot10^{-3}\cdot X_2-3.9\cdot10^{-3}\cdot X_3+$$
$$+6.3\cdot10^{-5}\cdot X_2\cdot X_3-8.3\cdot10^{-5}\cdot X_2^2$$

b)

$$y=0.935-6.7\cdot10^{-3}\cdot X_2-1.5\cdot10^{-3}\cdot X_3+$$
$$+3.8\cdot10^{-5}\cdot X_2\cdot X_3-5.4\cdot10^{-5}\cdot X_2^2-10^{-5}\cdot X_3^2$$

Figure 7. Surfaces of response functions:
a) Residual strength at X1 = 171 minutes; X4 = 6.3%;
b) Residual adhesion at X1 = 159 minutes; X4 = 6.1%.

4. Conclusion

Thus, the optimal parameters of the process of dismantling the EM windings using ultrasound are obtained. These parameters should be incorporated into the structural and electrical parts of the hardware complex for dismantling EM serial windings for the purpose of production tests of the developed method of their repair in electrical repair shops.

References

[1] Mankonen A, Kaikko J, Vakkilainen E, Sergeev V 2018 *MATEC Web of Conferences* **245** 07007
[2] Barsi D, Costa C, Satta F, Zunino P, Sergeev V 2018 *MATEC Web of Conferences* **245** 07008
[3] Barsi D, Perrone A, Qu Y, Ratto L, Ricci G, Sergeev V, Zunino P 2018 *Journal of Thermal Science* **27**(3) pp 259-69
[4] Gotovtsev Z B, Komova P M, Vasilov A V, Namsaraev R G 2018 *Renewable and Sustainable Energy Reviews* **81** pp 625-34
[5] Akhmetova I G, Chichirova N D 2017 *Journal of Physics: Conference Series* **891**(1) 012169
[6] Khomutov S O, Kobozev E V and Semichevsky P I 2009 Improving the efficiency of electric motor insulation restoration based on a comprehensive assessment of the influencing factors *Polzunovsky Vestnik* **1-2**.
[7] Sultanguzin I A, Fedyukhin A V, Kurzanov S Y, Tumanovsky V A, Titova D P 2015 *Thermal*

Engineering (English translation of Teploenergetika) **62**(5) A008 pp 359-64

[8] Politaeva N, Smyatskaya Y, Slugin V, Toumi A, Bouabdelli M 2018 *IOP Conference Series: Earth and Environmental Science* **115**(1) 012001

[9] Ulyanova V V, Sobgayda N A 2016 *Ecology and Industry of Russia* **20**(7) pp 4-9

[10] Sobgaida N A, Olshanskaya L N, Solodkova A B 2011 *Chemical and Petroleum Engineering,* **47**(3) pp 198-203

[11] Kashin A I, Aksenov D O and Kichigina G A 2016 Analysis of methods for dismantling the windings of an electric motor *XIV All-Russian Scientific Conference University science to the region* pp 7-9.

[12] Kashin A I and Nemirovsky A E 2016 Development of an ultrasonic bath for exploitation of an innovative method for dismantling the windings of electric motors *X Annual Scientific Session of graduate students and young scientists*.

[13] Kashin A I, Nemirovsky A E, Sergievskaya I Yu and Nikiforov O M 2017 The electrical part of the hardware complex for the repair of electric motors *Automation and energy saving of machine-building and metallurgical production, technology and reliability of machines, devices and equipment*.

[14] Nemirovsky A E, Petiforov V O and Sergievskaya I Yu 2017 Installation for studying the output mechanical characteristics of electric motors subjected to ultrasonic processing *Automation and energy saving of machine-building and metallurgical industries, technology and reliability of machines, devices and equipment*.

Thermal calculation of the installation for the moisture evaporation from petroleum products

Alisa Surikova[1*], Elena Sytsyanko[1], V F Kosmach[2], Yu F Titovec[2] and T R Akhmetov[3]

[1]Vologda State University, Department of Heat Gas Supply and Ventilation, 160000 Lenina st., 15, Russia
[2]Peter the Great St. Petersburg Polytechnic University, St. Petersburg, Russian Federation
[3]Kazan State Power Engineering University, Kazan, Russian Federation

* Corresponding author: alisa-surikova@bk.ru

Abstract. The article presents data on the verification thermal calculation of the existing installation for moisture evaporation from petroleum products. The measures to increase its performance and reduce specific fuel consumption are proposed.

Keywords: Sludge separation unit, evaporation plant, sludge, thermal efficiency

1. Relevance of the work

Currently there are a large number of various petroleum products containing moisture, not suitable for direct use [1-10]. Their processing in distillation towers at refineries is not economically feasible due to their relatively low yield, therefore, recently there appear installations for evaporation of moisture from petroleum products called oil sludge separation units (abbreviated OSSU). Oil sludge is usually an emulsion of mineral oil and water, sometimes containing solid impurities.

In the scientific and technical literature, there are no methods for thermal calculation of OSS units, since they appeared relatively recently and are not efficient.

2. Description of the installation

The object of the study was the OSSU-3 installation, in which, with a volume of loaded sludge of ~ 3 m^3 with an initial moisture content of 30%, the evaporation process up to a 1%-moisture content lasts about a day. The heat spent on the process of heating the sludge and evaporation of moisture from it, is supplied from the boiler with a nominal heat capacity of 95 kW. The boiler automation system does not allow water to heat to a temperature above 95°, while during the evaporation mode, as the operating experience has shown, the boiler works with long interruptions. From the heat balance it follows that if the nominal heat output of the boiler were used continuously, the evaporation process would last for about 6 hours, i.e. 4 times less than in the existing installation, while the productivity of the device would have increased by the same amount. This does not happen because, under the existing conditions, the heat exchange surface of the evaporator heater is insufficient to transfer the nominal heat output of the boiler to the sludge during the evaporation process. In addition, the heat of the steam leaving the

IOP Conf. Series: Earth and Environmental Science **337** (2019) 012072 doi:10.1088/1755-1315/337/1/012072

apparatus is not used at all and is lost in the heater, although numerically (with exception of losses) it is equal to the heat supplied to the oil sludge by heating water from the boiler.

The schematic diagram of the installation is shown in Figure 1.

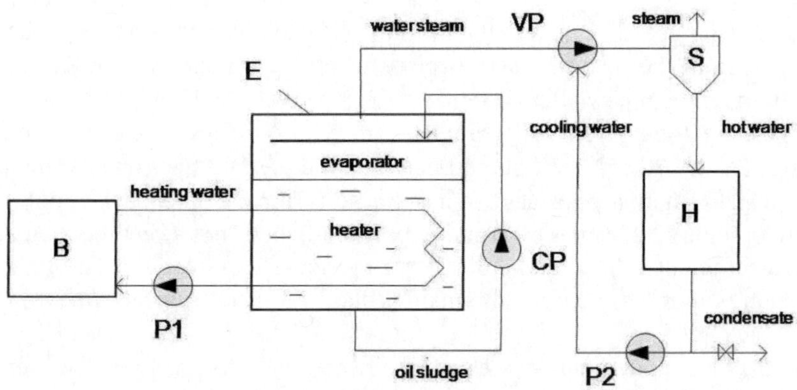

Figure 1. The schematic diagram of the OSSU installation.

First, the oil sludge is heated from a loading temperature of ~ 20°C to a temperature of 70÷80°C, at which the process of evaporation of moisture from the oil sludge begins. Evaporation takes place under vacuum created in the evaporator tank by a vacuum pump (VP). Thus, at water evaporation temperature of 75 °C, the absolute pressure inside the apparatus case should be 0.039 MPa (vacuum is about 65%). The heated sludge is taken from the bottom of the evaporator, and the circulation pump (CP) is fed to the shelves of the evaporator, located in the upper part of the apparatus, serving to increase the mass transfer surface of the sludge. With partial evaporation of water on the shelves of the evaporator, the sludge is cooled by several degrees and flows back to the lower part of the apparatus.

The water vapor produced in the apparatus is pumped out with a vacuum pump, in which it mixes with the cooling water and condenses. Heated water and non-condensable gases leave the vacuum pump and enter the separator (S) connected to the atmosphere. In the separator, gases are removed into the atmosphere, and the heated water enters the air heater (H), where it is cooled with atmospheric air, and is supplied by pump P2 to a vacuum pump. The resulting condensate is periodically removed from the system. When the process of evaporation of moisture from the sludge ends, the finished oil is poured out of the tank, and a new portion of sludge is poured into the tank. Thus, the evaporation process in the OSSU installation is periodic.

3. Scientific methodology for verification calculation

We consider the method of verification calculation of the OSSU-3 installation in the evaporation mode. The heating surface of the evaporator consists of two registers of smooth steel pipes Dy-40 mm, length l=4m, consisting of n = 9 tubes each, connected in series. The temperature of the heating water at the inlet to the apparatus is $t'_{liq1} = 95°C$, at the outlet it is $t''_{liq1} \cong 90°C$. The average temperature of oil sludge in the evaporation mode is $t_{liq2} = 75°C$.

Operating experience has shown that in the mode of evaporation of oil sludge with a moisture content of not more than 30%, the consumption of steam leaving the evaporator remains almost the same throughout the entire evaporation process, which means that at constant temperatures of oil sludge and heating water, the heat flux supplied to the oil sludge, remains constant. So, the heat transfer coefficient from the heating water to the oil sludge during evaporation also remains almost unchanged. This is possible if the thermophysical properties of oil sludge that affect heat transfer are practically independent on moisture content (if it does not exceed 30% by weight).

IOP Conf. Series: Earth and Environmental Science **337** (2019) 012072 doi:10.1088/1755-1315/337/1/012072

The surface area of the heat exchange evaporator, determined by the outer diameter of pipes, is F = 11.5 m^2. The heat transfer coefficient from the heating water to the sludge was determined by the expression, as that for a flat wall [11]:

$$k = \left(\frac{1}{\alpha_1} + \frac{\delta_c}{\lambda_c} + \frac{1}{\alpha_2}\right)^{-1}, \quad (1)$$

where α_1 is heat transfer coefficient from the heating water to the inner surface of the pipes; α_2 is heat transfer coefficient from the outer surface of pipes to the oil sludge, W/(m^2·K); λ_c is coefficient of thermal conductivity of the pipe wall; $\delta_c \cong 0{,}003$ is the pipe wall thickness, m.

For steel pipes, we took in our calculations λ_c = 51.5 W/(m·K), as for carbon steel 20 in the temperature range 20÷100°C [12]. We should note that the thermal resistance of the pipe wall δ_c/λ_c is many times less than the thermal resistance of heat transfer from the outer surface of pipe $1/\alpha_2$, therefore some inaccuracy in setting the values of δ_c and λ_c practically does not affect the accuracy of calculating the heat transfer coefficient k. In addition, the thermal resistance of heat transfer on the inner surface of a pipe $1/\alpha_1$ is many times less than the thermal resistance of heat transfer $1/\alpha_2$ on the outer surface, since $\alpha_1 >> \alpha_2$..

Therefore, in the first approximation, we can take $k \approx \alpha_2$. At the same time, the average temperatures on the inner and outer surfaces of pipes in the first approximation are equal to: $t_{c1} \approx t_{c2} \approx t_{liq1} = (t'_{liq1} + t''_{liq1})/2 = (95 + 90)/2 = 92.5$°C. Further, the values of k, t_{c1} and t_{c2} are defined more accurately, since $k < \alpha_2$, $t_{c2} < t_{c1} < t_{liq1}$.

The heat transfer coefficient α_1 from the heating water to the inner surface of pipes was determined by the well-known formula for the turbulent flow of liquid inside the pipes [13]:

$$Nu_{liq1} = 0.021 \cdot Re_{liq1}^{0.8} \cdot Pr_{liq1}^{0.43} \left(\frac{Pr_{liq}}{Pr_{c1}}\right)^{0.25}; \quad (2)$$

$$Nu_{li} = \alpha_1 \cdot \frac{\alpha_w}{\lambda_{liq}}; \quad (3)$$

$$Re_{li} = w \cdot \frac{d_{in}}{\nu_{liq}}; \quad (4)$$

where Pr_{liq1}, Pr_{c1} are Prandtl numbers for water at temperatures t_{liq1} and t_{c1}; w is water velocity inside pipes, m/s; d_{in} is internal diameter of pipes, m; λ_{liq1}, ν_{liq1} are coefficients of thermal conductivity and kinematic viscosity of water at temperature t_{liq1}.

According to the estimate, the water velocity in the heater tubes was 1.1 m/s, and the heat transfer coefficient from the heating water, calculated by the formula (2), is α_1 = 6245 W/(m^2·K).

Further we calculate the heat transfer coefficient α_2 on the outer surface of pipes. Since the cross-sectional area of the tank is large enough, the forced movement of oil sludge near heating pipes has almost no effect on convective heat transfer. The heat transfer coefficient to oil sludge α_2 is determined by free convection, and it can be calculated using the well-known formula for calculating heat transfer during free convection of liquid near horizontal pipes [13]:

$$Nu_{liq2} = 0.5 \cdot \left(Gr_{li} \cdot Pr_{liq2}\right)^{0.25} \left(\frac{Pr_{liq2}}{Pr_{c2}}\right)^{0.25}; \quad (5)$$

$$Nu_{liq2} = \alpha_2 \cdot \frac{d_{out}}{\lambda_{liq}} \quad (6)$$

$$Gr_{liq2} = \frac{g \cdot \beta_{liq2} \cdot (t_{c2} - t_{liq2}) \cdot d_{out}^3}{\nu_{liq2}^2}; \quad (7)$$

where d_{out} is outer diameter of pipes, m; Pr_{liq2}, Pr_{c2} are Prandtl numbers for oil sludge at temperatures t_{liq2} and t_{c2}; λ_{liq2}, β_{liq2}, ν_{liq2} are coefficients of heat transfer, thermal temperature volumetric expansion, kinematic viscosity of oil sludge at temperature t_{liq2}.

Since oil sludge is an oil-water emulsion in which water droplets are contained within the oil phase, the thermal sludge parameters affecting convective heat transfer are close to the properties of oil. The calculation will be carried out under the condition that the water content in the sludge is absent, which roughly corresponds to the end of the evaporation process. As an oil product, we choose transformer oil, its thermophysical properties were determined from tabular data [12].

At t_{liq2} = 75°C and t_{c2} = 92.5°C heat transfer coefficient α_2, calculated using the formula (5), is α_2 = 100.9 W/(m^2·K).

Total thermal resistance of heat transfer from water to oil sludge is $R = 1/\alpha_1 + \delta_c/\lambda_c + 1/\alpha_2 = 0.01$ m^2·K/W. The analysis shows that thermal resistance of heat transfer from water to the inner pipe surface $1/\alpha_1$ is 1.6% of the total thermal resistance R; thermal resistance of the pipe wall δ_c/λ_c is 0.75% of the total thermal resistance R; thermal resistance of heat transfer from outer wall of pipe to oil sludge $1/\alpha_2$ is 98.6% of the total thermal resistance R. Consequently, heat transfer from water to oil sludge is determined by the practical heat transfer from the outer pipe surface.

The heat transfer coefficient calculated using the formula (1), in the first approximation, is equal to: k = 99.5W/(m^2·K). We specify the values t_{c2}, α_2 and k. Heat flux density through the pipe wall: q = k·(t_{liq}1 t_{liq}2) = 1742 W/m^2. The temperature of the outer surface of the pipe: $t_{c2} = t_{liq1} + q/\alpha_2$ = 92.25°C. At t_{c2} = 92.25 °C, the heat transfer coefficient on the outer surface of the pipe is α_2=100.6 W/(m^2·K), and the heat transfer coefficient k = 99.2W/(m^2·K), which is slightly different from originally obtained values.

We determine the heat flux transferred from the heating water to the oil sludge through the heating surface according to the equation of heat transfer [11]:

$$Q = k\cdot(t_{liq1} - t_{liq2})\cdot F = 19990\ W \cong 20.0\ kW.$$

Thus, in the mode of evaporation of transformer oil, the thermal power transmitted through the existing heating surface is 19.96 kW, which is 4.7 times less than the nominal heat output of the boiler (95 kW). This means that the heat output of the boiler in the evaporation mode is largely unused.

We will check it using the heat balance method. The heat power supplied with the heating water in the evaporation mode is spent on steam formation of (if you do not take into account the small heat loss from the outer surface of the apparatus). The evaporation process of 2.8 m^3 of oil sludge with an initial water content by weight of 30% to a content of 1% lasts approximately $\tau \approx 24$ h (according to experimental data). With the oil density of 880 kg/m^3 and the water density of 998 kg/m^3 (at 20 °C), the average density of oil sludge with a water content of 0.3 is 912 kg/m^3, while the mass of 2.8 m^3 of loaded sludge is 2553 kg. During the process of evaporation, this amount of moisture is removed from the oil sludge: M = (0.3-0.01)·2553=740.3 kg. The evaporation of 1 kg of water at 75 °C consumes the heat of vaporization r = 2320 kJ/kg.

The average heat power consumed for evaporation of water is: $Q = M\cdot r/(3600\tau) = 740.3\cdot 2320/(3600\cdot 24) = 19.87$ kW.

Thus, the thermal power of the apparatus during evaporation, determined by the method of heat balance (19.87 kW), is slightly less than that from the heat transfer equation (20.0 kW). This suggests that the calculation was carried out quite correctly, and the assumptions made in the calculation of heat transfer to the oil sludge are correct.

The amount of moisture evaporated per unit of time in the evaporation mode, kg/s, can be determined from the heat balance equation:

$$G_n = \eta \cdot \frac{Q}{r} \tag{8}$$

where $\eta \le 1$ is efficiency of evaporator; Q is heat flow supplied to the oil sludge in the evaporation mode, kW; r is the heat of vaporization, kJ/kg.

So, at Q = 19.9 kW, r = 2320 kJ/kg and η = 0.98 the consumption of the produced steam is: Gp = 0.0086 kg/s = 30.9 kg/h = 741kg/day. If the heat output of the boiler of 95 kW were used continuously, then the productivity would be: Gn = 0.041 kg/s = 147.4 kg/h = 3537 kg/day.

We will carry out a verification thermal calculation of the existing evaporator for the regime of heating of oil sludge. When cold sludge is loaded into the apparatus with a temperature of t_{liq20} = 20÷25°C, it takes some time to warm it up to a temperature of 75°C, before the process of evaporation of moisture begins. Let us calculate the heat flux transferred from the heating water to the oil sludge in the heating mode using the considered method.

4. The results of the verification calculation

At the initial oil sludge temperature t_{liq20} = 20°C, the temperature pressure is (t_{liq1}-t_{liq20})= 92.5-25=67.5°C, which is 3.85 times more than the temperature pressure in the evaporation mode (t_{liq1}-t_{liq2}) = 92.5-75 = 17.5°C. The heat transfer coefficient on the outer surface at t_{liq20} = 25°C is α_2= 122.5 W/(m^2·K), and the heat transfer coefficient k = 120.5 W/ (m^2·K). The heat flux transferred from the water to the oil sludge at the heating surface F = 11.5 m^2 will be Q = 93.6 kW, which is close to the nominal power of boiler.

During the process of heating the sludge, its average temperature t_{liq2} increases, which leads to a decrease in the temperature head $\Delta t = t_{liq1}-t_{liq2}$, as a result of which the heat flux Q from water to oil sludge decreases, according to the expression: $Q = k\cdot\Delta t\cdot F$.

Table 1 shows the results of calculating Δt, α_2, k and Q at various temperatures of oil sludge t_{liq2}. The average temperature of heating water in tubes for calculation is assumed to be t_{liq1} = 92.5°C; the existing heating surface area F = 11.5 m^2.

Table 1. The results of the verification calculation for heating mode

tliq2, °C	Δt, °C	α2, W/(m2·K)	k, W/(m2·K)	Q, kW
25	67.5	122.5	120.5	93.6
30	62.5	121.7	119.6	90.0
40	52.5	128.1	125.9	76.0
50	42.5	122.6	120.6	58.9
60	32.5	116.2	114.3	42.7
70	22.5	107.0	105.5	27.3
75	17.5	101.0	99.7	20.0

As it is seen from Table 1, the boiler nominal capacity (95kW) is completely used in the existing evaporator unit only at the beginning of the sludge heating process, when the temperature difference between the heating water and sludge is maximum, after which the heat flux decreases continuously with increasing sludge temperature. At the oil sludge temperature t_{liq2} = 75°C, the heat flux is 20 kW.

We calculate the duration of the process of heating the sludge. Heat flow supplied to oil sludge is used to increase its temperature according to the equation:

$$Q = c_{liq2}\cdot G_{liq2}\frac{dt_{liq2}}{d\tau} \tag{9}$$

Where c_{liq2} is isobaric heat capacity of oil sludge, kJ/(kg·K), at temperature t_{liq2}; G_{liq2} is mass of the loaded oil sludge, kg; τ is time, s.

In expression (9), the values of Q and c_{liq2} depend on the oil sludge temperature t_{liq2}, therefore, this equation is solved numerically. The duration of the heating of the oil sludge from the temperature t_{liq20} to t_{liq2} is determined as a result of the integration of the latter expression:

$$\tau_{heat} = G_{liq2}\int_{t_{liq20}}^{t_{liq2}} c_{liq2}\cdot\frac{dt_{liq2}}{Q(t_{liq2})} \tag{10}$$

The relationship between the heat capacity of oil sludge and the temperature t_{liq2} in the range of 20÷75°C with a mass fraction of moisture of 30% can be described by the expression: $c_{liq2} = 2.38 + 0.00364\cdot t_{liq2}$, kJ/(kg·K).

The heat capacity of oil was selected according to the tables for transformer oil [12]. The heat capacity of water was assumed to be constant, equal to 4.19 kJ/(kg·K). The heat capacity of the oil sludge was determined taking into account the mass fractions of water and oil in the sludge.

The duration of heating of oil sludge of 2.8 m^3 volume and a mass G_{liq2} = 2553 kg with initial water content of 30% from a temperature of 20 °C to 75 °C, calculated by the formula (10), taking into account the data in Table. 1, will be τheat = 6990 s = 116.6 min. After the end of the heating mode, the evaporation mode begins.

IOP Conf. Series: Earth and Environmental Science **337** (2019) 012072 doi:10.1088/1755-1315/337/1/012072

5. Conclusions

The verification thermal calculation of the OSSU-3 installation showed that the boiler thermal power is used inefficiently, because only at the beginning of the oil sludge heating process all the nominal boiler power is transferred to the sludge, and during evaporation the heat flow from the heating water to the sludge decreases several times. To increase the capacity of the evaporator, it is necessary to increase the heater surface for several times, in addition, at the same time it is necessary to increase the cooling capacity of the heater for condensation of the increased steam consumption. In addition, the heat of the produced steam can be disposed in an additional evaporating stage, operating using the principle of the main apparatus, only instead of heating water, the additional stage will be heated with steam coming out of the main stage.

A method for thermal calculation of an evaporator for an oil sludge separation unit is proposed. Verification thermal calculation of the OSSU-3 evaporator was carried out. General recommendations for improving the thermal efficiency of this installation are given.

References

[1] Huang N, Zhao P, Ghosh S, Fedyukhin A 2019 *Applied Energy* **240** pp 882-92
[2] Barsi D, Costa C, Satta F, Zunino P, Sergeev V 2018 *MATEC Web of Conferences* **245** 07008
[3] Maksimova Y G, Vasil'ev D M, Zorina A S, Ovechkina G V, Maksimov A Y 2018 *Applied Biochemistry and Microbiology* **54**(2) pp 173-8
[4] Mingaleeva G R, Shamsutdinov E V, Ermolaev D V, Afanas'eva O V, Gil'Manov R M, Fedotov A I, Gal'keeva A A 2014 *Chemistry and Technology of Fuels and Oils* **49**(6) pp 459-68
[5] Shel E V, Paderin G V, Kabanova P K 2018 *Oil Industry* **12** pp 42-5
[6] Khasanov M M, Lezhnev K E, Pashkin V D, Roshchektaev A P 2018 *Oil Industry* **12** pp 63-7
[7] Nagorny A, Nagorny V, Tisenko V 2018 *International Russian Automation Conference, RusAutoCon 2018* 8501824
[8] Shel E, Paderin G, Kabanova P 2018 *Society of Petroleum Engineers - SPE Russian Petroleum Technology Conference 2018* RPTC 2018
[9] Antonov I D, Porubov A V, Bessonov N M 2018 *Materials Physics and Mechanics* **40**(1) pp 37-46
[10] Kizevetter D V, Reznik A S 2016 *Journal of Optical Technology (A Translation of Opticheskii Zhurnal)* **83**(5) pp 295-9
[11] Isachenko V L, Osipova V A. et al 1981 *Heat Transfer* **417** (Moscow: Energoizdat)
[12] Kazantsev E I 1975 *Industrial Furnaces* **368** (Moscow: Metallurgy)
[13] Lykov A V 1978 *Heat and Mass Transfer* **480** (Moscow: Energy)

IOP Conf. Series: Earth and Environmental Science **337** (2019) 012073 doi:10.1088/1755-1315/337/1/012073

Study of the temperature mode of a tunnel kiln using new Celsius© thermal sensors

Anton Sinitsyn[1*], Dmitry Belyanskiy[1], R P Kiyatkin[2], I I Loshchakov[2], Yu A Mamaev[2] and D A Zaripova[3]

[1]Vologda State University, Heat Gas Supply and Ventilation, 160000 Lenin st., 15, Russia
[2]Peter the Great St. Petersburg Polytechnic University, St. Petersburg, Russian Federation
[3]Kazan State Power Engineering University, Kazan, Russian Federation

* Corresponding author: sinitsyn.science@mail.ru

Abstract. The paper proposes a new approved system for control and monitoring of the working process parameters during production of building ceramic bricks for modern enterprises. We analyzed various devices for measuring the raw brick temperature, and identified the shortcomings of the existing instruments and measuring systems. To create high production efficiency, increase the competitiveness of products on the market, reduce the product cost as part of automation of long-term and energy-intensive technological processes, we proposed a method of organizing optimal control of drying process of ceramic bricks at the industrial plant using new Celsius © thermal sensors. A digital temperature sensor is used as a sensitive element, which eliminates the need for calibrating the primary transducer. The device can be installed on the object of examination using a threaded connection, magnetic mounting, or can be used to measure the temperature at the required depth due to remote primary measuring transducer. It is shown that device is universal and will be efficient for the needs of enterprises for which the control of thermodynamic parameters is the basis of the technological process.

1. Introduction

Development of construction industry is inextricably linked with increasing demand for building materials, and the requirements for the products quality are constantly increasing [1-7]. Therefore, the desire of managers to modernize existing production contributes to development of "healthy competition" among manufacturers. In order to increase the competitiveness of products on the market, first of all, it is necessary to automate long-term and energy-intensive technological processes, since the cost price of the product decreases [8-10].

One of the stages of production of building ceramic bricks is technological process of brick drying, which includes measuring and regulating parameters such as humidity, temperature and pressure [11].

In [12], drying refers to removal of excess moisture from any overwetted substance. During drying, water passes from a capillary or hygroscopic state to a vapor state. In this case, evaporation occurs under the condition that partial pressure or concentration of water vapor in the environment is less than the pressure of water vapor at the surface of the body to be dried. And the larger this difference is, the more intense is drying.

IOP Conf. Series: Earth and Environmental Science **337** (2019) 012073 doi:10.1088/1755-1315/337/1/012073

According to [13,14], drying of raw materials and products in ceramic industry is one of the important technological operations. It is applied to molded products to impart mechanical strength and reduce moisture, allowing to stack products in stacks during the subsequent firing or directly go to intensive firing of products without fear of cracking.

For each material and product, a certain drying mode is established, i.e. its permissible intensity, the highest heating temperature of material, parameters of drying agent (temperature, relative humidity and speed) and their change during different periods of drying. These parameters are individual for each technology, for example for production, processing of bricks or silicate or manufacturing of building ceramics.

Drying of brick occurs in drying chambers. Chamber dryers are loaded with bricks and they gradually change the temperature and humidity throughout the entire volume of dryer, in accordance with a given product drying curve and technology for manufacturing the building ceramics and brick production. During drying there is often a large reduction in volume of upper layers than the internal ones, which often leads to cracking of sample when it reaches a critical value.

Such a critical value is humidity corresponding to the lower limit of plasticity, i.e. the point where the clay mass during its drying goes from a plastic to a non-plastic state. In this regard, in the first moments of brick drying (especially at high temperatures and air circulation), the main task is to slow down the sudden evaporation of moisture from the surface.

Usually, the drying process in the brick production technology can be divided into three stages:

- Shrinkage period, in which shrinkage is proportional to the amount of evaporated water;
- Period of slow shrinkage;
- Period of steam water release.

The main interest is the period of intensive drying, during which the evaporation of shrink water occurs. Failure to comply with regimes of brick drying during its production can lead to cracking and warping of clay samples. One of the factors influencing the drying process of products is temperature and humidity of the surrounding air.

The main task during brick processing is selection of optimal operating modes of dryer, which ensures high quality at the maximum achievable productivity of a brick factory. Failure to comply with parameters of technological regime leads to failures in work. If at the beginning of dryer a lot of water vapor is formed in the raw material, then their pressure may exceed the tensile strength of raw material and a crack will appear. Therefore, the temperature and humidity in the first zone of dryer must be such that the water vapor pressure does not destroy the raw material needed for production of ceramic bricks.

If there is a decrease in the relative humidity of the coolant supplied to dryers, in the winter period below 65% in the first stages of drying the raw material, drying cracks appeared on it. Moreover, it is distinctive that when the raw material leaves the dryers with its residual moisture content of 3-4%, the cracks become invisible. However, during firing, the cracks open again and the output of the facing brick does not exceed 20%.

It was established that in order to eliminate formation of cracks in raw material during the initial drying period, the relative humidity of the coolant supplied to the dryers should be in the range of 70–75%, and its temperature at the raw entry positions in the dryers should not exceed the raw temperature by 5–7 °C.

Modern production of ceramic bricks is focused on the market, the main criterion of which is competitiveness. The competitiveness of products is higher, when its cost is lower and its quality is higher. One of the ways to improve the competitiveness of products is optimization of the production process [8].

2. Technology description

At each stage of drying bricks in order to reduce scrap and minimize costs, it is recommended to install a system for controlling and regulating humidity of both the coolant supplied and the products obtained. One solution could be in the use of humidity sensors and temperature sensors. For example, humidity and temperature sensors as a means of measurement are used to control the position of control valves

IOP Conf. Series: Earth and Environmental Science **337** (2019) 012073 doi:10.1088/1755-1315/337/1/012073

with parameters such as relative humidity, temperature, moisture content. Further, the measured signal can be supplied to the control units of ventilation systems (supply and recirculation of the drying agent) or control valve control units and transfer the direction of movement of the drying agent. Thus, the problem of organizing the optimal control of drying process of ceramic bricks is very important. To do this, one needs to know how the temperature and humidity of the product change during the entire drying process. Howcvcr, thc limited space in the drying oven does not allow performing any measurements using sensors that are directly connected to the computer.

To solve this task, the scientists of the "Heat and Gas Supply and Ventilation" Department of Vologda State University carried out a set of activities related to choice of a temperature measurement scheme at some local points of the raw brick preparation, the development of a temperature measurement system, and its introduction at an industrial enterprise.

A preliminary survey of the measurement technology market for temperature monitoring was conducted, as well as a review of patents for independent measuring devices. Among the enterprises which produce such systems, we can distinguish Rotronic, TermoChron Revisor and Dallas Semiconductor. However, their systems for measuring thermodynamic parameters of working fluids are very expensive for mass use, and also have low ranges of temperature measurements.

From the devices for autonomous collection of experimental data, we analyzed the simplest device [15] for continuous measurement and recording of temperature of the outer surface of pipes consisting of a base, a housing with a lid, a cable outlet and a sensing element made of a core with a wire thermistor. Temperature readings are recorded sequentially, at regular intervals, directly onto a computer via cable output, which in industrial environments makes little environmental pressure due to high temperatures and limited space, and the measurement system has a low level of reliability and durability.

A device for temperature control [16] was also analyzed, containing a sensitive element in the form of a temperature sensor and an ambient temperature sensor, the outputs of which are connected to the input of an analog-digital device (ADC). The output of the analog-digital conversion of ADC is connected to the input of the reprogramming unit, which is a kind of algorithm when recording the values of threshold temperature on a computer. The disadvantage of this device is the need to use preliminary mathematical calculations when moving this device from one production to another, or when replacing a sensitive element, as well as using thermoelectric converters that require compulsory preliminary calibration before operating in different temperature conditions.

Thus, the task was to develop our own system for measuring the working fluid temperature. The developed device [17] was aimed at expanding the arsenal of technical tools in this area, as well as at improving the usability of the measuring device by fixing it on the measurement object using a threaded or magnetic connection.

This is achieved due to the fact that the device casing is filled with heat-insulating material, which allows it to withstand large thermal loads in the temperature range from - 55 to 125 °C aimed at the electronic parts of the device. There are no elements in the device that are not protected from mechanical and thermal damage; all components are located inside the case. A digital temperature sensor is used as a sensitive element, which eliminates the need for calibrating the primary transducer. The device can be installed on the object of examination using a threaded connection, magnetic mounting, or can be used to measure the temperature at the required depth due to remote primary measuring transducer.

This set of measures ensures high reliability of the device, expands its application area, improves its usability and accuracy of measurement of temperature data.

The temperature measurement device is a self-sufficient system, which, after setting the user-selected setpoints by means of microcontroller, measures the temperature and records the results in a secure section of the built-in flash data memory. Recording is done at a user-defined rate using a programmable timer. The data is stored as sequential results of temperature values in the memory cells of buffer. The device allows one to save temperature values from 1 to 255 minutes at regular intervals. The device has a unique 64-bit laser-manufactured temperature registration identification number. The autonomy of the work is provided by the power source. The measurement medium can be gas, liquid and solid.

IOP Conf. Series: Earth and Environmental Science **337** (2019) 012073 doi:10.1088/1755-1315/337/1/012073

Information on temperature data can be transferred from Flash memory to a personal computer for further processing [17].

The device, if necessary, can be equipped with a threaded or magnetic connection, while the temperature sensor can be remote outside the device, which will provide an opportunity to measure the temperature in the recessed areas of the surface of the object under study.

To fix the temperature, the device is placed with the side with the temperature sensor on the measured surface. The measurement is carried out in accordance with a predetermined timer program with a record in the data storage buffer. Upon completion of the measurement, the user removes the device from the survey object and then extracts the data using the USB connector.

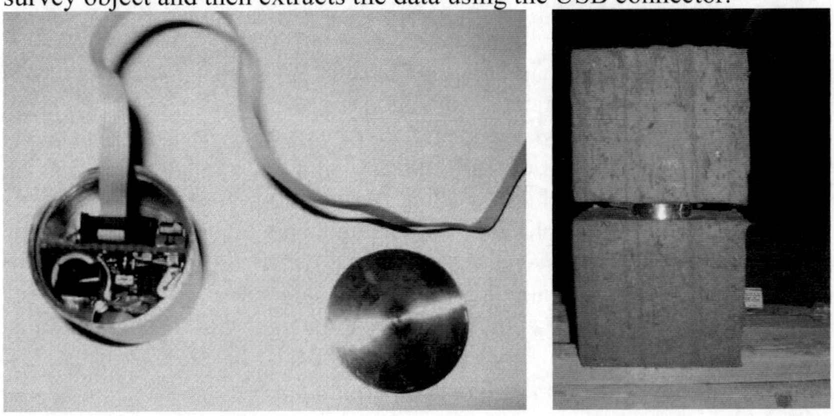

Figure 1. General view of the sensor (left) and the option of placing it in the raw brick (right).

3. Results

We consider a specific case, which is organization of optimal control of the process of drying ceramic bricks. In this case the device (Figure 1) is embedded in a prototype brick sample. The main task of such an autonomous system is to record the function of temperature inside the brick for the period of sample passing through the tunnel dryer (Figure 2). In accordance with the works [18, 19], the experiment plan was developed and the measurement procedure was performed. Results obtained from the instrument's flash memory are processed using standard Microsoft Office applications.

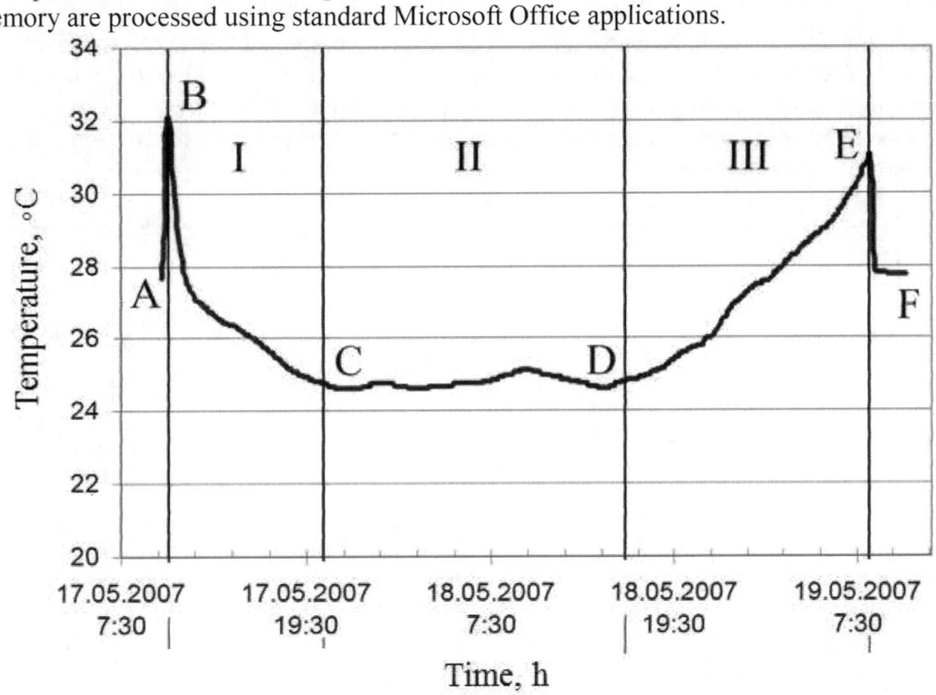

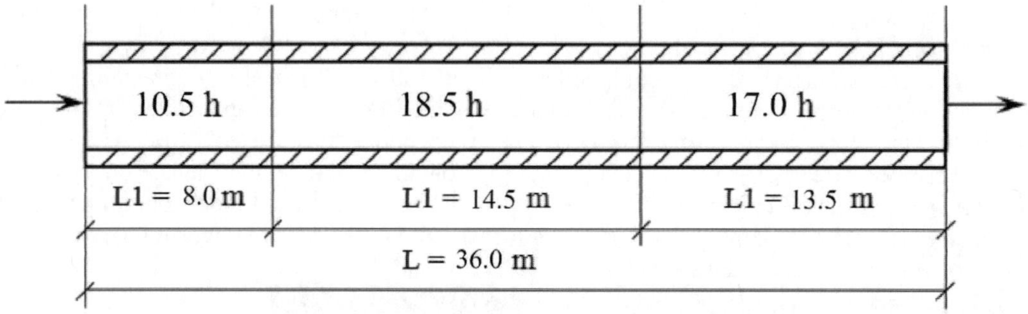

Figure 2. Temperature chart for a raw brick.

Convective drying of a raw brick is carried out in the tunnel kiln due to the heat of hot air (50 - 60 °C), which moves countercurrent to the direction of trolley movement. Comparing the experimental data on change in temperature and humidity parameters of a brick with the technology of moisture removal in such devices from [18] and [19], one can draw conclusions about the drying periods. For example, according to the schedule (Figure 2), the following conclusions were made: in the section AB, the sensor temperature rises from the ambient temperature to the temperature of raw brick; BCDE section is characterized by temperature change after loading the sample into the drying chamber; EF is cooling the brick up to the outside temperature. According to the experience, it can be said that the temperature inside the workpiece decreases at the CD section due to energy input of the heat supplied to evaporate moisture. After the brick is completely dried, its temperature rises (section DE).

4. Conclusions

We developed a unique system for measuring the temperature of the working fluid and a method for determining the drying efficiency by measuring the temperature and humidity characteristics of the raw brick during the entire process for the company Sokolstrom.

Thus, the use of possibilities of modern technology of receiving and transmitting information opens new innovative possibilities for managing the processes taking place while manufacturing the ceramic bricks.

After analyzing the existing scientific and technical developments, as well as assessing the competitiveness of the product, we can distinguish several of the most well-known devices: the Swiss company Rotronic and the American company Dallas Semiconductor (Table 1).

Table 1. Comparative analysis of the technical device characteristics with the foreign analogues

Device characteristics	Rotronic	iButton	Celsius ©
Dimensions	59 cm3	1.5 cm3	20 cm3
Amount of control points	from 1 to 2	1	from 1 to 32
PC connection via	USB interface for settings and data transmission	Data reading from the device is carried out using specialized device and software	1. USB interface for settings and data transmission 2. Bluetooth connection
Measurement range	from –25 to +80°C	from –55 to +125°C	from –55 to +125°C
Protection from dust, moisture and dynamic impact	The case is made of plastic	The case is stable to vibrations, shocks and other mechanical stress	The case is stable to vibrations, shocks and other mechanical stress

A comparative analysis showed the advantages of the Celsius© measuring complex over the existing analogues by several criteria: the number of control points of temperature and humidity can reach 32 pcs at the same time. In production, where access to the sensor is very difficult, it is possible to remotely

IOP Conf. Series: Earth and Environmental Science **337** (2019) 012073 doi:10.1088/1755-1315/337/1/012073

collect information via Bluetooth connection and control the device. The case of the device can be made of alloyed or stainless steel, capable of withstanding high loads.

The patented Celsius© measuring complex based on the developed temperature sensors and automated control system [17] can be used in various fields of science and production, where space limitations do not allow any measurements using sensors directly connected to a computer and where human presence is impossible for some time.

References

[1] Svatovskaya L, Sychov M, Sychova A and Gravit M 2016 New Geoecoprotective Properties of the Construction Materials for Underground Infrastructure Development *Procedia Engineering* **165** pp 1771–5

[2] Kalyutik A, Kiselev V and Rouzich E 2018 Impact of the Underground Metal Construction Cover Layer on the Cathode Protection Efficiency *IOP Conference Series: Materials Science and Engineering* **463**(2) p 02205

[3] Klyuev S, Klyuev A and Vatin N 2018 Fine-grained concrete with combined reinforcement by different types of fibers *MATEC Web of Conferences* **245** p 03006

[4] Musorina T, Katcay A, Selezneva A and Kamskov V 2018 Comparison of Thermal Stability of Dry High-strength Concrete and Wet High-strength Concrete *E3S Web of Conferences* **33** p 02048

[5] Cherkashin A V, Pykhtin K A, Begich Y E, Sherstobitova P A and Koltsova T S 2017 Mechanical properties of nanocarbon modified cement M*agazine of Civil Engineering* **72**(4) pp 54–61

[6] Pukharenko Yu V, Letenko D G, Nikitin V A and Morozov V I 2017 Obtaining the nanomodifier for cement composites based on the "DEALTOM" carbon nanotubes *Materials Physics and Mechanics* **31**(1-2) pp 59–62

[7] Batkov E, Tarasova D, Andreev K and Limarenko I 2016 Technologies, Equipment and Construction Materials for Underground Infrastructure Development: Shear Forces in Bolts of Semi-rigid Joints for Steel Constructions *Procedia Engineering* **165** pp 1595–603

[8] Jones J T 1995 *Ceramic Industry* **144**

[9] Kuladzhi T, Babkin A and Murtazaev S-A 2018 Matrix Tool for Efficiency Assessment of Production of Building Materials and Constructions in the Digital Economy *Advances in Intelligent Systems and Computing* **692** pp 1333–46

[10] Sychova A, Sychov M and Rusanova E 2017 A Method of Obtaining Geonoiseprotective Foam Concrete for Use on Railway Transport *Procedia Engineering* **189** pp 681–7

[11] Burkinshaw R F 2004 *Structural Survey* **22**

[12] Murugesan K, Seetharamu K N and Narayana P A A 1996 *Heat and Mass Transfer* **32**

[13] van der Zanden A J J and Wit M H 2012 *Drying Technology* **30**

[14] Suresh H N, Narayana P A A and Seetharamu K N 2001 *Heat and Mass Transfer* **37**

[15] Tsarenko G A, Serov N S, Isaev A A et al 2003 Pat. 2215271 RU, Int. Cl.7 G01K 7/16. Temperature sensor Pub. 27.10.2003

[16] Leonov S D, Maksimchuk A A and Troitsky Yu V 2004 Pat. 2405131 RU, Int. Cl.7 G01K 7/12. Temperature difference measuring device with thermistor sensors Pub. 20.10.2004

[17] Belyanskiy D A and Sinitsyn A A 2012 Pat. 119877 RU, Int. Cl.7 G01K 7/00. A device for controlling the temperature of the working fluid Pub. 27.08.2012

[18] Murugesan K, Suresh H N, Seetharamu K N and Narayana P A A 2001 *Int. J. of Heat and Mass Transfer* **44**

[19] Lykov A V 1968 *Theory of Drying* (Moscow: Energy)

IOP Conf. Series: Earth and Environmental Science **337** (2019) 012074 doi:10.1088/1755-1315/337/1/012074

Non-chemical water treatment in unified light and sound field and testing on waste water decontamination

O Lebedev[1], Yu Gavrilov[2*], A L Timkovskii[3], A G Bezrukova[3], E I Lezhnev[3], V A Polyanskii[3], A N Timofeev[3] and I G Akhmetova[4]

[1]OOO Novoteh-EKO, Vologda, Russia
[2]Vologda State University, Vologda, Russia
[3]Peter the Great St. Petersburg Polytechnic University, St. Petersburg, Russian Federation
[4]Kazan State Power Engineering University, Kazan, Russian Federation

* Corresponding author: gavr10@mail.ru

Abstract. The paper presents the results of testing of a high-performance method for decontaminating sewage, involving combination of ultrasonic and ultraviolet exposure at the final step of water purification aimed at destroying pathogenic microorganisms and thereby preventing the spread of infectious diseases.A variation of the method was developed and patented by the authors and passed the test at one of the most advanced water facilities in Russia which is the Vodokanal of St. Petersburg. The combined unit was manufactured at OOO Novotech-EKO and installed at the South-West Wastewater Treatment Plant (SWTP) of Vodokanal of St. Petersburg. The results of the test showed the efficiency of combined ultraviolet and ultrasonic wastewater treatment, both in terms of intensifying disinfection and in terms of the stability of the plant operation by preventing biofouling and salt deposition on the surface of the lamp casings. The latter fact was the basis for the pilot project of improvement of the existing system of UV wastewater disinfection at the SWTP.

1. Introduction

Wastewater of residential areas usually contains a significant amount of pathogenic microorganisms including agents of serious diseases. Their elimination before discharge into water bodies is an important final stage of water treatment preventing the spread of infectious diseases.

With this end in view, chlorine and ozone treatment technologies are used. They possess limited efficiency and can produce toxic materials in the water under treatment [1]. Besides, more effective and safe ultraviolet treatment is applied [2]. (Sometimes these technologies are combined.

One should note that the efficiency of ultraviolet decontamination decreases in high colour and turbid medium, which is characteristic of wastewater. What is more, according to the research, pathogenic microflora resistance to ultraviolet has increased 4 times for the recent 15-20 years [3]. Another disadvantage decreasing the efficiency of this process is salt deposition and biofouling of UV lamp casings [4], and as a result it requires regular cleaning the casing surfaces.

Development of powerful ultrasonic sources allowed their possible independent use in water decontamination. Special effects appearing under ultrasonic treatment in liquid media, which are high-speed microstreams > 1 km/s, powerful shock waves > 1 GPa, local heating areas > 1000 K, free radicals [5], being the result of cavitation bubbles collapse are undoubtedly useful for pathogenic flora

IOP Conf. Series: Earth and Environmental Science **337** (2019) 012074 doi:10.1088/1755-1315/337/1/012074

removal. However, the test showed that if the exposure time is little or the transmitter power is low the number of microorganisms even increases in a number of cases [6, 7]. That is why, ultrasonic treatment is to be applied only in combination with chemical dosing or, for instance, ultraviolet treatment [2].

The joint use of ultrasonic and ultraviolet water treatment is recommended by the authors of this article [8] and is considered the most promising decontamination method due to the combination of positive characteristics of both treatment ways. It has been proven that this combination leads to synergistic increase of decontamination efficiency due to the following ultrasonic effects:

1) crushing suspended particulates inside of which there can be microorganisms [9], as well as crushing microorganisms clusters [10] and breakdown of their cell structure [11, 12] and further accessibility which results in significant increase in microorganism sensitivity to ultraviolet (from this point of view ultrasonic treatment should precede ultraviolet one);
2) intensive water mixing which provides delivery of distant layers to the surface of lamp casings and as a result the whole volume of feed water is treated uniformly;
3) prevention of biofouling and salt deposition on the surface of lamp casings [3, 13] which provides the preserve of original radiation intensity during the whole service life.

The joint use of ultrasonic and ultraviolet water treatment results not only in increase of decontamination efficiency but also in the increase in lamp service life as well as in elimination of work pauses intended for casing cleaning [13]. It also eliminates the need for the use of acid solutions for quartz casing cleaning and etching which allows to diminish servicing costs and contributes to ecological safety. Some papers claim that combined decontamination is preferable economically [2, 3, 6, 13].

In recent years a number of experiments with the use of the mentioned above technology have been conducted [2, 3, 13-16] and showed its high efficiency. Moreover, in Russia OOO Novoteh-EKO (Vologda) and ZAO Svarog (Moscow) manufacture equipment for combined (ultraviolet and ultrasonic) water decontamination. The technology developed by OOO Novoteh-EKO involves using piezoceramic transducers. The number of equipment units using this technology has made up more than three hundred since 2006. The equipment operates both in Russia and CIS countries.

2. Methods

The UOV-SV-5 unit (Fig. 1), designed and manufactured by OOO Novoteh-EKO operates using the technology mentioned above. The unit is a pre-cavitator connected to a decontamination chamber and allows to treat waste water with capacity up to 9 m^3/h. With maximum waste water flow the unit consumes: in the ultraviolet treatment mode – 0,107 $kW \cdot h/m^3$, in the combined treatment mode with all transducers on – 0,196 $kW \cdot h/m^3$.

IOP Conf. Series: Earth and Environmental Science **337** (2019) 012074 doi:10.1088/1755-1315/337/1/012074

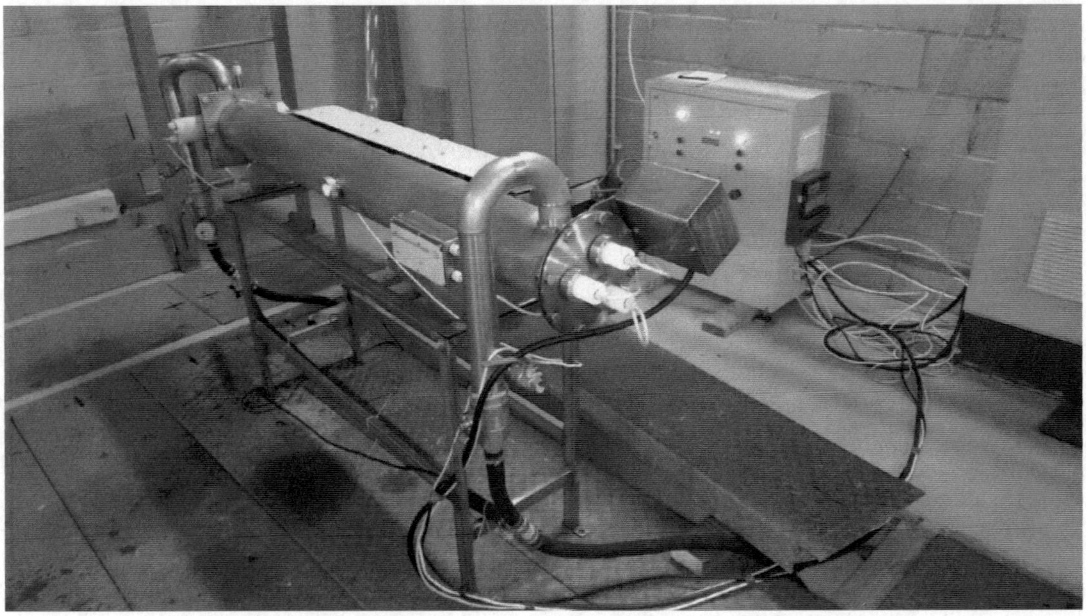

Figure 1. UOV-SV-5 unit installed in the testing facility.

UOV-SV-5 was tested at South-West Wastewater Treatment Plant (SWTP) of Vodokanal of St. Petersburg from October 2015 to December 2016. The unit was placed in the SWTP ultraviolet treatment facility which receives the waste water after primary and complete biological treatment. According to the chemical analysis results, suspended solids made up 5,0-10 mg/dm^3, and chemical oxygen demand was 20-40 mg/dm^3. (It should be noted that after the water treatment with the unit these measures did not change significantly).

The water before the ultraviolet treatment was taken from the canal basin with the pump, then it flew through the unit and was discharged back into the canal. The bacteriological analysis of water samples taken at the input and output of the unit was conducted by ZAO Water Research and Control Centre (St. Petersburg). The following factors were controlled: coliphages, total coliforms, E. coli, enterococci, staphylococci.

The tests were in three stages. At the first stage (I) the research was focused on different decontamination variants with low sewage flow rate through the unit (0,5-0,7 m^3/h). At the second stage (II) the sewage flow rate was high (8-10 m^3/h). Finally, the third stage (III) involved studying the intensity of biofouling and salt deposition on the UV lamp casings during long-term operation under the conditions of low sewage flow rate (0,5-0,7 m^3/h), as well as evaluation of the biofouling intensity decrease. The stage features are presented in Table 1.

At Stages I and II the water samples were collected in 45 minutes after the unit start in the appropriate mode. Thus, by the time of collecting the UV lamp casings were clean (without any traces of biofouling and salt deposition).

At Stage III the samples were collected after 30-50 days of uninterrupted operation in the appropriate mode.

Table 1. UOV-SV-5 operation modes at different testing stages.

Mode	Mode description	Stage		
		I	II	III
US	All ultrasonic transducers (8 pcs) are on (both in the pre-cavitator and in the decontamination chamber).	+		
US-c	Ultrasonic transducers are on only in the pre-cavitator (4 pcs).	+		
US+UV-1	Ultraviolet lamps and ultrasonic transducers are on only in the decontamination chamber (4 pcs).	+		

US+UV-2	Ultraviolet lamps and all ultrasonic transducers are on (8 pcs).	+	+	+
UV-c	Only ultraviolet lamps are on, all ultrasonic transducers are off.	+	+	+

3. Results

The results of Stage I tests are presented in table 2 where there are either values of logarithmic inactivation factor (LIF) or it is stated that complete cleaning was achieved for these indices. The LIF is calculated as a common logarithm of the living cells concentration ratio in the waste water volume unit before and after its treatment.

Table 2. Stage I test results (flow rate 0,5-0,7 m^3/h).

Index	Initial values interval	LIF or operation result in the mode				
		US	US-c	US+UV-1	US+UV-2	UV-c
Coliphages	400-1000 PFU/100 cm^3	0	0	Complete cleaning		
Total coliforms	32000-62000 CFU/100 cm^3	0	0	1,85	2,78	2,34
E. coli	29000-42000 CFU/100 cm^3	0	0,23	2,18	4,51	4,62
Enterococci	15000-32000 CFU/100 cm^3	0	0,32	Complete cleaning		
Staphylococci	45-150 CFU/100 cm^3	1,65	0	Complete cleaning		

It is obvious that ultrasonic treatment alone does not produce any significant decontaminating effect. Slight intensification of ultraviolet decontamination is noted under the use of ultrasound, and the biggest contribution was made by the pre-cavitator while the ultrasonic transducers in the decontamination chamber even produced some negative effect. Crushing suspended particulates and microorganisms clusters which deprives them from protection from ultraviolet radiation is likely to be a predominate mechanism. However, the low sewage flow rate through the unit predetermined high efficiency of ultraviolet decontamination on its own and small contribution of the ultrasonic treatment.

That is why at Stage II it was decided not to use pure ultrasonic modes (US, US-c) as well as the mode without the pre-cavitator (UV+US-1) and to increase the flow rate through the unit. The results are presented in Table 3 and Figure 2.

Table 3. Stage II test results (flow rate 8-10 m3/h).

Index	Initial values interval	LIF or operation result in the mode	
		US+UV-2	UV-c
Coliphages	330-1200 PFU/100 cm^3	Complete cleaning (> 3,08)	2,07
Total coliforms	30000-83000 CFU/100 cm^3	2,25	2,75
E. coli	29000-49000 CFU/100 cm^3	3,57	3,20
Enterococci	4700-25000 CFU/100 cm^3	4,03	3,06
Staphylococci	140-270 CFU/100 cm^3	Complete cleaning (>2,14)	2,09

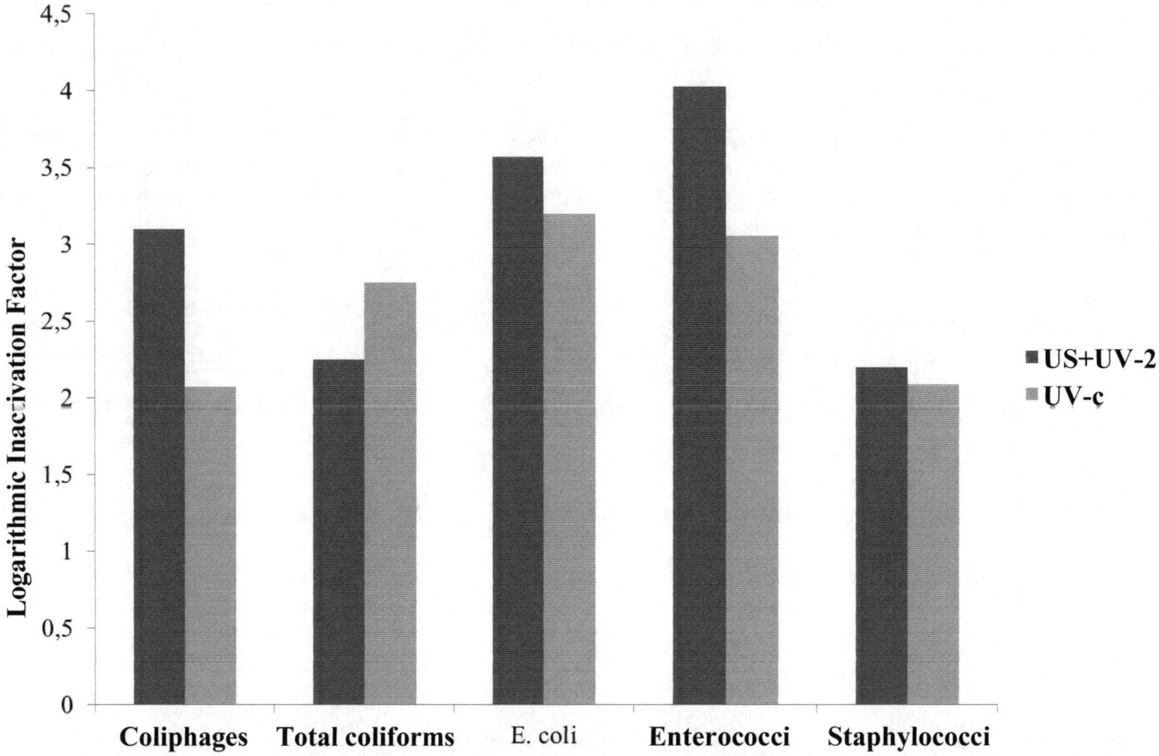

Figure 2. Stage II test results (flow rate 8-10 m^3/h).

It is obvious that ultrasonic waste water treatment going through the unit significantly (several-fold if absolute values are taken) increases decontaminating effect of ultraviolet radiation almost for all the indices under study while at Stage I ultrasound itself did not demonstrate disinfecting effect.

Stage III focused on ultrasound capability to prevent biofouling and salt deposition on the UV lamp casings. The flow rate was again decreased to 0,5-0,7 m^3/h, but the water samples were collected after 30-50 days of uninterrupted operation (see Table 4). Thus, the operational stability of the unit was controlled. The flow rate was decreased to ensure the integrity of the test to eliminate any possible washout of deposit and casing contaminants due to high linear flow rate. The results are presented in Table 4 and Figure 3.

Table 4. Stage III test results (flow rate 0,5-0,7 m^3/h).

Index	Initial values interval	LIF or operation result in the mode	
		US+UV-2 (50 days)	UV-c (30 days)
Coliphages	580-970 PFU/100 cm^3	Complete cleaning (>2,99)	Complete cleaning (>2,76)
Total coliforms	110000-190000 CFU/100 cm^3	4,11	3,70
E. coli	49000-72000 CFU/100 cm^3	4,90	3,70
Enterococci	22000-56000 CFU/100 cm^3	4,20	3,90
Staphylococci	45-270 CFU/100 cm^3	Complete cleaning (>2,43)	Complete cleaning (>1,65)

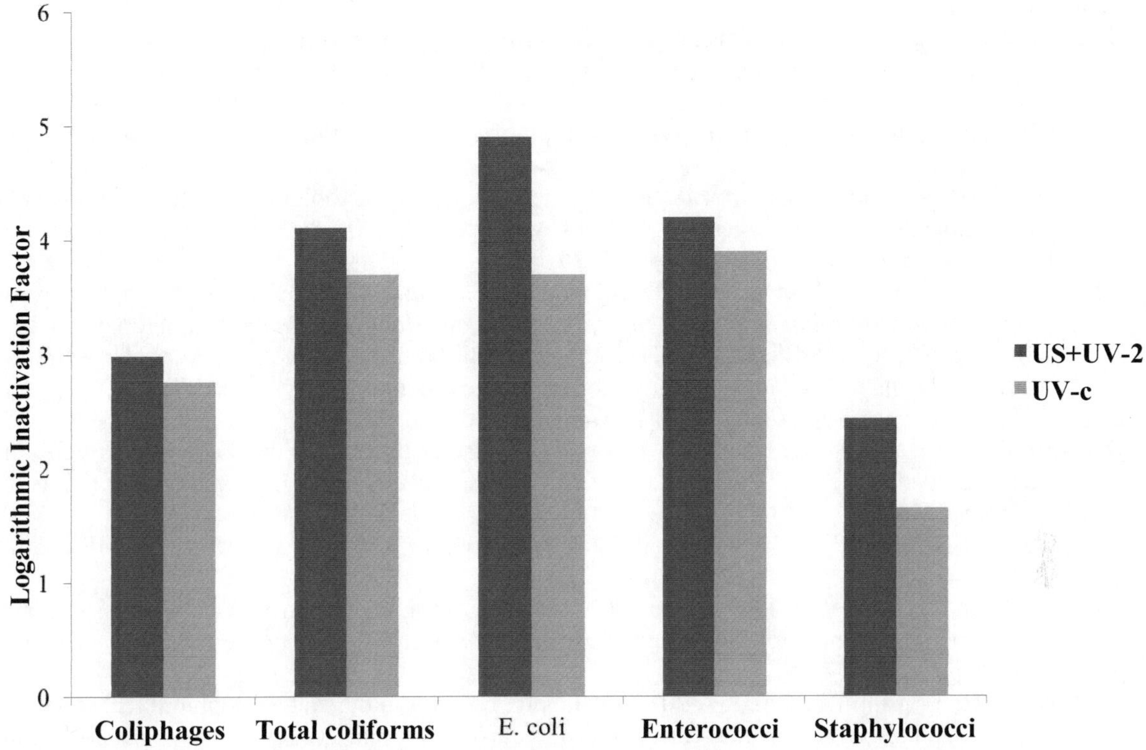

Figure 3. Stage III test results (flow rate 0,5-0,7 m^3/h, long-term uninterrupted operation in the appropriate mode).

The advantages of ultrasonic intensification were confirmed at Stage III too: decontamination efficiency increased a few times for all the microorganisms under study if absolute values are taken into account.

Moreover, the UV lamp casings were visually inspected after the long-term operation in each of the modes and the gravity study of deposition rate was conducted. The second index was measured in the following way: the deposits were removed from the casings by wiping their surface with polyester rayon fabric. The wiping was done after the determined operation period of the unit in the appropriate mode. Biofouling and salt deposition rate were estimated by the mass difference of the dry fabric before and after the wiping of the UV lamp quartz casings. This difference is proportional to the overall mass of contamination and biofouling products which deposited during the operation of the unit.

The total mass of air-dry deposits on three casings related to 1 m^3 of waste water cleaned in the UV+US-2 mode (all 3 UV lamps and 8 ultrasonic transducers are on) made up 45 $\mu g/m^3$, and in the UV-c mode (only 3 UV lamps are on) – 300 $\mu g/m^3$. Thus, the ultrasonic treatment allowed to decrease the rate of biofouling and saline deposition 6,7 times.

4. Conclusion

Thus, the test results proved the efficiency of the combined ultraviolet and ultrasonic waste water treatment both from the point of view of decontamination boost and in terms of the unit operational stability by preventing biological slime and salts from depositing on the lamp casing surfaces. The last fact was used as a basis for a pilot project for improving the existing system of UV waste water decontamination at the SWTP. The implementation of combined decontamination technology at the facility of the leading enterprise [17-19] will allow to recommend it for a wider use.

IOP Conf. Series: Earth and Environmental Science **337** (2019) 012074 doi:10.1088/1755-1315/337/1/012074

References:

[1] Degremont 2009 Water Treatment Technical Reference *Novyi zhurnal* **2**

[2] Sokolova N F 2013 Means and methods of water disinfection *Meditsinskii alfavit* **1** (5) pp 44–54

[3] Uliyanov A N 2009 Tekhnologiya 'L,azur' – new step in the disinfection of water and wastewater *Voda: khimiya i ekologiya* **3** pp 11–5

[4] Sevryukov I T and Afanasieva E V 2013 Some aspects of the problem of ensuring the safety of the population in emergency situations with pollution of the hydrosphere *Tekhnologii grazhdanskoi bezopasnosti* **10** 1 pp 22–25

[5] Osman H and Lim F et al 2016 Development of an ultrasonic resonato,r for ballast water disinfection *44th Annual Symposium of the Ultrasonic Industry Association* **87** pp 99–104

[6] Akhmedova O O and Stepanov S F et al 2009 Improving the efficiency of local wastewater treatment plants through the use of combined electrophysical methods of exposure *Sovremennye problem nauki i obrazovaniya* **5** pp 56–60

[7] Sesal N C and Kekec O 2014 Effects of pulsed ultrasound on Escherichia coli and Staphylococcus aureu *Transactions of the Royal Society of tropical medicine and hygiene* **108** pp 348–53

[8] Lebedev N M and Vasiliev A P et al 2008 Pat. 2332358 Rossiiskaya Federatsiya MPK C02F 1/36

[9] Jin X and Li Z F et al 2009 Improved wastewater ultraviolet disinfection by ultrasonic pre-treatment *Proc. of 2009 Beijing International environmental technology conference, Beiging, PRC* pp 498–505

[10] Bolotova Y V and Karelina K A 2015 Legionella bacteria in the water supply system. Methods of dealing with pathogenic microorganisms *Vestnik Permskogo natsionalnogo politekhnicheskogo universiteta. Prikladnaya ekologiya. Urbanistika* **3** pp 43–59

[11] Zhang S H and Wu C D 2009 Effect of Wastewater Ultraviolet Disinfection of Power Ultrasonic Enhancement *3rd International Conference on Bioinformatics and Biomedical Engineering, Beijing* **1-11** pp 4890–5

[12] Lebedev N M and Kazukov O V et al 2010 Experience of using installations for the disinfection of drinking water, pool water and wastewater by the combined effect of ultrasound and ultraviolet *Materialy VIII Mezhregionalnoi geologicheskoi nauchno-prakticheskoi konferentsii 'Geologiya, poleznye iskopaemye i problem Bashkortostana, Urala i sopredelnykh territorii,' Ufa* pp 275–6

[13] Sosnin E A and Lipatov E I 2016 The effect of ultraviolet radiation and ultrasonic vibrations on wastewater *Sovremennye nauchnye issledovaniya i innovatsii* **3** pp 125–31

[14] Lyamtsov A K and Kuzmichev A V et al 2015 Installation of water disinfection on livestock farms using ultrasound cavitation and ultraviolet radiation *Innovatsii v selskom khozyaistve* **3** pp 90–3

[15] Torres-Palma R A and Gibson J et al 2017 Surfactant-assisted sono-breakage of wastewater particles for improved UV disinfection *Water, Air and Soil Pollution* **228** (3) pp 106

[16] Grachev V and Karmazinov F et al 2010 Drinking water benchmarking *Novyi zhurnal* p 432

[17] Grachev V and Karmazinov F et al 2011 System channelization benchmarking *Novyi zhurnal* **2** p 528

[18] Grachev V and Karmazinov F et al 2012 System channelization benchmarking *Novyi zhurnal* **2** p 464

IOP Conf. Series: Earth and Environmental Science **337** (2019) 012075 doi:10.1088/1755-1315/337/1/012075

Influence of laser radiation on processes of electrochemical phytoremediation of copper from waste water by eichhornia

N A Politayeva[1*], L N Olshanskaya[1], E V Trukhina[1], A L Timkovskii[1], E I Lezhnev[1], V A Polyanskii[1], S A Kravchenko[1], D A Zaripova[2] and N L Bolotova[3]

[1]Peter the Great St. Petersburg Polytechnic University, St. Petersburg, Russian Federation
[2]Kazan State Power Engineering University, Kazan, Russian Federation
[3]Vologda State University, Department of biology and ecology, 160000 Lenin st., 15, Russia

*Corresponding author: conata07@list.ru

1. Introduction

The plant cell, a natural bioelectrochemical membrane reactor, is able to extract and utilize efficiently various contaminants, including heavy metals (TM). The method does not require expensive equipment and reagents, and there is no secondary water contamination, as the sorbent is natural biological material - higher aquatic plants. The costs of phytoradiation technologies do not exceed 20% of alternative methods with an average purification efficiency of 60-80%. The work done to increase the efficiency of purification by stimulating the phytoremediation process by the influence of physical factors: magnetic field, weak electric current, infrared and ultraviolet radiation gave a positive result [1-4].

The purpose of this work was to study the effect of laser influence on the process of copper ions extraction from sewage with the help of a higher aquatic plant - eichhornia.

2. Materials and methods

A laser, or an optical quantum generator, is a technical device emitting light in a very narrow spectral range in the form of a directed high-coherence, monochromatic, polarized beam, in other words in the form of a highly-ordered in space and time electromagnetic radiation of one color. The spectrum of laser application is very wide, sharp focusing in combination with high power of radiation is used in technological processes for cutting, welding, piercing holes in solid materials. They apply LR for monitoring studies on air pollution [5].

The effect of laser radiation (LR) on human body is widely studied and applied in the medical field. Lasers are used for surgical intervention. Laser scalpel allowed to shift to bloodless operations in medicine and open a new direction in eye microsurgery. LR is applied in immune system enhancement, in cosmetology and for therapeutic effects on the human body [6]. In agriculture, laser treatment accelerates the growth and development of plants, leading to an increase in crops yields [7, 8].

The purpose of this work was to study the effect of laser radiation duration on the process of extracting copper from sewage by electrochemical phytoremediation using eichhornia.

Despite the widespread use of LR and the variety of laser-related activities studied, its effect on the processes of phytoremediation has not been researched, thus novelty of this work is beyond doubt. The

IOP Conf. Series: Earth and Environmental Science **337** (2019) 012075 doi:10.1088/1755-1315/337/1/012075

LR source was a gas helium-neon laser (He-Ne) with the power of 2 mW, which operates in the visible red band (wavelength λ = 632.8 nm).

In order to study the effect of LR on the process of phytoremediation, copper sulfate based model solutions were prepared. For this purpose, a solution of copper sulfate with the concentration of 5 mg / l of Cu2 + was added to the conditionally clean sewage (after biological treatment at the sewage treatment facilities in the city of Engels, Saratov region). We stadied processes of accumulation and extraction of copper cations from one liter of the solution. Higher aquatic plants (eichhornia) of the same mass (20 g) and maturation period were placed in vessels. The plant was exposed to LR for 3, 5 and 10 min (LR-3, LR-5, LR-10, respectively). The analysis of water for the residual content of copper ions was carried out using the photoelectrocolorimetric method (device KFK 3-01) in accordance with GOST 4388-72 "Drinking water. Methods for determining the mass concentration of copper." To determine the final concentration in a 50 ml flask, 5 ml of the stock solution was placed and adjusted to 50 ml with distilled water. 1 to 2 drops of HCl (1: 1), 1 ml of Rochelle salt (1: 1), 5 ml of NH4OH (25%), 1 ml of starch (0.5) and 5 ml of sodium diethylene dithiocarb natate (0.1%) were then added to the solution. Thus a yellow-brown solution was obtained. When carrying out photocolorimetric analysis, a blue light filter with a wavelength λ = 430 nm and a cuvette with a working layer thickness of 50 ml was used.

The results of the change in the concentration of copper ions in solution under the influence of LR are shown in Figure 1.

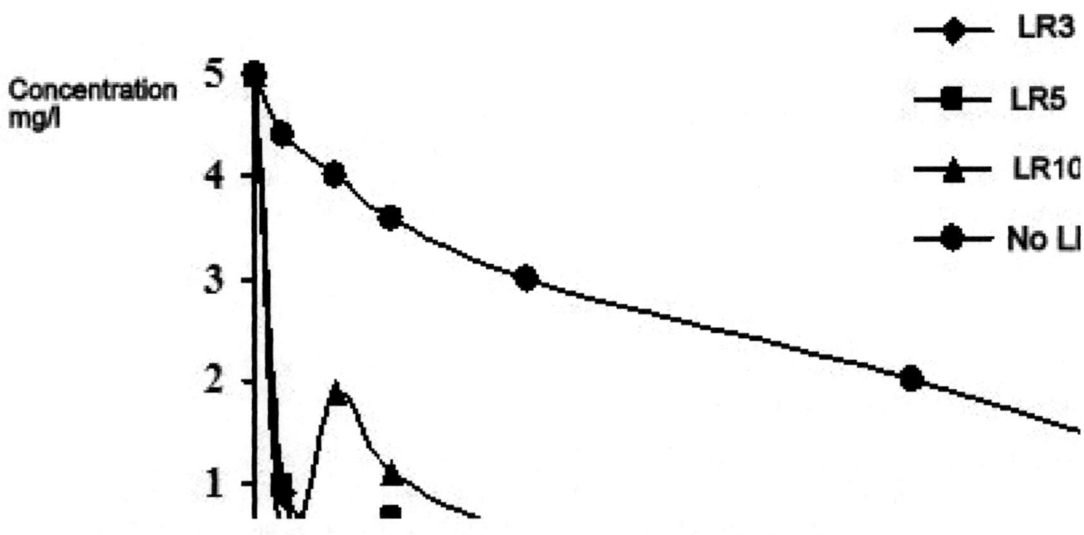

Figure 1. Change in the concentration of Cu2 + in solution during its extraction by eichhornia under the influence of laser radiation and without it.

3. Results and discussion

It can be seen from the figure and the table that laser treatment of eichhornia has a stimulating effect on the sorption of copper ions. It is known from the sources [9] that a low-intensity LR stimulates the metabolic activity of a cell. These processes are based on photophysical and photochemical reactions that arise in the body when exposed to laser radiation. Photophysical reactions are mainly due to the heating of the object (in the range 0.1-0.3 ° C) and the spread of the heat in the biotissues. The temperature difference is more pronounced on biological membranes, which leads to the outflow of Na + and K + ions, the opening of protein channels and the increase in the transport of molecules and ions. Photochemical reactions are due to the excitation of electrons in atoms, absorbing the light of matter. At the molecular level, this is expressed in the form of photoionization of the substance, its photoreduction, photooxidation, photodissociation of molecules, or in their rearrangement - photoisomerization

IOP Conf. Series: Earth and Environmental Science **337** (2019) 012075 doi:10.1088/1755-1315/337/1/012075

[9]. In this case, water absorbs visible light and red part of the spectrum. This changes the structural organization of the water layer in membranes and changes the function of the thermolabile channels of the membranes [10]. The plant cell is an electrochemically active biological membrane, with a working potassium - sodium pump (K + - Na + pump). The transport of charged particles, in particular, copper ions through the biological membrane of the plant cell, is carried out due to the bioelectric potential (Figure 2), which is created at theof border cell / electrolyte interface and its value can vary from 60 to 200 mV. Sorption of copper ions from the solution by eichornia takes place due to the diffusion-electrochemical mechanism of permeability of plant cell membranes. The plant cell is considered as a bioelectrochemical nanoreactor that accumulates and utilizes impurities,which is copper in our case.

Biological structures of the body own electromagnetic fields and free charges, which are redistributed under the influence of photons of radiation of GNL, which leads to direct "energy pumping" of the irradiated organism. Thus, at the basis of the mechanisms of low-power lasers influence on tissues in the visible and infrared spectrums are the processes occurring at the cellular and molecular levels [11].

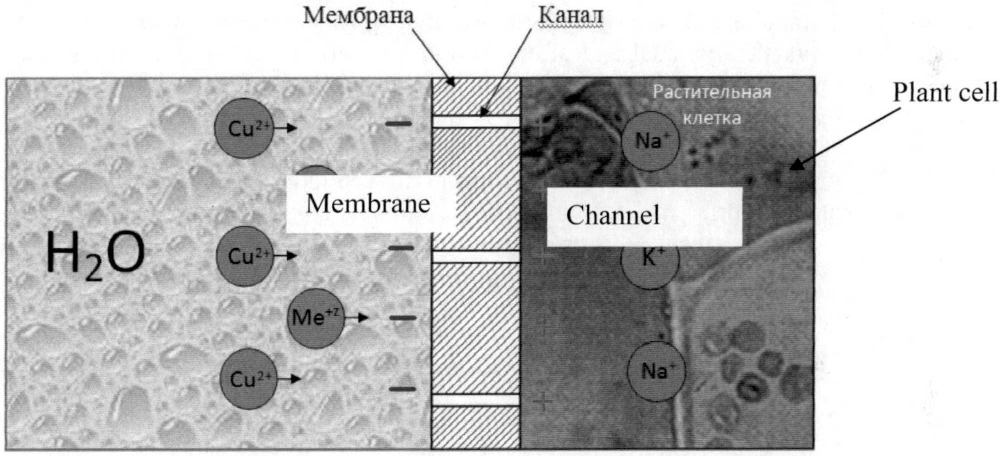

a)

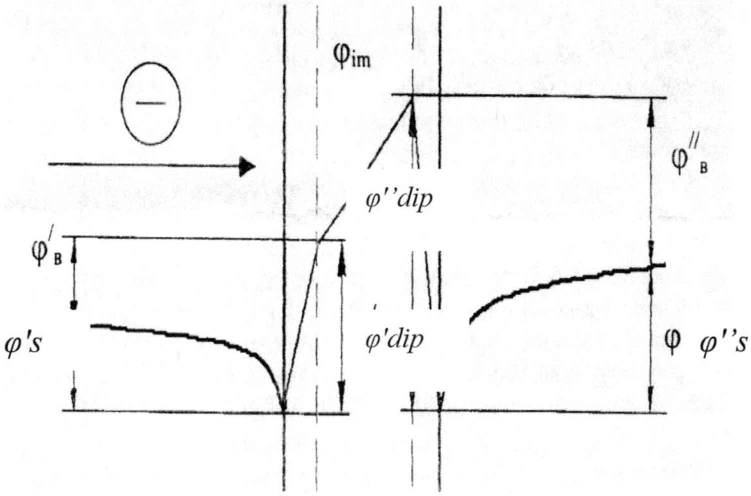

b)

Figure 2. (a) a schematic diagram of the work of the bioelectrochemical nanoreactor - a plant cell;

(b) the distribution of the potential on the cell membrane φ'b, φ'b - boundary potentials; φ's, φ's - surface jumps; φim-intramembrane potential; φ'dip, φ"di are dipole potentials.

As a result of the "energy swap" of the eichhornia due to the RI, a rapid increase in the rate of electro-sorption of copper ions from sewage takes place and the purification efficiency reaches 80-90% during the first hour of the phytoremediation process. In comparison with the purification process without LR, this effect is achieved only after a few days (see Figure 1).

It is known that photobiological effects depend on the parameters of laser radiation: wavelength, intensity of light energy flow, time of exposure on biological tissues. From the data obtained it can be seen that the processes of phytoremediation depend on the duration of the LR influence on the higher aquatic plant. So, when LR is applied for 10 min (LI-10) during the first hour, the maximum absorption of copper ions from the solution is achieved, and then the metal is ejected back into the solution. It is likely that all processes are accelerated in the plant cells that received the maximum push. The discharge of metal by the plant into the solution occurs with LR for 3 and 5 min (LI-3, LI-5), but with a lower intensity and speed. The speed and amount of copper ejection (peak height) decreases with decreasing the exposure time of LR in the series LR-10> LR-5> LR-3 (Figure 1, Table 1). After 10 hours, the concentrations of copper ions in the solutions of the samples under study at different durations of LR are practically equal and after 36 hours the efficiency of solution purification reaches 100% for all samples (Table). Consequently, for a positive effect of LR on the purification processes of copper ions by eichhornia, 1-3 minutes are sufficient, and it is economically unjustified and ecologically inexpedient to use energy for a longer exposure.

Table 1. The change in the concentration of copper (Concentration) and the cleaning efficiency (E, %) of solutions with eichhornia without and under the influence of Laser radiation (Initial concentration of Cu + 2 ions in CuSO4 solution, Cini = 5 mg / l).

Process time, t, h.	LR-3		LR-5		LR-10		No LR	
	Conc. mg/l	E, %	Conc. mg/l	E, %	Conc. mg/l	E, %	Conc. mg/l	E, %
0	5	0	5	0	5	0	5	0
1	0,9	82,0	1	80,0	0,4	92,0	4,4	12
3	0,13	97,4	0,4	92,0	1,9	82,0	4	20
5	0,3	94,0	0,65	87,0	1,1	78,0	3,6	28
10	0,43	91,4	0,29	94,2	0,55	89,0	3	40
24	0,08	98,4	0,25	95,0	0,1	98,0	2	60
36	0	100	0	100	0	100	1	80

4. Conclusions

On the basis of our studies we can draw the following conclusions:

1. It has been established that laser treatment of eichhornia has a stimulating effect on the sorption of copper ions by altering the metabolic activity of the plant cell.
2. The processes are based on photophysical and photochemical reactions occurring in the body under the action of laser radiation. The protein channels are opened in biological membranes, the outflow of Na + and K + ions, which increases the transport of copper ions.
3. The processes of phytoremediation depend on the duration of the LR effect on the higher aquatic plant. The process speed increased with decreasing radiation exposure time in the series LR-10 <LR-5 <LR-3.

References

[1] Ol'shanskaya L N, Tarushkina Yu A and Sobgayda N A 2008 Investigation of the dynamics of accumulation of zinc, copper and cadmium from highly concentrated solutions by aquatic plants *Ecology and industry of Russia* pp 32–3.

[2] Ol'shanskaya L N, Sobhayda N A, Tarushkina Yu A and Stoyanov A V 2008 Influence of the

magnetic field on the processes of electrochemical extraction of heavy metals from wastewater by duckweed *Chemical and Oil and Gas Engineering* **8** pp 41–4.

[3] Sobgaida N A, Olshanskaya L N and Stoyanov A V 2009 Influence of geomagnetic, magnetic and electric fields on the process of phytoremediation *Ecological problems of industrial cities: materials of the All-Russian Conf. Saratov April 4 - 6, 2009 Part 1* pp 62–5.

[4] Ol'shanskaya L N, Sobhayda N A and Stoyanov A V 2010 Influence of ultraviolet irradiation on the process of extraction of copper by eichhornia *Tatishchevsky readings: Actual problems of science and practice: materials of the VIIth International Scientific and Practical Conference in Tolyatti April 15-18, 2010* pp 64–6.

[5] Reggie J 1987 *Industrial use of lasers* (Moscow: Mir)

[6] Priezzhev A V, Tuchin V V and Shubochkin L P 1989 *Laser diagnostics in biology and medicine* (Moscow: Science)

[7] Belsky A I 2007 Quantum processing by laser irradiation in a magnetic field in the technology of annual growing of fruit tree seedlings *Plant growing* **2** pp 15–9.

[8] Zhurba P S, Zhurba T P and Zhurba E P 2003 Patent of the Russian Federation RU № 2240663 Method of industrial cultivation of agricultural crops using laser irradiation 2003.03.11

[9] Alexeev Yu V, Sokolov M E and Degrave T V 2001 Some aspects of biophotometry with LED radiation in the near infrared range *Materials of Intern. Confer. "Laser and Information Technologies in Medicine of the XXI Century". Part II.* pp 462–3.

[10] Fine S and Klein E 1968 *Biological action of laser radiation* trans. from English (Moscow: Mir)

[11] Gamaleya N F 1972 Some questions of biodynamics and bioelectronics of the organism in norm and pathology, biostimulation by laser radiation *Materials of the international conference. Krasnodar: KPI. 11-13 May 1972* pp 286–9.

Photobioreactors for microalga Chlorella Sorokiniana cultivation

N Politaeva[1*], Yu Smyatskaya[1], A L Timkovskii[1], E I Lezhnev[1], V A Polyanskii[1], A N Timofeev[1], A A Tulub[1], D A Zaripova[2] and O G Lopicheva[3]

[1]Peter the Great St. Petersburg Polytechnic University, St. Petersburg, Russian Federation
[2]Kazan State Power Engineering University, Kazan, Russian Federation
[3]Vologda State University, Department of biology and ecology, 160000 Lenin st., 15, Russia

* Corresponding author: conata07@list.ru

Abstract. This work presents various constructions of photobioreactors for microalgae Chlorella Sorokiniana cultivation. We studied dependence of biomass production rate on photobioreactor construction. We suggest special photobioreactor construction for pilot-scale biomass production. We explored the ways to use the obtained biomass as a source for lipids and carotinoids. We recommend to use waste biomass for obtaining of sorption materials for water purification. The spent sorption materials may be used as a co-substrate for fermentation of organic wastes.

Keywords: microalga, Chlorella Sorokiniana, photobioreactor, biomass, water purification

1. Introduction

Microalga Chlorella Sorokiniana (*C. Sorokiniana*) cultivation is carried out for synthesis of third-generation biofuel. High-speed synthesis of microalga biomass is carried out in special apparatus known as photobioreactor (PBR) [1-4].

The aim of this work was to develop photobioreactor construction and select appropriate light conditions for high-speed synthesis of *C. Sorokiniana* biomass.

Two types of photobioreactors are used for microalga *C. Sorokiniana* cultivation: open and closed. Open photobioreactors are artificial basins, open containers, ponds, vessels. The disadvantages of open-type PBR are the following: the possibility of other algae type reproduction, inflecting by pathogenic bacteria and water level reduction due to evaporation.

The closed-type PBR do not possess the above-mentioned disadvantages. But there are problems connected with biomass adhered to vessel walls, and long-term presence of microalgae at unlighted reactor zone. In order to achieve optimum cultivation conditions, one should provide equal lightning throughout the photobioreactor volume.

In case of heterotrophic organisms, uniform light access is provided throughout the whole microalgae cultivation period. Light factor is an exception for autotrophic cell population, which rather fast becomes a limiting one due to optical properties of the culture which is a medium of high light-absorption. Constant growth of cell amount and biomass per time unit provides rapid achievement of linear growth stage, despite the continuous cell amount growth. At this the culture density is various for reactors with various suspension width, however the surface cell concentration at the moment of this transition remains constant.

In dense culture which is in linear growth phase, the total photosynthesis productivity of the suspension is determined in each elementary time period as a work of cells, which are located at the irradiation zone. At this, a portion of cells at the irradiation zone carries out photosynthesis at the linear stage, another portion is at the plateau of photosynthesis light curve of a single cell. The other cells (outside the "photosynthesis area") are essentially ballast and do not make contribution to the general population photosynthesis, as the photosynthesis light reaction duration (about 10-3 s) is considerably shorter than cell migration period between transilluminated and non- transilluminated suspension zones (Figure 1), which is achieved by biomass mixture 6.

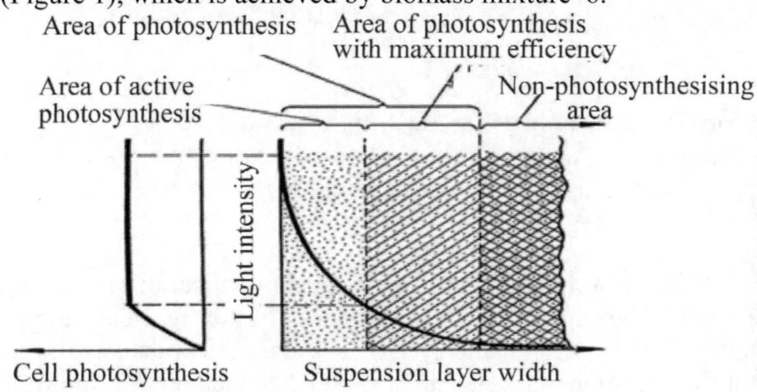

Figure 1. Optical and functional structure of aglae culture as a photosynthesising system: light distribution is on the right side, photosynthesis intensity of a single cell of microalgae population is on the left side.

2. Materials and methods

Nutrient broth composition for microalgae *C. Sorokiniana* cultivation was experimentally selected and is presented in Table 1.

Table 1. Nutrient broth composition.

Substance	Concentration, mg/l
$ZnSO_4 \cdot 7H_2O$	100
$CuSO_4 \cdot 5H_2O$	10
$CoSO_4 \cdot 7H_2O$	100
$MnCl_2 \cdot 4H_2O$	500
$H_3BO_3 \cdot WF$	50
$Na_2MoO_4 \cdot 2H_2O$	100
$FeCl_3 \cdot 6H_2O$	4
$Na_2EDTA \cdot 2H_2O$	6
KNO_3	1
KH_2PO_4	100
$MgSO_4 \cdot 7H_2O$	240

The prepared microalgae *C. Sorokiniana* suspension is placed in PBR of 500 ml. During the cultivation process an intermittent mixing takes place using magnetic mixing machine, located at the bottom of PBR. The mixing regime is as follows: 15± 2 min of mixing, 120 min of rest. Mixing rate was 500 rpm. Microalgae suspension is permanently aerated by air using bubbler throughout the whole cultivation period. This device consists of aerator pump and atmospheric air supply tube with 1.5±3 l/min consumption at 26 °C temperature. The cultivation was carried out at surface lightning with 2500±300 lx daylight fluorescent lamp (DFL) of 12 hours photoperiod. In order to increase power of luminous flux, which goes through biomass suspension, we put lamps of two types into the PBR.

In the first case (PBR-1) we used waterproof aquarium lamp of red visible light with wavelength λ =660 nm, which was directly put into the reactor (Figure 2). In the second case (PBR-2) the irradiation source was placed into protective glass casing with green-blue visible light λ =555 nm (Figure 3). PBR irradiation intensity with internal and external lamps was 4000 lx. The lamps were constantly turned on.

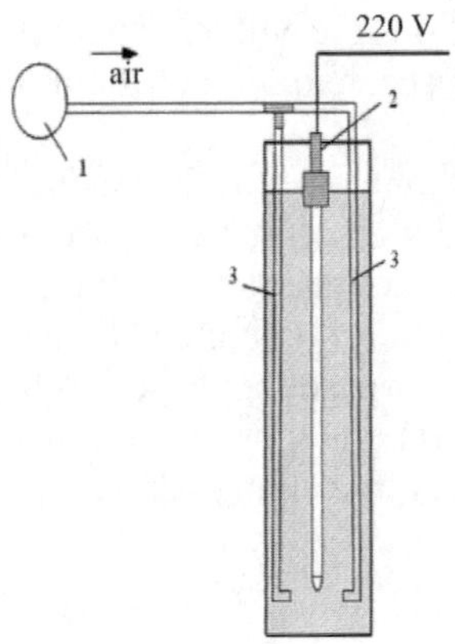

Figure 2. PBR-1: 1- aerator, 2 - waterproof lamp, 3 - aeration tubes.

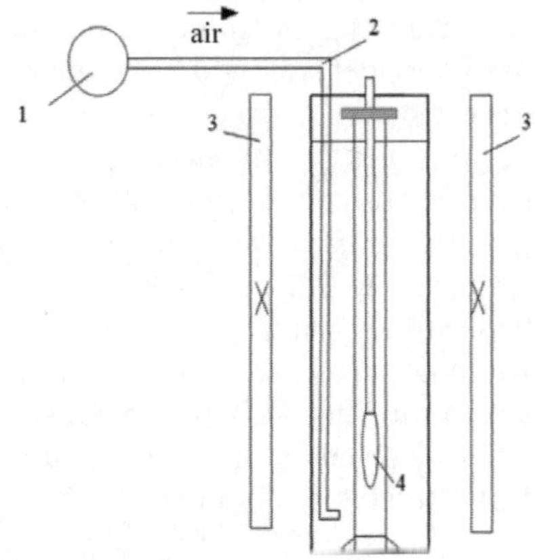

Figure 3. PBR-2: 1- aerator, 2 - aeration tube, 3- DFL, 4 - protective glass casing with DFL.

3. Results and discussion

Microalga *C. Sorokiniana* biomass growth was determined from the suspension optical density using spectrophotometer UNICO 1208 at 750 nm wavelength. Optical density was further recalculated for counted cell number in ml at hemocytometer (Goryaeva chamber) 7. The initial optical density of the *C. Sorokiniana* suspension equal to 0.120 corresponds to 1.87 mln/ml cell number. PBR without a submersible lamp was used as a control. The cultivation process was carried out within 10 days. The results of biomass growth are presented in Figure 3. Lamps without internal lightning were used as a control.

The highest biomass growth for *C. Sorokiniana* microalga was observed when cultivating at PBR-1, Figure 4. The additional lamp is placed into the photobioreactor with red light, wavelength is 660 nm.

The construction with submersible light source allows one to avoid non-photosynthesising areas.

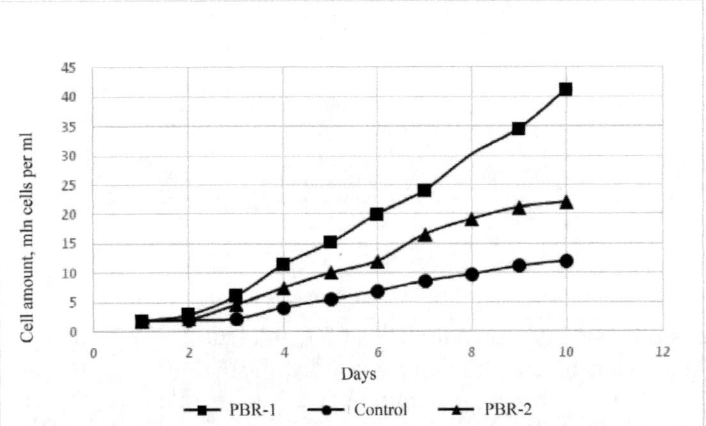

Figure 4. Dependence of cell number on time period (days) for PBR of various constructions.

Specially designed units PBR-3 (Figure 5) were used as pilot installations for biomass growth in laboratory conditions. These are narrowing down cylindrical vessels made of polymer. Sun and additional irradiation are used. Gas-air mixture supply is performed at the bottom of PBR-3, which provides aeration and suspension intermixing throughout the whole volume. PBR-3 volume is 100 l, it is equipped by sensors of medium pH, temperature and automatic system "day/night". PBR width is 10 cm, additional radiation is provided by day fluorescent lamps OSRAM DULUX S 11W/21-840, its' illumination intensity is 2200-2800 lx.

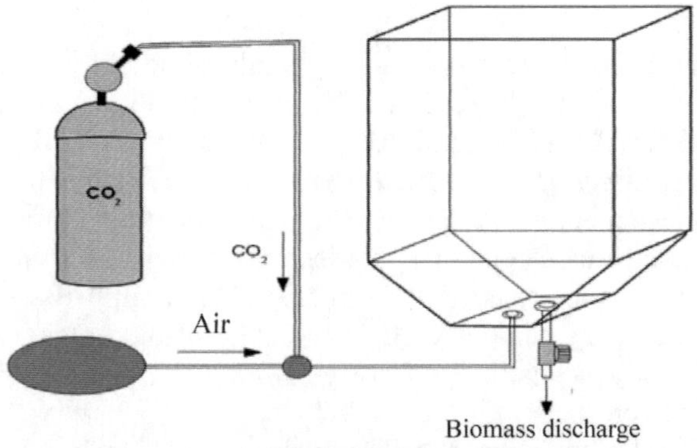

Figure 5. PBR-3.

We developed an economically efficient method for extraction of products with high value addition, such as lipids and carotenoids from the obtained *C. Sorokiniana* microalga biomass (Figure 6). A part of the residual biomass will be used as an adsorbent for industrial wastewater purification from heavy metals and oil products [8-12]. At the same time the microalga remained after extraction and spent sorption materials may be used as co-substrate for fermentation of organic wastes. Biogas, which generally consists of methane (CH_4) and carbon dioxide (CO_2) should be divided into components. Methane is recommended to use for housekeeping needs, for heat and power supply of the process [13-16]. Carbon dioxide is formed during the fermentation process and after methane combustion for energy obtaining. It will be used at microalga *C. Sorokiniana* cultivation stage as a source for carbon at photoautotrophic microalga biomass synthesis.

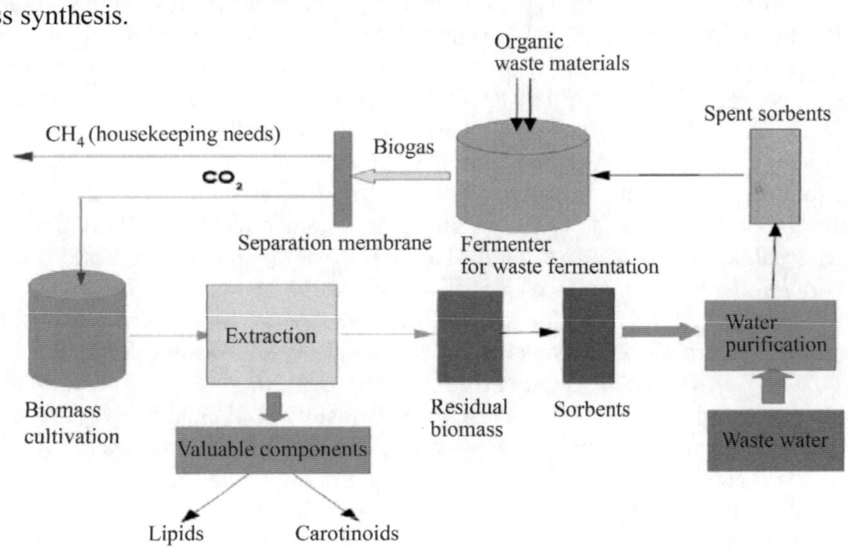

Figure 6. A scheme for *C. Sorokiniana* biomass processing.

4. Conclusions

During this work implementation we have selected photobioreactors for high-speed synthesis of *C. Sorokiniana* biomass and defined it zero waste management directions.

References

[1] Kshirsagar H H, Revankar M S, Kamat M Y and Lele S S 2002 Cultivation of Spirulina in Gas Induced Photobioreactor and Isolation of Phycobiliproteins Indian J Biotechnol. 1 (3) pp 255–62

[2] Kunjapur A M and Eldridge R B 2010 Photobioreactor design for commercial biofuel production

from microalgae *Ind Eng Chem Res* **49** (8) pp 3516–26. doi:10.1021/ie901459u
[3] Singh R N and Sharma S 2012 Development of suitable photobioreactor for algae production *Renew Sustain Energy Rev.* **16** (4) pp 2347–23 doi:10.1016/j.rser.2012.01.026
[4] Anderson G, Kommareddy A and Schipull M 2002 Photobioreactor design. *ASAE/CSAE North Central Intersectional Meeting Presentation MBSK02-216, Saskatoon, Canada*
[5] Nagornov S A and Meshcheryakova Yu V 2015 Investigation of cultivation conditions for chlorella microalgae at a tube reactor [Issledovanie uslovij kul'tivirovaniya mikrovodorosli chlorella v trubchatom fotobioreaktore] *Vestn TGTU* **21** (4) pp 653–9
[6] Vladimirova M G and Semenenko V E 1997 Mass microalgae cultivation [Massovoe kul'tivirovanie mikroskopicheskih vodoroslej] *Plant Life* vol 3 *[Zhizn' Rastenij tom 3]* (Moscow: Izd-vo «Prosveshchenie») pp 367–76
[7] Politaeva N, Kuznetsova T, Smyatskaya Y, Elena T and Ovchinnikov F 2017 Impact of various physical exposures on Chlorella Sorokiniana microalgae cultivation *Int J Appl Eng Res.* **12** (21) pp 11488–92
[8] Sobgajda N, Ol'shanskaya L N and Makarova Y A 2010 Waste water purification from oil products by composite filters based on production wastes [Ochistka stochnyh vod ot nefteproduktov kompozitnymi fil'trami na osnove othodov proizvodstv] *Him i Neftegazov Mashinostr* **3** p 37
[9] Sobgajda N A 2011 Sorption materials for waste and natural waters purification from oil produts [Sorbcionnye materialy dlya ochistki stochnyh i prirodnyh vod ot nefteproduktov] *Vestn Har'kovskogo Nac avtomobil'no-dorozhnogo Univ* **52** pp 120–4.
[10] Arakcheev E N, Brunman V E and Volkov A N et al 2016 Complex Electrolytic Unit for Producing Anode Liquor and Ferrate to Treat Waters Contaminated by Oil-Refining Products and Harmful Admixtures *Chem Pet Eng* **52** (1-2) pp 143–8 doi:10.1007/s10556-016-0164-y
[11] Das N, Vimala R and Karthika P 2008 Biosorption of heavy metals - an overview *Indian J Biotechnol* **7** (2) pp 159–69 doi:10.1016/j.is.2009.02.003
[12] Paknikar K M, Pethkar A V and Puranik P R 203 Bioremediation of Metalliferous Wastes and Products using Inactivated Microbial Biomass *Indian J Biotechnol* **2** (3) pp 426–43.
[13] Politaeva N, Kuznetsova T, Smyatskaya Y, Atamaniuk I and Trukhina E 2018 Chlorella Microalga Biomass Cultivation for Obtaining Energy in Climatic Conditions of St. Petersburg *Adv Intell Syst Comput* **692** pp 555–62 doi:10.1007/978-3-319-70987-1_59
[14] Kholodiakov A, Bondal A, Zubkova M, Chusov A and Stroganov A 2016 Methane component of the modern WWTP biogas as an alternative energy source *Solid State Phenom.* **871** pp 208–15 doi:10.4028/www.scientific.net/MSF.871.208
[15] Zubkova M, Maslikov V, Molodtsov D, Chusov A, Zhazhkov V and Stroganov A 2016 Technological Decision to Renewable Energy Usage Biogas for Off-grid Systems Consumption *MATEC Web of Conferences* **73** doi:10.1051/matecconf/20167304004
[16] Surov A V, Popov S D and Spodobin V A et al 2017 Powerful high-voltage AC plasma torches for plasma-chemical applications *J of Physics: Conference Series* **891**(1) doi:10.1088/1742-6596/891/1/012309

IOP Conf. Series: Earth and Environmental Science **337** (2019) 012077 doi:10.1088/1755-1315/337/1/012077

A new approach for recycling of spent activated sludge

N Politaeva[1*], V Prokhorov[1], E I Lezhnev[1], V A Polyanskii[1], B G Matisov[1], L R Mukhametova[2] and I A Mukhin[3]

[1]Peter the Great St. Petersburg Polytechnic University, St. Petersburg, Russian Federation
[2]Kazan State Power Engineering University, Kazan, Russian Federation
[3]Vologda State University, Department of biology and ecology, 160000 Lenin st., 15, Russia

* Corresponding author: conata07@list.ru

Abstract. The paper presents the basic methods of disposal and recycling of waste activated sludge. We propose a new technology of neutralization of waste activated sludge and subsequent disposal in the form of an adsorbent for cleaning oil-contaminated wastewater and production of complex fertilizers, allowing to intensify the process of phytoremediation of soils from oil.

Keywords: spent activated sludge, sorbents, fertilizers, sewage treatment.

1. Introduction

Chemical and petrochemical industry enterprises are the largest polluters of the environment, their activities produce a large number of polluted sewage. The main pollutants of sewage in chemical and petrochemical industry are heavy metal ions and oil products. Heavy metals are serious hazard in terms of their biological activity and toxic properties. Oil products getting into water affect ecological balance and make it impossible for biological systems to function normally.

In the world practice about 95% of chemical and petrochemical industries sewage is purified by biological methods, using activated sludge [1-13]. The main drawback of the classical scheme for cleaning sewage with activated sludge is the formation of a large amount of spent activated sludge as the result of the transformation of some of the initial contaminants into the active biomass. Utilization of spent activated sludge (SAS) is more labor-intensive than direct treatment of sewage, therefore, technologies that reduce the amount of this waste are economically viable and environmentally sound. There are a number of ways to recycle SAS: dropping into the seas and oceans, burning, burial in the soil, neutralization and use as organic fertilizers, as an additive in the preparation of various composts, etc.

Most often SAS is composted with household waste. Composting is a natural process of decomposition of organic waste under the influence of bacteria, insects, fungi and worms. The most widely this waste processing method is used in densely populated developed countries, where environmental problems are acute, and there is a shortage of natural resources. So, in the Netherlands 30-40% of household waste is processed for compost, in Austria and Belgium about 25%, in France 8%.

IOP Conf. Series: Earth and Environmental Science **337** (2019) 012077 doi:10.1088/1755-1315/337/1/012077

Studies have shown that adding sludge during to waste composting creates the conditions for the decomposition of the cellulosic components of the waste; in particular, it allows composting of waste containing large amounts of paper. At some composting plants in the United States addition of SAS allows to process containing up to 90% of paper waste to compost. In Germany semi-liquid precipitate with a moisture content of 92-96% (its share in the composted mass is 10-20%) and partly dewatered moisture 50-75% (its mass fraction is 14-34%) is used for this purpose.

Traditional way of field composting of waste in stacks under the open sky has not lost its importance. It is simple in technical terms, has low costs and high disinfecting effect. With this method compost is obtained from household waste and sewage sludge, which has a high agronomic value. There are two ways of composting: using so-called dynamic (with stirring of waste) and static (without stirring) stacks. Composting is carried out under the condition of forced aeration. Aerating improves the living conditions of microorganisms, thus significantly accelerating the process of waste rotting. , Recycling of household waste mixed with SAS sediment by the method of field composting is organized at many specialized enterprises. For example, in the USA 180 of the 200 composting enterprises are recycle using the method of field composting.

In Poland about 4,000 tons of compost per year is produced by field composting. The waste is stacked in three rows (the width of each row is about 2 m) with a distance between them of 2.5 m. After faeces are added the bulldozer on both sides levels the waste and forms a pile with a height of about 1.5 m. In one stack there are about 700 m^3 of waste, and in total 16,000 m^3 of garbage are laid annually at the plant. Faeces are added in proportion of 3 m^3 per 5 m^3 of waste. At the same time, the initial humidity is 60-65%, which is considered optimal for the process of fermentation and obtaining ready compost with a moisture content of at least 30%.

The largest garbage processing plant in Europe, composting household waste and SAS was built in Flensburg, Germany. Its productivity is 400 tons of compost per day. The plant processes all garbage and AAS sewage from Flensburg, which was previously dumped into the Baltic Sea. The composition of nutrients produced by compost is close to manure, and the amount of lime is superior to the latter.

Italian cities (Bologna, Ferrara, Madena, Bari, and others) have centers for waste dumping and composting. Waste of meat and fish industry, oil production, winemaking and sewage sludge - SAS - are added to urban waste. Due to this nitrogen content in composts rises to 4%, phosphorus - up to 3%, potassium - up to 2%. When composting waste in stacks, bacteria are added in the calculation of 700 thousand live cells per 1 g of compostable mass, 10-20% of which are actinomycetes and streptomycetes [14].

One of the most common methods of utilization of SAS is its use as an organomineral fertilizer (OMF), thus solving simultaneously a number of tasks: the need for storage (burial) is eliminated, soil fertility and crop yields are increased and surrounding environment is not polluted.

In the former USSR, the total annual volume of SAS for 1986 was 4-4.7 million tons of dry matter. By 1990 it had to increase to 9-10 million [15,16]. Unfortunately, at the present stage the level of SAS use in agriculture of CIS countries is still low. No more than 4-6% of sewage sludge is deposited into the soil from sewage treatment facilities in large cities. Most of the waste is taken to landfills that create dangerous foci of environmental pollution. At the same time, the useful components contained in the SAS are irretrievably lost.

At the beginning of the 21st century the problem of using SAS as fertilizers for soils attracted great attention.EU countries, using SAS in agriculture, agreed that the content of heavy metals in dry matter should not exceed the following values, mg/kg: 100 Ni, 100 Pb, 300 Cr, 600 Cu, 1500 Zn, 1 Hg, etc . In Russia usage of SAS in agriculture should comply with the norms of SanPiN 2.1.753-96 . According to these standards the content of HM in dry matter should be as follows, mg/kg, not more than 1000 Pb, 20 As, 15 Hg, 30 Cd, 400 Ni, 1200 Cr, 2000 Mn, 4000 Zn, 1500 Cu. For Russia there is a clear excess of the content of metals in soils3-4 times in terms of MACs and 10 times more than in the EU countries.

Many Russian and international scientists at the moment are developing technologies that allow the use of SAS as secondary raw material.

The authors of the patented technology [17] offer to process SAS containing heavy metals into fertilizers. For this purpose after thickening SAS is mixed with material containing poorly soluble calcium salts (CaCO3, CaSO4 · nH2O) in the form of natural minerals, at a ratio of 5-15 parts of the material per 100 parts of sludge, stirred for 3-6 hours at 6-300 °C, dividing solid and aqueous phase containing heavy metals. The mixing and stirring stages can be carried out prior to the thickening. HM are isolated from the aqueous phase by reagent precipitation, methods of ion exchange or adsorption.

To obtain fertilizer from SAS authors [18] proposed preliminary dehydration of SAS; mixing with sand from grit chambers; disinfection by means of a reagent and production of a recycling product by further neutralizing the sediments with a reagent. A complex powder is used as reagent: clay 40.0-60.0% by weight, lime 5.0-40.0% by weight, cement 5.0-40.0% by weight, complexing agent (mixture of metal oxides, ash, crushed slag, dolomite flour), ground limestone 5.0-10.0% by weight. The amount of reagent is 10.0-30.0% by weight of the mixture. The proposed invention provides neutralization of sewage sludge from various eco toxicants (heavy metals and their toxic complexes) and utilization of crushed and crushed waste, after mechanical cleaning.

Many scientists believe that in order to reduce soil contamination while using SAS as fertilizer it is essential to apply it in strictly controlled conditions, as soils react differently to its application. Its use should be excluded on clayly and compacted soils with very low or excessively high water resistance, and also in bare and poorly drained soils where periodic excessive moistening of the upper 50cm layer is possible, since soil pH has a significant effect on the degree of mobility of HM in it, increasing or reducing their absorption by plants. On acidic soils with pH less than 5.5 SAS should not be used at all. Soils with pH 5.5-6.5 should be preliminarily limed to a pH value exceeding 6.5.

Considering dry matter, SAS contains 37-52% of proteins, 20-35% of amino acids, fats, carbohydrates, as well as B vitamins. SAS is rich in nitrogen, phosphorus, copper, molybdenum and zinc. After special processing it can be used as a protein vitamin product for feeding animals, fish and birds. The protein is considered to be the most valuable raw material, millions of tons of which are practically not used. The processing mechanism is deep hydrolysis of SAS biomass into amino acids, followed by the synthesis of amino acid esters.

Various technological schemes for the production of a protein-vitamin product ("Belvitamil"), the production of a mixture of fodder yeast with silt and the production of technological vitamin B12 for the feed industry have been developed. The preparation of vitamin B12 consists of several stages of SAS processing: it is compacted to a moisture content of 95-96%, acidified with sulfuric acid to pH = 3 and sent to the reactor where it is heated up to 110 °C with steam. After cooling, solid particles are separated from the sludge in the centrifuge, which are then dried, crushed and used as a fertilizer, and the filtrate is treated with alkali in the coagulator to pH = 5. The mass is left to stand for 1 to 6 hours, the precipitate is separated in a centrifuge and processed into a fertilizer, and the filtrate is evaporated, then treated with alkali, dried and crushed and packaged.

To improve the quality of yeast, enriching them with the necessary amount of vitamin B12, yeast mixtures with SAS is prepared in a ratio of 10:1. The feed mixture is made by mixing condensed components of yeast with silt. SAS with the humidity of 96.5% from the secondary settling tanks is sent to the tank, from where it is fed into the mixer with concentrated yeast. The resulting mixture is dried.

Microbial protein can be obtained from SAS. To produce protein by extraction concentrated SAS is treated with hydrochloric acid so that its concentration after mixing is equal to the decinormal concentration. Under these conditions, the mixture is held for 24 hours with intermittent stirring. After separating the water the sludge is treated with 0.1 N NaOH solutions and held for 24 hours. After that the alkaline liquid fraction is separated from the sludge and heated to 80-900 °C and then cooled to 250 °C and the protein is precipitated by adding 10% hydrochloric acid to pH 4.6-4.7. The protein is dried and packaged.

According to another technology microbial protein can be extracted by alkaline or acid hydrolysis of cell membranes, transfer of intracellular proteins to the solution and by precipitating them at isoelectric point. The alkaline extraction is most effective at a temperature of 80-100 °C, followed by filtration and precipitation with acid at pH 4-6. At the same time 30-40% of the protein is extracted from the sludge.

To fulfill the requirement of reliable sterilization of the product, the temperature of the hydrolysis and extraction process was selected in the range of 100-135 °C. Heat treatment time is 20-30 min. During such a period cell membranes are sufficiently deeply destroyed, but intracellular proteins do not have time to hydrolyze into amino acids and polypeptides, not settling at the isoelectric point [19, 20].

The authors [21] proposed production of artificial ground mixtures on the basis of SAS after cleaning of household and industrial wastewater, for reclamation of slurry storage. The technical result is in the transformation of sludge-sediment into soil, characterized by ecological safety and low costs of production. The mixture for reclamation of slurry tanks contains OAI 100 parts by weight (w.p.), fly ash 15-20 w.p., fine-grained filler 15-30 w.p. (Clay or loam, alumino silicate rock 15-30 w.p.), and may additionally contain lime or Portland cement in the amount of 2-5 w.p. Mixture provides SAS cementing without its preliminary dehydration, which provides energy and labor saving, removal of sludge due to chemisorption of heavy metals contained in it, sterilization of the mixture - destruction of microorganisms, helminth eggs, etc.

The authors [22] propose technology for production of molded (briquetted, extruded or granulated) solid fuel from SAS. Molded fuel based on the dried mixture of crushed solid fuel and condensed SAS, containing, by weight: SAS with water content of 5-14% – 13-38, crushed solid fuel – the rest. The method for producing molded fuel comprises of mixing SAS and ground solid fuel, molding the mixture and subsequently drying the moldings. The mixture is formed at a pressure of 0.1-25 MPa. The moldings are dried at 50-180 °C for 1.5-0.4 hours or at an ambient temperature for 5-30 hours. Molded fuel possesses high mechanical strength, which allows reducing the cost of its storage, loading, unloading and transportation.

In Japan since 1981 there are about 500 units of final processing of waste in operation, while the amount of obtained SAS is about 24x10 m^3. SAS is burned and material is obtained, which consists of 80% of the dehydrated briquette and 11% of ash (resulting from combustion after dehydration) and other waste (dry or aggregated sludge) in an amount of 9%. These wastes (42%) are buried in the ground, dumped into the sea (36%), in the amount of 15% are effectively used as fertilizers for grassland and agricultural lands.

Perm State Technical University scientists developed technology for thermal energy neutralization of refineries SAS. According to this technology SAS is mixed with calcium oxide (1.0 to 1.5% dose), dehydrated in centrifuges (up to 40% moisture) with the removal of the fugate and dried (to 10% moisture), then fed to a continuous pyrolysis furnace. Pyrolysis gas at the outlet from the furnace enters the condenser, where the liquid phase of hydrocarbons is separated with a boiling point up to 400 °C. The liquid phase can be used as fuel or returned to the technological cycle of refining oil products. The primary heating of the furnace to the operating temperature is made by burning liquid fuel (fuel oil), after removing the pyrolysis furnace to the operating mode, its heating is carried out by burning pyrolysis gases. The fugate containing HM with a concentration of not more than 500 mg / l can be returned to the biochemical sewage treatment stage or processed for isolation of HM in the form of sparingly soluble compounds. Dyakov M.S., Glushankova I.S. used obtained organomineral composition after thermoenergetic neutralization of SAS as bio sorbents in oil and oil product spills on soils of podzolic soil [23-25].

2. Materials and Methods

Despite the development of numerous methods for utilizing spent activated sludge, this problem has not been completely solved. In industrial plants where sewage contains toxic elements, it is unacceptable to use spent activated sludge as a secondary raw material without preliminary neutralization. Therefore, work aimed at developing the technology for neutralizing spent activated sludge for the purpose of its further re-usage and solving environmental problems is very relevant and of great scientific and practical importance.

The authors of this article developed technology for neutralizing spent activated sludge with subsequent utilization as an adsorbent for cleaning oil-contaminated wastewaters and obtaining complex fertilizers that make it possible to intensify the processes of phytoremediation of soils from oil products.

IOP Conf. Series: Earth and Environmental Science **337** (2019) 012077 doi:10.1088/1755-1315/337/1/012077

Research was carried out on SAS, produced at sludge sites of JSC Saratov Oil Refinery (Saratov Refinery, Russia) after biological treatment of sewage waters of the enterprise.

3. Result and Discussion

Physicochemical studies of spent activated sludge from the sludge sites of "Saratov Refinery" were carried out on the X-ray fluorescence spectrometer "Spectroscan" and the gross composition of the elements was determined. It is shown that the spent activated sludge is characterized by high toxicity due to the presence of heavy metal impurities (Table 1).

Table 1. Average chemical composition of samples of spent activated sludge (In terms of dry matter).

Element	Concentration, mg / kg	Element	Concentration, mg / kg
Al	1.29	Cr	0.14
Si	2.74	Mn	3.97
P	3.46	Fe	27.80
S	4.58	Ni	0.14
Cl	0.70	Cu	1.04
K	1.09	Zn	1.01
Ca	8.97	Sr	0.34
Ti	0.40	B	0.57

The thermogravimetric analysis of the spent activated sludge on the derivatograph, at a heating rate of 10 deg/min, showed that during the combustion in the furnace the decomposition process proceeds in three stages: in the first stage (20-1500 °C), water and volatile organic substances are removed; at the second stage (150-500 °C) decomposition of organic substances proceeds; at the third stage, at temperatures above 500 °C, uncontrolled combustion of samples and the formation of a coke residue are observed.

Obtained physicochemical results showed the need to neutralize the spent activated sludge for secondary use, therefore it is necessary to carry out detoxification and thermal destruction of the spent activated sludge. Thermal destruction without access to oxygen makes it possible to obtain material containing amorphous carbon with adsorption properties and to use it as an effective adsorbent for extracting oil products from sewage. Detoxification from heavy metal ions from the spent activated sludge was carried out by treating the sludge with calcium-containing reagents (calcium hydroxide, calcium oxide, calcium carbonate, calcium sulfate, calcium chloride) in the amount of 1.0; 2.5 and 5.0% of the total mass of spent activated sludge. During the interaction of calcium-containing components with spent activated sludge, the ion exchange of heavy metal ions takes place on Ca^{+2} forming insoluble compounds of heavy metals that precipitate and are removed.

In order to remove pathogenic microorganisms, organic contaminants and active carbon formations, after detoxification and dehydration thermal destruction of spent activated sludge was carried out, at t = 105 °C. Conducted studies on the choice of heat treatment parameters: temperature (t = 350÷600 °C, step 50 °C) and time (τ = 20 · ÷ · 60 min, step 10 min) showed that the most rational is t = 500 °C and t = 30 minutes (Figure 1). Under these conditions are formed materials that ensure the highest purification efficiency (75%).

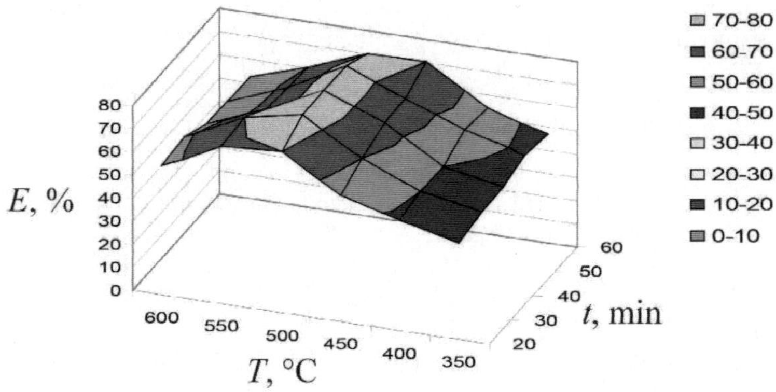

Figure 1. Relation between the efficiency of wastewater treatment from oil products with adsorbent neutralized spent activated sludge and temperature (°C) and time of destruction (min).

X-ray spectral microanalysis (Figure 2) of spent activated sludge after detoxification and thermal destruction showed the presence of reflection angles (0.330) characteristic of amorphous carbon. That allows us to recommend this material as an adsorbent.

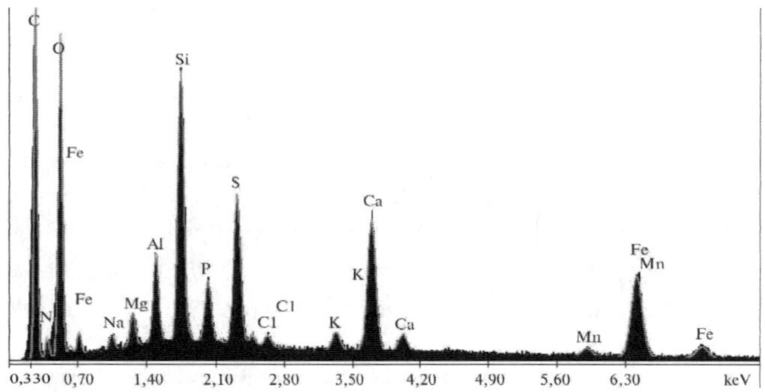

Figure 2. X-ray microanalysis of deactivated spent activated sludge.

The adsorbent obtained from the spent activated sludge was tested for the ability to extract oil products from the Saratov Refinery sewage water. Adsorption was carried out for 1 hour under static conditions in a ratio of 10 g of sorbent per 1 liter of wastewater with an initial concentration of NP - 32.6 mg / l. Then the adsorbent was extracted and the final concentration of oil products was determined on the AN-1 photometer in the accredited laboratory of Saratov Refinery (Table 2). Test results show that optimum ratio of calcium-containing reagents was 1% of the total mass of spent activated sludge.

Table 2. Change in the efficiency of purification (E) of oily sewage by adsorbents from the composition and quantity of reagent for detoxification of spent activated sludge under the selected thermal destruction conditions (t = 500 °C and τ = 30 min), (C_{ini} oil products = 32.6 mg / dm^3).

Detoxification reagent	Content of the reagent, % Wt.	Efficiency, %
Calcium carbonate, $CaCO_3$ + (T = 500 ° C, τ = 30 min)	1.0	65.0
	2.5	63.9
	5.0	53.4
Calcium sulphate, $CaSO_4$ + (T = 500 ° C, τ = 30 min)	1.0	61.6
	2.5	56.9
	5.0	50.1
Chlorine lime: 30% $Ca(OCl)_2$ + 35% $CaCl_2$ + 35% $Ca(OH)_2$) + (t = 500 ° C, τ = 30 min)	1.0	75.0
	2.5	69.2
	5.0	61.4
Calcium hydroxide, $Ca(OH)_2$ + (T = 500 °C, τ = 30 min)	1.0	74.0
	2.5	71.6

	5.0	66.6
	1.0	72.2
Calcium oxide, CaO + (t = 500 ° C, τ = 30 min)	2.5	68.2
	5.0	67.7

It has been established that the highest efficiency of treating sewage from oil products was shown by materials pretreated with chloric lime (E = 75%) and calcium hydroxide (E = 74%), taken in the amount of 1% of the total mass of spent activated sludge. The concentration of ions of heavy metals in the treatment of sludge with calcium-containing reagents in an amount of 1% is reduced significantly (Table 3).

Table 3. The content of heavy metal ions in the sludge mass before and after detoxification of spent activated sludge by various reagents.

Calcium-containing reagent, 1%	Concentration of metals, mg / kg				
	$Fe^{+2,+3}$	Cr^{+3}	Cu^{+2}	Zn^{+2}	Ni^{+2}
Initial OAI	27.80	0.14	1.04	1.01	0.14
Calcium carbonate ($CaCO_3$)	19.53	0.09	0.83	0.93	0.09
Calcium sulphate ($CaSO_4$)	21.21	0.11	0.95	0.92	0.10
Chlorine lime (30% $Ca(OCl)_2$ + 35% $CaCl_2$ + 35% $Ca(OH)_2$)	18.11	0.07	0.64	0.68	0.06
Calcium hydroxide, $Ca(OH)_2$	12.35	0.05	0.42	0.45	0.08
Calcium oxide, CaO	14.65	0.06	0.55	0.65	0.09

A technological scheme (Figure 3) for manufacturing and using an adsorbent based on spent activated sludge is proposed. Thermal processing of spent activated sludge produces flue gases of CO_2, SO_2, N_2O composition that must be to be trapped in order to prevent air pollution. Pyrolysis forms pyrolysis gases that can condense into the C11-C17 hydrocarbon fraction (M.S. Dyakov) with a boiling point up to 400 ° C, which are recommended to be captured and used as a liquid fuel to maintain the temperature in the furnace. Separated precipitate of insoluble heavy metals after additional processing can be used as pigment-fillers for paint.

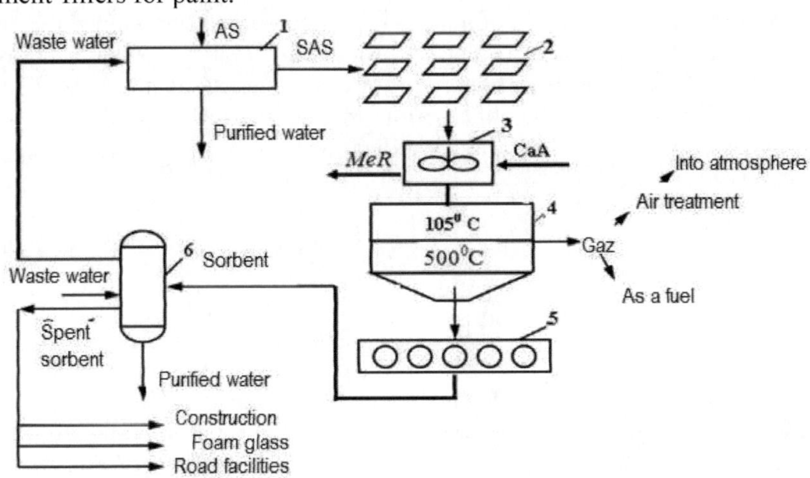

Figure 3. Technological scheme of manufacturing, use and utilization of adsorbent from spent activated sludge:
1 - aerotank;
2 - sludge areas;
3 - mixing chamber of reagents;
4 - muffle furnace;
5 - ball mill;
6 – adsorption.

The adsorption capacity of the adsorbent is determined primarily by the amount and size of pores of the material. Specific adsorption capacity (E, %) for iodine was 22.86%, indicating the presence of micropores with d = 1 nm, "E" for methylene blue was 115 mg / g, which is typical for mesopores with d = 1.5-1.7 nm. Since micro- and mesopores are present in the resulting material, it belongs to polyporous systems, which is confirmed by microstructural studies (Figure 4). It is shown that the dewatered spent activated sludge at a temperature of 105 °C has a microporous surface (Figure 4, a), while the adsorbent based on sludge after detoxification and thermal degradation has larger pores of tubular shape (Figure 4, b)

a) **b)**

Figure 4. Microstructure of spent activated sludge surface: a) dehydrated at t = 105 °C; b) after detoxification, dehydration and thermal degradation at 500 °C.

The permissible mass ratio of the adsorbent to the volume of sewage is 10 g / l. To determine the adsorption equilibrium time and the adsorption capacity, the adsorption of oil products from aqueous solutions by adsorbent (neutralized spent activated sludge) was carried out under static conditions with stirring and thermostating of the solution (t = 22 + 2 °C).

The spent adsorbent is proposed to be used as a burning additive in the production of ceramics and as a filler for construction, road construction, foam glass, asphalt and paving tiles.

The possibility of utilizing spent activated sludge as a complex fertilizer has also been studied. After detoxification with calcium-containing reagents (1% of the spent activated sludge mass) and thermal destruction at t = 350 °C, for 30 min. sludge was mixed with sawdust in various proportions (30:70, 50:50, 70:30%).

The resulting mixture was added into oil-contaminated soil and wheat was planted (Triticum). It was found that the use of a complex fertilizer on the basis of a mixture of neutralized activated sludge and sawdust in a ratio of 30:70% accelerates phytoremediation processes by 28% compared to soil purification technology without the application of fertilizers (Figure 5). The use of neutralized spent activated sludge as fertilizers will significantly reduce the permissible content of oil products in soils (MACs in soils are not established), solve the problem of its utilization and expand the production of organic fertilizers, that are of a very high demand.

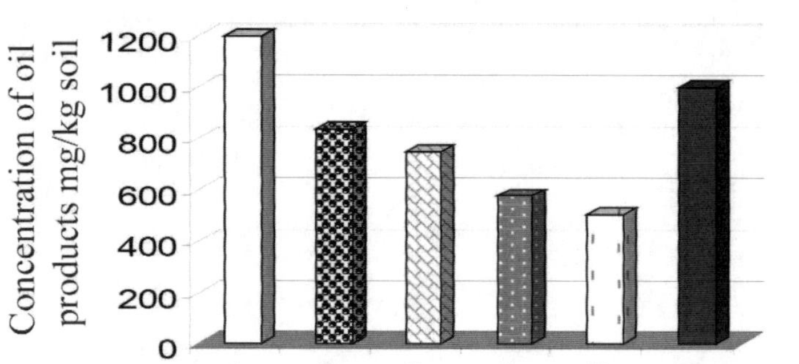

Figure 5. Concentration of oil products in the soils of the sludge collector after the phytoremediation process (wheat (Triticum)):
1 - without phyto-mediation processes;
2 - without the use of fertilizer;
3 - using fertilizers (70% of SAS: 30% of sawdust);
4 - using fertilizers (50% SAS: 50% of sawdust);
5 - using fertilizers (30% of SAS: 70% of sawdust);
6 - permissible level of oil products content in soil.

The calculation of ecological and economic indicators of production of adsorbent from SAS in the amount of 12 tons per year showed that its cost was 44.9 rubles per kg with the selling price of 56 rubles per kg, thus capital costs will pay off in 1.6 years. The calculated value of the prevented ecological and economic damage of Saratov Refinery water resources contaminated with oil products by using an adsorbent based on spent activated sludge is 2.3 million rubles.

4. Conclusion

1. Main technologies for obtaining secondary raw materials from SAS and ways of using it (composting, use as fertilizer, raw materials for protein and vitamin supplementation for animals and birds, amino acids, technical vitamin B12 for mixed fodders, etc.) were considered.
2. Physic-chemical studies of spent activated sludge of "Saratov Refinery" have been carried out, the technology of its neutralization with rational parameters of sludge neutralization processes: detoxification (1% of calcium-containing reagent - chloric lime) and thermal destruction (temperature T = 500 °C and time t = 30 min) was offered.
3. The possibility of using an adsorbent from spent activated sludge obtained with the established parameters for effective purification of oily wastewater (E = 70-75%) has been offered for the first time, and methods and directions for utilization of spent adsorbents have been proposed.
4. The technology of complex fertilizer production based on a mixture of spent activated sludge and sawdust was developed. It has been established that the application of complex fertilizer (spent activated sludge and sawdust in a ratio of 30:70%) to oil-contaminated soils accelerates the processes of phytoremediation of soils from petroleum products by 28%.

References

[1] Chusov A N, Bondarenko E A and Andrianova M J 2014 Study of electric conductivity of urban stream water polluted with municipal effluents *Appl. Mech. Mater* **641–642** pp 1172–5.

[2] Andrianova M and Bondarenko E 2016 Changing of Cations Concentrations in Waters of Polluted Urban River *MATEC Web of Conferences* **53**

[3] Arakcheev E N, Brunman V E, Brunman M V, Volkov A N, D'yachenko V A and Kochetkov A V 2016 Automated production system for reagents used in water and wastewater processing *Russ. Eng. Res.* **36** pp 745–50

[4] Didenko N, Skripnuk D and Mirolyubova O 2017 The effects of human behavior on fresh water resources *Int. Multidisciplinary Scientific GeoConference Surveying Geology and Mining Ecology Management, SGEM* **17** pp 901–10

[5] Andrianova M J, Molodkina L M and Chusov A N 2014 Changing of contaminants content and disperse state during treatment and transportation of drinking water *Appl. Mech. Mater* **587–589** pp 573–7

[6] Arakcheev E N, Brunman V E, Volkov A N, D'yachenko V A, Petkova A P and Sorokin A V 2016 Complex Electrolytic Unit for Producing Anode Liquor and Ferrate to Treat Waters Contaminated by Oil-Refining Products and Harmful Admixtures *Chem. Pet. Eng* **52** pp 143–8

[7] Pavlov S, Novikov A, Pavlov A, Skvortsova O, Nikonova O and Semanina E 2016 Methods of Improving Water Treatment Systems for Individual Residential Houses *MATEC Web of Conferences* **73**

[8] Sobgajda N A 2011 Sorption materials for waste and natural waters purification from oil products [Sorbcionnye materialy dlya ochistki stochnyh i prirodnyh vod ot nefteproduktov] *Vestn. Har'kovskogo Nac. avtomobil'no-dorozhnogo Univ.* **52** pp 120–4

[9] Sobgajda N A, Ol'shanskaya L N and Makarova Y A 2010 Waste water purification from oil products by composite filters based on production wastes Ochistka stochnyh vod ot nefteproduktov kompozitnymi fil'trami na osnove othodov proizvodstv. *Him. i Neftegazov. Mashinostr.* **3** pp 37

[10] Arakcheev E N, Petkova A P and Brunman M V 2016 Disinfection and water treatment using anode liquor and sodium ferrate and apparatus for their integrated production *Water Ecol. 2016* pp 26–36

[11] Osman M R, Azid A, Yunus K, Mustafa A D, Amran M A and Azaman F 2015 Assessment on bacteria in the heavy metal bioremediation *Malaysian J. Anal. Sci.* **19** 1405–14
[12] Azoddein A A M, Haris H B and Azli F A M 2015 Treatment of palm oil mill effluent (POME) using membrane bioreactor *Malaysian J. Anal. Sci.* **19** pp 463–71
[13] Syarifah Nazirah W I , Norhaniza Y, Farhana A and Nurasyikin M 2017 A review of oilfield wastewater treatment using membrane filtration over conventional technology *Malaysian J. Anal. Sci.* **21** pp 643–58
[14] Moffat A J, Matthews R W and Hall J E E 2005 The effects of sewage sludge on growth and foliar and soil chemistry in pole-stage Corsican pine at Ringwood Forest, Dorset, UK *Can. For. Res.* **21**
[15] Sunyaev K K, Zhmykhova E N and Chudinova S D 2004 Methodical aspects of utilization of municipal water from sludge drying beds *Proceedings of the international forum on the problems of science, technology and education* pp 110
[16] Sunyaev K K, Slipets A A and Tyutyunkova M V 2005 Technology and economics of soil way of municipal waste waters utilization from sludge drying beds *Collection of scientific works* **2** pp 41
[17] Panov V P, Zykova I V and Makashova T G 2001 Patent Russian Federation: A way for heavy metal removal from excess activated sludge
[18] Knatko V M, Shcherbakova E V, Knatko M V and Vladimirskaya N V 2007 Patent Russian Federation: A method for complex processing and utilization of municipal sludge
[19] Popova E E, Galyuzhnaya V I and Vikhoreva G A 2007 *Processing of Activated Sludge* (Moscow: High School)
[20] Repnikova A S, Yakovlev V I and Vikhoreva G A 2008 *Sewage and Activated Sludge* (Moscow: High School)
[21] Knat'ko V M, Knat'ko M V, Sherbakova E V and Maslennikova E S 1997 Mixture for recultivation
[22] Luriy V G 1999 Patent Russian Federation: Moulded fuel and methods for it's obtaining.
[23] Dyakov M S, Glushankova I S and Gulyaeva I S 2008 Complex processing of municipal sludge with obtainig of products *Environ. Prot. oil gas Sect.* **12** pp 29
[24] Dyakov M S , Vaysman Y I, Glushankova I S and Kalinina E V 2008 Thermal utilization of activated sludge of municipal purification facilities of Perm *Equatek-2008, collection of reports: International Congress "Water: Ecology and Technology"* pp 38
[25] Dyakov M S, Bikmansurova E K, Glushankova I S, Gulyaeva I S and Batrakov V A 2008 Application of organo-mineral composition as a structured sorbent at phytoremediation of sludge *Prot. Environ. oil gas Sect.* **12** pp 27

IOP Conf. Series: Earth and Environmental Science **337** (2019) 012078 doi:10.1088/1755-1315/337/1/012078

Lemna minor cultivation in the climatic conditions of Saint Petersburg

T Kuznetsova[1,*], N Politaeva[1], A R Ivanova[1], E I Lezhnev[1], V A Polyanskii[1], B G Matisov[1], L R Mukhametova[2] and E V Karmazina[3]

[1]Peter the Great St. Petersburg Polytechnic University, St. Petersburg, Russian Federation
[2]Kazan State Power Engineering University, Kazan, Russian Federation
[3]Vologda State University, Department of biology and ecology, 160000 Lenin st., 15, Russia

* Corresponding author: tano_lovely@mail.ru

Abstract. Cultivation of duckweed Lemna minor in conditions of St.Petersburg was carried out. Two cultivation variants were studied: 1- in a natural reservoir of Petergof; 2 - in an artificial reservoir at the yard of Peter the Great St. Petersburg Polytechnic University (Russia) during a period from May to July 2017. The following cultivation conditions were determined: lighting, temperature. The population growth intensity was estimated by increase of plant cover area of the pond. The most favourable conditions appeared to be at natural pond in Petergof. Daylight features, lightning intensity of the Leningrad Region are less favourable than that for south Russian regions, so biomass production rate is drastically lower. However, the climatic characteristics (moderate climate, transient from continental to marine) allows increasing L. minor cultivation period: from May to November inclusive. We carried out atomic-physiological investigation of fronds in summer and autumn, as well as epidermis and chlorenchyma cell analysis, which allow one to characterize photosynthesys system of the plants.

1. Introduction

Duckweed is quickly vegetatively divided in summer in fresh ponds with ditch-water, and form a green carpet on pond whole surface, it duplicates its' mass in 6 days [1]. L. minor is a typical species at middle-Russian area. It may produce 200 tons of biomass from a hectare during a season at multiple gathering. At Uzbekistan 276 tons were gathered from 1 hectare during 8 months [2].

Literature analysis have shown that the major directions of L. minor cultivation investigation are defined by further usage of its biomass:

- usage of L. minor in ecotoxicological investigations and tests [3–13];
- directed cultivation for obtaining valuable components in Lemna minor tissues. Duckweeds accumulate in fronds flavonoids, diterpenoids, tannins, B1, B2 and C vitamins, steroids, unsaturated fatty acids, including Omega-3, giving rise to its' medicinal properties and its' wide application in medicine [14].
- cultivation of L. minor with further usage as a lifestock feed (duckweed has a protein content up to 45% in dry weight) [15,16];

- L. minor cultivation for biomass production as a source for bioethanol and biogas obtaining. These technological cycles are based on enzymatic hydrolysis of starch, which is contained in duckweed [15,16];
- duckweeds are of interest as a potential biofactory for fused proteins production, such as vaccine proteins, serum albumin, hemoglobin, collagen; moreover, a genetic transformation of duckweeds might be made in order to enhance its' stability to pollutants, for example, heavy metals [17,18].

2. Materials and methods

2.1 Determination of intensity of L. minor population growth

The evaluation of L. minor growth intensity in natural conditions was carried out in accordance with pond surface incrustation area. For this purpose, square frames were cut from expanded polyethylene: the inner enclosure is 20×20 cm (400 cm^2), the external dimensions are 26×26 cm (676 cm^2), Figure 1.

In order to achieve high cultivation efficiency, one should use pure culture. Storage of L. *minor* stock culture is performed at 4-10°C, using MS-environment, pH is 5.7 [19].

Duckweed is gathered using fork or shovel devices with mesh-tip.

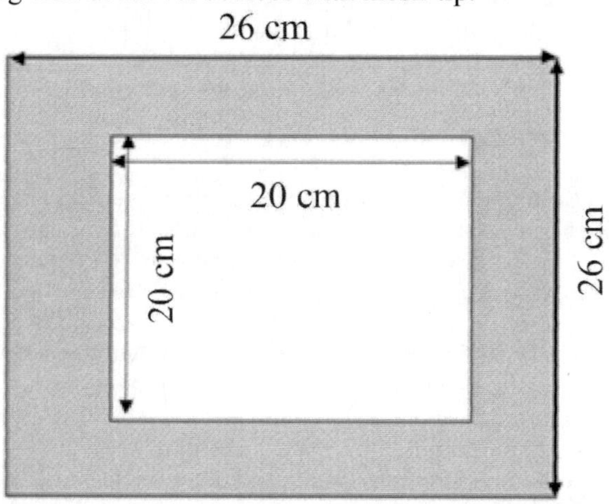

Figure 1. A scheme of enclosing frame.

The frame is placed at the pond surface. It should not sink. 10 plants are placed in the inner part of the frame. At intense population growth the frame is photo-recordered once a day (Figure 2).

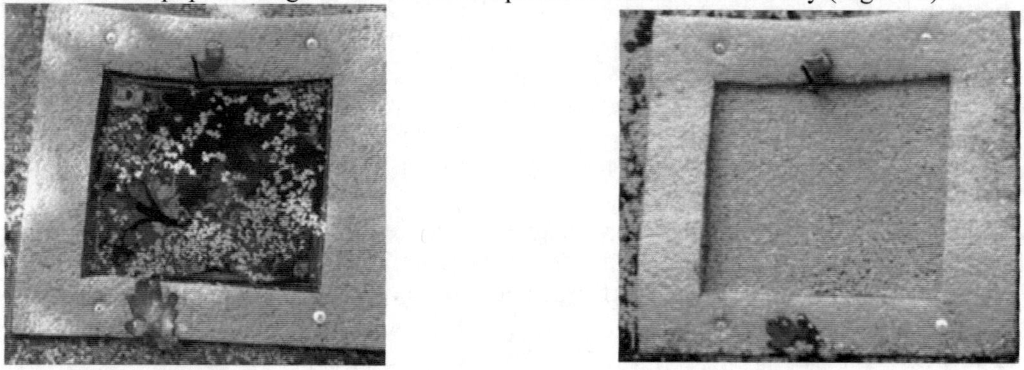

15 Days 34 Days

Figure 2. Photographs of L. *minor* population cultivated in enclosing frame.

The photographs are contrasted using Levenhuk ToupView software. The software instruments are used for total area estimation (S_{tot}) in pixels (px), which is enclosed by the inner frame. Also, it is used for determination of total area, which is occupied by plants (S_p) in pixels (px). Pond incrustation ratio (W, %) is determined according to the formula (1):

$$W = \frac{S_p}{S_{tot.}} 100, [\%] \quad (1)$$

Pond incrustation ratio is determined minimum in three frames and, using methods of statistical data processing, the average value of pond incrustation ratio and confidence range ($x_{av} \pm S_x$) are determined. Using the obtained data the cultivation curve is plotted: relation between pond surface incrustation area in enclosing frame (%) and cultivation time (days).

1.2 Morphological-anatomical investigation of L. minor

Morphological-anatomical study of L. minor are carried out by microscopy of intravital cells (colored and non-colored). Magnification power is 640 times. Epidermis investigation is carried out without its separation from parenchyma. In order to investigate chlorenchyma and chloroplast cells, we cut fronds by a thin blade at low magnification (×7).

For describing shape and dimensions of epidermis and parenchyma, and chloroplast dimensions, we made microphotograps using digital camera IS-500 and "Microanalysis FOTO" software for 10 fields of vision. Microphotographs analysis (colored and non-colored photographs) was performed using Levenguk software, which allows one to adjust contrast and brightness and additionally magnify the image.

In order to characterize major epidermis cells we determined its' area and perimeter, in order to characterize the degree of tortuosity of anclinal walls, we determined the coefficient (2):

$$K_a = \frac{P}{S} 100, [\mu m\text{-}1] \quad (2)$$

where P is anticlinal wall projection perimeter of major epidermis cells, μm;

where S is anticlinal wall projection area of major epidermis cells, μm2;

The diagnostic indicator which characterize metabolism (photosynthesis) is stomata condition: open stomata is a necessary condition for high metabolism intensity; the degree of stomata openness can be derived from formula (3):

$$\text{SOD} = \frac{S_2}{S_1} 100, [\%] \quad (3)$$

where SOD is stomata openness degree (%);

S^1 is stomata projection area, μm^2;

S^2 is stomata cleft area, μm^2;

Stomata projection area, and stomata cleft area can be determined using microphotographs analysing software, for example, Levenguk software. Chlorenchyma cell and chloroplast dimensions were determined according to its' projection area at frond cuts. Projection area was determined from the photographs by Levenguk software instruments.

3. Results and discussion

3.1 L. minor cultivation conditions

The experiment was conducted in Saint Petersburg (Russia) during the period from May, 29 to July, 3, Figure 3: variant 1 - in natural pond in southern Saint Petersburg (Petergof); variant 2 - in an artificial pond in northern Saint Petersburg (Saint Petersburg State Polytechnical University named after Peter the Great, SPBPU, Novorossiyskaya Str., 48) Colonies consisting of 2-4 green plates are taken from stock culture and moved into the pond for cultivation.

IOP Conf. Series: Earth and Environmental Science **337** (2019) 012078 doi:10.1088/1755-1315/337/1/012078

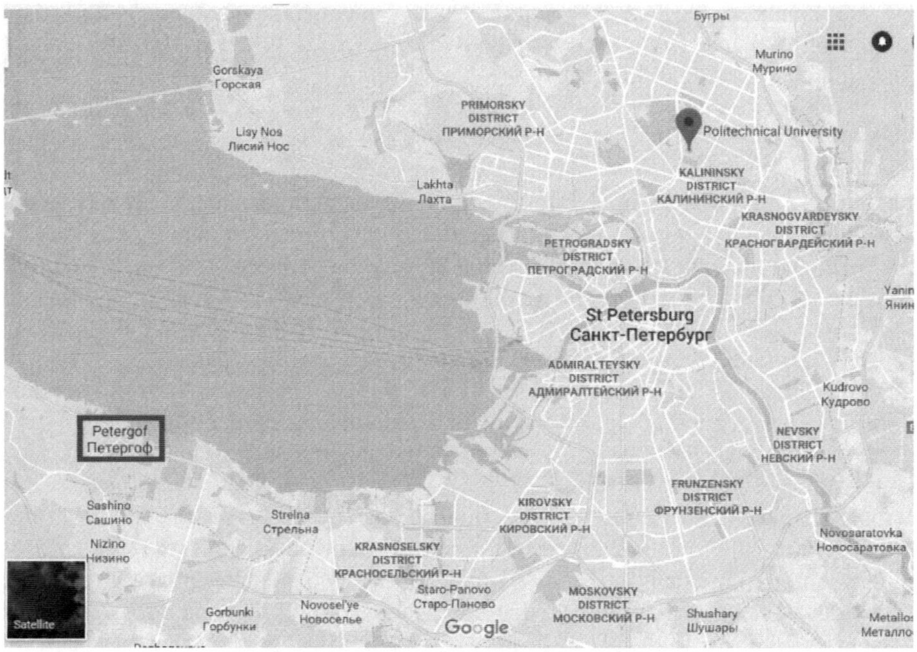

Figure 3. Map of Leningrad region.

In natural conditions of Leningrad region *L. minor* cultivation may be performed from April to November inclusive. The most efficient constant temperatures are 19-26 °C. The plants withstand short-term frosts. However, at long-term water temperature decrease below 3 °C we observed slowdown of plant growth rate. Day average temperatures of *Lemna minor* cultivation places are reliably equal, Figure 4.

Possible pH ranges in which *L. minor* is cultivated are 3.5-8.5 [20], at this optimal pH values are 6.3-8.0.

Cultivation took place at natural light. Lightning intensity in day time is the following: South of Saint Petersburg 405-1570 lx; North of Saint Petersburg 1350-4000 lx. Optimum range was from 405 to 9000 lx (3000 lx in average).

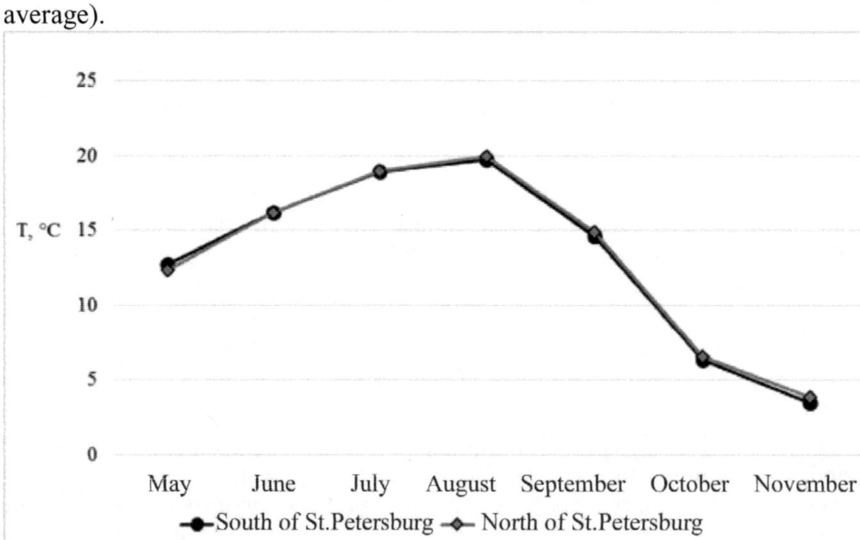

Figure 4. Day average temperatures of Saint Petersburg during Lemna minor cultivation.

3.2 L. minor cultivation curves

The most intensive growth of *L. minor* was observed in a natural reservoir (in pond in Petergof, Figure 5) in comparison with population in artificial pond (SPBPU). The time of *L. minor* population duplication is 20 days.

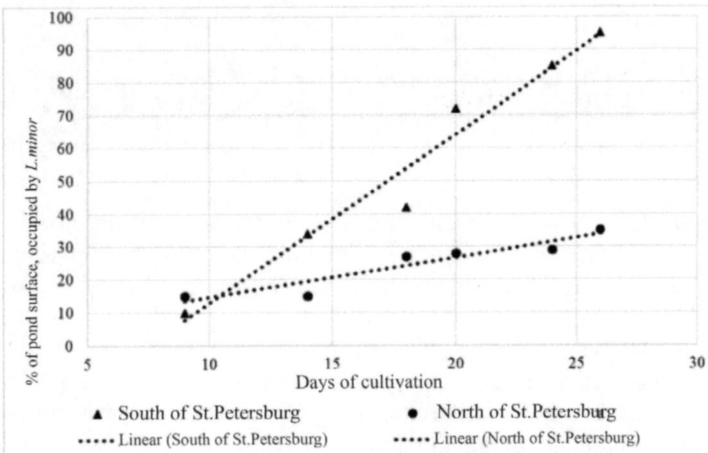

Figure 5. *L. minor* population growth at Saint Petersburg conditions (natural reservoir in Petergof, South of Saint Petersburg; artificial reservoir in SPBPU, North of Saint Petersburg).

Minimum sum of daily-average temperatures, which is necessary for population growth up to 100% of pond area in natural conditions is 406 °C at day lightning intensity 405-1570 lx.

3.3 Lemna minor plant characteristics at cultivation in Saint Petersburg conditions

Dry substance content in plant tissues (was determined by a commonly used method [21]) was 14.20±0.34%, water content was 85.79±0.34%.

Morphological characteristics were determined for *L. minor* population intensive-growth phase for natural conditions of Saint Petersburg (Table 1).

The features of fronds ultrastructure allow one to characterize the condition of photosynthesysing apparatus [21,22], which is directly relevant to accumulation of organic substances in tissues and biomass growth.

Table 1. Morphological characteristics of *L. minor* population in intensive-growth phase for natural conditions of Saint Petersburg, July 2017 (n=20).

Rootlet length, cm	Rootlet number	Diameter of frond rosette, cm	Frond number
1.64±0.13	7.20±0.68	0.89±0.06	3.55±0.39

The indicators of intensive metabolism are the following:

- large open stomata, its high concentration (which characterizes intensive gaseous exchange), Figure 6a.
- large chlorenchyma cells (accumulation of storage polysaccharides, intensive photosynthesis), Figure 6b.
- large chloroplasts (which characterize photosynthesis intensity), Figure 6b.

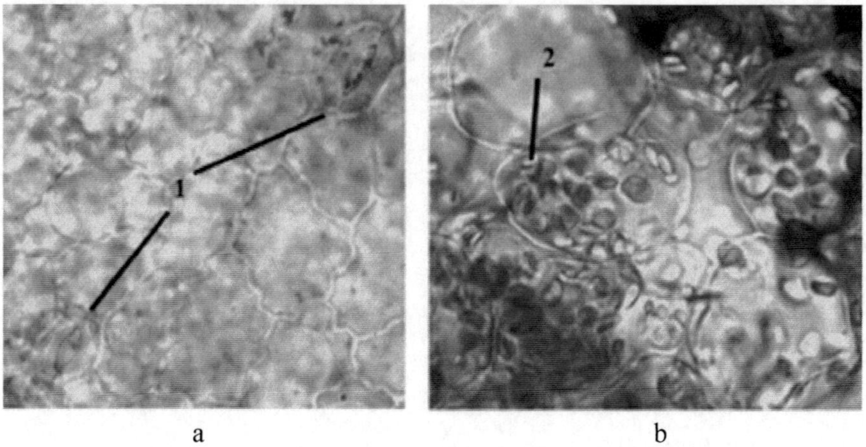

Figure 6. Microscopy image of L. minor fronds: a - epidermis (1- stoma); b - photosynthesising parenchyma cells (2 - chloroplasts).

We have carried out an anatomical investigation of *Lemna minor* fronds, which was cultivated in natural conditions of Saint Petersburg in July and October 2017. The population growth in July is intensive, while it slows down in October.

Chlorenchyma cell projection area in the phase of intensive growth is reliably equal to the data presented in literature [22], in autumn chlorenchyma cells increase for four times (Table 2), which is driven by intensive cell vacuolization (small vacuoles merge in a whole) and accumulation of reserved nutrients.

Table 2. Anatomical characteristics of *L. minor* fronds in natural St. Petersburg conditions, July-October 2017, n=50.

Cultivation period	S of chlorenchyma cell under epidermis	S of stoma projection	SOD, %	S of basic epidermis cells	Ka
summer	153.87±14.89	219.32±11.49	11.29±1.23	388.86±66.84	1.73±0.18
autumn	625.42±33.62	252.53±15.36	18.38±1.26	316.46±54.62	2.15±0.079

Note: SOD is stomata openness degree; Ka is coefficient of tortuosity of anticlinal walls of major epidermis cells.

Chloroplast dimensions in plant frond chlorenchyma at population intensive growth is somewhat less than literature data (Ekaterinburg, Russia) [22]. This might be connected with lower Sun intensity in St.Petersburg. However, at toxical phenolic impact we observed more intensive decrease of chloroplast dimensions.

Epidermis cells are more sensitive to changes of environmental factors, so its' characteristics are the most informative. Stomata of the upper epidermis are rather large and have a tendency to growth in autumn (Table 2).

Open stomata is a necessary condition of metabolism high intensity [23]. In autumn stomata dimensions and the degree of its' openness increase (Table 2), which states for high intensity of metabolic processes and accumulation of storage metabolites in tissues. This may be used for prediction of storage substances content in *L. minor* biomass.

By autumn major epidermis cells are formed smaller and with more tortuous anticlinal walls. Cell wall of young cells is rather elastic, so in such conditions cells can grow in size. At ageing or negative factors impact, induration of cell wall takes place, the cell stops growing, but non-uniform induration of cell tissue various regions results in edge tortuosity increase. Consequently, in autumn the major cells have stronger cell wall, which is driven by impact of unfavorable factors (low temperatures, short-term frosts) on *Lemna minor* fronds. These plants become less able to reproduction.

4. Conclusions

1. *Lemna minor* growth conditions appeared to be more favorable at natural pond of Petergof, rather than at artificial reservoir at the yard of Peter the Great St. Petersburg Polytechnic University.
2. Daylight features, lightning intensity of the Leningrad Region are less favourable than that for south Russian regions, so biomass production rate is drastically lower. However, the climatic characteristics (moderate climate, transient from continental to marine) allows increasing *L. minor* cultivation period: from May to November inclusive.
3. Anatomic-morphological characteristics of *L. minor* fronds are in good agreement with literature data (obtained for south regions of Russia). The most informative characteristics are the following: dimensions and openness degree of stomata, chlorenchyma cell dimensions, chloroplast dimensions. In autumn we observe a drastical increase of chlorenchyma cells, which is driven by vacuolizations of this cells and accumulation of nutrients. SOD remains high both at summer and at autumn, which is a necessary condition for intensive methabolic processes, including photosynthesis.

References

[1] Grudzinskaya I A 1982 Flowering plants *Plant life* vol 6 *[Zhizn' rastenij tom 6]* (Moscow: Izd-vo «Prosveshchenie») pp 493–500
[2] Spirulina & Duckweed. www.Istina1888.narod.ru
[3] Ol'shanskaya L N, Sobgajda N A and Tarushkina Y A 2011 *Phytoremediation Technologies for Protection of Hydrosphere: Monography [Fitoremediacionnye Tekhnologii v Zashchite Gidrosfery: Monografiya]* (Saratov: SGTU)
[4] Hurd N A and Sternberg S P K 2008 Bioremoval of aqueous lead using Lemna minor *Int. J. Phytoremediation* **10** pp 278–88
[5] Bokhari S H , Ahmad I , Mahmood-Ul-Hassan M and Mohammad A 2016 Phytoremediation potential of Lemna minor L. for heavy metals *Int. J. Phytoremediation* **18** pp 25–32
[6] Chaudhuri D, Majumder A, Misra A K and Bandyopadhyay K 2014 Cadmium Removal by Lemna minor and Spirodela polyrhiza *Int. J. Phytoremediation* **16** pp 1119–32
[7] Uysal Y and Taner F 2009 Effect of pH, temperature, and lead concentration on the bioremoval of lead from water using Lemna minor *Int. J. Phytoremediation* **11** pp 591–608
[8] Prasertsup P and Ariyakanon N 2011 Removal of chlorpyrifos by water lettuce (Pistia Stratiotes L.) And duckweed (Lemna Minor L.) *Int. J. Phytoremediation* **13** pp 383–95
[9] Basile A, Sorbo S, Conte B, Castaldo Cobianchi R, Trinchella F and Capasso C 2012 Toxicity, accumulation, and removal of heavy metals by three aquatic macrophytes *Int. J. Phytoremediation* **14** pp 374–87
[10] Lakatos G , Mészáros I, Bohátka S, Szabó S, Makádi M and Csatlós M 1993 Application of Lemna species in ecotoxicological studies of heavy metals and organic biocides *Sci. Total Environ.* **134** pp 773–8
[11] Fenske C, Daeschlein G, Günther B, Knauer A, Rudolph P and Schwahn C 2006 Comparison of different biological methods for the assessment of ecotoxicological risks *Int. J. Hyg. Environ. Health* **209** pp 275–84
[12] Radić S, Stipaničev D, Cvjetko P, Mikelić I L, Rajčić M M and Širac S 2010 Ecotoxicological assessment of industrial effluent using duckweed (Lemna minor L.) as a test organism *Ecotoxicology* **19** pp 216–22
[13] Khellaf N and Zerdaoui M 2009 Growth response of the duckweed lemna minor to heavy metal pollution *Iran. J. Environ. Heal. Sci. Eng.* **6** pp 161–6
[14] Duckweed – Uses, Benefits and Side Effects.

IOP Conf. Series: Earth and Environmental Science **337** (2019) 012078 doi:10.1088/1755-1315/337/1/012078

[15] Xu J, Zhao H, Stomp A-M and Cheng J J 2012 The production of duckweed as a source of biofuels *Biofuels* **3** pp 589–601.
[16] Cui W and Cheng J J 2015 Growing duckweed for biofuel production: a review. *Plant biology* **17** pp 16-23.
[17] Chhabra G, Chaudhary D, Sainger M and Jaiwal P K 2011 Genetic transformation of Indian isolate of Lemna minor mediated by Agrobacterium tumefaciens and recovery of transgenic plants *Physiol. Mol. Biol. Plants* **17** pp 129–36.
[18] Gaydykova S E 2011 *Development of system of genetic transformation of Lemna minor L. [Razrabotka sistemy geneticheskoj transformacii ryaski maloj (Lemna minor L.)]*. PhD Thesis.
[19] Kittiwongwattana C and Vuttipongchaikij S 2013 Effects of nutrient media on vegetative growth of Lemna minor and Landoltia punctata during in vitro and ex vitro cultivation *Maejo Int. J. Sci. Technol.* **7** pp 60–9.
[20] Khoan C K 2012 *Usage of aquatic macrophytes for water purification from heavy metals [Ispol'zovanie vodnyh makrofitov v ochishchenii vody ot tyazhelyh metallov]* PhD Thesis.
[21] Rondjina D A, Ivanov L A and Pyankov V I 2010 Chemical composition of leaf and structure of photosynthesising apparatus of higher aquatic plants *Plant Physiol.* **57** pp 389–97.
[22] Kislitsyna M N and Borisova G G 2016 Answer reactions of Lemna minor L. on the action of exogenyc phenolic compounds *Sci. Reports Petrozavodsk State Univ.* **4** pp 54–8.
[23] Kuznetsova T A 2015 Adaptation changes of leaf surface of Ribes nigrum L. depending on various conditions of water supply *Sci. News Belgorod State Univ. Nat. Sci.* **31** pp 29–36.

IOP Conf. Series: Earth and Environmental Science **337** (2019) 012079 doi:10.1088/1755-1315/337/1/012079

Flexible allocation of costs between electricity and heat as a factor for improving the CHP competitiveness

E Sukhareva[1*], A V Fedyukhin[1], E A Kikot[2], B G Matisov[2], N V Nikitkov[2], V V Izrantsev[2], V G Knorring[2], G A Kondrashkova[2], T R Akhmetov[3] and A A Sinitsyn[4]

[1]National Research University «Moscow Power Engineering Institute», Moscow, Russia Krasnokazarmennaya 14, Moscow, 111250 Russia
[2]Peter the Great St. Petersburg Polytechnic University, Polytechnicheskaya 29, Saint Petersburg, 195251 Russia
[3]Kazan State Power Engineering University, Kazan, Russian Federation
[4]Vologda State University, Department of biology and ecology, 160000 Lenin st., 15, Russia

* Corresponding author: evgeniyavs@mail.ru

Abstract. The review of cost allocation methods when forming cost of energy with the combined production was carried out. Their strengths and weaknesses were revealed. The article is devoted to the development of the distribution system costs in the form of energy cost.

Key words: combined heat and power (CHP); cost allocation method; method of separation of fuel consumption; cost of energy.

1. Introduction

The existing pricing rules for combined heat and power (CHP) plants participation in electricity market do not take into account major CHP function, which is combined production of heat and power [1,2]. During market transformations we observe an efficiency reduction of CHP production and commercial activities in comparison with that for condensation power plants (CPP). One of the main reasons for it is that electrical and thermal energy cost formations do not take into account modern competitive market requirements [3–5].

One of the main issues concerning bare cost formation for power produced at a CHP plant is development of economically feasible technique for total expenses allocation between electrical and thermal energies and calculation of bare cost of an energy unit [6]. Power enterprise management efficiency is greatly affected by cost allocation method which is used for bare cost formation.

Calculation and allocation of expenses at CHP plants are worldwide based only on market mechanisms and do not take into account production process technology and fuel consumption division [7].

In Germany the approach for thermal and electrical energy cost calculation at CHP plants is fully economic, in Russia it is called independent cost efficiency approach [8].

Two approaches are used in Denmark [8]. One of them is economical and is based on comparison of alternative schemes for thermal and electrical energy production. The other one states that electrical power of cogeneration turbine decreases when transition from condensation operation mode to heat supply from turbine bleed (for a predetermined hourly heat consumption and specified technical conditions) takes place. Further, for turbine unit the derating factor is determined. It is the electrical power decrease at transition to cogeneration mode for 1 GJ of heat, supplied from turbine bleed at constant steam consumption to the turbine unit head. This approach is used for analysis of power plant operation efficiency, while the first approach is chosen for calculation of cost indicators.

In France, the above mentioned approaches for cost allocation are used when electrical energy price is predetermined and after its sale income is estimated. Further, it is subtracted from total production costs, and the rest is charged to heat. This is known as thermal energy price [8]. In order to estimate the efficiency of fuel usage at CHP plant, the following approach is used: the heat generated at a CHP is put equal to the heat produced at the boiler house. This means that specific reference fuel consumption at CHP and boiler house accept to be equal. The rest fuel consumption is charged to electrical energy. Consequently, the full effect from combined generation is charged to electrical energy. This approach corresponds to Russian physical one.

Despite a large variety of cost allocation approaches for combined cycle, there is not any dispute about the way for CHP cost indicators determination. It is carried out solely basing on market mechanisms and it isn't connected with production process technology and fuel consumption division at a CHP [6].

It should be noted, that the major part of thermodynamic methods are based on the following principles:

1. Allocation of fuel consumption between generated electrical and thermal energies;
2. Allocation of relative-fixed costs is proportional to the chosen allocation base (fuel consumption, fuel cost, etc.)

The economical approaches are commonly based on CHP total costs allocation between generated energy types according to methods for cost accounting of 1kW·h electrical energy and 1GJ thermal energy [9].

The choice of cost allocation method between thermal and electrical energy has a great impact on the production bare costs [10,11,12].

Inefficiency of the described approaches is proved by the absence of the feasibility for CHP to overcome the economic crisis arisen from collapsing demand for the generated thermal energy.

CHP market success is defined by its competitiveness for heat and power. Let us consider flexible cost allocation between power and heat as one of the factor for CHP competitiveness improvement.

Now the cogeneration effect is charged to the both energy types for the existing cost allocation methods. However, now it is not possible to calculate the generated energy cost depending on the energy market environment and flexibly respond to its change.

The first step is devoted to consideration of the flexible cost allocation impact on the ability to function at the regional power market and local CHP heat market.

2. Experimental data and discussion

For several CHP plants of the Moscow region we have calculated a variety of variants for cost allocation between heat and power. Heat costs are charged from 0 to 100%, and, consequently, from 100 to 0% for the electrical energy, with 5% step. We have created a cost allocation diagram, which shows all possible variants of cost allocations expressed in "heat cost"-"electrical energy cost" coordinates (Figure 1). Bare costs for the introduced above approaches are market in points. It can be seen, that these data agree with theoretical curve of cost allocation.

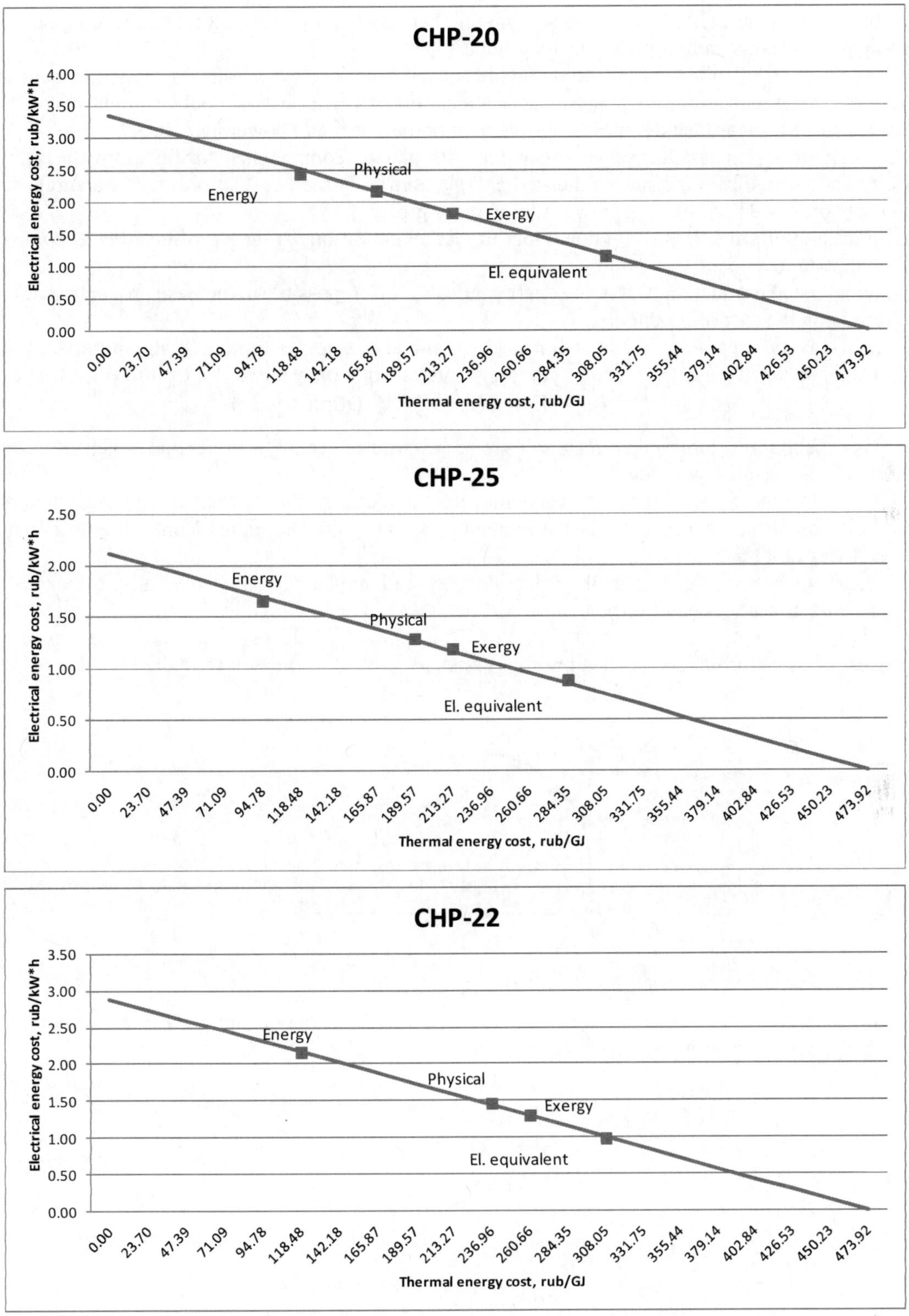

Figure 1. Heat and power cost allocation diagrams.

IOP Conf. Series: Earth and Environmental Science **337** (2019) 012079 doi:10.1088/1755-1315/337/1/012079

Using this diagram, the CHP plant can chose optimum bare costs for power and heat depending on price levels at power markets and set prices for its products.

Consequently, CHP plants could drive flexible pricing policy, which allows attracting new and retaining old costumers. An ability of free cost allocation between heat and power might become an additional source for it competitiveness improvement both at heat and power markets.

The profit of a cogeneration power plant consists of two components: profits from the sale of electricity and profits from the sale of thermal energy. Acting in market relations, manufacturers seek to maximize profits. There are several ways to increase profits:

1. Increasing of price. It is well known that in market conditions pricing is affected by the ratio of supply and demand.
2. Increasing of production. But in the energy industry it is impossible to increase production at will.
3. Reducing the cost of production.

Income (*I*) is an integrative indicator for production and business activities of an enterprise. It is a surplus value, obtained during production process, expressed in money terms (for example, rubles/year).

$$I = Rs - E = (P - C) \cdot V, \text{ (rubles/year)}$$

where *E* is expenditures, rubles/year; *Rs* is sales revenues, rubles/year; *P* is price, rubles/item; *C* is cost, rubles/item; *V* is overall input, items.

CHP plant income consists of two components: electrical energy and thermal energy sales income.

According to the market relations, the manufacturers tend to gain maximum income (Figure 2). There are several ways for this.

1. *Market price raising.* It is well-known, that demand and supply balance greatly influences the pricing at market conditions.

Price is a fundamental economical category, which is a product unit cost, expressed in money terms. The process of price formation is called pricing. Prices for power production are called tariffs.

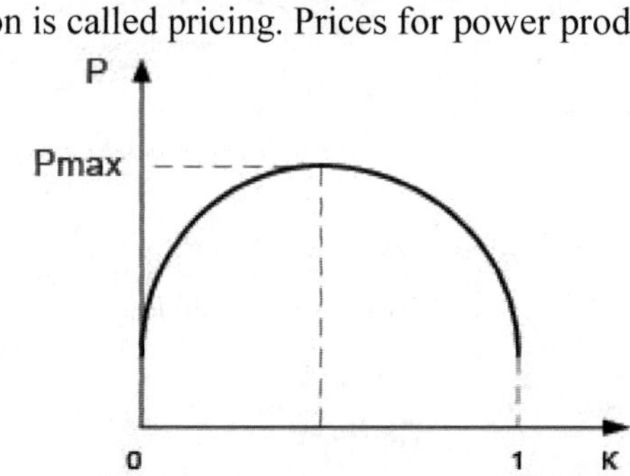

Figure 2. Income maximization condition.

Basing on the given above analysis we created an income factor model for CHP plant operating at thermal and electrical energy markets simultaneously (Figure 3).

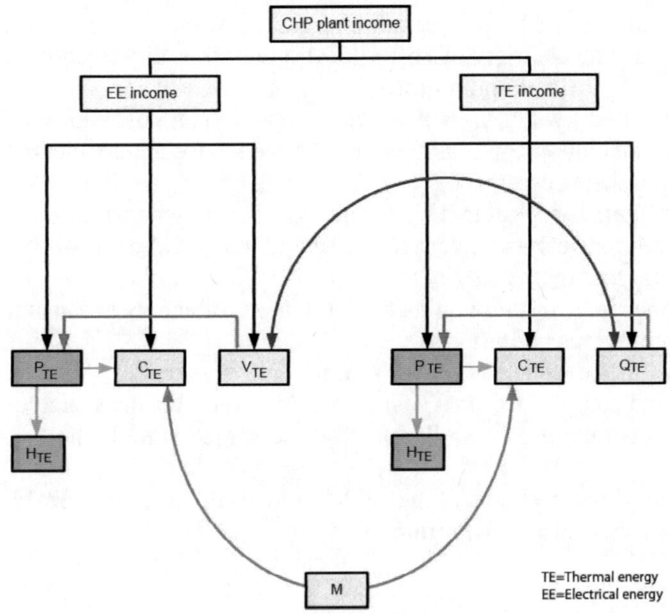

Figure 3. Income factor model.

Governing factors for enterprise production price making are the following:

- Production cost (it is influenced by cost allocation method for CHP plant);
- Projected income;
- Competitor prices;
- Financially reliability;
- Governing body and other social institutions requirements;
- Uniqueness of certain product characteristics.

The development of enterprise pricing strategy includes 6 steps, which are presented in Figure 4.

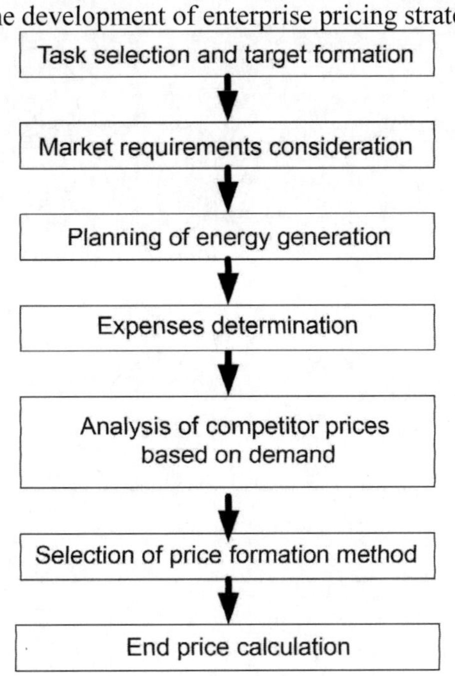

Figure 4. Pricing strategy development steps.

IOP Conf. Series: Earth and Environmental Science **337** (2019) 012079 doi:10.1088/1755-1315/337/1/012079

Figure 5 presents an algorithm for thermal and electrical energy cost forming at a CHP plant. The algorithm allows building a variety of cost allocation options, and selecting the cost distribution ratio, which allows obtaining the maximum profit.

1. Choose a step using which we will calculate a variety of cost allocation options.
2. For initial calculations, we assume that all costs relate to electricity, i.e., the distribution coefficient of heat costs = 0%.
3. Calculate the values of heat and electricity cost at a given cost ratio.
4. Choose prices for heat and electricity based on statistical data for the past relevant periods and analysis of prices on the energy markets.
5. Determine the planned sales volumes of both types of energy at a selected price based on demand graphs for previous periods.
6. We calculate the revenue and profit from the sale of energy.
7. We proceed to consider the next cost distribution ratio for the selected step.
8. Operations 3-6 are repeated until a profit is calculated with the heat cost distribution coefficient = 100%.
9. Thus, according to the results of the calculation of the maximum profit, the best value of the cost distribution coefficient is determined.

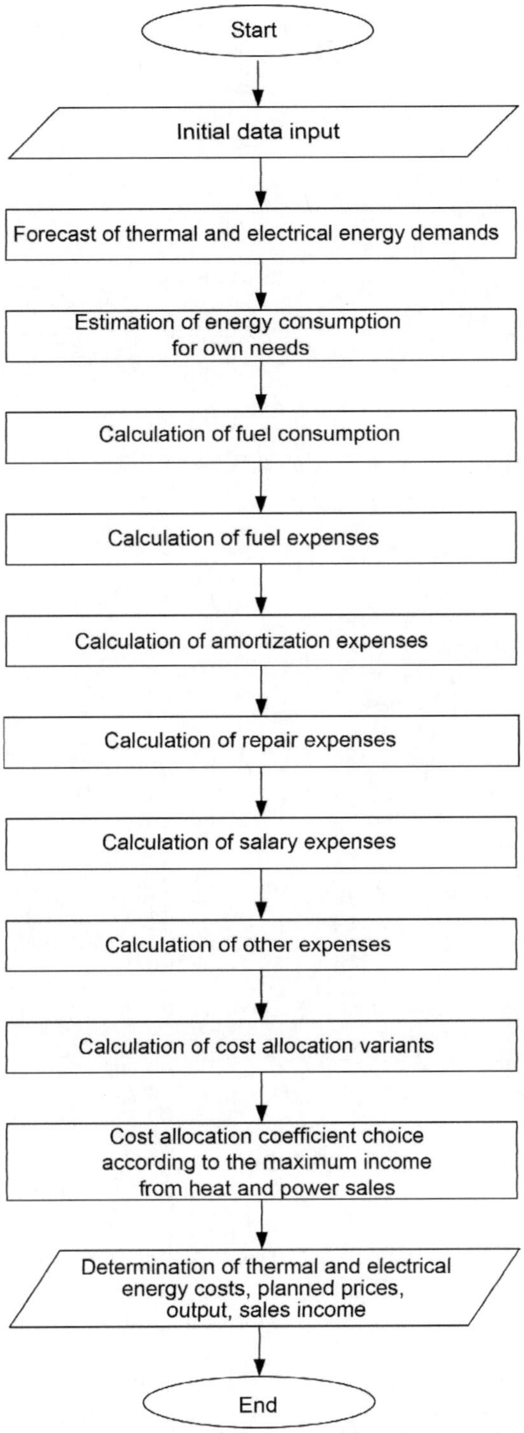

Figure 5. Algorithm for energy cost formation at a CHP plant.

Figure 6 presents a detailed algorithm for calculation of cost allocation variants.

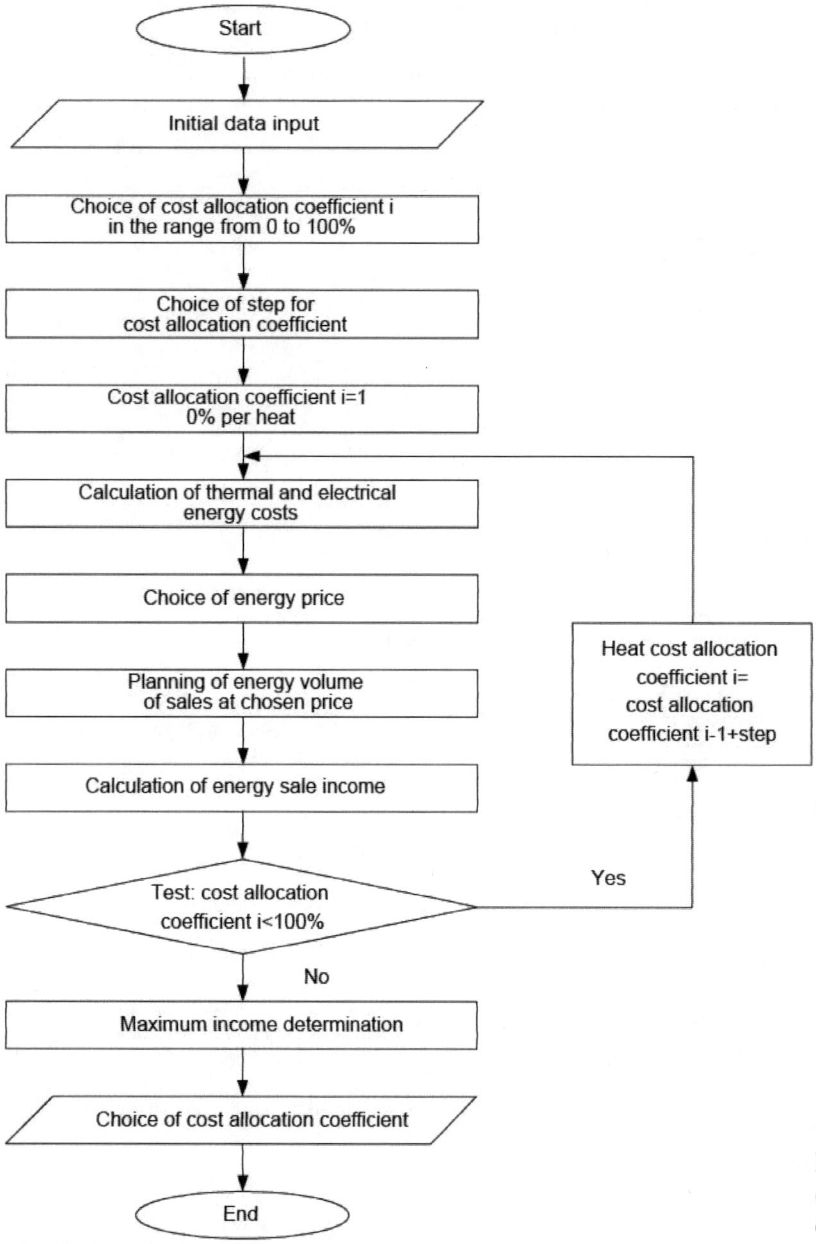

Figure 6. Algorithm for choice of cost allocation coefficient.

The designed algorithms were used for computation of CHP-20 characteristics using Mathcad and Excel.

The algorithm from Figure 6 was used to calculate a variety of variants for cost allocation with heat charging from 0 to 100% with 5% step. .

According to the algorithm from Figure 5 and factor model from Figure 3, we calculated income and costs for combined production of thermal and electrical energy. The calculated data was used to create a chart for CHP plant income dependence on cost allocation coefficient, Figure 7. It was performed under the condition of simultaneous operation of the cogeneration plant at electrical and thermal energy markets.

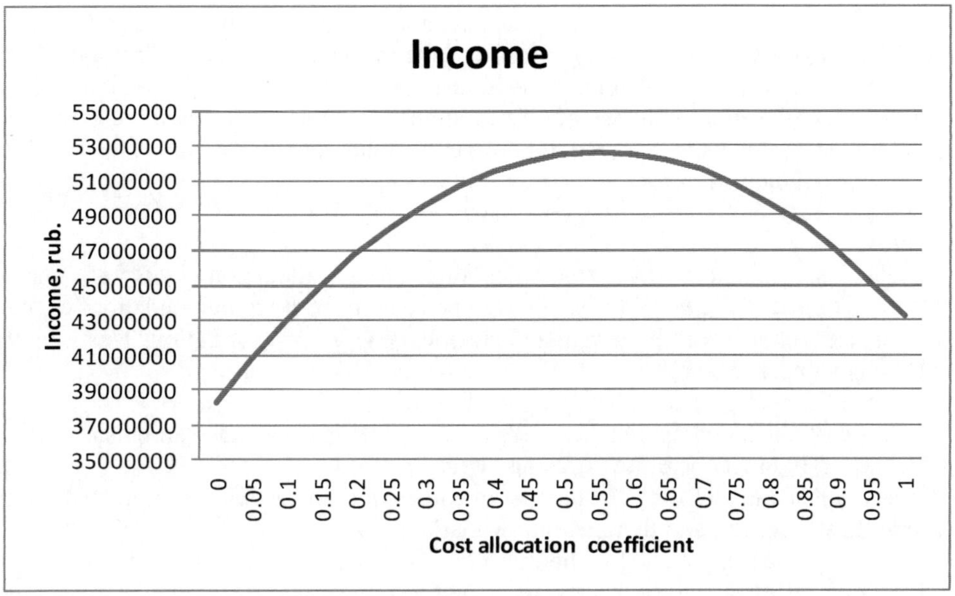

Figure 7. Relation between income and cost allocation coefficient by the example of CHP-20 (Mosenergo, Russia).

The obtained data show that the maximum income is reached at heat cost allocation coefficient, which is equal to 0.55.

As it is known, the cost allocation method is prescribed in the accounting policy of the enterprise. Accounting policy is adopted for the fiscal year. Thus, this algorithm can be used in operational accounting, when planning price bids for the day-ahead market. And in accounting there is a method which allows one to increase efficiency for the whole period.

The peculiarity of the Russian energy markets is that contracts for the supply of heat and electric power are concluded once a year, while electricity is sold every day. It is necessary to take into account the difference in the distribution of fixed and variable costs in order to avoid double counting that does not comply with the current legislation of our country.

In connection with the above stated, fixed costs are proposed to be allocated according to current methods (for example, physical method or ORGRES method). The proposed method of flexible cost allocation will be applicable only for variable costs, which will include fuel costs. Features of the distribution of costs are clearly presented in table 1. In connection with this division, at fixed costs, you can introduce the concept of average daily fixed costs for subsequent use in calculating production costs.

Table 1. Features of cost allocation.

Features	Thermal energy market	Electrical energy and power markets	
		Electrical energy market	Power market
Expenses	Alternating and constant components of expenses	Alternating component of expenses	Constant component of expenses
Period	For a year	For a day	For a year
Cost allocation	Constant expenses are allocated according to the existing method, alternating expenses are allocated according to the proposed method	Expenses are allocated according to the proposed method	Expenses are allocated according to the existing method

Thus, the proposed algorithm can be used not only for planning the price of electrical energy on the day-ahead when submitting a price bid, but also for taking into account the actual prices of electricity based on the results of the auction. In the first case, the application of the algorithm is an additional burden on the trader, which may entail the expansion of staff and the introduction of additional staff. In the second case, taking into account the actual prices, the adjustment of the planned indicators (sales volumes, prices, share of cost allocation with flexible distribution) for future periods can be carried out. These functions fall on the departments involved not only in planning, but also in analyzing the financial and economic activities of the enterprise.

3. Conclusions

The implemented investigation proves that calculation of cost allocation is of great importance for energy cost formation at CHP plant. We have developed a number of recommendations concerning cost allocation coefficient choice during thermal and electrical energy cost formation. This is aimed at CHP business efficiency improvement.

The main scientific and practical results are the following:

1. Various cost allocation approaches were analyzed, their advantages and disadvantages, which influence CHP plant operating efficiency were revealed.
2. We developed a model for CHP income maximization under the condition of its simultaneous operation at electrical and thermal energy markets.
3. We developed an approach for thermal and electrical energy cost formation, based on cost allocation coefficient choice in view of income maximization.
4. The developed approach was approved at an operating plant.

The approval results show that the implementation of the developed approach gives on opportunity to increase the total income of the CHP plant due to attraction of new consumers.

References

[1] Saari J, Sermyagina E, Kaikko J, Vakkilainen E and Sergeev V 2016 Integration of hydrothermal carbonization and a CHP plant: Part 2 – operational and economic analysis *Energy* **113** pp 574–85

[2] Fedyukhin A, Sultanguzin I, Gyul'Maliev A and Sergeev V 2017 Biomass pyrolysis and gasification comprehensive modeling for effective power generation at combined cycle power plant *Eurasian Chem. J.* **19** pp 245–53

[3] Russkov O and Saradgishvili S 2015 The electricity market prices forecast as energy efficient procedure for an industrial monotown enterprise *Procedia Engineering* **117** pp 309–16

[4] Russkov O V and Saradgishvili S E 2017 The method of planning the energy consumption for electricity market *IOP Conference Series: Earth and Environmental Science* **90**

[5] Khabachev L D, Plotkina U I, Bugaeva T M and Yurkova A B 2018 Assessment of systemic effects from integration of distributed generation facilities into regional energy systems *Reliability, Infocom Technologies and Optimization (Trends and Future Directions) (ICRITO), 2017 6th International Conference on* pp 188–93

[6] Korosteleva T S 2005 *Development of cost allocation procedure at CHP energy cost formation for market conditions.* Ph.D. Thesis in Economics (Samara, SSAU)

[7] Rawlings K J M 2017 Structure of energy industry in Zambia and Russia *Bull. Tomsk Polytech. Univ. Geo Assets Eng.* **328** pp 25–31

[8] Sultanov M M 2010 *Optimization of CHP operating modes in regards to energy efficiency.* Ph.D. Thesis in Engineering Sciences (Moscow Power Engineering Institute)

[9] Sukhareva E V and Lisin E M 2014 Analysis of CHP production and business efficiency reduction at power market conditions *Collect. Res. Pap. Bus. Soc. Responsib.* pp 268–77

[10] Sukhareva E V 2014 Modern situation for Russian power engineering *Transp. Bus. Russ.* **6** pp 157–9

[11] Russkov O V and Saradgishvili S E 2017 IT-Method for uneven energy consumption planning *2017 Int. Conf. Ind. Eng. Appl. Manuf. ICIEAM 2017 - Proc.*

[12] Aminov R Z and Garievsky M V. 2018 THE EFFICIENCY OF COMBINED-CYCLE CHP

PLANT WITH VARIABLE ELECTRIC LOADS, TAKING INTO ACCOUNT THE WEAR AND TEAR OF EQUIPMENT *Proc. High. Educ. institutions. ENERGY Sect. Probl.* **20** 10–22

IOP Conf. Series: Earth and Environmental Science **337** (2019) 012080 doi:10.1088/1755-1315/337/1/012080

Improvement of energy efficiency of motor fuel production at oil refinery

V Markov[1*], A Fedyukhin[2], I Sultanguzin[2], B G Matisov[3], N V Nikitkov[3], V V Izrantsev[3], V G Knorring[3], G A Konrashkova[3], L R Mukhametova[4]

[1]National Research University "Gubkin Russian State University of Oil and Gas"65 Leninsky Prospekt, Moscow, 119991, Russia
[2]National Research University "Moscow Power Engineering Institute", Krasnokazarmennaya 14, Moscow, 111250, Russia
[3]Peter the Great St. Petersburg Polytechnic University, St. Petersburg, Russia Polytechnicheskaya 29, Saint Petersburg, 195251, Russia
[4]Kazan State Power Engineering University, Kazan, Russian Federation

* Corresponding author: vitaliy_m_v_w@mail.ru

Abstract. This work presents a complex of technological solutions aimed at possible improvement of performance efficiency of oil refinery plants. These technological solutions are based on the control mechanisms of consumer material-heat balance for obtaining an optimum balance between fractionating accuracy of the obtained distillates, process energy characteristics and increase of the thermal energy integration at the plant. At this the proposed solutions do not provide the changes of the basic plant equipment, and no significant capital investments should be made for its implementation.

1. Introduction

Nowadays the issues concerning energy-efficiency and environmental sustainability are of vital importance for every area of human activities, and power engineering sector of industry is one of the highest-developing [1–11]. Oil refinery industry is one of the branches of huge fuel-and-energy complex, which also includes fuel extraction and production, energy distribution, as well as energy supply, etc. The major unfavorable external factors, which have a negative impact on oil refinery plant (ORP) efficiency, are instabilities of regional oil-energy balances and power supply [12–14]. The internal factors are the following: efficiency of energy-consuming and energy-generating equipment, energy capacity of technological processes, technological stream heat recuperation and regeneration degree. Energy-efficient and environmentally-friendly technologies are to be used here. Drastic economy of energy-resources may be obtained through implementation of energy-saving measures and advanced technologies into the most energy-consuming processes, as well as improvement of equipment and unit efficiencies[15–21]. At oil refinery plants the most energy-consuming are the processes of rectification, which consume a large amount of thermal energy in the form of steam, and electrical energy for cooling processes (Figure 1). To the greater extent rectification process is implemented at preliminary oil distillation units [22]. The example of this installation was used to study ways for improvement of energy-efficiency.

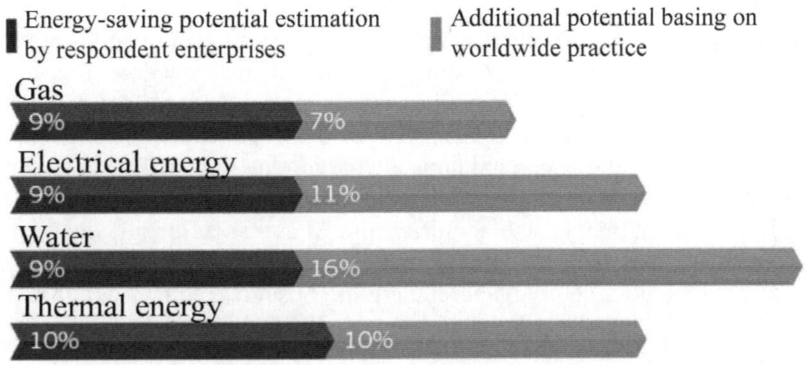

Figure 1. Estimation of energy-savings at modern oil refinery plants.

2. Physical-chemical basis of rectification process

The preliminary oil distillation units, in Russia known as ELOY-AVT (electric desalter unit with atmospheric-vacuum distillation) serve for removal of residual humidity and salts from oil with its further fractionation for obtaining of targeted product assortment (Figure 2).

The basic products which are obtained during oil division at AVT units are light distillates: straight-run gasoline (boiling range is 180 °C), jet fuel (150-220 °C), diesel fuel (180-360 °C), hydrocarbon gas and vacuum distillates (light and heavy vacuum gas oil) and vacuum distillation residue (gudron). Petroleum division for products is performed via rectification.

Rectification process can be organized periodically and continuously. Preliminary oil distillation units at oil refinery plants use nothing else but continuous rectification at various pressures: increased pressure is used for division of low-boiling hydro-carbons (crude oil stripping), near-atmospheric pressure is used for atmospheric distillation, vacuum is used for division of high-boiling hydrocarbons (vacuum distillation) [23].

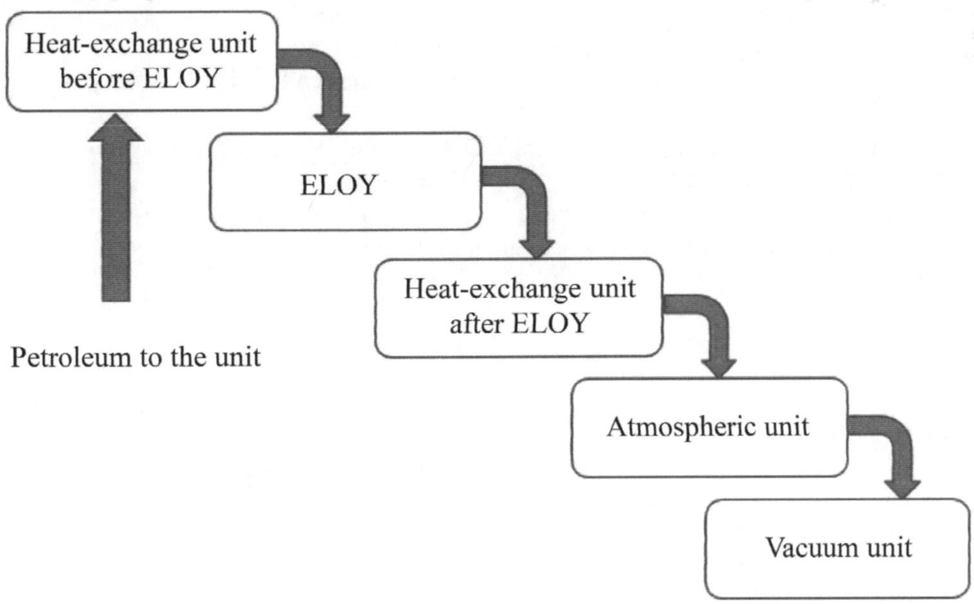

Figure 2. A schematic diagram for ELOY-AVT unit.

The quality of components, obtained at initial mixture division, is governed by its relative volatility (difference in boiling temperatures), relation between liquid and gas phases, and number of contact steps. The major technological parameters of rectification process are temperature mode, pressure, flow rates. Temperature of the rectifying column top defines rectification quality and is automatically maintained by upper delivery (reflux). Bottom temperature defines stripping degree from light-boiling components and is maintained by heat supply from pipe furnace. As for pressure, at its increase the fractionating accuracy deteriorates, as the relative volatility of the divided components is decreased.

Reflux supply to the upper part of the column is altered gradually in order not to cause reduction of its liquid level (flooding mode). At insufficient reflux supply, the top column temperature increases, and rectificate end-boiling temperature increases significantly. At excess reflux, the top column temperature decreases, rectificate end-boiling temperature decreases, its high-boiling components transform into residual. Increasing of reflux amount at relevant temperature raise of column bottom improves rectification sharpness.

3. Usage of energy potential of secondary energy-resources

The waste thermal energy is usually classified according to three temperature ranges: high-temperature (higher than 650 °C); medium-temperature (230÷650 °C); low-temperature (less than 230 °C). High- and medium-temperature thermal energies are used at oil refinery for steam generation of technological parameters, electrical energy, warming of air supplied at burner devices. Low-temperature thermal energy is used for heating of buildings, water and air. The main sources for thermal secondary energy-resources (SER) at oil refinery are the following;

- Flue gases of heating furnaces (300÷500 °C);
- Hot flows of liquid and gaseous oil products (125÷385 °C). After its' usage at technological cycle, the heat is removed at compressor-refrigerator (40÷90 °C);
- Thermal resources of light-fractions overhead (102÷125 °C), which are not used at Russian ORP.
- Thermal resources of combustion products, which are obtained during decoking of the catalyst processes (580÷600 °C), as well as whose failing to find complete usage for additional electrical energy generation at Russian installations of catalytic cracking;
- Thermal resources of condensate (70÷85 °C), exhaust steam (139 °C at P = 0.3 MPa) and hot water (45÷110 °C).
- The methods for waste heat utilization at ORP are the following:
- Direct utilization for heating of raw materials at the absence of heat-exchangers (so-called "heat tracers");
- Heat recuperation of production flows or exhaust gases at heat-exchangers or exhaust-heat boilers (EHB) for production of process steam or hot water;
- Regeneration of exhaust gases at air heaters or heat of product flows at combined units;
- Cogeneration with generation of electrical and thermal energies.

Regeneration of heat of warmed product flows is rather widely used at technological schemes at oil refineries. Excess heat of hot petroleum products at temperature 110÷160 °C and consumption more than 20 m^3/h is usually used for warming of chemically purified water (feeding of exhaust-heat boilers and water of industrial cogeneration), obtaining of freeze at absorption refrigeration units; while at temperature more than 160 °C it is used for steam obtaining. In order to cool-down the petroleum product flows with steam obtaining one use evaporators with steam chambers which are fed by condensate, collected at the unit. The majority steam consumers use only steam condensation heat and distribute condensate at saturation temperature [24,25].

Regenerative and recuperative air heaters and exhaust-heat boilers are widely used for utilization of flue gases thermal energy. It's a common practice to utilize heat of furnaces of efficiency more than 25 GJ/h at exhaust gases temperature higher than 220 °C. It should be noted that due to mismatch between real and design temperatures of exhaust gases at operating units, the exhaust-heat boiler efficiency is less than passport one. Special aspect of exhaust-heat boiler exploitation at ORP is instability of heating gas consumption, which leads to significant load fluctuations at exhaust-heat boilers [25,26].

IOP Conf. Series: Earth and Environmental Science **337** (2019) 012080 doi:10.1088/1755-1315/337/1/012080

4. Development of ELOY-AVT model

Computer model development was performed at Aspen HYSYS environment, which is widely used for various technological process modeling [27,28]. In order to calculate physical-chemical properties of petroleum we used design data on composition and properties of narrow fractions for typical Russian installation ELOY-AVT-7 (annual production of crude oil is 7 mln tones per year). As a thermodynamic package we chose Peng-Robinson equation of state. It allows one to calculate with high accuracy equilibrium and thermodynamic parameters of hydrocarbon systems at pressures lower than 30 MPa [29].

It was established that atmospheric column furnaces H0801A/B set limits for unit efficiency improvement. When they are loaded for 120%, temperature of pipe coil walls is close to maximum allowable one (650 °C), flue gases temperature is more than 300 °C, which states for low thermal efficiency of the furnace.

In order to reduce thermal load at H0801A/B/C furnaces, we suggest to warm up stripped oil, supplied from C0401 column cube by high-potential gudron flow from vacuum column C0501 (Figure 3). Gudron flow is the highest-potential technological flow at this installation, its temperature is 362 °C. We chose gudron for oil warming up as it has rather high ratio of hot/cold coolant due to its high content in petroleum (about 25%). For implementation of this solution we suggest to use the existing heat-exchanger for oil warming up before ELOY E-104A with surface area of 427.4 m^2.

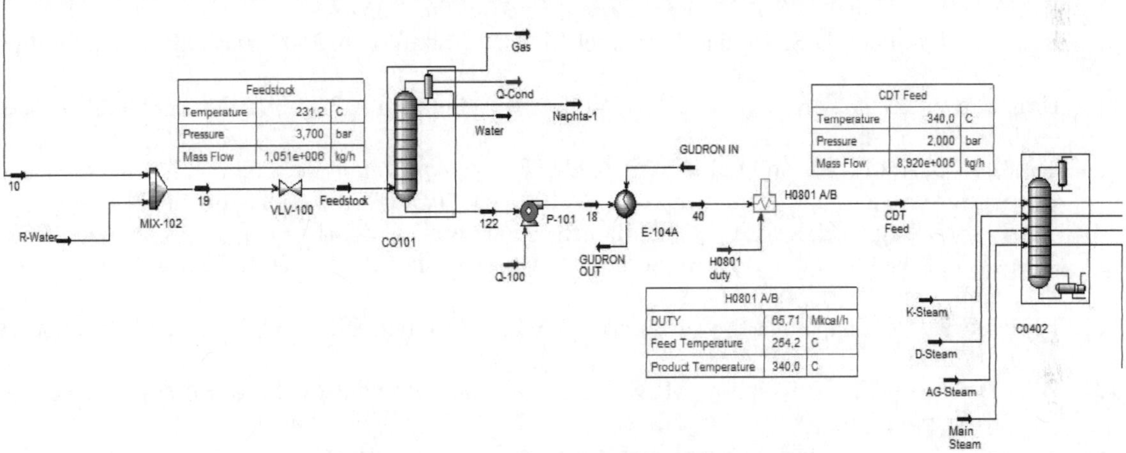

Figure 3. ELOY-AVT unit modeling in Aspen HYSYS environment.

After stripped oil warming up by gudron, its temperature before the furnace H0801 A/B increased from 230 to 254 °C, while thermal load decreased from 79.36 to 65.71 Gcal, which is 17%.

After stripped oil warming before the atmospheric furnaces H0801 A/B/C a drastic reduction of gudron temperature took place (up to 280 °C). To compensate heat, directed at reduction of H0801A/B/C furnaces load, one should increase heat recuperation degree at heat-exchange unit before and after ELOY. For this purpose we redistributed heat balance of the main atmospheric column C0402: recovery of higher-potential 2nd and 3rd pumparound (PA) was increased, recovery of low-potential 1st PA was decreased. This resulted in 1st PA load reduction from 21.82 to15.42 Gcal/h, 2nd and 3rd PA load increase, which allows increasing of recuperation degree at oil warming unit after ELOY. Load redistribution between pumparounds was performed in such a way that provided maximum recovery increase of high-potential 2nd and 3rd PA. At this the quality indicators were maintained within the specification limits.

5. Conclusions

This work presents a variant for ELOY-AVT unit modernization. Here we suggest warming up of a stripped oil flow part before the H0801A/B furnace by gudrone from vacuum column C0501. Using the computer modeling we obtained the following results:

- Change of configuration and connection circuit of the heat-consuming equipment allowed reducing natural gas consumption by H0801 A/B furnace for 15%.
- In order to increase heat recuperation degree we redistributed heat balance at atmospheric column C0402 and changed the arrangement of material warming up unit.
- During material flow optimization we succeeded in improvement of thermal energy recuperation degree at the unit and maintaining of normative oil temperature before the atmospheric block.

References

[1] Staritcyna A, Pollock E, Sokolova E and Martynenko E 2016 *MATEC Web of Conferences* **73**
[2] Zadvinskaya T O and Gorshkov A S 2014 *Adv. Mater. Res.* **953–954** pp 1570–7
[3] Baranova D, Sovetnikov D, Semashkina D and Borodinecs A 2017 *Procedia Engineering* **205** pp 503–10
[4] Vatin N and Gamayunova O 2014 *Appl. Mech. Mater.* **633–634** pp 972–6
[5] Krolin A A and Guzhov S V 2015 *Reg. energy energy Effic.* pp 4–5
[6] Raafat T, El-Gendy N S, Farahat L, Kamel M and El-Shafy E A 2007 *Eurasian Chem. J.* **9** pp 153–62
[7] Brunman V E, Vataev A S, Volkov A N, Volkov E A, Petkova A P and Kochanzhi F I 2017 *Russ. Eng. Res.* **37** pp 378–82
[8] Tarasova D, Andreev K and Lakić S 2016 *MATEC Web of Conferences* **53**
[9] Borovkov V M, Zysin L V and Sergeev V V 2002 *Izv. Akad. Nauk. Energ.* pp 13–24
[10] Saari J, Sermyagina E, Kaikko J, Vakkilainen E and Sergeev V 2016 *Energy* **113** pp 574–85
[11] Zysin L V, Koshkin N L, Orlov E I, Sergeev V V and Steshenkov L P 2002 *Therm. Eng. (English Transl. Teploenerg)* **49** pp 14–9
[12] Pogodaeva T V, Zhaparova D V, Rudenko D Y and Skripnuk D F 2015 *Mediterr. J. Soc. Sci.* **6** pp 129–35
[13] A.N. Dmitrievskii, N.I. Komkov, M. V Krotova and V.S. Romantsov, Stud. Russ. Econ. Dev. 27 (2016), pp. 21–33
[14] Melnikov A V, Nadezhina O S and Rudskaya I A 2016 pp 2201–09
[15] Arkharova I V and Zaporotskova I V 2015 *Eurasian Chem. J.* **17** pp 295–300
[16] Polonsky V L and Tyurin A P 2015 *Chem. Pet. Eng.* **51** pp 37–40
[17] Arutyunov V S, Strekova L N and Nikitin A V 2013 *Eurasian Chem. J.* **15** pp 265–73
[18] Sergeev V V and Aleshina A S 2011 *Therm. Eng.* **58** pp 268–70
[19] Arutyunov V S, Savchenko V I, Sedov I V, Nikitin A V, Troshin K Y and Borisov A A 2017 *Eurasian Chem. J.* **19** pp 265–71
[20] Alimzhanova M B, Kenessov B N, Nauryzbayev M K and Koziel J A 2012 *Eurasian Chem. J.* **14** pp 331–35
[21] E.S. Dremicheva and E.R. Zvereva, STUDY CORROSION PROCESSES OF OIL EQUIPMENT, «Izvestiya Vyss. uchebnykh Zaved. Probl. Energ. 20 (2018), pp. 138–143
[22] «2009 Tekhnologicheskij reglament. Ustanovka motornyh topliv (UMT)" TR-6100-20367-02-2009 (SPb.: Lengiproneftekhim)»
[23] Lastovkin L A, Radchenko E D and Rudin M G 1986 *Oil-Refinery, Spravochnik Neftepererabotchika [Worker Textbook]* (Leningrad: Himiya)
[24] Stepanov A V, Sul'zhik N I and Goryunov V S 1989 *Racional'noe Ispol'zovanie Syr'evyh I Ehnergeticheskih Resursov Pri Pererabotke Uglevodorodov [Rational Usage of Raw Materials and Energy Resources and Hydro-Carbon Processing]* (Kiev: Tekhnika)
[25] Vagin G Y, Dudnikova L V, Zenyutich E A et al 2001 *Ekonomiya Ehnergoresursov v Promyshlennyh Tekhnologiyah. Spravochno-Metodicheskoe Posobie [Economy of Energy-*

Resources at Industrial Technologies. Reference-Guidance Manual] (N. Novgorod: NGTU, NICEH)
[26] Linnhoff B, Dunford H and Smith R 1983 *Chem Eng Sci* **38** pp 1175–88
[27] Kemp I C 2007 *Pinch Analysis and Process Integration. A User Guide on Process Integration for the Efficient Use of Energy, 2nd Edition* (Amsterdam: Elsevier)
[28] Murzin D Y, Garcia S, Russo V, Kilpiö T, Godina L I and Tokarev A V 2017 *Ind. Eng. Chem. Res.* **56** pp 13240–53
[29] Fedyukhin A, Sultanguzin I, Gyul'Maliev A and Sergeev V 2017 *Eurasian Chem. J.* **19** pp 245–53
[30] Fedyuhin A V, Sultanguzin I A, Kurzanov S Y, Belov R V, Bakulin A V and Shomova T P 2016 *Primenenie Prikladnyh Programmnyh Sredstv Dlya Resheniya Zadach Promyshlennoj Teploehnergetiki: Uchebnoe Posobie [Application of Software for Solution of Industrial Thermal-Energetics Tasks: Textbook]* (Moscow: Izdatel'stvo MEI)

IOP Conf. Series: Earth and Environmental Science **337** (2019) 012081 doi:10.1088/1755-1315/337/1/012081

Egypt and nuclear energy: aspects, reasons and future

D A Cauich-López[1*], L F Barrera Payan[1], A M N Abdelhalim[1], V V Izrantsev[1], V G Knorring[1], G A Kondrashkova[1], Yu M Pankratov[1], L R Mukhametova[2] and A A Sinitsyn[3]

[1]Peter the Great St. Petersburg Polytechnic University, Power Plant Engineering, Saint Petersburg, Russia
[2]Kazan State Power Engineering University, Kazan, Russian Federation
[3]Vologda State University, Heat Gas Supply and Ventilation, 160000 Lenin st., 15, Russia

* Corresponding author: daniel.cauich@gmail.com

Abstract. This article tries to give an analysis of the current energy situation in Egypt and to the Egyptian nuclear program giving a spot light on the history of the nuclear program and the reasons that made the nuclear energy a possible and applicable solution to the energy problems in Egypt. Also, it shows the Russian – Egyptian cooperation in the field of nuclear energy and the role of the Russian side in development of the Egyptian nuclear program

1. Introduction

Energy is defined as the ability to do work, and it is one of the most fundamental parts of our universe. Everything we do in our life requires energy, moving, thinking, sports, etc.

The cornerstone of the development of each country is to be able to meet its demand of energy. That's why each country started to look for various types of energy and to develop techniques and methods to increase the amount of energy production and to find new resources of energy in order to be able to satisfy its needs of energy. In particular electrical energy is playing a vital role in countries economy, while it's considered basic unit in achieving sustainable development.

Egypt has the third-largest population and the third highest gross national income (GNI), in Africa, according to the World Bank.

Over the last decade, Egypt's energy policies and energy sector have witnessed fundamental changes. Accordingly, several characteristics of the country's energy sector could be identified. Egypt's growing population and economic (industrial) development led to a significant rise in the demand for energy products in all sectors.

After 2011 revolution the country experienced a sharp decline in tourism revenue and foreign direct investment, according to the International Monetary Fund (IMF). According to the IMF, financial support from some Persian Gulf countries has helped Egypt meet its domestic energy demand.

Nevertheless, starting from late 2014 and shortly after new Egyptian president Abdel Fattah El-Sisi implemented a new strategy to improve the energy sector and to be able to face the increase in energy demand by increasing number of energy production units and expanding the energy infrastructure and find more energy resources and increase the share of renewables, also the strategy included future plans to depend on the nuclear energy.

IOP Conf. Series: Earth and Environmental Science **337** (2019) 012081 doi:10.1088/1755-1315/337/1/012081

2. Energy Resources in Egypt

2.1 Oil, Petroleum and Natural gas:
According to U.S. Energy information administration report about Egypt, Egypt is the largest non-OPEC (Organization of the Petroleum Exporting Countries) oil producer in Africa and is the largest consumer of oil on the continent [1] . The country's use of oil accounts for 20 percent of Africa's total consumption. Moreover, Egypt has Africa's largest oil refinery capacity.

Egypt's total primary energy consumption was 1.7 million barrels per day (b/d) of oil equivalent in 2013, according to the BP 2014 Statistical Review of World Energy [2]. Natural gas and oil are the primary fuels used to meet Egypt's energy needs.

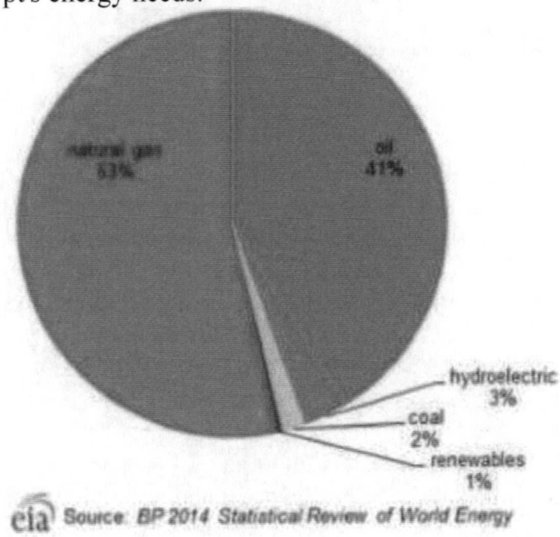

Figure 1. Primary Energy consumption by fuel in Egypt, 2013 [2].

While being a major producer, Egypt is also an importer of oil because the country's consumption of oil has been increasing by approximately 3 percent per year due to economic and population growth.

Egypt also is accounting for about 20% of petroleum and other liquids consumption and 40% of dry natural gas consumption in Africa in 2013.

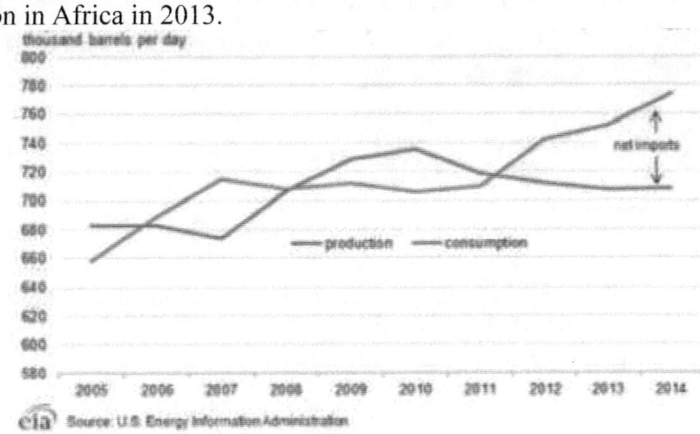

Figure 2. Petroleum and other liquid production in Egypt [1].

In 2014, Egypt's petroleum and other liquids production averaged 708,000 b/d, unchanged from the previous year. About half of Egypt's oil production comes from the Western Desert and the remainder comes from the Gulf of Suez, Eastern Desert, Sinai, Mediterranean Sea, Nile Delta, and Upper Egypt. Most of Egypt's production is derived from relatively small fields that are connected to larger regional production systems.

IOP Conf. Series: Earth and Environmental Science **337** (2019) 012081 doi:10.1088/1755-1315/337/1/012081

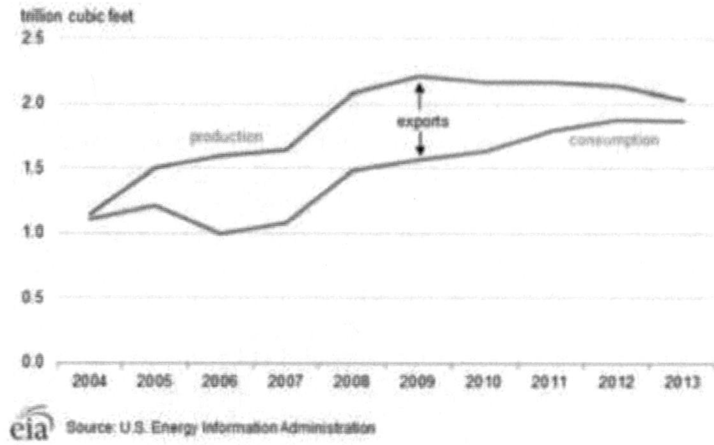

Figure 3. Dry natural gas production and consumption in Egypt [1].

Most of the natural gas consumed in Egypt is used to fuel electric power plants. The Egyptian government encourages households, businesses, and the industrial sector to consider natural gas as a substitute for petroleum products and coal.

Egyptian consumption of natural gas has been increasing by approximately 7 percent per year over the past decade.

According to the American security background report about Egypt, with the increasing level of consumption combined with the decreasing level of production Egypt will begin to import liquefied natural gas (LNG). Cairo signed an agreement with Algeria in December 2014 to import LNG between April and September 2015. Egypt signed a contract in February 2015 to import natural gas from an Israeli offshore gas field in the future. Cairo also signed an agreement with Gazprom to import LNG from Russia.

2.2 Hydro-power:

According to the Annual report of the ministry of Electricity and renewable energy [3], Hydropower is Egypt's third-largest energy source after natural gas and oil. Egypt generated Power of hydroelectricity, accounting for about 9% of Egypt's total power generation. Most of the country's hydroelectricity comes from the Aswan High Dam and the Aswan Reservoir Dams across the Nile River

Plant	12/13	13/14	Variance %
High Dam	9113	9304	2.1
Aswan Dam 1	1502	1559	3.8
Aswan Dam 2	1596	1503	(5.8)
Esna	473	535	13
Naga Hamady	437	451	3.2
Total	13121	13352	1.8

Figure 4. Hydropower generation in Egypt 2013/2014, taken from the Egyptian electricity holding company annual report.

According to the American security background report about Egypt [4], In January 2015, the Egyptian government announced that it plans to build additional hydroelectric power stations with the help of China; however the new stations are not likely to make a big change in Egypt's hydroelectric capacity

2.3 Solar and wind energy:

According to the Annual report of the ministry of Electricity and renewable energy for 2013/2014 [3], Egypt's first solar-thermal power plant, located in Kuraymat just south of Cairo, has the capacity to generate 140 MW of solar-thermal energy, and it has been operational since June 2011.

Also, Egypt is planning to establish a 140 MW solar power plant along the Red Sea coast and a 100 MW plant in Kom Ombo. Egypt plans to make an increase in the solar generation capacity by 3,500 MW by 2027.

Egypt generates wind power mainly from the 545 MW Zafarana wind farm and the 5 MW Hurghada wind farm.

According to the Annual report of the ministry of Electricity and renewable energy [3], the energy council approved the strategy that helps to increase the amount of generated energy by renewables to 20% by 2020, at which 12% will be from wind energy. The plan includes the establishment of new shared wind projects with the private sector to increase the total capacity to 7200 MW by 2020

3. Energy situation in Egypt

According to the reports of the ministry of electricity and renewable resources and the nuclear power plants authority in 2015 [5], the electrical energy generation in Egypt can be shown as follows:

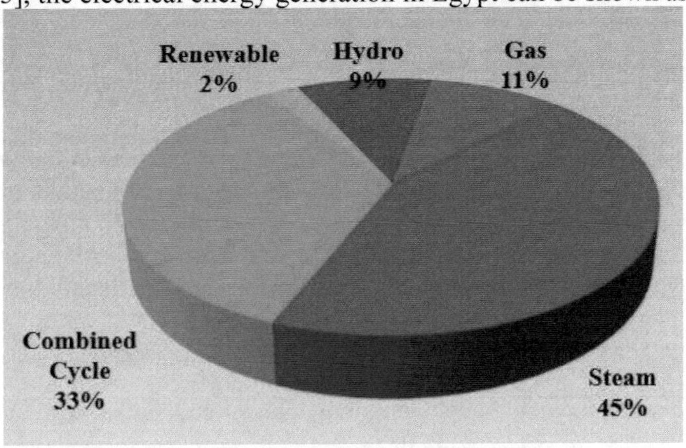

Figure 5. Distribution of energy generation in Egypt - Taken from Nuclear power plants authority report (January 2015) [5]

According to the annual report of the ministry of electricity and renewable energy for year 2013/2014 [3], the annual rate of growth for the peak load in Egypt was 3.5% in the period between 2009 and 2014.

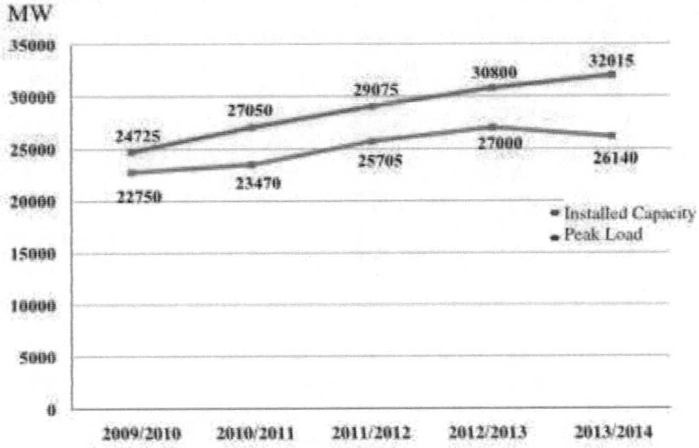

Figure 6. Electricity peak load curve [3].

In May 2015 and according to U.S. Energy information administration report about Egypt and Middle East economic survey, Egypt's generation capacity of electricity were 31.45 GW while the expected peak demand was 30 GW and according to research done by the ministry of electricity and renewable energy in 2015, the peak load is expected to reach 61 GW by year 2027.

In the same report it was mentioned that 70% of Egypt's electricity consumption is fueled by Natural gas.

IOP Conf. Series: Earth and Environmental Science **337** (2019) 012081 doi:10.1088/1755-1315/337/1/012081

The shortage of natural gas and problems in distribution and generation due to aging infrastructure, Egypt suffered from several blackouts.

4. Why Nuclear energy in Egypt?

For the above mentioned reasons, the Egyptian government is trying to increase the capacity of electrical generation to be able to match the increase of electricity demand expected in the future. The government started to boost the renewable energy in form of solar and wind energy; these plans are in parallel with the strategic plan to increase power generation using fossil fuels resources.

The Egyptian government started to discuss the usage of nuclear energy and building nuclear power plants to improve electricity production and generation and to ensure the ability to match the demand in the future.

The Egyptian Government realized that Nuclear energy can be considered a key solution to the energy problems in Egypt and found that it has a lot of advantages and benefits.

In the report issued by the nuclear power plants authority in 2015 [5], nuclear power was considered one of the best possible ways to generate electricity and meet the demand in the future.

Egypt has consumptions of fossil fuels higher than the amount the government can supply knowing that most of the natural gas consumed in Egypt is used to fuel electric power plants. With these limited resources and high consumption, Egypt is forced to find another source of energy to meet its demand and since nuclear power doesn't depend on the fossil fuels so it' considered the best substitute for energy generation.

Also, the market of fossil energy resources isn't stable and prices vary according to the dollar price and political circumstances which makes it hard to be imported.

Nuclear energy is considered a clean source of energy, The World Nuclear Association carried out a review of over twenty studies assessing the greenhouse gas emissions produced by different forms of electricity generation [6–8]. The results was shown in the chart below show that generating electricity from fossil fuels cause greenhouse gas emissions higher than when using nuclear energy or using renewable energy.

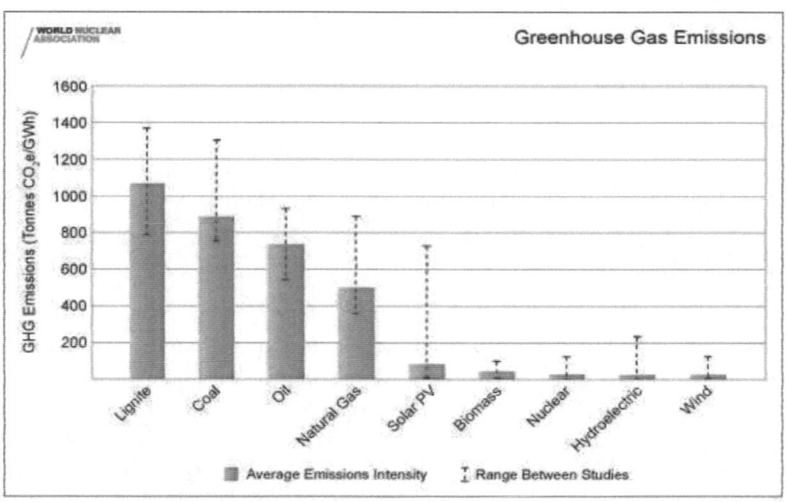

Figure 7. Greenhouse gas emissions for different types of fuel used for electricity generation - World Nuclear Association.

The nuclear energy has almost full utilization of hydropower. A power plant generates electricity by boiling water to steam, and uses the steam to turn the turbine and then the turbine is connected to the generator which generates the electric power. Cooling systems are required to cool the steam back to water so the cycle can continue.

Nuclear energy has a low operation cost, after construction of the nuclear power plant the operation cost and maintenance cost is very low compared to fossil fuels [9].

Operating costs include operating and maintenance (O&M) and fuel. Fuel cost figures include used fuel management and final waste disposal. These costs, while usually external for other technologies, are internal for nuclear power (i.e. they have to be paid or set aside securely by the utility generating the power, and the cost passed on to the customer in the actual tariff).

Uranium has the advantage of being a highly concentrated source of energy which is easily and cheaply transportable. The quantities necessary for operating the nuclear power plant are very much less than for coal or oil.

5. Egyptian Nuclear Program

Egypt started to think about nuclear energy long time ago and the Egyptian nuclear program has been an on-and-off process due to several political and economic circumstances.

The Egyptian program started officially in 1955 with the establishment of the atomic energy committee. In the same year, Egypt took the first step in the nuclear program by signing an agreement with the Soviet Union to acquire Light water reactor of capacity 2 MW as a research reactor. Egypt started the construction of the reactor in 1958 and in 1961 the reactor was operated.

The Egyptian Nuclear program was disturbed for long time due to the political circumstances and the war till 1975 when USA agreed to support the Egyptian nuclear program, but the conditions that USA introduced to Egypt in 1976 were unacceptable to the Egyptian government.

By the end of 1976, Egypt and France reached an agreement to develop the research reactor, but under the pressure from the U.S government, France changed the terms of agreement with Egypt and the co-operation between the two countries collapsed.

As a part of new regulation and steps to enhance the management and planning for the nuclear program, the Egyptian government established the Nuclear power plants authority in 1976.

From 1981 to 1984, Egypt signed several cooperation agreements with USA, France, Germany and Switzerland in the field of civil usage of nuclear energy and nuclear power plants due to political circumstances.

In 1986, Egypt suspended its nuclear plans after the disaster of Chernobyl Nuclear power plant, but in 1992 Egypt signed an agreement with to Argentinean Company INVAP to construct light water research reactor of capacity of 22 MW. The construction of the plant started in the same year and it was first operated in 1997.

After 10 years of silence about the Egyptian nuclear program, Egypt and Russian agreed on starting cooperation in the field of nuclear energy during to the visit of President Mubarak to Russia in 2008.

The Egyptian government announced the creation of new regulations to support the nuclear program in 2009 and these regulations were discussed and finally issued in 2010 [10]. In the same year the IAEA had approved (El-Dabaa site – near Alexandria) an acceptable site for Egypt to build its nuclear power plant.

The Nuclear program stopped again after that due to the political circumstances in Egypt and the Egyptian revolution in January 2011. Egypt approached the Russian side to renew the agreement signed in 2008 concerning the Egyptian nuclear program in April 2013.

6. The Egyptian Russian Nuclear reactor

The Egyptian Russian cooperation has a long history since the revolution of July 1952, building the 1st research reactor in 1958 and the Russian support to Egypt during the 1973 war and the building of the High Dam in Aswan.

As mentioned before that in April 2013, Egypt approached Russia to renew the agreement of nuclear cooperation. The Russian foreign minister announced in November 2013 that Russia is ready to finance and support the Egyptian Nuclear program and the building of the nuclear power plant. This became possible as Russian scientists and economists have a significant experience in nuclear power plant operating technology development [11–22].

In February 2015, The Russian and Egyptian presidents announced the Russian Egyptian cooperation in building Egypt's first nuclear power plant [23]. The ministry of electricity and Rosatom signed a primary agreement to establish the nuclear power plant that will be formed from 4 units of VVER 1200 Nuclear power plants with total capacity of 4800 MW [24,25]

The agreement includes building nuclear fuel storage facilities to supply the nuclear plant with the required nuclear fuel along the 60-year operational period. Rosatom will also manage nuclear waste, maintenance and operation, and training Egyptian experts.

The station will be completely safe, especially since Rosatom will provide them with the VVER-1200 safety protocols to protect against internal and external factors and influences.

7. Future of Nuclear energy in Egypt

The success of the First NPP will be built in Egypt is the key to start an ambitious program to build more NPPs in Egypt to insure that the amount of energy generated is more than the peak demand in the future.

The ambitious of the Egyptian government to build a competitive industrial infrastructure will help in pushing forward the nuclear program to insure the stability of these projects. The plan is to build 6 more NPPs in case of the success of the first project.

8. Conclusion

Egypt nuclear program is based on infrastructure with two research reactors and regulatory mechanisms for the planning and control of nuclear development. Due to the increasing domestic demand for electricity to improve the energy infrastructure which will help in the development of the industrial sector too, Egypt has strong interest in nuclear energy and building nuclear power plants. As its current energy industry relies mainly on oil and gas, with some wind power, the nuclear power will be considered a possible and applicable approach to help in the development of the energy sector.

The Russian – Egyptian cooperation in the development of the Egyptian nuclear program will increase the chances of the success of the program and the possibility of the construction of more nuclear power plants due to the high technology offered from the Russian side and training for Egyptian experts and engineers.

References

[1] U.S.E. information Administration 2014 *International energy data and analysis – Egypt (June 2014)*
[2] BP 2014 *BP 2014 Statistical Review of World Energy*
[3] The Ministry of Electricity and renewable energy 2014 *Annual Report (2013/2014)*
[4] American Security Project report 2015 *Energy In Egypt – Background and issues*
[5] Nuclear power plants authority report 2015 *An Overview of the Egyptian Nuclear Power Project*
[6] *Electricity generation - what are the options?* Available at http://www.world-nuclear.org/nuclearbasics/electricitygeneration-what-are-the-options.aspx
[7] World Nuclear Association report *Nuclear Energy and Climate Change*
[8] World Nuclear Association report *The Economics of Nuclear Power*
[9] Rogner H-H 2008 *Nuclear Power: Benifts and Risks*
[10] N.P.P.A. (NPPA) and WorleyParsons 2010 *Egyptian Nuclear Power Plant Program History and Status* p 18
[11] Arefiev N, Mikhalev M, Zotov D, Zotov K, Vatin N and Nikonova O 2015 Physical modeling of suspended sediment deposition in marine intakes of nuclear power plants *Procedia engineering* **117** pp 32–8
[12] Legkostupova V V and Sudakov A V 2015 Estimated and experimental research of separation part of SPP-500-1 *MATEC Web of Conferences* **23**
[13] Malinina T and Murina V 2015 National policy in the area of spent fuel management in Russia *Proceedings of the 8th International Scientific Symposium on Electrical Power Engineering, ELEKTROENERGETIKA* pp. 425–6
[14] Davydov V V, Dudkin V I, Velichko E N and Karseev A Y 2015 Fiber-optic system for simulating

IOP Conf. Series: Earth and Environmental Science **337** (2019) 012081 doi:10.1088/1755-1315/337/1/012081

accidents in the cooling circuits of a nuclear power plant *J. Opt. Technol.* **82** pp 132–5
[15] Obraztsov E P, Belikov A N, Menshikov N N and Popkov E N 2016 Electrical calculation code relecs in css “veb” for npp designs **2** pp 1131–40
[16] Skvortsova O, Dashkina A, Petrovskaia E, Terleev V, Nikonorov A and Badenko V 2016 The Classification of Accidental Situations’ Scenarios on Hydropower Plants *MATEC Web of Conferences* **53**
[17] A.M. Pankin and N. V Korovkin, Diagnostics Indicators in Monitoring the Technical State of Direct Charge Sensors in Nuclear Reactors, At. Energy 122 (2017), pp. 284–9
[18] Laaksonen J 2012 Profitable nuclear power generation *Nucl. Plant J.* **30** pp 32–7
[19] Kuznetsov Y N 2006 A simple & passively safe reactor *Nucl. Plant J.* **24** pp 36–7
[20] Kuznetsov Y N, Romenkov A A, Yarmolenko O A and Alekseev A I 2001 Prospects for the development of innovative simplified boiling-water reactor VK-300 *Nucl. Plant J.* **19** pp 31,33,35–36
[21] Gabaraev B A, Kuznetsov Y N, Moskin V A , Reshetov V A and Rylov I I 2002 Leasing of nuclear power plants *Nucl. Plant J.* **20** pp 37–40
[22] Oshkanov N 2005 Plant profile: Beloyarsk NPP, Russia *Nucl. Plant J.* **23** pp 48–50
[23] Agreement between Russian federation government and The Egyptian government for construction of Nuclear power plant in Egypt, 2015
[24] Shay S 2015 *The Egyptian Russian Nuclear Deal* (IDC Herzliya, Institute for Policy and Strategy, Lauder School of Government, Diplomacy and Strategy)
[25] Egypt’s Nuclear dream realized with Russian technology after 60 years delay. Available at http://www.dailynewsegypt.com/author/m-farag/

Investigation of gas fuel combustion efficiency at magnetic field treatment

A Sinitsyn[1*], D Rusakov[1], V V Izrantsev[2], S A Kravchenko[2] and G A Konrashkova[2]

[1]Vologda State University, Lenina Street 15, Vologda, 160000, Russia
[2]Peter the Great St. Petersburg Polytechnic University,
Polytechnicheskaya 29, Saint Petersburg, 195251 Russia

* Corresponding author: sinitsyn.science@mail.ru

Abstract. This paper presents an investigation of gas fuel combustion efficiency at magnetic field treatment. Equipment used for magnetic fuel activation is analyzed. Basing on the conducted analysis, the authors developed a laboratory setup and special technique for investigation of magnetic field impact on gas fuel characteristics during its combustion. The obtained experimental results are presented.

Keywords: magnetic activation, gas fuel, ortho-modifier, neodymium-iron-boron.

1. Introduction

Nowadays decreasing of oil and gas fuel consumption is of urgent market and scientific interest, due to its cost growth and decreasing of world reserves. Studying of regularities and mechanism of various types of fuel combustion, including ecological aspects, are presented in a vast number of paper, but a lot of problems still remain unsolved [1-17]. The surveys concerning efficiency increase for organic fuel combustion presented in [18,19] are dealing only with the problem of initial air-fuel mixture treatment. Magnetic fuel activation is one of the new technologies in this research area, whereas it is extensively used in other science fields [20-22].

In 1950s S. Raskin found that there was a special construction of fuel activators, which resulted in improvement of internal combustion engine characteristics. In order to increase intensity of hydrocarbon chain breakage in fuel, he used a special magnetic-neodymium alloy. This alloy is used at magnetic water treatment device described in [23, 24]. Further, a number of scientists [25-27] improved his results for magnetic resonance theory. They used the special neodymium composition as a catalytic material for hydrocarbon chain breakage in fuel. They have proved that magnetic fuel treatment might change the molecular fuel structure. While addressing the issue concerning pollution abatement and hazardous emission decreasing for fuel combustion at car engines, a SMF product was invented. According to scientific classification, SMF is a magnetic-frequency resonance device. These devices are widely used in various scientific areas, but magnetic resonance principles used in SMF for hydrocarbons ionization are unique.

Scientific group of Kherson State Marine Academy carried out some tests of magnetic fuel activators for 6ChN25/34 engine. The tests were carried out using operating process monitoring and diagnostic system D4.0H (DEPAS). The object under test was steady-state operating mode ($n=411$ min^{-1})

IOP Conf. Series: Earth and Environmental Science **337** (2019) 012082 doi:10.1088/1755-1315/337/1/012082

with constant cyclic fuel supply, with various magnetic activators. In addition, the magnetic activators were placed in various parts of engine fuel system in order to achieve maximum efficiency. The maximum effect was achieved when the five activators were placed before fuel pump of high pressure [28]. The similar principle of discrete-gradient fuel treatment is described in the Ukrainian patent [29].

The known devices are "Powermag", "Soyuzintellect", inventions are "Ecomag-10G", "Stat-7", "Pomazkin Apparatus". However, all patents for these devices became expired and are not supported now. These devices are based on medium changes which occur during its movement in magnetic field. According to the manufacturer's data, the total engine cleaning from soot and coke takes place, carbon formation on the valves, piston rings, cylinder walls is avoided, toxic exhaust emission to the atmosphere reduces, noise and vibration levels decrease. However, despite the deceptive simplicity of device design and treatment availability, the announced positive effect of exploitation is difficult to obtain.

Ortho-modifier OMT-5 developed by LLC "Innova-Orto" (Russia) is one of the modern devices for magnetic fuel treatment. A number of Russian companies have tested this device. According to test reports, the following results were obtained: the engine start and warm-up time in winter period was decreased; the accumulator launch-load became less; engine operation at low temperature and high humidity was improved; fuel consumption was decreased for 15%, oil consumption was decreased for 12%; engine smoking was decreased; operating productivity was increased for 7%.

Magnetic modifier "ortho-modifier OMT-5" by LLC "Innova-Orto" satisfies the requirements of Russian Specifications 4591-005-90449293-2011 and is classified as fit for exploitation. However during the Specifications review the general guidelines and standards for the magnetic activator were not revealed. This says that nowadays there are no existing normative documents, which standardize the device and its operation. Thus, the further investigations and development of the similar devices should be aimed at formation of general specifications.

2. Materials and methods

We investigated impact of magnetic field on gas characteristics during its combustion through the example of fuel activator ortho-modifier OMT-5.

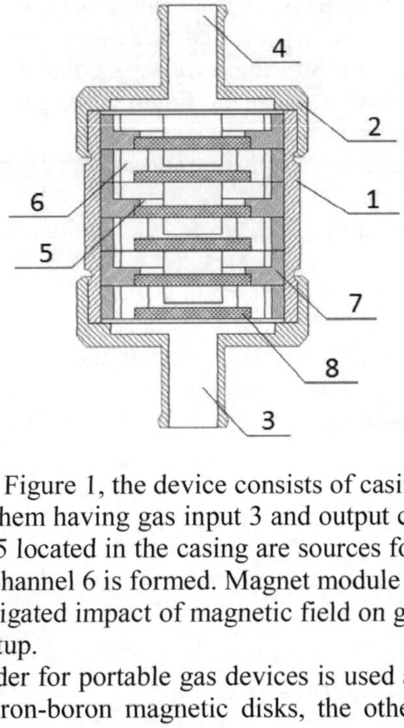

Figure 1. A device for magnetic medium treatment ortho-modifier OMT-5:
1 - casing,
2 - cap,
3 - gas input connection,
4 - gas output connection,
5 - magnet module,
6 - zigzag flow channel,
7 - washer,
8 - magnet.

According to Figure 1, the device consists of casing 1, caps 2 are connected from both sides of the casing, each of them having gas input 3 and output connections 4 from modifier of 8 mm diameter. Magnet modules 5 located in the casing are sources for magnetic field. Due to specific modules location, a zigzag flow channel 6 is formed. Magnet module 5 consists of washer 7 and magnet disk 8.

We investigated impact of magnetic field on gas characteristics during its combustion at a designed laboratory setup.

Gas cylinder for portable gas devices is used as a fuel source. One cylinder is intended for tests of neodymium-iron-boron magnetic disks, the other one is for testing of copper-nickel non-magnetic disks. Gas composition was the following: propane, isobutane, butane.

It corresponds to Russian State Standard GOST 20448-90 "Hydrocarbon liquid fuel gases for domestic consumption", European Standard EN 417, meets the requirements ISO 9001, ISO 14001.

Figure 2 presents laboratory setup for investigation of magnetic field impact on gas characteristics during its combustion.

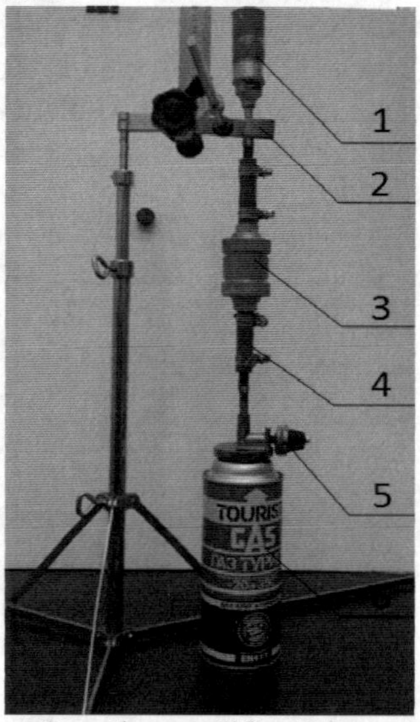

Figure 2. The laboratory setup for investigation of magnetic field impact on gas characteristics:
1 - gas torch,
2 - mount pillar,
3 - ortho-modifier OMT-5,
4 - gas hose,
5 - tap with a clamp,
6 - gas cylinder.

According to Figure 2, ortho-modifier 3 is connected using hose 4 to the gas cylinder 6 from one side, and to the gas torch 1 from the other side. Mount pillar 2 serves as a support for the setup as a whole.

The setup operates in the following way: The flow of the treated gas from gas cylinder enters the casing through the input connection. In the casing the initial flow is divided into two. The flows move towards each other and intensively collide with each other. Further, the flow turns at a set angle α. Consequently, the treated gas moves in a zigzag way along the flow channel perpendicularly to magnetic field force lines. Further, the flow is turbulized by force of hydrodynamic resistance, caused by the shape of the flow channel. In a turbulent flow the groups of molecules are rubbed and collide with each other. This results in breakage of several groups, viscosity reduction, gas volume ratio increase and, finally, the fuel activity in oxidation reactions increases. Magnetic field-treated gas comes out of the casing through the output connection and enters the torch [25].

Work sequence at the laboratory setup for investigation of magnetic field impact on gas characteristics is as follows:

1. Installation of neodymium-iron-boron magnetic disks (copper-nickel non-magnetic disks) at washers of the ortho-modifier OMT-5.
2. Formation of modules with flow channel and their fixation at the modifier.
3. Caps connection to the modifier casing.
4. Tap with a clamp connection to the modifier with the help of two clamps and a hose segment.
5. Gas torch connection to the modifier with the help of two clamps and a hose segment.
6. Weighting of the gas cylinder using electronic balance.
7. Writing the obtained data in Table.
8. Gas cylinder connection.
9. Gas torch mounting on the mount pillar with target height.
10. Opening of valve with fixing unit.
11. Simultaneous ignition and timer turn on.
12. Combustion time is three minutes.
13. Valve closing.
14. Gas cylinder dismounting and weighting.
15. Writing the obtained data in Table.

The tests are performed several times. Further, the magnetic disks at modifier modules are replaced by the non-magnetic ones, the tests are repeated.

3. Results and discussion

Figure 3 presents a chart for gas cylinder mass change when using neodymium-iron-boron magnetic disks, and copper-nickelnon-magnetic disks at washers of the ortho-modifier OMT-5.

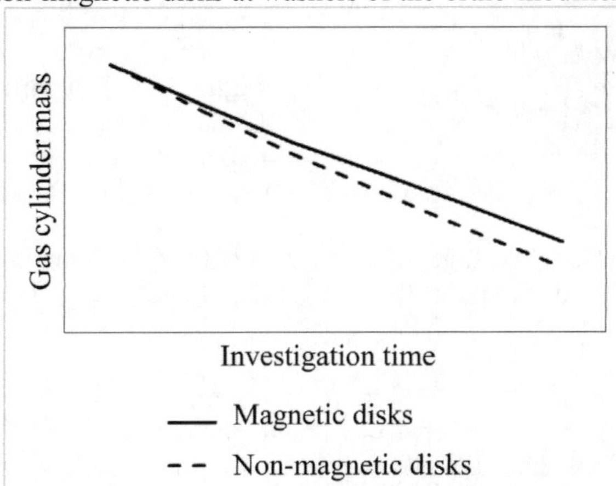

Figure 3. Gas cylinder mass change chart.

The performed tests showed that the examined ortho-modifier OMT-5 is more efficient for gas combustion when using neodymium-iron-boron magnetic disks, at this gas consumption reduces for 12%.

We used the following free torch schemes for describing the shape and character of the flame: laminar and turbulent [25]. Figure 4a presents laminar flare in comparison to torch flare. The torch flare is from modifier with neodymium-iron-boron magnetic disks.

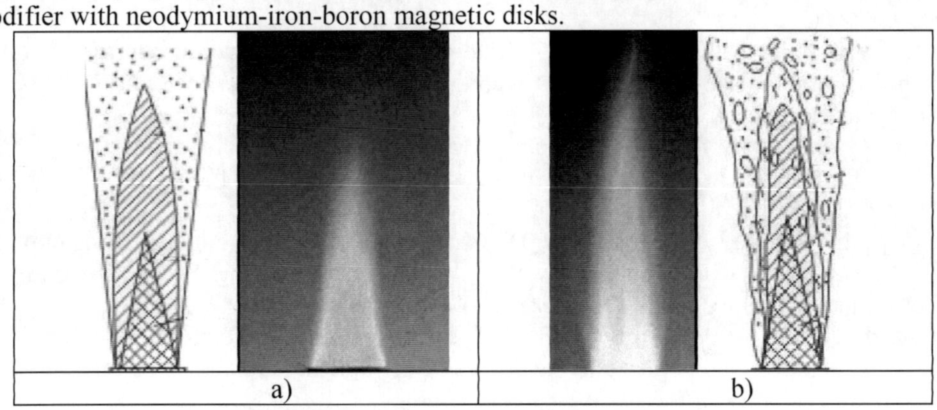

Figure 4. Comparison between flame shapes:
a) laminar flare with non-magnetic disks
b) turbulent flare with magnetic disks.

According to Figure 4a, the flame has an increased brightness and noticeable conic front. Also the flame has distinct boundaries which are common for laminar shape.

Figure 4b presents turbulent flare in comparison to torch flare. The torch flare is from modifier with neodymium-iron-boron magnetic disks. According to Figure 4b, there is no noticeable conic front of flame, at this separate flare particles are observed, divided by flame impulses. The combustion process occurs throughout the volume, an increased intensity of flame is observed, by analogy to results from [26, 27]. Figure 5 presents comparison between torch flames when using magnetic and non-magnetic disks.

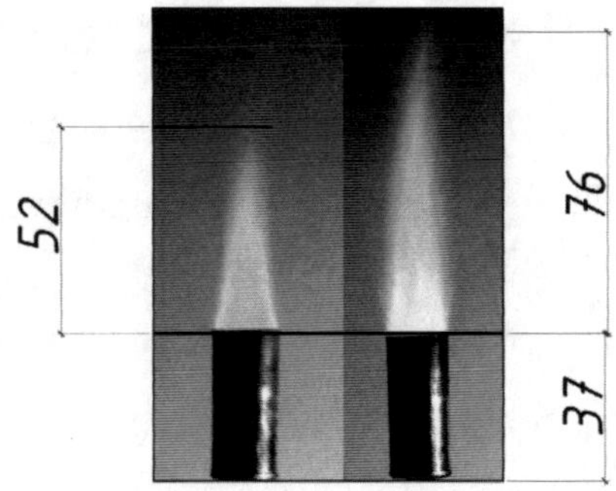

Figure 5. Comparison between torch flames when using magnetic and non-magnetic disks.

4. Conclusions

It was experimentally shown that usage of neodymium-iron-boron magnetic disks in the ortho-modifier cassette reduces gas consumption for 12%, at this the magnetic field has an impact on flame character.

References

[1] Trinchenko A, Paramonov A, Kadyrov M and Koryabkin A 2017 Numerical research of reburning-process of burning of coal-dust torch *IOP Conference Series: Earth and Environmental Science* **90** (1) p 012012

[2] Betelin V B, Nikitin V F, Smirnov N N, Smirnova M N, Stamov L I and Tyurenkova V V 2017 Using GPUs for solving problems of combustion and physicochemical transformations *Mathematical Models and Computer Simulations* **9** (6) pp 727–41

[3] Snegirev A, Markus E, Kuznetsov E, Harris J and Wu T 2017 On soot and radiation modeling in buoyant turbulent diffusion flames *Heat and Mass Transfer/Waerme- und Stoffuebertragung* pp 1–19

[4] Korshunov G I and Solnitsev R I 2017 System catalytic neutralization control of combustion engines waste gases in mining technologies *IOP Conference Series: Earth and Environmental Science* **87** (4) p 042008

[5] Zubkova M, Stroganov A, Chusov A and Molodtsov D 2017 Hydrogenous fuel as an energy material for efficient operation of tandem system based on fuel cells *Key Engineering Materials* **723** pp 616–21

[6] Zubkova M, Maslikov V, Molodtsov D, Chusov A, Zhazhkov V and Stroganov A 2016 Technological Decision to Renewable Energy Usage Biogas for Off-grid Systems Consumption *MATEC Web of Conferences* **73** p 04004

[7] Kolodyazhnyi D Y and Nagornyi V S 2016 Electric charge imposition on hydrocarbon fuel drops in sharply inhomogeneous electric field of injector modules *Russian Aeronautics* **59** (3) pp 402–7

[8] Konakov S A, Dzyubanenko S V and Krzhizhanovskaya V V 2016 Computer Simulation Approach in Development of Propane-air Combustor Microreactor *Procedia Computer Science* **101** pp 76–85

[9] Poletaev N I, Shevchuk V G and Khlebnikova M E 2016 Energy and technological aspects of the combustion of ionized gas-dispersed systems *Eurasian Chemico-Technological J* **18** (3) pp 215–22

[10] Kolodyazhnyi D Y and Nagornyi V S 2015 Experimental study of the influence of electric field on parameters of kerosene-air mixture co-mbustion *Russian Aeronautics* **58** (4) pp 438–42

[11] Gus'kov S Y, Zmitrenko N V, Il'in D V and Sherman V E 2015 Fast ignition of an inertial fusion target with a solid noncryogenic fuel by an ion beam *Plasma Physics Reports* **41** (9) pp 725–36

[12] Ponomarev D A, Mettee H D and Miller J 2015 Empirically estimated heats of combustion of

oxygenated hydrocarbon bio-type oils *BioResources* **10** (2) pp 3638–56
[13] Snegirev A Yu 2014 Generalized approach to model pyrolysis of flammable materials. *Thermochimica Acta* **590** pp 242–50
[14] Smirnov N, Phylippov Y, Nikitin V and Silnikov M 2014 Modeling of combustion in engines fed by hydrogen *WSEAS Transactions on Fluid Mechanics* **9** pp 154–67
[15] Braun-Unkhoff M, Goos E, Kathrotia T, Kick T, Naumann C, Slavinskaya N and Riedel U 2014 The importance of detailed chemical mechanisms in gas turbine combustion simulations *Eurasian Chemico-Technological J* **16** (2-3) pp 179–94
[16] Mamytov K Z, Beisenbayev O K, Shvets V F and Syrmanova K K 2012 The multifunctional automobile gasoline additive on the basis of amino-aromatic hydrocarbons and oxygen-containing compounds *Eurasian Chemico-Technological J* **14** (3) pp 249–52
[17] Mansurov Z A 2011 Synthesis of carbon nanomaterials in flames *Eurasian Chemico-Technological J* **13** (1-2) pp 5–16
[18] Sinitsyn A A and Igonin V I 2006 Influence of vibrational characteristics on pulse combustion boiler operation *Engineering Journal* **12** (117) pp 53–7
[19] Sinitsyn A A 2007 *Pulsed combustion Energy: economics, technology, ecology* (Moscow: Nayka)
[20] Krivosheev S I, Shneerson G A, Platonov V V, Selemir V D, Tatsenko O M, Filippov A V and Bychkova E A 2016 Effect of strong magnetic fields on gas adsorption *Technical Physics* **61** (1) pp 125–9
[21] Romanov N G, Tolmachev D O, Gurin A S, Uspenskaya Yu A, Asatryan H R, Badalyan A G, Baranov P G, Wieczorek H and Ronda C 2015 Dramatic impact of the giant local magnetic fields on spin-dependent recombination processes in gadolinium based garnets *Applied Physics Letters* **106** (26) p 262404
[22] Sivasubramanaiam R, Caruana D J, Kuznetsov M V, Pankhurst Q A and Parkin I P 2011 Combustion reactions of some "Metal-Oxide" systems under conditions of zero and applied magnetic fields: Thermal imaging experiments *Eurasian Chemico-Technological J* **13** (3-4) pp 225–34
[23] Yurchenko V N, Boyko I N and Dykachev I A 2012 Russian Patent RU№2452691 A device for magnetic water treatment. №2011101482/05. Declared 17.01.2011 Publ. 10.06.2012
[24] Powell M R 1998 Magnetic Water and Fuel Treatment: Myth, Magic, or Mainstream Science? *Skeptical inquirer* **22** (1)
[25] Friedrich H and Wintgen H 1989 The hydrogen atom in- a uniform magnetic field: An example of chaos/resonance *Physics Reports* **183** (2) pp 37–79
[26] Gershenfeld N A and Chuang I L 1997 Bulk Spin Resonance *Science* **275** pp 350–6
[27] Slichter C P 1992 *Principles of Magnetic Resonance. Third edition* (Verlag: Springer)
[28] Malygin B V, Pogorletskiy D S, Vasil'chenko G Yu and Sapronov A A 2011 Methods for improvement of ecological safety during magnetic treatment of hydrocarbon fuels for internal combustion engines *Bulletin of the Kherson State Marine Academy Naykovii J* **2** (5) pp 130–9
[29] Malygin B V, Ben' A P, Blakh I V, Konovalov M Yu and Klimenko V V 2006 A method for magnetic activation of organic energy

IOP Conf. Series: Earth and Environmental Science **337** (2019) 012083 doi:10.1088/1755-1315/337/1/012083

Implementation of noise data into building information model (BIM) to reduce noise in the environment and at workplace

Marina Butorina[1,*], Lyudmila Drozdova[1],Denis Kuklin[1], Alexander Sharkov[2], Konstantin Aref'ev[2], Sergey Sopozhnikov[2], Grigoriy Topazh[2], Boris Lyamaev[2], Vladimir Nagornyy[2], Anatoliy Simonov[2], Liliya Muhametova[3]

[1]Baltic State Technical University "VOENMEH", 1st Krasnoarmeyskaya Str., 190005 Saint-Petersburg, Russia
[2]Peter the Great St.Petersburg Polytechnic University, 195251, Polytechnicheskaya, 29, St. Petersburg, Russia
[3]Kazan State Power Engineering University, Kazan, Russian Federation
*Corresponding author: marina_butorina@inbox.ru

Abstract. Noise mapping is the best way to present information on the acoustic pollution. To design noise protection measures, authors use modern software package SoundPLAN, which is based on the up-to-date normative documentation and scientific researches. The program allows importing the results of calculations to Autodesk tools. Improving the efficiency of the noise protection design process is executed through the implementation of building information modeling (BIM). Data exchange between SoundPLAN, AutoCAD 3D or Revit and Navisworks is used in the development of building information model. In the article we present a new approach to noise reduction through implementation of noise data into the overall information model of the infrastructure or building project. BIM helps to track the collision of calculated noise levels with residential buildings, protected premises and workplaces. It lays a basis for the sound proof allocation of workplaces and development of protection measures. BIM also serves as a support tool for the design process of sound protection barriers, since it helps to avoid interconnection of noise barriers basement with engineering networks or paste a barrier in the proper place to provide its efficiency.

1. Introduction

Noise is the main source of pollution in cities. It comes from automobile and rail transport and industrial sources. The results of noise mapping in Europe show that 33% of European population suffer from increased environmental noise levels. About 56 millions of people are affected by increased noise levels coming from roads, 9 million suffer of railroad noise and about 1 million is exposed to the noise of industrial sources [1].

Industrial noise is a cause of professional diseases that affect workers at the workplace during the production process. About 7% of workers suffer from work related hearing difficulties [2].

So, at the design of infrastructure projects, such as roads or railroads, and design of industrial buildings the noise factor should be evaluated. Noise levels should be decreased to the limits to avoid noise caused diseases.

Evaluation of the noise impact is executed using up-to-date calculation methods that take into account all main parameters of traffic and buildings. At the calculation of noise propagation one should also consider distance, atmospheric attenuation, sound absorption and screening [3]. On the basis of the

IOP Conf. Series: Earth and Environmental Science **337** (2019) 012083 doi:10.1088/1755-1315/337/1/012083

calculation results, the noise protection measures should be developed. The most effective way to design noise protection measures is to use specialized software [4].

At the environmental impact assessment, development of noise maps and noise protection measures the researches and practitioners use modern software systems based on updated methods established according to [5] in Europe and according to [6] in Russia. All these methods are implemented in SoundPLAN noise calculation program that we use at the design process. SoundPLAN imports the data necessary for the calculation, and exports the results to formats that allow data exchange between SoundPLAN and other programs intended for design [7].

Increasing the effectiveness of noise design is implemented through the introduction of building information model or BIM technology. The essence of this concept is that in the design process, we use the information model that acts as a general source of data on the object, providing optimal decisions at all stages of its life cycle. The fundamental difference between BIM and the traditional approach based on the preparation of project drawings is that BIM creates, processes and uses computable information about the object, which can be automatically reconciled, matched and internally regulated. BIM contains geometry, spatial relationships, geographic information, the amount and characteristics of building elements, cost estimates, inventories and project lead times. BIM provides close interaction in the design, involving and integrating the various parties involved in the project development. The joint work of the project team allows to identify potential problems before they occur during the construction process [8, 9, 10].

In the process of implementation of noise data into BIM we use data exchange between Geo-Information System (GIS), SoundPLAN, AutoCAD 3D or Revit and Autodesk Navisworks.

2. Algorithm of development of information model on noise pollution

The algorithm for entering data on noise pollution into BIM is executed with following steps:

1. *Import of layouts and information on heights and elevation lines from Geo-Information Systems or Civil 3D as *.dxf/*.dbf files into SoundPLAN.*
2. *Creation of digital ground model in SoundPlan software* (Figure 1). Using information on the heights and elevations the digital ground model of the terrain is automatically calculated in SoundPLAN.

Figure 1. 3D visualization of the terrain and buildings in SoundPLAN.

3. *Entering of parameters of noise sources and buildings in SoundPLAN.* For infrastructure linear project such as road or railroad the necessary information include data on intensity and speed of traffic, percentage of different types of vehicles or trains, type of road pavement or rails, bridges, overpasses, slope of road and other parameters that should be taken into account according to the calculation method. For the recipient buildings we should enter their shapes, height – either in meters or amount of floors, usage and address if necessary.

IOP Conf. Series: Earth and Environmental Science **337** (2019) 012083 doi:10.1088/1755-1315/337/1/012083

The same approach is used to implement noise information into the BIM of any production plant. For the industrial building we should also enter the layout of premises at each level of building, allocation of noise sources and their power levels, thickness and material or sound isolation of walls, allocation of workplaces, sound absorption surfaces, etc. [11,12,13]. These data are used for the calculation of noise maps inside the building. On Figure 2 a model of the industrial building with calculated noise levels is presented.

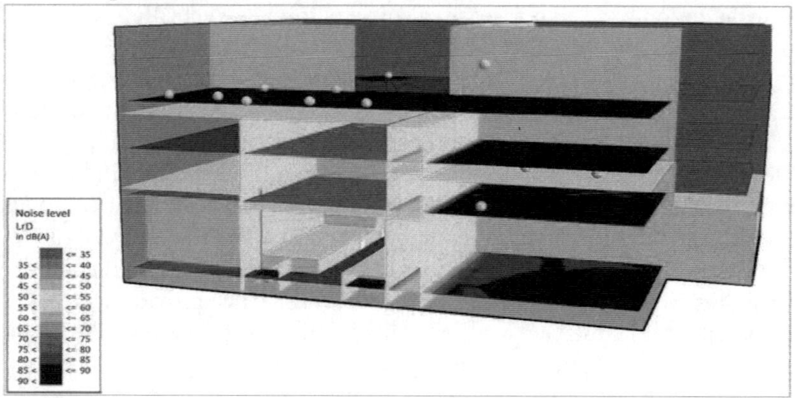

Figure 2. 3D model of industrial building combined with noise maps.

4. *Calculation of noise levels and creation of a noise map.*

With the purpose to implement the results of calculation into building information model we should develop a noise map. The noise map is the most visible way of presenting information on the acoustic situation. Noise maps are also an effective tool in the design and implementation of noise protection measures.

To calculate noise propagation, the calculated area is divided on a grid with a certain step, that is the most appropriate to the purposes of mapping, i.e. 5 m, 10 m but not more than 30 m. After determining the noise levels at the grid points, the points with equal sound levels are connected by isolines, resulting in equal-sound lines. The most convenient step for presentation of sound levels is 5 dBA, which corresponds to the sanitary noise limits [14].

On Figure 3 the noise map for the territory adjoined the rail road is presented. The noise levels at the territory are presented as the isolines and the areas between them are marked according to the provided color scale. The buildings with different usage are marked out with different colors to assign noise limits.

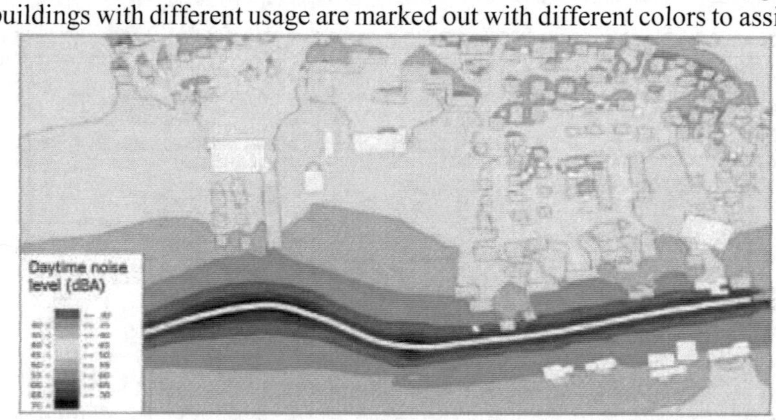

Figure 3. Noise map of the territory adjoined the rail road.

5. *Export of isolines of equal noise levels from SoundPLAN to Civil 3D.* The most spread format of data *.dxf is used to allow exchange of data between different programs.

6. *Creation of surfaces of equal noise levels in Civil 3D at different heights.* Normally, a plain surface at every floor of building as it is shown on Figure 4 is enough to evaluate a necessity of implementation of noise measures.
7. *Creation of 3D models of buildings with use of Civil 3D, Revit and Nawisworks.* SoundPLAN does not allow to evaluate the amount of premises (or windows) that are exposed to the increased noise levels. Civil 3D allows to make different layers with windows of different usage within the same building. Navisworks includes a useful tool that helps to stretch a real photo of the existing building on its virtual model that helps to understand the usage of premises.
8. *Export of equal noise levels' surfaces and models of buildings from Civil 3D into Navisworks using *.nwc files.* The result of import is shown on Figure 4. The exported noise levels' surfaces and models of buildings with different premises (or windows) are used to trace the collisions, i.e. the intersection of the surface of sound level of 55 dBA (the limit noise level) with premises intended for residential use. In Figure 4 the surface with a sound level of 55 dBA crosses a number of rooms that have a residential purpose. Therefore, it is necessary to develop a complex of noise protection measures.

Figure 4. Imported limit noise level surface and a model of building [15].

3. Results of noise information model development and protection measures

3.1 Analysis of BIM for infrastructure project

The most spread measures to reduce road noise are installation of noise barriers and usage of noise-proof glazing [14, 16]. For the case presented on Figure 4 the use of a noise barrier was proposed. The barriers are installed with account of guidance provided in [17,18]. The noise levels after installation of the barrier were calculated using SoundPLAN. The results are shown on Figure 5.

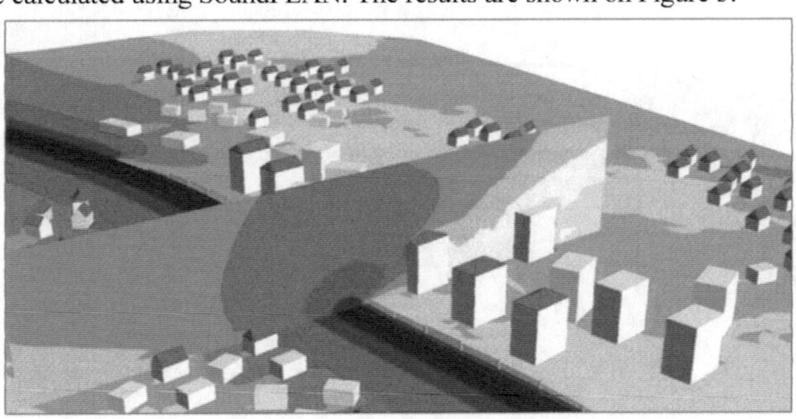

IOP Conf. Series: Earth and Environmental Science **337** (2019) 012083 doi:10.1088/1755-1315/337/1/012083

Figure 5. Noise map after installation of a noise barrier.

The calculated noise levels after the barrier installation were imported into the Navisworks program using the algorithm described above. In the Navisworks program, collisions were tracked again. It was found that the excess of the noise levels is observed at heights of more than 10 m, therefore, additional use of noise-proof glazing is recommended for residential premises located above the third floor.

BIM is also useful when developing working documentation for noise barriers. It contains a digital terrain model which helps to determine the exact location of the screen to provide its efficiency. By tracking collisions in the Navisworks, it is possible to assess whether the pile foundation of a barrier or a barrier itself crosses the engineering networks or other road construction elements, such as lighting supports or road fencing, that are also parts of BIM of a road construction project (Figure 6).

Figure 6. Tracking crossing of pipes and piles [15].

3.2 Analysis of BIM for building project

The same procedure of creation of noise information model is used for industrial premises to reduce noise levels at workplaces. On Figure 7 a) the noise map for the production plant of the industrial building presented on Figure 2 is shown. In the result of calculation, we have obtained that the noise levels are exceeded by 10-15 dBA at the majority of workplaces.

For the reduction of noise levels inside the production plant the sound absorption with baffles installed at the ceiling and noise curtains are proposed according provisions of [19-26]. The results of calculations of noise levels with sound absorption (Figure 7 b) show that the noise levels are decreased by 2-14 dBA at the workplaces No. 5, 6 and 7. For the workplaces No. 1 and 4 a rational planning should be used, i.e. the permanent workplaces should be shifted from the zone with excess levels to the quitter zones. Workers No. 2 and 3 should use personal protection measures such as earphones.

The limit noise line should be pasted into the BIM of the plant to redevelop the workplaces [27-32]. The noise protection constructions should be imported into BIM to avoid collision with ceiling constructions and to allow the cost of sound absorption in the project budget documentation.

a)

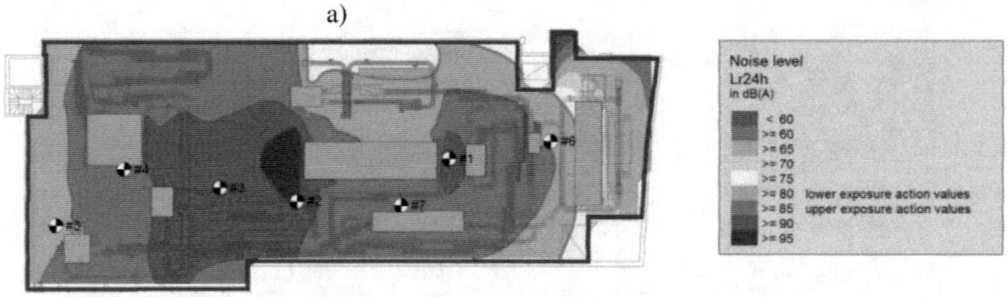

IOP Conf. Series: Earth and Environmental Science **337** (2019) 012083 doi:10.1088/1755-1315/337/1/012083

b)

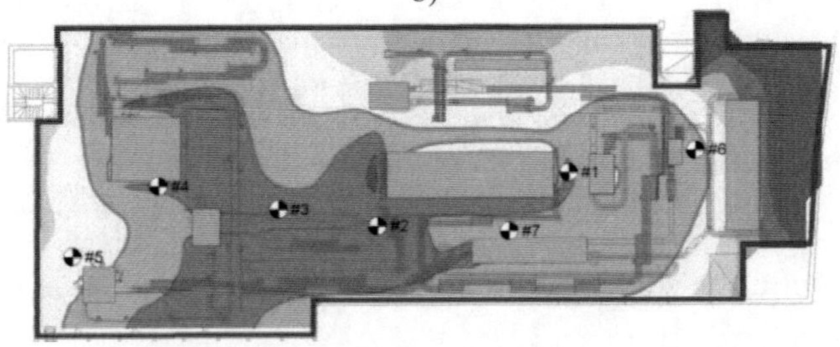

Figure 7. Noise maps of a plant: a) without noise protection; b) with noise protection.

Thus, the use of BIM is an effective tool of noise protection measures' design. BIM allows designers to take into account the results obtained at the development of related sections, promptly eliminate emerging problems and allow protection measures in the budget documentation. The joint work of the project team allows to identify potential problems before they occur at the construction or even at the exploitation of the object designed.

4. Conclusions

1. Noise is one of the main factors affecting population and workers. It can be minimized with proper evaluation and introduction of protection measures.
2. The design of noise protection measures is performed using specialized software calculation tools.
3. At the implementation of noise information to BIM, data is exchanged between SoundPLAN, AutoCAD 3D and Navisworks software.
4. Using BIM, it is possible to track the excess of noise levels in residential premises and at workplaces, the models of which are developed in Civil 3D or Revit.
5. BIM allows designers to take into account the results of noise evaluation, eliminate problems and allow protection measures in the budget documentation.

References

[1] Directive E U 2002 Directive 2002/49/EC of the European Parliament and the Council of 25 June 2002 relating to the assessment and management of environmental noise.

[2] Eurostat 2004 *Work and health in the EU: a statistical portrait.*

[3] ISO 9613-2 1996 *Acoustics - Attenuation of sound during propagation outdoors - Part 2: General method of calculation.*

[4] European Commission Working Group Assessment of Exposure to Noise (WG-AEN) 2006 *Good Practice Guide for Strategic Noise Mapping and the Production of Associated Data on Noise Exposure. Position paper. Final Draft, Version 2*

[5] Commission Directive (EU) 2015/996 establishing common noise assessment methods according to Directive 2002/49/EC of the European Parliament and of the Council. *Official Journal of the European Union.*

[6] SP 276.1325800.2016 2016 *Buildings and territories. Protection design rules from traffic noise* (Moscow: SC "Technorma")

[7] SoundPLAN GmbH 2015 *SoundPLAN manual*

[8] 2014 Otchet o NIR po teme «Razrabotka rekomendatsiy po vnedreniyu innovatsionnykh tekhnologiy na vsekh etapakh zhiznennogo tsikla avtomobil'nykh dorog». (Moscow: OOO «Avtodor-Inzhiniring»).

[9] 2015 *Corporate BIM Standard For Building/Construction Projects.* (Moscow: Autodesk, Inc.) p 118.

[10] 2015 *Corporate BIM Standard For Building/Construction Projects.* (Moscow: Autodesk, Inc.) p 54.

[11] EN 12354-5:2009 2009 *Building acoustics - Estimation of acoustic performance of buildings from*

the performance of elements - Part 5: Sounds levels due to the service equipment.
[12] EN 12354-4:2000 2000 *Building acoustics - Estimation of acoustic performance of buildings from the performance of elements - Part 4: Transmission of indoor sound to the outside.*
[13] EN 12354-6:2004 2004 *Building acoustics - Estimation of acoustic performance of buildings from the performance of elements - Part 6: Sound absorption in enclosed spaces.*
[14] Butorina M and Minina N 2006 *13th International Congress on Sound and Vibration 2006, ICSV 2006* **7** pp 5761–5768.
[15] Anisimov E P, Butorina M V, Turina N V and Kuklin D A 2016 *Bulletin of building technics* **6** (982) pp 18-21.
[16] Schaal J 2015 *Proceedings of 5 th International Conference on Acoustics & Vibration*
[17] 2003 Recommandation de la Commission du 6 aout 2003 relative aux lignes directrices sur les provisoires de calcul du bruit industriel, du bruit des avions, du bruit du trafic routier et du bruit des trains, ainsi qu'aux correspondantes, Journal official L 212, pp 49-64
[18] 2018 Part 1 HA 65/94 Design Guide for Environmental Barriers. Section 5 Environmental barriers. Part 1, Highways England
[19] Maekawa Z, Rindel J H and Lord P 2011 *Environmental and Architectural Acoustics*. Second edition (London: Spon Press)
[20] Ivanov N 2016 *Engineering acoustics: Theory and practice of noise control.* Handbook (Moscow: Logos)
[21] Bobrova T V and Panchenko P M 2017 *Magazine of Civil Engineering* **76** pp 84–97 doi:10.18720/MCE.76.8
[22] Giyasov B I, Ledenyov V I and Matveeva I V 2018 *Magazine of Civil Engineering* **77** pp 13–22 doi:10.18720/MCE.77.2
[23] Burdonov A E 2014 *Magazine of Civil Engineering* **47** pp 9–16 doi:10.5862/MCE.47.1
[24] Kiesewetter D 2015 *2015 International Siberian Conference on Control and Communications, SIBCON 2015 - Proceedings* doi:10.1109/SIBCON.2015.7147096
[25] Petritchenko M R, Kotov E V, Nemova D V, Tarasova D S and Sergeev V V 2018 *Magazine of Civil Engineering* **77** (1)
[26] Bolotin A and Bakayev V 2017 *Journal of Human Sport and Exercise* **12** pp 405–13 doi:10.14198/jhse.2017.122.17
[27] Anikina I D, Sergeyev V V, Amosov N T and Luchko M G 2017 *Int. J. of Hydrogen Energy* **42** (1)
[28] Drozhzhov K A and Ivanov S I 2018 *Proceedings of the 2018 IEEE Conference of Russian Young Researchers in Electrical and Electronic Engineering, ElConRus 2018* pp 1071–1074 doi:10.1109/EIConRus.2018.8317275
[29] Martínez-Lera P, Kucukcoskun K, Tournour M, Shur M and Travin A 2016 *22nd AIAA/CEAS Aeroacoustics Conference, 2016* doi:10.2514/6.2016-2796
[30] Vatin N I, Pestryakov I I, Kiski S S and Teplova Z S 2014 *Applied Mechanics and Materials* pp 584–6, 1381–7 doi:10.4028/www.scientific.net/AMM.584-586.1381
[31] Petritchenko M R, Subbotina S A, Khairutdinova F F, Reich E V, Nemova D V, Olshevskiy V Ya and Sergeev V V 2017 *Magazine of Civil Engineering* **73** (5)
[32] Pavlov M, Lukin S and Derevianko O 2018 *MATEC Web of Conferences* **193** p 03006.

IOP Conf. Series: Earth and Environmental Science **337** (2019) 012084 doi:10.1088/1755-1315/337/1/012084

The study of the properties of ABS plastics and polyamide products, made by 3D printing for use in turbine construction

V. Barskov[1,*], S. Besedin[1], V. Rassokhin[1], A. Smetankin[1], K. Besedina[2], N. Lavrov[2], D. Panfilov[2], Y. Smirnov[1], Y. Lypar[1], L R Mukhametova[3]

[1]Peter the Great St.Petersburg Polytechnic University, 195251 Polytechnicheskaya, 29, St.Petersburg, Russia
[2]Saint-Petersburg State Institute of Technology, 190013 Moskovsky prospect 26, St.Petersburg, Russia
[3]Kazan State Power Engineering University, Kazan, Russian Federation
*Corresponding author: viktorbarskov@ntcmtt.ru

Abstract. Physical and mechanical tests of samples made from ABS plastic and polyamide obtained by 3D printing were carried out. Based on the results of experimental studies, the possibility of using additive technologies in the manufacture of low-power turbo generator impellers has been shown.

1. Introduction

Technical solutions utilized in turbo machines for centralized electric power generation are not optimal for turbo machines designated for distributed generation. In distributed generation, low-power turbo machines are allocated to a special class due to the essential effect of engine size on net power, parameters selection and design of the engine, which in turn determines the overall economic efficiency. Alternative technological solutions should be applied to achieve the economic efficiency of distributed turbo-machines for distributed power. All over the globe R&D groups are investigating various alternative technological solutions and materials. In mechanical engineering research often focuses on application of powder metals and ceramics. Industrial enterprises are looking for application of new technology to distributed power generation in order to decrease costs-labor, material consumption and usage of manufacturing equipment. When developing low-capacity turbomachines, it is important to manufacture their flow parts from polymeric materials (PM) instead of metals. The use of additive manufacturing [1, 2] is a promising trend in the development of industrial scale manufacturing of products with complex configuration from PM. Polycarbonate, polystyrene, ABS plastics, polyamides, polymethyl methacrylate, polylactide and photopolymers [1-3] are used as PM in additive manufacturing. After analyzing the requirements for materials to be used in the design of flow parts of turbomachines, and details of operating parameters, ABS plastics and polyamides have been chosen [4]. Despite the variety of additive technologies, for each of various types of PM a certain type of manufacturing technology should be chosen. For parts made out of ABS plastic, FDM technology is used, for polyamides - SLS technology. FDM technology (Fused Deposition Modeling), or modeling by fusing, is suitable not only for ABS, but also for polylactide [3, 4], as well as for some other thermoplastics. SLS technology (Selective Laser Sintering) is employed for polyamides, as well as for sintering ceramic and metal powders [3, 4].

Additive manufacturing is widely discussed in various papers with information on applications of 3D-printing to various industries [5–19]. But there is almost no information on how properties of PM

IOP Conf. Series: Earth and Environmental Science **337** (2019) 012084 doi:10.1088/1755-1315/337/1/012084

change while processed by these methods, and whether and how they are correlate with data in relevant reference books [22-28].

2. Methods

For the purpose of present research we employed the ABS-plastic REC and polyamide PA-12. Sample preparation was performed by 3D printing. Samples of ABS plastic have been prepared with FDM technology using modes in the middle of the recommended range (melt temperature 228 °C, melt deposition rate 0.05 m/s). Samples of polyamide were obtained using SLS technology as recommended by the manufacturer. The shape and size of the samples corresponded to GOST, 3-5 samples were used to define each parameter. The tests were carried out using the equipment of the company SHIMADZU (universal tensile testing machine of the AG-X plus series with an accuracy of 0.5%). Hardness tests were carried out in accordance with GOST 4670-70 using Brinell method.

3. Results and Discussion

Table 1 presents the results for experimental tests with samples made from ABS plastic using the 3D printing method, as well as the properties of ABS plastics given in professional literature. While reference books do not indicate a method for samples production, for thermoplastics it is usually the injection moulding method.

Table 1. ABS-plastic properties.

Parameter	Manufacturing method	
	3D-printing	Injection moulding
Destructive stresses under tension (σ_t), MPa	44	36-60
Relative elongation at break (Δl),%	3	1.0-3.0
Destructive stress under compression (σ_c), MPa	71	50-87
Bending stress (σ_b), MPa	73	46-80
Toughness (A), kJ/m^2 (kgf*cm)	337	80-100
Brinell hardness (HB), MPa	46	100
Material density, kg/m^3	997-1020	1040

If the properties presented in Table 1 were compared, it could be noted that the properties of samples produced by the method of 3D printing fall into the range of values indicated by referenced publications for most of the parameters. Density of the resulting product is also close to the range provided. When 3D printing is used, the product is formed without pressure and one could expect decline in the density of polymer product and in it's strength properties. But this has not been observed, apparently due to the fact that in the selected temperature range, ABS plastic used has low viscosity, which makes it possible to obtain non-porous products without internal stresses as a result of 3D printing. However, the experimental data of toughness and hardness differ significantly, which can be explained by the different composition of plastics. It is known that in ABS plastics the composition may vary. Apparently, a higher content of butadiene in the copolymer used for 3D printing led to increased impact strength and decreased hardness. Table 2 presents the results of testing samples of polyamide brand PA-12, manufactured by selective laser sintering (SLS) technology. For comparison, the indicators characterizing values of physical and mechanical properties of injection-moulded polyamide PA-12L are provided below in Table 2.

Table 2. Polyamid product properties.

Parameter	Manufacturing method	
	3D-printing	Injection moulding
Destructive stresses under tension (σ_t), MPa	37	50
Relative elongation at break (Δl),%	-	200-280
Destructive stress under compression (σ_c), MPa	68	60
Bending stress (σ_b), MPa	53	60
Toughness (A), kJ/m^2 (kgf*cm)	72.9	80-90

Brinell scale (BHN), MPa	36,1	75
Material density, kg/m^3	945	1020

From Table 2 it can be concluded that most of the indicators of polyamide, obtained by 3D printing, have lower values than those given by literature, while ABS samples, obtained by 3D printing, have properties close to the properties of samples obtained by injection molding. The reason is that to obtain products from ABS plastics FDM technology is used, in which polymer is applied to substrate in a melted form — the product design process is similar to welding PM, and as a result products gain in density. SLS technology is used to manufacture polyamide products, in which sintering of polymer particles allows to produce products with pores. Variations have been observed in properties of products from PM obtained by different methods. For example, it was previously shown [20-28] that properties of PM products obtained by rotational molding, also having a porous structure, differ from the properties of materials claimed by the manufacturer.

In addition to the reduction of some strength properties and apparent density, products made of polyamide, obtained by SLS technology, have a rough surface. A study was conducted on the possibility of using parts made of ABS plastics and polyamides in an experimental turbine installation; for this purpose, a strength analysis was performed using the experimentally determined physical and mechanical properties of PA-12 and ABS plastic. Since the rotating part was chosen as an object, calculations were performed at a maximum rotation speed of 2093 rad/sec (20.000 rpm). Figure 1 shows the stress state of a full disk. Figure 2 shows the results of the calculation of the segment in equivalent stresses and strains by von-Mesis, performed by the FEM method in the ANSYSWB program.

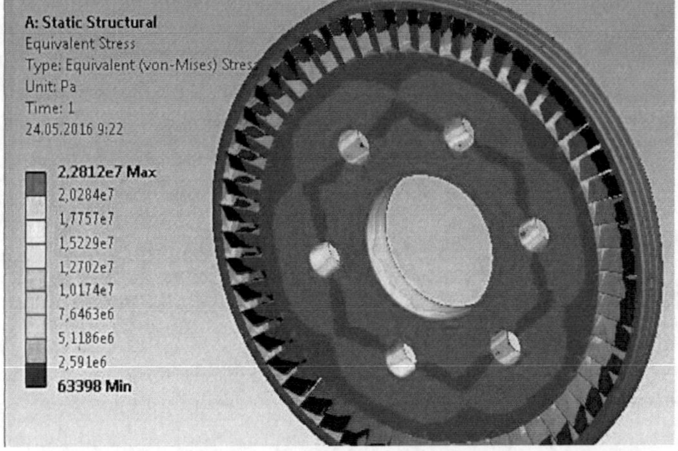

Figure 1. The stress state of a plastic disc at rotation speed of 20.000 rpm (view on the output edges of the blades).

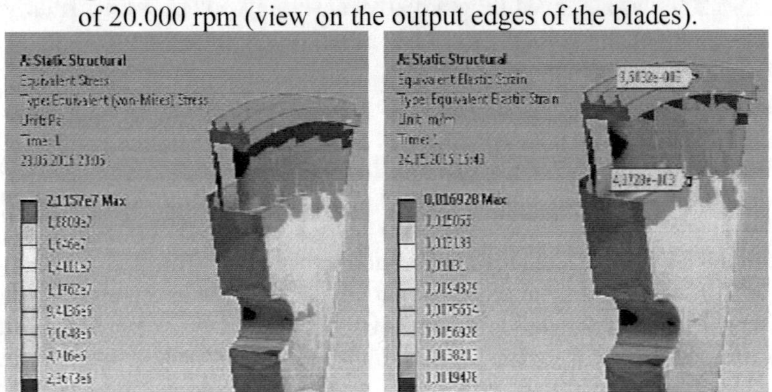

a) b)

Figure 2. Stressed (a) and deformation (b) state of a plastic disc segment (view on the output edges of the blades).

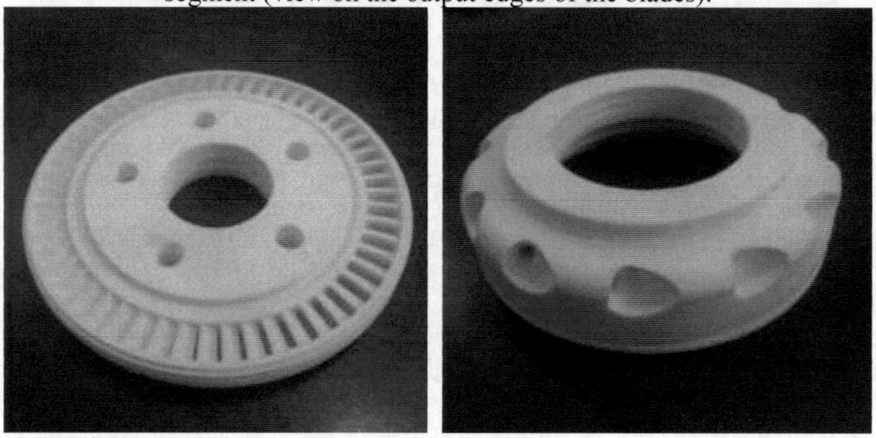

Figure 3. Turbogenerator flow part made of PA-12 polyamide according to SLS technology.

The experimental prototype of a turbo-generator unit with electrical power of 1 kW (commercial name MDG-1) has been developed and manufactured at the production site of LLC "NTC Microturbine Technologies"; the photographs of the assembled unit are shown on Figure 4.

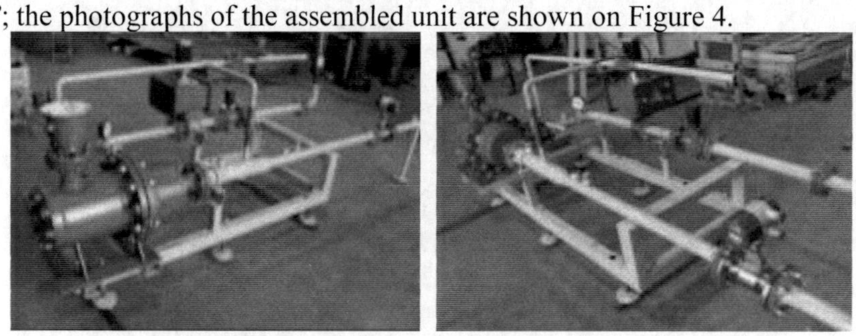

Figure 4. Photo of an experimental sample of a turbogenerator installation with an electrical output of 1 kW; left and right views.

The tests were carried out in accordance with the "Testing Program and Methods", jointly developed by the department of "Turbines, hydraulic machines and aircraft engines" and the team from LLC "STC Microturbine Technologies". Tests were conducted with a special "experimental" stand. The maximum active power load of 1 kW was achieved by gradually increasing power incrementally by 0.2 kW with a 1 minute delay. Testing of electrical energy quality effect on the power load was performed during 72 hours. Samples successfully passed the examination.

4. Conclusions

In Russia, many research groups are investigating the application of polymeric materials. Currently, polymers are used in almost all areas of production. They are used in toys, building materials, implants, fabrics, medical drugs, machine lubricants, protective masks and goggles, optical glass, awnings and windows, furniture fabrics and fillers, leatherette and processed natural leather, rubber, packaging materials, promotional products, instrument cases, artificial and synthetic fabrics and fibers, films for various purposes, construction materials, materials for the electrical and radio engineering industry, decoration, ion exchange and epoxy resins, plastics with extreme properties (heat-resistant and cold-resistant, high hardness, fireproof etc.). Polymers are actively used in additive technologies, as an alternative to metals and ceramics. This is due to the diversity of polymers and a wide range of physical and mechanical properties of these materials.

According to calculations, it was shown that a single plastic disk made of PA-12 and ABS plastic can withstand a rotation speed of 20.000 rpm with a margin of 1.8 for maximum stresses. Maximum stresses in the RK disk do not exceed 14 MPa. As a result of calculations and tests, it was concluded that polyamide and ABS plastic can be used in manufacturing of certain parts of turbomachines.

The advantages of using additive technologies within engineering industry:

1. A smaller amount of waste than in traditional production: according to statistics, at machinery manufacturing enterprises the utilization rate of metals during mechanical processing is less than 0.2
2. The price of the parts produced, since the cost of plastic is several times less than the cost of the metal, which will significantly reduce the cost of finished products.
3. Affordable price for 3D printers, which varies from 30 to 1500 US dollars and could be easily purchased on the Internet.

At LLC "STC Microturbine Technologies", authors of present paper continue their research on polymer-based additive technologies for turbine engineering.

References

[1] Burlov B V, Spirov K N and Kryzhanovsky V K 2016 *Izvestia SPbSTI (TU)* **36** pp 52-7
[2] Besedina K S, Lavrov N A and Barskov V V 2018 *Izvestia SPbSTI (TU)* **70** p 44
[3] Lavrov N A, Igumenov M S and Talk K S 2014 *Plast. Masses.* **56** (1-2)
[4] Barskov V, Besedin S, Besedina K, Rassokhin V, Smetankin A and Nikiforova O 2018 *J. of Adv Research in Dynamical & Control Systems* **10 p** 348
[5] Kulhavý P, Fliegel V, Lepšík P and Srb P 2018 *Experimental Stress Analysis - 56th International Scientific Conference* pp 227-33
[6] Oral M, Ersoy O and Serhatli E 2018 *J. of Plastic Film and Sheeting* 8756087918768348
[7] Zhang H, Cai L, Golub M, Zhang Y, Yang X, Schlarman K and Zhang J 2018 *J. of Materials Engineering and Performance* **27** pp 57-62
[8] Woern A, Byard D, Oakley R, Fiedler M, Snabes S and Pearce J 2018 *Materials* **11** p 1413
[9] Duty C, Ajinjeru C, Kishore V, Compton B, Hmeidat N, Chen X, Liu P, Hassen A, Lindahl J and Kunc V 2018 *J. of Manufacturing Processes* **35** pp 526-37
[10] Sonsalla T, Moore A, Meng W, Radadia A and Weiss L 2018 *Polymer Testing* **70** pp 398-395
[11] Osman M and Atia M 2018 *Rapid Prototyping J.* **24** pp 1067-75
[12] Kumar R, Singh R and Ahuja I 2018 *Measurement: J. of the International Measurement Confederation* **120** pp 11-20
[13] Balderrama-Armendariz C O, MacDonald E, Espalin D, Cortes-Saenz D, Wicker R and Maldonado-Macias A 2018 *Int. J. of Advanced Manufacturing Technology* **96** pp 307-17
[14] Balzereit S, Proes F, Altstädt V and Emmelmann C 2018 *Additive Manufacturing* **23** pp 347-54
[15] Walter R, Friedrich K and Gurka M 2018 *AIP Conference Proceedings* **1981** p 020033
[16] Tiwari S, Pande S and Kumar S 2018 *Procedia Manufacturing* **21** pp 630-7
[17] Lai C, Chen S, Chen M, Chen H, Hsiao H, Liu L and Chen C 2019 *J. of Molecular Structure* **1175** pp 836-43
[18] Liu Z, Shi J and Wang Y 2018 *ASME 2018 13th International Manufacturing Science and Engineering Conference* **1**
[19] Craft G, Nussbaum J, Crane N and Harmon J 2018 *Additive Manufacturing* **22** 800-6
[20] Barskov V, Besedin S, Besedina K, Zabelin N, Matveev Yu, Rassokhin V, Lavrov N and Fokin G 2017 *Int. J. of Advanced Biotechnology and Research* **8** 1708-15

[21] Besedina K, Lavrov N and Barskov V 2018 *St. Petersburg State University of Film and Television* pp 26-27

[22] Bolotin A and Bakayev V 2017 *J. of Physical Therapy Science* **29** pp 1092–4 doi:10.1589/jpts.29.1092

[23] Rakov G L, Rassokhin V A, Zabelin N A, Olennikov S Y, Sebelev A A, Sukhanov A I and Schislyaev S M 2016 *Int. J. of Environmental and Science Education* **11** pp 11721–33

[24] Zobacheva A Y, Nemov A S, Borovkov A I, Novokshenov A D, Khovaiko M V and Ermolenko N A 2017 *Materials Physics and Mechanics* **34** pp 51–8 doi:10.18720/MPM.3412017_6

[25] Perevoznik D, Kurselis K, Kiyan R, Nepomnyashchaya E K, Aksenov E T, Velichko E N and Chichkov B N 2017 *Progress in Electromagnetics Research Symposium* pp 2313–5 doi:10.1109/PIERS.2017.8262137

[26] Konnikov E A, Pogrebova O A, Maskova Y R and Glukhov V V 2018 *2017 6th International Conference on Reliability, Infocom Technologies and Optimization: Trends and Future Directions, ICRITO 2017* pp 541–7 doi:10.1109/ICRITO.2017.8342487

[27] Zobacheva A U, Nemov A S and Borovkov A I 2017 *Materials Physics and Mechanics* **32** pp 74–82

[28] Popovich A A, Sufiiarov V S, Borisov E V, Polozov I A and Masaylo D V 2018 *Materials Physics and Mechanics* **38** 1–10 doi:10.18720/MPM.3812018_1

IOP Conf. Series: Earth and Environmental Science **337** (2019) 012085 doi:10.1088/1755-1315/337/1/012085

Aluminum-based composite reinforced with fullerene soot

F. Yunusov[1], E. Bobrynina[1], T. J. Ma[1], T. Larionova[1], M. Okrepilov[1], V. Mazin[1], S. Kravchenko[1], V. Yakovlev[1], S. Chulkin[1], D Zaripova[2]

[1]Peter the Great St.Petersburg Polytechnic University Polytechnicheskaya 29, St.-Petersburg, 195029, Russian Federation
[2]Kazan State Power Engineering University, Kazan, Russian Federation

Abstract. Al-based composites with fullerene soot have been successfully prepared using powder metallurgy route. The selected parameters of the powder preparation for extrusion are discussed. The microstructure and phase composition are analyzed by optical microscopy and XRD analysis. It is shown that carbon reacts with Al with very fine carbide particles formation, resulting in micro stress of the aluminum matrix, and providing high strength of the composite. The obtained composites have excellent strength characteristic (660MPa) and satisfactory ductility (4.5%). It was shown that the developed Al-based composites can successfully compete with the best aluminum alloys used for electrical wire production.

1. Introduction

Aluminum alloys used in electrical wire production are subject to a number of requirements for their physical and mechanical properties. Among the main is the requirement for a combination of high strength, satisfactory plasticity, and high electrical conductivity. Pure aluminum possesses high conductivity of about 64% of IACS, but very low strength of about 60MPa [1]. Alloying allows to increase the strength by an order of magnitude of pure aluminum, but leads to a loss of conductivity because of solid solution formation. Nevertheless, development of Al alloys for electro technical application continues rather successfully; to the present the main tendency is addition of low soluble in aluminum elements, such as REM, Zr, Sc. However, such alloying considerably increases the cost of the material, what limits their commercial attractiveness. Another requirement that is imposed on the properties of materials to be used in electro technical application is their thermal stability up to 300°C. Ordinary aluminum alloys lose their properties at heating either due to recovery of the cold deformed structure or due to coarsening and dissolution of strengthening phases [2, 3].

A new class of materials able to meet the mentioned above targets are composites consisting of pure metal matrix and undissolved in it nanoparticles. This approach allows to combine the high conductivity of pure aluminum with elevated strength caused by nanoparticles. Besides, the ultrafine homogeneously distributed chemically stable particles retain their structure at heating and hinder the self-diffusion of aluminum preventing its recrystallization.

Nano-carbon materials are known to be one of the attractive reinforcing agents due to their excellent physic-mechanical properties [4] and an abundance of carbon in nature. Carbon does not dissolve in aluminum, but reacts with it forming Al_4C_3, which has a high decomposition temperature about 2500°C. To the present there are rather high level of mechanical properties of the aluminum-nanocarbon composite has been reported in literature [5-15]. According to the different authors the effect of the strengthening was attributed to different phenomena, such as solid solution strengthening [10, 15], grain boundary refinement [7,9], dispersion strengthening [7,9] and good load transfer [5,7-9,12].

As the reinforcing agent carbon nanotubes [6-9], graphene [12-14], and fullerene [10,11,15] have been used. All of these nanocarbon materials are expensive because of complex, high technology production process. In the present paper we would like to consider a possibility to manufacture Al-based composite reinforced with nanoparticle of fullerene soot, estimate strengthening effect and thermal stability.

2. Materials and method

Aluminum matrix composites were produced using an aluminum powder of 98% purity, particle size 80-125 μm, and fullerene soot supplied by Suzhou Dade Carbon Nanotechnology Co. Stearic acid was added to prevent the agglomeration of the fullerene soot, and a mix of NaCl, KCl and CaF2 was added as a flux to improve the contact between the Al particles. The powders were prepared varying fullerene soot concentration up to 3wt.%. Ball milling was performed in 80ml stainless steel chambers with 5mm diameter stainless balls with ball to powder ratio of 10:1 using Fritsch planetary ball mill Pulverisette 7 premium line. The milling was performed in two stages, the first one at 200 rpm for 60 min and the second one - at 600 rpm for two hours. To prevent excessive powder heating the milling was performed with 20 min interruptions after each 15 min of milling. All the procedures were conducted in an Ar atmosphere.

For extrusion the ball-milled composite powder was diluted with aluminum powder in a ratio of 2:1, 1:1, and 1:2 correspondingly. The diluted powders were either mix in a tilting drum blender for 2 h or additionally ball milled at 200 rpm for 15 min and at 600 rpm for 15 min in an Ar atmosphere. Aftervthat the powder was pre-pressed at 50 MPa in a cylindrical container of 30 mm in diameter, 100 mm in height, and 2 mm thick made of 5056 Al alloy. Then the prepared capsule was heated to 400°C and hot extruded into 10 mm diameter rods with an extrusion ratio of 9:1. Finally, the rods were hot rolled at 380°C in calibrated profile rolls. The rolling was conducted three times; as a result, there were samples of a square section with a thickness of 5, 3, and 2 mm. Annealing was performed in resistance furnace in Ar atmosphere at 470°C.

For optical observation the specimens had been polished and etched with 10% NaOH. The observation was fulfilled on Carl Zeiss Observer D1m microscope.

Vickers microhardness was tested with ZWICK ZHU under a load of 100N and a dwelling time of 10c. The 3-point bend test was carried out on the samples of 60 mm length, with a distance between the supports of 26mm. The test was carried out on the universal testing machine Instron 5965 with a tool for 3-point bending (radius of the supports 10 mm, radius of the loading tip 10 mm) at a strain rate of 10mm/min.

The phase composition was examined with the Bruker D8 Advance diffractometer under Cu Kα radiation. The aluminum solid solution lattice parameter, internal microstresses and size of the coherent scattering region were determined on the base of XRD patterns, which had been collected in the range from 20° to 140° 2Θ with a speed of 2°/min. The definition of the lattice parameter was calculated using Nelson-Riley [16] extrapolation function in order to eliminate the systematic error after the subtraction of K2 doublet. Internal microstresses and size of the coherent scattering region have been determined by line-broadening analysis using Williamson-Hall plot [17].

3. Results and Discussion

SEM image of the composite powder produced by high energy ball milling are presented in Figure 1. As it is seen, the milled particles are equiaxial dense agglomerates of size about 20- 40μm. Figure 2 shows photographs of the as-extruded and rolled samples, as well as the samples after bending test.

IOP Conf. Series: Earth and Environmental Science **337** (2019) 012085 doi:10.1088/1755-1315/337/1/012085

Figure 1. SEM microphotographs of the composite powder after milling.

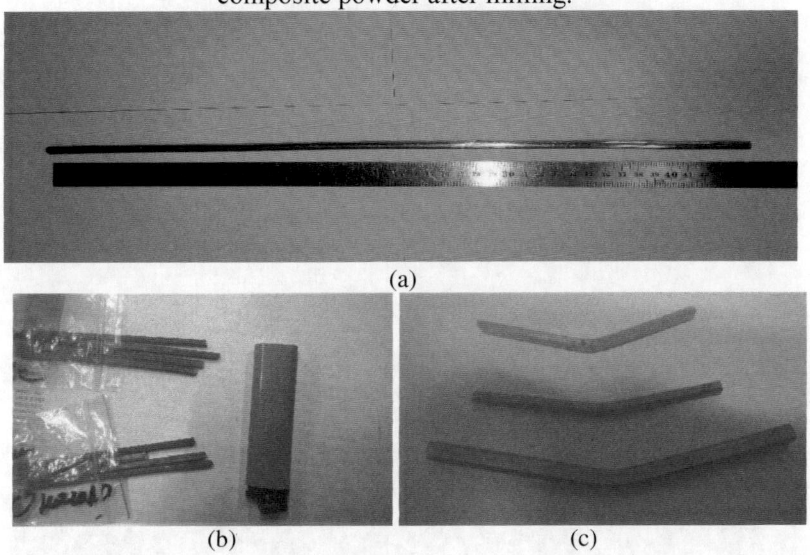

Figure 2. Photographs of as-extruded (a) and rolled (b) samples, (c) – the rolled samples after bending test.

For extrusion the ball milled powder were diluted with pure aluminum powder. The dilution was undertaken with an aim to stimulate plastic deformation of the composite during extrusion and prevent an excessive equipment wear. The parameters of the dilution and mechanical properties of the samples are presented in Table 1. The corresponding stress-strain curves are shown in Figure 3. All the samples had been compacted by the same way: through extrusion and rolling to 3 mm of thickness.

The samples cut from the extruded rode prepared from powder diluted in the blender are very different from each other, it may be explained by the fact that mixing in a blender does not allow to achieve homogeneous mixture of aluminum and composite particles because of their different morphology.

Table 1. Parameters of the dilution and mechanical properties of the samples.

No	Mode of dilution	Initial concentration of the ball milled composite powder, wt.%C	Dilution ratio X/Y, X-composite powder, Y – pure aluminum	Resultant concentration, wt.%C	HV	σ_b, MPa	ε, %

IOP Conf. Series: Earth and Environmental Science **337** (2019) 012085 doi:10.1088/1755-1315/337/1/012085

1	Tilting drum blender	3	1:2	1	80	250-390	4-10
2	Tilting drum blender	3	2:1	2	120	610	0,5
3	Ball mill	2	1:1	1	125	660	4,5
4	Ball mill	3	1:2	1	95	450	3,8

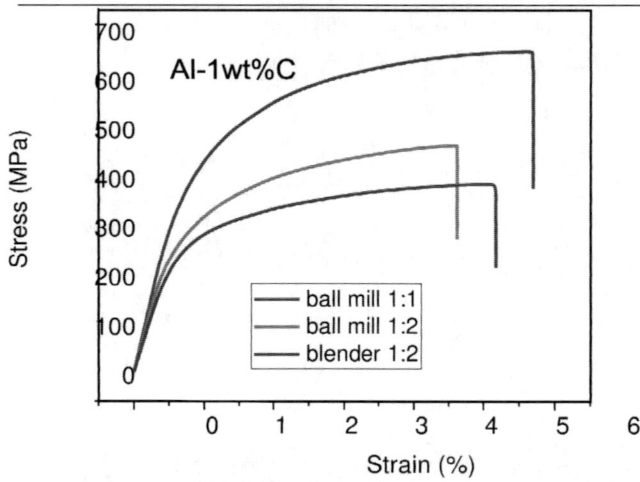

Figure 3. Bending deformation curves at for the samples prepared with equal content of carbon, but according to various dilution parameters.

Microstructure of the extruded and rolled samples in depending on the dilution parameters are presented in Figure 4.

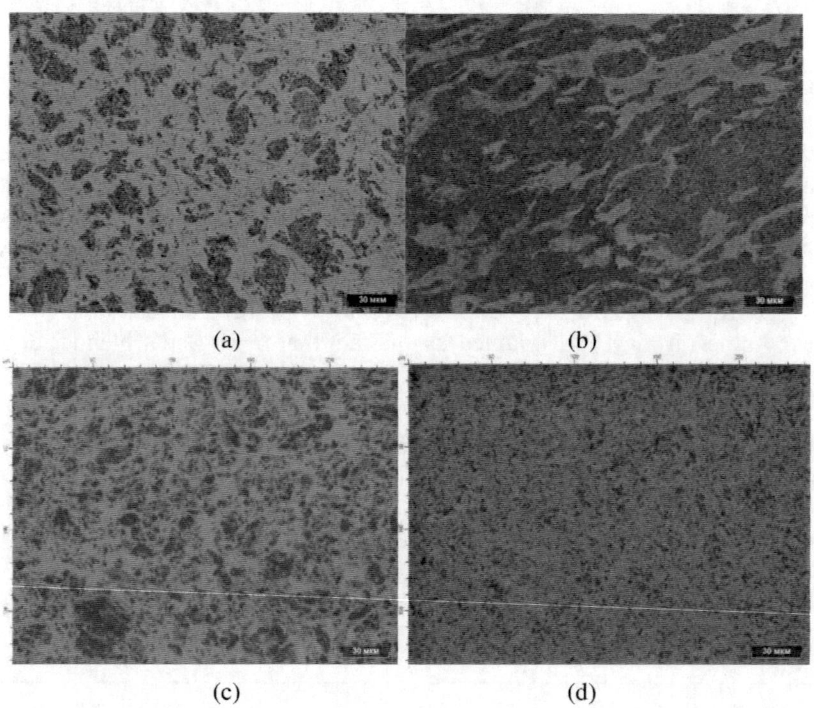

Figure 4. Optical images of the samples microstructures in depending on the dilution parameters (Table 1): a) – 1; b) 2; c) – 3; d) – 4.

All the microstructures contain light and dark constituents; the first one corresponds to the soft pure aluminum matrix and the second - to the hard composite particles distributed in the latter. Ball milling leads to the more homogeneous and fine structure compared to that obtained after mixing in a drum blender. During ball milling soft aluminium particles envelope the harder composite ones leading to a disintegration of the composite agglomerates and promoting the homogeneous distribution. The microstructure of the sample obtained by the dilution of Al-2%C in a ball mill to Al-1%C (sample number 4 in Table 1 and Figure 4,d) has the best appearance, it is very dense, fine, and homogeneous. Owing to the achieved microstructure this sample has the best combination of strength and ductility. Annealing at 470°C does not lead to notable structural change.

XRD patterns of the powders, as well as extruded and annealed samples are presented in Figure 5. Lattice parameter, crystallite size and microstress evaluated from XRD data are summarized in Table 2.

Ball milling leads to a strong broadening of the Al XRD peaks compared to the initial powder. The XRD peaks broadening means a decrease of the crystallite size and microstress growth caused by the high concentration of the introduced defects and crystallite disintegration. After compacting and annealing the crystallite size increases due to recrystallization, however, the microstress does not relax nor during compacting neither annealing.

In the insertion of Figure 6 is the region of the XRD pattern where peaks of Al_4C_3 may be observed. It is seen that the carbide appears on a stage of compacting obviously because of heating. The peaks belonging to Al_4C_3 are very wide, what is caused by a very low size of this precipitates. Annealing does not change the form and location of the Al_4C_3 peak.

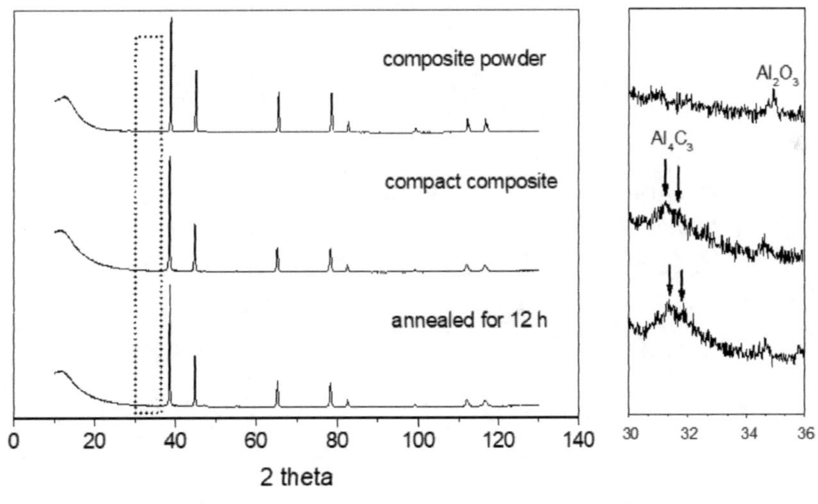

Figure 5. XRD pattern of the Al-C composite powder and the compact sample after extrusion and annealing at 470°C for 12 h.

Table 2. Aluminum lattice parameter, microstress and crystallite size (size of coherent scattering region) determined by XRD.

Composition	Microstress	Crystallite size, nm	Lattice parameter, nm
Al initial powder	-	-	4.0478±0.0004
Composite powder	0.20	48	4.0474±0.0005
Compact sample	0.12	79	4.0500±0.0005
Annealed sample	0.11	137	4.0507±0.0005

IOP Conf. Series: Earth and Environmental Science **337** (2019) 012085 doi:10.1088/1755-1315/337/1/012085

According to Al-C equilibrium phase diagram [18] carbon does not dissolve in Al crystal lattice. However, high energy mechanical milling is believed [19] to be possible to significantly expand solid solution composition limits. Taking into account a radius of Al atom equaled to 143 pm, for interstitial solid solution formation a size of octahedral interstitial site should be no less than the radius of a C atom equaled to 70 pm, it should lead to an increase of Al lattice parameter for more than 20%. As found by XRD Al lattice parameter has not grown after milling, i.e. aluminum lattice did not comprise carbon atoms [21-27]. An increase of the Al lattice parameter after heating along with simultaneously growing microstress (Table 2) may be explained by an elastic stress caused by a formation of ultrafine carbide phase coherently bound to aluminum matrix. Change of microhardness measured for the compacted and annealed at 470 °C for 4, 8, 12 h samples does not exceed 10%, what confirms the stability of the obtained structure.

Figure 6 compares the mechanical properties obtained for the selective samples in the present work with those for traditional and new Al-based materials. The obtained composites have excellent strength characteristic and satisfactory ductility. Figure 7. shows location of the composites in the strength – resistivity diagram among the advanced alloyscontaining REM, Zr, Sc. It is seen that the Al-based composite reinforced with carbon nanoparticles can successfully compete with the best aluminum alloys.

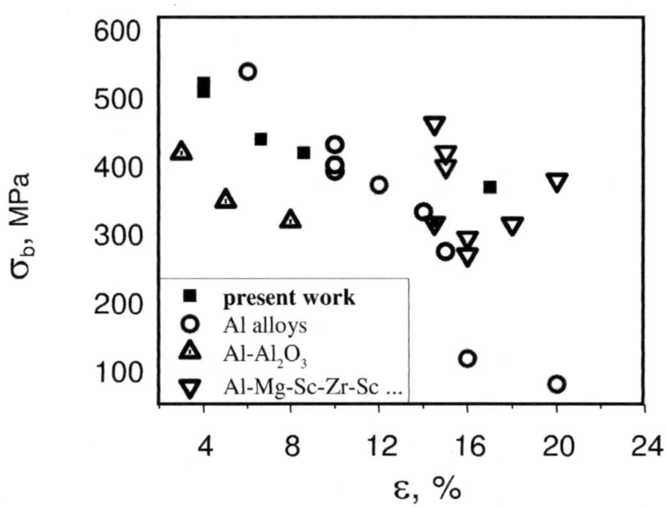

Figure 6. Comparison of the mechanical properties (tensile) of the obtained materials with existing ones.

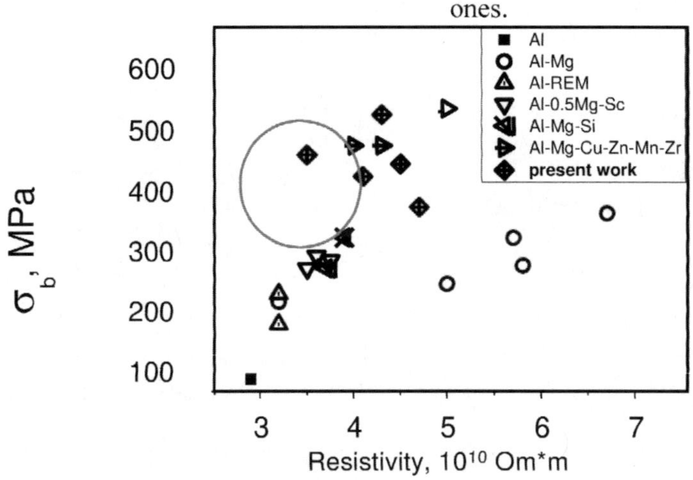

Figure 7. Comparison of the properties for the developed material with Al alloys used for electrical wire production.

4. Conclusion

Al-based composites with fullerene soot have been successfully prepared using powder metallurgy route. The microstructure and phase composition had been analyzed. It was shown that carbon does not dissolve in aluminum, but at compacting forms very fine carbide particles, resulting in micro stress of the aluminum matrix, and providing high strength of the composite. The obtained composites have excellent strength characteristic and satisfactory ductility. It was shown that the developed Al-based composites can successfully compete with the best aluminum alloys used for electrical wire production.

References

[1] Davis J R 2001 *Aluminum and Aluminum Alloys*
[2] Polmear I J 2005 *Light alloy. From Traditional Alloys to Nanocrystals, 4th ed*
[3] Weiland H 2017 *IOP Conf. Ser.: Mater. Sci. Eng.* **25** (2)
[4] Bakshi S R, Lahiri D and Argawal A 2010 *Intern Mater Rev* **55** (1)
[5] Iacob G, Ghica V G, Buzatu M, Buzatu T and Petrescu M I 2015 *Compos Part B* **69**
[6] Xu R, Tan Z, Xiong D, Fan G, Zhang Q-G J, Su Y, Li Z and Zhang D 2017 *Compos. Part A* **96**
[7] Li Ch, Qiu R, Luan B and Li Zh 2017 *Mater Sci Eng A* **704**
[8] Zhu X, Zhao Y-G, Wu M, Wang H-Y, Jiang Q-C 2016 *J. Alloys Comp.* **674**
[9] Nam D H, Cha S I, Lim B K, Park H M, Han D S and Hong S H 2012 *Carbon* **50**
[10] Choi H J, Shin J H and Bae D H 2010 *Carbon* **48**
[11] Tokunaga T, Kaneko K, Sato K and Horita Z 2008 *Scr Mater.* **58** (9)
[12] Shin S E, Choi H Y, Hwang J Y and Bae D H 2015 *Sci. rep.* **5** (16114)
[13] Wang J, Li Z, Fan G, Pan H, Chen Z and Zhang D 2012 *Scr. Mater.* **66**
[14] Bartolucci S F, Paras J, Rafiee M A, Rafiee J, Lee S, Kapoor D and Koratkar N 2011 *Mater Sci Eng A* **528**
[15] Shin J, Choi K, Shiko S, Choi H and Bae D 2015 *Compos. Part B* **77**
[16] Nelson J B and Riley D P 1945 *Proc Phys Soc* **57**
[17] Williamson G K and Hall W H 1953 *Acta Metall* **1**
[18] Massalski T B, Subramanian P R, Okamoto H and Kacprzak L 1990 *Binary Alloy Phase Diagrams, 2nd ed, (*Ohio:ASM International, Materials Park)
[19] Suryanarayana C 2001 *Prog Mater Sci.* **46**
[20] Zobacheva A U, Nemov A S and Borovkov A I 2017 *Materials Physics and Mechanics* **32** 74–82
[21] Seyedkavoosi S, Zaytsev D, Drach B, Panfilov P, Gutkin M Yu and Sevostianov I 2017 *Int. J. of Engineering Science* **111** pp 52–60 doi:10.1016/j.ijengsci.2016.11.005
[22] Bakaev V V, Bolotin A E and Aganov S S 2016 *Teoriya i Praktika Fizicheskoy Kultury* **2016** pp 6–8
[23] Nekliudova E A, Semenov A S, Melnikov B E and Semenov S G 2014 *Magazine of Civil Engineering* **47** pp 25–39 doi:10.5862/MCE.47.3
[24] Rudskoy A I, Tolochko O V, Kol'tsova T S and Nasibulin A G 2014 *Metal Science and Heat Treatment* **55** pp 564–8 doi:10.1007/s11041-014-9670-8
[25] Ovid'ko I A and Sheinerman A G 2014 *J. of Physics D: Applied Physics* **47** doi:10.1088/0022-3727/47/49/495302
[26] Nizina T A and Balykov A S 2016 *Magazine of Civil Engineering* **62** pp 13–25 doi:10.5862/MCE.62.2
[27] Klyuev S V, Klyuev A V, Abakarov A D, Shorstova E S and Gafarova N G 2017 *Magazine of Civil Engineering* **75** pp 66–75 doi:10.18720/MCE.75.6

IOP Conf. Series: Earth and Environmental Science **337** (2019) 012086 doi:10.1088/1755-1315/337/1/012086

Local buckling of box-shaped beams due to skew bending

K. Manzhula[1], A. Naumov[1,*], S. Sokolov[1], M. Okrepilov[1], S. Kravchenko[1], G. Kondrashkova[1], V. Shashikhin[1], V. Yakovlev[1], S. Chulkin[1], L Mukhametova[1]

[1]Peter the Great St. Petersburg Polytechnic University, St.Petersburg,195251,Russia
[2]Kazan State Power Engineering University, Kazan, Russian Federation
*Corresponding author: anaumov@nissan.ru

Abstract. Local buckling efficiency of different types of beams was reviewed. For beams with non-linear walls and parameterized geometry local buckling analysis was made in ANSYS. Based on the results of ANSYS calculation analytic formula was evaluated in MathCAD using the least square method. Analysis of the influence of different geometrical parameters on local buckling stress value for beams with non-linear walls was completed.

1. Introduction

There are different structures of beams which are used in metal structures. The simplest of these are standard elements. These kinds of beams are usually included in simple structures of buildings and other standard constructions. More complex structures, like box-shaped beams, are usually used in complicated facilities. Such beams have upper and lower flanges, one or two walls and different rigidity-increasing elements such as longitudinal ribs and cross members. Cross members have cut-outs to avoid interference with longitudinal ribs (Figure 1, a). This design makes production of beams more complex and expensive. Simple metal structures do not have box-shaped beams. Usually engineers prefer beams with corrugated walls for these needs [1-4]. Such type of beam is mostly applicable to cargos and facilities. Beams with corrugated walls are not popular solutions for cranes and bridges [5]. It is economically profitable to use corrugated walls in mass production manufacturing, for instance in freight cars (Figure 1, b). A corrugated wall itself is not enough when the load-to-structure is intricate. In this case longitudinal ribs are required (Figure 1, c). However, welding of corrugated walls to longitudinal ribs makes production more complicated and leads to increased labor intensity. Due to this fact it is better to use box-shaped beams for low volume products. Wall cross members and longitudinal beams for box-shaped beams are made from standard straight metal panels. This increases the mass of the beam and quantity of welding seams which results in an increase in production cost. Moreover, hard mode usage of box-shaped beams raises the chances of deformations in welding seams [6].

a) b) c)

Figure 1. Types of metal structures: a) box-shaped beam; b) freight cars; c) beam with corrugated wall.

Load is usually applied to the upper flange of the box-shaped beam so the load can make maximum impact on each point of the entire length of the beam. As a result, local buckling can be raised in any area of the beam's wall. However, there are three calculating schemes for local buckling calculation (Figure 2). In these schemes, local buckling stress depends on normal stress, shear stress, or a combination of normal and shear stresses [7].

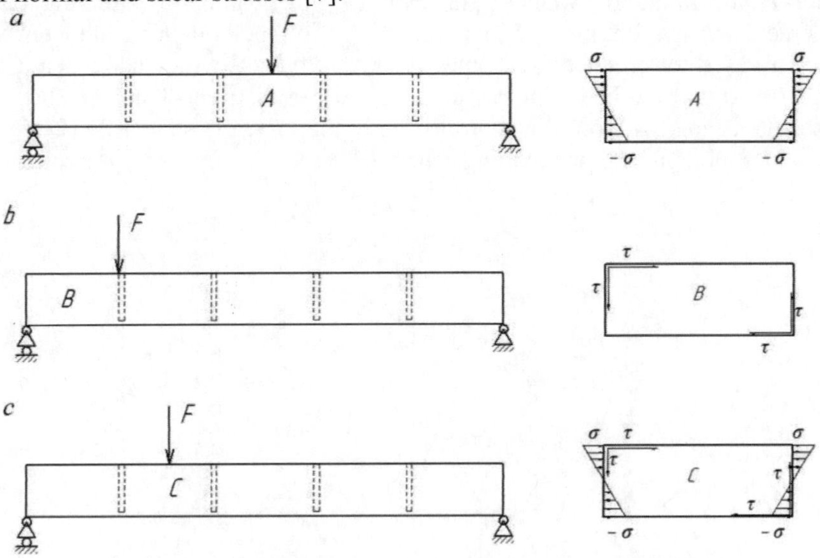

Figure 2. Calculation schemes of local buckling stress.

Calculation of local buckling stress for linear walls can be done based on the following formulas [8]:

$$\sqrt{\left(\frac{\sigma}{\sigma_{cr}}\right)^2+\left(\frac{\tau}{\tau_{cr}}\right)^2}\leq\frac{1}{n}; \tag{1}$$

$$\sigma\leq\frac{\sigma_{cr}}{n=[\sigma_{cr}]}; \tag{2}$$

$$\tau\leq\frac{\tau_{cr}}{n=[\tau_{cr}]}; \tag{3}$$

σ and τ - normal and shear stresses in a beam, σ_{cr} and τ_{cr}- critical normal and shear stresses (when a beam is buckled), n - safety factor.

Formula (1) is used when a load is applied to ¼ of the beam's length (Figure 2, c). When a load is applied to the center of the beam (Figure 2, a) only normal stress makes an impact on the local buckling stress. Formula (2) should be used in this case. Formula (3) is used when only shear stress affects due to the load on the ends of the beam (Figure 2, b).

The critical stress is expressed by the formula $\sigma_{cr}=k_sK_{so}(t/d)^2$, where: k_s is the buckling factor reflecting the effect of fixing the conditions and distribution of stress across the width of the plate; K_{so}- the factor having the dimension MPa; t - plate thickness; and, d - height of the plate. The length of the plate is included in the determination of k_s.

This calculating scheme was analyzed in the previous article [13]. In that article the influence of the flection radius on local buckling was reviewed. The formula for calculating the local buckling stress value was found.

The second formula applies the load to the ends of the beam near the fixation areas (Figure 4, b). In this case the local buckling depends on the shear. The stability of the plate at stress is represented by the equation $\tau\leq\frac{\tau_{cr}}{n=[\tau_{cr}]}$, where: τ - the maximum stress in the plate; τ_{cr} - the critical stress of stability; and,

IOP Conf. Series: Earth and Environmental Science **337** (2019) 012086 doi:10.1088/1755-1315/337/1/012086

n - the safety factor. The critical stress is expressed by the formula $\tau_{cr} = k_q K_{so}(t/d)^2$, where: k_q is the buckling factor reflecting the effect of fixing the conditions and distribution of shearing stress across the width of the plate; K_{so}- the factor having the dimension MPa, t - plate thickness; and, d - height of the plate. The length of the plate is included in the determination of k_q. In this case a combination of normal stress and shear stress affects local buckling properties [18]. The second calculation scheme will be analyzed in this article.

Flection radius application to the walls of the beams (Figure 3) under the bending load is a solution to exclude cross members and longitudinal ribs. The production method of metal plates with a radius is not new. It is always used for production of pipes. There are patents which show beam structures with flection radius [9-11]. However, there is no research on how radius affects local buckling stress and how to calculate it for the beams with non-linear walls. Only the LLOYD standard [12] describes how to calculate stress values, but only for one loading scheme.

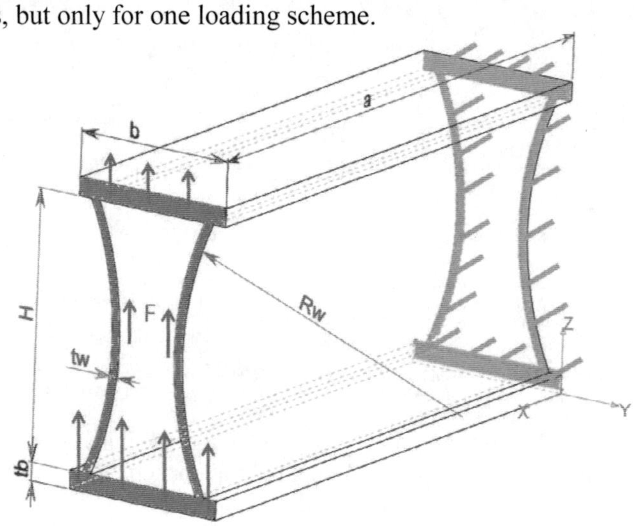

Figure 3. Box-shaped beam with non-linear walls.

Papers [13-17] show that beams with non-linear walls have a better local buckling performance than box-shaped beams with standard walls under a pressure and a bending load

2. Methods

Previous dependencies (1)-(3) are not applicable to beams with non-linear walls because they do not take into the account the curvature of the walls. An assessment of the local buckling of the beam wall was made in ANSYS Workbench. To do this, a cross-section of the beam was developed. Also, variable geometric parameters were selected (table 1) [19].

After that, a project in ANSYS Workbench was created. Structure of calculation program includes static analysis and linear buckling analysis (Figure 4). In this project the same 3D model with geometric parameters as in the table 1 was created. Previously discussed parameters were selected on the created 3D model. It provided an opportunity to calculate all possible variants of beams geometry [20-28].

Table 1. Geometric parameters

Geometric parameter	Unit of measurement	Value
H	mm	1000, 1600, 2000
t_b	mm	10, 16, 20, 24
t_w	mm	6, 10, 12, 16
b	mm	400, 600, 800
R_w	mm	2000, 4000, 8000
a	mm	1000, 2000, 4000

The second step was to select beam material. It was performed in Ansys material manager. As a result, structural steel from Ansys engineering data was selected as the material of the beam. Tensile yield strength of this steel is 250 mPa and tensile ultimate strength is 460 mPa. The last step was to input the type of fixation of the beam (highlighted in pink on (Figure 3)), select the load method (F on (Figure 3)) and define the maximum size of the grid elements. Stress and displacement values were also entered as preferable results of the calculation. After that an evaluation of all geometry combinations of beams was completed.

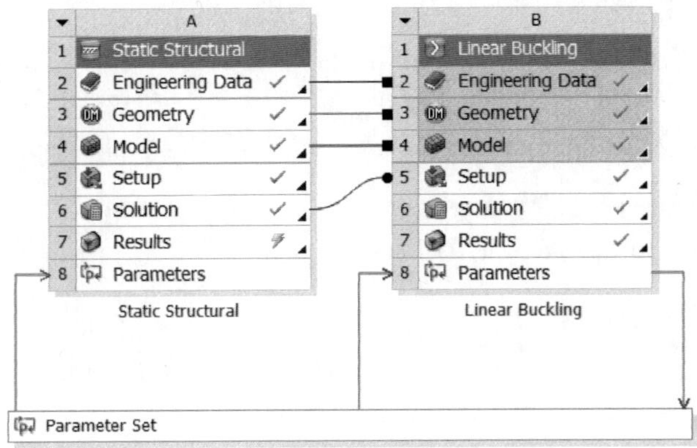

Figure 4. Project schematic of buckling analysis in Ansys.

To analyze the results of the calculation in ANSYS, the received data can be approximated with the following formula

$$\tau_{cr} = A_0 (X_1)^{a_1} (X_2)^{a_2} (X_3)^{a_3} (X_4)^{a_4} (X_5)^{a_5} \quad (4)$$

To do it, value of parameters A_o, a_i should be found. Firstly, the logarithm of formula (4) should be taken

$$log\,\tau = log\left(A_0 \prod_{i=1}^{N} X_i^{a_i}\right) = log\,A_0 + \sum_{i=1}^{N} a_i \cdot log\,X_i \quad (5)$$

Where X_i is geometric parameter of beams, for example

$$X_1 = \frac{t_b}{H}, X_2 = \frac{t_w}{H}, X_3 = \frac{R_w}{H}, X_4 = \frac{a}{H}, X_5 = \frac{b}{H}.$$

Secondly, representing$log\,Y = \tilde{y}$,$log\,A_0 = a_0$, $log\,X_i = \tilde{x}_i$ formula (5) can be rewritten as:

$$\tilde{y} = a_0 + a_1\tilde{x}_1 + a_2\tilde{x}_2 + \ldots + a_N\tilde{x}_N = a_0 + \sum_{i=1}^{N} a_i\tilde{x}_i \quad (6)$$

Now, equation (6) can be solved with the least square method. This method is based on the fact that the difference between values of critical stresses y_m, which were found in Ansys Workbench, and values of critical stresses, which can be found by solving the equation (6), should be approached to 0:

$$\Delta = \frac{1}{n}\sum_{m=1}^{n}(y_m - \tilde{y}_m)^2 = \frac{1}{n}\sum_{m=1}^{n}\left(y_m - \left(a_0 + \sum_{i=1}^{N} a_i\tilde{x}_{im}\right)\right)^2 \rightarrow min \quad (7)$$

where n is the quantity of beams variants calculated in Ansys Workbench. Solution of the formula (7) is partial derivative, which should be alike 0:

$$\frac{\partial F}{\partial a_{0,i}} = 0.$$

Application of this solution to all parameters ai can be found from a set of equations:

$$\begin{cases} \frac{\partial F}{\partial a_0}=\sum_{m=1}^{n}\left(y_m-\left(a_0+\sum_{i=1}^{N} a_i\tilde{x}_{im}\right)\right)=0 \\ \frac{\partial F}{\partial a_1}=\sum_{m=1}^{n}\left(\left(y_m-\left(a_0+\sum_{i=1}^{N} a_i\tilde{x}_{im}\right)\right)\cdot\tilde{x}_{1m}\right)=0 \\ \dots\dots \\ \frac{\partial F}{\partial a_i}=\sum_{m=1}^{n}\left(\left(y_m-\left(a_0+\sum_{i=1}^{N} a_i\tilde{x}_{im}\right)\right)\cdot\tilde{x}_{im}\right)=0 \\ \dots\dots \\ \& \frac{\partial F}{\partial a_N}=\sum_{m=1}^{n}\left(\left(y_m-\left(a_0+\sum_{i=1}^{N} a_i x_{im}\right)\right)\cdot\tilde{x}_{Nm}\right)=0 \end{cases} \quad (8)$$

Set of equations can be represented as:

$$\begin{cases} \sum_{m=1}^{n} y_m=a_0\sum_{m=1}^{n} 1+a_1\sum_{m=1}^{n}\tilde{x}_{1m}+a_2\sum_{m=1}^{n}\tilde{x}_{2m}+\dots+a_N\sum_{m=1}^{n}\tilde{x}_{Nm} \\ \sum_{m=1}^{n} y_m\tilde{x}_{1m}=a_0\sum_{m=1}^{n}\tilde{x}_{1m}+a_1\sum_{m=1}^{n}\tilde{x}_{1m}\tilde{x}_{1m}+a_2\sum_{m=1}^{n}\tilde{x}_{2m}\tilde{x}_{1m}+\dots+a_N\sum_{m=1}^{n}\tilde{x}_{Nm}\tilde{x}_{1m} \\ \dots\dots \\ \sum_{m=1}^{n} y_m\tilde{x}_{im}=a_0\sum_{m=1}^{n}\tilde{x}_{im}+a_1\sum_{m=1}^{n}\tilde{x}_{1m}\tilde{x}_{im}+a_2\sum_{m=1}^{n}\tilde{x}_{2m}\tilde{x}_{im}+\dots+a_N\sum_{m=1}^{n}\tilde{x}_{Nm}\tilde{x}_{im} \\ \dots\dots \\ \sum_{m=1}^{n} y_m\tilde{x}_{Nm}=a_0\sum_{m=1}^{n}\tilde{x}_{Nm}+a_1\sum_{m=1}^{n}\tilde{x}_{1m}\tilde{x}_{Nm}+a_2\sum_{m=1}^{n}\tilde{x}_{2m}\tilde{x}_{Nm}+\dots+a_N\sum_{m=1}^{n}\tilde{x}_{Nm}\tilde{x}_{Nm} \end{cases} \quad (9)$$

and rewritten in the matrix form $\tilde{Y} = A \cdot \tilde{X}$, where

$$\tilde{Y} = \begin{pmatrix} \sum_{m=1}^{n} y_m \\ \sum_{m=1}^{n} y_m\tilde{x}_{1m} \\ \sum_{m=1}^{n} y_m\tilde{x}_{2m} \\ \dots \\ \sum_{m=1}^{n} y_m\tilde{x}_{Nm} \end{pmatrix} \qquad A = \begin{pmatrix} a_0 \\ a_1 \\ a_2 \\ \dots \\ a_N \end{pmatrix},$$

$$\tilde{X} = \begin{pmatrix} \sum_{m=1}^{n} 1 & \sum_{m=1}^{n}\tilde{x}_{1m} & \sum_{m=1}^{n}\tilde{x}_{2m} & \dots & \sum_{m=1}^{n}\tilde{x}_{Nm} \\ \sum_{m=1}^{n}\tilde{x}_{1m} & \sum_{m=1}^{n}\tilde{x}_{1m}\tilde{x}_{1m} & \sum_{m=1}^{n}\tilde{x}_{1m}\tilde{x}_{2m} & \dots & \sum_{m=1}^{n}\tilde{x}_{1m}\tilde{x}_{Nm} \\ \sum_{m=1}^{n}\tilde{x}_{2m} & \sum_{m=1}^{n}\tilde{x}_{2m}\tilde{x}_{1m} & \sum_{m=1}^{n}\tilde{x}_{2m}\tilde{x}_{2m} & \dots & \sum_{m=1}^{n}\tilde{x}_{2m}\tilde{x}_{1m} \\ \dots & \dots & \dots & \dots & \dots \\ \sum_{m=1}^{n}\tilde{x}_{Nm} & \sum_{m=1}^{n}\tilde{x}_{Nm}\tilde{x}_{1m} & \sum_{m=1}^{n}\tilde{x}_{Nm}\tilde{x}_{2m} & \dots & \sum_{m=1}^{n}\tilde{x}_{Nm}\tilde{x}_{Nm} \end{pmatrix}.$$

From the matrix equation $\tilde{Y} = A \cdot \tilde{X}$ parameters a_i can be found as

$$A = \tilde{X}^{-1} \cdot \tilde{Y} \quad (10)$$

3. Results and Discussion

In ANSYS, 240 beams with different geometrical parameters were calculated. Diagrams of local buckling have the same shape: two big beams on both walls (Figure 5). Beams with linear walls have the same shape. However, beams with non-linear walls have a higher value of critical buckling stress.

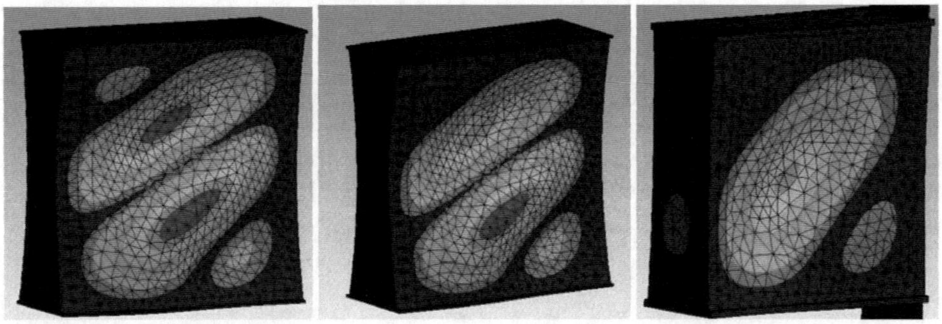

Figure 5. Movements diagrams with loss of local stability for different beam geometry.

Evaluations based on the least square method from the previous article section can be used to find the analytic formula for local buckling stress. To do it, calculations were done in MathCad software. Solutions of parameters ai from the formula (7) are shown in table 2.

Table 2. Values of parameters ai.

a_0	6.035
a_1	-0.926
a_2	2.32

a_3	-0.385
a_4	-0.41
a_5	0.483

Finally, all parameters a_i can be applied to the formula (4) and it can be represented as

$$\tau_{cr} = \frac{1.086 \cdot 10^6 \cdot \left(\frac{b}{H}\right)^{0.483} \cdot \left(\frac{t_w}{H}\right)^{2.32}}{\left(\frac{R_w}{H}\right)^{0.385} \cdot \left(\frac{a}{H}\right)^{0.41} \cdot \left(\frac{t_b}{H}\right)^{0.926}} \quad (11)$$

Also, constant K_{so}=0.759 MPa for structural steels can be added to the formula (7) [8]

$$\tau_{cr} = \frac{1.43 \cdot 10^6 \cdot K_{SO} \cdot \left(\frac{b}{H}\right)^{0.483} \cdot \left(\frac{t_w}{H}\right)^{2.332}}{\left(\frac{R_w}{H}\right)^{0.385} \cdot \left(\frac{a}{H}\right)^{0.41} \cdot \left(\frac{t_b}{H}\right)^{0.926}} \quad (12)$$

The assessed value of formula (11) error can be found as

$$\delta = \frac{\tau_{ansys} - \tau_{cr}}{\tau_{ansys}} \cdot 100\% \quad (13)$$

Average value of formula (11) error is 8.1%. Distribution of formula (12) results to calculate Ansys stress is represented on Figure 6.

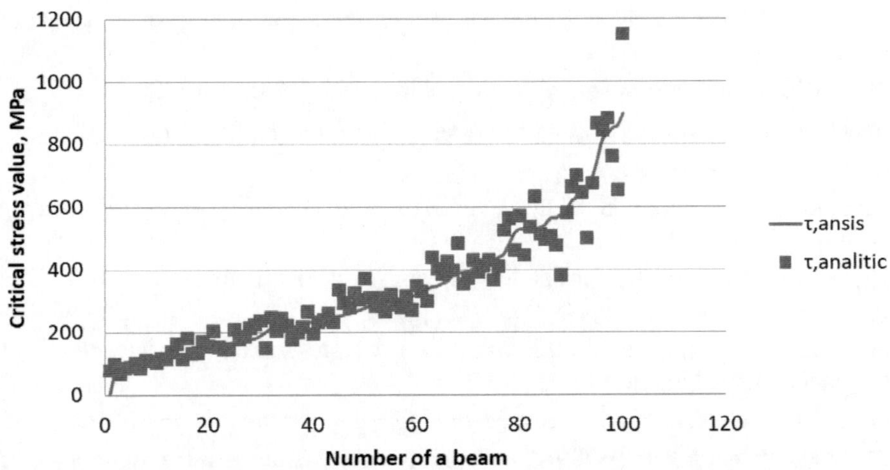

Figure 6. Graphic of ansys and analytic formula (12) buckling stress.

A graph was created to understand how radius makes an impact on the value of buckling critical stress (Figure 7).

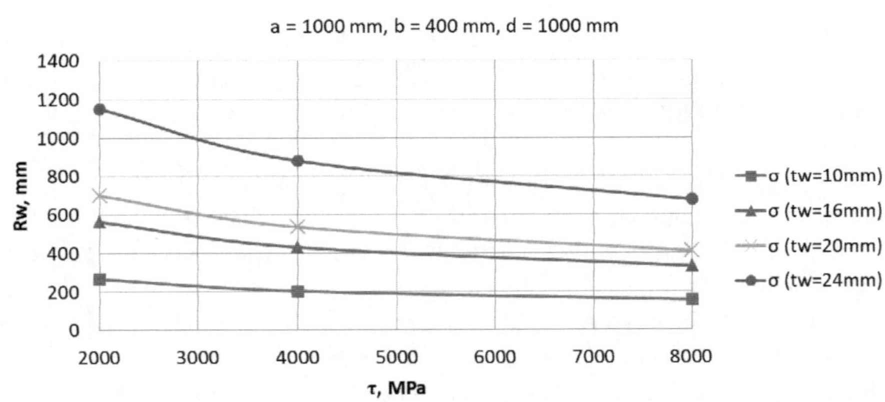

Figure 7. Impact of a radius on a stress value for different wall thickness.

4. Conclusions

In this study, we analyzed box-shaped beams with non-linear walls. This type of beam is not frequently used in metals structures but there are patents that describe the layout and geometry of such beams. The advantage of such beams is a larger value of local buckling stress. This property is important for main beams in traveling cranes, so that beams with non-linear walls can improve the robustness of current structures. This is the reason why it is essential to study beams with non-linear walls. Only a finite element method can be used to calculate the stress value of beams with non-linear walls. Mechanic properties variations of these beams are not studied due to the flection radius increase.

Also, it was defined that local buckling critical force of beams with non-linear walls was bigger than the critical force of the standard box-shaped beams. The value of the critical force increases with the gain of flections' radius. The most important result of the study is the development of an analytic formula for local buckling stress. It can be used in earlier stages of metal structure engineering to get estimated values of stress and geometric parameters.

References

[1] Pasternak H and Kubieniec G 2010 *J. of Civil Engineering and Management* **16 (2)** (2010)

[2] Fainstein A 2007 *Steel beams with minimal weight.* Publishing house of Polytechnic university **95**

[3] Balzannikov M, Holopov I, Soloviev A and Lukin A 2013 *Vestnik MGSU* **11**

[4] Azhermachev S and Semenov P 2005 *Building and technogenetic safety*

[5] Kulkova N N 2004 *Stress-strain behavior aspects of crans beams*

[6] Manzhula K P and Petinov S V 2001 *Strength and durability of metal structures.* (Saint-Petersburg, SPBGTU)

[7] EN 1993-1-5 (2006) (English): Eurocode 3 - Design of steel structures - Part 1-5: Plated structural elements

[8] Sokolov S 2011 *Construction mechanics and metal structures of machines,* (Spb: SPB Politechnika publishing house)

[9] Watson G 1935 Patent. 1989834 USA. *Method for reinforcement beams* published. 5.02.1935

[10] Aust R 2001 Patent EP 0293789 A1 Germany. *Als Kastentrager ausgebildeter Krantrager* published. 10.10.2001

[11] Belyi A 2011 Patent 2478557, Russia. *Travaling beam* published 1.06.2011

[12] Germanischer Lloyd SE 2011 *LLOYD Rules for Classification and Construction IV Industrial Services*

[13] Manzhula K P and Naumov A V 2017 *Int. Review of Mechanical Engineering (I.RE.M.E.)* **11** (5)

[14] Manzhula K P and Naumov A V 2017 *Modern mechanic engineering: science and education, 6th international scientific-practical conference* (SPb: Publishing house of Polytechnic university)

[15] Manzhula K P and Naumov A V 2016 *Interstoymech-2016* p 244

[16] Manzhula K P and Naumov A V 2016 *Modern machine building: science and education: materials of the 5th International Scientific and Practical Conference* (SPb: Publishing house of Polytechnic university)

[17] Manzhula K P and Naumov A V 2016 *Science Week SPbPU. Materials of the scientific conference with international participation. Best reports* **70**

[18] Tovstik P E 2005 *Local buckling of plates and shapes.* (Moscow: RAN)

[19] Sidnyaev N I and Vilisova N T 2011 *Theory of experiment planning* (Moscow: MGTU)

[20] Basov K A 2005 *Guide for ANSYS* (Moscow: DMK-PROSS)

[21] Vatin N, Sinelnikov A, Garifullin M and Trubina D 2014 *Appl.Mechanics and Materials* **633–4** pp 1037–41 doi:10.4028/www.scientific.net/AMM.633-634.1037

[22] Lalin V, Rybakov V and Sergey A 2014 *Appl. Mechanics and Materials* **578–9** pp 858–63 doi:10.4028/www.scientific.net/AMM.578-579.858

[23] Beyer B, Sholukha V, Dugailly P M, Rooze M, Moiseev F, Feipel V and Jan S V S 2014 *Clinical Biomechanics* **29** 434–8 doi:10.1016/j.clinbiomech.2014.01.007

[24] Freidin A B, Vilchevskaya E N and Korolev I K 2014 *Int. J. of Engineering Science* **83** pp 57–75 doi:10.1016/j.ijengsci.2014.03.008

[25] Alexander B and Vladislav B 2016 *J. of Physical Education and Sport* **16** pp 910–3 doi:10.7752/jpes.2016.03143

[26] Gorshkov A S, Rymkevich P P and Vatin N I 2014 *Magazine of Civil Engineering* **52** doi:10.5862/MCE.52.5

[27] Nazmeeva T V and Vatin N I 2016 *Magazine of Civil Engineering* **62** 92–101 doi:10.5862/MCE.62.9

[28] Vatin N I, Nazmeeva T and Guslinscky R 2014 Problems of cold-bent notched c-shaped profile members *Advanced Materials Research* **941** pp 944

IOP Conf. Series: Earth and Environmental Science **337** (2019) 012087 doi:10.1088/1755-1315/337/1/012087

Nanocomposite antifriction molybdenum disulfide (filler) – silicon oxide (matrix) coatings

S. Alexandrov[1], K. Tyurikov[1,*], A. Breki[1], G. Kondrashkova[1], V. Shashikhin[1], B. Matisov[1], Y. Mamaev[1], A. Timofeev[1], S. Burdakov[1], D A Zaripova[2]

[1]Peter the Great St. Petersburg Polytechnic University, 195251, Politekhnicheskaya 29, St. Petersburg, Russia
[2]Kazan State Power Engineering University, Kazan, Russian Federation
*Corresponding author: kirill.tyurikov@spbstu.ru

Abstract. It was demonstrated experimentally that the spatial separation of two processes of chemical vapor deposition, one of which provides synthesis of filler (MoS2) nanoparticles and the other yields the matrix (SiO2) of the nanocomposite coating, performed in a common reactor, enables an independent control over two process rates and makes it possible to widely vary the composition of the films deposited in this way. The deposition was performed in a double-zone vertical tubular quartz reactor. Molybdenum disulfide particles were produced by pyrolysis of aerosols of ammonium thiomolybdate solutions in dimethylformamide in the upper zone of the reactor, and the plasma-chemical deposition of a nanocomposite coating occurred in the lower zone into which MoS2 nanoparticles were transported by the gas flow and tetraethoxysilane was delivered. It was shown that the nanocomposite coatings composed of molybdenum disulfide (filler) and silicon oxide (matrix) possess improve the antifriction properties as compared with the matrix (SiO2 layers), these properties being determined by the relative amounts of MoS2 nanoparticles in the layer and by their average size.

1. Introduction

Nanocomposite materials have unique combinations of physical and chemical properties and therefore have attracted considerable interest [1-4]. Methods for forming coatings from nanocomposites are of particular interest, since in many practical cases it is necessary to provide only particular properties of the surface of products, which itself can be made of low cost material. Obviously, processes carried out at atmospheric pressure and allowing deposition of coatings at relatively low temperatures are the most attractive because the use of expensive vacuum equipment is not necessary and they can be employed for layer deposition on thermally unstable materials.

The results of our analysis [5] of the most common methods for depositing nanocomposite coatings at atmospheric pressure have shown that among the various processes, chemical vapor deposition (CVD) with various activation methods is one of the technologies allowing to form not only high-quality layers but also synthesize various types of nanomaterials, and even nanocomposites [6]. Only a few papers have been published on the deposition of nanocomposite coatings using CVD technologies [7-10]. Some of them are based on the use of a combination of physical methods of sputtering (PVD) targets for the formation of nanoparticles with CVD processes for coating formation [8]. It should be noted that with this approach it is necessary to provide the same total pressure in PVD and CVD reactors, which is a technically difficult task, since normally PVD processes are carried out at much lower pressures than conventional CVD processes. Nanocomposite layers can be formed by chemical deposition from the gas phase while simultaneously introducing into the reaction zone the reagents required for the synthesis of the filler and the matrix of the composite [9], or using reagents of complex chemical composition [10, 11]. The main disadvantage of these approaches is the serious limitations in the variation of the nanocomposite composition within a wide range. Recently, a very interesting and promising approach to the CVD technologies of nanocomposite coatings has been developed, in which the pre-synthesized nanoparticles of the filler and the reagents necessary for the formation of the matrix layer are simultaneously introduced into the reaction chamber [12, 13]. Despite the difficulties associated with storing and transporting nanoparticles, the authors succeeded in obtaining nanocomposite coatings consisting of siloxane (matrix) and carbon (nanoparticles of filler), as well as layers of siloxane - an aromatic resin [12] and films representing a glassy SixOy matrix with polytetrafluoroethylene (PTFE) nanoparticles distributed therein [13].

Obviously, the most attractive from the viewpoint of the possibility of controlled variation in the composition of the resulting composite are the CVD processes in which the nanoscale filler particles are homogeneously synthesized in one zone of the reactor and then transported by the gas stream together with additional reagents to another zone where the nanocomposite coating is heterogeneously deposited. The results of the preliminary experimental studies carried out by us for the first time [5] indicate that, in fact, the separation in the space of a single reactor of two CVD processes (i.e, the synthesis of nanoparticles and the deposition of coatings) provides an independent control of the rates of the two processes and therefore one can vary composition of deposited layers over a wide range.

The purpose of this work was the experimental testing of a two-stage plasma enhanced CVD process to form nanocomposite coatings, which are a matrix layer of silicon dioxide containing nanoparticles of molybdenum disulfide, and possessing improved antifriction properties. The choice of the components of the composite was made on the basis that the silicon dioxide layers are often used as hardening coatings on the polymer surface [14], and the MoS2 nanoparticles have unique antifriction properties [15]. It was of interest to ascertain the nature of the influence of the layers deposition conditions on their antifriction properties.

2. Experimental

The design of the quartz reactor used to carry out the two-step CVD process for the deposition of nanocomposite coatings in the "MoS_2 (filler) -SiO_2 (matrix)" system is described in detail in [5]. The reactor consisted of two main parts: an upper zone intended for the synthesis of molybdenum disulfide nanoparticles and a lower zone used for plasma enhanced CVD of nanocomposite coatings at atmospheric pressure by maintaining a low-frequency corona discharge (27 kHz). MoS_2 nanoparticles were synthesized in a gas stream by pyrolyzing the aerosol particles of a $(NH_4)_2MoS_4$ solution in dimethylformamide [16] and then transported by a gas stream to a deposition zone where the substrates were placed on a grounded steel substrate holder and tetraethoxysilane (TEOS) vapor - the reagent necessary for precipitation of silicon dioxide - was injected. The upper zone of the reactor was a quartz tube equipped with two resistive heaters. The first heater designed to evaporate the solvent from aerosol particles was heated to temperatures (T_1) 150-750±10°C. The second heater was used to heat the pyrolysis zone of ammonium thiomolybdate to 800±10°C (T_2). The aerosol of a solution of ammonium thiomolybdate in dimethylformamide created by a piezoelectric nebulizer operating at 2.4 MHz was transported to the reactor with a carrier gas (helium of high purity) at a rate of about 0.3 slm. The concentration of the solution of ammonium thiomolybdate was in the range 0.0025-0.01 $mol*L^{-1}$. The lower part of the reactor was used for plasma enhanced CVD of nanocomposite coatings by co-deposition of MoS_2 particles coming from the upper part of the reactor and a layer of silicon dioxide formed from tetraethoxysilane vapors. A high-voltage electrode made of a sharp tungsten rod was placed 25-30 mm above the grounded substrate holder heated to 300°C. Corona discharge at atmospheric pressure was created in this zone by means of a low frequency power supply (27 kHz, 1.8-2.2 kV). Typical values of the discharge power were 40-90W. The vapors of tetraethoxysilane were fed to the reactor from a quartz evaporator stabilized at 70°C, the design of which provided an unchanged level of the evaporated liquid. The carrier gas helium saturated with TEOS vapor was injected into the upper part of the deposition zone approximately 30 mm downstream of the first zone through the injection nozzle. The total flow of helium through the evaporator was changed within the range 0-1 slm. The flow rate of the diluent helium was varied within the same limits so that the total flow rate was 1 slm. Substrates of silicon and polished stainless steel were placed on a grounded substrate holder.

The composition of the obtained samples and the particle size characteristics were determined using a Zeiss SUPRA 55VP scanning electron microscope (SEM) equipped with an x-ray fluorescence microanalysis unit Oxford Instruments X-Max. The structure of the samples was studied using a Jeol JEM-2100F TEM with an accelerating voltage of 200 kV and a point resolution of 0.19 nm. The composition of the layer was evaluated using a IR-Fourier spectrometer FSM 1201. The thickness of the layers of the nanocomposite was estimated from the images of the cross sections of composite layers obtained with the use of SEM.

To estimate the antifriction properties, the sliding friction force was measured. As a rotating counter body, a cylindrical sample (roller) was used from the bearing steel, with a diameter of 10 mm. During the tests, the counter body with its flat part pressed against the surface of the sample, which was a nanocomposite coating with thickness of about 1μm, deposited on a substrate of stainless steel with a polished surface. The clamp fixed to the bearing assembly with the help of a cable and a tensile beam in the course of the experiment was kept from the rotation imparted by the movable roller. Data on the frictional force and the acting load were transmitted from the tensile beams to the computer. In all the experiments, the load on the friction pair was 390 N, and the speed of the shaft in which the steel roller was clamped was 200 rpm.

3. Results and discussion

Taking into account the results of preliminary studies of the synthesis of MoS_2 particles by the pyrolysis of aerosols [16] and the plasma enhanced CVD of silicon dioxide layers [5], a series of experiments on the deposition of nanocomposite coatings were carried out. The main process parameters for these experiments were fixed (Table 1), except for the concentration of ammonium thiomolybdate in dimethylformamide, which varied in the range 0.0025-0.01 mol^{-1}, and the ratio of the helium-carrier and helium-diluent vapours of tetraethoxysilane.

Table 1. Deposition parameters of nanocomposite coatings.

Parameter	Value
Temperature of the first zone of the reactor, °C	800
Temperature of the second zone of the reactor, °C	800
Substrate temperature, °C	300
TEOS evaporator temperature, °C	70
Flow rate of the aerosol carrier gas, L min-1	0.3
Flow rate of the TEOS vapor diluent gas, L min-1	0.1–1

The results of the study of samples using SEM indicate that the deposited layers contain spherical particles characterized by a statistical distribution in the coating (Figure 1, a) and are characterized by a rough surface, probably due to the joint deposition of nanoparticles on the surface of the growing layer. The thickness of the coating was estimated from the SEM image of the cross-section of the coating (Figure 1, *b)*.

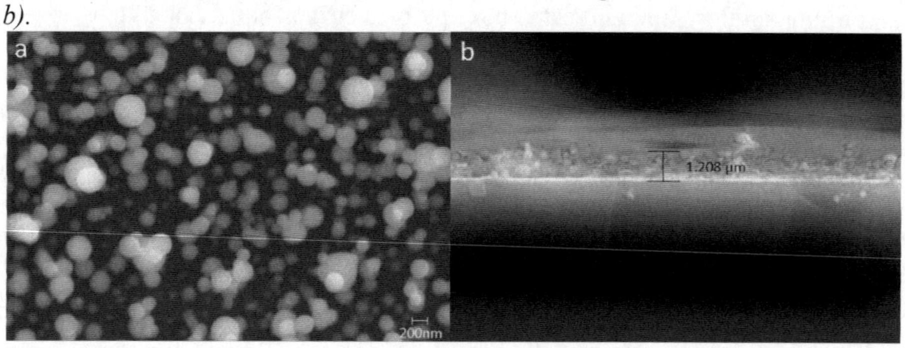

IOP Conf. Series: Earth and Environmental Science **337** (2019) 012087 doi:10.1088/1755-1315/337/1/012087

Figure 1. Plan-view (a) and cross-sectional SEM images (b) of the nanocomposite coating constituted by silicon dioxide (matrix) and MoS_2 nanoparticles (filler).

The growth rate of the layers did not depend on the presence of aerosol particles in the deposition zone and was approximately the same as for plasma enhanced CVD of silica (15-20 nm*min^{-1}). It is interesting to note that the average nanoparticle size in the layers depends on the concentration of $(NH_4)_2MoS_4$ in the solution used to make the aerosol and decreases with the dilution of the solution (Figure 2), as in the case of aerosol synthesis of nanoparticles without simultaneous deposition of SiO_2.

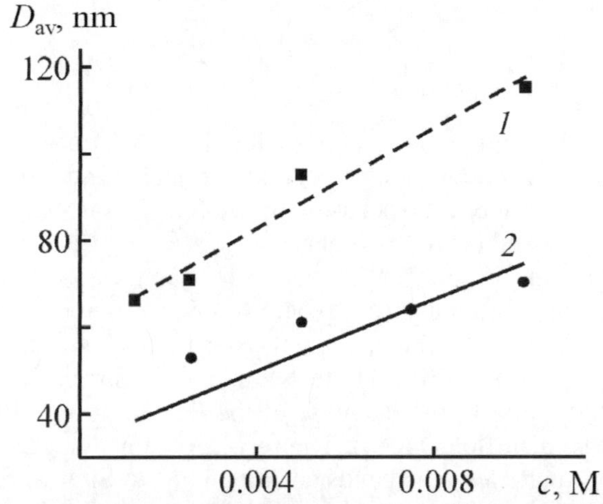

Figure 2. Average diameter D_{av} of MoS_2 nanoparticles (*1*) distributed in a nanocomposite coating and (*2*) collected on an electrostatic filter without deposition of SiO_2 vs. the $(NH_4)_2MoS_4$ concentration *c* in solution.

The increase in the average size of nanoparticles of molybdenum disulphide in the case of nanocomposite production is due to deposition of silicon dioxide layers on them (Figure 3).

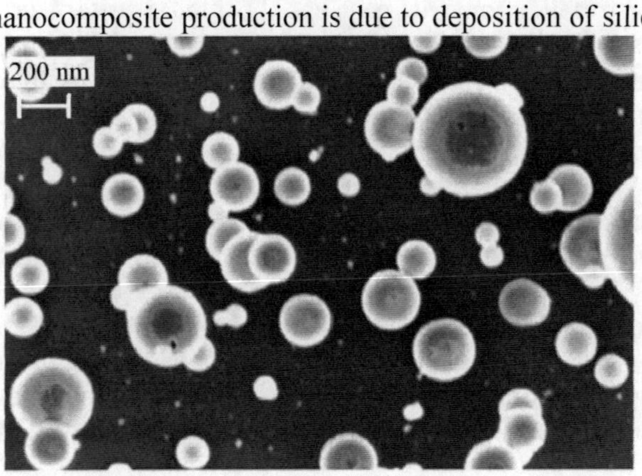

Figure 3. SEM image of MoS_2 nanoparticles on the surface of a nanocomposite coating, covered with a SiO_2 fi lm.

The composition of the deposited composite layers was studied using Raman spectroscopy. The spectra of four samples obtained using an 1800-line per mm grid and a spectral resolution of 1 cm^{-1} are shown in Figure 4.

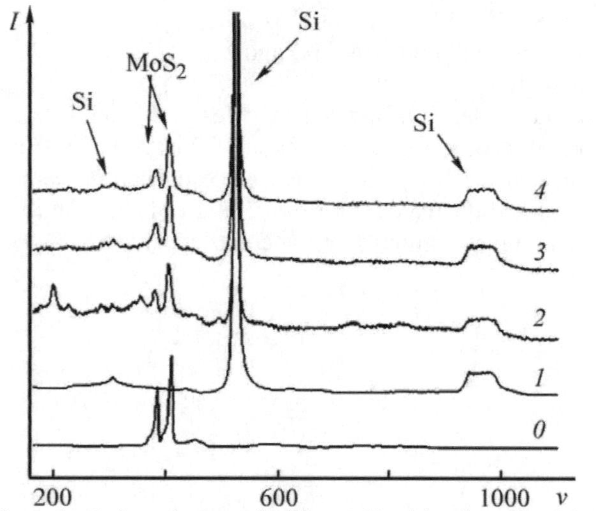

Figure 4. (*1–4*) Raman spectra of nanocomposite coating samples and (0) reference spectrum of MoS_2. (*I*) Intensity and (v) wavenumber.

Sample 1 is a layer of silicon dioxide deposited by the plasma enhanced CVD. Samples 2-4 are composite layers deposited using solutions with different $(NH_4)_2MoS_4$ concentrations in dimethylformamide (2 - 0.01M, *3* - 0.005M and *4* - 0.0025M). As can be seen, the spectra of all samples contain the Si (304, 520, 940-980 cm^{-1}) bands [17] due to the silicon substrate. Samples 2-4 also revealed two additional bands (384 and 409 cm^{-1}), which, according to the database of the Raman spectra of RRUFF [18], are due to the presence of molybdenum disulfide (MoS_2). The results obtained suggest that the samples of the composite layers contain MoS_2 particles. The bands corresponding to SiO_2 were not detected in the Raman spectra of all deposited coatings because of the amorphous nature of the matrix layer. However, the results of IR spectroscopy of all composite coatings show that in all spectra there are three main absorption bands with maxima of about 450, 800 and about 1075 cm^{-1}, corresponding to different Si-O-Si bond vibrations [19]. In addition, a relatively small peak with a maximum at 965-975 cm-1 associated with deformation vibrations of the Si-CH_3 bonds was observed, with its relative normalized intensity increasing by 10-15% compared to deposition of pure silicon dioxide, probably due to higher concentrations of various hydrocarbons in the nanocomposite coating deposition zone. Thus, the analysis results of the Raman and IR spectra show that the deposited coatings are indeed a composite material consisting of MoS_2 nanoparticles statistically distributed in a silica matrix.

For detailed determination of the structure of composite coatings they were investigated using high resolution transmission electron microscopy. For this purpose, the coatings were deposited on a monocrystalline NaCl, which was then removed by dissolution in water. In Figure 5 shows an image of a fragment of a small area of coverage and a map of the distribution of the main elements (Mo, Si, S and O) obtained with the help of a XRD analysis.

IOP Conf. Series: Earth and Environmental Science **337** (2019) 012087 doi:10.1088/1755-1315/337/1/012087

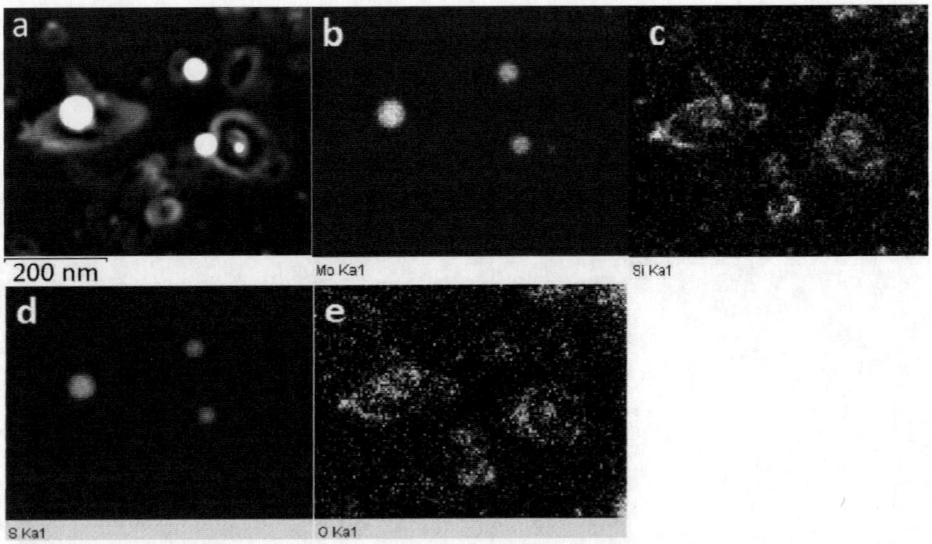

Figure 5. (a) TEM image of a part of the composite layer and (b–e) EDX element distribution maps for the layer: (b) Mo, (c) Si, (d) S, (e) O.

The presented results give a direct proof of the nanocompositional nature of deposited coatings consisting of MoS_2 nanoparticles statistically distributed in a matrix of silicon dioxide.

The evaluation of the antifriction properties was carried out based on the results of the measurement of the frictional force produced by the contact of the rotating counter body with the surface of the coatings deposited to the polished stainless steel substrates, the selected loading regimes were corresponding to the most severe operating conditions of friction pairs in various equipment. In Figure 6 is a plot of the frictional force versus the test time of a sample on which a silica layer and a nanocomposite coating containing nanoparticles of molybdenum disulfide were applied.

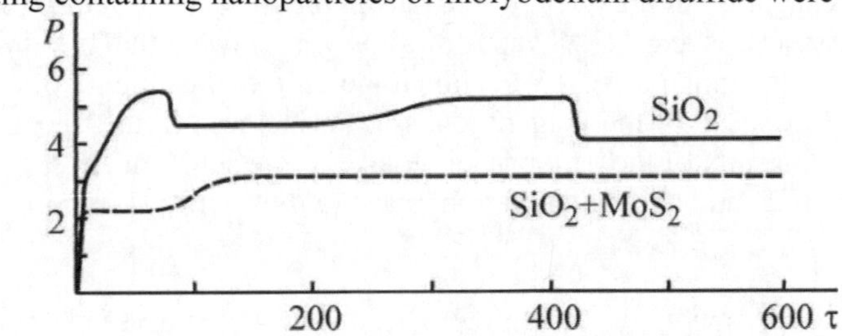

Figure 6. Spinning friction force P vs. test duration τ.

As can be seen from the presented data, the coefficient of friction for the nanocomposite coating was significantly lower over the whole range of the test time (almost twice at the initial stage). It should be noted that the frictional force measured on a sample with a SiO_2 layer has always been characterized by a large scatter of values compared to samples with deposited nanocomposite layers. The nature of the time dependence of the frictional force is due to the complex influence of several factors, the main of which are the following: the formation of wear products of the counter body, the destruction of the coating layer, which, on the one hand, is accompanied by the entry of friction particles of SiO_2 into the friction region and the other hand by the parts of antifriction particles MoS_2, and heating of the friction area. It is possible that these phenomena are more important for a silicon dioxide layer, and in the case of a nanocomposite coating, judging by the smooth nature of the dependence, they play a smaller role. In connection with these reasons, for a more objective comparative evaluation of the frictional force of the samples of nanocomposite coatings obtained under different conditions, the values of the frictional force were chosen in the steady-state part of the time dependence, which corresponds to test times exceeding 200 s.

In the first series of experiments, the effect of ammonium thiomolybdate concentration in a solution of dimethylformamide, used to create aerosols, on the antifriction properties of coatings was investigated. Figure 7 shows the data characterizing the effect of this factor on the frictional force of rotation.

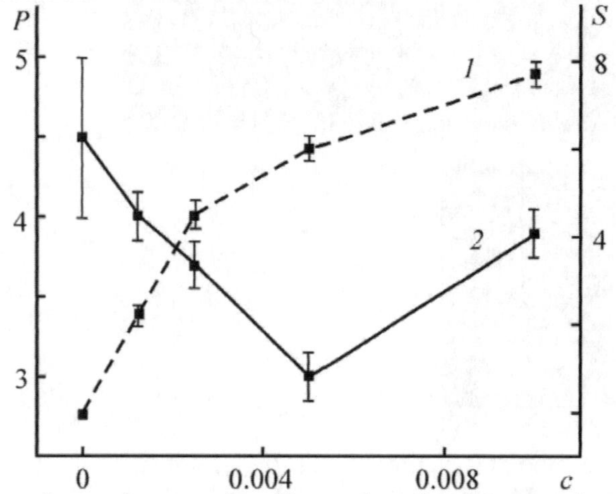

Figure 7. Spinning friction force P and surface density of nanoparticles vs. the $(NH_4)_2MoS_4$ concentration c in the dimethylformamide solution used to produce aerosols.

As can be seen, for all samples of nanocomposite coatings the frictional force has lower values than for SiO2 layers. Reduction of the frictional force with increasing concentration $(NH_4)_2MoS_4$ to 0.005 mol*l^{-1}, is most likely due to an increase in the relative amount of MoS_2 nanoparticles in the coating, as evidenced indirectly by an increase in their surface density determined by SEM. With increasing thiomolybdate concentration, the average size of synthesized molybdenum disulfide nanoparticles increases, and bearing in mind the fact [20-27] that an increase in the size of the nanoparticles of the lubricant material leads to a deterioration in its antifriction properties, this circumstance apparently causes an increase in the frictional force with a further increase in concentration.

The results of the investigation of the effect of the partial pressure of tetraethoxysilane, varied due to a change in the ratio of carrier gas and TEOS vapor dilution gas, showed that the sliding frictional force increased insignificantly in the range of 10-15% with an increase in the amount of reagent. Most likely, an increase in the partial pressure of the reagent leads to an increase in the deposition rate of silicon dioxide and at a constant intensity of the nanoparticle intake to the substrate, a smooth decrease in their relative amount in the coating, and accordingly an increase in the frictional force.

4. Conclusions

It has been demonstrated experimentally that the separation of two chemical vapor deposition processes in the space, one of which provides the synthesis of the nanoparticles of the filler (MoS_2), and the other, the formation of the nanocomposite coating matrix (SiO_2) in one reactor, provides independent control of the two process speeds and allows to vary the composition of deposited films within a wide range.

It is shown that the nanocomposite coatings of molybdenum disulphide (filler)-silicon (matrix) have improved antifriction properties in comparison with the matrix (SiO_2 layers), which are determined by the relative amount of MoS_2 nanoparticles in the layer and their average size.

Acknowledgments
The authors are grateful to the Russian Scientific Foundation for the financial support of the study (contract No. 15-13-00045).

References

[1] Binions R and Parkin I P 2011 *Advances in Nanocomposites — Synthesis, Characterization and Industrial Applications*

[2] Milionis A, Dang K and Prato M 2015 *J. Mater., Chem. A* **3** pp 12880-9

[3] Liu Y and Kumar S 2014 *ACS Appl. Mater. Interfaces* **6**(9) pp 6069-87

[4] Zhang S, Sun D, Fu Y and Du H 2003 *Surf, Coat. Technol.* **167** pp 113-9

[5] Alexandrov S E, Tyurikov K S and Kirilenko D A 2017 *Adv. Mater. Interfaces* **4** p 1700241

[6] Kruis F E, Fissan H and Peled A 1998 *J. Aerosol Sci.* **5-6** p 511-35

[7] Barreca D, Comini E and Ferrucci A P 2007 *Chem. Mater.* **19** p 5642-9

[8] Simon Q, Barreca D and Bekermann D 2011 *Int. J. Hydrogen Energy* **36** pp 15527-37

[9] Mungkalasiri J, Bedel L, Emieux F 2010 *Chem. Vap. Deposition* **16** pp 35-41

[10] Cao Y, Su Q and Che R 2012 *Synth. Met.* **11-12** pp 968-73

[11] Shelemin A, Choukourov A and Kousal J 2014 *Plasma Process. Polym.* **11** pp 864-77

[12] Zuber K, Evans D R, McClure S and Murphy P J 2015 *Adv. Eng. Mater.* **17** pp 1547-55

[13] Zuber K, Murphy P and Evans D 2015 *Adv. Eng. Mater.* **17** pp 474-82

[14] Leterrier Y, Boogh L, Andersons J and Manson J-A 1997 *J. Polym., Sci., Part B: Polym. Phys.* **35** pp 1449-61

[15] Lansdown A R 1999 *Molybdenum Disulphide Lubrication* (Amsterdam: Elsevier)

[16] Aleksandrov S E, Filatov K D and Tyurikov K S 2016 *Russ. J. Appl. Chem.* **89** pp 714-8

[17] Temple P A and Hathaway C E 1973 *Phys. Rev. B* **7** pp 3685-97

[18] Downs R *Database of Raman spectroscopy, X-ray diffraction and chemistry of minerals*, http://rruff.info.

[19] Rubio F, Rubio J, Oteo J L 1998 *Spectrosc. Lett.* **31** pp 199-219

[20] Gunda R K and Narala S K 2016 *Surf. And Coat. Technol.* **308** pp 203-12

[21] Kolosko A G, Popov E O, Filippov S V and Romanov P A 2014 *Technical Digest - 2014 27th International Vacuum Nanoelectronics Conference, IVNC 2014* pp 186–7 doi:10.1109/IVNC.2014.6894803

[22] Kolosko A G, Popov E O, Filippov S V, Romanov P A and Terukov E I 2015 *IVNC 2015 - Technical Digest: 28th International Vacuum Nanoelectronics Conference* pp 40–41 doi:10.1109/IVNC.2015.7225523

[23] Alexander B and Vladislav B 2016 *J. of Physical Education and Sport* **16** pp 910–3 doi:10.7752/jpes.2016.03143

[24] Atkarskaya A B, Nartzev V M, Privalov V E and Shemanin V G 2017 *Optical Memory and Neural Networks (Information Optics)* **26** pp 34–9 doi:10.3103/S1060992X17010027

[25] Ilkiv I, Kotlyar K, Amel'chuk D, Lebedev S, Cirlin G and Bouravleuv A 2017 *Acta Physica Polonica A* **132** pp 366–8 doi:10.12693/APhysPolA.132.366

[26] Maslov V V, Yurkinskii V P and Egorov N A 2014 *Russian J. of Applied Chemistry* **87** 1700–6 doi:10.1134/S1070427214110214

[27] Il'inskii A V, Moshnikov V A, Pashkevich M E, Permyakov N V and Shadrin E B 2015 *Technical Physics* **60** pp 123–7 doi:10.1134/S1063784215010132

Aerosol-assisted chemical deposition of nanocomposite material in the "iron-carbon-oxygen" system

I. Tyurikova[1], S. Alexandrov[1], K. Tyurikov[1,*], S. Murashkin[1], Y. Baron[1], V. Kosmach[1], I. Loshchakov[1], Y. Pankratov[1], S. Burdakov[1], L R Mukhametova[2]

[1]Peter the Great St. Petersburg Polytechnic University, 195251, Politekhnicheskaya 29, St. Petersburg, Russia
[2]Kazan State Power Engineering University, Kazan, Russian Federation*e-mail: Corresponding author: kirill.tyurikov@spbstu.ru

Abstract. The process based on aerosol chemical deposition of a composite with a carbon matrix and iron oxide nanoparticles incorporated therein was developed. O-xylene is proposed as the source of carbon, and stabilized and unstabilized magnetic fluid based on water or dimethylformamide - source of nanoscale iron oxide. The proposed method allows to obtain composite materials of various types with several directions of growth.

1. Introduction

Nanocomposite materials based on nanoparticles (NPs) of iron oxide and carbon are of considerable interest in connection with the possibility of their practical application in various fields of industry, for example, as an anode material of lithium-ion batteries [1,2, 3], adsorbents [4, 5, 6], electrochemical sensors [7], for the magnetoswitchable control of electron-transfer reactions [8], supercapacitors material [9, 10], and etc.

Currently, there are various methods of forming such composite materials. For example, chemical deposition of oxide particles on carbon nanotubes [11, 12, 13] or into porous carbon [14, 15] from solutions, formation of iron oxide nanoparticles directly in the pores of the carbon matrix [16], ion exchange [17]. These methods for the formation of nanocomposite materials with the preliminary preparation of a carbon matrix and the further formation of iron oxide nanoparticles are mainly used, which, obviously, can lead to the formation of nanoparticles with a wide and uncontrolled size distribution. Thus a method providing formation of a nanocomposite material by pre-synthesizing iron oxide nanoparticles and introducing them into the area of carbon matrix synthesis can be considered as promising.

2. Experimental

The preparation of the nanocomposite material was carried out in a vertical quartz tubular reactor with an inner diameter of about 15.5 mm equipped with two successively located resistive-type heaters. Reagents necessary for the formation of a nanocomposite material were introduced into the reactor in the form of an aerosol. The source of magnetic nanoparticles was an aqueous magnetic liquid (ML) prepared according to the method described in [18] as well as a MF based on dimethylformamide (DMF) with a stabilizer and without a stabilizer. The formation of the carbon matrix was carried out by pyrolysis of an ortho-xylene aerosol. The first heater of the reactor was used to evaporate o-xylene aerosol droplets, and its temperature T1 varied in the range from 200°C to the temperature of the second heater

(T2) used to heat the pyrolysis region from 600 to 800°C. The accuracy of temperature measurements in the heated zones was 2°C.

Dimensional analysis of the obtained materials was carried out using a Zeiss SUPRA 55VP electron microscope. The composition of the products was studied by X-ray photoelectron spectroscopy on the ESCA platform of the Nanofab-25 complex. Statistical analysis of the dimensional characteristics of the synthesized particles was carried out by image processing using the Digimizer software package for at least 200 measurements. X-ray phase analysis was performed using a Bruker Advance D8 X-ray diffractometer. To determine the nature of chemical bonds in the deposited product and the qualitative evaluation of the composition of the reaction gas medium, Fourier transform infrared spectroscopy (IR Fourier spectrometer FSM-1201) was used.

3. Results and discussion

The results of preliminary experiments on the pyrolysis of ortho-xylene (experimental conditions: the first zone temperature is 250°C, the second zone temperature is 700°C, the carrier gas flow rate is 0.3 l/min) show that spherical black particles are formed, and 95% of which have the diametral size of about 50–300 nm (Fig. 1).

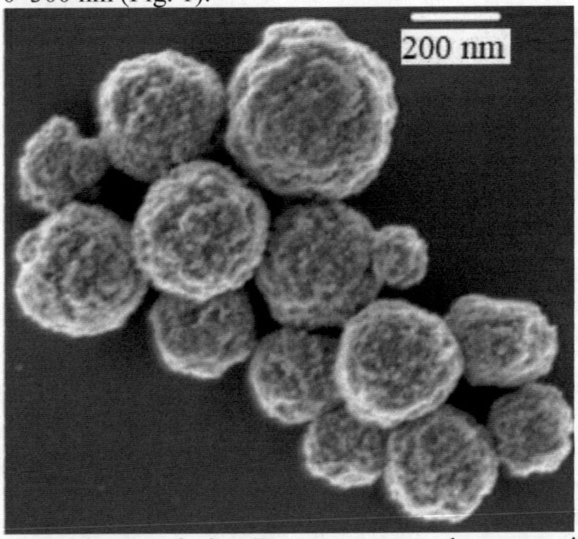

Figure 1. SEM image of carbon-containing particles obtained at T_1=250°C, T_2=700°C, carrier gas flow rate 0.3 l/min.

Using the transmission IR spectroscopy, the composition of the obtained powder was estimated. Figure 2 shows the typical FTIR spectrum of the obtained powder sample and the reference spectrum of ortho-xylene [19].

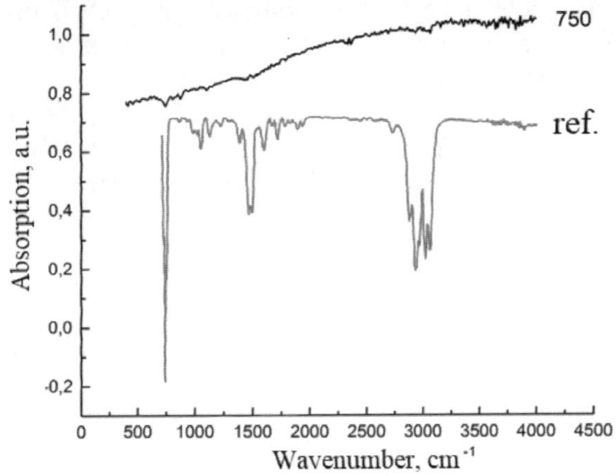

Figure 2. IR spectrum of the sample of the powder obtained during the preliminary experiment at 750°C, and the reference spectrum of o-xylene.

IOP Conf. Series: Earth and Environmental Science **337** (2019) 012088 doi:10.1088/1755-1315/337/1/012088

It can be seen that the absorption of IR radiation by the powder sample is very low and comparing the spectrum of the sample with the reference spectrum of ortho-xylene, one can assume that the resulting black powder contains only impurities of undecomposed reagent. In addition, it is possible to suggest that there are also unidentifiable hydrocarbon pyrolysis products while in general the lack of significant absorption of IR radiation suggests that the powder consists of pure carbon which cannot be detected by IR spectroscopy.

The influence of temperature of deposition process on the properties of the resulting powders (composition and dimensional characteristics) was investigated.

The results from a FTIR study of the composition of the samples deposited in powder form at different temperatures of the preheating zone in the range from 200 to 800°C (the temperature of the second zone was 700°C) show that variation of temperature in the preheating zone does not lead to a significant change in the composition of the products obtained.

The composition of deposited carbon-containing powders was studied at various temperatures of o-xylene pyrolysis in the range of 600-800°C. Figure 3 shows the IR transmission spectra of powders deposited in the temperature range of 600-800°C, and also the reference IR spectrum of o-xylene.

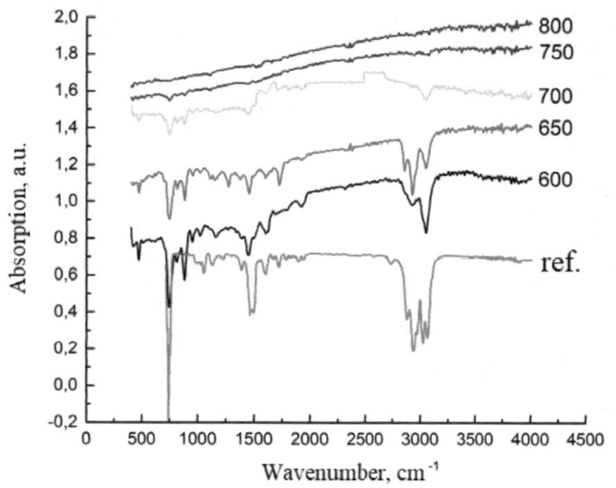

Figure 3. IR spectra of o-xylene and carbon-containing powder obtained from o-xylene at pyrolysis temperatures of 600-800°C in a deposition zone.

From the FTIR spectra presented on figure 3 it can be seen that the composition of the powders contains undecomposed o-xylene and the products of its incomplete pyrolysis. It should be noted that with an increase in the pyrolysis zone temperature the content of the by-products such as various hydrocarbon radicals in the powders decreases significantly and their traces disappeared at pyrolysis temperature higher than 750°C.

Using ESCA it was found the content of C–H bonds in the deposits decreases from 12 to about 7% with increasing temperature from 700 to 800°C. The data obtained by X-ray photoelectron spectroscopy qualitatively and quantitatively confirm the presence of amorphous carbon. It was also established using scanning electron microscopy that the temperature of the pyrolysis zone influences the dimensional characteristics of the particles produced. Formation at 700°C of carbon particles with a wide particle size distribution in the range of 50–300 nm is observed (see Figure 4). With an increase in pyrolysis temperature up to 750°C the average particle size decreases till 150–200 nm. Further increase in the temperature pyrolysis does not influence the sizes of the particles.

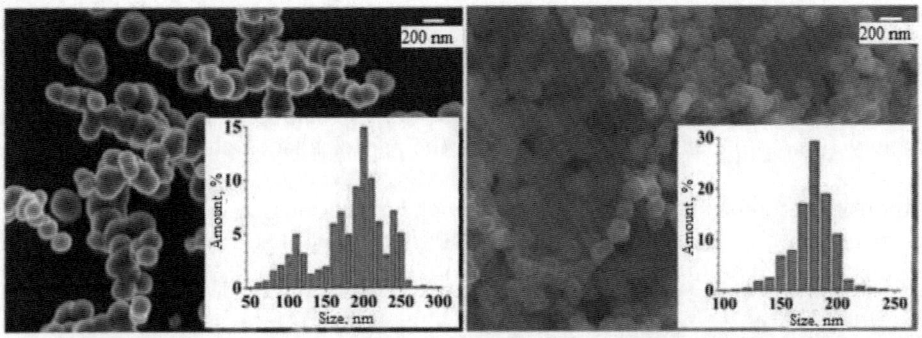

Figure 4. Carbon particles obtained at 700°C (left) and 800°C (right).

For a more detailed study of the role of the process temperature the composition of the reaction gas phase was studied at the exit from the reaction volume using IR Fourier spectroscopy at various temperatures of pyrolysis. It was found that with an increase in the pyrolysis temperature the methane content in the gas phase increased. This fact allows to suggest that there is a correlation between the composition of the deposited powder and the composition of the gas phase. If we assume that the amount of hydrogen in the reaction volume to be constant then an increase in the amount of methane with increasing temperature correlates with the decrease in the content of C-H groups in the deposited powder. An increase in the methane content also indicates a more complete pyrolysis of the reagents or intermediates and this leads to the decrease of the C-H groups content in the deposited powders.

Magnetic nanoparticles of iron oxide were obtained by drop-by-drop introduction of an aqueous solution of iron (II) and (III) salts into an aqueous solution of ammonia. The sources of iron ions were double iron salts: iron ammonium alum and Mohr salt, which are more stable in air and less susceptible to oxidation than simple iron salts. This allows to estimate more accurately the ratio of iron (III) to iron (II) in the resulting solution and help to improve reproducibility of results. The use of mannitol as a stabilizer for nanoparticles allows to obtain MNPs of Fe_3O_4, which has an excellent set of desired characteristics allowing surpass the obtained MNPs with the use of an oleic acid stabilizer. As a result of the work, MLs of Fe_3O_4 were obtained with the average size about 7 nm and a narrow size distribution (Fig. 5).

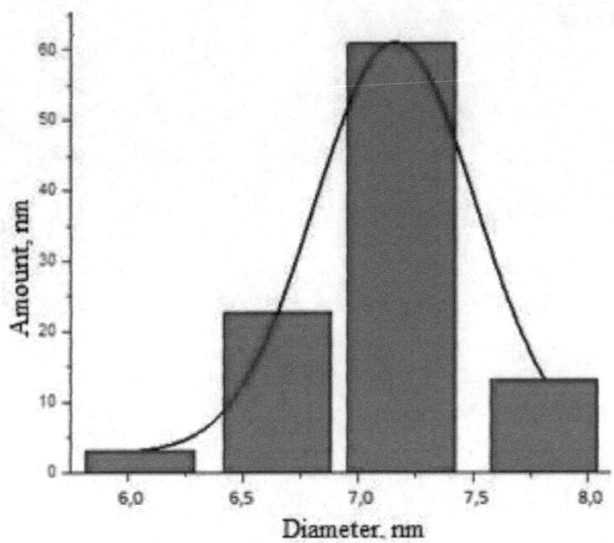

Figure 5. Fe3O4 magnetic nanoparticles size distribution.

Fe_3O_4 ML was prepared for the formation of nanocomposite material with the use of mannitol stabilizer and without it. Mannitol solutions in water of various concentrations from 5% to 20% were used for

IOP Conf. Series: Earth and Environmental Science **337** (2019) 012088 doi:10.1088/1755-1315/337/1/012088

stabilization. The experiment showed that the concentration of 5% was not enough to stabilize the MNPs and large agglomerates were formed while the use of very high concentrations at room temperature is hampered by the precipitation of a solid phase in saturated aqueous solutions of mannitol. In general, the size of the particles is significantly reduced with the use of a stabilizer and the typical average value is approximately 10 nm. Intensive agglomeration of the particles takes place in an unstabilized liquid [20-26].

As a result of nanocomposite deposition processes the formation of a carbon shell on the surface of stabilized particles proceeds less actively. On the initial stage of pyrolysis carbon particles with spherical shape are formed with a plurality of flat faces (Fig. 6).

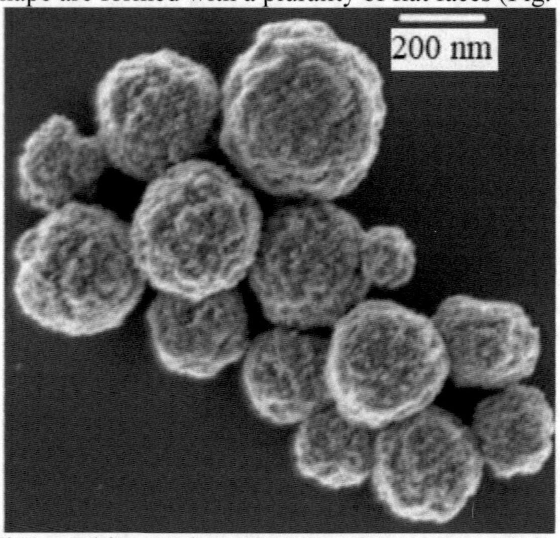

Figure 6. The image of carbon-containing particles formed at the initial stage of pyrolysis, obtained by SEM.

The resulting carbon layer on the surface of the nanoparticles protects them from oxidation and contributes to the formation of iron-containing long whisker structures, which growth can be controlled mainly by varying the technological parameters of the experiment. Fig. 7 shows images of individual iron-containing particles and combined in chains distributed in large carbon particulates.

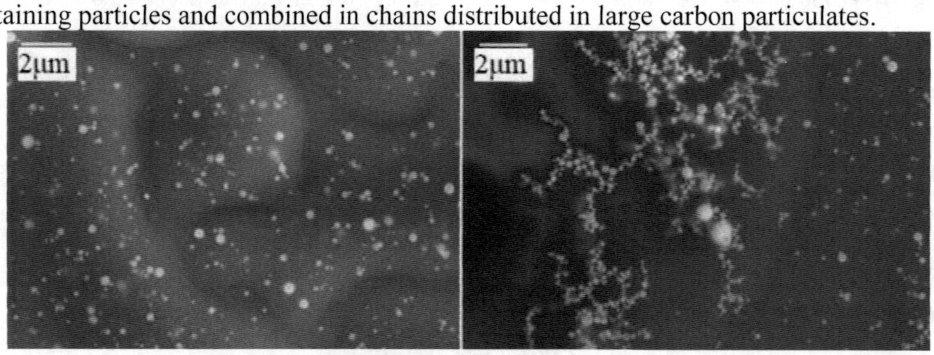

Figure 7. Micrographs of carbon particles with iron-containing inclusions.

The results from the test of a MLs using DMF show that chemical vapor deposition processes form composites consisting of nanotubes (NT) growing at the active centers which presumably are iron NPs (Fig. 8). The limiting growth factors are the residence time of particles in the pyrolysis zone and the size of the particles from which NTs begin to grow. The carbon layer on the surface of the NPs is actively subjected to etching due to the presence of a large amount of oxygen which is part of mannitol. This leads to formation of a porous surface and NTs starts to grow from the pores. Therefore, with the use of mannitol, significantly more product is formed, NTs are thinner than in the case of unstabilized NPs, and have a narrow diametral size distribution.

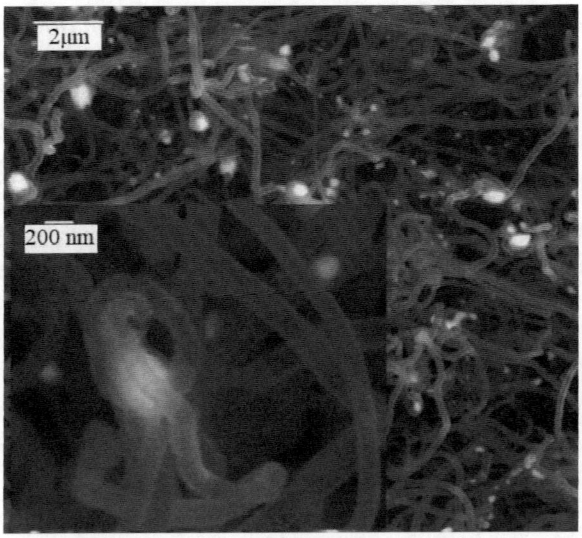

Figure 8. Images obtained on SEM, a composite material containing carbon nanotubes growing from iron nanoparticles.

4. Conclusions

The principal possibility of the formation of a carbon-iron oxide nanocomposite material in a single process with the introduction of magnetic nanoparticles into the region of the carbon matrix formation is shown; the influence of the process parameters on the characteristics of the composite components is described.

References

[1] Latorre-Sanchez M 2012 *J. Mater. Chem.* **22** pp 21373-5
[2] Yang Z, Shen J, Archer L A 2011 *J. Mater. Chem.* **21** pp 11092-7
[3] Ju H, Zhang X, Wang J 2011 *NanoBiosensing. Principles, Development and Application* (New York : Springer)
[4] Gong J-L 2009 *J. Hazard Mater* **164** pp 1517–22
[5] Ahmad R, Kumar R 2010 *J. Korean, Chem. Soc.* **54** pp 125-30
[6] Gupta V K 2011 *Water Res.* **45** pp 2207-12
[7] Qu S, Wang J, Kong J, Yang P and Chen G 2007 *Talanta* **71** pp 1096–102
[8] Musameh M, J Wang 2005 *Langmuir* **21** pp 8565-8
[9] Lin Y 2014 *Chem., Mater.* **26** pp 2128–37
[10] Zhi M 2013 *Nanoscale* **5** pp 72-88
[11] Peng X 2005 *Carbon* **43** pp 855–94
[12] Chen C 2009 *J. Hazard, Mater.* **164** pp 923-8
[13] Chen C L 2009 *Environ., Sci. Technol.* **43** pp 2362-7
[14] Yang N 2008 *Mater. Lett.* **62** pp 645–7
[15] Nagao M 2011 *J., Power Sources* **196** pp 4741-6
[16] Wu Z X 2012 *Adv. Mater.* **24** pp 485-91
[17] Huwe H 2003 *Microporous Mesoporous Mater.* **60** pp 151-8
[18] Tyurikova I A and Demidov A I 2017 *Inorg., Mater.* **53** pp 413–8
[19] Smith A L 1982 *The Coblentz Society Desk Book of Infrared Spectra*
[20] Song H 2001 *J. Am, Chem. Soc.* **127** pp 9992–3

[21] Khudolii O M, Ivashchenko O V, Iermakov S S and Rumba O G 2016 *Science of Gymnastics J.* **8** pp 215–28

[22] Micheli D, Giusti A, Delfini A, Pastore R, Vricella A, Marchetti M, Tolochko O and Vasilyeva E 2016 *Proceedings of the International Astronautical Congress, IAC*

[23] Maslov V V, Yurkinskii V P, Egorov N A 2016 *Russian J. of Applied Chemistry* **89** pp 120–4 doi:10.1134/S1070427216010195

[24] Semenova I V, Belashov A V, Garbuzov F E, Samsonov A M and Semenov A A 2017 *Proceedings of SPIE - The International Society for Optical Engineering* p 10329 doi:10.1117/12.2270375

[25] Konakov V G, Archakov I Y, Solovieva E N, Golubev S N and Orlov A V 2017 *Materials Physics and Mechanics* **30** pp 75–85

[26] Feklistov E G, Plotnikov A P and Tsobkallo E S 2017 *Proceedings of the 2017 IEEE Russia Section Young Researchers in Electrical and Electronic Engineering Conference, ElConRus 2017* pp 1148–51 doi:10.1109/EIConRus.2017.7910761

AUTHOR INDEX

AUTHOR INDEX

AUTHOR INDEX

AUTHOR INDEX